STATA MULTIVARIATE STATISTICS REFERENCE MANUAL

RELEASE 12

A Stata Press Publication
StataCorp LP
College Station, Texas

Published by Stata Press, 4905 Lakeway Drive, College Station, Texas 77845
Typeset in TEX
Printed in the United States of America
10 9 8 7 6 5 4 3 2 1

ISBN-10: 1-59718-090-4
ISBN-13: 978-1-59718-090-0

The suggested citation for this software is

StataCorp. 2011. *Stata: Release 12*. Statistical Software. College Station, TX: StataCorp LP.

Table of contents

Cross-referencing the documentation

When reading this manual, you will find references to other Stata manuals. For example,

[U] **26 Overview of Stata estimation commands**
[R] **regress**
[D] **reshape**

The first example is a reference to chapter 26, *Overview of Stata estimation commands*, in the *User's Guide*; the second is a reference to the `regress` entry in the *Base Reference Manual*; and the third is a reference to the `reshape` entry in the *Data-Management Reference Manual.*

All the manuals in the Stata Documentation have a shorthand notation:

[GSM] *Getting Started with Stata for Mac*
[GSU] *Getting Started with Stata for Unix*
[GSW] *Getting Started with Stata for Windows*
[U] *Stata User's Guide*
[R] *Stata Base Reference Manual*
[D] *Stata Data-Management Reference Manual*
[G] *Stata Graphics Reference Manual*
[XT] *Stata Longitudinal-Data/Panel-Data Reference Manual*
[MI] *Stata Multiple-Imputation Reference Manual*
[MV] *Stata Multivariate Statistics Reference Manual*
[P] *Stata Programming Reference Manual*
[SEM] *Stata Structural Equation Modeling Reference Manual*
[SVY] *Stata Survey Data Reference Manual*
[ST] *Stata Survival Analysis and Epidemiological Tables Reference Manual*
[TS] *Stata Time-Series Reference Manual*
[I] *Stata Quick Reference and Index*

[M] *Mata Reference Manual*

Detailed information about each of these manuals may be found online at

http://www.stata-press.com/manuals/

Title

intro — Introduction to multivariate statistics manual

Description

This entry describes this manual and what has changed since Stata 11.

Remarks

This manual documents Stata's multivariate analysis features and is referred to as the [MV] manual in cross-references.

Following this entry, [MV] **multivariate** provides an overview of the multivariate analysis features in Stata and Stata's multivariate analysis commands. The other parts of this manual are arranged alphabetically.

Stata is continually being updated, and Stata users are always writing new commands. To find out about the latest multivariate analysis features, type `search multivariate analysis` after installing the latest official updates; see [R] **update**.

What's new

This section is intended for previous Stata users. If you are new to Stata, you may as well skip it.

1. **Structural equation modeling (SEM)**. New command `sem` estimates multivariate linear models that can include observed and latent variables. `sem` can perform confirmatory factor analysis, multivariate regression, path analysis, and much more; see the new *Stata Structural Equation Modeling Reference Manual.*

2. **Contrasts**, which is to say, tests of linear hypotheses involving factor variables and their interactions from the most recently fit model. Tests include ANOVA-style tests of main effects, simple effects, interactions, and nested effects. Effects can be decomposed into comparisons with reference categories, comparisons of adjacent levels, comparisons with the grand mean, and more. New commands `contrast` and `margins, contrast` are available after `manova`; see [MV] **manova postestimation**. Also see [R] **contrast** and [R] **margins, contrast**.

3. **Pairwise comparisons** of means, estimated cell means, estimated marginal means, predictive margins of linear and nonlinear responses, intercepts, and slopes. In addition to ANOVA-style comparisons, comparisons can be made of population averages. New commands `pwcompare` and `margins, pwcompare` are available after `manova`; see [MV] **manova postestimation**. New command `pwmean` computes all pairwise differences of means. See [R] **pwcompare**, [R] **margins, pwcompare**, and [R] **pwmean**.

4. **Graphs of margins, marginal effects, contrasts, and pairwise comparisons**. Margins and effects can be obtained from linear or nonlinear (for example, probability) responses. New command `marginsplot` is available after `manova`; see [MV] **manova postestimation**. Also see [R] **marginsplot**.

5. **Option addplot() now places added graphs above or below**. Graph commands that allow option `addplot()` can now place the added plots above or below the command's plots. Affected by this are the multivariate commands `screeplot` and `cluster dendrogram`; see [MV] **screeplot** and [MV] **cluster dendrogram**.

For a complete list of all the new features in Stata 12, see [U] **1.3 What's new**.

Also see

[U] **1.3 What's new**

[R] **intro** — Introduction to base reference manual

Title

multivariate — Introduction to multivariate commands

Description

The *Multivariate Reference Manual* organizes the commands alphabetically, which makes it easy to find individual command entries if you know the name of the command. This overview organizes and presents the commands conceptually, that is, according to the similarities in the functions that they perform. The table below lists the manual entries that you should see for additional information.

Cluster analysis.
These commands perform cluster analysis on variables or the similarity or dissimilarity values within a matrix. An introduction to cluster analysis and a description of the `cluster` and `clustermat` subcommands is provided in [MV] **cluster** and [MV] **clustermat**.

Discriminant analysis.
These commands provide both descriptive and predictive linear discriminant analysis (LDA), as well as predictive quadratic discriminant analysis (QDA), logistic discriminant analysis, and *k*th-nearest-neighbor (KNN) discriminant analysis. An introduction to discriminant analysis and the `discrim` command is provided in [MV] **discrim**.

Factor analysis and principal component analysis.
These commands provide factor analysis of a correlation matrix and principal component analysis (PCA) of a correlation or covariance matrix. The correlation or covariance matrix can be provided directly or computed from variables.

Rotation.
These commands provide methods for rotating a factor or PCA solution or for rotating a matrix. Also provided is Procrustean rotation analysis for rotating a set of variables to best match another set of variables.

Multivariate analysis of variance and related techniques.
These commands provide canonical correlation analysis, multivariate regression, multivariate analysis of variance (MANOVA), and comparison of multivariate means. Also provided are multivariate tests on means, covariances, and correlations, and tests for multivariate normality.

Structural equation modeling.
These commands provide multivariate linear models that can include observed and latent variables. These models include confirmatory factor analysis, multivariate regression, path analysis, mediator analysis, and more; see the new *Stata Structural Equation Modeling Reference Manual*.

Multidimensional scaling and biplots.
These commands provide classic and modern (metric and nonmetric) MDS and two-dimensional biplots. MDS can be performed on the variables or on proximity data in a matrix or as proximity data in long format.

Correspondence analysis.
These commands provide simple correspondence analysis (CA) on the cross-tabulation of two categorical variables or on a matrix and multiple correspondence analysis (MCA) and joint correspondence analysis (JCA) on two or more categorical variables.

Cluster analysis

[MV] **cluster**	Introduction to cluster-analysis commands
[MV] **clustermat**	Introduction to clustermat commands
[MV] **matrix dissimilarity**	Compute similarity or dissimilarity measures; may be used by `clustermat`

Discriminant analysis

[MV] **discrim**	Introduction to discriminant-analysis commands
[MV] **discrim lda**	Linear discriminant analysis (LDA)
[MV] **discrim lda postestimation**	Postestimation tools for discrim lda
[MV] **candisc**	Canonical (descriptive) linear discriminant analysis
[MV] **discrim qda**	Quadratic discriminant analysis (QDA)
[MV] **discrim qda postestimation**	Postestimation tools for discrim qda
[MV] **discrim logistic**	Logistic discriminant analysis
[MV] **discrim logistic postestimation**	Postestimation tools for discrim logistic
[MV] **discrim knn**	*k*th-nearest-neighbor (KNN) discriminant analysis
[MV] **discrim knn postestimation**	Postestimation tools for discrim knn
[MV] **discrim estat**	Postestimation tools for discrim

Factor analysis and principal component analysis

[MV] **factor**	Factor analysis
[MV] **factor postestimation**	Postestimation tools for factor and factormat
[MV] **pca**	Principal component analysis
[MV] **pca postestimation**	Postestimation tools for pca and pcamat
[MV] **rotate**	Orthogonal and oblique rotations after factor and pca
[MV] **screeplot**	Scree plot
[MV] **scoreplot**	Score and loading plots

Rotation

[MV] **rotate**	Orthogonal and oblique rotations after factor and pca
[MV] **rotatemat**	Orthogonal and oblique rotation of a Stata matrix
[MV] **procrustes**	Procrustes transformation
[MV] **procrustes postestimation**	Postestimation tools for procrustes

Multivariate analysis of variance and related techniques

[MV] **canon**	Canonical correlations
[MV] **canon postestimation**	Postestimation tools for canon
[R] **mvreg**	Multivariate regression
[R] **mvreg postestimation**	Postestimation tools for mvreg
[MV] **manova**	Multivariate analysis of variance and covariance
[MV] **manova postestimation**	Postestimation tools for manova
[MV] **hotelling**	Hotelling's T-squared generalized means test
[MV] **mvtest**	Multivariate tests on means, covariances, correlations, and of normality

Structural equation modeling

[SEM] *Stata Structural Equation Modeling Reference Manual*

Multidimensional scaling and biplots

[MV] **mds**	Multidimensional scaling for two-way data
[MV] **mds postestimation**	Postestimation tools for mds, mdsmat, and mdslong
[MV] **mdslong**	Multidimensional scaling of proximity data in long format
[MV] **mdsmat**	Multidimensional scaling of proximity data in a matrix
[MV] **biplot**	Biplots

Correspondence analysis

[MV] **ca**	Simple correspondence analysis
[MV] **ca postestimation**	Postestimation tools for ca and camat
[MV] **mca**	Multiple and joint correspondence analysis
[MV] **mca postestimation**	Postestimation tools for mca

Remarks

Remarks are presented under the following headings:

Cluster analysis
Discriminant analysis
Factor analysis and principal component analysis
Rotation
Multivariate analysis of variance and related techniques
Structural equation modeling
Multidimensional scaling and biplots
Correspondence analysis

Cluster analysis

Cluster analysis is concerned with finding natural groupings, or clusters. Stata's cluster-analysis commands provide several hierarchical and partition clustering methods, postclustering summarization methods, and cluster-management tools. The hierarchical clustering methods may be applied to the data with the `cluster` command or to a user-supplied dissimilarity matrix with the `clustermat` command. See [MV] **cluster** for an introduction to cluster analysis and the `cluster` and `clustermat` suite of commands.

A wide variety of similarity and dissimilarity measures are available for comparing observations; see [MV] ***measure_option***. Dissimilarity matrices, for use with `clustermat`, are easily obtained using the `matrix dissimilarity` command; see [MV] **matrix dissimilarity**. This provides the building blocks necessary for clustering variables instead of observations or for clustering using a dissimilarity not automatically provided by Stata; [MV] **clustermat** provides examples.

Discriminant analysis

Discriminant analysis may be used to describe differences between groups and to exploit those differences in allocating (classifying) observations to the groups. These two purposes of discriminant analysis are often called descriptive discriminant analysis and predictive discriminant analysis.

`discrim` has both descriptive and predictive LDA; see [MV] **discrim lda**. The `candisc` command computes the same thing as `discrim lda`, but with output tailored for the descriptive aspects of the discrimination; see [MV] **candisc**.

The remaining `discrim` subcommands provide alternatives to linear discriminant analysis for predictive discrimination. [MV] **discrim qda** provides quadratic discriminant analysis. [MV] **discrim logistic** provides logistic discriminant analysis. [MV] **discrim knn** provides *k*th-nearest-neighbor discriminant analysis.

Postestimation commands provide classification tables (confusion matrices), error-rate estimates, classification listings, and group summarizations. In addition, postestimation tools for LDA and QDA include display of Mahalanobis distances between groups, correlations, and covariances. LDA postestimation tools also include discriminant-function loading plots, discriminant-function score plots, scree plots, display of canonical correlations, eigenvalues, proportion of variance, likelihood-ratio tests for the number of nonzero eigenvalues, classification functions, loadings, structure matrix, standardized means, and ANOVA and MANOVA tables. See [MV] **discrim estat**, [MV] **discrim lda postestimation**, and [MV] **discrim qda postestimation**.

Factor analysis and principal component analysis

Factor analysis and principal component analysis (PCA) have dual uses. They may be used as a dimension-reduction technique, and they may be used in describing the underlying data.

In PCA, the leading eigenvectors from the eigen decomposition of the correlation or covariance matrix of the variables describe a series of uncorrelated linear combinations of the variables that contain most of the variance. For data reduction, a few of these leading components are retained. For describing the underlying structure of the data, the magnitudes and signs of the eigenvector elements are interpreted in relation to the original variables (rows of the eigenvector).

`pca` uses the correlation or covariance matrix computed from the dataset. `pcamat` allows the correlation or covariance matrix to be directly provided. The `vce(normal)` option provides standard errors for the eigenvalues and eigenvectors, which aids in their interpretation. See [MV] **pca** for details.

Factor analysis finds a few common factors that linearly reconstruct the original variables. Reconstruction is defined in terms of prediction of the correlation matrix of the original variables, unlike PCA, where reconstruction means minimum residual variance summed across all variables. Factor loadings are examined for interpretation of the structure of the data.

`factor` computes the correlation from the dataset, whereas `factormat` is supplied the matrix directly. They both display the eigenvalues of the correlation matrix, the factor loadings, and the "uniqueness" of the variables. See [MV] **factor** for details.

To perform factor analysis or PCA on binary data, compute the tetrachoric correlations and use these with `factormat` or `pcamat`. Tetrachoric correlations are available with the `tetrachoric` command; see [R] **tetrachoric**.

After factor analysis and PCA, a suite of commands are available that provide for rotation of the loadings; generation of score variables; graphing of scree plots, loading plots, and score plots; display of matrices and scalars of interest such as anti-image matrices, residual matrices, Kaiser–Meyer–Olkin measures of sampling adequacy, squared multiple correlations; and more. See [MV] **factor postestimation**, [MV] **pca postestimation**, [MV] **rotate**, [MV] **screeplot**, and [MV] **scoreplot** for details.

Rotation

Rotation provides a modified solution that is rotated from an original multivariate solution such that interpretation is enhanced. Rotation is provided through three commands: `rotate`, `rotatemat`, and `procrustes`.

rotate works directly after pca, pcamat, factor, and factormat. It knows where to obtain the component- or factor-loading matrix for rotation, and after rotating the loading matrix, it places the rotated results in e() so that all the postestimation tools available after pca and factor may be applied to the rotated results. See [MV] **rotate** for details.

Perhaps you have the component or factor loadings from a published source and want to investigate various rotations, or perhaps you wish to rotate a loading matrix from some other multivariate command. rotatemat provides rotations for a specified matrix. See [MV] **rotatemat** for details.

A large selection of orthogonal and oblique rotations are provided for rotate and rotatemat. These include varimax, quartimax, equamax, parsimax, minimum entropy, Comrey's tandem 1 and 2, promax power, biquartimax, biquartimin, covarimin, oblimin, factor parsimony, Crawford–Ferguson family, Bentler's invariant pattern simplicity, oblimax, quartimin, target, and weighted target rotations. Kaiser normalization is also available.

The procrustes command provides Procrustean analysis. The goal is to transform a set of source variables to be as close as possible to a set of target variables. The permitted transformations are any combination of dilation (uniform scaling), rotation and reflection (orthogonal and oblique transformations), and translation. Closeness is measured by the residual sum of squares. See [MV] **procrustes** for details.

A set of postestimation commands are available after procrustes for generating fitted values and residuals; for providing fit statistics for orthogonal, oblique, and unrestricted transformations; and for providing a Procrustes overlay graph. See [MV] **procrustes postestimation** for details.

Multivariate analysis of variance and related techniques

The first canonical correlation is the maximum correlation that can be obtained between a linear combination of one set of variables and a linear combination of another set of variables. The second canonical correlation is the maximum correlation that can be obtained between linear combinations of the two sets of variables subject to the constraint that these second linear combinations are orthogonal to the first linear combinations, and so on.

canon estimates these canonical correlations and provides the loadings that describe the linear combinations of the two sets of variables that produce the correlations. Standard errors of the loadings are provided, and tests of the significance of the canonical correlations are available. See [MV] **canon** for details.

Postestimation tools are available after canon for generating the variables corresponding to the linear combinations underlying the canonical correlations. Various matrices and correlations may also be displayed. See [MV] **canon postestimation** for details.

In canonical correlation, there is no real distinction between the two sets of original variables. In multivariate regression, however, the two sets of variables take on the roles of dependent and independent variables. Multivariate regression is an extension of regression that allows for multiple dependent variables. See [R] **mvreg** for multivariate regression, and see [R] **mvreg postestimation** for the postestimation tools available after multivariate regression.

Just as analysis of variance (ANOVA) can be formulated in terms of regression where the categorical independent variables are represented by indicator (sometimes called dummy) variables, multivariate analysis of variance (MANOVA), a generalization of ANOVA that allows for multiple dependent variables, can be formulated in terms of multivariate regression where the categorical independent variables are represented by indicator variables. Multivariate analysis of covariance (MANCOVA) allows for both continuous and categorical independent variables.

The `manova` command fits MANOVA and MANCOVA models for balanced and unbalanced designs, including designs with missing cells, and for factorial, nested, or mixed designs, or designs involving repeated measures. Four multivariate test statistics—Wilks' lambda, Pillai's trace, the Lawley–Hotelling trace, and Roy's largest root—are computed for each term in the model. See [MV] **manova** for details.

Postestimation tools are available after `manova` that provide for univariate Wald tests of expressions involving the coefficients of the underlying regression model and that provide for multivariate tests involving terms or linear combinations of the underlying design matrix. Linear combinations of the dependent variables are also supported. Also available are marginal means, predictive margins, marginal effects, and average marginal effects. See [MV] **manova postestimation** for details.

Related to MANOVA is Hotelling's T-squared test of whether a set of means is zero or whether two sets of means are equal. It is a multivariate test that reduces to a standard t test if only one variable is involved. The `hotelling` command provides Hotelling's T-squared test; see [MV] **hotelling**, but also see [MV] **mvtest means** for more extensive multivariate means testing.

A suite of `mvtest` commands perform assorted multivariate tests. `mvtest means` performs one-sample and multiple-sample multivariate tests on means, assuming multivariate normality. `mvtest covariances` performs one-sample and multiple-sample multivariate tests on covariances, assuming multivariate normality. `mvtest correlations` performs one-sample and multiple-sample tests on correlations, assuming multivariate normality. `mvtest normality` performs tests for univariate, bivariate, and multivariate normality. See [MV] **mvtest**.

Structural equation modeling

Structural equation modeling (SEM) is a flexible estimation method for fitting a variety of multivariate models, and it allows for latent (unobserved) variables. See the *Stata Structural Equation Modeling Reference Manual.*

Multidimensional scaling and biplots

Multidimensional scaling (MDS) is a dimension-reduction and visualization technique. Dissimilarities (for instance, Euclidean distances) between observations in a high-dimensional space are represented in a lower-dimensional space (typically two dimensions) so that the Euclidean distance in the lower-dimensional space approximates the dissimilarities in the higher-dimensional space.

The `mds` command provides classical and modern (metric and nonmetric) MDS for dissimilarities between observations with respect to the variables; see [MV] **mds**. A wide variety of similarity and dissimilarity measures are allowed (the same ones available for the `cluster` command); see [MV] ***measure_option***.

`mdslong` and `mdsmat` provide MDS directly on the dissimilarities recorded either as data in long format (`mdslong`) or as a dissimilarity matrix (`mdsmat`); see [MV] **mdslong** and [MV] **mdsmat**.

Postestimation tools available after `mds`, `mdslong`, and `mdsmat` provide MDS configuration plots and Shepard diagrams; generation of the approximating configuration or the disparities, dissimilarities, distances, raw residuals and transformed residuals; and various matrices and scalars, such as Kruskal stress (loss), quantiles of the residuals per object, and correlations between disparities or dissimilarities and approximating distances. See [MV] **mds postestimation** for details.

Biplots are two-dimensional representations of data. Both the observations and the variables are represented. The observations are represented by marker symbols, and the variables are represented by arrows from the origin. Observations are projected to two dimensions so that the distance between the observations is approximately preserved. The cosine of the angle between arrows approximates the correlation between the variables. A biplot aids in understanding the relationship between the variables, the observations, and the observations and variables jointly. The `biplot` command produces biplots; see [MV] **biplot**.

Correspondence analysis

Simple correspondence analysis (CA) is a technique for jointly exploring the relationship between rows and columns in a cross-tabulation. It is known by many names, including dual scaling, reciprocal averaging, and canonical correlation analysis of contingency tables.

`ca` performs CA on the cross-tabulation of two integer-valued variables or on two sets of crossed (stacked) integer-valued variables. `camat` performs CA on a matrix with nonnegative entries—perhaps from a published table. See [MV] **ca** for details.

A suite of commands are available following `ca` and `camat`. These include commands for producing CA biplots and dimensional projection plots; for generating fitted values, row coordinates, and column coordinates; and for displaying distances between row and column profiles, individual cell inertia contributions, χ^2 distances between row and column profiles, and the fitted correspondence table. See [MV] **ca postestimation** for details.

`mca` performs multiple (MCA) or joint (JCA) correspondence analysis on two or more categorical variables and allows for crossing (stacking). See [MV] **mca**.

Postestimation tools available after `mca` provide graphing of category coordinate plots, dimensional projection plots, and plots of principal inertias; display of the category coordinates, optionally with column statistics; the matrix of inertias of the active variables after JCA; and generation of row scores. See [MV] **mca postestimation**.

Also see

[R] **intro** — Introduction to base reference manual

[MV] **Glossary**

Title

biplot — Biplots

Syntax

`biplot` *varlist* [*if*] [*in*] [, *options*]

options	Description
Main	
`rowover(`*varlist*`)`	identify observations from different groups of *varlist*; may not be combined with `separate` or `norow`
`dim(`*# #*`)`	two dimensions to be displayed; default `dim(2 1)`
`std`	use standardized instead of centered variables
`alpha(`*#*`)`	row weight = #; column weight = 1 − #; default is 0.5
`stretch(`*#*`)`	stretch the column (variable) arrows
`mahalanobis`	approximate Mahalanobis distance; implies `alpha(0)`
`xnegate`	negate the data relative to the x axis
`ynegate`	negate the data relative to the y axis
`autoaspect`	adjust aspect ratio on the basis of the data; default aspect ratio is 1
`separate`	produce separate plots for rows and columns; may not be combined with `rowover()`
`nograph`	suppress graph
`table`	display table showing biplot coordinates
Rows	
`rowopts(`*row_options*`)`	affect rendition of rows (observations)
`row#opts(`*row_options*`)`	affect rendition of rows (observations) in the # group of *varlist* defined in `rowover()`; available only with `rowover()`
`rowlabel(`*varname*`)`	specify label variable for rows (observations)
`norow`	suppress row points; may not be combined with `rowover()`
`generate(`$newvar_x$ $newvar_y$`)`	store biplot coordinates for observations in variables $newvar_x$ and $newvar_y$
Columns	
`colopts(`*col_options*`)`	affect rendition of columns (variables)
`negcol`	include negative column (variable) arrows
`negcolopts(`*col_options*`)`	affect rendition of negative columns (variables)
`nocolumn`	suppress column arrows
Y axis, X axis, Titles, Legend, Overall	
twoway_options	any options other than `by()` documented in [G-3] ***twoway_options***

row_options	Description
marker_options	change look of markers (color, size, etc.)
marker_label_options	change look or position of marker labels
`nolabel`	remove the default row (variable) label from the graph
`name(`*name*`)`	override the default name given to rows (observations)

col_options	Description
pcarrow_options	affect the rendition of paired-coordinate arrows
`nolabel`	remove the default column (variable) label from the graph
`name(`*name*`)`	override the default name given to columns (variables)

See [G-2] **graph twoway pcarrow**.

Menu

Statistics > Multivariate analysis > Biplot

Description

`biplot` displays a two-dimensional biplot of a dataset. A biplot simultaneously displays the observations (rows) and the relative positions of the variables (columns). Marker symbols (points) are displayed for observations, and arrows are displayed for variables. Observations are projected to two dimensions such that the distance between the observations is approximately preserved. The cosine of the angle between arrows approximates the correlation between the variables.

Options

Main

`rowover(`*varlist*`)` distinguishes groups among observations (rows) by highlighting observations on the plot for each group identified by equal values of the variables in *varlist*. By default, the graph contains a legend that consists of group names. `rowover()` may not be combined with `separate` or `norow`.

`dim(`*# #*`)` identifies the dimensions to be displayed. For instance, `dim(3 2)` plots the third dimension (vertically) versus the second dimension (horizontally). The dimension numbers cannot exceed the number of variables. The default is `dim(2 1)`.

`std` produces a biplot of the standardized variables instead of the centered variables.

`alpha(`*#*`)` specifies that the variables be scaled by $\lambda^{\#}$ and the observations by $\lambda^{(1-\#)}$, where λ are the singular values. It is required that $0 \leq \# \leq 1$. The most common values are 0, 0.5, and 1. The default is `alpha(0.5)` and is known as the symmetrically scaled biplot or symmetric factorization biplot. The result with `alpha(1)` is the principal-component biplot, also called the row-preserving metric (RPM) biplot. The biplot with `alpha(0)` is referred to as the column-preserving metric (CPM) biplot.

`stretch(`*#*`)` causes the length of the arrows to be multiplied by *#*. For example, `stretch(1)` would leave the arrows the same length, `stretch(2)` would double their length, and `stretch(0.5)` would halve their length.

`mahalanobis` implies `alpha(0)` and scales the positioning of points (observations) by $\sqrt{n-1}$ and positioning of arrows (variables) by $1/\sqrt{n-1}$. This additional scaling causes the distances between observations to change from being approximately proportional to the Mahalanobis distance to instead being approximately equal to the Mahalanobis distance. Also, the inner products between variables approximate their covariance.

`xnegate` specifies that dimension-1 (x axis) values be negated (multiplied by -1).

`ynegate` specifies that dimension-2 (y axis) values be negated (multiplied by -1).

`autoaspect` specifies that the aspect ratio be automatically adjusted based on the range of the data to be plotted. This option can make some biplots more readable. By default, `biplot` uses an aspect ratio of one, producing a square plot. Some biplots will have little variation in the y-axis direction, and using the `autoaspect` option will better fill the available graph space while preserving the equivalence of distance in the x and y axes.

As an alternative to `autoaspect`, the *twoway_option* `aspectratio()` can be used to override the default aspect ratio. `biplot` accepts the `aspectratio()` option as a suggestion only and will override it when necessary to produce plots with balanced axes; that is, distance on the x axis equals distance on the y axis.

twoway_options, such as `xlabel()`, `xscale()`, `ylabel()`, and `yscale()`, should be used with caution. These *axis_options* are accepted but may have unintended side effects on the aspect ratio. See [G-3] ***twoway_options***.

`separate` produces separate plots for the row and column categories. The default is to overlay the plots. `separate` may not be combined with `rowover()`.

`nograph` suppresses displaying the graph.

`table` displays a table with the biplot coordinates.

Rows

`rowopts(`*row_options*`)` affects the rendition of the points plotting the rows (observations). This option may not be combined with `rowover()`. The following *row_options* are allowed:

marker_options affect the rendition of markers drawn at the plotted points, including their shape, size, color, and outline; see [G-3] ***marker_options***.

marker_label_options specify the properties of marker labels; see [G-3] ***marker_label_options***. `mlabel()` in `rowopts()` may not be combined with the `rowlabel()` option.

`nolabel` removes the default row label from the graph.

`name(`*name*`)` overrides the default name given to rows.

`row#opts(`*row_options*`)` affects rendition of the points plotting the rows (observations) in the #th group identified by equal values of the variables in *varlist* defined in `rowover()`. This option requires specifying `rowover()`. See `rowopts()` above for the allowed *row_options*, except `mlabel()` is not allowed with `row#opts()`.

`rowlabel(`*varname*`)` specifies label variable for rows (observations).

`norow` suppresses plotting of row points. This option may not be combined with `rowover()`.

`generate(`$newvar_x$ $newvar_y$`)` stores biplot coordinates for rows in variables $newvar_x$ and $newvar_y$.

Columns

`colopts(`*col_options*`)` affects the rendition of the arrows and points plotting the columns (variables). The following *col_options* are allowed:

pcarrow_options affect the rendition of paired-coordinate arrows; see [G-2] **graph twoway pcarrow**.

`nolabel` removes the default column label from the graph.

`name(`*name*`)` overrides the default name given to columns.

`negcol` includes negative column (variable) arrows on the plot.

`negcolopts(`*col_options*`)` affects the rendition of the arrows and points plotting the negative columns (variables). The *col_options* allowed are given above.

`nocolumn` suppresses plotting of column arrows.

Y axis, X axis, Titles, Legend, Overall

twoway_options are any of the options documented in [G-3] ***twoway_options***, excluding `by()`. These include options for titling the graph (see [G-3] ***title_options***) and for saving the graph to disk (see [G-3] ***saving_option***). See `autoaspect` above for a warning against using options such as `xlabel()`, `xscale()`, `ylabel()`, and `yscale()`.

Remarks

The `biplot` command produces what Cox and Cox (2001) refer to as the "classic biplot". Biplots were introduced by Gabriel (1971); also see Gabriel (1981). Gower and Hand (1996) discuss extensions and generalizations to biplots and place many of the well-known multivariate techniques into a generalized biplot framework extending beyond the classic biplot implemented by Stata's `biplot` command. Cox and Cox (2001), Jolliffe (2002), Gordon (1999), Jacoby (1998), Rencher (2002), and Seber (1984) discuss the classic biplot. Kohler (2004) provides a Stata implementation of biplots.

Let $\mathbf{X}$ be the centered (or standardized if the `std` option is specified) data. A biplot splits the information in $\mathbf{X}$ into a portion related to the observations (rows of $\mathbf{X}$) and a portion related to the variables (columns of $\mathbf{X}$)

$$\mathbf{X} \approx (\mathbf{U}_2\, \mathbf{\Lambda}_2^{\alpha})(\mathbf{V}_2\, \mathbf{\Lambda}_2^{1-\alpha})'$$

where $0 \leq \alpha \leq 1$; see *Methods and formulas* for details. $\mathbf{U}_2\, \mathbf{\Lambda}_2^{\alpha}$ contains the plotting coordinates corresponding to observations (rows), and $\mathbf{V}_2\, \mathbf{\Lambda}_2^{1-\alpha}$ contains the plotting coordinates corresponding to variables (columns). In a biplot, the row coordinates are plotted as symbols, and the column coordinates are plotted as arrows from the origin.

The commonly used values for α are 0, 0.5, and 1. The default is 0.5. The `alpha()` option allows you to set α.

Biplots with an α of 1 are also called principal-component biplots because $\mathbf{U}_2\, \mathbf{\Lambda}_2$ contains the principal-component scores and $\mathbf{V}_2$ contains the principal-component coefficients. Euclidean distance between points in this kind of biplot approximates the Euclidean distance between points in the original higher-dimensional space.

Using an α of 0, Euclidean distances in the biplot are approximately proportional to Mahalanobis distances in the original higher-dimensional space. Also, the inner product of the arrows is approximately proportional to the covariances between the variables.

When you set α to 0 and specify the `mahalanobis` option, the Euclidean distances are not just approximately proportional but are approximately equal to Mahalanobis distances in the original space. Likewise, the inner products of the arrows are approximately equal (not just proportional) to the covariances between the variables. This means that the length of an arrow is approximately equal to the standard deviation of the variable it represents. Also, the cosine of the angle between two arrows is approximately equal to the correlation between the two variables.

A biplot with an α of 0.5 is called a symmetric factorization biplot or symmetrically scaled biplot. It often produces reasonable looking biplots where the points corresponding to observations and the arrows corresponding to variables are given equal weight. Using an α of 0 (or 1) causes the points (or the arrows) to be bunched tightly around the origin while the arrows (or the points) are predominant in the graph. Here many authors recommend picking a scaling factor for the arrows to bring them back into balance. The `stretch()` option allows you to do this.

Regardless of your choice of α, the position of a point in relation to an arrow indicates whether that observation is relatively large, medium, or small for that variable. Also, although the special conditions mentioned earlier may not strictly hold for all α, the biplot still aids in understanding the relationship between the variables, the observations, and the observations and variables jointly.

▷ Example 1

Gordon (1999, 176) provides a simple example of a biplot based on data having five rows and three columns.

```
. input v1 v2 v3
              v1          v2          v3
  1.    60    80 -240
  2. -213    66  180
  3.   123 -186  180
  4.    -9    38  -60
  5.    39     2  -60
  6. end
. biplot v1 v2 v3
Biplot of 5 observations and 3 variables
    Explained variance by component 1  0.6283
    Explained variance by component 2  0.3717
             Total explained variance  1.0000
```

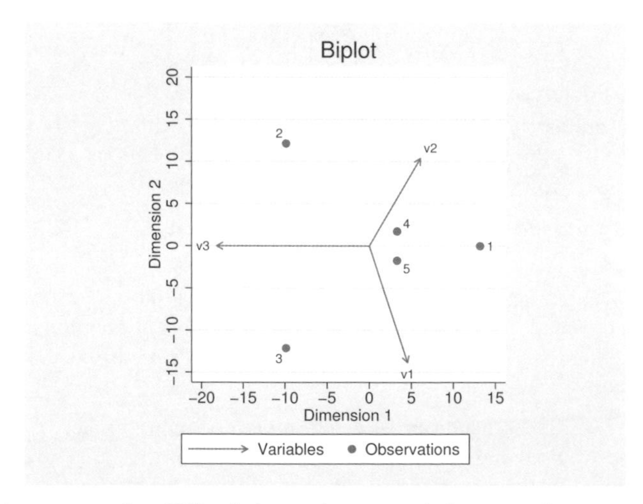

The first component accounts for 63% of the variance, and the second component accounts for the remaining 37%. All the variance is accounted for because, here, the 5-by-3 data matrix is only of rank 2.

Gordon actually used an α of 0 and performed the scaling to better match Mahalanobis distance. We do the same using the options `alpha(0)` and `mahalanobis`. (We could just use `mahalanobis` because it implies `alpha(0)`.) With an α of 0, Gordon decided to scale the arrows by a factor of

.01. We accomplish this with the `stretch()` option and add options to provide a title and subtitle in place of the default title obtained previously.

```
. biplot v1 v2 v3, alpha(0) mahalanobis stretch(.01) title(Simple biplot)
> subtitle(See figure 6.10 of Gordon (1999))
Biplot of 5 observations and 3 variables

    Explained variance by component 1  0.6283
    Explained variance by component 2  0.3717
              Total explained variance  1.0000
```

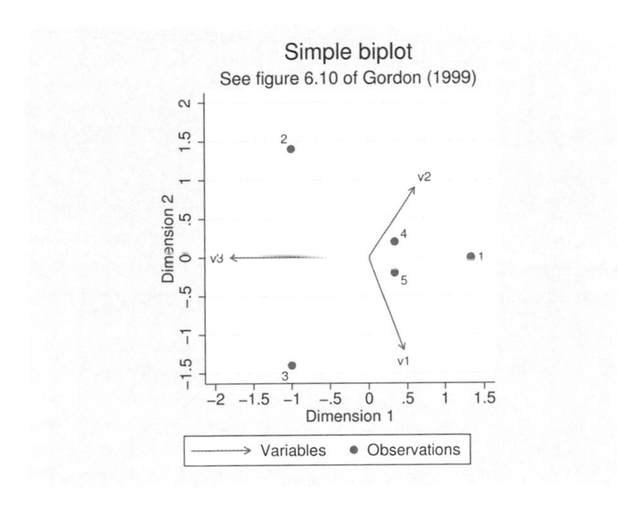

The outcome is little changed between the first and second biplot except for the additional titles and the scale of the x and y axes.

`biplot` allows you to highlight observations belonging to different groups by using option `rowover()`. Suppose our data come from two groups defined by variable `group`, `group=1` and `group=2`.

```
. generate byte group = cond(_n<3, 1, 2)
. list
```

	v1	v2	v3	group
1.	60	80	-240	1
2.	-213	66	180	1
3.	123	-186	180	2
4.	-9	38	-60	2
5.	39	2	-60	2

Here is the previous biplot with group-specific markers:

```
. biplot v1 v2 v3, alpha(0) mahalanobis stretch(.01) title(Simple biplot)
> subtitle(Grouping variable group) rowover(group)
> row1opts(name("Group 1") msymbol(O) nolabel)
> row2opts(name("Group 2") msymbol(T) nolabel)
Biplot of 5 observations and 3 variables

    Explained variance by component 1  0.6283
    Explained variance by component 2  0.3717
              Total explained variance  1.0000
```

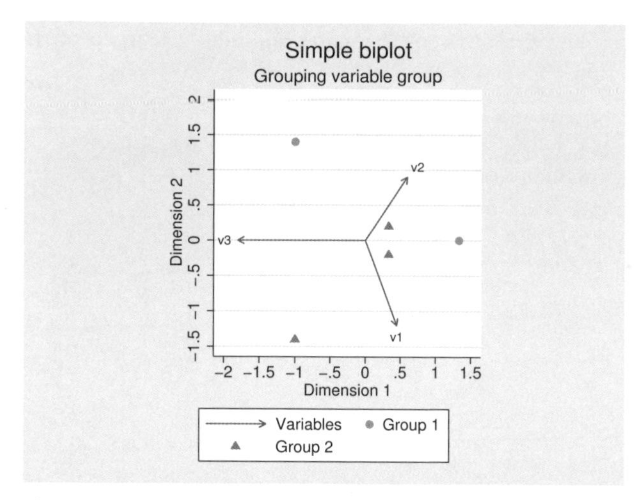

In the above example, groups are defined by a single variable `group` but you can specify multiple variables with `rowover()`. The rendition of group markers is controlled by options `row1opts()` and `row2opts()`. The marker labels are disabled by using the `nolabel` option.

◁

▷ Example 2

Table 7.1 of Cox and Cox (2001) provides the scores of 10 Renaissance painters on four attributes using a scale from 0 to 20, as judged by Roger de Piles in the 17th century.

```
. use http://www.stata-press.com/data/r12/renpainters, clear
(Scores by Roger de Piles for Renaissance Painters)
. list, abbrev(12)
```

	painter	composition	drawing	colour	expression
1.	Del Sarto	12	16	9	8
2.	Del Piombo	8	13	16	7
3.	Da Udine	10	8	16	3
4.	Giulio Romano	15	16	4	14
5.	Da Vinci	15	16	4	14
6.	Michelangelo	8	17	4	8
7.	Fr. Penni	0	15	8	0
8.	Perino del Vaga	15	16	7	6
9.	Perugino	4	12	10	4
10.	Raphael	17	18	12	18

```
. biplot composition-expression, alpha(1) stretch(10) table rowopts(name(Painters))
> rowlabel(painter) colopts(name(Attributes)) title(Renaissance painters)
Biplot of 10 painters and 4 attributes

    Explained variance by component 1  0.6700
    Explained variance by component 2  0.2375
             Total explained variance  0.9075

Biplot coordinates

        Painters      dim1       dim2

      Del Sarto     1.2120     0.0739
      Del Piombo   -4.5003     5.7309
       Da Udine    -7.2024     7.5745
    Giulio Rom~o    8.4631    -2.5503
       Da Vinci     8.4631    -2.5503
    Michelangelo    0.1284    -5.9578
       Fr Penni   -11.9449    -5.4510
    Perino del~a    2.2564    -0.9193
       Perugino    -7.8886    -0.8757
        Raphael    11.0131     4.9251

      Attributes      dim1       dim2

     composition    6.4025     3.3319
         drawing    2.4952    -3.3422
          colour   -2.4557     8.7294
      expression    6.8375     1.2348
```

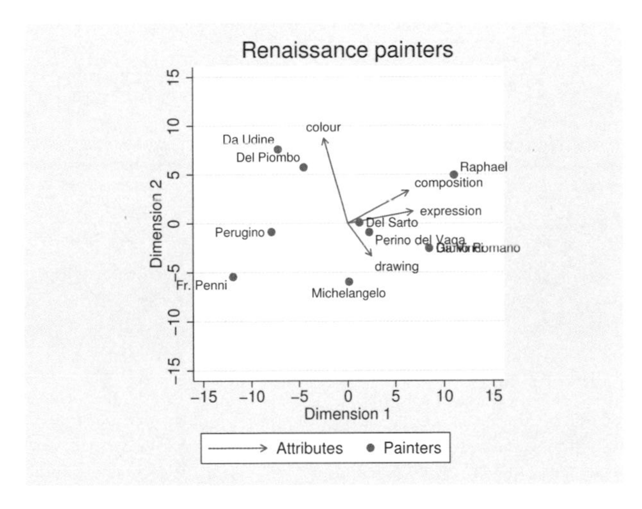

`alpha(1)` gave us an α of 1. `stretch(10)` made the arrows 10 times longer. `table` requested that the biplot coordinate table be displayed. `rowopts()` and `colopts()` affected the rendition of the rows (observations) and columns (variables). The `name()` suboption provided a name to use instead of the default names “Observations” and “Variables” in the graph legend and in the biplot coordinate table. The `rowlabel(painter)` option requested that the variable `painter` be used to label the row points (observations) in both the graph and table. The `title()` option was used to override the default title.

The default is to produce a square graphing region. Because the x axis containing the first component has more variability than the y axis containing the second component, there are often no observations or arrows appearing in the upper and lower regions of the graph. The `autoaspect` option sets the aspect ratio and the x-axis and y-axis scales so that more of the graph region is used while maintaining the equivalent interpretation of distance for the x and y axes.

Here is the previous biplot with the omission of the `table` option and the addition of the `autoaspect` option. We also add the `ynegate` option to invert the orientation of the data in the y-axis direction to match the orientation shown in figure 7.1 of Cox and Cox (2001). We add the `negcol` option to include column (variable) arrows pointing in the negative directions, and the rendition of these negative columns (variables) is controlled by `negcolopts()`.

```
. biplot composition-expression, autoaspect alpha(1) stretch(10) ynegate
> rowopts(name(Painters)) rowlabel(painter) colopts(name(Attributes))
> title(Renaissance painters) negcol negcolopts(name(-Attributes))
Biplot of 10 painters and 4 attributes

    Explained variance by component 1  0.6700
    Explained variance by component 2  0.2375
              Total explained variance  0.9075
```

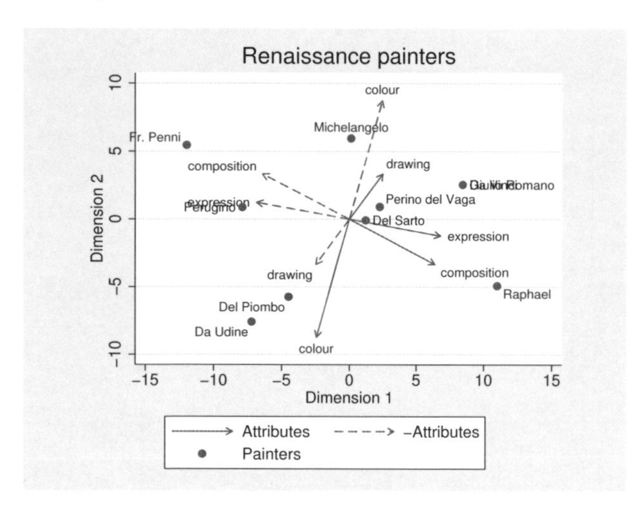

◁

Saved results

`biplot` saves the following in `r()`:

Scalars

`r(rho1)`	explained variance by component 1
`r(rho2)`	explained variance by component 2
`r(rho)`	total explained variance
`r(alpha)`	value of `alpha()` option

Matrices

`r(U)`	biplot coordinates for the observations; stored only if the row dimension does not exceed Stata's maximum matrix size; as an alternative, use `generate()` to store biplot coordinates for the observations in variables
`r(V)`	biplot coordinates for the variables
`r(Vstretch)`	biplot coordinates for the variables times `stretch()` factor

Methods and formulas

`biplot` is implemented as an ado-file.

Let $\mathbf{X}$ be the centered (standardized if `std` is specified) data with N rows (observations) and p columns (variables). A biplot splits the information in $\mathbf{X}$ into a portion related to the observations (rows of $\mathbf{X}$) and a portion related to the variables (columns of $\mathbf{X}$). This task is done using the singular value decomposition (SVD).

$$\mathbf{X} = \mathbf{U}\boldsymbol{\Lambda}\mathbf{V}'$$

The biplot formula is derived from this SVD by first splitting $\boldsymbol{\Lambda}$, a diagonal matrix, into

$$\boldsymbol{\Lambda} = \boldsymbol{\Lambda}^{\alpha}\,\boldsymbol{\Lambda}^{1\quad\alpha}$$

and then retaining the first two columns of $\mathbf{U}$, the first two columns of $\mathbf{V}$, and the first two rows and columns of $\boldsymbol{\Lambda}$. Using the subscript 2 to denote this, the biplot formula is

$$\mathbf{X} \approx \mathbf{U}_2\,\boldsymbol{\Lambda}_2^{\alpha}\,\boldsymbol{\Lambda}_2^{1-\alpha}\,\mathbf{V}_2'$$

where $0 \le \alpha \le 1$. This is then written as

$$\mathbf{X} \approx (\mathbf{U}_2\,\boldsymbol{\Lambda}_2^{\alpha})(\mathbf{V}_2\,\boldsymbol{\Lambda}_2^{1-\alpha})'$$

$\mathbf{U}_2\,\boldsymbol{\Lambda}_2^{\alpha}$ contains the plotting coordinates corresponding to observations (rows) and $\mathbf{V}_2\,\boldsymbol{\Lambda}_2^{1-\alpha}$ contains the plotting coordinates corresponding to variables (columns). In a biplot, the row coordinates are plotted as symbols and the column coordinates are plotted as arrows from the origin.

Let λ_i be the ith diagonal of $\boldsymbol{\Lambda}$. The explained variance for component 1 is

$$\rho_1 = \left\{\sum_{i=1}^{p}\lambda_i^2\right\}^{-1}\lambda_1^2$$

and for component 2 is

$$\rho_2 = \left\{\sum_{i=1}^{p}\lambda_i^2\right\}^{-1}\lambda_2^2$$

The total explained variance is

$$\rho = \rho_1 + \rho_2$$

Acknowledgment

Several `biplot` options were based on the work of Ulrich Kohler (2004) from WZB, Berlin.

Kuno Ruben Gabriel (1929–2003) was born in Germany, raised in Israel, and studied at the London School of Economics and the Hebrew University of Jerusalem, earning a doctorate in demography in 1957. After several years on the faculty at the Hebrew University, he moved to the University of Rochester in 1975. His wide-ranging statistical interests spanned applications in meteorology, including weather-modification experiments, and medicine. Gabriel's best-known contribution is the biplot.

References

Cox, T. F., and M. A. A. Cox. 2001. *Multidimensional Scaling*. 2nd ed. Boca Raton, FL: Chapman & Hall/CRC.

Gabriel, K. R. 1971. The biplot graphical display of matrices with application to principal component analysis. *Biometrika* 58: 453–467.

——. 1981. Biplot display of multivariate matrices for inspection of data and diagnosis. In *Interpreting Multivariate Data*, ed. V. Barnett, 571–572. Chichester, UK: Wiley.

Gordon, A. D. 1999. *Classification*. 2nd ed. Boca Raton, FL: Chapman & Hall/CRC.

Gower, J. C., and D. J. Hand. 1996. *Biplots*. London: Chapman & Hall.

Hall, W. J. 2003. Ruben Gabriel, 1929–2003. *IMS Bulletin* 32(4): 9.

Jacoby, W. G. 1998. *Statistical Graphics for Visualizing Multivariate Data*. Thousand Oaks, CA: Sage.

Jolliffe, I. T. 2002. *Principal Component Analysis*. 2nd ed. New York: Springer.

Kohler, U. 2004. Biplots, revisited. 2004 German Stata Users Group meeting. http://www.stata.com/meeting/2german/.

Kohler, U., and M. Luniak. 2005. Data inspection using biplots. *Stata Journal* 5: 208–223.

Rencher, A. C. 2002. *Methods of Multivariate Analysis*. 2nd ed. New York: Wiley.

Seber, G. A. F. 1984. *Multivariate Observations*. New York: Wiley.

Also see

[MV] **ca** — Simple correspondence analysis

[MV] **mds** — Multidimensional scaling for two-way data

[MV] **pca** — Principal component analysis

Title

ca — Simple correspondence analysis

Syntax

Simple correspondence analysis of two categorical variables

ca *rowvar colvar* [*if*] [*in*] [*weight*] [, *options*]

Simple correspondence analysis with crossed (stacked) variables

ca *row_spec col_spec* [*if*] [*in*] [*weight*] [, *options*]

Simple correspondence analysis of an $n_r \times n_c$ *matrix*

camat *matname* [, *options*]

where *spec* = *varname* | (*newvar* : *varlist*)

options	Description
Model 2	
dimensions(*#*)	number of dimensions (factors, axes); default is dim(2)
normalize(*nopts*)	normalization of row and column coordinates
rowsupp(*matname*$_r$)	matrix of supplementary rows
colsupp(*matname*$_c$)	matrix of supplementary columns
rowname(*string*)	label for rows
colname(*string*)	label for columns
missing	treat missing values as ordinary values (ca only)
Codes (ca only)	
report(variables)	report coding of crossing variables
report(crossed)	report coding of crossed variables
report(all)	report coding of crossing and crossed variables
length(min)	use minimal length unique codes of crossing variables
length(*#*)	use # as coding length of crossing variables
Reporting	
ddimensions(*#*)	number of singular values to be displayed; default is ddim(.)
norowpoints	suppress table with row category statistics
nocolpoints	suppress table with column category statistics
compact	display tables in a compact format
plot	plot the row and column coordinates
maxlength(*#*)	maximum number of characters for labels; default is maxlength(12)

nopts	Description
symmetric	symmetric coordinates (canonical); the default
standard	row and column standard coordinates
row	row principal, column standard coordinates
column	column principal, row standard coordinates
principal	row and column principal coordinates
#	power $0 \leq \# \leq 1$ for row coordinates; seldom used

bootstrap, by, jackknife, rolling, and statsby are allowed with ca; see [U] **11.1.10 Prefix commands**. However, bootstrap and jackknife results should be interpreted with caution; identification of the ca parameters involves data-dependent restrictions, possibly leading to badly biased and overdispersed estimates (Milan and Whittaker 1995).

Weights are not allowed with the bootstrap prefix; see [R] **bootstrap**.

aweights are not allowed with the jackknife prefix; see [R] **jackknife**.

fweights, aweights, and iweights are allowed with ca; see [U] **11.1.6 weight**.

See [U] **20 Estimation and postestimation commands** for more capabilities of estimation commands.

Menu

ca

Statistics > Multivariate analysis > Correspondence analysis > Two-way correspondence analysis (CA)

camat

Statistics > Multivariate analysis > Correspondence analysis > Two-way correspondence analysis of a matrix

Description

ca performs a simple correspondence analysis (CA) of the cross-tabulation of the integer-valued variables *rowvar* and *colvar* with n_r and n_c categories with n_r, $n_c \geq 2$. CA is formally equivalent to various other geometric approaches, including dual scaling, reciprocal averaging, and canonical correlation analysis of contingency tables (Greenacre 1984, chap. 4).

camat performs a simple CA of an $n_r \times n_c$ matrix *matname* having nonnegative entries and strictly positive margins. The correspondence table need not contain frequencies. The labels for the row and column categories are obtained from the matrix row and column names.

Optionally, a CA biplot may be produced. The biplot displays the row and column coordinates within the same two-dimensional graph.

Results may be replayed using ca or camat; there is no difference.

Options

Model 2

dimensions(*#*) specifies the number of dimensions (= factors = axes) to be extracted. The default is dimensions(2). If you may specify dimensions(1), the row and column categories are placed on one dimension. *#* should be strictly smaller than the number of rows and the number of columns, counting only the active rows and columns, excluding supplementary rows and columns (see options rowsupp() and colsupp()).

CA is a hierarchical method, so extracting more dimensions does not affect the coordinates and decomposition of inertia of dimensions already included. The percentages of inertia accounting for the dimensions are in decreasing order as indicated by singular values. The first dimension accounts for the most inertia, followed by the second dimension, and then the third dimension, etc.

`normalize(`*nopt*`)` specifies the normalization method, that is, how the row and column coordinates are obtained from the singular vectors and singular values of the matrix of standardized residuals. See *Normalization and interpretation of correspondence analysis* in *Remarks* for a discussion of these different normalization methods.

`symmetric`, the default, distributes the inertia equally over rows and columns, treating the rows and columns symmetrically. The symmetric normalization is also known as the standard, or canonical, normalization. This is the most common normalization when making a biplot. `normalize(symmetric)` is equivalent to `normalize(0.5)`. `canonical` is a synonym for `symmetric`.

`standard` specifies that row and column coordinates should be in standard form (singular vectors divided by the square root of mass). This normalization method is not equivalent to `normalize(`*#*`)` for any *#*.

`row` specifies principal row coordinates and standard column coordinates. This option should be chosen if you want to compare row categories. Similarity of column categories should not be interpreted. The biplot interpretation of the relationship between row and column categories is appropriate. `normalize(row)` is equivalent to `normalize(1)`.

`column` specifies principal column coordinates and standard row coordinates. This option should be chosen if you want to compare column categories. Similarity of row categories should not be interpreted. The biplot interpretation of the relationship between row and column categories is appropriate. `normalize(column)` is equivalent to `normalize(0)`.

`principal` is the normalization to choose if you want to make comparisons among the row categories and among the column categories. In this normalization, comparing row and column points is not appropriate. Thus a biplot in this normalization is best avoided. In the principal normalization, the row and column coordinates are obtained from the left and right singular vectors, multiplied by the singular values. This normalization method is not equivalent to `normalize(`*#*`)` for any *#*.

#, $0 \leq \# \leq 1$, is seldom used; it specifies that the row coordinates are obtained as the left singular vectors multiplied by the singular values to the power *#*, whereas the column coordinates equal the right singular vectors multiplied by the singular values to the power $1 - \#$.

`rowsupp(`$matname_r$`)` specifies a matrix of supplementary rows. $matname_r$ should have n_c columns. The row names of $matname_r$ are used for labeling. Supplementary rows do not affect the computation of the dimensions and the decomposition of inertia. They are, however, included in the plots and in the table with statistics of the row points. Because supplementary points do not contribute to the dimensions, their entries under the column labeled `contrib` are left blank.

`colsupp(`$matname_c$`)` specifies a matrix of supplementary columns. $matname_c$ should have n_r rows. The column names of $matname_c$ are used for labeling. Supplementary columns do not affect the computation of the dimensions and the decomposition of inertia. They are, however, included in the plots and in the table with statistics of the column points. Because supplementary points do not contribute to the dimensions, their entries under the column labeled `contrib` are left blank.

`rowname(`*string*`)` specifies a label to refer to the rows of the matrix. The default is `rowname(rowvar)` for `ca` and `rowname(rows)` for `camat`.

`colname(`*string*`)` specifies a label to refer to the columns of the matrix. The default is `colname(colvar)` for `ca` and `colname(columns)` for `camat`.

`missing`, allowed only with `ca`, treats missing values of *rowvar* and *colvar* as ordinary categories to be included in the analysis. Observations with missing values are omitted from the analysis by default.

Codes

`report(`*opt*`)` displays coding information for the crossing variables, crossed variables, or both. `report()` is ignored if you do not specify at least one crossed variable.

`report(variables)` displays the coding schemes of the crossing variables, that is, the variables used to define the crossed variables.

`report(crossed)` displays a table explaining the value labels of the crossed variables.

`report(all)` displays the codings of the crossing and crossed variables.

`length(`*opt*`)` specifies the coding length of crossing variables.

`length(min)` specifies that the minimal-length unique codes of crossing variables be used.

`length(`*#*`)` specifies that the coding length # of crossing variables be used, where # must be between 4 and 32.

Reporting

`ddimensions(`*#*`)` specifies the number of singular values to be displayed. The default is `ddimensions(.)`, meaning all.

`norowpoints` suppresses the table with row point (category) statistics.

`nocolpoints` suppresses the table with column point (category) statistics.

`compact` specifies that the table with point statistics be displayed multiplied by 1,000 as proposed by Greenacre (2007), enabling the display of more columns without wrapping output. The compact tables can be displayed without wrapping for models with two dimensions at line size 79 and with three dimensions at line size 99.

`plot` displays a plot of the row and column coordinates in two dimensions. With row principal normalization, only the row points are plotted. With column principal normalization, only the column points are plotted. In the other normalizations, both row and column points are plotted. You can use `cabiplot` directly if you need another selection of points to be plotted or if you want to otherwise refine the plot; see [MV] **ca postestimation**.

`maxlength(`*#*`)` specifies the maximum number of characters for row and column labels in plots. The default is `maxlength(12)`.

Note: the reporting options may be specified during estimation or replay.

Remarks

Remarks are presented under the following headings:

Introduction

Correspondence analysis (CA) offers a geometric representation of the rows and columns of a two-way frequency table that is helpful in understanding the similarities between the categories of variables and the association between the variables. For an informal introduction to CA and related metric approaches, see Weller and Romney (1990). Greenacre (2007) provides a much more thorough introduction with few mathematical prerequisites. More advanced treatments are given by Greenacre (1984) and Gower and Hand (1996).

In some respects, CA can be thought of as an analogue to principal components for nominal variables. It is also possible to interpret CA in reciprocal averaging (Greenacre 1984, 96–102; Cox and Cox 2001, 193–200), in optimal scaling (Greenacre 1984, 102–108), and in canonical correlations (Greenacre 1984, 108–116; Gower and Hand 1996, 183–185). Scaling refers to the assignment of scores to the categories of the row and column variables. Different criteria for the assignment of scores have been proposed, generally with different solutions. If the aim is to maximize the correlation between the scored row and column, the problem can be formulated in terms of CA. The optimal scores are the coordinates on the first dimension. The coordinates on the second and subsequent dimensions maximize the correlation between row and column scores subject to orthogonality constraints. See also [MV] **ca postestimation**.

A first example

▷ Example 1

We illustrate CA with an example of smoking behavior by different ranks of personnel. This example is often used in the CA literature (for example, Greenacre 1984, 55; Greenacre 2007, 66), so you have probably encountered these (artificial) data before. By using these familiar data, we make it easier to relate the literature on CA to the output of the ca command.

```
. use http://www.stata-press.com/data/r12/ca_smoking
. tabulate rank smoking
```

rank	smoking intensity none	light	medium	heavy	Total
senior_mngr	4	2	3	2	11
junior_mngr	4	3	7	4	18
senior_empl	25	10	12	4	51
junior_empl	18	24	33	13	88
secretary	10	6	7	2	25
Total	61	45	62	25	193

ca displays the results of a CA on two categorical variables in a multipanel format.

```
. ca rank smoking
Correspondence analysis                      Number of obs     =       193
                                             Pearson chi2(12)  =     16.44
                                             Prob > chi2       =    0.1718
                                             Total inertia     =    0.0852
    5 active rows                            Number of dim.    =         2
    4 active columns                         Expl. inertia (%) =     99.51

                 singular     principal                                  cumul
    Dimension      value       inertia            chi2    percent    percent

        dim 1   .2734211      .0747591           14.43      87.76      87.76
        dim 2   .1000859      .0100172            1.93      11.76      99.51
        dim 3   .0203365      .0004136            0.08       0.49     100.00

        total                 .0851899           16.44        100

Statistics for row and column categories in symmetric normalization

                        overall                       dimension_1
    Categories     mass  quality  %inert      coord    sqcorr  contrib

  rank
    senior mngr   0.057    0.893   0.031      0.126    0.092    0.003
    junior mngr   0.093    0.991   0.139     -0.495    0.526    0.084
    senior empl   0.264    1.000   0.450      0.728    0.999    0.512
    junior empl   0.456    1.000   0.308     -0.446    0.942    0.331
      secretary   0.130    0.999   0.071      0.385    0.865    0.070

  smoking
           none   0.316    1.000   0.577      0.752    0.994    0.654
          light   0.233    0.984   0.083     -0.190    0.327    0.031
         medium   0.321    0.983   0.148     -0.375    0.982    0.166
          heavy   0.130    0.995   0.192     -0.562    0.684    0.150

                        dimension_2
    Categories     coord   sqcorr  contrib

  rank
    senior mngr    0.612    0.800    0.214
    junior mngr    0.769    0.465    0.551
    senior empl    0.034    0.001    0.003
    junior empl   -0.183    0.058    0.152
      secretary   -0.249    0.133    0.081

  smoking
           none    0.096    0.006    0.029
          light   -0.446    0.657    0.463
         medium   -0.023    0.001    0.002
          heavy    0.625    0.310    0.506
```

The order in which we specify the variables is mostly immaterial. The first variable (`rank`) is also called the row variable, and the second (`smoking`) is the column variable. This ordering is important only as far as the interpretation of some options and some labeling of output are concerned. For instance, the option `norowpoints` suppresses the table with row points, that is, the categories of rank. `ca` requires two integer-valued variables. The rankings of the categories and the actual values used to code categories are not important. Thus, rank may be coded 1, 2, 3, 4, 5, or 0, 1, 4, 9, 16, or −2, −1, 0, 1, 2; it does not matter. We do suggest assigning value labels to the variables to improve the interpretability of tables and plots.

Correspondence analysis seeks to offer a low-dimensional representation describing how the row and column categories contribute to the inertia in a table. ca reports Pearson's test of independence, just like `tabulate` with the `chi2` option. Inertia is Pearson's chi-squared statistic divided by the sample size, $16.44/193 = 0.0852$. Pearson's chi-squared test has significance level $p = 0.1718$, casting doubt on any association between rows and columns. Still, given the prominence of this example in the CA literature, we will continue.

The first panel produced by ca displays the decomposition of total inertia in orthogonal dimensions—analogous to the decomposition of the total variance in principal component analysis (see [MV] **pca**). The first dimension accounts for 87.76% of the inertia; the second dimension accounts for 11.76% of the inertia. Because the dimensions are orthogonal, we may add the contributions of the two dimensions and say that the two leading dimensions account for $87.76\% + 11.76\% = 99.52\%$ of the total inertia. A two-dimensional representation seems in order. The remaining output is discussed later.

◁

How many dimensions?

▷ Example 2

In the first example with the smoking data, we displayed coordinates and statistics for a two-dimensional approximation of the rows and columns. This is the default. We can specify more or fewer dimensions with the option `dimensions()`. The maximum number is $\min(n_r - 1, n_c - 1)$. At this maximum, the chi-squared distances between the rows and columns are exactly represented by CA; 100% of the inertia is accounted for. This is called the saturated model; the fitted values of the CA model equal the observed correspondence table.

The minimum number of dimensions is one; the model with zero dimensions would be a model of independence of the rows and columns. With one dimension, the rows and columns of the table are identified by points on a line, with distance on the line approximating the chi-squared distance in the table, and a biplot is no longer feasible.

```
. ca rank smoking, dim(1)
Correspondence analysis                         Number of obs     =        193
                                                Pearson chi2(12)  =      16.44
                                                Prob > chi2       =     0.1718
                                                Total inertia     =     0.0852
    5 active rows                               Number of dim.    =          1
    4 active columns                            Expl. inertia (%) =      87.76

                 |  singular    principal                            cumul
       Dimension |   value       inertia         chi2    percent    percent
    -------------+----------------------------------------------------------
           dim 1 |  .2734211    .0747591        14.43      87.76      87.76
           dim 2 |  .1000859    .0100172         1.93      11.76      99.51
           dim 3 |  .0203365    .0004136         0.08       0.49     100.00
    -------------+----------------------------------------------------------
           total |              .0851899        16.44        100
```

```
Statistics for row and column categories in symmetric normalization

                        overall                   dimension_1
   Categories     mass  quality  %inert     coord   sqcorr  contrib

rank
  senior mngr    0.057    0.092   0.031     0.126    0.092    0.003
  junior mngr    0.093    0.526   0.139    -0.495    0.526    0.084
  senior empl    0.264    0.999   0.450     0.728    0.999    0.512
  junior empl    0.456    0.942   0.308    -0.446    0.942    0.331
    secretary    0.130    0.865   0.071     0.385    0.865    0.070

smoking
         none    0.316    0.994   0.577     0.752    0.994    0.654
        light    0.233    0.327   0.083    -0.190    0.327    0.031
       medium    0.321    0.982   0.148    -0.375    0.982    0.166
        heavy    0.130    0.684   0.192    -0.562    0.684    0.150
```

The first panel produced by ca does not depend on the number of dimensions extracted; thus, we will always see all singular values and the percentage of inertia explained by the associated dimensions. In the second panel, the only thing that depends on the number of dimensions is the overall quality of the approximation. The overall quality is the sum of the quality scores on the extracted dimensions and so increases with the number of extracted dimensions. The higher the quality, the better the chi-squared distances with other rows (columns) are represented by the extracted number of dimensions. In a saturated model, the overall quality is 1 for each row and column category.

So, how many dimensions should we retain? It is common for researchers to extract the minimum number of dimensions in a CA to explain at least 90% of the inertia, analogous to similar heuristic rules on the number of components in principal component analysis. We could probably also search for a scree, the number of dimensions where the singular values flatten out (see [MV] **screeplot**). A screeplot of the singular values can be obtained by typing

```
. screeplot e(Sv)
  (output omitted)
```

where e(Sv) is the name where ca has stored the singular values.

◁

Statistics on the points

▷ Example 3

We now turn our attention to the second panel. The overall section of the panel lists the following statistics:

- The mass of the category, that is, the proportion in the marginal distribution. The masses of all categories of a variable add up to 1.
- The quality of the approximation for a category, expressed as a number between 0 (very bad) and 1 (perfect). In a saturated model, quality is 1.
- The percentage of inertia contained in the category. Categories are divided through by the total inertia; the inertias of the categories of a variable add up to 100%.

For each of the dimensions, the panel lists the following:

- The coordinate of the category.

- The squared residuals between the profile and the categories. The sum of the squared residuals over the dimensions adds up to the quality of the approximation for the category.
- The contribution made by the categories to the dimensions. These add up to 1 over all categories of a variable.

The table with point statistics becomes pretty large, especially with more than two dimensions. ca can also list the second panel in a more compact form, saving space by multiplying all entries by 1,000; see Greenacre (2007).

```
. ca rank smoking, dim(2) compact
Correspondence analysis                        Number of obs     =        193
                                               Pearson chi2(12)  =      16.44
                                               Prob > chi2       =     0.1718
                                               Total inertia     =     0.0852
    5 active rows                              Number of dim.    =          2
    4 active columns                           Expl. inertia (%) =      99.51

                  singular    principal                          cumul
    Dimension      value       inertia        chi2    percent   percent

        dim 1    .2734211     .0747591       14.43      87.76     87.76
        dim 2    .1000859     .0100172        1.93      11.76     99.51
        dim 3    .0203365     .0004136        0.08       0.49    100.00

        total                 .0851899       16.44        100
Statistics for row and column categories in symmetric norm. (x 1000)
                    overall           dimension 1          dimension 2
    Categories  mass qualt %inert  coord sqcor contr  coord sqcor contr

  rank
   senior mngr    57   893    31    126    92     3    612   800   214
   junior mngr    93   991   139   -495   526    84    769   465   551
   senior empl   264  1000   450    728   999   512     34     1     3
   junior empl   456  1000   308   -446   942   331   -183    58   152
     secretary   130   999    71    385   865    70   -249   133    81

  smoking
          none   316  1000   577    752   994   654     96     6    29
         light   233   984    83   -190   327    31   -446   657   463
        medium   321   983   148   -375   982   166    -23     1     2
         heavy   130   995   192   -562   684   150    625   310   506
```

◁

Normalization and interpretation of correspondence analysis

The normalization method used in CA determines whether and how the similarity of the row categories, the similarity of the column categories, and the relationship (association) between the row and column variables can be interpreted in terms of the row and column coordinates and the origin of the plot.

How does one compare row points—provided that the normalization method allows such a comparison? Formally, the Euclidean distance between the row points approximates the chi-squared distances between the corresponding row profiles. Thus in the biplot, row categories mapped close together have similar row profiles; that is, the distributions on the column variable are similar. Row categories mapped widely apart have dissimilar row profiles. Moreover, the Euclidean distance between a row point and the origin approximates the chi-squared distance from the row profile and the row centroid, so it indicates how different a category is from the population.

An analogous interpretation applies to column points.

For the association between the row and column variables: in the CA biplot, you should not interpret the distance between a row point r and a column point c as the relationship of r and c. Instead, think in terms of the vectors origin to r (OR) and origin to c (OC). Remember that CA decomposes scaled deviations $d(r,c)$ from independence and $d(r,c)$ is approximated by the inner product of OR and OC. The larger the absolute value of $d(r,c)$, the stronger the association between r and c. In geometric terms, $d(r,c)$ can be written as the product of the length of OR, the length of OC, and the cosine of the angle between OR and OC.

What does this mean? First, consider the effects of the angle. The association in (r,c) is strongly positive if OR and OC point in roughly the same direction; the frequency of (r,c) is much higher than expected under independence, so r tends to flock together with c—if the points r and c are close together. Similarly, the association is strongly negative if OR and OC point in opposite directions. Here the frequency of (r,c) is much lower than expected under independence, so r and c are unlikely to occur simultaneously. Finally, if OR and OC are roughly orthogonal (angle = ± 90), the deviation from independence is small.

Second, the association of r and c increases with the lengths of OR and OC. Points far from the origin tend to have large associations. If a category is mapped close to the origin, all its associations with categories of the other variable are small: its distribution resembles the marginal distribution.

Here are the interpretations enabled by the main normalization methods as specified in the `normalize()` option.

Normalization method	Similarity row cat.	Similarity column cat.	Association row vs. column
`symmetric`	No	No	Yes
`principal`	Yes	Yes	No
`row`	Yes	No	Yes
`column`	No	Yes	Yes

If we say that a comparison between row categories or between column categories is not possible, we really mean that the chi-squared distance between row profiles or column profiles is actually approximated by a weighted Euclidean distance between the respective plots in which the weights depend on the inertia of the dimensions rather than on the standard Euclidean distance.

You may want to do a CA in principal normalization to study the relationship between the categories of a variable and do a CA in symmetric normalization to study the association of the row and column categories.

Plotting the points

▷ Example 4

In our discussion of normalizations, we stated that CA offers simple geometric interpretations to the similarity of categories and the association of the variables. We may specify the option `plot` with `ca` during estimation or during replay.

```
. ca, norowpoint nocolpoint plot
Correspondence analysis                    Number of obs      =       193
                                           Pearson chi2(12)   =     16.44
                                           Prob > chi2        =    0.1718
                                           Total inertia      =    0.0852
    5 active rows                          Number of dim.     =         2
    4 active columns                       Expl. inertia (%)  =     99.51

                 |  singular    principal                          cumul
       Dimension |   value       inertia       chi2    percent    percent
    -------------+-----------------------------------------------------
           dim 1 |  .2734211    .0747591      14.43      87.76      87.76
           dim 2 |  .1000859    .0100172       1.93      11.76      99.51
           dim 3 |  .0203365    .0004136       0.08       0.49     100.00
    -------------+-----------------------------------------------------
           total |              .0851899      16.44        100
```

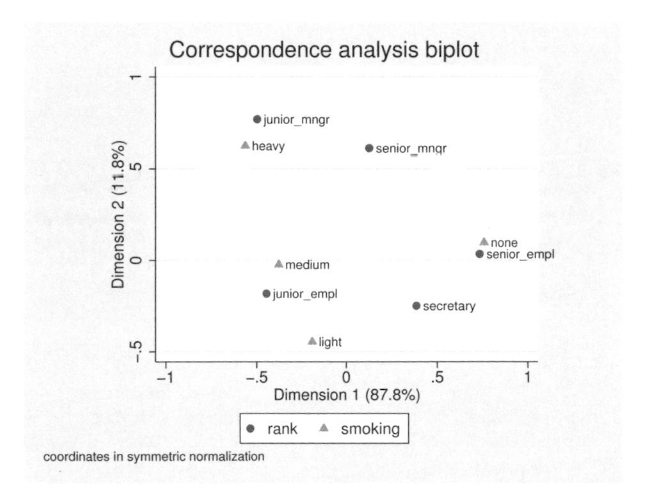

The options `norowpoint` and `nocolpoint` suppress the large tables of statistics for the rows and columns. If we did not request the plot during estimation, we can still obtain it with the `cabiplot` postestimation command. Unlike requesting the plot at estimation time, `cabiplot` allows us to fine-tune the plot; see [MV] **ca postestimation**.

The horizontal dimension seems to distinguish smokers from nonsmokers, whereas the vertical dimensions can be interpreted as intensity of smoking. Because the orientations from the origin to `none` and from the origin to `senior_empl` are so close, we conclude that senior employees tend not to smoke. Similarly, junior managers tend to be heavy smokers, and junior employees tend to be medium smokers.

◁

Supplementary points

A useful feature of CA is the ability to locate supplementary rows and columns in the space generated by the "active" rows and columns (see Greenacre [1984, 70–74]; Greenacre [2007, chap. 12], for an extensive discussion). Think of supplementary rows and columns as having mass 0; therefore, supplementary points do not influence the approximating space—their contribution values are zero.

▷ Example 5

In our example, we want to include the national distribution of smoking intensity as a supplementary row.

ca requires that we define the supplementary row distributions as rows of a matrix. In this example, we have only one supplementary row, with the percentages of the smoking categories in a national sample. The matrix should have one row per supplementary row category and as many columns as there are active columns. We define the row name to obtain appropriately labeled output.

```
. matrix S_row = ( 42, 29, 20, 9 )
. matrix rowname S_row = national
```

Before we show the CA analysis with the supplementary row, we also include two supplementary columns for the rank distribution of alcoholic beverage drinkers and nondrinkers. It will be interesting to see where smoking is located relative to drinking and nondrinking.

```
. matrix S_col = (  0, 11 \
                    1, 19 \
                    5, 44 \
                   10, 78 \
                    7, 18 )
. matrix colnames S_col = nondrink drink
```

We now invoke ca, specifying the names of the matrices with supplementary rows and columns with the options rowsupp() and colsupp().

```
. ca rank smoking, rowsupp(S_row) colsupp(S_col) plot
Correspondence analysis                           Number of obs     =       193
                                                  Pearson chi2(12)  =     16.44
                                                  Prob > chi2       =    0.1718
                                                  Total inertia     =    0.0852
    5 active + 1 supplementary rows               Number of dim.    =         2
    4 active + 2 supplementary columns            Expl. inertia (%) =     99.51

                 singular   principal                            cumul
    Dimension      value      inertia        chi2    percent    percent

        dim 1   .2734211    .0747591        14.43      87.76      87.76
        dim 2   .1000859    .0100172         1.93      11.76      99.51
        dim 3   .0203365    .0004136         0.08       0.49     100.00

        total               .0851899        16.44        100
```

Statistics for row and column categories in symmetric normalization

Categories	overall mass	overall quality	overall %inert	dimension_1 coord	dimension_1 sqcorr	dimension_1 contrib
rank						
senior mngr	0.057	0.893	0.031	0.126	0.092	0.003
junior mngr	0.093	0.991	0.139	-0.495	0.526	0.084
senior empl	0.264	1.000	0.450	0.728	0.999	0.512
junior empl	0.456	1.000	0.308	-0.446	0.942	0.331
secretary	0.130	0.999	0.071	0.385	0.865	0.070
suppl_rows						
national	0.518	0.761	0.644	0.494	0.631	
smoking						
none	0.316	1.000	0.577	0.752	0.994	0.654
light	0.233	0.984	0.083	-0.190	0.327	0.031
medium	0.321	0.983	0.148	-0.375	0.982	0.166
heavy	0.130	0.995	0.192	-0.562	0.684	0.150
suppl_cols						
nondrink	0.119	0.439	0.460	0.220	0.040	
drink	0.881	0.838	0.095	-0.082	0.202	

Categories	dimension_2 coord	dimension_2 sqcorr	dimension_2 contrib
rank			
senior mngr	0.612	0.800	0.214
junior mngr	0.769	0.465	0.551
senior empl	0.034	0.001	0.003
junior empl	-0.183	0.058	0.152
secretary	-0.249	0.133	0.081
suppl_rows			
national	-0.372	0.131	
smoking			
none	0.096	0.006	0.029
light	-0.446	0.657	0.463
medium	-0.023	0.001	0.002
heavy	0.625	0.310	0.506
suppl_cols			
nondrink	-1.144	0.398	
drink	0.241	0.636	

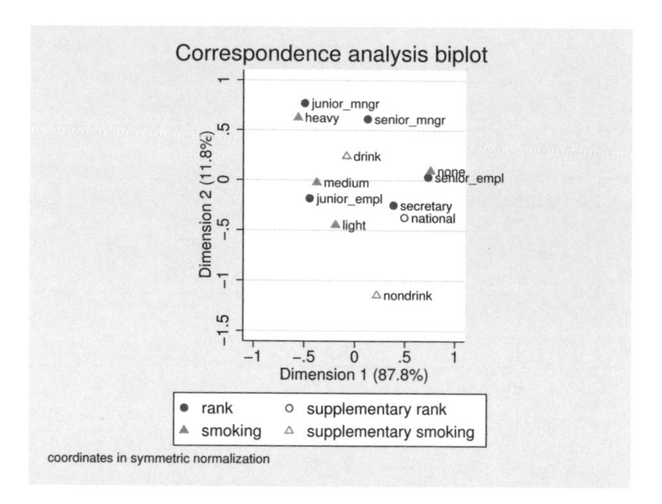

The first panel and the information about the five active rows and the four active columns have not changed—the approximating space is fully determined by the active rows and columns and is independent of the location of the supplementary rows and columns.

The table with statistics for the row and column categories now also contains entries for the supplementary rows and columns. The `contrib` entries for the supplementary points are blank. Supplementary points do not “contribute to” the location of the dimensions—their contribution is 0.000, but displaying blanks makes the point more clearly. All other columns for the supplementary points are informative. The inertia of supplementary points is the chi-squared distance to the respective centroid. The coordinates of supplementary points are obtained by applying the transition equations of the CA. Correlations of the supplementary profiles with the dimensions are also well defined. Finally, we may consider the quality of the two-dimensional approximation for the supplementary points. These are lower than for the active points, which will be the case in most applications—the active points exercise influence on the dimensions to improve their quality, whereas the supplementary points simply have to accept the dimensions as determined by the active points.

If we look at the biplot, the supplementary points are shown along with the active points. We may interpret the supplementary points just like the active points. Secretaries are close to the national sample in terms of smoking. Drinking alcohol is closer to the smoking categories than to nonsmoking, indicating that alcohol consumption and smoking are similar behaviors—but concluding that the *same* people smoke and drink is not possible because we do not have three-way data.

◁

Matrix input

▷ Example 6

If we want to do a CA of a published two-way frequency table, we typically do not have immediate access to the data in the form of a dataset. We could enter the data with frequency weights.

```
. input rank smoking freq
  1.        1         1    4
  2.        1         2    2
  3.        1         3    3
 (output omitted)
 19.        5         3    7
 20.        5         4    2
 21.   end
. label define vl_rank  1  "senior_mngr" ...
. label value rank vl_rank
. label define vl_smoke 1  "none" ...
. label value smoke vl_smoke
. ca rank smoking [fw=freq]
 (output omitted)
```

Or we may enter the data as a matrix and use `camat`. First, we enter the frequency matrix with proper column and row names and then list the matrix for verification.

```
. matrix F = ( 4,2,3,2 \ 4,3,7,4 \ 25,10,12,4 \ 18,24,33,13 \ 10,6,7,2 )
. matrix colnames F = none light medium heavy
. matrix rownames F = senior_mngr junior_mngr senior_empl junior_empl secretary
. matlist F, border
```

	none	light	medium	heavy
senior_mngr	4	2	3	2
junior_mngr	4	3	7	4
senior_empl	25	10	12	4
junior_empl	18	24	33	13
secretary	10	6	7	2

We can use `camat` on `F` to obtain the same results as from the raw data. We use the `compact` option for a more compact table.

```
. camat F, compact
Correspondence analysis                         Number of obs     =       193
                                                Pearson chi2(12)  =     16.44
                                                Prob > chi2       =    0.1718
                                                Total inertia     =    0.0852
    5 active rows                               Number of dim.    =         2
    4 active columns                            Expl. inertia (%) =     99.51

                 singular    principal                                cumul
    Dimension     value       inertia          chi2     percent    percent

        dim 1    .2734211    .0747591          14.43      87.76      87.76
        dim 2    .1000859    .0100172           1.93      11.76      99.51
        dim 3    .0203365    .0004136           0.08       0.49     100.00

        total                .0851899          16.44        100
Statistics for row and column categories in symmetric norm. (x 1000)

                        overall          dimension 1         dimension 2
    Categories    mass qualt %inert   coord sqcor contr   coord sqcor contr

    rows
     senior mngr    57   893     31     126    92     3     612   800   214
     junior mngr    93   991    139    -495   526    84     769   465   551
     senior empl   264  1000    450     728   999   512      34     1     3
     junior empl   456  1000    308    -446   942   331    -183    58   152
       secretary   130   999     71     385   865    70    -249   133    81

    columns
            none   316  1000    577     752   994   654      96     6    29
           light   233   984     83    -190   327    31    -446   657   463
          medium   321   983    148    -375   982   166     -23     1     2
           heavy   130   995    192    -562   684   150     625   310   506
```

◁

▷ Example 7

The command camat may also be used for a CA of nonfrequency data. The data should be nonnegative, with strictly positive margins. An example are the compositional data on the distribution of government R&D funds over 11 areas in five European countries in 1989; the data are listed in Greenacre (1993, 82). The expenditures are scaled to 1,000 within country, to focus the analysis on the intranational distribution policies. Moreover, with absolute expenditures, small countries, such as The Netherlands, would have been negligible in the analysis.

We enter the data as a Stata matrix. The command matrix input (see [P] **matrix define**) allows us to input row entries separated by blanks, rather than by commas; rows are separated by the backward slash (\).

```
. matrix input RandD = (
     18  19  14  14   6 \
     12  34   4  15  31 \
     44  33  36  58  25 \
     37  88  67 101  40 \
     42  20  36  28  43 \
     90 156 107 224 176 \
     28  50  59  88  28 \
    165 299 120 303 407 \
     48 128 147  62 103 \
    484 127 342  70  28 \
     32  46  68  37 113 )
```

```
. matrix colnames RandD = Britain West_Germany France Italy Netherlands

. matrix rownames RandD = earth_exploration pollution human_health
                          energy agriculture industry space university
                          nonoriented defense other
```

We perform a CA, suppressing the voluminous row- and column-point statistics. We want to show a biplot, and therefore we select symmetric normalization.

```
. camat RandD, dim(2) norm(symm) rowname(source) colname(country) norowpoints
> nocolpoints plot

Correspondence analysis                          Number of obs     =      5000
                                                 Pearson chi2(40)  =   1321.55
                                                 Prob > chi2       =    0.0000
                                                 Total inertia     =    0.2643
    11 active rows                               Number of dim.    =         2
    5 active columns                             Expl. inertia (%) =     89.08

                 |   singular    principal                              cumul
       Dimension |      value      inertia        chi2     percent    percent
    -------------+------------------------------------------------------------
           dim 1 |    .448735     .2013631     1006.82       76.18      76.18
           dim 2 |   .1846219     .0340852      170.43       12.90      89.08
           dim 3 |   .1448003     .0209671      104.84        7.93      97.01
           dim 4 |   .0888532     .0078949       39.47        2.99     100.00
    -------------+------------------------------------------------------------
           total |                .2643103     1321.55        100
```

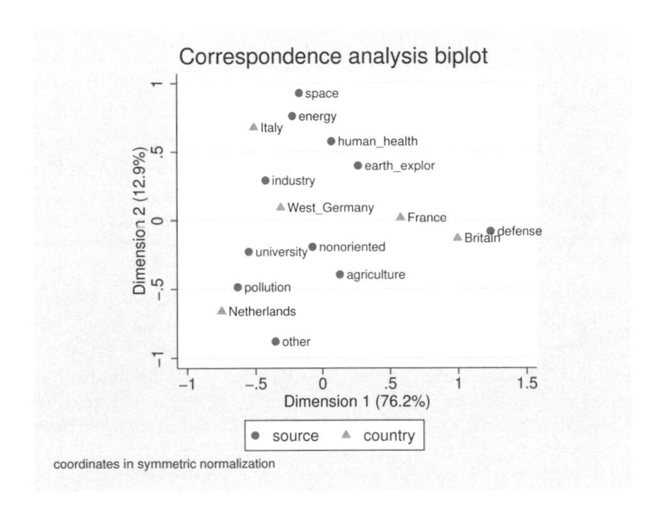

The two dimensions account for 89% of the inertia in this example, justifying an interpretation of the biplot. Let us focus on the position of The Netherlands. The orientation of The Netherlands from the origin is in the same direction as the orientation of pollution and university from the origin, indicating that The Netherlands spends more on academic research and on research to reduce environmental pollution than the average country. Earth exploration and human health are in the opposite direction, indicating investments much lower than average in these areas. Industry and agriculture are approximately orthogonal to the orientation of The Netherlands, indicating average investments by The Netherlands in these areas. Britain and France have big military investments, whereas Germany and Italy have more of an industrial orientation.

◁

❑ Technical note

The interpretation of the biplot is not fully in line with what we easily see in the row and column profiles—surprisingly, Greenacre does not seem to feel the need to comment on this. Why is this the case? The clue is in the statistics we did not show. Although the two dimensions account for 90% of the total inertia, this does not mean that all rows and columns are approximated to this extent. There are some row and column categories that are not well described in two dimensions. For instance, the quality of the source categories nonoriented, agriculture, and earth_exploration are only 0.063, 0.545, and 0.584, respectively, indicating that these rows are poorly represented in a two-dimensional space. The quality of West_Germany is also rather low at 0.577. Adding a third dimension improves the quality of the category nonoriented but hardly affects the other two problematic categories. This effect can be seen only from the squared correlations between the third dimension and the profiles of the row and column categories—these correlations are small for all categories but nonoriented. Thus, nonoriented does not seem to really belong with the other categories and should probably be omitted from the analysis.

❑

Crossed variables

ca can include interactions between variables in the analysis; variables that contain interactions are called crossed or stacked variables, whereas the variables that make them up are the crossing or stacking variables.

▷ Example 8

We illustrate crossed variables with ca by using the ISSP (1993) data from [MV] **mca**, which explores attitudes toward science and the environment. We are interested in whether responses to item A differ with education and gender. The item asks for a response to the statement "We believe too often in science, and not enough in feelings or faith," with a 1 indicating strong agreement and a 5 indicating strong disagreement. We are interested in how education and gender influence response. We cross the variables sex and edu into one demographic variable labeled demo to explore this question.

```
. use http://www.stata-press.com/data/r12/issp93
(Selection from ISSP (1993))
. tabulate A edu
```

too much science, not enough feelings&faith	education (6 categories) primary i	primary c	secondary	secondary	Total
agree strongly	7	59	29	11	119
agree	15	155	84	27	322
neither agree nor dis	7	84	65	18	204
disagree	8	68	54	26	178
disagree strongly	1	12	10	12	48
Total	38	378	242	94	871

too much science, not enough feelings&faith	education (6 categories) tertiary	tertiary	Total
agree strongly	5	8	119
agree	20	21	322
neither agree nor dis	11	19	204
disagree	8	14	178
disagree strongly	5	8	48
Total	49	70	871

We notice immediately the long labels for variable A and on edu. We replace these labels with short labels that can be abbreviated, so that in our analysis we will easily be able to identify categories. We use the length(2) option to ca to ensure that labels from each of the crossing variables are restricted to two characters.

```
. label define response 1 "++" 2 "+" 3 "+/-" 4 "-" 5 "--"
. label values A response
. label define education 1 "-pri" 2 "pri" 3 "-sec" 4 "sec" 5 "-ter" 6 "ter"
. label values edu education
. ca A (demo: sex edu), norowpoints nocolpoints length(2) plot norm(symmetric)
Correspondence analysis                        Number of obs     =        871
                                               Pearson chi2(44)  =      72.52
                                               Prob > chi2       =     0.0043
                                               Total inertia     =     0.0833
    5 active rows                              Number of dim.    =          2
    12 active columns                          Expl. inertia (%) =      80.17

                 |  singular    principal                                cumul
      Dimension  |     value      inertia        chi2    percent       percent
    -------------+--------------------------------------------------------------
          dim 1  |  .2108455     .0444558       38.72      53.39         53.39
          dim 2  |    .14932     .0222965       19.42      26.78         80.17
          dim 3  |  .1009876     .0101985        8.88      12.25         92.42
          dim 4  |  .0794696     .0063154        5.50       7.58        100.00
    -------------+--------------------------------------------------------------
          total  |               .0832662       72.52        100
```

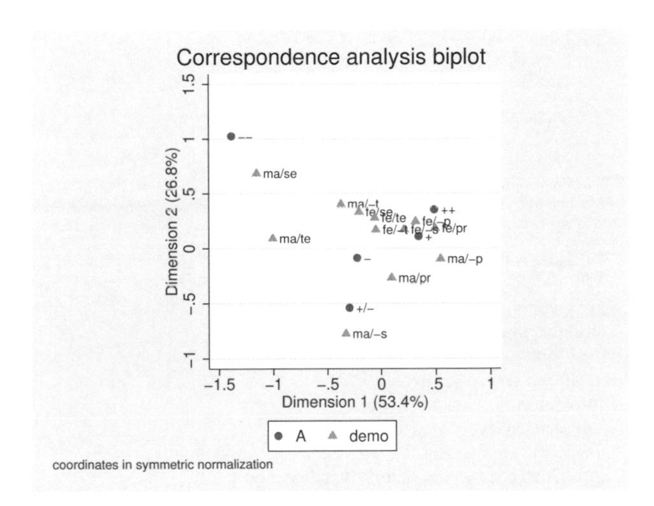

We see clearly that the responses of the males vary more widely than those of the females. Strong agreement with item A is most closely associated with females with little education, and strong disagreement is most closely associated with males with a secondary or tertiary education. Educated males are more closely associated with a negative response than educated females are, and females with little education are more closely associated with a positive response than males with little education are.

◁

Saved results

Let r be the number of rows, c be the number of columns, and f be the number of retained dimensions. ca and camat save the following in e():

Scalars

e(N)	number of observations
e(f)	number of dimensions (factors, axes); maximum of $\min(r-1,c-1)$
e(inertia)	total inertia = e(X2)/e(N)
e(pinertia)	inertia explained by e(f) dimensions
e(X2)	χ^2 statistic
e(X2_df)	degrees of freedom $(r-1)(c-1)$
e(X2_p)	p-value for e(X2)

Macros

e(cmd)	ca (even for camat)
e(cmdline)	command as typed
e(Rcrossvars)	row crossing variable names (ca only)
e(Ccrossvars)	column crossing variable names (ca only)
e(varlist)	the row and column variable names (ca only)
e(wtype)	weight type (ca only)
e(wexp)	weight expression (ca only)
e(title)	title in estimation output
e(ca_data)	variables or crossed
e(Cname)	name for columns
e(Rname)	name for rows
e(norm)	normalization method
e(sv_unique)	1 if the singular values are unique, 0 otherwise
e(properties)	nob noV eigen
e(estat_cmd)	program used to implement estat
e(predict)	program used to implement predict
e(marginsnotok)	predictions disallowed by margins

Matrices

e(Ccoding)	column categories $(1\times c)$ (ca only)
e(Rcoding)	row categories $(1\times r)$ (ca only)
e(GSC)	column statistics $(c\times 3(1+f))$
e(GSR)	row statistics $(r\times 3(1+f))$
e(TC)	normalized column coordinates $(c\times f)$
e(TR)	normalized row coordinates $(r\times f)$
e(Sv)	singular values $(1\times f)$
e(C)	column coordinates $(c\times f)$
e(R)	row coordinates $(r\times f)$
e(c)	column mass (margin) $(c\times 1)$
e(r)	row mass (margin) $(r\times 1)$
e(P)	analyzed matrix $(r\times c)$
e(GSC_supp)	supplementary column statistics
e(GSR_supp)	supplementary row statistics
e(PC_supp)	principal coordinates supplementary column points
e(PR_supp)	principal coordinates supplementary row points
e(TC_supp)	normalized coordinates supplementary column points
e(TR_supp)	normalized coordinates supplementary row points

Functions

e(sample)	marks estimation sample (ca only)

Methods and formulas

ca and camat are implemented as ado-files.

Our presentation of simple CA follows that of Greenacre (1984, 83–125); see also Blasius and Greenacre (1994) and Rencher (2002, 514–530). See Greenacre and Blasius (1994) for a concise presentation of CA from a computational perspective. Simple CA seeks a geometric representation of the rows and column of a (two mode) matrix with nonnegative entries in a common low-dimensional space so that chi-squared distances between the rows and between the columns are well approximated by the Euclidean distances in the common space.

Let $\mathbf{N}$ be an $I \times J$ matrix with nonnegative entries and strictly positive margins. $\mathbf{N}$ may be frequencies of a two-way cross-tabulation, but this is not assumed in most of CA. Let $n = N_{++}$ be the overall sum of N_{ij} ("number of observations"). Define the *correspondence table* as the matrix $\mathbf{P}$ where $P_{ij} = N_{ij}/n$, so the overall sum of P_{ij} is $P_{++} = 1$. Let $\mathbf{r} = \mathbf{P}\,\mathbf{1}$ be the row margins, also known as the *row masses*, with elements $r_i > 0$. Similarly, $\mathbf{c} = \mathbf{P}'\mathbf{1}$ contains the column margins, or *column masses*, with elements $c_j > 0$.

CA is defined in terms of the generalized singular value decomposition (GSVD) of $\mathbf{P} - \mathbf{rc}'$ with respect to the inner products normed by $\mathbf{D}_r^{-1}$ and $\mathbf{D}_c^{-1}$, where $\mathbf{D}_r = \text{diag}(\mathbf{r})$ and $\mathbf{D}_c = \text{diag}(\mathbf{c})$. The GSVD can be expressed in terms of the orthonormal (or standard) SVD of the standardized residuals

$$\mathbf{Z} = \mathbf{D}_r^{-\frac{1}{2}}(\mathbf{P} - \mathbf{rc}')\mathbf{D}_c^{-\frac{1}{2}} \quad \text{with elements} \quad Z_{ij} = \frac{P_{ij} - r_i c_j}{\sqrt{r_i c_j}}$$

Denote by $\mathbf{Z} = \mathbf{R\Lambda C}'$ the SVD of $\mathbf{Z}$ with $\mathbf{R}'\mathbf{R} = \mathbf{C}'\mathbf{C} = \mathbf{I}$ and $\mathbf{\Lambda}$ a diagonal matrix with singular values in decreasing order. `ca` displays a warning message if $\mathbf{Z}$ has common singular values.

The *total principal inertia* of the correspondence table $\mathbf{P}$ is defined as $\chi^2/n = \sum_{i,j} Z_{ij}^2$, where χ^2 is Pearson's chi-squared statistic. We can express the inertia of $\mathbf{P}$ in terms of the singular values of $\mathbf{Z}$:

$$\text{inertia} = \frac{1}{n}\chi^2 = \sum_{k=1}^{\min(I-1,J-1)} \lambda_k^2$$

The inertia accounted for by d dimensions is $\sum_{k=1}^{d} \lambda_k^2$. The fraction of inertia accounted for (explained) by the d dimensions is defined as

$$\text{explained inertia} = \frac{\sum_{k=1}^{d} \lambda_k^2}{\sum_{k=1}^{\min(I-1,J-1)} \lambda_k^2}$$

Principal row ($\widetilde{R}_{ik}$) and principal column ($\widetilde{C}_{jk}$) coordinates are defined as

$$\widetilde{R}_{ik} = \frac{R_{ik}\lambda_k}{\sqrt{r_i}} = (\mathbf{D}_r^{-\frac{1}{2}}\mathbf{R\Lambda})_{ik} \qquad \widetilde{C}_{jk} = \frac{C_{jk}\lambda_k}{\sqrt{c_j}} = (\mathbf{D}_c^{-\frac{1}{2}}\mathbf{C\Lambda})_{jk}$$

The α-normalized row and column coordinates are defined as

$$R_{ik}^{(\alpha)} = \frac{R_{ik}\lambda_k^{\alpha}}{\sqrt{r_i}} \qquad C_{jk}^{(\alpha)} = \frac{C_{jk}\lambda_k^{1-\alpha}}{\sqrt{c_j}}$$

The row principal coordinates are obtained with $\alpha = 1$. The column principal coordinates are obtained with $\alpha = 0$. The symmetric coordinates are obtained with $\alpha = 1/2$.

Decomposition of inertia by rows ($\text{In}^{(r)}$) and by columns ($\text{In}^{(c)}$) is defined as

$$\text{In}_i^{(r)} = \sum_{j=1}^{J} Z_{ij}^2 \qquad \text{In}_j^{(c)} = \sum_{i=1}^{I} Z_{ij}^2$$

Quality of subspace approximations for the row and column categories are defined as

$$Q_i^{(r)} = \frac{r_i}{\text{In}_i^{(r)}} \sum_{k=1}^{d} \widetilde{R}_{ik}^2 \qquad Q_j^{(c)} = \frac{c_j}{\text{In}_j^{(c)}} \sum_{k=1}^{d} \widetilde{C}_{jk}^2$$

If $d = \min(I-1, J-1)$, the quality index satisfies $Q_i^{(r)} = Q_j^{(c)} = 1$.

CA provides several diagnostics for a more detailed analysis of inertia: what do the categories contribute to the inertia explained by the dimensions, and what do the dimensions contribute to the inertia explained for the categories?

The relative contributions of row i ($G_{ik}^{(r)}$) and of column j ($G_{jk}^{(c)}$) to the inertia of principal dimension k are defined as

$$G_{ik}^{(r)} = \frac{r_i \widetilde{R}_{ik}}{\lambda_k^2} \qquad G_{jk}^{(c)} = \frac{c_j \widetilde{C}_{jk}}{\lambda_k^2}$$

$G_{+k}^{(r)} = G_{+k}^{(c)} = 1$.

The correlations $H_{ik}^{(r)}$ of the ith row profile and kth principal row dimension and, analogously, $H_{jk}^{(c)}$ for columns are

$$H_{ik}^{(r)} = \frac{r_i}{\text{In}_i^{(r)}} \widetilde{R}_{ik}^2 \qquad H_{jk}^{(c)} = \frac{c_j}{\text{In}_j^{(c)}} \widetilde{C}_{jk}^2$$

We now define the quantities returned by the `estat` subcommands after `ca`. The row profiles are $\mathbf{U} = \mathbf{D}_r^{-1}\mathbf{P}$. The chi-squared distance between rows i_1 and i_2 of $\mathbf{P}$ is defined as the Mahalanobis distance between the respective row profiles $\mathbf{U}_{i_1}$ and $\mathbf{U}_{i_2}$ with respect to $\mathbf{D}_c$,

$$(\mathbf{U}_{i_1} - \mathbf{U}_{i_2})\mathbf{D}_c^{-1}(\mathbf{U}_{i_1} - \mathbf{U}_{i_2})'$$

The column profiles and the chi-squared distances between columns are defined analogously. The chi-squared distances for the approximated correspondence table are defined analogously in terms of $\widehat{\mathbf{P}}$.

The fitted or reconstructed values $\widehat{P}_{ij}$ are

$$\widehat{P}_{ij} = r_i c_j \left(1 + \lambda_k^{-1} \sum_{k=1}^{d} \widetilde{R}_{ik} \widetilde{C}_{jk}\right)$$

References

Blasius, J., and M. J. Greenacre. 1994. Computation of correspondence analysis. In *Correspondence Analysis in the Social Sciences*, ed. M. Greenacre and J. Blasius. London: Academic Press.

Cox, T. F., and M. A. A. Cox. 2001. *Multidimensional Scaling*. 2nd ed. Boca Raton, FL: Chapman & Hall/CRC.

Gower, J. C., and D. J. Hand. 1996. *Biplots*. London: Chapman & Hall.

Greenacre, M. J. 1984. *Theory and Applications of Correspondence Analysis*. London: Academic Press.

——. 1993. *Correspondence Analysis in Practice*. London: Academic Press.

——. 2007. *Correspondence Analysis in Practice*. 2nd ed. Boca Raton, FL: Chapman & Hall/CRC.

Greenacre, M. J., and J. Blasius, ed. 1994. *Correspondence Analysis in the Social Sciences*. London: Academic Press.

ISSP. 1993. International Social Survey Programme: Environment. http://www.issp.org.

Milan, L., and J. Whittaker. 1995. Application of the parametric bootstrap to models that incorporate a singular value decomposition. *Applied Statistics* 44: 31–49.

Rencher, A. C. 2002. *Methods of Multivariate Analysis*. 2nd ed. New York: Wiley.

Van Kerm, P. 1998. sg78: Simple and multiple correspondence analysis in Stata. *Stata Technical Bulletin* 42: 32–37. Reprinted in *Stata Technical Bulletin Reprints*, vol. 7, pp. 210–217. College Station, TX: Stata Press.

Weller, S. C., and A. K. Romney. 1990. *Metric Scaling: Correspondence Analysis*. Newbury Park, CA: Sage.

Also see

[MV] **ca postestimation** — Postestimation tools for ca and camat

[MV] **mca** — Multiple and joint correspondence analysis

[R] **tabulate twoway** — Two-way tables of frequencies

[U] **20 Estimation and postestimation commands**

Title

ca postestimation — Postestimation tools for ca and camat

Description

The following postestimation commands are of special interest after `ca` and `camat`:

Command	Description
`cabiplot`	biplot of row and column points
`caprojection`	CA dimension projection plot
`estat coordinates`	display row and column coordinates
`estat distances`	display χ^2 distances between row and column profiles
`estat inertia`	display inertia contributions of the individual cells
`estat loadings`	display correlations of profiles and axes
`estat profiles`	display row and column profiles
† `estat summarize`	estimation sample summary
`estat table`	display fitted correspondence table
`screeplot`	plot singular values

† `estat summarize` is not available after `camat`.

For information about these commands, except for `screeplot`, see below. For information about `screeplot`, see [MV] **screeplot**.

The following standard postestimation commands are also available:

Command	Description
* `estimates`	cataloging estimation results
† `predict`	fitted values, row coordinates, or column coordinates

* All `estimates` subcommands except `table` and `stats` are available.

† `predict` is not available after `camat`.

See the corresponding entries in the *Base Reference Manual* for details.

Special-interest postestimation commands

`cabiplot` produces a plot of the row points or column points, or a biplot of the row and column points. In this plot, the (Euclidean) distances between row (column) points approximates the χ^2 distances between the associated row (column) profiles if the CA is properly normalized. Similarly, the association between a row and column point is approximated by the inner product of vectors from the origin to the respective points (see [MV] **ca**).

`caprojection` produces a line plot of the row and column coordinates. The goal of this graph is to show the ordering of row and column categories on each principal dimension of the analysis. Each principal dimension is represented by a vertical line; markers are plotted on the lines where the row and column categories project onto the dimensions.

estat coordinates displays the row and column coordinates.

estat distances displays the χ^2 distances between the row profiles and between the column profiles. Also, the χ^2 distances between the row and column profiles to the respective centers (marginal distributions) are displayed. Optionally, the fitted profiles rather than the observed profiles are used.

estat inertia displays the inertia (χ^2/N) contributions of the individual cells.

estat loadings displays the correlations of the row and column profiles and the axes, comparable to the loadings of principal component analysis.

estat profiles displays the row and column profiles; the row (column) profile is the conditional distribution of the row (column) given the column (row). This is equivalent to specifying the row and column options with the tabulate command; see [R] **tabulate twoway**.

estat summarize displays summary information about the row and column variables over the estimation sample.

estat table displays the fitted correspondence table. Optionally, the observed "correspondence table" and the expected table under independence are displayed.

Syntax for predict

predict [*type*] *newvar* [*if*] [*in*] [, *statistic*]

statistic	Description
Main	
fit	fitted values; the default
rowscore(#)	row score for dimension #
colscore(#)	column score for dimension #

predict is not available after camat.

Menu

Statistics > Postestimation > Predictions, residuals, etc.

Options for predict

Main

fit specifies that fitted values for the correspondence analysis model be computed. fit displays the fitted values p_{ij} according to the correspondence analysis model. fit is the default.

rowscore(#) generates the row score for dimension #, that is, the appropriate elements from the normalized row coordinates.

colscore(#) generates the column score for dimension #, that is, the appropriate elements from the normalized column coordinates.

Syntax for estat

Display row and column coordinates

estat coordinates [, norow nocolumn format(%*fmt*)]

Display chi-squared distances between row and column profiles

estat distances [, norow nocolumn approx format(%*fmt*)]

Display inertia contributions of cells

estat inertia [, total noscale format(%*fmt*)]

Display correlations of profiles and axes

estat loadings [, norow nocolumn format(%*fmt*)]

Display row and column profiles

estat profiles [, norow nocolumn format(%*fmt*)]

Display summary information

estat summarize [, labels noheader noweights]

Display fitted correspondence table

estat table [, fit obs independence noscale format(%*fmt*)]

options	Description
norow	suppress display of row results
nocolumn	suppress display of column results
format(%*fmt*)	display format; default is format(%9.4f)
approx	display distances between fitted (approximated) profiles
total	add row and column margins
noscale	display χ^2 contributions; default is inertias $= \chi^2/N$ (with estat inertia)
labels	display variable labels
noheader	suppress the header
noweights	ignore weights
fit	display fitted values from correspondence analysis model
obs	display correspondence table ("observed table")
independence	display expected values under independence
noscale	suppress scaling of entries to 1 (with estat table)

Menu

Statistics > Postestimation > Reports and statistics

Options for estat

`norow`, an option used with `estat coordinates`, `estat distances`, and `estat profiles`, suppresses the display of row results.

`nocolumn`, an option used with `estat coordinates`, `estat distances`, and `estat profiles`, suppresses the display of column results.

`format(%`*fmt*`)`, an option used with many of the subcommands of `estat`, specifies the display format for the matrix, for example, `format(%8.3f)`. The default is `format(%9.4f)`.

`approx`, an option used with `estat distances`, computes distances between the fitted profiles. The default is to compute distances between the observed profiles.

`total`, an option used with `estat inertia`, adds row and column margins to the table of inertia or χ^2 (χ^2/N) contributions.

`noscale`, as an option used with `estat inertia`, displays χ^2 contributions rather than inertia ($= \chi^2/N$) contributions. (See below for the description of `noscale` with `estat table`.)

`labels`, an option used with `estat summarize`, displays variable labels.

`noheader`, an option used with `estat summarize`, suppresses the header.

`noweights`, an option used with `estat summarize`, ignores the weights, if any. The default when weights are present is to perform a weighted `summarize` on all variables except the weight variable itself. An unweighted `summarize` is performed on the weight variable.

`fit`, an option used with `estat table`, displays the fitted values for the correspondence analysis model. `fit` is implied if `obs` and `independence` are not specified.

`obs`, an option used with `estat table`, displays the observed table with nonnegative entries (the "correspondence table").

`independence`, an option used with `estat table`, displays the expected values p_{ij} assuming independence of the rows and columns, $p_{ij} = r_i c_j$, where r_i is the mass of row i and c_j is the mass of column j.

`noscale`, as an option used with `estat table`, normalizes the displayed tables to the sum of the original table entries. The default is to scale the tables to overall sum 1. (See above for the description of `noscale` with `estat inertia`.)

Syntax for cabiplot

`cabiplot` [, *options*]

options	Description
Main	
dim(*# #*)	the two dimensions to be displayed; default is dim(2 1)
norow	suppress row coordinates
nocolumn	suppress column coordinates
xnegate	negate the data relative to the x axis
ynegate	negate the data relative to the y axis
maxlength(*#*)	maximum number of characters for labels; default is maxlength(12)
origin	display the origin on the plot
originlopts(*line_options*)	affect rendition of origin axes
Rows	
rowopts(*row_opts*)	affect rendition of rows
Columns	
colopts(*col_opts*)	affect rendition of columns
Y axis, X axis, Titles, Legend, Overall	
twoway_options	any options other than by() documented in [G-3] ***twoway_options***

row_opts and *col_opts*	Description
plot_options	change look of markers (color, size, etc.) and look or position of marker labels
suppopts(*plot_options*)	change look of supplementary markers and look or position of supplementary marker labels

plot_options	Description
marker_options	change look of markers (color, size, etc.)
marker_label_options	add marker labels; change look or position

Menu

Statistics > Multivariate analysis > Correspondence analysis > Postestimation after CA > Biplot of row and column points

Options for cabiplot

Main

dim(*# #*) identifies the dimensions to be displayed. For instance, dim(3 2) plots the third dimension (vertically) versus the second dimension (horizontally). The dimension number cannot exceed the number of extracted dimensions. The default is dim(2 1).

norow suppresses plotting of row points.

`nocolumn` suppresses plotting of column points.

`xnegate` specifies that the x-axis values are to be negated (multiplied by -1).

`ynegate` specifies that the y-axis values are to be negated (multiplied by -1).

`maxlength(#)` specifies the maximum number of characters for row and column labels; the default is `maxlength(12)`.

`origin` specifies that the origin be displayed on the plot. This is equivalent to adding the options `xline(0, lcolor(black) lwidth(vthin)) yline(0, lcolor(black) lwidth(vthin))` to the `cabiplot` command.

`originlopts(`*line_options*`)` affects the rendition of the origin axes; see [G-3] ***line_options***.

Rows

`rowopts(`*row_opts*`)` affects the rendition of the rows. The following *row_opts* are allowed:

plot_options affect the rendition of row markers, including their shape, size, color, and outline (see [G-3] ***marker_options***) and specify if and how the row markers are to be labeled (see [G-3] ***marker_label_options***).

`suppopts(`*plot_options*`)` affects supplementary markers and supplementary marker labels; see above for description of *plot_options*.

Columns

`colopts(`*col_opts*`)` affects the rendition of columns. The following *col_opts* are allowed:

plot_options affect the rendition of column markers, including their shape, size, color, and outline (see [G-3] ***marker_options***) and specify if and how the column markers are to be labeled (see [G-3] ***marker_label_options***).

`suppopts(`*plot_options*`)` affects supplementary markers and supplementary marker labels; see above for description of *plot_options*.

Y axis, X axis, Titles, Legend, Overall

twoway_options are any of the options documented in [G-3] ***twoway_options***, excluding `by()`. These include options for titling the graph (see [G-3] ***title_options***) and for saving the graph to disk (see [G-3] ***saving_option***).

`cabiplot` automatically adjusts the aspect ratio on the basis of the range of the data and ensures that the axes are balanced. As an alternative, the *twoway_option* `aspectratio()` can be used to override the default aspect ratio. `cabiplot` accepts the `aspectratio()` option as a suggestion only and will override it when necessary to produce plots with balanced axes; that is, distance on the x axis equals distance on the y axis.

twoway_options, such as `xlabel()`, `xscale()`, `ylabel()`, and `yscale()` should be used with caution. These *axis_options* are accepted but may have unintended side effects on the aspect ratio. See [G-3] ***twoway_options***.

Syntax for caprojection

`caprojection` [, *options*]

options	Description
Main	
dim(*numlist*)	dimensions to be displayed; default is all
norow	suppress row coordinates
nocolumn	suppress column coordinates
alternate	alternate labels
maxlength(#)	number of characters displayed for labels; default is maxlength(12)
combine_options	affect the rendition of the combined column and row graphs
Rows	
rowopts(*row_opts*)	affect rendition of rows
Columns	
colopts(*col_opts*)	affect rendition of columns
Y axis, X axis, Titles, Legend, Overall	
twoway_options	any options other than by() documented in [G-3] ***twoway_options***

row_opts and *col_opts*	Description
plot_options	change look of markers (color, size, etc.) and look or position of marker labels
suppopts(*plot_options*)	change look of supplementary markers and look or position of supplementary marker labels

plot_options	Description
marker_options	change look of markers (color, size, etc.)
marker_label_options	add marker labels; change look or position

Menu

Statistics > Multivariate analysis > Correspondence analysis > Postestimation after CA > Dimension projection plot

Options for caprojection

Main

dim(*numlist*) identifies the dimensions to be displayed. By default, all dimensions are displayed.

norow suppresses plotting of rows.

nocolumn suppresses plotting of columns.

alternate causes adjacent labels to alternate sides.

`maxlength(#)` specifies the maximum number of characters for row and column labels; the default is `maxlength(12)`.

combine_options affect the rendition of the combined plot; see [G-2] **graph combine**. *combine_options* may not be specified with either `norow` or `nocolumn`.

Rows

`rowopts(`*row_opts*`)` affects the rendition of rows. The following *row_opts* are allowed:

plot_options affect the rendition of row markers, including their shape, size, color, and outline (see [G-3] ***marker_options***) and specify if and how the row markers are to be labeled (see [G-3] ***marker_label_options***).

`suppopts(`*plot_options*`)` affects supplementary markers and supplementary marker labels; see above for description of *plot_options*.

Columns

`colopts(`*col_opts*`)` affects the rendition of columns. The following *col_opts* are allowed:

plot_options affect the rendition of column markers, including their shape, size, color, and outline (see [G-3] ***marker_options***) and specify if and how the column markers are to be labeled (see [G-3] ***marker_label_options***).

`suppopts(`*plot_options*`)` affects supplementary markers and supplementary marker labels; see above for description of *plot_options*.

Y axis, X axis, Titles, Legend, Overall

twoway_options are any of the options documented in [G-3] ***twoway_options***, excluding `by()`. These include options for titling the graph (see [G-3] ***title_options***) and for saving the graph to disk (see [G-3] ***saving_option***).

Remarks

Remarks are presented under the following headings:

Postestimation statistics
Postestimation graphs
Predicting new variables

Postestimation statistics

After you conduct a correspondence analysis, there are several additional tables to help you understand and interpret your results. Some of these tables resemble tables produced by other Stata commands but are provided as part of the `ca` postestimation suite of commands for a unified presentation style.

▷ Example 1

We continue with the classic example of correspondence analysis, namely, the data on smoking in organizations. We extract only one dimension.

```
. use http://www.stata-press.com/data/r12/ca_smoking
. ca rank smoking, dim(1)
Correspondence analysis                        Number of obs      =       193
                                               Pearson chi2(12)   =     16.44
                                               Prob > chi2        =    0.1718
                                               Total inertia      =    0.0852
    5 active rows                              Number of dim.     =         1
    4 active columns                           Expl. inertia (%)  =     87.76

            |   singular   principal                                  cumul
  Dimension |    value      inertia         chi2     percent        percent
  ----------+---------------------------------------------------------------
      dim 1 |   .2734211   .0747591        14.43       87.76          87.76
      dim 2 |   .1000859   .0100172         1.93       11.76          99.51
      dim 3 |   .0203365   .0004136         0.08        0.49         100.00
  ----------+---------------------------------------------------------------
      total |              .0851899        16.44         100

Statistics for row and column categories in symmetric normalization

             |           overall          |          dimension_1
  Categories |  mass  quality    %inert   |   coord    sqcorr   contrib
  -----------+----------------------------+-----------------------------
  rank       |                            |
   senior mngr | 0.057   0.092    0.031   |   0.126    0.092    0.003
   junior mngr | 0.093   0.526    0.139   |  -0.495    0.526    0.084
   senior empl | 0.264   0.999    0.450   |   0.728    0.999    0.512
   junior empl | 0.456   0.942    0.308   |  -0.446    0.942    0.331
     secretary | 0.130   0.865    0.071   |   0.385    0.865    0.070
  -----------+----------------------------+-----------------------------
  smoking    |                            |
        none | 0.316   0.994    0.577     |   0.752    0.994    0.654
       light | 0.233   0.327    0.083     |  -0.190    0.327    0.031
      medium | 0.321   0.982    0.148     |  -0.375    0.982    0.166
       heavy | 0.130   0.684    0.192     |  -0.562    0.684    0.150
  -------------------------------------------------------------------
```

CA analyzes the similarity of row and of column categories by comparing the row profiles and the column profiles—some may prefer to talk about conditional distributions for a two-way frequency distribution, but CA is not restricted to this type of data.

```
. estat profiles
Row profiles (rows normalized to 1)

             |     none      light     medium      heavy  |      mass
  -----------+--------------------------------------------+----------
  senior mngr|   0.3636     0.1818     0.2727     0.1818  |    0.0570
  junior mngr|   0.2222     0.1667     0.3889     0.2222  |    0.0933
  senior empl|   0.4902     0.1961     0.2353     0.0784  |    0.2642
  junior empl|   0.2045     0.2727     0.3750     0.1477  |    0.4560
    secretary|   0.4000     0.2400     0.2800     0.0800  |    0.1295
  -----------+--------------------------------------------+----------
         mass|   0.3161     0.2332     0.3212     0.1295  |

Column profiles (columns normalized to 1)

             |     none      light     medium      heavy  |      mass
  -----------+--------------------------------------------+----------
  senior mngr|   0.0656     0.0444     0.0484     0.0800  |    0.0570
  junior mngr|   0.0656     0.0667     0.1129     0.1600  |    0.0933
  senior empl|   0.4098     0.2222     0.1935     0.1600  |    0.2642
  junior empl|   0.2951     0.5333     0.5323     0.5200  |    0.4560
    secretary|   0.1639     0.1333     0.1129     0.0800  |    0.1295
  -----------+--------------------------------------------+----------
         mass|   0.3161     0.2332     0.3212     0.1295  |
```

The tables also include the row and column masses—marginal probabilities. Two row categories are similar to the extent that their row profiles (that is, their distribution over the columns) are the same. Similar categories could be collapsed without distorting the information in the table. In CA, similarity or dissimilarity of the row categories is expressed in terms of the χ^2 distances between the rows. These are sums of squares, weighted with the inverse of the column masses. Thus a difference is counted "heavier" (inertia!) the smaller the respective column mass. In the table, we also add the χ^2 distances of the rows to the row centroid, that is, to the marginal distribution. This allows us to easily see which row categories are similar to each other as well as which row categories are similar to the population.

```
. estat distances, nocolumn
Chi2 distances between the row profiles
```

rank	junior_~r	senior_~l	junior_~l	secretary	center
senior_mngr	0.3448	0.3721	0.3963	0.3145	0.2166
junior_mngr		0.6812	0.3044	0.5622	0.3569
senior_empl			0.6174	0.2006	0.3808
junior_empl				0.4347	0.2400
secretary					0.2162

We see that senior employees are especially dissimilar from junior managers in terms of their smoking behavior but are rather similar to secretaries. Also the senior employees are least similar to the average staff member among all staff categories.

One of the goals of CA is to come up with a low-dimensional representation of the rows and columns in a common space. One way to see the adequacy of this representation is to inspect the implied approximation for the χ^2 distances—are the similarities between the row categories and between the column categories adequately represented in lower dimensions?

```
. estat distances, nocolumn approx
Chi2 distances between the dim=1 approximations of the row profiles
```

rank	junior_~r	senior_~l	junior_~l	secretary	center
senior_mngr	0.3247	0.3148	0.2987	0.1353	0.0658
junior_mngr		0.6396	0.0260	0.4600	0.2590
senior_empl			0.6135	0.1795	0.3806
junior_empl				0.4340	0.2330
secretary					0.2011

Some of the row distances are obviously poorly approximated, whereas the quality of other approximations is hardly affected. The dissimilarity in smoking behavior between junior managers and junior employees is particularly poorly represented in one dimension. From the CA with two dimensions, the second dimension is crucial to adequately represent the senior managers and the junior managers. By itself, this does not explain where the one-dimensional approximation fails; for this, we would have to take a closer look at the representation of the smoking categories as well.

A correspondence analysis can also be seen as equivalent to fitting the model

$$P_{ij} = r_i c_j (1 + R_{i1} C_{j1} + R_{i2} C_{j2} + \cdots)$$

to the correspondence table $\mathbf{P}$ by some sort of least squares, with parameters r_i, c_j, R_{ij}, and C_{jk}. We may compare the (observed) table $\mathbf{P}$ with the fitted table $\widehat{\mathbf{P}}$ to assess goodness of fit informally. Here we extract only one dimension, and so the fitted table is

$$\widehat{P}_{ij} = r_i c_j (1 + \widehat{R}_{i1} \widehat{C}_{j1})$$

with **R** and **C** the coordinates in symmetric (or row principal or column principal) normalization. We display the observed and fitted tables.

```
. estat table, fit obs
```

Correspondence table (normalized to overall sum = 1)

	none	light	medium	heavy
senior_mngr	0.0207	0.0104	0.0155	0.0104
junior_mngr	0.0207	0.0155	0.0363	0.0207
senior_empl	0.1295	0.0518	0.0622	0.0207
junior_empl	0.0933	0.1244	0.1710	0.0674
secretary	0.0518	0.0311	0.0363	0.0104

Approximation for dim = 1 (normalized to overall sum = 1)

	none	light	medium	heavy
senior_mngr	0.0197	0.0130	0.0174	0.0069
junior_mngr	0.0185	0.0238	0.0355	0.0154
senior_empl	0.1292	0.0531	0.0617	0.0202
junior_empl	0.0958	0.1153	0.1710	0.0738
secretary	0.0528	0.0280	0.0356	0.0132

Interestingly, some categories (for example, the junior employees, the nonsmokers, and the medium smokers) are very well represented in one dimension, whereas the quality of the fit of other categories is rather poor. This can, of course, also be inferred from the quality column in the `ca` output. We would consider the fit unsatisfactory and would refit the model with a second dimension.

◁

❑ Technical note

If the data are two-way cross-classified frequencies, as with `ca`, it may make sense to assume that the data are multinomial distributed, and the parameters can be estimated by maximum likelihood. The estimator has well-established properties in contrast to the estimation method commonly used in CA. One advantage is that sampling variability, for example, in terms of standard errors of the parameters, can be easily assessed. Also, the likelihood-ratio test against the saturated model may be used to select the number of dimensions to be extracted. See Van der Heijden and de Leeuw (1985).

❑

Postestimation graphs

In example 4 of [MV] **ca**, we showed that plots can be obtained simply by specifying the `plot` option during estimation (or replay). If the default plot is not exactly what you want, the **`cabiplot`** postestimation command provides control over the appearance of the plot.

▷ Example 2

For instance, if we constructed a CA in row principal normalization, we would want to look only at the (points for the) row categories, omitting the column categories. In this normalization, the Euclidean distances between the row points approximate the χ^2 distances between the corresponding row profiles, but the Euclidean distances between the column categories are a distortion of the χ^2 distances of the column profiles. We can use `cabiplot` with the `nocolumn` option to suppress the graphing of the column points.

```
. quietly ca rank smoking, norm(principal)
. cabiplot, nocolumn legend(on label(1 rank))
```

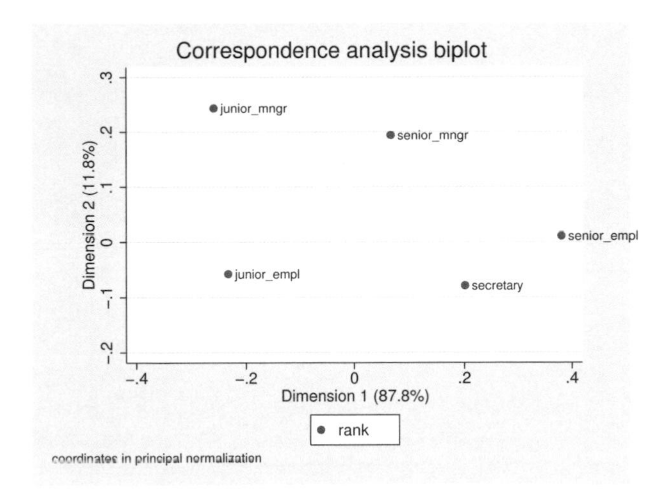

The default graph would not have provided a legend, so we included `legend(on label(1 rank))` to produce one. We see that secretaries have smoking behavior that is rather similar to that of senior employees but rather dissimilar to that of the junior managers, with the other two ranks taking intermediate positions. Because we actually specified the principal normalization, we may also interpret the distances between the smoking categories as approximations to χ^2 distances.

```
. cabiplot, norow legend(on label(1 smoking))
```

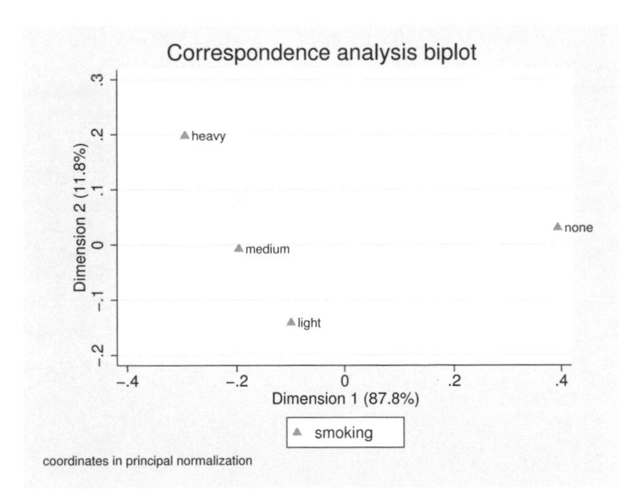

You may not like the orientation of the dimensions. For instance, in this plot, the smokers are on the left and the nonsmokers are on the right. It is more natural to locate the nonsmokers on the left and the smokers on the right so that smoking increases from left to right. This is accomplished with the `xnegate` option.

```
. cabiplot, xnegate norow legend(on label(1 smoking))
```

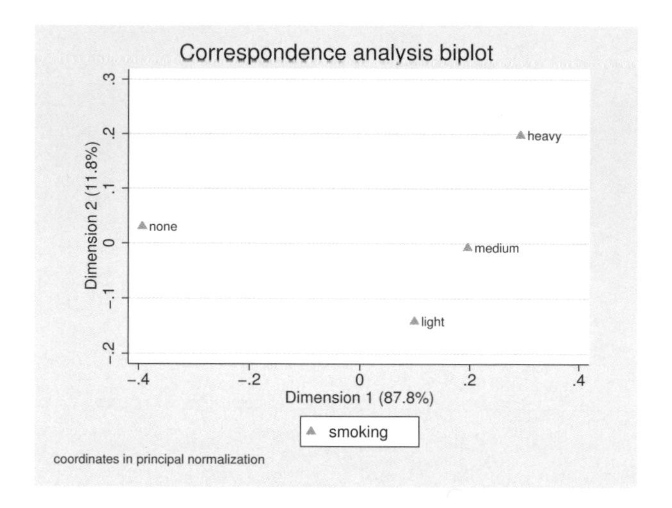

◁

❑ Technical note

To see that negating is possible, think in terms of the fitted values

$$\widehat{P}_{ij} = r_i c_j (1 + \widehat{R}_{i1}\widehat{C}_{j1} + \widehat{R}_{i2}\widehat{C}_{j2} + \cdots)$$

If the sign of the first column of $\mathbf{R}$ and $\mathbf{C}$ is changed at the same time, the fitted values are not affected. This is true for all CA statistics, and it holds true for other columns of $\mathbf{R}$ and $\mathbf{C}$ as well.

❑

▷ Example 3

Using the symmetric normalization allows us to display a biplot where row categories may be compared with column categories. We execute `ca` again, with the `normalize(symmetric)` option, but suppress the output. This normalization somewhat distorts the interpretation of the distances between row points (or column points) as approximations to χ^2 distances. Thus the similarity of the staff categories (or smoking categories) cannot be adequately assessed. However, this plot allows us to study the association between smoking and rank.

```
. quietly ca rank smoking, normalize(symmetric) dim(2)
. cabiplot, origin
```

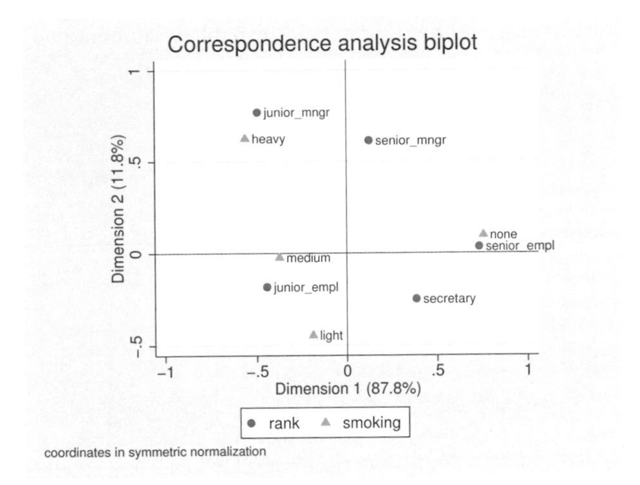

With this symmetric normalization, we do not interpret the distances between categories of `smoking` and `rank`. Rather, we have to think in terms of vectors from the origin. The inner product of vectors approximates the residuals from a model of independence of the rows and columns. The inner product depends on the lengths of the vectors and the (cosine of the) angle between the vectors. If the vectors point in the same direction, the residuals are positive—these row and column categories tend to occur together. In our example, we see that senior employees tend to be nonsmokers. If the vectors point in opposite directions, the residuals are negative—these row and column categories tend to be exclusive. In our example, senior managers tend not to be light smokers. Finally, if the vectors are orthogonal ($\pm$90 degrees), the residuals tend to be small; that is, the observed frequencies correspond to what we expect under independence. For instance, junior managers have an average rate of light smoking.

Using various graph options, we can enhance the look of the plot.

```
. cabiplot, origin subtitle("Fictitious data, N = 193")
> legend(pos(2) ring(0) col(1) lab(1 Employee rank) lab(2 Smoking status))
```

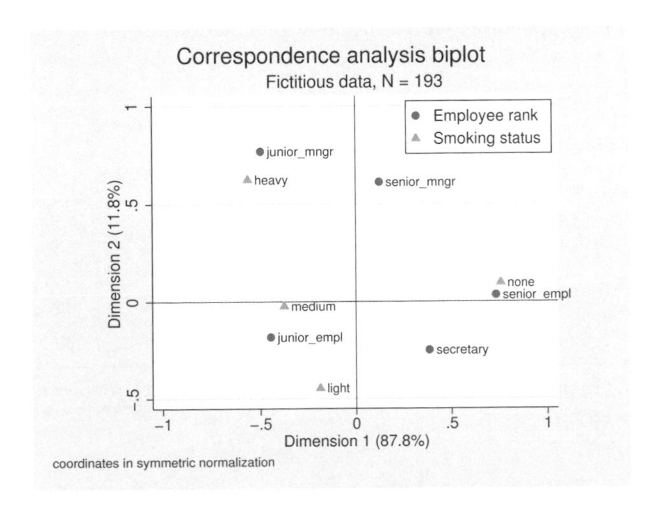

◁

▷ Example 4

`caprojection` produces a projection plot of the row and column coordinates after `ca` or `camat` and is especially useful if we think of CA as optimal scaling of the categories of the variables to maximize the correlations between the row and column variables. We continue where we left off with our previous example.

```
. caprojection
```

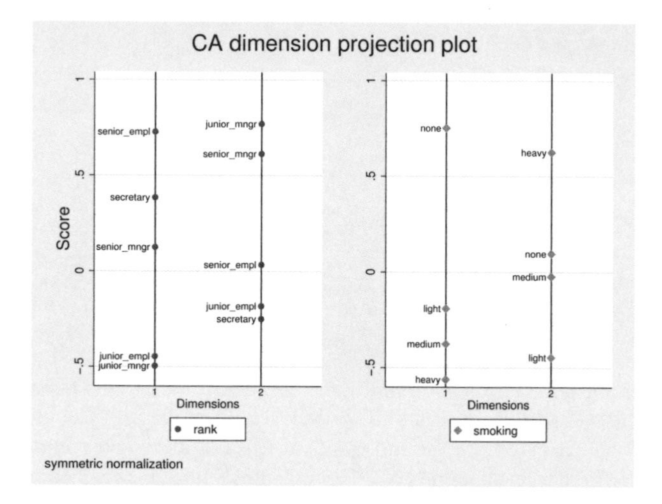

This example has relatively few categories, so we could visualize the orderings of the rows and columns from the previous biplots. However, CA is often used with larger problems, and in those cases, a projection plot is a useful presentation device.

◁

Predicting new variables

If you use `ca` to obtain the optimal scaling positions for the rows and columns, you may use `predict` to obtain the corresponding scores in the normalization used.

▷ Example 5

First, we obtain scores for the first dimension.

```
. quietly ca rank smoking, normalize(symmetric) dim(2)
. predict r1, row(1)
. predict c1, col(1)
. describe r1 c1
                storage  display     value
variable name     type   format      label      variable label
-------------------------------------------------------------------------
r1               float   %9.0g                  rank score(1) in symmetric norm.
c1               float   %9.0g                  smoking score(1) in symmetric
                                                  norm.
```

```
. correlate r1 c1
(obs=193)

             |       r1       c1
-------------+------------------
          r1 |   1.0000
          c1 |   0.2734   1.0000
```

The correlation of `r1` and `c1` is 0.2734, which equals the first singular value reported in the first panel by `ca`. In the same way, we may obtain scores for the second dimension.

```
. predict r2, row(2)

. predict c2, col(2)

. correlate r1 r2 c1 c2
(obs=193)

             |       r1       r2       c1       c2
-------------+------------------------------------
          r1 |   1.0000
          r2 |  -0.0000   1.0000
          c1 |   0.2734   0.0000   1.0000
          c2 |   0.0000   0.1001   0.0000   1.0000
```

The correlation between the row and column scores `r2` and `c2` for the second dimension is 0.1001, which is the same as the second singular value. Moreover, the row scores for dimensions 1 and 2 are not correlated, nor are the column scores.

◁

Obtaining the fitted values of the CA model is also possible,

$$\pi_{ij} = r_i c_j (1 + R_{i1} C_{i1} + R_{i2} C_{i2})$$

where **R** and **C** are the row and column scales in symmetric normalization. These may be used, say, to compute fit measures, for instance, from the Cressie–Read power family to analyze the fit of the CA model (Weesie 1997).

Saved results

`estat distances` saves the following in `r()`:

Matrices

`r(Dcolumns)`	χ^2 distances between the columns and between the columns and the column center
`r(Drows)`	χ^2 distances between the rows and between the rows and the row center

`estat inertia` saves the following in `r()`:

Matrices

`r(Q)`	matrix of (squared) inertia (or χ^2) contributions

`estat loadings` saves the following in `r()`:

Matrices

`r(LC)`	column loadings
`r(LR)`	row loadings

`estat profiles` saves the following in `r()`:

Matrices

`r(Pcolumns)`	column profiles (columns normalized to 1)
`r(Prows)`	row profiles (rows normalized to 1)

`estat table` saves the following in `r()`:

Matrices

`r(Fit)`	fitted (reconstructed) values
`r(Fit0)`	fitted (reconstructed) values, assuming independence of row and column variables
`r(Obs)`	correspondence table

Methods and formulas

All postestimation commands listed above are implemented as ado-files. See *Methods and formulas* in [MV] **ca** for information.

References

Van der Heijden, P. G. M., and J. de Leeuw. 1985. Correspondence analysis used complementary to loglinear analysis. *Psychometrika* 50: 429–447.

Weesie, J. 1997. sg68: Goodness-of-fit statistics for multinomial distributions. *Stata Technical Bulletin* 36: 26–28. Reprinted in *Stata Technical Bulletin Reprints*, vol. 6, pp. 183–186. College Station, TX: Stata Press.

Also see *References* in [MV] **ca**.

Also see

[MV] **ca** — Simple correspondence analysis

[MV] **screeplot** — Scree plot

Title

candisc — Canonical linear discriminant analysis

Syntax

`candisc` *varlist* [*if*] [*in*] [*weight*], `group(`*groupvar*`)` [*options*]

options	Description
Model	
*`group(`*groupvar*`)`	variable specifying the groups
`priors(`*priors*`)`	group prior probabilities
`ties(`*ties*`)`	how ties in classification are to be handled
Reporting	
`notable`	suppress resubstitution classification table
`lootable`	display leave-one-out classification table
`nostats`	suppress display of canonical statistics
`nocoef`	suppress display of standardized canonical discriminant function coefficients
`nostruct`	suppress display of canonical structure matrix
`nomeans`	suppress display of group means on canonical variables

priors	Description
`equal`	equal prior probabilities; the default
`proportional`	group-size-proportional prior probabilities
matname	row or column vector containing the group prior probabilities
matrix_exp	matrix expression providing a row or column vector of the group prior probabilities

ties	Description
`missing`	ties in group classification produce missing values; the default
`random`	ties in group classification are broken randomly
`first`	ties in group classification are set to the first tied group

*`group()` is required.

`statsby` and `xi` are allowed; see [U] **11.1.10 Prefix commands**.

`fweight`s are allowed; see [U] **11.1.6 weight**.

See [U] **20 Estimation and postestimation commands** for more capabilities of estimation commands.

Menu

Statistics > Multivariate analysis > Discriminant analysis > Canonical linear discriminant analysis

Description

candisc performs canonical linear discriminant analysis (LDA). What is computed is the same as with [MV] **discrim lda**. The difference is in what is presented. See [MV] **discrim** for other discrimination commands.

Options

Model

group(*groupvar*) is required and specifies the name of the grouping variable. *groupvar* must be a numeric variable.

priors(*priors*) specifies the prior probabilities for group membership. The following *priors* are allowed:

priors(equal) specifies equal prior probabilities. This is the default.

priors(proportional) specifies group-size-proportional prior probabilities.

priors(*matname*) specifies a row or column vector containing the group prior probabilities.

priors(*matrix_exp*) specifies a matrix expression providing a row or column vector of the group prior probabilities.

ties(*ties*) specifies how ties in group classification will be handled. The following *ties* are allowed:

ties(missing) specifies that ties in group classification produce missing values. This is the default.

ties(random) specifies that ties in group classification are broken randomly.

ties(first) specifies that ties in group classification are set to the first tied group.

Reporting

notable suppresses the computation and display of the resubstitution classification table.

lootable displays the leave-one-out classification table.

nostats suppresses the display of the table of canonical statistics.

nocoef suppresses the display of the standardized canonical discriminant function coefficients.

nostruct suppresses the display of the canonical structure matrix.

nomeans suppresses the display of group means on canonical variables.

Remarks

See [MV] **discrim** for background on discriminant analysis (classification) and see [MV] **discrim lda** for more information on linear discriminant analysis. What candisc displays by default with

```
. candisc x y z, group(group)
```

you can also obtain with the following sequence of discrim commands and estat postestimation commands.

```
. discrim x y z, group(group) notable
. estat canontest
. estat loadings
. estat structure
. estat grmeans, canonical
. estat classtable
```

The candisc command will appeal to those performing descriptive LDA.

▷ Example 1

Example 2 of [MV] **discrim knn** introduces a head-measurement dataset from Rencher (2002) that has six discriminating variables and three groups. The three groups are high school football players, college football players, and nonplayers. The data were collected as a preliminary step in determining the relationship between helmet design and neck injuries.

Descriptive discriminant analysis allows us to explore the relationship in this dataset between head measurements and the separability of the three groups.

```
. use http://www.stata-press.com/data/r12/head
(Table 8.3 Head measurements -- Rencher (2002))

. candisc wdim circum fbeye eyehd earhd jaw, group(group)

Canonical linear discriminant analysis

                                              Like-
         Canon.   Eigen-      Variance        lihood
  Fcn    Corr.    value     Prop.   Cumul.    Ratio       F       df1     df2  Prob>F

    1    0.8107  1.91776  0.9430  0.9430    0.3071  10.994      12     164  0.0000 e
    2    0.3223  .115931  0.0570  1.0000    0.8961  1.9245       5      83  0.0989 e

  Ho: this and smaller canon. corr. are zero;                    e = exact F

Standardized canonical discriminant function coefficients

                function1   function2

         wdim    .6206412    .9205834
       circum   -.0064715   -.0009114
        fbeye   -.0047581    -.021145
        eyehd   -.7188123    .5997882
        earhd   -.3965116   -.3018196
          jaw   -.5077218   -.9368745

Canonical structure

                function1   function2

         wdim    .1482946    .3766581
       circum   -.2714134    .1305383
        fbeye   -.1405813    -.061071
        eyehd    -.824502    .5363578
        earhd   -.5177312    .1146999
          jaw   -.2119042   -.3895934

Group means on canonical variables

        group   function1   function2

  high school   -1.910378   -.0592794
      college     1.16399   -.3771343
    nonplayer    .7463888    .4364137
```

```
Resubstitution classification summary

  +---------+
  | Key     |
  |---------|
  | Number  |
  | Percent |
  +---------+

             | Classified
  True group | high school      college    nonplayer |      Total
-------------+-------------------------------------+-----------
 high school |          26            1            3 |         30
             |       86.67         3.33        10.00 |     100.00
             |                                     |
     college |           1           20            9 |         30
             |        3.33        66.67        30.00 |     100.00
             |                                     |
   nonplayer |           2            8           20 |         30
             |        6.67        26.67        66.67 |     100.00
-------------+-------------------------------------+-----------
       Total |          29           29           32 |         90
             |       32.22        32.22        35.56 |     100.00
             |                                     |
      Priors |      0.3333       0.3333       0.3333 |
```

As seen in the canonical-correlation table, the first linear discriminant function accounts for almost 95% of the variance. The standardized discriminant function coefficients (loadings) indicate that two of the variables, `circum` (head circumference) and `fbeye` (front-to-back measurement at eye level), have little discriminating ability for these three groups. The first discriminant function is contrasting `wdim` (head width at widest dimension) to a combination of `eyehd` (eye-to-top-of-head measurement), `earhd` (ear-to-top-of-head measurement), and `jaw` (jaw width).

The canonical structure coefficients, which measure the correlation between the discriminating variables and the discriminant function, are also shown. There is controversy on whether the standardized loadings or the structure coefficients should be used for interpretation; see Rencher (2002, 291) and Huberty (1994, 262–264).

The group means on the canonical variables are shown, giving some indication of how the groups are separated. The means on the first function show the high school group separated farthest from the other two groups.

The resubstitution classification table, also known as a confusion matrix, indicates how many observations from each group are classified correctly or misclassified into the other groups. The college and nonplayer groups appear to have more misclassifications between them, indicating that these two groups are harder to separate.

All the postestimation tools of `discrim lda` are available after `candisc`; see [MV] **discrim lda postestimation**. For example, `estat grsummarize` can produce discriminating-variable summaries for each of our three groups.

```
. estat grsummarize

Estimation sample candisc
Summarized by group
```

Mean	group high school	college	nonplayer	Total
wdim	15.2	15.42	15.58	15.4
circum	58.937	57.37967	57.77	58.02889
fbeye	20.10833	19.80333	19.81	19.90722
eyehd	13.08333	10.08	10.94667	11.37
earhd	14.73333	13.45333	13.69667	13.96111
jaw	12.26667	11.94333	11.80333	12.00444
N	30	30	30	90

A score plot graphs observation scores from the first two discriminant functions; see [MV] **scoreplot**. After `candisc`, `scoreplot` automatically labels the points with the value labels assigned to the groups. The value labels for our three groups are long—the resulting graph is too crowded.

To overcome this, we create a new label language (see [D] **label language**), define one letter labels for the groups, assign this label to our group variable, and then call `scoreplot`. We then reset the label language back to the default containing the longer, more descriptive value labels.

```
. label language short, new
(language short now current language)
. label define fball 1 "H" 2 "C" 3 "X"
. label values group fball
. scoreplot, msymbol(i) aspect(.625)
. label language default
```

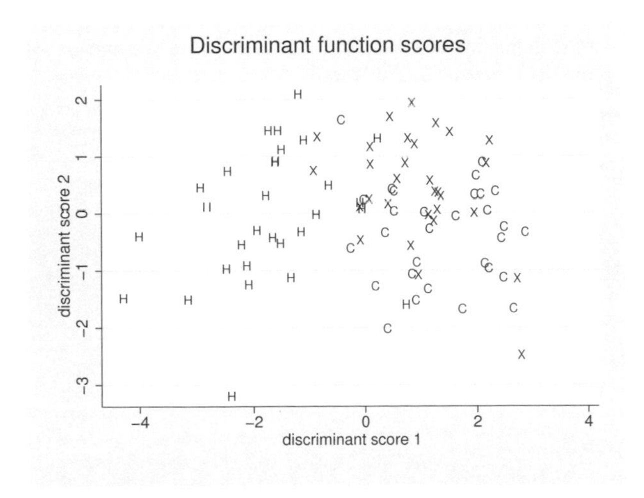

The score plot illustrates the separation due to the first and second canonical linear discriminant functions. As expected from our examination of the earlier descriptive output, the high school group (labeled `H`) is reasonably well separated from the college (labeled `C`) and nonplayer (labeled `X`) groups. There is some separation in the second dimension between the college and nonplayer groups, but with substantial overlap.

A loading plot provides a graphical way of looking at the standardized discriminant function coefficients (loadings) that we previously examined in tabular form.

```
. loadingplot
```

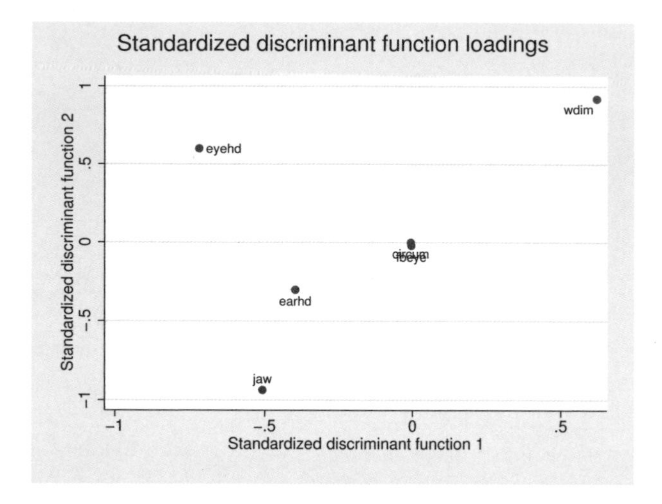

circum and fbeye are near the origin, indicating that they provide almost no discriminating ability in comparison to the other discriminating variables. The relative locations of the remaining variables indicate their contribution to the discriminant functions.

◁

Saved results

candisc saves the same items in e() as [MV] **discrim lda** with the exception that e(subcmd) is not set and the following e() results are different:

Macros

e(cmd)	candisc
e(title)	Canonical linear discriminant analysis

Methods and formulas

candisc is implemented as an ado-file.

See *Methods and formulas* in [MV] **discrim lda** for information.

References

Huberty, C. J. 1994. *Applied Discriminant Analysis*. New York: Wiley.

Rencher, A. C. 2002. *Methods of Multivariate Analysis*. 2nd ed. New York: Wiley.

Also see

[MV] **discrim lda postestimation** — Postestimation tools for discrim lda

[MV] **discrim lda** — Linear discriminant analysis

Title

canon — Canonical correlations

Syntax

`canon (`*varlist*$_1$`) (`*varlist*$_2$`)` [*if*] [*in*] [*weight*] [, *options*]

options	Description
Model	
lc(#)	calculate the linear combinations for canonical correlation #
first(#)	calculate the linear combinations for the first # canonical correlations
noconstant	do not subtract means when calculating correlations
Reporting	
stdcoef	output matrices of standardized coefficients
stderr	display raw coefficients and conditionally estimated standard errors
level(#)	set confidence level; default is level(95)
test(*numlist*)	display significance tests for the specified canonical correlations
notests	do not display tests
format(%*fmt*)	numerical format for coefficient matrices; default is format(%8.4f)

by and statsby are allowed; see [U] **11.1.10 Prefix commands**.
aweights and fweights are allowed; see [U] **11.1.6 weight**.
See [MV] **canon postestimation** for features available after estimation.

Menu

Statistics > Multivariate analysis > MANOVA, multivariate regression, and related > Canonical correlations

Description

canon estimates canonical correlations and provides the coefficients for calculating the appropriate linear combinations corresponding to those correlations.

canon typed without arguments redisplays previous estimation results.

Options

Model

lc(#) specifies that linear combinations for canonical correlation # be calculated. By default, all are calculated.

first(#) specifies that linear combinations for the first # canonical correlations be calculated. By default, all are calculated.

noconstant specifies that means not be subtracted when calculating correlations.

Reporting

`stdcoef` specifies that the first part of the output contain the standard coefficients of the canonical correlations in matrix form. The default is to present the raw coefficients of the canonical correlations in matrix form.

`stderr` specifies that the first part of the output contains the raw coefficients of the canonical correlations, the conditionally estimated standard errors, and the conditionally estimated confidence intervals in the standard estimation table. The default is to present the raw coefficients of the canonical correlations in matrix form.

`level(`*#*`)` specifies the confidence level, as a percentage, for confidence intervals of the coefficients. The default is `level(95)` or as set by `set level`; see [U] **20.7 Specifying the width of confidence intervals**. These "confidence intervals" are the result of an approximate calculation; see the technical note later in this entry.

`test(`*numlist*`)` specifies that significance tests of the canonical correlations in the *numlist* be displayed. Because of the nature of significance testing, if there are three canonical correlations, `test(1)` will test the significance of all three correlations, `test(2)` will test the significance of canonical correlations 2 and 3, and `test(3)` will test the significance of the third canonical correlation alone.

`notests` specifies that significance tests of the canonical correlation not be displayed.

`format(%`*fmt*`)` specifies the display format for numbers in coefficient matrices; see [D] **format**. `format(%8.4f)` is the default. `format()` may not be specified with `stderr`.

Remarks

Canonical correlations attempt to describe the relationships between two sets of variables. Given two sets of variables, $\mathbf{X} = (x_1, x_2, \dots, x_K)$ and $\mathbf{Y} = (y_1, y_2, \dots, y_L)$, the goal is to find linear combinations of $\mathbf{X}$ and $\mathbf{Y}$ so that the correlation between the linear combinations is as high as possible. That is, letting $\widehat{x}_1$ and $\widehat{y}_1$ be the linear combinations,

$$\widehat{x}_1 = \beta_{11}x_1 + \beta_{12}x_2 + \cdots + \beta_{1K}x_K$$

$$\widehat{y}_1 = \gamma_{11}y_1 + \gamma_{12}y_2 + \cdots + \gamma_{1L}y_L$$

you wish to find the maximum correlation between $\widehat{x}_1$ and $\widehat{y}_1$ as functions of the β's and the γ's. The second canonical correlation coefficient is defined as the ordinary correlation between

$$\widehat{x}_2 = \beta_{21}x_1 + \beta_{22}x_2 + \cdots + \beta_{2K}x_K \quad \text{and}$$

$$\widehat{y}_2 = \gamma_{21}y_1 + \gamma_{22}y_2 + \cdots + \gamma_{2L}y_L$$

This correlation is maximized subject to the constraints that $\widehat{x}_1$ and $\widehat{x}_2$, along with $\widehat{y}_1$ and $\widehat{y}_2$, are orthogonal and that $\widehat{x}_1$ and $\widehat{y}_2$, along with $\widehat{x}_2$ and $\widehat{y}_1$, are also orthogonal. The third and further correlations are defined similarly. There are $m = \min(K, L)$ such correlations.

Canonical correlation analysis originated with the work of Hotelling (1935, 1936). For an introduction, see Rencher (2002) or Johnson and Wichern (2007).

Example 1

Consider two scientists trying to describe how "big" a car is. The first scientist takes physical measurements—the length, weight, headroom, and trunk space—whereas the second takes mechanical measurements—the engine displacement, mileage rating, gear ratio, and turning circle. Can they agree on a conceptual framework?

```
. use http://www.stata-press.com/data/r12/auto
(1978 Automobile Data)

. canon (length weight headroom trunk) (displ mpg gear_ratio turn)

Canonical correlation analysis                        Number of obs =       74

Raw coefficients for the first variable set

             |        1          2          3          4
-------------+--------------------------------------------
      length |   0.0095     0.1441     0.0329     0.0212
      weight |   0.0010    -0.0037    -0.0010     0.0007
    headroom |   0.0351    -0.3701     1.5361    -0.0440
       trunk |  -0.0023    -0.0343    -0.2135    -0.3253
----------------------------------------------------------

Raw coefficients for the second variable set

             |        1          2          3          4
-------------+--------------------------------------------
displacement |   0.0054    -0.0125     0.0191    -0.0005
         mpg |  -0.0461    -0.0413     0.0683     0.2478
  gear_ratio |   0.0330     1.0280     3.6596    -1.0311
        turn |   0.0794     0.3113     0.0033     0.2240
----------------------------------------------------------

----------------------------------------------------------------------------
Canonical correlations:
  0.9476  0.3400  0.0634  0.0447

----------------------------------------------------------------------------
Tests of significance of all canonical correlations

                          Statistic      df1      df2            F     Prob>F
         Wilks' lambda     .0897314      16   202.271      15.1900    0.0000 a
        Pillai's trace      1.01956      16       276       5.9009    0.0000 a
Lawley-Hotelling trace      8.93344      16       258      36.0129    0.0000 a
    Roy's largest root      8.79667       4        69     151.7426    0.0000 u
----------------------------------------------------------------------------
                            e = exact, a = approximate, u = upper bound on F
```

By default, canon presents the raw coefficients of the canonical correlations in matrix form, reports the canonical correlations, and finally reports the tests of significance of all canonical correlations. The two views on car size are closely related: the best linear combination of the physical measurements is correlated at almost 0.95 with the best linear combination of the mechanical measurements. All the tests are significant.

To see the standardized coefficients instead of the raw coefficients, we can use the `stdcoef` option on replay, which gives the standardized coefficients in matrix form. We specify the `notests` option to suppress the display of tests this time.

```
. canon, stdcoef notests

Canonical correlation analysis                          Number of obs =        74
Standardized coefficients for the first variable set
             |        1          2          3          4
-------------+--------------------------------------------
      length |   0.2110     3.2095     0.7334     0.4714
      weight |   0.7898    -2.8469    -0.7448     0.5308
    headroom |   0.0297    -0.3131     1.2995    -0.0373
       trunk |  -0.0098    -0.1466    -0.9134    -1.3914
----------------------------------------------------------

Standardized coefficients for the second variable set
             |        1          2          3          4
-------------+--------------------------------------------
displacement |   0.4932    -1.1525     1.7568    -0.0493
         mpg |  -0.2670    -0.2388     0.3954     1.4337
  gear_ratio |   0.0150     0.4691     1.6698    -0.4705
        turn |   0.3493     1.3694     0.0145     0.9857
----------------------------------------------------------

------------------------------------------------------------------------------
Canonical correlations:
  0.9476  0.3400  0.0634  0.0447
```

◁

❑ Technical note

canon, with the stderr option, reports standard errors for the coefficients in the linear combinations; most other software does not. You should view these standard errors as lower bounds for the true standard errors. It is based on the assumption that the coefficients for one set of measurements are correct for calculating the coefficients and standard errors of the other relationship on the basis of a linear regression.

After canon, if you predict a canonical variate and regress it on the other variable set, the variance you get from the regression will be the variance you get from canon multiplied by the square of the corresponding canonical correlation.

❑

Saved results

canon saves the following in e():

Scalars

e(N)	number of observations
e(df_r)	residual degrees of freedom
e(df)	degrees of freedom
e(df1)	numerator degrees of freedom for significance tests
e(df2)	denominator degrees of freedom for significance tests
e(n_lc)	the linear combination calculated
e(n_cc)	number of canonical correlations calculated
e(rank)	rank of e(V)

Macros

e(cmd)	canon
e(cmdline)	command as typed
e(wtype)	weight type
e(wexp)	weight expression
e(properties)	b V
e(estat_cmd)	program used to implement estat
e(predict)	program used to implement predict
e(marginsnotok)	predictions disallowed by margins

Matrices

e(b)	coefficient vector
e(stat_#)	statistics for canonical correlation #
e(stat_m)	statistics for overall model
e(canload11)	canonical loadings for $varlist_1$
e(canload22)	canonical loadings for $varlist_2$
e(canload12)	correlation between $varlist_1$ and the canonical variates from $varlist_2$
e(canload21)	correlation between $varlist_2$ and the canonical variates from $varlist_1$
e(rawcoef_var1)	raw coefficients for $varlist_1$
e(rawcoef_var2)	raw coefficients for $varlist_2$
e(stdcoef_var1)	standardized coefficients for $varlist_1$
e(stdcoef_var2)	standardized coefficients for $varlist_2$
e(ccorr)	canonical correlation coefficients
e(corr_var1)	correlation matrix for $varlist_1$
e(corr_var2)	correlation matrix for $varlist_2$
e(corr_mixed)	correlation matrix between $varlist_1$ and $varlist_2$
e(V)	variance–covariance matrix of the estimators

Functions

e(sample)	marks estimation sample

Methods and formulas

canon is implemented as an ado-file.

Let the covariance matrix between the two sets of variables be

$$\begin{pmatrix} \mathbf{S_{yy}} & \mathbf{S_{yx}} \\ \mathbf{S_{xy}} & \mathbf{S_{xx}} \end{pmatrix}$$

Here $\mathbf{y}$ indicates the first variable set and $\mathbf{x}$ indicates the second variable set.

The squared canonical correlations are the eigenvalues of $\mathbf{V} = \mathbf{S}_{\mathbf{yy}}^{-1}\mathbf{S}_{\mathbf{yx}}\mathbf{S}_{\mathbf{xx}}^{-1}\mathbf{S}_{\mathbf{xy}}$ or $\mathbf{W} = \mathbf{S}_{\mathbf{xx}}^{-1}\mathbf{S}_{\mathbf{xy}}\mathbf{S}_{\mathbf{yy}}^{-1}\mathbf{S}_{\mathbf{yx}}$ (either will work), which are both nonsymmetric matrices (Rencher 1998, 312–317; Rencher 2002, 361–364). Let the eigenvalues of $\mathbf{V}$ (and $\mathbf{W}$) be called r_k, the eigenvectors of $\mathbf{V}$ be called $\mathbf{a}_k$, and the eigenvectors of $\mathbf{W}$ be called $\mathbf{b}_k$. These eigenvectors are the raw coefficients for calculating the canonical variates, which are the linear combinations for the two sets of variables with maximal correlation. The eigenvalue equation for $\mathbf{V}$ is

$$\mathbf{S}_{\mathbf{yy}}^{-1}\mathbf{S}_{\mathbf{yx}}\mathbf{S}_{\mathbf{xx}}^{-1}\mathbf{S}_{\mathbf{xy}}\mathbf{a}_k - r_k^2\mathbf{a}_k = 0$$

Premultiplying by $\mathbf{S}_{\mathbf{xx}}^{-1}\mathbf{S}_{\mathbf{xy}}$, we see that

$$(\mathbf{S}_{\mathbf{xx}}^{-1}\mathbf{S}_{\mathbf{xy}}\mathbf{S}_{\mathbf{yy}}^{-1}\mathbf{S}_{\mathbf{yx}})(\mathbf{S}_{\mathbf{xx}}^{-1}\mathbf{S}_{\mathbf{xy}}\mathbf{a}_k) - r_k^2\mathbf{S}_{\mathbf{xx}}^{-1}\mathbf{S}_{\mathbf{xy}}\mathbf{a}_k = 0$$

so the $\mathbf{b}_k$ are proportional to $\mathbf{S}_{\mathbf{xx}}^{-1}\mathbf{S}_{\mathbf{xy}}\mathbf{a}_k$. Eigenvectors are determined up to a scale factor, and we choose the eigenvectors to give canonical variates with variance one. The canonical variates with correlation r_k are given by

$$\mathbf{u}_k = \mathbf{a}_k\mathbf{x} \qquad \text{and} \qquad \mathbf{v}_k = \mathbf{b}_k\mathbf{y}$$

In fact

$$\mathbf{b}_k = \frac{1}{r_k}\mathbf{S}_{\mathbf{xx}}^{-1}\mathbf{S}_{\mathbf{xy}}\mathbf{a}_k$$

To calculate lower bounds for the standard errors in this form, assume that the eigenvectors $\mathbf{a}_k$ are fixed. The formula relating $\mathbf{a}_k$ and $\mathbf{b}_k$ is given above. The coefficients given by $\mathbf{b}_k$ have covariance matrix

$$\frac{1 - r_k^2}{r_k^2(n-k-1)}\mathbf{S}_{\mathbf{xx}}^{-1}$$

Here n is the number of observations and k is the number of variables in the set $\mathbf{x}$.

Likewise, we can let the correlation matrix between the two sets of variables be

$$\begin{pmatrix} \mathbf{R}_{\mathbf{yy}} & \mathbf{R}_{\mathbf{yx}} \\ \mathbf{R}_{\mathbf{xy}} & \mathbf{R}_{\mathbf{xx}} \end{pmatrix}$$

That is, $\mathbf{R}_{\mathbf{yy}}$ is the correlation matrix of the first set of variables with themselves, $\mathbf{R}_{\mathbf{xx}}$ is the correlation matrix of the second set of variables with themselves, and $\mathbf{R}_{\mathbf{yx}}$ (and $\mathbf{R}_{\mathbf{xy}}$) contains the cross-correlations.

Using correlation matrices, the squared canonical correlations are the eigenvalues of $\widetilde{\mathbf{V}} = \mathbf{R}_{\mathbf{yy}}^{-1}\mathbf{R}_{\mathbf{yx}}\mathbf{R}_{\mathbf{xx}}^{-1}\mathbf{R}_{\mathbf{xy}}$ or $\widetilde{\mathbf{W}} = \mathbf{R}_{\mathbf{xx}}^{-1}\mathbf{R}_{\mathbf{xy}}\mathbf{R}_{\mathbf{yy}}^{-1}\mathbf{R}_{\mathbf{yx}}$ (Rencher 1998, 318–319; Rencher 2002, 365). The corresponding eigenvectors are the standardized coefficients for determining the canonical variates from the centered and standardized original variables (mean 0 and variance 1). Eigenvectors are determined only up to a scale factor; we choose the scale to give the canonical variates in standardized (variance 1) form.

If the eigenvalues are $r_1, r_2, \ldots, r_m$ where m is the number of canonical correlations, we test the hypothesis that there is no (linear) relationship between the two variable sets. This is equivalent to the statement that none of the correlations $r_1, r_2, \ldots, r_m$ is significant.

Wilks' (1932) lambda statistic is

$$\Lambda_1 = \prod_{i=1}^{m}(1 - r_i^2)$$

and is a likelihood-ratio statistic. This statistic is distributed as the Wilks Λ-distribution. Rejection of the null hypothesis is for small values of Λ_1.

Pillai's (1955) trace for canonical correlations is

$$V^{(m)} = \sum_{i=1}^{m} r_i^2$$

and the Lawley–Hotelling trace (Lawley 1938 and Hotelling 1951) is

$$U^{(m)} = \sum_{i=1}^{m} \frac{r_i^2}{1 - r_i^2}$$

Roy's (1939) largest root is given by

$$\theta = r_1^2$$

Rencher (2002) has tables providing critical values for these statistics and discussion on significance testing for canonical correlations.

Canonical loadings, the correlation between a variable set and its corresponding canonical variate set, are calculated by `canon` and used in [MV] **canon postestimation**.

For a note about Harold Hotelling, see [MV] **hotelling**.

Acknowledgment

Significance testing of canonical correlations is based on the `cancor` package originally written by Philip B. Ender, UCLA.

References

Hotelling, H. 1935. The most predictable criterion. *Journal of Educational Psychology* 26: 139–142.

——. 1936. Relations between two sets of variates. *Biometrika* 28: 321–377.

——. 1951. A generalized t^2 test and measurement of multivariate dispersion. *Proceedings of the Second Berkeley Symposium on Mathematical Statistics and Probability* 1: 23–41.

Johnson, R. A., and D. W. Wichern. 2007. *Applied Multivariate Statistical Analysis*. 6th ed. Englewood Cliffs, NJ: Prentice Hall.

Lawley, D. N. 1938. A generalization of Fisher's z-test. *Biometrika* 30: 180–187.

Pillai, K. C. S. 1955. Some new test criteria in multivariate analysis. *Annals of Mathematical Statistics* 26: 117–121.

Rencher, A. C. 1998. *Multivariate Statistical Inference and Applications*. New York: Wiley.

——. 2002. *Methods of Multivariate Analysis*. 2nd ed. New York: Wiley.

Roy, S. N. 1939. p-statistics or some generalizations in analysis of variance appropriate to multivariate problems. *Sankhyā* 4: 381–396.

Wilks, S. S. 1932. Certain generalizations in the analysis of variance. *Biometrika* 24: 471–494.

——. 1962. *Mathematical Statistics*. New York: Wiley.

Also see

[MV] **canon postestimation** — Postestimation tools for canon

[MV] **factor** — Factor analysis

[MV] **pca** — Principal component analysis

[R] **correlate** — Correlations (covariances) of variables or coefficients

[R] **mvreg** — Multivariate regression

[R] **pcorr** — Partial and semipartial correlation coefficients

[R] **regress** — Linear regression

[U] **20 Estimation and postestimation commands**

Title

canon postestimation — Postestimation tools for canon

Description

The following postestimation commands are of special interest after canon:

Command	Description
estat correlations	show correlation matrices
estat loadings	show loading matrices
estat rotate	rotate raw coefficients, standard coefficients, or loading matrices
estat rotatecompare	compare rotated and unrotated coefficients or loadings
screeplot	plot canonical correlations

For information about estat correlation, estat loadings, estat rotate, and estat rotatecompare, see below.
For information about screeplot, see [MV] **screeplot**.

The following standard postestimation commands are also available:

Command	Description
estat	VCE and estimation sample summary
estimates	cataloging estimation results
lincom	point estimates, standard errors, testing, and inference for linear combinations of coefficients
nlcom	point estimates, standard errors, testing, and inference for nonlinear combinations of coefficients
predict	predictions, residuals, influence statistics, and other diagnostic measures
predictnl	point estimates, standard errors, testing, and inference for generalized predictions
test	Wald tests of simple and composite linear hypotheses
testnl	Wald tests of nonlinear hypotheses

See the corresponding entries in the *Base Reference Manual* for details.

Special-interest postestimation commands

estat correlations displays the correlation matrices calculated by canon for $varlist_1$ and $varlist_2$ and between the two lists.

estat loadings displays the canonical loadings computed by canon.

estat rotate performs orthogonal varimax rotation of the raw coefficients, standard coefficients, or canonical loadings. Rotation is calculated on the canonical loadings regardless of which coefficients or loadings are actually rotated.

estat rotatecompare displays the rotated and unrotated coefficients or loadings and the most recently rotated coefficients or loadings. This command may be used only if estat rotate has been performed first.

Syntax for predict

predict [*type*] *newvar* [*if*] [*in*], *statistic** [correlation(#)]

*statistic**	Description
Main	
u	calculate linear combination of $varlist_1$
v	calculate linear combination of $varlist_2$
stdu	calculate standard error of the linear combination of $varlist_1$
stdv	calculate standard error of the linear combination of $varlist_2$

* There is no default statistic; you must specify one *statistic* from the list.

These statistics are available both in and out of sample; type `predict ... if e(sample) ...` if wanted only for the estimation sample.

Menu

Statistics > Postestimation > Predictions, residuals, etc.

Options for predict

Main

u and v calculate the linear combinations of $varlist_1$ and $varlist_2$, respectively. For the first canonical correlation, u and v are the linear combinations having maximal correlation. For the second canonical correlation, specified in predict with the correlation(2) option, u and v have maximal correlation subject to the constraints that u is orthogonal to the u from the first canonical correlation, and v is orthogonal to the v from the first canonical correlation. The third and higher correlations are defined similarly. Canonical correlations may be chosen either with the lc() option to canon or by specifying the correlation() option to predict.

stdu and stdv calculate the standard errors of the respective linear combinations.

correlation(#) specifies the canonical correlation for which the requested statistic is to be computed. The default value for correlation() is 1. If the lc() option to canon was used to calculate a particular canonical correlation, then only this canonical correlation is in the estimation results. You can obtain estimates for it either by specifying correlation(1) or by omitting the correlation() option.

Syntax for estat

Display the correlation matrices

estat correlations [, format(%*fmt*)]

Display the canonical loadings

estat loadings [, format(%*fmt*)]

Perform orthogonal varimax rotation

estat rotate [, rawcoefs stdcoefs loadings format(%*fmt*)]

Display the rotated and unrotated coefficients or loadings

estat rotatecompare [, format(%*fmt*)]

Menu

Statistics > Postestimation > Reports and statistics

Option for estat

format(%*fmt*) specifies the display format for numbers in matrices; see [D] **format**. format(%8.4f) is the default.

rawcoefs, an option for estat rotate, requests the rotation of raw coefficients. It is the default.

stdcoefs, an option for estat rotate, requests the rotation of standardized coefficients.

loadings, an option for estat rotate, requests the rotation of the canonical loadings.

Remarks

In addition to the coefficients presented by canon in computing canonical correlations, several other matrices may be of interest.

▷ Example 1

Recall from canon the example of two scientists trying to describe how "big" a car is. One took physical measurements—the length, weight, headroom, and trunk space—whereas the second took mechanical measurements—engine displacement, mileage rating, gear ratio, and turning radius. We discovered that these two views are closely related, with the best linear combination of the two types of measurements, the largest canonical correlation, at 0.9476. We can prove that the first canonical correlation is correct by calculating the two linear combinations and then calculating the ordinary correlation.

```
. use http://www.stata-press.com/data/r12/auto
(1978 Automobile Data)
. quietly canon (length weight headroom trunk) (displ mpg gear_ratio turn)
. predict physical, u corr(1)
. predict mechanical, v corr(1)
. correlate mechanical physical
(obs=74)

             | mechan~l physical
-------------+------------------
  mechanical |   1.0000
    physical |   0.9476   1.0000

. drop mechanical physical
```

◁

▷ Example 2

Researchers are often interested in the canonical loadings, the correlations between the original variable lists and their canonical variates. The canonical loadings are used to interpret the canonical variates. However, as shown in the technical note later in this entry, Rencher (1988; 1992; 1998, sec. 8.6.3; 2002, sec. 8.7.3) has shown that there is no information in these correlations about how one variable list contributes jointly to canonical correlation with the other. Loadings are still often discussed, and `estat loadings` reports these as well as the cross-loadings or correlations between *varlist*$_1$ and the canonical variates for *varlist*$_2$ and the correlations between *varlist*$_2$ and the canonical variates for *varlist*$_1$. The loadings and cross-loadings are all computed by `canon`.

```
. estat loadings
```

Canonical loadings for variable list 1

	1	2	3	4
length	0.9664	0.2481	0.0361	-0.0566
weight	0.9972	-0.0606	-0.0367	0.0235
headroom	0.5140	-0.1295	0.7134	-0.4583
trunk	0.6941	0.0644	-0.0209	-0.7167

Canonical loadings for variable list 2

	1	2	3	4
displacement	0.9404	-0.3091	0.1050	0.0947
mpg	-0.8569	-0.1213	0.1741	0.4697
gear_ratio	-0.7945	0.3511	0.4474	-0.2129
turn	0.9142	0.3286	-0.0345	0.2345

Correlation between variable list 1 and canonical variates from list 2

	1	2	3	4
length	0.9158	0.0844	0.0023	-0.0025
weight	0.9449	-0.0206	-0.0023	0.0011
headroom	0.4871	-0.0440	0.0452	-0.0205
trunk	0.6577	0.0219	-0.0013	-0.0320

Correlation between variable list 2 and canonical variates from list 1

	1	2	3	4
displacement	0.8912	-0.1051	0.0067	0.0042
mpg	-0.8120	-0.0413	0.0110	0.0210
gear_ratio	-0.7529	0.1194	0.0284	-0.0095
turn	0.8663	0.1117	-0.0022	0.0105

```
. mat load2 = r(canload22)
```

◁

▷ Example 3

In example 2, we saved the loading matrix for *varlist*$_2$, containing the mechanical variables, and we wish to verify that it is correct. We predict the canonical variates for *varlist*$_2$ and then find the canonical correlations between the canonical variates and the original mechanical variables as a means of getting the correlation matrices, which we then display using `estat correlations`. The mixed correlation matrix is the same as the loading matrix that we saved.

```
. predict mechanical1, v corr(1)
. predict mechanical2, v corr(2)
. predict mechanical3, v corr(3)
. predict mechanical4, v corr(4)
. quietly canon (mechanical1-mechanical4) (displ mpg gear_ratio turn)
. estat correlation
Correlations for variable list 1

             | mechan~1  mechan~2  mechan~3  mechan~4
-------------+----------------------------------------
 mechanical1 |   1.0000
 mechanical2 |  -0.0000    1.0000
 mechanical3 |  -0.0000    0.0000    1.0000
 mechanical4 |  -0.0000   -0.0000   -0.0000    1.0000
------------------------------------------------------

Correlations for variable list 2

             | displa~t       mpg  gear_r~o      turn
-------------+----------------------------------------
displacement |   1.0000
         mpg |  -0.7056    1.0000
  gear_ratio |  -0.8289    0.6162    1.0000
        turn |   0.7768   -0.7192   -0.6763    1.0000
------------------------------------------------------

Correlations between variable lists 1 and 2

             | mechan~1  mechan~2  mechan~3  mechan~4
-------------+----------------------------------------
displacement |   0.9404   -0.3091    0.1050    0.0947
         mpg |  -0.8569   -0.1213    0.1741    0.4697
  gear_ratio |  -0.7945    0.3511    0.4474   -0.2129
        turn |   0.9142    0.3286   -0.0345    0.2345
------------------------------------------------------

. matlist load2, format(%8.4f) border(bottom)

             |        1         2         3         4
-------------+----------------------------------------
displacement |   0.9404   -0.3091    0.1050    0.0947
         mpg |  -0.8569   -0.1213    0.1741    0.4697
  gear_ratio |  -0.7945    0.3511    0.4474   -0.2129
        turn |   0.9142    0.3286   -0.0345    0.2345
------------------------------------------------------
```

◁

▷ Example 4

Here we observe the results of rotation of the canonical loadings, via the Kaiser varimax method outlined in Cliff and Krus (1976). This observation is often done for interpretation of the results; however, rotation destroys several fundamental properties of canonical correlation.

```
. quietly canon (length weight headroom trunk) (displ mpg gear_ratio turn)
. estat rotate, loadings
    Criterion          varimax
    Rotation class     orthogonal
    Normalization      none
```

Rotated canonical loadings

	1	2	3	4
length	0.3796	0.7603	0.4579	0.2613
weight	0.6540	0.5991	0.3764	0.2677
headroom	0.0390	0.1442	0.3225	0.9347
trunk	0.1787	0.2052	0.8918	0.3614
displacement	0.7638	0.4424	0.2049	0.4230
mpg	-0.3543	-0.4244	-0.8109	-0.1918
gear_ratio	-0.9156	-0.3060	-0.2292	0.1248
turn	0.3966	0.8846	0.2310	0.0832

Rotation matrix

	1	2	3	4
1	0.5960	0.6359	0.3948	0.2908
2	-0.6821	0.6593	0.1663	-0.2692
3	-0.3213	0.1113	-0.3400	0.8768
4	0.2761	0.3856	-0.8372	-0.2724

```
. estat rotatecompare
```

Rotated canonical loadings — orthogonal varimax

	1	2	3	4
length	0.3796	0.7603	0.4579	0.2613
weight	0.6540	0.5991	0.3764	0.2677
headroom	0.0390	0.1442	0.3225	0.9347
trunk	0.1787	0.2052	0.8918	0.3614
displacement	0.7638	0.4424	0.2049	0.4230
mpg	-0.3543	-0.4244	-0.8109	-0.1918
gear_ratio	-0.9156	-0.3060	-0.2292	0.1248
turn	0.3966	0.8846	0.2310	0.0832

Unrotated canonical loadings

	1	2	3	4
length	0.9664	0.2481	0.0361	-0.0566
weight	0.9972	-0.0606	-0.0367	0.0235
headroom	0.5140	-0.1295	0.7134	-0.4583
trunk	0.6941	0.0644	-0.0209	-0.7167
displacement	0.9404	-0.3091	0.1050	0.0947
mpg	-0.8569	-0.1213	0.1741	0.4697
gear_ratio	-0.7945	0.3511	0.4474	-0.2129
turn	0.9142	0.3286	-0.0345	0.2345

◁

❑ Technical note

estat loadings reports the canonical loadings or correlations between a *varlist* and its corresponding canonical variates. It is widely claimed that the loadings provide a more valid interpretation of the canonical variates. Rencher (1988; 1992; 1998, sec. 8.6.3; 2002, sec. 8.7.3) has shown that a weighted sum of the correlations between an $x_j \in varlist_1$ and the canonical variates from $varlist_1$ is equal to the squared multiple correlation between x_j and the variables in $varlist_2$. The correlations do not give new information on the importance of a given variable in the context of the others. Rencher (2002, 373) notes, "The researcher who uses these correlations for interpretation is unknowingly reducing the multivariate setting to a univariate one."

❑

Saved results

estat correlations saves the following in r():

Matrices

r(corr_var1)	correlations for $varlist_1$
r(corr_var2)	correlations for $varlist_2$
r(corr_mixed)	correlations between $varlist_1$ and $varlist_2$

estat loadings saves the following in r():

Matrices

r(canload11)	canonical loadings for $varlist_1$
r(canload22)	canonical loadings for $varlist_2$
r(canload21)	correlations between $varlist_2$ and the canonical variates for $varlist_1$
r(canload12)	correlations between $varlist_1$ and the canonical variates for $varlist_2$

estat rotate saves the following in r():

Macros

r(coefficients)	coefficients rotated
r(class)	rotation classification
r(criterion)	rotation criterion

Matrices

r(AT)	rotated coefficient matrix
r(T)	rotation matrix

Methods and formulas

All postestimation commands listed above are implemented as ado-files.

Cliff and Krus (1976) state that they use the Kaiser varimax method with normalization for rotation. The loading matrix, the correlation matrix between the original variables and their canonical variates, is already normalized. Consequently, normalization is not required, nor is it offered as an option.

Rotation after canonical correlation is a subject fraught with controversy. Although some researchers wish to rotate coefficients and loadings for greater interpretability, and Cliff and Krus (1976) have shown that some properties of canonical correlations are preserved by orthogonal rotation, rotation does destroy some of the fundamental properties of canonical correlation. Rencher (1992, 2002) and Thompson (1996) both contribute on the topic. Rencher speaks starkly against rotation. Thompson explains why rotation is desired as well as why it is at odds with the principles of canonical correlation analysis.

The researcher is encouraged to consider carefully his or her goals in canonical correlation analysis and these references when evaluating whether rotation is an appropriate tool to use.

Harris (2001) gives an amusing critique on the misuse of canonical loadings in the interpretation of canonical correlation analysis results. As mentioned, Rencher (1988; 1992; 1998, sec. 8.6.3; 2002, sec. 8.7.3) critiques the use of canonical loadings.

References

Cliff, N., and D. J. Krus. 1976. Interpretation of canonical analysis: Rotated vs. unrotated solutions. *Psychometrika* 41: 35–42.

Harris, R. J. 2001. *A Primer of Multivariate Statistics.* 3rd ed. Mahwah, NJ: Lawrence Erlbaum.

Rencher, A. C. 1988. On the use of correlations to interpret canonical functions. *Biometrika* 75: 363–365.

——. 1992. Interpretation of canonical discriminant functions, canonical variates, and principal components. *American Statistician* 46: 217–225.

——. 1998. *Multivariate Statistical Inference and Applications.* New York: Wiley.

——. 2002. *Methods of Multivariate Analysis.* 2nd ed. New York: Wiley.

Thompson, B. 1996. *Canonical Correlation Analysis: Uses and Interpretation.* Thousand Oaks, CA: Sage.

Also see

[MV] **canon** — Canonical correlations

[MV] **rotatemat** — Orthogonal and oblique rotations of a Stata matrix

[MV] **screeplot** — Scree plot

Title

cluster — Introduction to cluster-analysis commands

Syntax

Cluster analysis of data

`cluster` *subcommand* ...

Cluster analysis of a dissimilarity matrix

`clustermat` *subcommand* ...

Description

Stata's cluster-analysis routines provide several hierarchical and partition clustering methods, postclustering summarization methods, and cluster-management tools. This entry presents an overview of cluster analysis, the `cluster` and `clustermat` commands (also see [MV] **clustermat**), as well as Stata's cluster-analysis management tools. The hierarchical clustering methods may be applied to the data by using the `cluster` command or to a user-supplied dissimilarity matrix by using the `clustermat` command.

The `cluster` command has the following *subcommands*, which are detailed in their respective manual entries.

Partition-clustering methods for observations

`kmeans`	[MV] **cluster kmeans and kmedians**	Kmeans cluster analysis
`kmedians`	[MV] **cluster kmeans and kmedians**	Kmedians cluster analysis

Hierarchical clustering methods for observations

`singlelinkage`	[MV] **cluster linkage**	Single-linkage cluster analysis
`averagelinkage`	[MV] **cluster linkage**	Average-linkage cluster analysis
`completelinkage`	[MV] **cluster linkage**	Complete-linkage cluster analysis
`waveragelinkage`	[MV] **cluster linkage**	Weighted-average linkage cluster analysis
`medianlinkage`	[MV] **cluster linkage**	Median-linkage cluster analysis
`centroidlinkage`	[MV] **cluster linkage**	Centroid-linkage cluster analysis
`wardslinkage`	[MV] **cluster linkage**	Ward's linkage cluster analysis

Postclustering commands

`stop`	[MV] **cluster stop**	Cluster-analysis stopping rules
`dendrogram`	[MV] **cluster dendrogram**	Dendrograms for hierarchical cluster analysis
`generate`	[MV] **cluster generate**	Generate summary or grouping variables from a cluster analysis

User utilities

notes	[MV] **cluster notes**	Place notes in cluster analysis
dir	[MV] **cluster utility**	Directory list of cluster analyses
list	[MV] **cluster utility**	List cluster analyses
drop	[MV] **cluster utility**	Drop cluster analyses
use	[MV] **cluster utility**	Mark cluster analysis as most recent one
rename	[MV] **cluster utility**	Rename cluster analyses
renamevar	[MV] **cluster utility**	Rename cluster-analysis variables

Programmer utilities

	[MV] **cluster programming subroutines**	Add cluster-analysis routines
query	[MV] **cluster programming utilities**	Obtain cluster-analysis attributes
set	[MV] **cluster programming utilities**	Set cluster-analysis attributes
delete	[MV] **cluster programming utilities**	Delete cluster-analysis attributes
parsedistance	[MV] **cluster programming utilities**	Parse (dis)similarity measure names
measures	[MV] **cluster programming utilities**	Compute (dis)similarity measures

The clustermat command has the following *subcommands*, which are detailed along with the related cluster command manual entries. Also see [MV] **clustermat**.

Hierarchical clustering methods for matrices

singlelinkage	[MV] **cluster linkage**	Single-linkage cluster analysis
averagelinkage	[MV] **cluster linkage**	Average-linkage cluster analysis
completelinkage	[MV] **cluster linkage**	Complete-linkage cluster analysis
waveragelinkage	[MV] **cluster linkage**	Weighted-average linkage cluster analysis
medianlinkage	[MV] **cluster linkage**	Median-linkage cluster analysis
centroidlinkage	[MV] **cluster linkage**	Centroid-linkage cluster analysis
wardslinkage	[MV] **cluster linkage**	Ward's linkage cluster analysis

Also, the clustermat stop postclustering command has syntax similar to that of the cluster stop command; see [MV] **cluster stop**. For the remaining postclustering commands and user utilities, you may specify either cluster or clustermat—it does not matter which.

If you are new to Stata's cluster-analysis commands, we recommend that you first read this entry and then read the following:

[MV] ***measure_option***	Option for similarity and dissimilarity measures
[MV] **clustermat**	Cluster analysis of a dissimilarity matrix
[MV] **cluster kmeans and kmedians**	Kmeans and kmedians cluster analysis
[MV] **cluster linkage**	Hierarchical cluster analysis
[MV] **cluster dendrogram**	Dendrograms for hierarchical cluster analysis
[MV] **cluster stop**	Cluster-analysis stopping rules
[MV] **cluster generate**	Generate summary or grouping variables from a cluster analysis

Remarks

Remarks are presented under the following headings:

Introduction to cluster analysis

Cluster analysis attempts to determine the natural groupings (or clusters) of observations. Sometimes this process is called "classification", but this term is used by others to mean discriminant analysis, which is related but is not the same; see [MV] **discrim**. To avoid confusion, we will use "cluster analysis" or "clustering" when referring to finding groups in data. Defining cluster analysis is difficult (maybe impossible). Kaufman and Rousseeuw (1990) start their book by saying, "Cluster analysis is the art of finding groups in data." Everitt et al. (2011, 7) use the terms "cluster", "group", and "class" and say, concerning a formal definition for these terms, "In fact it turns out that such formal definition is not only difficult but may even be misplaced."

Everitt et al. (2011) and Gordon (1999) provide examples of the use of cluster analysis, such as in refining or redefining diagnostic categories in psychiatry, detecting similarities in artifacts by archaeologists to study the spatial distribution of artifact types, discovering hierarchical relationships in taxonomy, and identifying sets of similar cities so that one city from each class can be sampled in a market research task. Also, the activity now called "data mining" relies extensively on cluster-analysis methods.

We view cluster analysis as an exploratory data-analysis technique. According to Everitt, "Many cluster-analysis techniques have taken their place alongside other exploratory data-analysis techniques as tools of the applied statistician. The term exploratory is important here because it explains the largely absent 'p-value', ubiquitous in many other areas of statistics. . . . Clustering methods are intended largely for generating rather than testing hypotheses" (1993, 10).

Although some have said that there are as many cluster-analysis methods as there are people performing cluster analysis. This is a gross understatement! There exist infinitely more ways to perform a cluster analysis than people who perform them.

There are several general types of cluster-analysis methods, each having many specific methods. Also, most cluster-analysis methods allow a variety of distance measures for determining the similarity or dissimilarity between observations. Some of the measures do not meet the requirements to be called a distance metric, so we use the more general term "dissimilarity measure" in place of distance. Similarity measures may be used in place of dissimilarity measures. There are an infinite number of similarity and dissimilarity measures. For instance, there are an infinite number of Minkowski distance metrics, with the familiar Euclidean, absolute-value, and maximum-value distances being special cases.

In addition to cluster method and dissimilarity measure choice, if you are performing a cluster analysis, you might decide to perform data transformations and/or variable selection before clustering. Then you might need to determine how many clusters there really are in the data, which you can do using stopping rules. There is a surprisingly large number of stopping rules mentioned in the literature. For example, Milligan and Cooper (1985) compare 30 different stopping rules.

Looking at all these choices, you can see why there are more cluster-analysis methods than people performing cluster analysis.

Stata's cluster-analysis system

Stata's `cluster` and `clustermat` commands were designed to allow you to keep track of the various cluster analyses performed on your data. The main clustering subcommands—`singlelinkage`, `averagelinkage`, `completelinkage`, `waveragelinkage`, `medianlinkage`, `centroidlinkage`, `wardslinkage` (see [MV] **cluster linkage**), `kmeans`, and `kmedians` (see [MV] **cluster kmeans and kmedians**)—create named Stata cluster objects that keep track of the variables these methods create and hold other identifying information for the cluster analysis. These cluster objects become part of your dataset. They are saved with your data when your data are `saved` and are retrieved when you again `use` your dataset; see [D] **save** and [D] **use**.

Post–cluster-analysis subcommands are available with the `cluster` and `clustermat` commands so that you can examine the created clusters. Cluster-management tools are provided that allow you to add information to the cluster objects and to manipulate them as needed. The main clustering subcommands, postclustering subcommands, and cluster-management tools are discussed in the following sections.

Stata's clustering methods fall into two general types: partition and hierarchical. These two types are discussed below. There exist other types, such as fuzzy partition (where observations can belong to more than one group). Stata's `cluster` command is designed so that programmers can extend it by adding more methods; see [MV] **cluster programming subroutines** and [MV] **cluster programming utilities** for details.

❑ Technical note

If you are familiar with Stata's large array of estimation commands, be careful to distinguish between cluster analysis (the `cluster` command) and the `vce(cluster` *clustvar*`)` option (see [R] ***vce_option***) allowed with many estimation commands. Cluster analysis finds groups in data. The `vce(cluster` *clustvar*`)` option allowed with various estimation commands indicates that the observations are independent across the groups defined by the option but are not necessarily independent within those groups. A grouping variable produced by the `cluster` command will seldom satisfy the assumption behind the use of the `vce(cluster` *clustvar*`)` option.

❑

Data transformations and variable selection

Stata's `cluster` command has no built-in data transformations, but because Stata has full data-management and statistical capabilities, you can use other Stata commands to transform your data before calling the `cluster` command. Standardizing the variables is sometimes important to keep a variable with high variability from dominating the cluster analysis. In other cases, standardizing variables hides the true groupings present in the data. The decision to standardize or perform other data transformations depends on the type of data and the nature of the groups.

Data transformations (such as standardization of variables) and the variables selected for use in clustering can also greatly affect the groupings that are discovered. These and other cluster-analysis data issues are covered in Milligan and Cooper (1988) and Schaffer and Green (1996) and in many of the cluster-analysis texts, including Anderberg (1973); Gordon (1999); Everitt et al. (2011); and Späth (1980).

Similarity and dissimilarity measures

Several similarity and dissimilarity measures have been implemented for Stata's clustering commands for both continuous and binary variables. For information, see [MV] ***measure_option***.

Partition cluster-analysis methods

Partition methods break the observations into a distinct number of nonoverlapping groups. Stata has implemented two partition methods, kmeans and kmedians.

One of the more commonly used partition clustering methods is called kmeans cluster analysis. In kmeans clustering, the user specifies the number of clusters, k, to create using an iterative process. Each observation is assigned to the group whose mean is closest, and then based on that categorization, new group means are determined. These steps continue until no observations change groups. The algorithm begins with k seed values, which act as the k group means. There are many ways to specify the beginning seed values.

A variation of kmeans clustering is kmedians clustering. The same process is followed in kmedians as in kmeans, except that medians, instead of means, are computed to represent the group centers at each step. See [MV] **cluster kmeans and kmedians** for the details of the `cluster kmeans` and `cluster kmedians` commands.

These partition-clustering methods will generally be quicker and will allow larger datasets than the hierarchical clustering methods outlined next. However, if you wish to examine clustering to various numbers of clusters, you will need to execute `cluster` many times with the partition methods. Clustering to various numbers of groups by using a partition method typically does not produce clusters that are hierarchically related. If this relationship is important for your application, consider using one of the hierarchical methods.

Hierarchical cluster-analysis methods

Hierarchical clustering creates hierarchically related sets of clusters. Hierarchical clustering methods are generally of two types: agglomerative or divisive.

Agglomerative hierarchical clustering methods begin with each observation's being considered as a separate group (N groups each of size 1). The closest two groups are combined ($N - 1$ groups, one of size 2 and the rest of size 1), and this process continues until all observations belong to the same group. This process creates a hierarchy of clusters.

In addition to choosing the similarity or dissimilarity measure to use in comparing 2 observations, you can choose what to compare between groups that contain more than 1 observation. The method used to compare groups is called a linkage method. Stata's `cluster` and `clustermat` commands provide several hierarchical agglomerative linkage methods, which are discussed in the next section.

Unlike hierarchical agglomerative clustering, divisive hierarchical clustering begins with all observations belonging to one group. This group is then split in some fashion to create two groups. One of these two groups is then split to create three groups; one of these three is then split to create four groups, and so on, until all observations are in their own separate group. Stata currently has no divisive hierarchical clustering commands. There are relatively few mentioned in the literature, and they tend to be particularly time consuming to compute.

To appreciate the underlying computational complexity of both agglomerative and divisive hierarchical clustering, consider the following information paraphrased from Kaufman and Rousseeuw (1990). The first step of an agglomerative algorithm considers $N(N-1)/2$ possible fusions of observations to find the closest pair. This number grows quadratically with N. For divisive hierarchical clustering, the first step would be to find the best split into two nonempty subsets, and if all possibilities were considered, it would amount to $2^{(N-1)}-1$ comparisons. This number grows exponentially with N.

Agglomerative methods

Stata's `cluster` and `clustermat` commands provide the following hierarchical agglomerative linkage methods: single, complete, average, Ward's method, centroid, median, and weighted average. There are others mentioned in the literature, but these are the best-known methods.

Single-linkage clustering computes the similarity or dissimilarity between two groups as the similarity or dissimilarity between the closest pair of observations between the two groups. Complete-linkage clustering, on the other hand, uses the farthest pair of observations between the two groups to determine the similarity or dissimilarity of the two groups. Average-linkage clustering uses the average similarity or dissimilarity of observations between the groups as the measure between the two groups. Ward's method joins the two groups that result in the minimum increase in the error sum of squares. The other linkage methods provide alternatives to these basic linkage methods.

The `cluster singlelinkage` and `clustermat singlelinkage` commands implement single-linkage hierarchical agglomerative clustering; see [MV] **cluster linkage** for details. Single-linkage clustering suffers (or benefits, depending on your point of view) from what is called chaining. Because the closest points between two groups determine the next merger, long, thin clusters can result. If this chaining feature is not what you desire, consider using one of the other methods, such as complete linkage or average linkage. Because of special properties that can be computationally exploited, single-linkage clustering is faster and uses less memory than the other linkage methods.

Complete-linkage hierarchical agglomerative clustering is implemented by the `cluster completelinkage` and `clustermat completelinkage` commands; see [MV] **cluster linkage** for details. Complete-linkage clustering is at the other extreme from single-linkage clustering. Complete linkage produces spatially compact clusters, so it is not the best method for recovering elongated cluster structures. Several sources, including Kaufman and Rousseeuw (1990), discuss the chaining of single linkage and the clumping of complete linkage.

Kaufman and Rousseeuw (1990) indicate that average linkage works well for many situations and is reasonably robust. The `cluster averagelinkage` and `clustermat averagelinkage` commands provide average-linkage clustering; see [MV] **cluster linkage**.

Ward (1963) presented a general hierarchical clustering approach where groups were joined to maximize an objective function. He used an error-sum-of-squares objective function to illustrate. Ward's method of clustering became synonymous with using the error-sum-of-squares criteria. Kaufman and Rousseeuw (1990) indicate that Ward's method does well with groups that are multivariate normal and spherical but does not do as well if the groups are of different sizes or have unequal numbers of observations. The `cluster wardslinkage` and `clustermat wardslinkage` commands provide Ward's linkage clustering; see [MV] **cluster linkage**.

At each step of the clustering, centroid linkage merges the groups whose means are closest. The centroid of a group is the componentwise mean and can be interpreted as the center of gravity for the group. Centroid linkage differs from average linkage in that centroid linkage is concerned with the distance between the means of the groups, whereas average linkage looks at the average distance between the points of the two groups. The `cluster centroidlinkage` and `clustermat centroidlinkage` commands provide centroid-linkage clustering; see [MV] **cluster linkage**.

Weighted-average linkage and median linkage are variations on average linkage and centroid linkage, respectively. In both cases, the difference is in how groups of unequal size are treated when merged. In average linkage and centroid linkage, the number of elements of each group is factored into the computation, giving correspondingly larger influence to the larger group. These two methods are called unweighted because each observation carries the same weight. In weighted-average linkage and median linkage, the two groups are given equal weighting in determining the combined group, regardless of the number of observations in each group. These two methods are said to be weighted because observations from groups with few observations carry more weight than observations from groups with many observations. The `cluster waveragelinkage` and `clustermat waveragelinkage` commands provide weighted-average linkage clustering. The `cluster medianlinkage` and `clustermat medianlinkage` commands provide median linkage clustering; see [MV] **cluster linkage**.

Lance and Williams' recurrence formula

Lance and Williams (1967) developed a recurrence formula that defines, as special cases, most of the well-known hierarchical clustering methods, including all the hierarchical clustering methods found in Stata. Anderberg (1973); Jain and Dubes (1988); Kaufman and Rousseeuw (1990); Gordon (1999); Everitt et al. (2011); and Rencher (2002) discuss the Lance–Williams formula and how most popular hierarchical clustering methods are contained within it.

From the notation of Everitt et al. (2011, 78), the Lance–Williams recurrence formula is

$$d_{k(ij)} = \alpha_i d_{ki} + \alpha_j d_{kj} + \beta d_{ij} + \gamma |d_{ki} \quad d_{kj}|$$

where d_{ij} is the distance (or dissimilarity) between cluster i and cluster j; $d_{k(ij)}$ is the distance (or dissimilarity) between cluster k and the new cluster formed by joining clusters i and j; and α_i, α_j, β, and γ are parameters that are set based on the particular hierarchical cluster-analysis method.

The recurrence formula allows, at each new level of the hierarchical clustering, the dissimilarity between the newly formed group and the rest of the groups to be computed from the dissimilarities of the current grouping. This approach can result in a large computational savings compared with recomputing at each step in the hierarchy from the observation-level data. This feature of the recurrence formula allows `clustermat` to operate on a similarity or dissimilarity matrix instead of the data.

The following table shows the values of α_i, α_j, β, and γ for the hierarchical clustering methods implemented in Stata. n_i, n_j, and n_k are the number of observations in group i, j, and k, respectively.

Clustering linkage method	α_i	α_j	β	γ
Single	$\frac{1}{2}$	$\frac{1}{2}$	0	$-\frac{1}{2}$
Complete	$\frac{1}{2}$	$\frac{1}{2}$	0	$\frac{1}{2}$
Average	$\frac{n_i}{n_i+n_j}$	$\frac{n_j}{n_i+n_j}$	0	0
Weighted average	$\frac{1}{2}$	$\frac{1}{2}$	0	0
Centroid	$\frac{n_i}{n_i+n_j}$	$\frac{n_j}{n_i+n_j}$	$-\alpha_i\alpha_j$	0
Median	$\frac{1}{2}$	$\frac{1}{2}$	$-\frac{1}{4}$	0
Ward's	$\frac{n_i+n_k}{n_i+n_j+n_k}$	$\frac{n_j+n_k}{n_i+n_j+n_k}$	$\frac{-n_k}{n_i+n_j+n_k}$	0

For information on the use of various similarity and dissimilarity measures in hierarchical clustering, see the next two sections.

Dissimilarity transformations and the Lance and Williams formula

The Lance–Williams formula, which is used as the basis for computing hierarchical clustering in Stata, is designed for use with dissimilarity measures. Before performing hierarchical clustering, Stata transforms similarity measures, both continuous and binary, to dissimilarities. After cluster analysis, Stata transforms the fusion values (heights at which the various groups join in the hierarchy) back to similarities.

Stata's `cluster` command uses

$$\text{dissimilarity} = 1 - \text{similarity}$$

to transform from a similarity to a dissimilarity measure and back again; see Kaufman and Rousseeuw (1990, 21). Stata's similarity measures range from either 0 to 1 or -1 to 1. The resulting dissimilarities range from 1 down to 0 and from 2 down to 0, respectively.

For continuous data, Stata provides both the `L2` and `L2squared` dissimilarity measures, as well as both the `L(#)` and `Lpower(#)` dissimilarity measures. Why have both an `L2` and `L2squared` dissimilarity measure, and why have both an `L(#)` and `Lpower(#)` dissimilarity measure?

For single- and complete-linkage hierarchical clustering (and for kmeans and kmedians partition clustering), there is no need for the additional `L2squared` and `Lpower(#)` dissimilarities. The same cluster solution is obtained when using `L2` and `L2squared` (or `L(#)` and `Lpower(#)`), except that the resulting heights in the dendrogram are raised to a power.

However, for the other hierarchical clustering methods, there is a difference. For some of these other hierarchical clustering methods, the natural default for dissimilarity measure is `L2squared`. For instance, the traditional Ward's (1963) method is obtained by using the `L2squared` dissimilarity option.

Warning concerning similarity or dissimilarity choice

With hierarchical centroid, median, Ward's, and weighted-average linkage clustering, Lance and Williams (1967); Anderberg (1973); Jain and Dubes (1988); Kaufman and Rousseeuw (1990); Everitt et al. (2011); and Gordon (1999) give various levels of warnings about using many of the similarity and dissimilarity measures ranging from saying that you should never use anything other than the default squared Euclidean distance (or Euclidean distance) to saying that the results may lack a useful interpretation.

Example 2 of [MV] **cluster linkage** illustrates part of the basis for this warning. The simple matching coefficient is used on binary data. The range of the fusion values for the resulting hierarchy is not between 1 and 0, as you would expect for the matching coefficient. The conclusions from the cluster analysis, however, agree well with the results obtained in other ways.

Stata does not restrict your choice of similarity or dissimilarity. If you are not familiar with these hierarchical clustering methods, use the default dissimilarity measure.

Synonyms

Cluster-analysis methods have been developed by researchers in many different disciplines. Because researchers did not always know what was happening in other fields, many synonyms for the different hierarchical cluster-analysis methods exist.

Blashfield and Aldenderfer (1978) provide a table of equivalent terms. Jain and Dubes (1988) and Day and Edelsbrunner (1984) also mention some of the synonyms and use various acronyms. Here is a list of synonyms:

Single linkage
- Nearest-neighbor method
- Minimum method
- Hierarchical analysis
- Space-contracting method
- Elementary linkage analysis
- Connectedness method

Complete linkage
- Furthest-neighbor method
- Maximum method
- Compact method
- Space-distorting method
- Space-dilating method
- Rank-order typal analysis
- Diameter analysis

Average linkage
- Arithmetic-average clustering
- Unweighted pair-group method using arithmetic averages
- UPGMA
- Unweighted clustering
- Group-average method
- Unweighted group mean
- Unweighted pair-group method

Weighted-average linkage
- Weighted pair-group method using arithmetic averages
- WPGMA
- Weighted group-average method

Centroid linkage
- Unweighted centroid method
- Unweighted pair-group centroid method
- UPGMC
- Nearest-centroid sorting

Median linkage
- Gower's method
- Weighted centroid method
- Weighted pair-group centroid method
- WPGMC
- Weighted pair method
- Weighted group method

Ward's method
- Minimum-variance method
- Error-sum-of-squares method
- Hierarchical grouping to minimize tr(W)
- HGROUP

Reversals

Unlike the other hierarchical methods implemented in Stata, centroid linkage and median linkage (see [MV] **cluster linkage**) can (and often do) produce reversals or crossovers; see Anderberg (1973), Jain and Dubes (1988), Gordon (1999), and Rencher (2002). Normally, the dissimilarity or clustering criterion increases monotonically as the agglomerative hierarchical clustering progresses from many to few clusters. (For similarity measures, it monotonically decreases.) The dissimilarity value at which $k+1$ clusters form will be larger than the value at which k clusters form. When the dissimilarity does not increase monotonically through the levels of the hierarchy, it is said to have reversals or crossovers.

The word *crossover*, in this context, comes from the appearance of the resulting dendrogram (see [MV] **cluster dendrogram**). In a hierarchical clustering without reversals, the dendrogram branches extend in one direction (increasing dissimilarity measure). With reversals, some of the branches reverse and go in the opposite direction, causing the resulting dendrogram to be drawn with crossing lines (crossovers).

When reversals happen, Stata still produces correct results. You can still generate grouping variables (see [MV] **cluster generate**) and compute stopping rules (see [MV] **cluster stop**). However, the `cluster dendrogram` command will not draw a dendrogram with reversals; see [MV] **cluster dendrogram**. In all but the simplest cases, dendrograms with reversals are almost impossible to interpret visually.

Hierarchical cluster analysis applied to a dissimilarity matrix

What if you want to perform a cluster analysis using a similarity or dissimilarity measure that Stata does not provide? What if you want to cluster variables instead of observations? The `clustermat` command gives you the flexibility to do either; see [MV] **clustermat**.

User-supplied dissimilarities

There are situations where the dissimilarity between objects is evaluated subjectively (perhaps on a scale from 1 to 10 by a rater). These dissimilarities may be entered in a matrix and passed to the `clustermat` command to perform hierarchical clustering. Likewise, if Stata does not offer the dissimilarity measure you desire, you may compute the dissimilarities yourself and place them in a matrix and then use `clustermat` to perform the cluster analysis. [MV] **clustermat** illustrates both of these situations.

Clustering variables instead of observations

Sometimes you want to cluster variables rather than observations, so you can use the `cluster` command. One approach to clustering variables in Stata is to use `xpose` (see [D] **xpose**) to transpose the variables and observations and then to use `cluster`. Another approach is to use the `matrix dissimilarity` command with the `variables` option (see [MV] **matrix dissimilarity**) to produce a dissimilarity matrix for the variables. This matrix is then passed to `clustermat` to obtain the hierarchical clustering. See [MV] **clustermat**.

Postclustering commands

Stata's `cluster stop` and `clustermat stop` commands are used to determine the number of clusters. Two stopping rules are provided, the Caliński and Harabasz (1974) pseudo-F index and the Duda, Hart, and Stork (2001, sec. 10.10) Je(2)/Je(1) index with associated pseudo-T-squared. You can easily add stopping rules to the `cluster stop` command; see [MV] **cluster stop** for details.

The `cluster dendrogram` command presents the dendrogram (cluster tree) after a hierarchical cluster analysis; see [MV] **cluster dendrogram**. Options allow you to view the top portion of the tree or the portion of the tree associated with a group. These options are important with larger datasets because the full dendrogram cannot be presented.

The `cluster generate` command produces grouping variables after hierarchical clustering; see [MV] **cluster generate**. These variables can then be used in other Stata commands, such as those that tabulate, summarize, and provide graphs. For instance, you might use `cluster generate` to create a grouping variable. You then might use the `pca` command (see [MV] **pca**) to obtain the first two principal components of the data. You could follow that with a graph (see *Stata Graphics Reference Manual*) to plot the principal components, using the grouping variable from the `cluster generate` command to control the point labeling of the graph. This method would allow you to get one type of view into the clustering behavior of your data.

Cluster-management tools

You may add notes to your cluster analysis with the `cluster notes` command; see [MV] **cluster notes**. This command also allows you to view and delete notes attached to the cluster analysis.

The `cluster dir` and `cluster list` commands allow you to list the cluster objects and attributes currently defined for your dataset. `cluster drop` lets you remove a cluster object. See [MV] **cluster utility** for details.

Cluster objects are referred to by name. If no name is provided, many of the `cluster` commands will, by default, use the cluster object from the most recently performed cluster analysis. The `cluster use` command tells Stata which cluster object to use. You can change the name attached to a cluster object with the `cluster rename` command and the variables associated with a cluster analysis with the `cluster renamevar` command. See [MV] **cluster utility** for details.

You can exercise fine control over the attributes that are stored with a cluster object; see [MV] **cluster programming utilities**.

References

Anderberg, M. R. 1973. *Cluster Analysis for Applications*. New York: Academic Press.

Blashfield, R. K., and M. S. Aldenderfer. 1978. The literature on cluster analysis. *Multivariate Behavioral Research* 13: 271–295.

Caliński, T., and J. Harabasz. 1974. A dendrite method for cluster analysis. *Communications in Statistics* 3: 1–27.

Day, W. H. E., and H. Edelsbrunner. 1984. Efficient algorithms for agglomerative hierarchical clustering methods. *Journal of Classification* 1: 7–24.

Duda, R. O., P. E. Hart, and D. G. Stork. 2001. *Pattern Classification*. 2nd ed. New York: Wiley.

Everitt, B. S. 1993. *Cluster Analysis*. 3rd ed. London: Arnold.

Everitt, B. S., S. Landau, M. Leese, and D. Stahl. 2011. *Cluster Analysis*. 5th ed. Chichester, UK: Wiley.

Gordon, A. D. 1999. *Classification*. 2nd ed. Boca Raton, FL: Chapman & Hall/CRC.

Jain, A. K., and R. C. Dubes. 1988. *Algorithms for Clustering Data*. Englewood Cliffs, NJ: Prentice Hall.

Kaufman, L., and P. J. Rousseeuw. 1990. *Finding Groups in Data: An Introduction to Cluster Analysis*. New York: Wiley.

Lance, G. N., and W. T. Williams. 1967. A general theory of classificatory sorting strategies: 1. Hierarchical systems. *Computer Journal* 9: 373–380.

Milligan, G. W., and M. C. Cooper. 1985. An examination of procedures for determining the number of clusters in a dataset. *Psychometrika* 50: 159–179.

——. 1988. A study of standardization of variables in cluster analysis. *Journal of Classification* 5: 181–204.

Raciborski, R. 2009. Graphical representation of multivariate data using Chernoff faces. *Stata Journal* 9: 374–387.

Rencher, A. C. 2002. *Methods of Multivariate Analysis*. 2nd ed. New York: Wiley.

Rohlf, F. J. 1982. Single-link clustering algorithms. In Vol. 2 of *Handbook of Statistics*, ed. P. R. Krishnaiah and L. N. Kanal, 267–284. Amsterdam: North-Holland.

Schaffer, C. M., and P. E. Green. 1996. An empirical comparison of variable standardization methods in cluster analysis. *Multivariate Behavioral Research* 31: 149–167.

Sibson, R. 1973. SLINK: An optimally efficient algorithm for the single-link cluster method. *Computer Journal* 16: 30–34.

Späth, H. 1980. *Cluster Analysis Algorithms for Data Reduction and Classification of Objects*. Chichester, UK: Ellis Horwood.

Ward, J. H., Jr. 1963. Hierarchical grouping to optimize an objective function. *Journal of the American Statistical Association* 58: 236–244.

Also see

[MV] **clustermat** — Introduction to clustermat commands

[MV] **cluster programming subroutines** — Add cluster-analysis routines

[MV] **cluster programming utilities** — Cluster-analysis programming utilities

[MV] **discrim** — Discriminant analysis

Title

clustermat — Introduction to clustermat commands

Syntax

`clustermat` *linkage matname* ...

linkage	Description
`singlelinkage`	single-linkage cluster analysis
`averagelinkage`	average-linkage cluster analysis
`completelinkage`	complete-linkage cluster analysis
`waveragelinkage`	weighted-average linkage cluster analysis
`medianlinkage`	median-linkage cluster analysis
`centroidlinkage`	centroid-linkage cluster analysis
`wardslinkage`	Ward's linkage cluster analysis

See [MV] **cluster linkage**.

`clustermat stop` has similar syntax to that of `cluster stop`; see [MV] **cluster stop**. For the remaining postclustering subcommands and user utilities, you may specify either `cluster` or `clustermat`—it does not matter which.

Description

`clustermat` performs hierarchical cluster analysis on the dissimilarity matrix *matname*. `clustermat` is part of the `cluster` suite of commands; see [MV] **cluster**. All Stata hierarchical clustering methods are allowed with `clustermat`. The partition-clustering methods (`kmeans` and `kmedians`) are not allowed because they require the data.

See [MV] **cluster** for a listing of all the `cluster` and `clustermat` commands. The `cluster dendrogram` command (see [MV] **cluster dendrogram**) will display the resulting dendrogram, the `clustermat stop` command (see [MV] **cluster stop**) will help in determining the number of groups, and the `cluster generate` command (see [MV] **cluster generate**) will produce grouping variables. Other useful `cluster` subcommands include `notes`, `dir`, `list`, `drop`, `use`, `rename`, and `renamevar`; see [MV] **cluster notes** and [MV] **cluster utility**.

Remarks

If you are clustering observations by using one of the similarity or dissimilarity measures provided by Stata, the `cluster` command is what you need. If, however, you already have a dissimilarity matrix or can produce one for a dissimilarity measure that Stata does not provide, or if you want to cluster variables instead of observations, the `clustermat` command is what you need.

▷ Example 1

Table 6 of Kaufman and Rousseeuw (1990) provides a subjective dissimilarity matrix among 11 sciences. Fourteen postgraduate economics students from different parts of the world gave subjective dissimilarities among these 11 sciences on a scale from 0 (identical) to 10 (very different). The final dissimilarity matrix was obtained by averaging the results from the students.

We begin by creating a label variable and a shorter version of the label variable corresponding to the 11 sciences. Then we create a row vector containing the lower triangle of the dissimilarity matrix.

```
. input str13 science

             science
  1. Astronomy
  2. Biology
  3. Chemistry
  4. Computer sci.
  5. Economics
  6. Geography
  7. History
  8. Mathematics
  9. Medicine
 10. Physics
 11. Psychology
 12. end

. gen str4 shortsci = substr(science,1,4)

. matrix input D = (
0.00
7.86 0.00
6.50 2.93 0.00
5.00 6.86 6.50 0.00
8.00 8.14 8.21 4.79 0.00
4.29 7.00 7.64 7.71 5.93 0.00
8.07 8.14 8.71 8.57 5.86 3.86 0.00
3.64 7.14 4.43 1.43 3.57 7.07 9.07 0.00
8.21 2.50 2.93 6.36 8.43 7.86 8.43 6.29 0.00
2.71 5.21 4.57 4.21 8.36 7.29 8.64 2.21 5.07 0.00
9.36 5.57 7.29 7.21 6.86 8.29 7.64 8.71 3.79 8.64 0.00 )
```

There are several ways that we could have stored the dissimilarity information in a matrix. To avoid entering both the upper and lower triangle of the matrix, we entered the dissimilarities as a row vector containing the lower triangular entries of the dissimilarity matrix, including the diagonal of zeros (although there are options that would allow us to omit the diagonal of zeros). We typed `matrix input D = ...` instead of `matrix D = ...` so that we could omit the commas between entries; see [P] **matrix define**.

We now perform a complete-linkage cluster analysis on these dissimilarities. The `name()` option names the cluster analysis. We will name it `complink`. The `shape(lower)` option is what signals that the dissimilarity matrix is stored as a row vector containing the lower triangle of the dissimilarity matrix, including the diagonal of zeros. The `add` option indicates that the resulting cluster information should be added to the existing dataset. Here the existing dataset consists of the `science` label variable and the shortened version `shortsci`. See [MV] **cluster linkage** for details concerning these options. The short labels are passed to `cluster dendrogram` so that we can see which subjects were most closely related when viewing the dendrogram; see [MV] **cluster dendrogram**.

```
. clustermat completelinkage D, shape(lower) add name(complink)
. cluster dendrogram complink, labels(shortsci)
                    title(Complete-linkage clustering)
                    ytitle("Subjective dissimilarity"
                           "0=Same, 10=Very different")
```

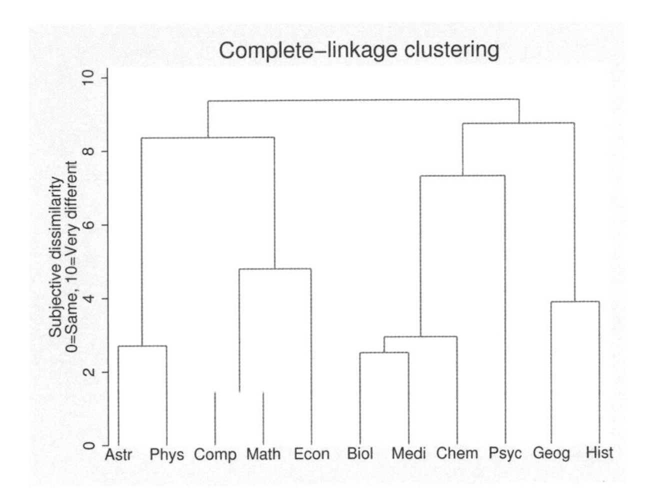

From the dendrogram, we see that mathematics and computer science were deemed most similar and that the economists most closely related their field of study to those two disciplines.

◁

▷ Example 2

Stata does not provide the Bray and Curtis (1957) dissimilarity measure first described by Odum (1950). Using the same notation as that found in [MV] ***measure_option***, we find that the Bray–Curtis dissimilarity between observations i and j is

$$\frac{\sum_{a=1}^{p} |x_{ia} - x_{ja}|}{\sum_{a=1}^{p} (x_{ia} + x_{ja})}$$

Stata does not provide this measure because of the many cases where the measure is undefined (because of dividing by zero). However, when the data are positive the Bray–Curtis dissimilarity is well behaved.

Even though Stata does not automatically provide this measure, it is easy to obtain it and then use it with `clustermat` to perform hierarchical clustering. The numerator of the Bray–Curtis dissimilarity measure is the `L1` (absolute value) distance. We use the `matrix dissimilarity` command (see [MV] **matrix dissimilarity**) to obtain the `L1` dissimilarity matrix and then divide the elements of that matrix by the appropriate values to obtain the Bray–Curtis dissimilarity.

Fisher (1936) presented data, originally from Anderson (1935), on three species of iris. Measurements of the length and width of the sepal and petal were obtained for 50 samples of each of the three iris species. We obtained the data from Morrison (2005). Here we demonstrate average-linkage clustering of these 150 observations.

```
. use http://www.stata-press.com/data/r12/iris, clear
(Iris data)

. summarize seplen sepwid petlen petwid

    Variable |       Obs        Mean    Std. Dev.       Min        Max
-------------+--------------------------------------------------------
      seplen |       150    5.843333    .8280661        4.3        7.9
      sepwid |       150    3.057333    .4358663          2        4.4
      petlen |       150       3.758    1.765298          1        6.9
      petwid |       150    1.199333    .7622377         .1        2.5

. matrix dissimilarity irisD = seplen sepwid petlen petwid, L1

. egen rtot = rowtotal(seplen sepwid petlen petwid)

. forvalues a = 1/150 {
  2.         forvalues b = 1/150 {
  3.                 mat irisD[`a',`b'] = irisD[`a',`b']/(rtot[`a']+rtot[`b'])
  4.         }
  5. }

. matlist irisD[1..5,1..5]

             |      obs1       obs2       obs3       obs4       obs5
-------------+-------------------------------------------------------
        obs1 |         0
        obs2 |   .035533          0
        obs3 |  .0408163    .026455          0
        obs4 |  .0510204    .026455   .0212766          0
        obs5 |  .0098039    .035533   .0408163   .0510204          0
```

The `egen rowtotal()` function provided the row totals used in the denominator of the Bray–Curtis dissimilarity measure; see [D] **egen**. We listed the dissimilarities between the first 5 observations.

We now compute the average-linkage cluster analysis on these 150 observations (see [MV] **cluster linkage**) and examine the Caliński–Harabasz pseudo-*F* index and the Duda–Hart Je(2)/Je(1) index (cluster stopping rules; see [MV] **cluster stop**) to try to determine the number of clusters.

```
. clustermat averagelink irisD, name(iris) add

. clustermat stop, variables(seplen sepwid petlen petwid)
```

Number of clusters	Calinski/ Harabasz pseudo-F
2	502.82
3	299.96
4	201.58
5	332.89
6	288.61
7	244.61
8	252.39
9	223.28
10	268.47
11	241.51
12	232.61
13	233.46
14	255.84
15	273.96

```
. clustermat stop, variables(seplen sepwid petlen petwid) rule(duda)
```

Number of clusters	Duda/Hart Je(2)/Je(1)	pseudo T-squared
1	0.2274	502.82
2	0.8509	17.18
3	0.8951	5.63
4	0.4472	116.22
5	0.6248	28.23
6	0.9579	2.55
7	0.5438	28.52
8	0.8843	5.10
9	0.5854	40.37
10	0.0000	.
11	0.8434	6.68
12	0.4981	37.28
13	0.5526	25.91
14	0.6342	16.15
15	0.6503	3.23

The stopping rules are not conclusive here. From the Duda–Hart pseudo-T-squared (small values) you might best conclude that there are three, six, or eight natural clusters. The Caliński and Harabasz pseudo-F (large values) indicates that there might be two, three, or five groups.

With the iris data, we know the three species. Let's compare the average-linkage hierarchical cluster solutions with the actual species. The `cluster generate` command (see [MV] **cluster generate**) will generate grouping variables for our hierarchical cluster analysis.

```
. cluster generate g = groups(2/6)
. tabulate g2 iris
```

g2	Iris species Setosa	Versicolo	Virginica	Total
1	50	0	0	50
2	0	50	50	100
Total	50	50	50	150

```
. tabulate g3 iris
```

g3	Iris species Setosa	Versicolo	Virginica	Total
1	50	0	0	50
2	0	46	50	96
3	0	4	0	4
Total	50	50	50	150

```
. tabulate g4 iris
```

g4	Iris species Setosa	Versicolo	Virginica	Total
1	49	0	0	49
2	1	0	0	1
3	0	46	50	96
4	0	4	0	4
Total	50	50	50	150

```
. tabulate g5 iris

           |            Iris species
        g5 |    Setosa  Versicolo  Virginica |     Total
-----------+---------------------------------+----------
         1 |        49          0          0 |        49
         2 |         1          0          0 |         1
         3 |         0         45         15 |        60
         4 |         0          1         35 |        36
         5 |         0          4          0 |         4
-----------+---------------------------------+----------
     Total |        50         50         50 |       150

. tabulate g6 iris

           |            Iris species
        g6 |    Setosa  Versicolo  Virginica |     Total
-----------+---------------------------------+----------
         1 |        41          0          0 |        41
         2 |         8          0          0 |         8
         3 |         1          0          0 |         1
         4 |         0         45         15 |        60
         5 |         0          1         35 |        36
         6 |         0          4          0 |         4
-----------+---------------------------------+----------
     Total |        50         50         50 |       150
```

The two-group cluster solution splits *Iris setosa* from *Iris versicolor* and *Iris virginica*. The three- and four-group cluster solutions appear to split off some outlying observations from the two main groups. The five-group solution finally splits most of *Iris virginica* from the *Iris versicolor* but leaves some overlap.

Though this is not shown here, cluster solutions that better match the known species can be found by using dissimilarity measures other than Bray–Curtis.

◁

▷ Example 3

The `cluster` command clusters observations. If you want to cluster variables, you have two choices. You can use `xpose` (see [D] **xpose**) to transpose the variables and observations, or you can use `matrix dissimilarity` with the `variables` option (see [MV] **matrix dissimilarity**) and then use `clustermat`.

In example 2 of [MV] **cluster kmeans and kmedians**, we introduce the women's club data. Thirty women were asked 35 yes–no questions. In [MV] **cluster kmeans and kmedians**, our interest was in clustering the 30 women for placement at luncheon tables. Here our interest is in understanding the relationship among the 35 variables. Which questions produced similar response patterns from the 30 women?

```
. use http://www.stata-press.com/data/r12/wclub, clear
. describe
Contains data from http://www.stata-press.com/data/r12/wclub.dta
  obs:            30
 vars:            35                          1 May 2011 16:56
 size:         1,050

              storage  display     value
variable name   type   format      label      variable label

bike            byte   %8.0g                  enjoy bicycle riding Y/N
bowl            byte   %8.0g                  enjoy bowling Y/N
swim            byte   %8.0g                  enjoy swimming Y/N
jog             byte   %8.0g                  enjoy jogging Y/N
hock            byte   %8.0g                  enjoy watching hockey Y/N
foot            byte   %8.0g                  enjoy watching football Y/N
base            byte   %8.0g                  enjoy baseball Y/N
bask            byte   %8.0g                  enjoy basketball Y/N
arob            byte   %8.0g                  participate in aerobics Y/N
fshg            byte   %8.0g                  enjoy fishing Y/N
dart            byte   %8.0g                  enjoy playing darts Y/N
clas            byte   %8.0g                  enjoy classical music Y/N
cntr            byte   %8.0g                  enjoy country music Y/N
jazz            byte   %8.0g                  enjoy jazz music Y/N
rock            byte   %8.0g                  enjoy rock and roll music Y/N
west            byte   %8.0g                  enjoy reading western novels Y/N
romc            byte   %8.0g                  enjoy reading romance novels Y/N
scif            byte   %8.0g                  enjoy reading sci. fiction Y/N
biog            byte   %8.0g                  enjoy reading biographies Y/N
fict            byte   %8.0g                  enjoy reading fiction Y/N
hist            byte   %8.0g                  enjoy reading history Y/N
cook            byte   %8.0g                  enjoy cooking Y/N
shop            byte   %8.0g                  enjoy shopping Y/N
soap            byte   %8.0g                  enjoy watching soap operas Y/N
sew             byte   %8.0g                  enjoy sewing Y/N
crft            byte   %8.0g                  enjoy craft activities Y/N
auto            byte   %8.0g                  enjoy automobile mechanics Y/N
pokr            byte   %8.0g                  enjoy playing poker Y/N
brdg            byte   %8.0g                  enjoy playing bridge Y/N
kids            byte   %8.0g                  have children Y/N
hors            byte   %8.0g                  have a horse Y/N
cat             byte   %8.0g                  have a cat Y/N
dog             byte   %8.0g                  have a dog Y/N
bird            byte   %8.0g                  have a bird Y/N
fish            byte   %8.0g                  have a fish Y/N

Sorted by:
```

The `matrix dissimilarity` command allows us to compute the Jaccard similarity measure (the `Jaccard` option), comparing variables (the `variables` option) instead of observations, saving one minus the Jaccard measure (the `dissim(oneminus)` option) as a dissimilarity matrix.

```
. matrix dissimilarity clubD = , variables Jaccard dissim(oneminus)
. matlist clubD[1..5,1..5]

             |      bike       bowl       swim        jog       hock
-------------+-------------------------------------------------------
        bike |         0
        bowl |  .7333333          0
        swim |     .5625       .625          0
         jog |        .6   .8235294   .5882353          0
        hock |  .8461538         .6         .8   .8571429          0
```

We pass the `clubD` matrix to `clustermat` and ask for a single-linkage cluster analysis. We need to specify the `clear` option to replace the 30 observations currently in memory with the 35 observations containing the cluster results. Using the `labelvar()` option, we also ask for a label variable, `question`, to be created from the `clubD` matrix row names. To see the resulting cluster analysis, we call `cluster dendrogram`; see [MV] **cluster dendrogram**.

```
. clustermat singlelink clubD, name(club) clear labelvar(question)
obs was 0, now 35

. describe

Contains data
  obs:            35
 vars:             4
 size:           490
--------------------------------------------------------------------------------
              storage  display     value
variable name   type   format      label      variable label
--------------------------------------------------------------------------------
club_id         byte   %8.0g
club_ord        byte   %8.0g
club_hgt        double %10.0g
question        str4   %9s
--------------------------------------------------------------------------------
Sorted by:
     Note:  dataset has changed since last saved

. cluster dendrogram club, labels(question)
>                   xlabel(, angle(90) labsize(*.75))
>                   title(Single-linkage clustering)
>                   ytitle(1 - Jaccard similarity, suffix)
```

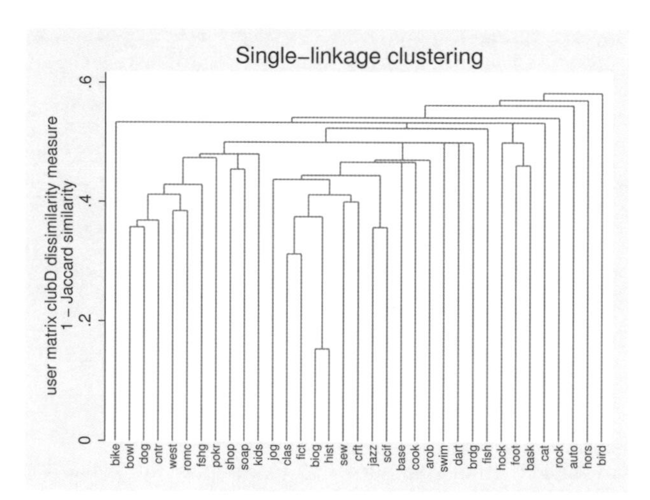

From these 30 women, we see that the `biog` (enjoy reading biographies) and `hist` (enjoy reading history) questions were most closely related. `auto` (enjoy automobile mechanics), `hors` (have a horse), and `bird` (have a bird) seem to be the least related to the other variables. These three variables, in turn, merge last into the supergroup containing the remaining variables.

◁

References

Anderson, E. 1935. The irises of the Gaspe Peninsula. *Bulletin of the American Iris Society* 59: 2–5.

Bray, R. J., and J. T. Curtis. 1957. An ordination of the upland forest communities of southern Wisconsin. *Ecological Monographs* 27: 325–349.

Fisher, R. A. 1936. The use of multiple measurements in taxonomic problems. *Annals of Eugenics* 7: 179–188.

Kaufman, L., and P. J. Rousseeuw. 1990. *Finding Groups in Data: An Introduction to Cluster Analysis*. New York: Wiley.

Morrison, D. F. 2005. *Multivariate Statistical Methods*. 4th ed. Belmont, CA: Duxbury.

Odum, E. P. 1950. Bird populations of the Highlands (North Carolina) plateau in relation to plant succession and avian invasion. *Ecology* 31: 587–605.

Also see

[MV] **cluster programming subroutines** — Add cluster-analysis routines

[MV] **cluster programming utilities** — Cluster-analysis programming utilities

[MV] **cluster** — Introduction to cluster-analysis commands

Title

cluster dendrogram — Dendrograms for hierarchical cluster analysis

Syntax

`cluster dendrogram` [*clname*] [*if*] [*in*] [, *options*]

option	Description
Main	
`quick`	do not center parent branches
`labels(`*varname*`)`	name of variable containing leaf labels
`cutnumber(`*#*`)`	display top # branches only
`cutvalue(`*#*`)`	display branches above # (dis)similarity measure only
`showcount`	display number of observations for each branch
`countprefix(`*string*`)`	prefix the branch count with *string*; default is "`n=`"
`countsuffix(`*string*`)`	suffix the branch count with *string*; default is empty string
`countinline`	put branch count in line with branch label
`vertical`	orient dendrogram vertically (default)
`horizontal`	orient dendrogram horizontally
Plot	
line_options	affect rendition of the plotted lines
Add plots	
`addplot(`*plot*`)`	add other plots to the dendrogram
Y axis, X axis, Titles, Legend, Overall	
twoway_options	any options other than `by()` documented in [G-3] ***twoway_options***

Note: `cluster tree` is a synonym for `cluster dendrogram`.

In addition to the restrictions imposed by `if` and `in`, the observations are automatically restricted to those that were used in the cluster analysis.

Menu

Statistics > Multivariate analysis > Cluster analysis > Postclustering > Dendrograms

Description

`cluster dendrogram` produces dendrograms (also called cluster trees) for a hierarchical clustering. See [MV] **cluster** for a discussion of cluster analysis, hierarchical clustering, and the available `cluster` commands.

Dendrograms graphically present the information concerning which observations are grouped together at various levels of (dis)similarity. At the bottom of the dendrogram, each observation is considered its own cluster. Vertical lines extend up for each observation, and at various (dis)similarity values, these lines are connected to the lines from other observations with a horizontal line. The observations continue to combine until, at the top of the dendrogram, all observations are grouped together.

The height of the vertical lines and the range of the (dis)similarity axis give visual clues about the strength of the clustering. Long vertical lines indicate more distinct separation between the groups. Long vertical lines at the top of the dendrogram indicate that the groups represented by those lines are well separated from one another. Shorter lines indicate groups that are not as distinct.

Options

Main

`quick` switches to a different style of dendrogram in which the vertical lines go straight up from the observations instead of the default action of being recentered after each merge of observations in the dendrogram hierarchy. Some people prefer this representation, and it is quicker to render.

`labels(`*varname*`)` specifies that *varname* be used in place of observation numbers for labeling the observations at the bottom of the dendrogram.

`cutnumber(`*#*`)` displays only the top # branches of the dendrogram. With large dendrograms, the lower levels of the tree can become too crowded. With `cutnumber()`, you can limit your view to the upper portion of the dendrogram. Also see the `cutvalue()` option.

`cutvalue(`*#*`)` displays only those branches of the dendrogram that are above the # (dis)similarity measure. With large dendrograms, the lower levels of the tree can become too crowded. With `cutvalue()`, you can limit your view to the upper portion of the dendrogram. Also see the `cutnumber()` option.

`showcount` requests that the number of observations associated with each branch be displayed below the branches. `showcount` is most useful with `cutnumber()` and `cutvalue()` because, otherwise, the number of observations for each branch is one. When this option is specified, a label for each branch is constructed by using a prefix string, the branch count, and a suffix string.

`countprefix(`*string*`)` specifies the prefix string for the branch count label. The default is `countprefix(n=)`. This option implies the use of the `showcount` option.

`countsuffix(`*string*`)` specifies the suffix string for the branch count label. The default is an empty string. This option implies the use of the `showcount` option.

`countinline` requests that the branch count be put in line with the corresponding branch label. The branch count is placed below the branch label by default. This option implies the use of the `showcount` option.

`vertical` and `horizontal` specify whether the x and y coordinates are to be swapped before plotting—`vertical` (the default) does not swap the coordinates, whereas `horizontal` does.

Plot

line_options affect the rendition of the lines; see [G-3] ***line_options***.

Add plots

`addplot(`*plot*`)` allows adding more `graph twoway` plots to the graph; see [G-3] ***addplot_option***.

Y axis, X axis, Titles, Legend, Overall

twoway_options are any of the options documented in [G-3] ***twoway_options***, excluding `by()`. These include options for titling the graph (see [G-3] ***title_options***) and for saving the graph to disk (see [G-3] ***saving_option***).

Remarks

Examples of the `cluster dendrogram` command can be found in [MV] **cluster linkage**, [MV] **clustermat**, [MV] **cluster stop**, and [MV] **cluster generate**. Here we illustrate some of the additional options available with `cluster dendrogram`.

▷ Example 1

Example 1 of [MV] **cluster linkage** introduces a dataset with 50 observations on four variables. Here we show the dendrogram for a complete-linkage analysis:

```
. use http://www.stata-press.com/data/r12/labtech
. cluster completelinkage x1 x2 x3 x4, name(L2clnk)
. cluster dendrogram L2clnk, labels(labtech) xlabel(, angle(90) labsize(*.75))
```

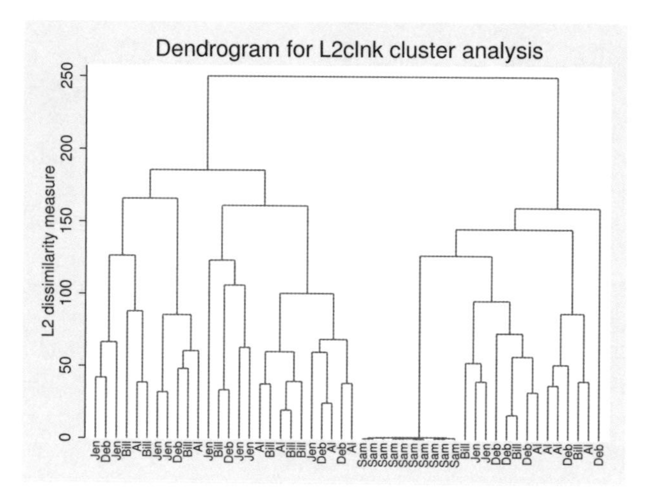

The same dendrogram can be rendered in a slightly different format by using the `quick` option:

```
. cluster dendrogram L2clnk, quick labels(labtech)
          xlabel(, angle(90) labsize(*.75))
```

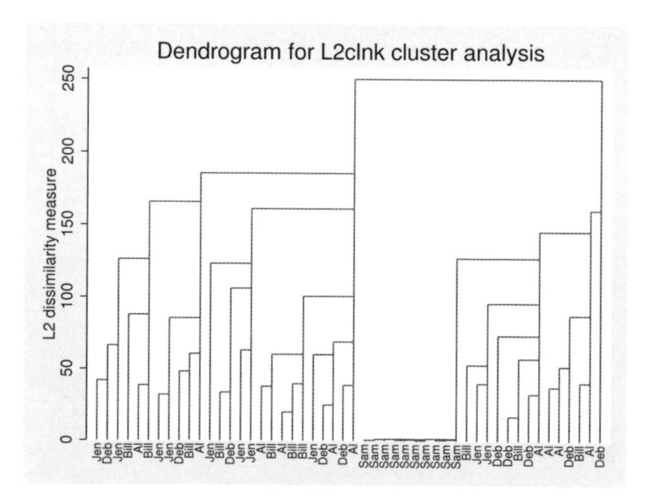

Some people prefer this style of dendrogram. As the name implies, this style of dendrogram is quicker to render.

You can use the `if` and `in` conditions to restrict the dendrogram to the observations for one subgroup. This task is usually accomplished with the `cluster generate` command, which creates a grouping variable; see [MV] **cluster generate**.

Here we show the third of three groups in the dendrogram by first generating the grouping variable for three groups and then using `if` in the command for `cluster dendrogram` to restrict it to the third of those three groups.

```
. cluster gen g3 = group(3)
. cluster tree if g3==3
```

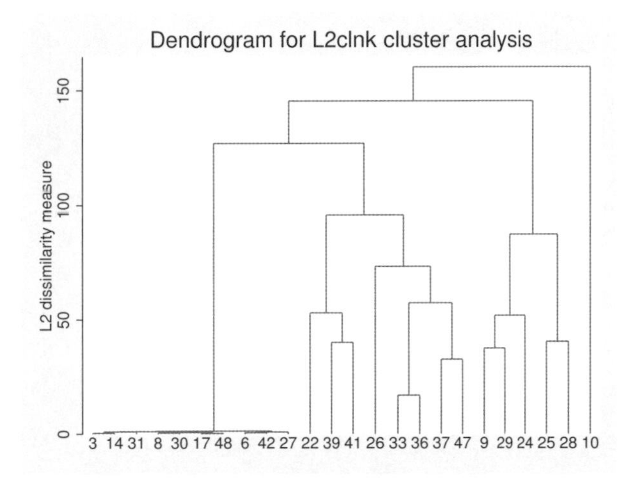

Because we find it easier to type, we used the synonym `tree` instead of `dendrogram`. We did not specify the cluster name, allowing it to default to the most recently performed cluster analysis. We also omitted the `labels()` and `xlabel()` options, which brings us back to the default action of showing, horizontally, the observation numbers.

This example has only 50 observations. When there are many observations, the dendrogram can become too crowded. You will need to limit which part of the dendrogram you display. One way to view only part of the dendrogram is to use `if` and `in` to limit to one particular group, as we did above.

The other way to limit your view of the dendrogram is to specify that you wish to view only the top portion of the tree. The `cutnumber()` and `cutvalue()` options allow you to do this:

```
. cluster tree, cutn(15) showcount
```

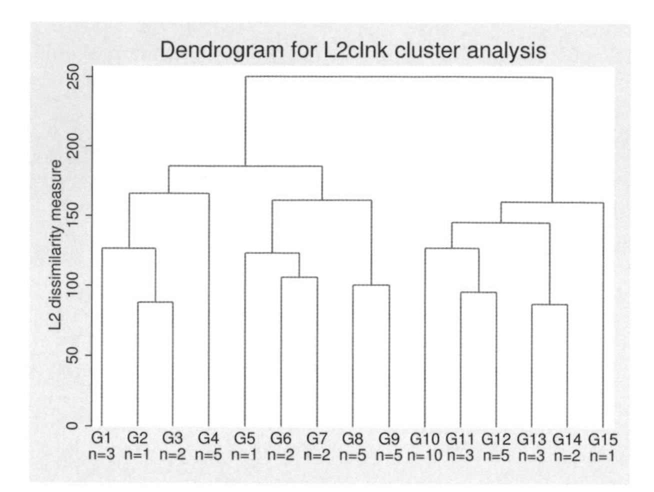

We limited our view to the top 15 branches of the dendrogram with `cutn(15)`. By default, the 15 branches were labeled `G1`–`G15`. The `showcount` option provided, below these branch labels, the number of observations in each of the 15 groups.

The `cutvalue()` option provides another way to limit the view to the top branches of the dendrogram. With this option, you specify the similarity or dissimilarity value at which to trim the tree.

```
. cluster tree, cutvalue(75.3)
        countprefix("(") countsuffix(" obs)") countinline
        ylabel(, angle(0)) horizontal
```

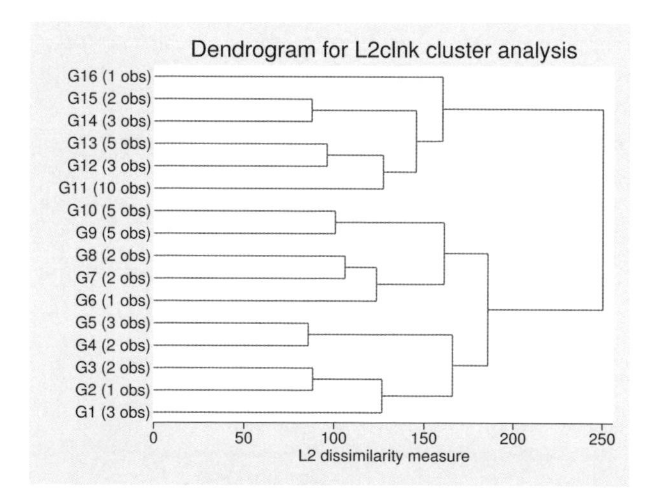

This time, we limited the dendrogram to those branches with dissimilarity greater than 75.3 by using the `cutvalue(75.3)` option. There were 16 branches (groups) that met that restriction. We used the `countprefix()` and `countsuffix()` options to display the number of observations in each branch as "(# obs)" instead of "n=#". The `countinline` option puts the branch counts in line with the branch labels. We specified the `horizontal` option and the `angle(0)` suboption of `ylabel()` to get a horizontal dendrogram with horizontal branch labels.

◁

❑ Technical note

Programmers can control the graphical procedure executed when `cluster dendrogram` is called. This ability will be helpful to programmers adding new hierarchical clustering methods that require a different dendrogram algorithm. See [MV] **cluster programming subroutines** for details.

❑

Methods and formulas

`cluster dendrogram` is implemented as an ado-file.

Reference

Falcaro, M., and A. Pickles. 2010. riskplot: A graphical aid to investigate the effect of multiple categorical risk factors. *Stata Journal* 10: 61–68.

Also see

[MV] **cluster** — Introduction to cluster-analysis commands

[MV] **clustermat** — Introduction to clustermat commands

Title

cluster generate — Generate summary or grouping variables from a cluster analysis

Syntax

Generate grouping variables for specified numbers of clusters

cluster generate {*newvar* | *stub*} = groups(*numlist*) [, *options*]

Generate grouping variable by cutting the dendrogram

cluster generate *newvar* = cut(#) [, name(*clname*)]

option	Description
name(*clname*)	name of cluster analysis to use in producing new variables
ties(error)	produce error message for ties; default
ties(skip)	ignore requests that result in ties
ties(fewer)	produce results for largest number of groups smaller than your request
ties(more)	produce results for smallest number of groups larger than your request

Menu

Statistics > Multivariate analysis > Cluster analysis > Postclustering > Summary variables from cluster analysis

Description

The cluster generate command generates summary or grouping variables from a hierarchical cluster analysis. The result depends on the function. See [MV] **cluster** for information on available cluster-analysis commands.

The groups(*numlist*) function generates grouping variables, giving the grouping for the specified numbers of clusters from a hierarchical cluster analysis. If one number is given, *newvar* is produced with group numbers going from 1 to the number of clusters requested. If more than one number is specified, a new variable is generated for each number by using the provided *stub* name appended with the number. For instance,

```
cluster gen xyz = groups(5/7), name(myclus)
```

creates variables xyz5, xyz6, and xyz7, giving groups 5, 6, and 7 obtained from the cluster analysis named myclus.

The cut(#) function generates a grouping variable corresponding to cutting the dendrogram (see [MV] **cluster dendrogram**) of a hierarchical cluster analysis at the specified (dis)similarity value.

More cluster generate functions may be added; see [MV] **cluster programming subroutines**.

Options

`name(`*clname*`)` specifies the name of the cluster analysis to use in producing the new variables. The default is the name of the cluster analysis last performed, which can be reset by using the `cluster use` command; see [MV] **cluster utility**.

`ties(error | skip | fewer | more)` indicates what to do with the `groups()` function for ties. A hierarchical cluster analysis has ties when multiple groups are generated at a particular (dis)similarity value. For example, you might have the case where you can uniquely create two, three, and four groups, but the next possible grouping produces eight groups because of ties.

`ties(error)`, the default, produces an error message and does not generate the requested variables.

`ties(skip)` specifies that the offending requests be ignored. No error message is produced, and only the requests that produce unique groupings will be honored. With multiple values specified in the `groups()` function, `ties(skip)` allows the processing of those that produce unique groupings and ignores the rest.

`ties(fewer)` produces the results for the largest number of groups less than or equal to your request. In the example above with `groups(6)` and using `ties(fewer)`, you would get the same result that you would by using `groups(4)`.

`ties(more)` produces the results for the smallest number of groups greater than or equal to your request. In the example above with `groups(6)` and using `ties(more)`, you would get the same result that you would by using `groups(8)`.

Remarks

Examples of how to use the `groups()` function of `cluster generate` can be found in [MV] **cluster dendrogram**, [MV] **cluster linkage**, and [MV] **cluster stop**. More examples of the `groups()` and `cut()` functions of `cluster generate` are provided here.

You may find it easier to understand these functions by looking at a dendrogram from a hierarchical cluster analysis. The `cluster dendrogram` command produces dendrograms (cluster trees) from a hierarchical cluster analysis; see [MV] **cluster dendrogram**.

▷ Example 1

Example 1 of [MV] **cluster linkage** examines a dataset with 50 observations with four variables. Here we use complete-linkage clustering and use the `groups()` function of `cluster generate` to produce a grouping variable, splitting the data into two groups.

```
. use http://www.stata-press.com/data/r12/labtech
. cluster completelinkage x1 x2 x3 x4, name(L2clnk)
. cluster dendrogram L2clnk, xlabel(, angle(90) labsize(*.75))
  (graph omitted)
. cluster generate g2 = group(2), name(L2clnk)
. codebook g2
```

```
g2                                                          (unlabeled)

                  type:  numeric (byte)

                 range:  [1,2]                       units:  1
         unique values:  2                       missing .:  0/50

            tabulation:  Freq.  Value
                            26  1
                            24  2
```

```
. by g2, sort: summarize x*

-------------------------------------------------------------------------------
-> g2 = 1

    Variable |       Obs        Mean    Std. Dev.       Min        Max
-------------+--------------------------------------------------------
          x1 |        26        91.5    37.29432       17.4        143
          x2 |        26    74.58077    41.19319        4.8      142.1
          x3 |        26    101.0077    36.95704       16.3      147.9
          x4 |        26    71.77308    43.04107        6.6      146.1

-------------------------------------------------------------------------------
-> g2 = 2

    Variable |       Obs        Mean    Std. Dev.       Min        Max
-------------+--------------------------------------------------------
          x1 |        24        18.8    23.21742          0         77
          x2 |        24    30.05833    37.66979          0      143.6
          x3 |        24    18.54583    21.68215         .2       69.7
          x4 |        24    41.89167    43.62025         .1      130.9
```

The `group()` function of `cluster generate` created a grouping variable named `g2`, with ones indicating the 26 observations that belong to the left main branch of the dendrogram and twos indicating the 24 observations that belong to the right main branch of the dendrogram. The summary of the `x` variables used in the cluster analysis for each group shows that the second group is characterized by lower values.

We could have obtained the same grouping variable by using the `cut()` function of `cluster generate`.

```
. cluster gen g2cut = cut(200)

. table g2 g2cut

----------------------
          |   g2cut
       g2 |    1     2
----------+-----------
        1 |   26
        2 |         24
----------------------
```

We did not need to specify the `name()` option because this was the latest cluster analysis performed, which is the default. The `table` output shows that we obtained the same result with `cut(200)` as with `group(2)` for this example.

How many groups are produced if we cut the tree at the value 105.2?

```
. cluster gen z = cut(105.2)
. codebook z, tabulate(20)
```

```
---------------------------------------------------------------------------------
z                                                                     (unlabeled)
---------------------------------------------------------------------------------

                  type:  numeric (byte)

                 range:  [1,11]                       units:  1
         unique values:  11                       missing .:  0/50

            tabulation:  Freq.  Value
                             3  1
                             3  2
                             5  3
                             1  4
                             2  5
                             2  6
                            10  7
                            10  8
                             8  9
                             5  10
                             1  11
```

The codebook command (see [D] **codebook**) shows that the result of cutting the dendrogram at the value 105.2 produced 11 groups ranging in size from 1 to 10 observations.

The group() function of cluster generate may be used to create multiple grouping variables with one call. Here we create the grouping variables for groups of size 3–12:

```
. cluster gen gp = gr(3/12)
. summarize gp*
```

Variable	Obs	Mean	Std. Dev.	Min	Max
gp3	50	2.26	.8033095	1	3
gp4	50	3.14	1.030356	1	4
gp5	50	3.82	1.438395	1	5
gp6	50	3.84	1.461897	1	6
gp7	50	3.96	1.603058	1	7
gp8	50	4.24	1.911939	1	8
gp9	50	5.18	2.027263	1	9
gp10	50	5.94	2.385415	1	10
gp11	50	6.66	2.781939	1	11
gp12	50	7.24	3.197959	1	12

Here we used abbreviations for generate and group(). The group() function takes a numlist; see [U] **11.1.8 numlist**. We specified 3/12, indicating the numbers 3–12. gp, the stub name we provide, is appended with the number as the variable name for each group variable produced.

◁

▷ Example 2

Example 2 of [MV] **cluster linkage** shows the following dendrogram from the single-linkage clustering of 30 observations on 60 variables. In that example, we used the group() function of cluster generate to produce a grouping variable for three groups. What happens when we try to obtain four groups from this clustering?

```
. use http://www.stata-press.com/data/r12/homework, clear
. cluster singlelinkage a1-a60, measure(matching)
cluster name: _clus_1
. cluster tree
```

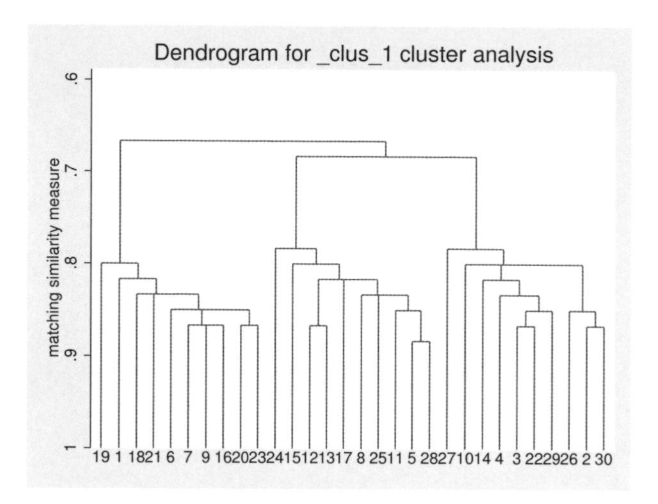


```
. cluster gen g4 = group(4)
cannot create 4 groups because of ties
r(198);
```

Stata complains that it cannot create four groups from this cluster analysis.

The `ties()` option gives us control over this situation. We just need to decide whether we want more groups or fewer groups than we asked for when faced with ties. We demonstrate both ways.

```
. cluster gen more4 = gr(4), ties(more)
. cluster gen less4 = gr(4), ties(fewer)
. summarize more4 less4
```

Variable	Obs	Mean	Std. Dev.	Min	Max
more4	30	2.933333	1.638614	1	5
less4	30	2	.8304548	1	3

For this cluster analysis, `ties(more)` with `group(4)` produces five groups, whereas `ties(fewer)` with `group(4)` produces three groups.

The `ties(skip)` option is convenient when we want to produce a range of grouping variables.

```
. cluster gen group = gr(4/20), ties(skip)
. summarize group*
```

Variable	Obs	Mean	Std. Dev.	Min	Max
group5	30	2.933333	1.638614	1	5
group9	30	4.866667	2.622625	1	9
group13	30	7.066667	3.92106	1	13
group18	30	9.933333	5.419844	1	18

With this cluster analysis, the only unique groupings available are 5, 9, 13, and 18 within the range 4–20.

◁

Methods and formulas

`cluster generate` is implemented as an ado-file.

Also see

[D] **egen** — Extensions to generate

[D] **generate** — Create or change contents of variable

[MV] **cluster** — Introduction to cluster-analysis commands

[MV] **clustermat** — Introduction to clustermat commands

Title

cluster kmeans and kmedians — Kmeans and kmedians cluster analysis

Syntax

Kmeans cluster analysis

cluster kmeans [*varlist*] [*if*] [*in*] , k(*#*) [*options*]

Kmedians cluster analysis

cluster kmedians [*varlist*] [*if*] [*in*] , k(*#*) [*options*]

option	Description
Main	
k(#*)	perform cluster analysis resulting in # groups
measure(*measure*)	similarity or dissimilarity measure; default is L2 (Euclidean)
name(*clname*)	name of resulting cluster analysis
Options	
start(*start_option*)	obtain *k* initial group centers by using *start_option*; see *Options* for details
keepcenters	append the *k* final group means or medians to the data
Advanced	
generate(*groupvar*)	name of grouping variable
iterate(*#*)	maximum number of iterations; default is iterate(10000)

k(#*) is required.

Menu

cluster kmeans

Statistics > Multivariate analysis > Cluster analysis > Cluster data > Kmeans

cluster kmedians

Statistics > Multivariate analysis > Cluster analysis > Cluster data > Kmedians

Description

cluster kmeans and cluster kmedians perform kmeans and kmedians partition cluster analysis, respectively. See [MV] **cluster** for a general discussion of cluster analysis and a description of the other cluster commands.

Options

Main

k(#) is required and indicates that # groups are to be formed by the cluster analysis.

measure(*measure*) specifies the similarity or dissimilarity measure. The default is measure(L2), Euclidean distance. This option is not case sensitive. See [MV] ***measure_option*** for detailed descriptions of the supported measures.

name(*clname*) specifies the name to attach to the resulting cluster analysis. If name() is not specified, Stata finds an available cluster name, displays it for your reference, and attaches the name to your cluster analysis.

Options

start(*start_option*) indicates how the k initial group centers are to be obtained. The available *start_options* are

krandom[(*seed#*)], the default, specifies that k unique observations be chosen at random, from among those to be clustered, as starting centers for the k groups. Optionally, a random-number seed may be specified to cause the command set seed *seed#* (see [R] **set seed**) to be applied before the k random observations are chosen.

firstk[, exclude] specifies that the first k observations from among those to be clustered be used as the starting centers for the k groups. With the exclude option, these first k observations are not included among the observations to be clustered.

lastk[, exclude] specifies that the last k observations from among those to be clustered be used as the starting centers for the k groups. With the exclude option, these last k observations are not included among the observations to be clustered.

random[(*seed#*)] specifies that k random initial group centers be generated. The values are randomly chosen from a uniform distribution over the range of the data. Optionally, a random-number seed may be specified to cause the command set seed *seed#* (see [R] **set seed**) to be applied before the k group centers are generated.

prandom[(*seed#*)] specifies that k partitions be formed randomly among the observations to be clustered. The group means or medians from the k groups defined by this partitioning are to be used as the starting group centers. Optionally, a random-number seed may be specified to cause the command set seed *seed#* (see [R] **set seed**) to be applied before the k partitions are chosen.

everykth specifies that k partitions be formed by assigning observations $1, 1+k, 1+2k, \ldots$ to the first group; assigning observations $2, 2+k, 2+2k, \ldots$ to the second group; and so on, to form k groups. The group means or medians from these k groups are to be used as the starting group centers.

segments specifies that k nearly equal partitions be formed from the data. Approximately the first N/k observations are assigned to the first group, the second N/k observations are assigned to the second group, and so on. The group means or medians from these k groups are to be used as the starting group centers.

group(*varname*) provides an initial grouping variable, *varname*, that defines k groups among the observations to be clustered. The group means or medians from these k groups are to be used as the starting group centers.

keepcenters specifies that the group means or medians from the k groups that are produced be appended to the data.

Advanced

`generate(`*groupvar*`)` provides the name of the grouping variable to be created by `cluster kmeans` or `cluster kmedians`. By default, this will be the name specified in `name()`.

`iterate(`*#*`)` specifies the maximum number of iterations to allow in the kmeans or kmedians clustering algorithm. The default is `iterate(10000)`.

Remarks

Two examples are presented, one using `cluster kmeans` with continuous data and the other using `cluster kmeans` and `cluster kmedians` with binary data. Both commands work similarly with the different types of data.

▷ Example 1

You have measured the flexibility, speed, and strength of the 80 students in your physical education class. You want to split the class into four groups, based on their physical attributes, so that they can receive the mix of flexibility, strength, and speed training that will best help them improve.

Here is a summary of the data and a matrix graph showing the data:

```
. use http://www.stata-press.com/data/r12/physed
. summarize flex speed strength
    Variable |       Obs        Mean    Std. Dev.       Min        Max
-------------+--------------------------------------------------------
 flexibility |        80    4.402625    2.788541        .03       9.97
       speed |        80    3.875875    3.121665        .03       9.79
    strength |        80    6.439875    2.449293        .05       9.57
. graph matrix flex speed strength
```

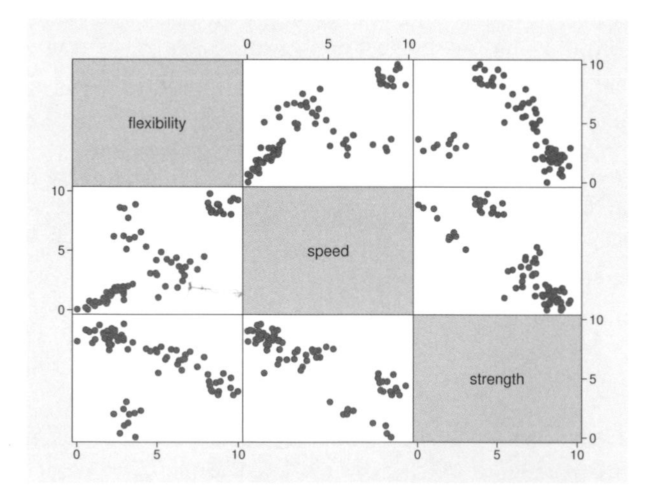

As you expected, based on what you saw the first day of class, the data indicate a wide range of levels of performance for the students. The graph seems to indicate that there are some distinct groups, which leads you to believe that your plan will work well.

You decide to perform a cluster analysis to create four groups, one for each of your class assistants. You have had good experience with kmeans clustering in the past and generally like the behavior of the absolute-value distance.

You do not really care what starting values are used in the cluster analysis, but you do want to be able to reproduce the same results if you ever decide to rerun your analysis. You decide to use the `krandom()` option to pick k of the observations at random as the initial group centers. You supply a random-number seed for reproducibility. You also add the `keepcenters` option so that the means of the four groups will be added to the bottom of your dataset.

```
. cluster k flex speed strength, k(4) name(g4abs) s(kr(385617)) mea(abs) keepcen
. cluster list g4abs
g4abs  (type: partition,  method: kmeans,  dissimilarity: L1)
      vars: g4abs (group variable)
     other: cmd: cluster kmeans flex speed strength, k(4) name(g4abs)
                 start(kr(385617)) measure(abs) keepcen
            varlist: flexibility speed strength
            k: 4
            start: krandom(385617)
            range: 0 .
. table g4abs
```

g4abs	Freq.
1	15
2	20
3	35
4	10

```
. list flex speed strength in 81/L, abbrev(12)
```

	flexibility	speed	strength
81.	8.852	8.743333	4.358
82.	5.9465	3.4485	6.8325
83.	1.969429	1.144857	8.478857
84.	3.157	6.988	1.641

```
. drop in 81/L
(4 observations deleted)
```

```
. tabstat flex speed strength, by(g4abs) stat(min mean max)
Summary statistics: min, mean, max
  by categories of: g4abs

   g4abs |  flexib~y     speed  strength
---------+------------------------------
       1 |      8.12      8.05      3.61
         |     8.852  8.743333     4.358
         |      9.97      9.79      5.42
---------+------------------------------
       2 |      4.32      1.05      5.46
         |    5.9465    3.4485    6.8325
         |      7.89      5.32      7.66
---------+------------------------------
       3 |       .03       .03      7.38
         |  1.969429  1.144857  8.478857
         |      3.48      2.17      9.57
---------+------------------------------
       4 |      2.29      5.11       .05
         |     3.157     6.988     1.641
         |      3.99      8.87      3.02
---------+------------------------------
   Total |       .03       .03       .05
         |  4.402625  3.875875  6.439875
         |      9.97      9.79      9.57
----------------------------------------
```

After looking at the last 4 observations (which are the group means because you specified `keepcenters`), you decide that what you really wanted to see was the minimum and maximum values and the mean for the four groups. You remove the last 4 observations and then use the `tabstat` command to view the desired statistics.

Group 1, with 15 students, is already doing well in flexibility and speed but will need extra strength training. Group 2, with 20 students, needs to emphasize speed training but could use some improvement in the other categories as well. Group 3, the largest, with 35 students, has serious problems with both flexibility and speed, though they did well in the strength category. Group 4, the smallest, with 10 students, needs help with flexibility and strength.

Because you like looking at graphs, you decide to view the matrix graph again but with group numbers used as plotting symbols.

```
. graph matrix flex speed strength, m(i) mlabel(g4abs) mlabpos(0)
```

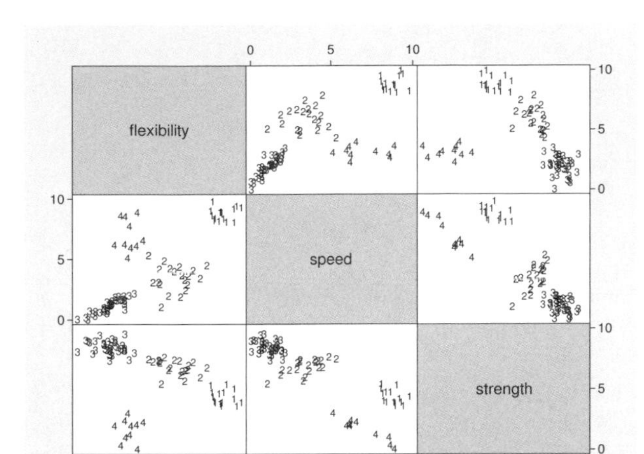

The groups, as shown in the graph, do appear reasonably distinct. However, you had hoped to have groups that were about the same size. You are curious what clustering to three or five groups would produce. For no good reason, you decide to use the first k observations as initial group centers for clustering to three groups and random numbers within the range of the data for clustering to five groups.

```
. cluster k flex speed strength, k(3) name(g3abs) start(firstk) measure(abs)
. cluster k flex speed strength, k(5) name(g5abs) start(random(33576))
> measure(abs)
. table g3abs g4abs, col
```

g3abs	g4abs 1	2	3	4	Total
1				10	10
2		18	35		53
3	15	2			17

```
. table g5abs g4abs, col
```

g5abs	g4abs 1	2	3	4	Total
1		20			20
2	15				15
3				6	6
4				4	4
5			35		35

With three groups, the unequal-group-size problem gets worse. With five groups, the smallest group gets split. Four groups seem like the best option for this class. You will try to help the assistant assigned to group 3 in dealing with the larger group.

You might want to investigate the results of using different random seeds in the command used to generate the 4 groups earlier in this example. Because these data do not have clearly defined, well-separated clusters, there is a good chance that clusters based on different starting values will be different.

◁

▷ Example 2

You have just started a women's club. Thirty women from throughout the community have sent in their requests to join. You have them fill out a questionnaire with 35 yes–no questions relating to sports, music, reading, and hobbies. A description of the 35 variables is found in example 3 of [MV] **clustermat**.

In planning the first meeting of the club, you want to assign seats at the five lunch tables on the basis of shared interests among the women. You really want people placed together who share the same positive interests, not who share dislikes. From all the available binary similarity measures, you decide to use the Jaccard coefficient as the binary similarity measure because it does not include jointly zero comparisons in its formula; see [MV] ***measure_option***. The Jaccard coefficient is also easy to understand.

You decide to examine the groupings produced by kmeans and kmedians clustering.

```
. use http://www.stata-press.com/data/r12/wclub, clear
. cluster kmeans bike-fish, k(5) measure(Jaccard) st(firstk) name(gr5)
. cluster kmed bike-fish, k(5) measure(Jaccard) st(firstk) name(kmedian5)
. cluster list kmedian5
kmedian5  (type: partition,  method: kmedians,  similarity: Jaccard)
      vars: kmedian5 (group variable)
     other: cmd: cluster kmedians bike-fish, k(5) measure(Jaccard) st(firstk)
                 name(kmedian5)
            varlist: bike bowl swim jog hock foot base bask arob fshg dart clas
                 cntr jazz rock west romc scif biog fict hist cook shop soap
                 sew crft auto pokr brdg kids hors cat dog bird fish
            k: 5
            start: firstk
            range: 1 0
```

You used the first *k* observations as starting centers for both kmeans and kmedians—the `st(firstk)` option.

What size groups did each method produce, and how closely did the results agree?

```
. table gr5 kmedian5, row col
```

	kmedian5					
gr5	1	2	3	4	5	Total
1	7					7
2	1	6				7
3			5			5
4				5		5
5	1		1		4	6
Total	9	6	6	5	4	30

There is reasonably strong agreement between the results from `cluster kmeans` and `cluster kmedians`. Because the tables can seat only eight comfortably, the grouping produced by `cluster kmeans` will be used because the group sizes range from five to seven, whereas the groups from `cluster kmedians` range from four to nine.

◁

Methods and formulas

`cluster kmeans` and `cluster kmedians` are implemented as ado-files.

Kmeans cluster analysis and its variant, kmedians cluster analysis, are discussed in most cluster-analysis books; see *References* in [MV] **cluster**. [MV] **cluster** also provides a general discussion of cluster analysis, including kmeans and kmedians clustering, and discusses the available `cluster` subcommands.

Kmeans and kmedians clustering are iterative procedures that partition the data into k groups or clusters. The procedure begins with k initial group centers. Observations are assigned to the group with the closest center. The mean or median of the observations assigned to each of the groups is computed, and the process is repeated. These steps continue until all observations remain in the same group from the previous iteration.

To avoid endless loops, an observation will be reassigned to a different group only if it is closer to the other group center. For a tied distance between an observation and two or more group centers, the observation is assigned to its current group if that is one of the closest and to the lowest numbered group otherwise.

The `start()` option provides many ways to specify the beginning group centers. These include methods that specify the actual starting centers, as well as methods that specify initial partitions of the data from which the beginning centers are computed.

Some kmeans clustering algorithms recompute the group centers after each reassignment of an observation. Other algorithms, including Stata's `cluster kmeans` and `cluster kmedians` commands, recompute the group centers only after a complete pass through the data. A disadvantage of this method is that orphaned group centers—one that has no observations that are closest to it—can occur. The advantage of recomputing means only at the end of each pass through the data is that the sort order of the data does not potentially change your result.

Stata deals with orphaned centers by finding the observations that are farthest from the centers and using them as new group centers. The observations are then reassigned to the closest groups, including these new centers.

Continuous or binary data are allowed with `cluster kmeans` and `cluster kmedians`. The mean of a group of binary observations for a variable is the proportion of ones for that group of observations and variable. The median of a group of binary observations for a variable is almost always either zero or one. However, if there are an equal number of zeros and ones for a group, the median is 0.5. The binary similarity measures can accommodate the comparison of a binary observation to a proportion. See [MV] ***measure_option*** for details on this subject and for the formulas for all the available (dis)similarity measures.

Also see

[MV] **cluster notes** — Place notes in cluster analysis

[MV] **cluster stop** — Cluster-analysis stopping rules

[MV] **cluster utility** — List, rename, use, and drop cluster analyses

[MV] **cluster** — Introduction to cluster-analysis commands

[MV] **clustermat** — Introduction to clustermat commands

Title

cluster linkage — Hierarchical cluster analysis

Syntax

Cluster analysis of data

cluster *linkage* [*varlist*] [*if*] [*in*] [, *cluster_options*]

Cluster analysis of a dissimilarity matrix

clustermat *linkage* *matname* [*if*] [*in*] [, *clustermat_options*]

linkage	Description
singlelinkage	single-linkage cluster analysis
averagelinkage	average-linkage cluster analysis
completelinkage	complete-linkage cluster analysis
waveragelinkage	weighted-average linkage cluster analysis
medianlinkage	median-linkage cluster analysis
centroidlinkage	centroid-linkage cluster analysis
wardslinkage	Ward's linkage cluster analysis

cluster_options	Description
Main	
measure(*measure*)	similarity or dissimilarity measure
name(*clname*)	name of resulting cluster analysis
Advanced	
generate(*stub*)	prefix for generated variables; default prefix is *clname*

clustermat_options	Description
Main	
shape(*shape*)	shape (storage method) of *matname*
add	add cluster information to data currently in memory
clear	replace data in memory with cluster information
labelvar(*varname*)	place dissimilarity matrix row names in *varname*
name(*clname*)	name of resulting cluster analysis
Advanced	
force	perform clustering after fixing *matname* problems
generate(*stub*)	prefix for generated variables; default prefix is *clname*

shape	*matname* is stored as a
`full`	square symmetric matrix; the default
`lower`	vector of rowwise lower triangle (with diagonal)
`llower`	vector of rowwise strict lower triangle (no diagonal)
`upper`	vector of rowwise upper triangle (with diagonal)
`uupper`	vector of rowwise strict upper triangle (no diagonal)

Menu

cluster singlelinkage

Statistics > Multivariate analysis > Cluster analysis > Cluster data > Single linkage

cluster averagelinkage

Statistics > Multivariate analysis > Cluster analysis > Cluster data > Average linkage

cluster completelinkage

Statistics > Multivariate analysis > Cluster analysis > Cluster data > Complete linkage

cluster waveragelinkage

Statistics > Multivariate analysis > Cluster analysis > Cluster data > Weighted-average linkage

cluster medianlinkage

Statistics > Multivariate analysis > Cluster analysis > Cluster data > Median linkage

cluster centroidlinkage

Statistics > Multivariate analysis > Cluster analysis > Cluster data > Centroid linkage

cluster wardslinkage

Statistics > Multivariate analysis > Cluster analysis > Cluster data > Ward's linkage

Description

Stata's `cluster` and `clustermat` commands provide the following hierarchical agglomerative linkage methods: single, complete, average, Ward's method, centroid, median, and weighted average. There are others mentioned in the literature, but these are the best-known methods.

The `clustermat` linkage commands perform hierarchical agglomerative linkage cluster analysis on the dissimilarity matrix *matname*. See [MV] **clustermat** for a general discussion of cluster analysis of dissimilarity matrices and a description of the other `clustermat` commands.

After a `cluster` *linkage* or `clustermat` *linkage* command, the `cluster dendrogram` command (see [MV] **cluster dendrogram**) displays the resulting dendrogram, the `cluster stop` or `clustermat stop` command (see [MV] **cluster stop**) helps determine the number of groups, and the `cluster generate` command (see [MV] **cluster generate**) produces grouping variables.

Options for cluster linkage commands

Main

`measure(`*measure*`)` specifies the similarity or dissimilarity measure. The default for `averagelinkage`, `completelinkage`, `singlelinkage`, and `waveragelinkage` is `L2` (synonym `Euclidean`). The default for `centroidlinkage`, `medianlinkage`, and `wardslinkage` is `L2squared`. This option is not case sensitive. See [MV] ***measure_option*** for a discussion of these measures.

Several authors advise using the `L2squared` *measure* exclusively with centroid, median, and Ward's linkage. See *Dissimilarity transformations and the Lance and Williams formula* and *Warning concerning similarity or dissimilarity choice* in [MV] **cluster** for details.

`name(`*clname*`)` specifies the name to attach to the resulting cluster analysis. If `name()` is not specified, Stata finds an available cluster name, displays it for your reference, and attaches the name to your cluster analysis.

Advanced

`generate(`*stub*`)` provides a prefix for the variable names created by `cluster` *linkage*. By default, the variable name prefix will be the name specified in `name()`. Three variables with the suffixes `_id`, `_ord`, and `_hgt` are created and attached to the cluster-analysis results. Users generally will not need to access these variables directly.

Centroid linkage and median linkage can produce reversals or crossovers; see [MV] **cluster** for details. When reversals happen, `cluster centroidlinkage` and `cluster medianlinkage` also create a fourth variable with the suffix `_pht`. This is a pseudoheight variable that is used by some postclustering commands to properly interpret the `_hgt` variable.

Options for clustermat linkage commands

Main

`shape(`*shape*`)` specifies the storage mode of *matname*, the matrix of dissimilarities. `shape(full)` is the default. The following shapes are allowed:

`full` specifies that *matname* is an $n \times n$ symmetric matrix.

`lower` specifies that *matname* is a row or column vector of length $n(n+1)/2$, with the rowwise lower triangle of the dissimilarity matrix including the diagonal of zeros.

$$\mathrm{D}_{11}\ \mathrm{D}_{21}\ \mathrm{D}_{22}\ \mathrm{D}_{31}\ \mathrm{D}_{32}\ \mathrm{D}_{33}\ \ldots\ \mathrm{D}_{n1}\ \mathrm{D}_{n2}\ \ldots\ \mathrm{D}_{nn}$$

`llower` specifies that *matname* is a row or column vector of length $n(n-1)/2$, with the rowwise lower triangle of the dissimilarity matrix excluding the diagonal.

$$\mathrm{D}_{21}\ \mathrm{D}_{31}\ \mathrm{D}_{32}\ \mathrm{D}_{41}\ \mathrm{D}_{42}\ \mathrm{D}_{43}\ \ldots\ \mathrm{D}_{n1}\ \mathrm{D}_{n2}\ \ldots\ \mathrm{D}_{n,n-1}$$

`upper` specifies that *matname* is a row or column vector of length $n(n+1)/2$, with the rowwise upper triangle of the dissimilarity matrix including the diagonal of zeros.

$$\mathrm{D}_{11}\ \mathrm{D}_{12}\ \ldots\ \mathrm{D}_{1n}\ \mathrm{D}_{22}\ \mathrm{D}_{23}\ \ldots\ \mathrm{D}_{2n}\ \mathrm{D}_{33}\ \mathrm{D}_{34}\ \ldots\ \mathrm{D}_{3n}\ \ldots\ \mathrm{D}_{nn}$$

`uupper` specifies that *matname* is a row or column vector of length $n(n-1)/2$, with the rowwise upper triangle of the dissimilarity matrix excluding the diagonal.

$$D_{12}\ D_{13}\ \dots\ D_{1n}\ D_{23}\ D_{24}\ \dots\ D_{2n}\ D_{34}\ D_{35}\ \dots\ D_{3n}\ \dots\ D_{n-1,n}$$

`add` specifies that `clustermat`'s results be added to the dataset currently in memory. The number of observations (selected observations based on the `if` and `in` qualifiers) must equal the number of rows and columns of *matname*. Either `clear` or `add` is required if a dataset is currently in memory.

`clear` drops all the variables and cluster solutions in the current dataset in memory (even if that dataset has changed since the data were last saved) before generating `clustermat`'s results. Either `clear` or `add` is required if a dataset is currently in memory.

`labelvar(`*varname*`)` specifies the name of a new variable to be created containing the row names of matrix *matname*.

`name(`*clname*`)` specifies the name to attach to the resulting cluster analysis. If `name()` is not specified, Stata finds an available cluster name, displays it for your reference, and attaches the name to your cluster analysis.

Advanced

`force` allows computations to continue when *matname* is nonsymmetric or has nonzeros on the diagonal. By default, `clustermat` will complain and exit when it encounters these conditions. `force` specifies that `clustermat` operate on the symmetric matrix $(matname * matname')/2$, with any nonzero diagonal entries treated as if they were zero.

`generate(`*stub*`)` provides a prefix for the variable names created by `clustermat`. By default, the variable name prefix is the name specified in `name()`. Three variables are created and attached to the cluster-analysis results with the suffixes `_id`, `_ord`, and `_hgt`. Users generally will not need to access these variables directly.

Centroid linkage and median linkage can produce reversals or crossovers; see [MV] **cluster** for details. When reversals happen, `clustermat centroidlinkage` and `clustermat medianlinkage` also create a fourth variable with the suffix `_pht`. This is a pseudoheight variable that is used by some of the postclustering commands to properly interpret the `_hgt` variable.

Remarks

▷ Example 1

As the senior data analyst for a small biotechnology firm, you are given a dataset with four chemical laboratory measurements on 50 different samples of a particular plant gathered from the rain forest. The head of the expedition that gathered the samples thinks, based on information from the natives, that an extract from the plant might reduce the negative side effects associated with your company's best-selling nutritional supplement.

While the company chemists and botanists continue exploring the possible uses of the plant and plan future experiments, the head of product development asks you to look at the preliminary data and to report anything that might be helpful to the researchers.

Although all 50 plants are supposed to be of the same type, you decide to perform a cluster analysis to see if there are subgroups or anomalies among them. You arbitrarily decide to use single-linkage clustering with the default Euclidean distance.

```
. use http://www.stata-press.com/data/r12/labtech
. cluster singlelinkage x1 x2 x3 x4, name(sngeuc)
. cluster list sngeuc
sngeuc  (type: hierarchical,  method: single,  dissimilarity: L2)
      vars: sngeuc_id (id variable)
            sngeuc_ord (order variable)
            sngeuc_hgt (height variable)
     other: cmd: cluster singlelinkage x1 x2 x3 x4, name(sngeuc)
            varlist: x1 x2 x3 x4
            range: 0 .
```

The `cluster singlelinkage` command generated some variables and created a cluster object with the name `sngeuc`, which you supplied as an argument. `cluster list` provides details about the cluster object; see [MV] **cluster utility**.

What you really want to see is the dendrogram for this cluster analysis; see [MV] **cluster dendrogram**.

```
. cluster dendrogram sngeuc, xlabel(, angle(90) labsize(*.75))
```

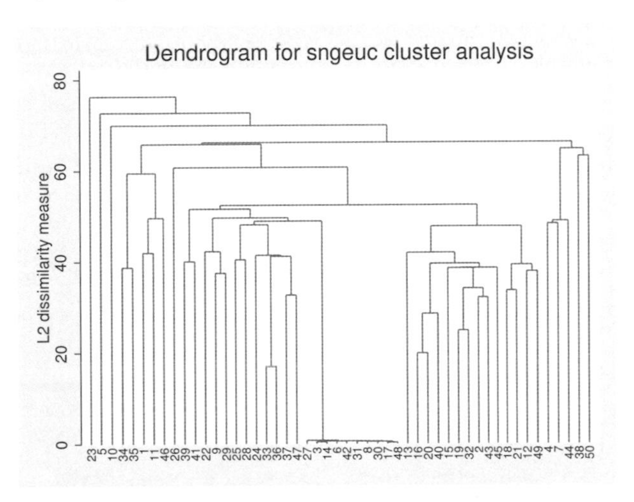

From your experience looking at dendrograms, two things jump out at you about this cluster analysis. The first is the observations showing up in the middle of the dendrogram that are all close to each other (short vertical bars) and are far from any other observations (the long vertical bar connecting them to the rest of the dendrogram). Next you notice that if you ignore those 10 observations, the rest of the dendrogram does not indicate strong clustering, as shown by the relatively short vertical bars in the upper portion of the dendrogram.

You start to look for clues why these 10 observations are so peculiar. Looking at scatterplots is usually helpful, so you examine the matrix of scatterplots.

```
. graph matrix x1 x2 x3 x4
```

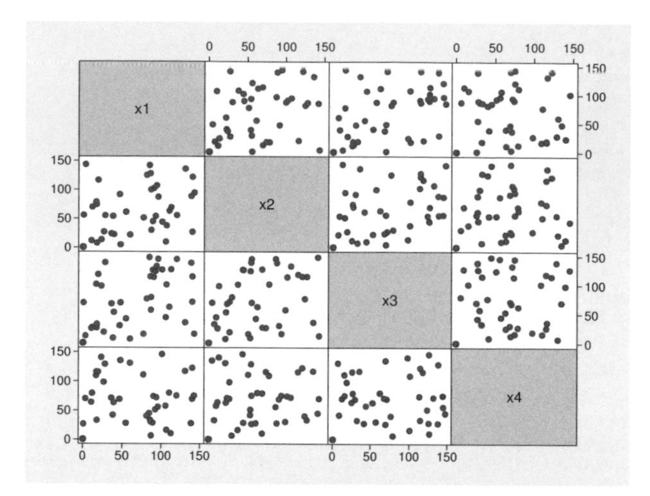

Unfortunately, these scatterplots do not indicate what might be going on.

Suddenly, from your past experience with the laboratory technicians, you have an idea of what to check next. Because of past data mishaps, the company started the policy of placing within each dataset a variable giving the name of the technician who produced the measurement. You decide to view the dendrogram, using the technician's name as the label instead of the default observation number.

```
. cluster dendrogram sngeuc, labels(labtech) xlabel(, angle(90) labsize(*.75))
```

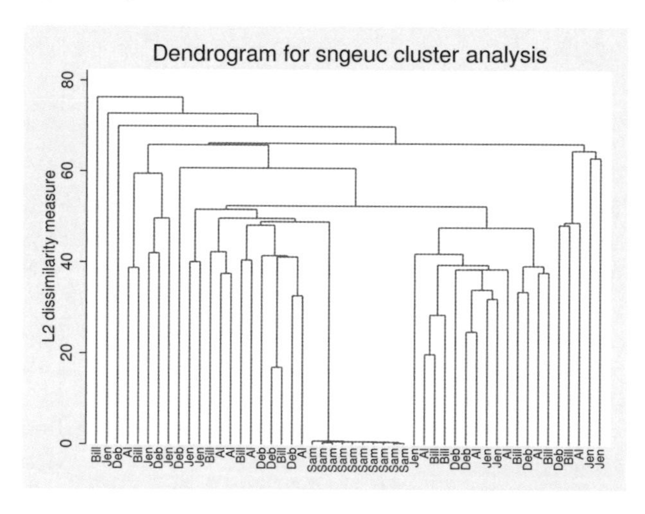

Your suspicions are confirmed. Sam, one of the laboratory technicians, has messed up again. You list the data and see that all his observations are between zero and one, whereas the other four technicians' data range up to about 150, as expected. It looks like Sam forgot, once again, to calibrate his sensor before analyzing his samples. You decide to save a note of your findings with this cluster analysis (see [MV] **cluster notes** for the details) and to send the data back to the laboratory to be fixed.

◁

▷ Example 2

The sociology professor of your graduate-level class gives, as homework, a dataset containing 30 observations on 60 binary variables, with the assignment to tell him something about the 30 subjects represented by the observations. You think that this assignment is too vague, but because your grade depends on it, you get to work trying to figure something out.

Among the analyses you try is the following cluster analysis. You decide to use single-linkage clustering with the simple matching binary coefficient because it is easy to understand. Just for fun, though it makes no difference to you, you specify the `generate()` option to force the generated variables to have `zstub` as a prefix. You let Stata pick a name for your cluster analysis by not specifying the `name()` option.

```
. use http://www.stata-press.com/data/r12/homework, clear
. cluster s a1-a60, measure(matching) gen(zstub)
cluster name: _clus_1
. cluster list
_clus_1  (type: hierarchical,  method: single,  similarity: matching)
      vars: zstub_id (id variable)
            zstub_ord (order variable)
            zstub_hgt (height variable)
     other: cmd: cluster singlelinkage a1-a60, measure(matching) gen(zstub)
            varlist: a1 a2 a3 a4 a5 a6 a7 a8 a9 a10 a11 a12 a13 a14 a15 a16 a17
                 a18 a19 a20 a21 a22 a23 a24 a25 a26 a27 a28 a29 a30 a31 a32
                 a33 a34 a35 a36 a37 a38 a39 a40 a41 a42 a43 a44 a45 a46 a47
                 a48 a49 a50 a51 a52 a53 a54 a55 a56 a57 a58 a59 a60
            range: 1 0
```

Stata selected `_clus_1` as the cluster name and created the variables `zstub_id`, `zstub_ord`, and `zstub_hgt`.

You display the dendrogram by using the `cluster tree` command, which is a synonym for `cluster dendrogram`. Because Stata uses the most recently performed cluster analysis by default, you do not need to type the name.

```
. cluster tree
```

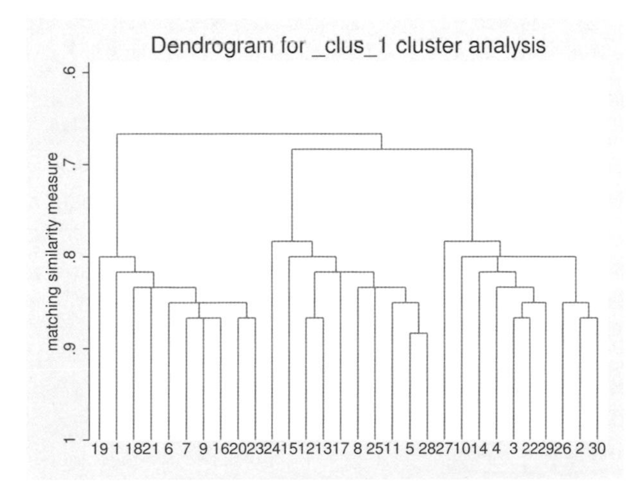

The dendrogram seems to indicate the presence of three groups among the 30 observations. You decide that this is probably the structure your teacher wanted you to find, and you begin to write up your report. You want to examine the three groups further, so you use the `cluster generate` command (see [MV] **cluster generate**) to create a grouping variable to make the task easier. You examine various summary statistics and tables for the three groups and finish your report.

After the assignment is turned in, your professor gives you the same dataset with the addition of one more variable, `truegrp`, which indicates the groupings he thinks are in the data. You do a cross-tabulation of the `truegrp` and `grp3`, your grouping variable, to see if you are going to get a good grade on the assignment.

```
. cluster gen grp3 = group(3)
. table grp3 truegrp
```

grp3	truegrp 1	2	3
1		10	
2			10
3	10		

Other than the numbers arbitrarily assigned to the three groups, both you and your professor agree. You rest easier that night knowing that you may survive one more semester.

In addition to examining single-linkage clustering of these data, you decide to see what median-linkage clustering shows. As with the single-linkage clustering, you pick the simple matching binary coefficient to measure the similarity between groups. The `name()` option is used to attach the name `medlink` to the cluster analysis. `cluster list` displays the details; see [MV] **cluster utility**.

```
. cluster median a1-a60, measure(match) name(medlink)
. cluster list medlink
medlink  (type: hierarchical,  method: median,  similarity: matching)
      vars: medlink_id (id variable)
            medlink_ord (order variable)
            medlink_hgt (real_height variable)
            medlink_pht (pseudo_height variable)
     other: cmd: cluster medianlinkage a1-a60, measure(match) name(medlink)
            varlist: a1 a2 a3 a4 a5 a6 a7 a8 a9 a10 a11 a12 a13 a14 a15 a16 a17
                 a18 a19 a20 a21 a22 a23 a24 a25 a26 a27 a28 a29 a30 a31 a32
                 a33 a34 a35 a36 a37 a38 a39 a40 a41 a42 a43 a44 a45 a46 a47
                 a48 a49 a50 a51 a52 a53 a54 a55 a56 a57 a58 a59 a60
            range: 1 0
```

You attempt to use the `cluster dendrogram` command to display the dendrogram, but because this particular cluster analysis produced reversals, `cluster dendrogram` refuses to produce the dendrogram. You realize that with reversals, the resulting dendrogram would not be easy to interpret anyway.

You use the `cluster generate` command (see [MV] **cluster generate**) to create a three-group grouping variable, based on your median-linkage clustering, to compare with `truegrp`.

```
. cluster gen medgrp3 = group(3)
. table medgrp3 truegrp
```

medgrp3	truegrp 1	2	3
1		10	
2	10		
3			10

Because you were unable to view a dendrogram by using median-linkage clustering, you turn to Ward's linkage clustering method.

```
. cluster ward a1-a60, measure(match) name(wardlink)
. cluster list wardlink
wardlink  (type: hierarchical,  method: wards,  similarity: matching)
      vars: wardlink_id (id variable)
            wardlink_ord (order variable)
            wardlink_hgt (height variable)
     other: cmd: cluster wardslinkage a1-a60, measure(match) name(wardlink)
            varlist: a1 a2 a3 a4 a5 a6 a7 a8 a9 a10 a11 a12 a13 a14 a15 a16 a17
                 a18 a19 a20 a21 a22 a23 a24 a25 a26 a27 a28 a29 a30 a31 a32
                 a33 a34 a35 a36 a37 a38 a39 a40 a41 a42 a43 a44 a45 a46 a47
                 a48 a49 a50 a51 a52 a53 a54 a55 a56 a57 a58 a59 a60
            range: 1 0
. cluster tree wardlink
```

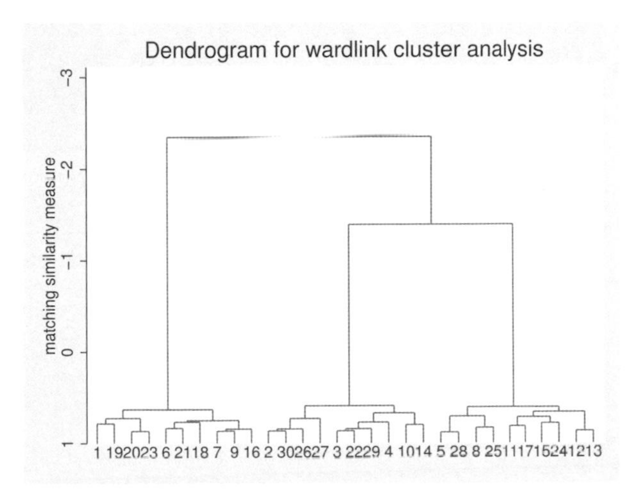

As with single-linkage clustering, the dendrogram from Ward's linkage clustering seems to indicate the presence of three groups among the 30 observations. However, notice the y-axis range for the resulting dendrogram. How can the matching similarity coefficient range from 1 to less than -2? By definition, the matching coefficient is bounded between 1 and 0. This is an artifact of the way Ward's linkage clustering is defined, and it underscores the warning mentioned in the discussion of the choice of *measure*. Also see *Dissimilarity transformations and the Lance and Williams formula* and *Warning concerning similarity or dissimilarity choice* in [MV] **cluster** for more details.

A cross-tabulation of `truegrp` and `wardgrp3`, a three-group grouping variable from this cluster analysis, is shown next.

```
. cluster generate wardgrp3 = group(3)
. table wardgrp3 truegrp
```

wardgrp3	truegrp 1	2	3
1		10	
2	10		
3			10

Other than the numbers arbitrarily assigned to the three groups, your teacher's conclusions and the results from the Ward's linkage clustering agree. So, despite the warning against using something other than squared Euclidean distance with Ward's linkage, you were still able to obtain a reasonable cluster-analysis solution with the matching similarity coefficient.

◁

▷ Example 3

The wclub dataset contains answers from 30 women to 35 yes–no questions. The variables are described in example 3 of [MV] **clustermat**. We are interested in seeing how weighted-average linkage clustering will cluster the 35 variables (instead of the observations).

We use the `matrix dissimilarity` command to produce a dissimilarity matrix equal to one minus the Jaccard similarity; see [MV] **matrix dissimilarity**.

```
. use http://www.stata-press.com/data/r12/wclub, clear
. matrix dissimilarity clubD = , variables Jaccard dissim(oneminus)
. clustermat waverage clubD, name(clubwav) clear labelvar(question)
obs was 0, now 35
. cluster dendrogram clubwav, labels(question)
                    xlabel(, angle(90) labsize(*.75))
                    title(Weighted-average linkage clustering)
                    ytitle(1 - Jaccard similarity, suffix)
```

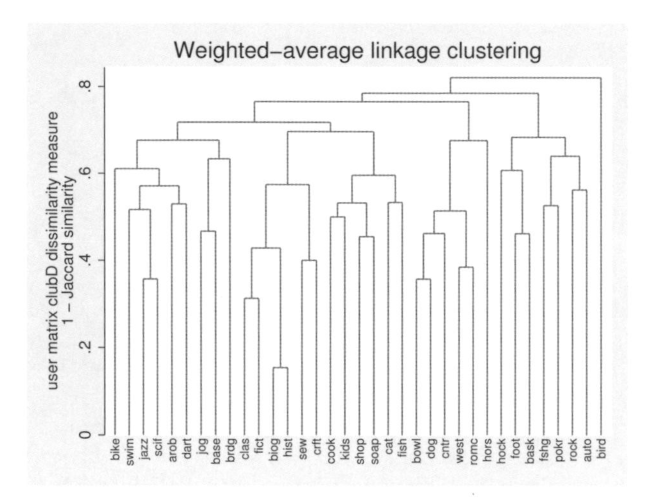

From these 30 women, we see that the `biog` (enjoy reading biographies) and `hist` (enjoy reading history) questions were most closely related. `bird` (have a bird) seems to be the least related to the other variables. It merges last into the supergroup containing the remaining variables.

◁

❑ Technical note

`cluster` commands require a significant amount of memory and execution time. With many observations, the execution time may be significant.

❑

Methods and formulas

All `cluster` *linkage* and `clustermat` *linkage* commands listed above are implemented as ado-files.

[MV] **cluster** discusses and compares the hierarchical clustering methods.

Conceptually, hierarchical agglomerative linkage clustering proceeds as follows. The N observations start out as N separate groups, each of size one. The two closest observations are merged into one group, producing $N-1$ total groups. The closest two groups are then merged so that there are $N-2$ total groups. This process continues until all the observations are merged into one large group, producing a hierarchy of groupings from one group to N groups. The difference between the various hierarchical-linkage methods depends on how they define "closest" when comparing groups.

For single-linkage clustering, the closest two groups are determined by the closest observations between the two groups.

In complete linkage, the closest two groups are determined by the farthest observations between the two groups.

For average-linkage clustering, the closest two groups are determined by the average (dis)similarity between the observations of the two groups.

The Lance–Williams formula provides the basis for extending the well-known Ward's method of clustering into the general hierarchical-linkage framework that allows a choice of (dis)similarity measures.

Centroid linkage merges the groups whose means are closest.

Weighted-average linkage clustering is similar to average-linkage clustering, except that it gives each group of observations equal weight. Average linkage gives each observation equal weight.

Median linkage is a variation on centroid linkage in that it treats groups of unequal size differently. Centroid linkage gives each observation equal weight. Median linkage, however, gives each group of observations equal weight, meaning that with unequal group sizes, the observations in the smaller group will have more weight than the observations in the larger group.

The linkage clustering algorithm produces two variables that together act as a pointer representation of a dendrogram. To this, Stata adds a third variable used to restore the sort order, as needed, so that the two variables of the pointer representation remain valid. The first variable of the pointer representation gives the order of the observations. The second variable has one less element and gives the height in the dendrogram at which the adjacent observations in the order variable join.

When reversals happen, a fourth variable, called a pseudoheight, is produced and is used by postclustering commands with the height variable to properly interpret the ordering of the hierarchy.

See [MV] ***measure_option*** for the details and formulas of the available *measures*, which include (dis)similarity measures for continuous and binary data.

Joe H. Ward, Jr. (1926–) obtained degrees in mathematics and educational psychology from the University of Texas. He worked as a personnel research psychologist for the U.S. Air Force Human Resources Laboratory, applying educational psychology, statistics, and computers to a wide variety of problems.

Also see

[MV] **cluster dendrogram** — Dendrograms for hierarchical cluster analysis

[MV] **cluster generate** — Generate summary or grouping variables from a cluster analysis

[MV] **cluster notes** — Place notes in cluster analysis

[MV] **cluster stop** — Cluster-analysis stopping rules

[MV] **cluster utility** — List, rename, use, and drop cluster analyses

[MV] **cluster** — Introduction to cluster-analysis commands

[MV] **clustermat** — Introduction to clustermat commands

Title

cluster notes — Place notes in cluster analysis

Syntax

Add a note to a cluster analysis

`cluster notes` *clname* : *text*

List all cluster notes

`cluster notes`

List cluster notes associated with specified cluster analyses

`cluster notes` *clnamelist*

Drop cluster notes

`cluster notes drop` *clname* [`in` *numlist*]

Menu

Statistics > Multivariate analysis > Cluster analysis > Postclustering > Cluster analysis notes

Description

The `cluster notes` command attaches notes to a previously run cluster analysis. The notes become part of the data and are saved when the data are saved and retrieved when the data are used; see [D] **save**.

To add a note to a cluster analysis, type `cluster notes`, the cluster-analysis name, a colon, and the text.

Typing `cluster notes` by itself lists all cluster notes associated with all defined cluster analyses. `cluster notes` followed by one or more cluster names lists the notes for those cluster analyses.

`cluster notes drop` allows you to drop cluster notes. If `in` *numlist* argument is omitted, all notes for *clname* are dropped.

See [MV] **cluster** for information on the available cluster analysis commands.

Remarks

The cluster-analysis system in Stata has many features that allow you to manage the various cluster analyses that you perform. See [MV] **cluster** for information on all the available cluster-analysis commands; see [MV] **cluster utility** for other `cluster` commands, including `cluster list`, that help you manage your analyses. The `cluster notes` command is modeled after Stata's `notes` command (see [D] **notes**), but they are different systems and do not interact.

▷ Example 1

We illustrate the `cluster notes` command starting with three cluster analyses that have already been performed. The `cluster dir` command shows us the names of all the existing cluster analyses; see [MV] **cluster utility**.

```
. cluster dir
sngeuc
sngabs
kmn3abs
. cluster note sngabs : I used single linkage with absolute value distance
. cluster note sngeuc : Euclidean distance and single linkage
. cluster note kmn3abs : This has the kmeans cluster results for 3 groups
. cluster notes
sngeuc
     notes:   1. Euclidean distance and single linkage
sngabs
     notes:   1. I used single linkage with absolute value distance
kmn3abs
     notes:   1. This has the kmeans cluster results for 3 groups
```

After adding a note to each of the three cluster analyses, we used the `cluster notes` command without arguments to list all the notes for all the cluster analyses.

The `*` and `?` characters may be used when referring to cluster names; see [U] **11.2 Abbreviation rules**.

```
. cluster note k* : Verify that observation 5 is correct.  I am suspicious that
> there was a typographical error or instrument failure in recording the
> information.
. cluster notes kmn3abs
kmn3abs
     notes:   1. This has the kmeans cluster results for 3 groups
              2. Verify that observation 5 is correct. I am suspicious that
                 there was a typographical error or instrument failure in
                 recording the information.
```

`cluster notes` expanded `k*` to `kmn3abs`, the only cluster name that begins with a `k`. Notes that extend to multiple lines are automatically wrapped when displayed. When entering long notes, you just continue to type until your note is finished. Pressing *Return* signals that you are done with that note.

After examining the dendrogram (see [MV] **cluster dendrogram**) for the `sngeuc` single-linkage cluster analysis and seeing one small group of data that split off from the main body of data at a very large distance, you investigate further and find data problems. You decide to add some notes to the `sngeuc` analysis.

```
. cluster note *euc : All of Sam's data look wrong to me.
. cluster note *euc : I think Sam should be fired.
. cluster notes sng?*
sngeuc
     notes:   1. Euclidean distance and single linkage
              2. All of Sam's data look wrong to me.
              3. I think Sam should be fired.
sngabs
     notes:   1. I used single linkage with absolute value distance
```

Sam, one of the lab technicians, who happens to be the owner's nephew and is paid more than you, really messed up. After adding these notes, you get second thoughts about keeping the notes attached to the cluster analysis (and the data). You decide you really want to delete those notes and to add a more politically correct note.

```
. cluster note sngeuc : Ask Jennifer to help Sam reevaluate his data.
. cluster note sngeuc
sngeuc
     notes:   1. Euclidean distance and single linkage
              2. All of Sam's data looks wrong to me.
              3. I think Sam should be fired.
              4. Ask Jennifer to help Sam reevaluate his data.
. cluster note drop sngeuc in 2/3
. cluster notes kmn3abs s*
kmn3abs
     notes:   1. This has the kmeans cluster results for 3 groups
              2. Verify that observation 5 is correct. I am suspicious that
                 there was a typographical error or instrument failure in
                 recording the information.
sngeuc
     notes:   1. Euclidean distance and single linkage
              2. Ask Jennifer to help Sam reevaluate his data.
sngabs
     notes:   1. I used single linkage with absolute value distance
```

Just for illustration purposes, the new note was added before deleting the two offending notes. `cluster notes drop` can take an `in` argument followed by a list of note numbers. The numbers correspond to those shown in the listing provided by the `cluster notes` command. After the deletions, the note numbers are reassigned to remove gaps. So `sngeuc` note 4 becomes note 2 after the deletion of notes 2 and 3 as shown above.

Without an `in` argument, the `cluster notes drop` command drops all notes associated with the named cluster.

◁

Remember that the cluster notes are stored with the data and, as with other updates you make to the data, the additions and deletions are not permanent until you save the data; see [D] **save**.

❑ Technical note

Programmers can access the notes (and all the other cluster attributes) by using the `cluster query` command; see [MV] **cluster programming utilities**.

❑

Methods and formulas

`cluster notes` is implemented as an ado-file.

Also see

[MV] **cluster programming utilities** — Cluster-analysis programming utilities

[MV] **cluster utility** — List, rename, use, and drop cluster analyses

[D] **notes** — Place notes in data

[D] **save** — Save datasets

[MV] **cluster** — Introduction to cluster-analysis commands

[MV] **clustermat** — Introduction to clustermat commands

Title

cluster programming subroutines — Add cluster-analysis routines

Description

This entry describes how to extend Stata's `cluster` command; see [MV] **cluster**. Programmers can add subcommands to `cluster`, add functions to `cluster generate` (see [MV] **cluster generate**), add stopping rules to `cluster stop` (see [MV] **cluster stop**), and set up an alternative command to be executed when `cluster dendrogram` is called (see [MV] **cluster dendrogram**).

The `cluster` command also provides utilities for programmers; see [MV] **cluster programming utilities** to learn more.

Remarks

Remarks are presented under the following headings:

Adding a cluster subroutine

You add a `cluster` subroutine by creating a Stata program with the name `cluster_`*subcmdname*. For example, to add the subcommand `xyz` to `cluster`, create `cluster_xyz.ado`. Users could then execute the `xyz` subcommand with

```
cluster xyz ...
```

Everything entered on the command line after `cluster xyz` is passed to the `cluster_xyz` command.

You can add new clustering methods, new cluster-management tools, and new postclustering programs. The `cluster` command has subcommands that can be helpful to cluster-analysis programmers; see [MV] **cluster programming utilities**.

▷ Example 1

We will add a `cluster` subroutine by writing a simple postcluster-analysis routine that provides a cross-tabulation of two cluster-analysis grouping variables. The syntax of the new command will be

`cluster mycrosstab` *clname1 clname2* [, *tabulate_options*]

Here is the program:

```
program cluster_mycrosstab
        version 12
        gettoken clname1 0 : 0 , parse(" ,")
        gettoken clname2 rest : 0 , parse(" ,")
        cluster query `clname1'
        local groupvar1 `r(groupvar)'
        cluster query `clname2'
        local groupvar2 `r(groupvar)'
        tabulate `groupvar1' `groupvar2' `rest'
end
```

See [P] **gettoken** for information on the `gettoken` command, and see [R] **tabulate twoway** for information on the `tabulate` command. The `cluster query` command is one of the cluster programming utilities that is documented in [MV] **cluster programming utilities**.

We can demonstrate `cluster mycrosstab` in action. This example starts with two cluster analyses, `cl1` and `cl2`. The dissimilarity measure and the variables included in the two cluster analyses differ. We want to see how closely the two cluster analyses match.

```
. use http://www.stata-press.com/data/r12/auto
(1978 Automobile Data)
. cluster kmeans gear head tr, L1 k(5) name(cl1) start(krandom(55234))
> gen(cl1gvar)
. cluster kmeans tr tu mpg, L(1.5) k(5) name(cl2) start(krandom(22132))
> gen(gvar2)
. cluster list, type method dissim var
cl2  (type: partition,  method: kmeans,  dissimilarity: L(1.5))
      vars: gvar2 (group variable)
cl1  (type: partition,  method: kmeans,  dissimilarity: L1)
      vars: cl1gvar (group variable)
. cluster mycrosstab cl1 cl2, chi2
```

	gvar2					
cl1gvar	1	2	3	4	5	Total
1	0	0	10	0	0	10
2	1	4	0	5	6	16
3	8	0	0	4	1	13
4	9	0	8	4	0	21
5	0	8	0	0	6	14
Total	18	12	18	13	13	74

```
          Pearson chi2(16) =  98.4708   Pr = 0.000
```

The `chi2` option was included to demonstrate that we were able to exploit the existing options of `tabulate` with little programming effort. We just pass along to `tabulate` any of the extra arguments received by `cluster_mycrosstab`.

◁

Adding a cluster generate function

Programmers can add functions to the `cluster generate` command (see [MV] **cluster generate**) by creating a command called `clusgen_`*name*. For example, to add a function called `abc()` to `cluster generate`, you could create `clusgen_abc.ado`. Users could then execute

`cluster generate` *newvar* `= abc(` ... `)` ...

Everything entered on the command line following `cluster generate` is passed to `clusgen_abc`.

▷ Example 2

Here is the beginning of a `clusgen_abc` program that expects an integer argument and has one option called `name(`*clname*`)`, which gives the name of the cluster. If `name()` is not specified, the name defaults to that of the most recently performed cluster analysis. We will assume, for illustration purposes, that the cluster analysis must be `hierarchical` and will check for this in the `clusgen_abc` program.

```
program clusgen_abc
        version 12
        // we use gettoken to work our way through the parsing
        gettoken newvar 0 : 0 , parse(" =")
        gettoken temp 0 : 0 , parse(" =")
        if `"`temp'"' != "=" {
                error 198
        }
        gettoken temp 0 : 0 , parse(" (")
        if `"`temp'"' != "abc" {
                error 198
        }
        gettoken funcarg 0 : 0 , parse(" (") match(temp)
        if `"`temp'"' != "(" {
                error 198
        }
        // funcarg holds the integer argument to abc()
        confirm integer number `funcarg'
        // we can now use syntax to parse the option
        syntax [, Name(str) ]
        // cluster query will give us the list of cluster names
        if `"`name'"' == "" {
                cluster query
                local clnames `r(names)'
                if "`clnames'" == "" {
                        di as err "no cluster solutions defined"
                        exit 198
                }
                // first name in the list is the latest clustering
                local name : word 1 of `clnames'
        }
        // cluster query followed by name will tell us the type
        cluster query `name'
        if "`r(type)'" != "hierarchical" {
                di as err "only allowed with hierarchical clustering"
                exit 198
        }
        /*
           you would now pull more information from the call of
                     cluster query `name'
           and do your computations and generate `newvar'
        */
        ...
end
```

See [MV] **cluster programming utilities** for details on the `cluster query` command.

◁

Adding a cluster stopping rule

Programmers can add stopping rules to the `rule()` option of the `cluster stop` command (see [MV] **cluster stop**) by creating a Stata program with the name `clstop_`*name*. For example, to add a stopping rule named `mystop` so that `cluster stop` would now have a `rule(mystop)` option, you could create `clstop_mystop.ado` defining the `clstop_mystop` program. Users could then execute

`cluster stop` [*clname*] `, rule(mystop)` ...

The `clstop_mystop` program is passed the cluster name (*clname*) provided by the user (or the name of the current cluster result if no name is specified), followed by a comma and all the options entered by the user except for the `rule(mystop)` option.

▷ Example 3

We will add a `rule(stepsize)` option to `cluster stop`. This option implements the simple step-size stopping rule (see Milligan and Cooper 1985), which computes the difference in fusion values between levels in a hierarchical cluster analysis. (A fusion value is the similarity or dissimilarity measure at which clusters are fused or split in the hierarchical cluster structure.) Large values of the step-size stopping rule indicate groupings with more distinct cluster structure.

Examining cluster dendrograms (see [MV] **cluster dendrogram**) to visually determine the number of clusters is equivalent to using a visual approximation to the step-size stopping rule.

Here is the `clstop_stepsize` program:

```
program clstop_stepsize, sortpreserve rclass
        version 12
        syntax anything(name=clname) [, Depth(integer -1) ]
        cluster query `clname'
        if "`r(type)'" != "hierarchical" {
                di as error ///
                    "rule(stepsize) only allowed with hierarchical clustering"
                exit 198
        }
        if "`r(pseudo_heightvar)'" != "" {
                di as error "dendrogram reversals encountered"
                exit 198
        }
        local hgtvar `r(heightvar)'
        if `"`r(similarity)'"' != "" {
                sort `hgtvar'
                local negsign "-"
        }
        else if `"`r(dissimilarity)'"' != "" {
                gsort -`hgtvar'
        }
        else {
                di as error "dissimilarity or similarity not set"
                exit 198
        }
        quietly count if !missing(`hgtvar')
        local depth = cond(`depth'<=1, r(N), min(`depth',r(N)))
        tempvar diff
        qui gen double `diff'=`negsign'(`hgtvar'-`hgtvar'[_n+1]) if _n<`depth'
        di
        di as txt "Depth" _col(10) "Stepsize"
        di as txt "{hline 17}"
        forvalues i = 1/`= `depth'-1' {
                local j = `i' + 1
                di as res `j' _col(10) %8.0g `diff'[`i']
                return scalar stepsize_`j' = `diff'[`i']
        }
        return local rule "stepsize"
end
```

See [P] **syntax** for information about the `syntax` command, [P] **forvalues** for information about the `forvalues` looping command, and [P] **macro** for information about the `‘= ... ’` macro function. The `cluster query` command is one of the cluster programming utilities that is documented in [MV] **cluster programming utilities**.

With this program, users can obtain the step-size stopping rule. We demonstrate this process by using an average-linkage hierarchical cluster analysis on the data found in the second example of [MV] **cluster linkage**. The dataset contains 30 observations on 60 binary variables. The simple matching coefficient is used as the similarity measure in the average-linkage clustering.

```
. use http://www.stata-press.com/data/r12/homework, clear
. cluster a a1-a60, measure(match) name(alink)
. cluster stop alink, rule(stepsize) depth(15)
Depth     Stepsize

2         .065167
3         .187333
4          .00625
5         .007639
6         .002778
7         .005952
8         .002381
9         .008333
10        .005556
11        .002778
12              0
13              0
14        .006667
15            .01
```

In the `clstop_stepsize` program, we included a `depth()` option. `cluster stop`, when called with the new `rule(stepsize)` option, can also have the `depth()` option. Here we specified that it stop at a depth of 15.

The largest step size, .187, happens at the three-group level of the hierarchy. This number, .187, represents the difference between the matching coefficient created when two groups are formed and that created when three groups are formed in this hierarchical cluster analysis.

The `clstop_stepsize` program could be enhanced by using a better output table format. An option could also be added that saves the results to a matrix.

◁

Applying an alternate cluster dendrogram routine

Programmers can change the behavior of the `cluster dendrogram` command (alias `cluster tree`); see [MV] **cluster dendrogram**. This task is accomplished by using the `other()` option of the `cluster set` command (see [MV] **cluster programming utilities**) with a *tag* of `treeprogram` and with *text* giving the name of the command to be used in place of the standard Stata program for `cluster dendrogram`. For example, if you had created a new hierarchical cluster-analysis method for Stata that needed a different algorithm for producing dendrograms, you would use the command

`cluster set` *clname*`, other(treeprogram` *progname*`)`

to set *progname* as the program to be executed when `cluster dendrogram` is called.

▷ Example 4

If we were creating a new hierarchical cluster-analysis method called `myclus`, we could create a program called `cluster_myclus` (see *Adding a cluster subroutine*). If `myclus` needed a different dendrogram routine from the standard one used within Stata, we could include the following line in `cluster_myclus.ado` at the point where we set the cluster attributes.

```
cluster set `clname', other(treeprogram myclustree)
```

We could then create a program called `myclustree` in a file called `myclustree.ado` that implements the particular dendrogram program needed by `myclus`.

◁

Reference

Milligan, G. W., and M. C. Cooper. 1985. An examination of procedures for determining the number of clusters in a dataset. *Psychometrika* 50: 159–179.

Also see

[MV] **cluster** — Introduction to cluster-analysis commands

[MV] **clustermat** — Introduction to clustermat commands

[MV] **cluster programming utilities** — Cluster-analysis programming utilities

Title

cluster programming utilities — Cluster-analysis programming utilities

Syntax

Obtain various attributes of a cluster analysis

cluster query [*clname*]

Set various attributes of a cluster analysis

cluster set [*clname*] [, *set_options*]

Delete attributes from a cluster analysis

cluster delete *clname* [, *delete_options*]

Check similarity and dissimilarity measure name

cluster parsedistance *measure*

Compute similarity and dissimilarity measure

cluster measures *varlist* [*if*] [*in*] , compare(*numlist*) generate(*newvarlist*) [*measures_options*]

set_options	Description
addname	add *clname* to the master list of cluster analyses
type(*type*)	set the cluster type for *clname*
method(*method*)	set the name of the clustering method for the cluster analysis
similarity(*measure*)	set the name of the similarity measure used for the cluster analysis
dissimilarity(*measure*)	set the name of the dissimilarity measure used for the cluster analysis
var(*tag varname*)	set *tag* that points to *varname*
char(*tag charname*)	set *tag* that points to *charname*
other(*tag text*)	set *tag* with *text* attached to the tag marker
note(*text*)	add a note to the *clname*

delete_options	Description
zap	delete all possible settings for *clname*
delname	remove *clname* from the master list of current cluster analyses
type	delete the cluster type entry from *clname*
method	delete the cluster method entry from *clname*
similarity	delete the similarity entries from *clname*
dissimilarity	delete the dissimilarity entries from *clname*
notes(*numlist*)	delete the specified numbered notes from *clname*
allnotes	remove all notes from *clname*
var(*tag*)	remove *tag* from *clname*
allvars	remove all the entries pointing to variables for *clname*
varzap(*tag*)	same as var(), but also delete the referenced variable
allvarzap	same as allvars, but also delete the variables
char(*tag*)	remove *tag* that points to a Stata characteristic from *clname*
allchars	remove all entries pointing to Stata characteristics for *clname*
charzap(*tag*)	same as char(), but also delete the characteristic
allcharzap	same as allchars, but also delete the characteristics
other(*tag*)	delete *tag* and its associated text from *clname*
allothers	delete all entries from *clname* that have been set using other()

measures_options	Description
* compare(*numlist*)	use *numlist* as the comparison observations
* generate(*newvarlist*)	create *newvarlist* variables
measure	(dis)similarity measure; see *Options for cluster measures* for available measures; default is L2
propvars	interpret observations implied by if and in as proportions of binary observations
propcompares	interpret comparison observations as proportions of binary observations

*compare(*numlist*) and generate(*newvarlist*) are required.

Description

The cluster query, cluster set, cluster delete, cluster parsedistance, and cluster measures commands provide tools for programmers to add their own cluster-analysis subroutines to Stata's cluster command; see [MV] **cluster** and [MV] **cluster programming subroutines**. These commands make it possible for the new command to take advantage of Stata's cluster-management facilities.

cluster query provides a way to obtain the various attributes of a cluster analysis in Stata. If *clname* is omitted, cluster query returns in r(names) a list of the names of all currently defined cluster analyses. If *clname* is provided, the various attributes of the specified cluster analysis are returned in r(). These attributes include the type, method, (dis)similarity used, created variable names, notes, and any other information attached to the cluster analysis.

`cluster set` allows you to set the various attributes that define a cluster analysis in Stata, including naming your cluster results and adding the name to the master list of currently defined cluster results. With `cluster set`, you can provide information on the type, method, and (dis)similarity measure of your cluster-analysis results. You can associate variables and Stata characteristics (see [P] **char**) with your cluster analysis. `cluster set` also allows you to add notes and other specified fields to your cluster-analysis result. These items become part of the dataset and are saved with the data.

`cluster delete` allows you to delete attributes from a cluster analysis in Stata. This command is the inverse of `cluster set`.

`cluster parsedistance` takes the similarity or dissimilarity *measure* name and checks it against the list of those provided by Stata, taking account of allowed minimal abbreviations and aliases. Aliases are resolved (for instance, `Euclidean` is changed into the equivalent `L2`).

`cluster measures` computes the similarity or dissimilarity *measure* between the observations listed in the `compare()` option and the observations included based on the `if` and `in` conditions and places the results in the variables specified by the `generate()` option. See [MV] **matrix dissimilarity** for the `matrix dissimilarity` command that places (dis)similarities in a matrix.

Stata also provides a method for programmers to extend the `cluster` command by providing subcommands; see [MV] **cluster programming subroutines**.

Options for cluster set

`addname` adds *clname* to the master list of currently defined cluster analyses. When *clname* is not specified, the `addname` option is mandatory, and here, `cluster set` automatically finds a cluster name that is not currently in use and uses this as the cluster name. `cluster set` returns the name of the cluster in `r(name)`. If `addname` is not specified, the *clname* must have been added to the master list previously (for instance, through a previous call to `cluster set`).

`type(`*type*`)` sets the cluster type for *clname*. `type(hierarchical)` indicates that the cluster analysis is hierarchical-style clustering, and `type(partition)` indicates that it is a partition-style clustering. You are not restricted to these types. For instance, you might program some kind of fuzzy partition-clustering analysis, so you then use `type(fuzzy)`.

`method(`*method*`)` sets the name of the clustering method for the cluster analysis. For instance, Stata uses `method(kmeans)` to indicate a kmeans cluster analysis and uses `method(single)` to indicate single-linkage cluster analysis. You are not restricted to the names currently used within Stata.

`similarity(`*measure*`)` and `dissimilarity(`*measure*`)` set the name of the similarity or dissimilarity measure used for the cluster analysis. For example, Stata uses `dissimilarity(L2)` to indicate the L2 or Euclidean distance. You are not restricted to the names currently used within Stata. See [MV] ***measure_option*** and [MV] **cluster** for a listing and discussion of (dis)similarity measures.

`var(`*tag varname*`)` sets a marker called *tag* in the cluster analysis that points to the variable *varname*. For instance, Stata uses `var(group` *varname*`)` to set a grouping variable from a kmeans cluster analysis. With single-linkage clustering, Stata uses `var(id` *idvarname*`)`, `var(order` *ordervarname*`)`, and `var(height` *hgtvarname*`)` to set the `id`, `order`, and `height` variables that define the cluster-analysis result. You are not restricted to the names currently used within Stata. Up to 10 `var()` options may be specified with a `cluster set` command.

`char(`*tag charname*`)` sets a marker called *tag* in the cluster analysis that points to the Stata characteristic named *charname*; see [P] **char**. This characteristic can be either an `_dta[]` dataset characteristic or a variable characteristic. Up to 10 `char()` options may be specified with a `cluster set` command.

`other(`*tag text*`)` sets a marker called *tag* in the cluster analysis with *text* attached to the *tag* marker. Stata uses `other(k` *#*`)` to indicate that `k` (the number of groups) was # in a kmeans cluster analysis. You are not restricted to the names currently used within Stata. Up to 10 `other()` options may be specified with a `cluster set` command.

`note(`*text*`)` adds a note to the *clname* cluster analysis. The `cluster notes` command (see [MV] **cluster notes**) is the command to add, delete, or view cluster notes. The `cluster notes` command uses the `note()` option of `cluster set` to add a note to a cluster analysis. Up to 10 `note()` options may be specified with a `cluster set` command.

Options for cluster delete

`zap` deletes all possible settings for cluster analysis *clname*. It is the same as specifying the `delname`, `type`, `method`, `similarity`, `dissimilarity`, `allnotes`, `allcharzap`, `allothers`, and `allvarzap` options.

`delname` removes *clname* from the master list of current cluster analyses. This option does not affect the various settings that make up the cluster analysis. To remove them, use the other options of `cluster delete`.

`type` deletes the cluster type entry from *clname*.

`method` deletes the cluster method entry from *clname*.

`similarity` and `dissimilarity` delete the similarity and dissimilarity entries, respectively, from *clname*.

`notes(`*numlist*`)` deletes the specified numbered notes from *clname*. The numbering corresponds to the returned results from the `cluster query` *clname* command. The `cluster notes drop` command (see [MV] **cluster notes**) drops a cluster note. It, in turn, calls `cluster delete`, using the `notes()` option to drop the notes.

`allnotes` removes all notes from the *clname* cluster analysis.

`var(`*tag*`)` removes from *clname* the entry labeled *tag* that points to a variable. This option does not delete the variable.

`allvars` removes all the entries pointing to variables for *clname*. This option does not delete the corresponding variables.

`varzap(`*tag*`)` is the same as `var()` and actually deletes the variable in question.

`allvarzap` is the same as `allvars` and actually deletes the variables.

`char(`*tag*`)` removes from *clname* the entry labeled *tag* that points to a Stata characteristic (see [P] **char**). This option does not delete the characteristic.

`allchars` removes all the entries pointing to Stata characteristics for *clname*. This option does not delete the characteristics.

`charzap(`*tag*`)` is the same as `char()` and actually deletes the characteristics.

`allcharzap` is the same as `allchars` and actually deletes the characteristics.

`other(`*tag*`)` deletes from *clname* the *tag* entry and its associated text, which were set by using the `other()` option of the `cluster set` command.

`allothers` deletes all entries from *clname* that have been set using the `other()` option of the `cluster set` command.

Options for cluster measures

`compare(`*numlist*`)` is required and specifies the observations to use as the comparison observations. Each of these observations will be compared with the observations implied by the `if` and `in` conditions, using the specified (dis)similarity *measure*. The results are stored in the corresponding new variable from the `generate()` option. There must be the same number of elements in *numlist* as there are variable names in the `generate()` option.

`generate(`*newvarlist*`)` is required and specifies the names of the variables to be created. There must be as many elements in *newvarlist* as there are numbers specified in the `compare()` option.

measure specifies the similarity or dissimilarity measure. The default is `L2` (synonym `Euclidean`). This option is not case sensitive. See [MV] ***measure_option*** for detailed descriptions of the supported measures.

`propvars` is for use with binary measures and specifies that the observations implied by the `if` and `in` conditions be interpreted as proportions of binary observations. The default action with binary measures treats all nonzero values as one (excluding missing values). With `propvars`, the values are confirmed to be between zero and one, inclusive. See [MV] ***measure_option*** for a discussion of the use of proportions with binary measures.

`propcompares` is for use with binary measures. It indicates that the comparison observations (those specified in the `compare()` option) are to be interpreted as proportions of binary observations. The default action with binary measures treats all nonzero values as one (excluding missing values). With `propcompares`, the values are confirmed to be between zero and one, inclusive. See [MV] ***measure_option*** for a discussion of the use of proportions with binary measures.

Remarks

▷ Example 1

Programmers can determine which cluster solutions currently exist by using the `cluster query` command without specifying a cluster name to return the names of all currently defined clusters.

```
. use http://www.stata-press.com/data/r12/auto
(1978 Automobile Data)
. cluster k gear turn trunk mpg displ, k(6) name(grpk6L2) measure(L2) gen(g6l2)
. cluster k gear turn trunk mpg displ, k(7) name(grpk7L2) measure(L2) gen(g7l2)
. cluster kmed gear turn trunk mpg displ, k(6) name(grpk6L1) measure(L1) gen(g6l1)
. cluster kmed gear turn trunk mpg displ, k(7) name(grpk7L1) measure(L1) gen(g7l1)
. cluster dir
grpk7L1
grpk6L1
grpk7L2
grpk6L2
. cluster query
. return list
macros:
              r(names) : "grpk7L1 grpk6L1 grpk7L2 grpk6L2"
```

Here there are four cluster solutions. A programmer can further process the `r(names)` returned macro. For example, to determine which current cluster solutions used kmeans clustering, we would loop through these four cluster solution names and, for each one, call `cluster query` to determine its properties.

```
. local clusnames `r(names)'
. foreach cname of local clusnames {
  2.         cluster query `cname'
  3.         if "`r(method)'" == "kmeans" {
  4.                 local kmeancls `kmeancls' `cname'
  5.         }
  6. }
. di "{tab}Cluster analyses using kmeans: `kmeancls'"
        Cluster analyses using kmeans: grpk7L2 grpk6L2
```

Here we examined `r(method)`, which records the name of the cluster-analysis method. Two of the four cluster solutions used kmeans.

◁

▷ Example 2

We interactively demonstrate `cluster set`, `cluster delete`, and `cluster query`, though in practice these would be used within a program.

First, we add the name `myclus` to the master list of cluster analyses and, at the same time, set the type, method, and similarity.

```
. cluster set myclus, addname type(madeup) method(fake) similarity(who knows)
. cluster query
. return list
macros:
             r(names) : "myclus grpk7L1 grpk6L1 grpk7L2 grpk6L2"
. cluster query myclus
. return list
macros:
              r(name) : "myclus"
        r(similarity) : "who knows"
            r(method) : "fake"
              r(type) : "madeup"
```

`cluster query` shows that `myclus` was successfully added to the master list of cluster analyses and that the attributes that were `cluster set` can also be obtained.

Now we add a reference to a variable. We will use the word `group` as the *tag* for a variable `mygrpvar`. We also add another item called `xyz` and associate some text with the `xyz` item.

```
. cluster set myclus, var(group mygrpvar) other(xyz some important info)
. cluster query myclus
. return list
macros:
              r(name) : "myclus"
            r(o1_val) : "some important info"
            r(o1_tag) : "xyz"
          r(groupvar) : "mygrpvar"
           r(v1_name) : "mygrpvar"
            r(v1_tag) : "group"
        r(similarity) : "who knows"
            r(method) : "fake"
              r(type) : "madeup"
```

The cluster query command returned the mygrpvar information in two ways. The first way is with r(v#_tag) and r(v#_name). Here there is only one variable associated with myclus, so we have r(v1_tag) and r(v1_name). This information allows the programmer to loop over all the saved variable names without knowing beforehand what the *tags* might be or how many there are. You could loop as follows:

```
local i 1
while "`r(v`i'_tag)'" != "" {
        ...
        local ++i
}
```

The second way the variable information is returned is in an r() result with the *tag* name appended by var, r(*tag*var). In our example, this is r(groupvar). This second method is convenient when, as the programmer, you know exactly which *varname* information you are seeking.

The same logic applies to characteristic attributes that are cluster set.

Now we continue with our interactive example:

```
. cluster delete myclus, method var(group)
. cluster set myclus, note(a note) note(another note) note(a third note)
. cluster query myclus
. return list
macros:
              r(name) : "myclus"
             r(note3) : "a third note"
             r(note2) : "another note"
             r(note1) : "a note"
            r(o1_val) : "some important info"
            r(o1_tag) : "xyz"
        r(similarity) : "who knows"
              r(type) : "madeup"
```

We used cluster delete to remove the method and the group variable we had associated with myclus. Three notes were then added simultaneously by using the note() option of cluster set. In practice, users will use the cluster notes command (see [MV] **cluster notes**) to add and delete cluster notes. The cluster notes command is implemented with the cluster set and cluster delete programming commands.

We finish our interactive demonstration of these commands by deleting more attributes from myclus and then eliminating myclus. In practice, users would remove a cluster analysis with the cluster drop command (see [MV] **cluster utility**), which is implemented with the zap option of the cluster delete command.

```
. cluster delete myclus, allnotes similarity
. cluster query myclus
. return list
macros:
              r(name) : "myclus"
            r(o1_val) : "some important info"
            r(o1_tag) : "xyz"
              r(type) : "madeup"
. cluster delete myclus, zap
. cluster query
. return list
macros:
             r(names) : "grpk7L1 grpk6L1 grpk7L2 grpk6L2"
```

The cluster attributes that are `cluster set` become a part of the dataset. They are saved with the dataset when it is `saved` and are available again when the dataset is `used`; see [D] **save**.

◁

❑ Technical note

You may wonder how Stata's cluster-analysis data structures are implemented. Stata data characteristics (see [P] **char**) hold the information. The details of the implementation are not important, and in fact, we encourage you to use the `set`, `delete`, and `query` subcommands to access the cluster attributes. This way, if we ever decide to change the underlying implementation, you will be protected through Stata's version-control feature.

❑

▷ Example 3

The `cluster parsedistance` programming command takes as an argument the name of a similarity or dissimilarity measure. Stata then checks this name against those that are implemented within Stata (and available to you through the `cluster measures` command). Uppercase or lowercase letters are allowed, and minimal abbreviations are checked. Some of the measures have aliases, which are resolved so that a standard measure name is returned. We demonstrate the `cluster parsedistance` command interactively:

```
. cluster parsedistance max
. sreturn list
macros:
            s(drange) : "0 ."
             s(dtype) : "dissimilarity"
              s(unab) : "maximum"
              s(dist) : "Linfinity"
. cluster parsedistance Eucl
. sreturn list
macros:
            s(drange) : "0 ."
             s(dtype) : "dissimilarity"
              s(unab) : "Euclidean"
              s(dist) : "L2"
. cluster parsedistance correl
. sreturn list
macros:
            s(drange) : "1 -1"
             s(dtype) : "similarity"
              s(unab) : "correlation"
              s(dist) : "correlation"
. cluster parsedistance jacc
. sreturn list
macros:
            s(drange) : "1 0"
            s(binary) : "binary"
             s(dtype) : "similarity"
              s(unab) : "Jaccard"
              s(dist) : "Jaccard"
```

cluster parsedistance returns s(dtype) as either similarity or dissimilarity. It returns s(dist) as the standard Stata name for the (dis)similarity and returns s(unab) as the unabbreviated standard Stata name. s(drange) gives the range of the measure (most similar to most dissimilar). If the measure is designed for binary variables, s(binary) is returned with the word binary, as seen above.

See [MV] ***measure_option*** for a listing of the similarity and dissimilarity measures and their properties.

◁

▷ Example 4

cluster measures computes the similarity or dissimilarity measure between each comparison observation and the observations implied by the if and in conditions (or all the data if no if or in conditions are specified).

We demonstrate with the auto dataset:

```
. use http://www.stata-press.com/data/r12/auto, clear
. cluster measures turn trunk gear_ratio in 1/10, compare(3 11) gen(z3 z11) L1
. format z* %8.2f
. list turn trunk gear_ratio z3 z11 in 1/11
```

	turn	trunk	gear_r~o	z3	z11
1.	40	11	3.58	6.50	14.30
2.	40	11	2.53	6.55	13.25
3.	35	12	3.08	0.00	17.80
4.	40	16	2.93	9.15	8.65
5.	43	20	2.41	16.67	1.13
6.	43	21	2.73	17.35	2.45
7.	34	10	2.87	3.21	20.59
8.	42	16	2.93	11.15	6.65
9.	43	17	2.93	13.15	4.65
10.	42	13	3.08	8.00	9.80
11.	44	20	2.28	.	.

Using the three variables turn, trunk, and gear_ratio, we computed the L1 (or absolute value) distance between the third observation and the first 10 observations and placed the results in the variable z3. The distance between the 11th observation and the first 10 was placed in variable z11.

There are many measures designed for binary data. Below we illustrate cluster measures with the matching coefficient binary similarity measure. We have 8 observations on 10 binary variables, and we will compute the matching similarity measure between the last 3 observations and all 8 observations.

```
. use http://www.stata-press.com/data/r12/clprogxmpl1, clear
. cluster measures x1-x10, compare(6/8) gen(z6 z7 z8) matching
. format z* %4.2f
. list
```

	x1	x2	x3	x4	x5	x6	x7	x8	x9	x10	z6	z7	z8
1.	1	0	0	0	1	1	0	0	1	1	0.60	0.80	0.40
2.	1	1	1	0	0	1	0	1	1	0	0.70	0.30	0.70
3.	0	0	1	0	0	0	1	0	0	1	0.60	0.40	0.20
4.	1	1	1	1	0	0	0	1	1	1	0.40	0.40	0.60
5.	0	1	0	1	1	0	1	0	0	1	0.20	0.60	0.40
6.	1	0	1	0	0	1	0	0	0	0	1.00	0.40	0.60
7.	0	0	0	1	1	1	0	0	1	1	0.40	1.00	0.40
8.	1	1	0	1	0	1	0	1	0	0	0.60	0.40	1.00

Stata treats all nonzero observations as one (except missing values, which are treated as missing values) when computing these binary measures.

When the similarity measure between binary observations and the means of groups of binary observations is needed, the `propvars` and `propcompares` options of `cluster measures` provide the solution. The mean of binary observations is a proportion. The value 0.2 would indicate that 20% of the values were one and 80% were zero for the group. See [MV] ***measure_option*** for a discussion of binary measures. The `propvars` option indicates that the main body of observations should be interpreted as proportions. The `propcompares` option specifies that the comparison observations be treated as proportions.

We compare 10 binary observations on five variables to 2 observations holding proportions by using the `propcompares` option:

```
. use http://www.stata-press.com/data/r12/clprogxmpl2, clear
. cluster measures a* in 1/10, compare(11 12) gen(c1 c2) matching propcompare
. list
```

	a1	a2	a3	a4	a5	c1	c2
1.	1	1	1	0	1	.6	.56
2.	0	0	1	1	1	.36	.8
3.	1	0	1	0	0	.76	.56
4.	1	1	0	1	1	.36	.44
5.	1	0	0	0	0	.68	.4
6.	0	0	1	1	1	.36	.8
7.	1	0	1	0	1	.64	.76
8.	1	0	0	0	1	.56	.6
9.	0	1	1	1	1	.32	.6
10.	1	1	1	1	1	.44	.6
11.	.8	.4	.7	.1	.2	.	.
12.	.5	0	.9	.6	1	.	.

◁

Saved results

`cluster query` with no arguments saves the following in `r()`:

Macros

`r(names)`	cluster solution names

`cluster query` with an argument saves the following in `r()`:

Macros

`r(name)`	cluster name
`r(type)`	type of cluster analysis
`r(method)`	cluster-analysis method
`r(similarity)`	similarity measure name
`r(dissimilarity)`	dissimilarity measure name
`r(note#)`	cluster note number #
`r(v#_tag)`	variable tag number #
`r(v#_name)`	varname associated with `r(v#_tag)`
`r(`*tag*`var)`	varname associated with *tag*
`r(c#_tag)`	characteristic tag number #
`r(c#_name)`	characteristic name associated with `r(c#_tag)`
`r(c#_val)`	characteristic value associated with `r(c#_tag)`
`r(`*tag*`char)`	characteristic name associated with *tag*
`r(o#_tag)`	other tag number #
`r(o#_val)`	other value associated with `r(o#_tag)`

`cluster set` saves the following in `r()`:

Macros

`r(name)`	cluster name

`cluster parsedistance` saves the following in `s()`:

Macros

`s(dist)`	(dis)similarity measure name
`s(unab)`	unabbreviated (dis)similarity measure name (before resolving alias)
`s(darg)`	argument of (dis)similarities that take them, such as `L(#)`
`s(dtype)`	`similarity` or `dissimilarity`
`s(drange)`	range of measure (most similar to most dissimilar)
`s(binary)`	`binary` if the measure is for binary observations

`cluster measures` saves the following in `r()`:

Macros

`r(generate)`	variable names from the `generate()` option
`r(compare)`	observation numbers from the `compare()` option
`r(dtype)`	`similarity` or `dissimilarity`
`r(distance)`	the name of the (dis)similarity measure
`r(binary)`	`binary` if the measure is for binary observations

Methods and formulas

All `cluster` commands listed above are implemented as ado-files.

Also see

[MV] **cluster** — Introduction to cluster-analysis commands

[MV] **clustermat** — Introduction to clustermat commands

[MV] **cluster programming subroutines** — Add cluster-analysis routines

Title

cluster stop — Cluster-analysis stopping rules

Syntax

Cluster analysis of data

`cluster stop` [*clname*] [, *options*]

Cluster analysis of a dissimilarity matrix

`clustermat stop` [*clname*] , `variables(`*varlist*`)` [*options*]

options	Description
`rule(calinski)`	use Caliński–Harabasz pseudo-F index stopping rule; the default
`rule(duda)`	use Duda–Hart Je(2)/Je(1) index stopping rule
`rule(`*rule_name*`)`	use *rule_name* stopping rule; see *Options* for details
`groups(`*numlist*`)`	compute stopping rule for specified groups
`matrix(`*matname*`)`	save results in matrix *matname*
* `variables(`*varlist*`)`	compute the stopping rule using *varlist*

* `variables(`*varlist*`)` is required with a `clustermat` solution and optional with a `cluster` solution.

`rule(`*rule_name*`)` is not shown in the dialog box. See [MV] **cluster programming subroutines** for information on how to add stopping rules to the `cluster stop` command.

Menu

Statistics > Multivariate analysis > Cluster analysis > Postclustering > Cluster analysis stopping rules

Description

Cluster-analysis stopping rules are used to determine the number of clusters. A stopping-rule value (also called an index) is computed for each cluster solution (for example, at each level of the hierarchy in a hierarchical cluster analysis). Larger values (or smaller, depending on the particular stopping rule) indicate more distinct clustering. See [MV] **cluster** for background information on cluster analysis and on the `cluster` and `clustermat` commands.

The `cluster stop` and `clustermat stop` commands currently provide two stopping rules, the Caliński and Harabasz (1974) pseudo-F index and the Duda–Hart (2001, sec. 10.10) Je(2)/Je(1) index. For both rules, larger values indicate more distinct clustering. Presented with the Duda–Hart Je(2)/Je(1) values are pseudo-T-squared values. Smaller pseudo-T-squared values indicate more distinct clustering.

clname specifies the name of the cluster analysis. The default is the most recently performed cluster analysis, which can be reset using the `cluster use` command; see [MV] **cluster utility**.

More `stop` rules may be added; see [MV] **cluster programming subroutines**, which illustrates this ability by showing a program that adds the step-size stopping rule.

Options

`rule(calinski` | `duda` | *rule_name*`)` indicates the stopping rule. `rule(calinski)`, the default, specifies the Caliński–Harabasz pseudo-F index. `rule(duda)` specifies the Duda–Hart Je(2)/Je(1) index.

`rule(calinski)` is allowed for both hierarchical and nonhierarchical cluster analyses. `rule(duda)` is allowed only for hierarchical cluster analyses.

You can add stopping rules to the `cluster stop` command (see [MV] **cluster programming subroutines**) by using the `rule(`*rule_name*`)` option. [MV] **cluster programming subroutines** illustrates how to add stopping rules by showing a program that adds a `rule(stepsize)` option, which implements the simple step-size stopping rule mentioned in Milligan and Cooper (1985).

`groups(`*numlist*`)` specifies the cluster groupings for which the stopping rule is to be computed. `groups(3/20)` specifies that the measure be computed for the three-group solution, the four-group solution, ..., and the 20-group solution.

With `rule(duda)`, the default is `groups(1/15)`. With `rule(calinski)` for a hierarchical cluster analysis, the default is `groups(2/15)`. `groups(1)` is not allowed with `rule(calinski)` because the measure is not defined for the degenerate one-group cluster solution. The `groups()` option is unnecessary (and not allowed) for a nonhierarchical cluster analysis.

If there are ties in the hierarchical cluster-analysis structure, some (or possibly all) of the requested stopping-rule solutions may not be computable. `cluster stop` passes over, without comment, the `groups()` for which ties in the hierarchy cause the stopping rule to be undefined.

`matrix(`*matname*`)` saves the results in a matrix named *matname*.

With `rule(calinski)`, the matrix has two columns, the first giving the number of clusters and the second giving the corresponding Caliński–Harabasz pseudo-F stopping-rule index.

With `rule(duda)`, the matrix has three columns: the first column gives the number of clusters, the second column gives the corresponding Duda–Hart Je(2)/Je(1) stopping-rule index, and the third column provides the corresponding pseudo-T-squared values.

`variables(`*varlist*`)` specifies the variables to be used in the computation of the stopping rule. By default, the variables used for the cluster analysis are used. `variables()` is required for cluster solutions produced by `clustermat`.

Remarks

Everitt et al. (2011) and Gordon (1999) discuss the problem of determining the number of clusters and describe several stopping rules, including the Caliński–Harabasz (1974) pseudo-F index and the Duda–Hart (2001, sec. 10.10) Je(2)/Je(1) index. There are many cluster stopping rules. Milligan and Cooper (1985) evaluate 30 stopping rules, singling out the Caliński–Harabasz index and the Duda–Hart index as two of the best rules.

Large values of the Caliński–Harabasz pseudo-F index indicate distinct clustering. The Duda–Hart Je(2)/Je(1) index has an associated pseudo-T-squared value. A large Je(2)/Je(1) index value and a small pseudo-T-squared value indicate distinct clustering. See *Methods and formulas* at the end of this entry for details.

Example 2 of [MV] **clustermat** shows the use of the `clustermat stop` command.

Some stopping rules such as the Duda–Hart index work only with a hierarchical cluster analysis. The Caliński–Harabasz index, however, may be applied to both nonhierarchical and hierarchical cluster analyses.

▷ Example 1

Previously, you ran kmeans cluster analyses on data where you measured the flexibility, speed, and strength of the 80 students in your physical education class; see example 1 of [MV] **cluster kmeans and kmedians**. Your original goal was to split the class into four groups, though you also examined the three- and five-group kmeans cluster solutions as possible alternatives.

Now out of curiosity, you wonder what the Caliński–Harabasz stopping rule shows for the three-, four-, and five-group solutions from a kmedian clustering of this dataset.

```
. use http://www.stata-press.com/data/r12/physed
. cluster kmed flex speed strength, k(3) name(kmed3) measure(abs) start(lastk)
. cluster kmed flex speed strength, k(4) name(kmed4) measure(abs) start(kr(11736))
. cluster kmed flex speed strength, k(5) name(kmed5) measure(abs) start(prand(8723))
. cluster stop kmed3
```

Number of clusters	Calinski/ Harabasz pseudo-F
3	132.75

```
. cluster stop kmed4
```

Number of clusters	Calinski/ Harabasz pseudo-F
4	337.10

```
. cluster stop kmed5
```

Number of clusters	Calinski/ Harabasz pseudo-F
5	300.45

The four-group solution with a Caliński–Harabasz pseudo-F value of 337.10 is largest, indicating that the four-group solution is the most distinct compared with the three-group and five-group solutions.

The three-group solution has a much lower stopping-rule value of 132.75. The five-group solution, with a value of 300.45, is reasonably close to the four-group solution.

Though you do not think it will change your decision on how to split your class into groups, you are curious to see what a hierarchical cluster analysis might produce. You decide to try an average-linkage cluster analysis using the default Euclidean distance; see [MV] **cluster linkage**. You examine the resulting cluster analysis with the `cluster tree` command, which is an easier-to-type alias for the `cluster dendrogram` command; see [MV] **cluster dendrogram**.

```
. cluster averagelink flex speed strength, name(avglnk)
. cluster tree avglnk, xlabel(, angle(90) labsize(*.75))
```

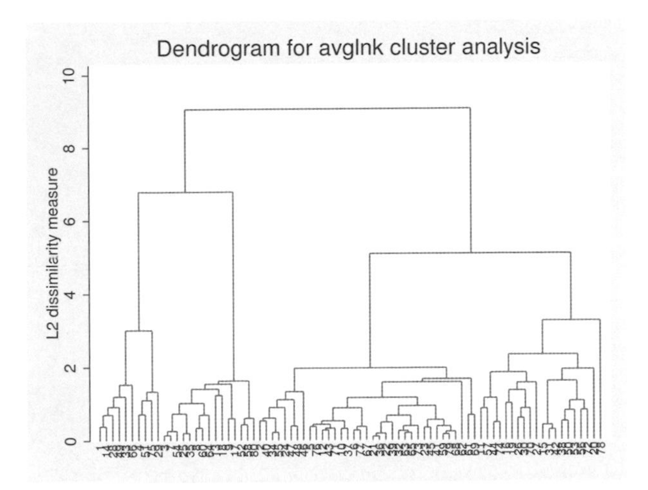

You are curious to see how the four- and five-group solutions from this hierarchical cluster analysis compare with the four- and five-group solutions from the kmedian clustering.

```
. cluster gen avgg = groups(4/5), name(avglnk)
. table kmed4 avgg4
```

	avgg4			
kmed4	1	2	3	4
1			35	
2		15		
3				20
4	10			

```
. table kmed5 avgg5
```

	avgg5				
kmed5	1	2	3	4	5
1		15			
2				19	1
3			20		
4	10				
5			15		

The four-group solutions are identical, except for the numbers used to label the groups. The five-group solutions are different. The kmedian clustering split the 35-member group into subgroups having 20 and 15 members. The average-linkage clustering instead split one member off from the 20-member group.

Now you examine the Caliński–Harabasz pseudo-F stopping-rule values associated with the kmedian hierarchical cluster analysis.

```
. cluster stop avglnk, rule(calinski)
```

Number of clusters	Calinski/ Harabasz pseudo-F
2	131.86
3	126.62
4	337.10
5	269.07
6	258.40
7	259.37
8	290.78
9	262.86
10	258.53
11	249.93
12	247.85
13	247.53
14	236.98
15	226.51

Because `rule(calinski)` is the default, you could have obtained this same table by typing

```
. cluster stop avglnk
```

or, because `avglnk` was the most recent cluster analysis performed, by typing

```
. cluster stop
```

You did not specify the number of groups to examine from the hierarchical cluster analysis, so it defaulted to examining up to 15 groups. The highest Caliński–Harabasz pseudo-F value is 337.10 for the four-group solution.

What does the Duda–Hart stopping rule produce for this hierarchical cluster analysis?

```
. cluster stop avglnk, rule(duda) groups(1/10)
```

Number of clusters	Duda/Hart Je(2)/Je(1)	pseudo T-squared
1	0.3717	131.86
2	0.1349	147.44
3	0.2283	179.19
4	0.8152	4.08
5	0.2232	27.85
6	0.5530	13.74
7	0.5287	29.42
8	0.6887	3.16
9	0.4888	8.37
10	0.7621	7.80

This time, we asked to see the results for one to 10 groups. The largest Duda–Hart Je(2)/Je(1) stopping-rule value is 0.8152, corresponding to four groups. The smallest pseudo-T-squared value is 3.16 for the eight-group solution, but the pseudo-T-squared value for the four-group solution is also low, with a value of 4.08.

Distinct clustering is characterized by large Caliński–Harabasz pseudo-F values, large Duda–Hart Je(2)/Je(1) values, and small Duda–Hart pseudo-T-squared values.

The conventional wisdom for deciding the number of groups based on the Duda–Hart stopping-rule table is to find one of the largest Je(2)/Je(1) values that corresponds to a low pseudo-T-squared value that has much larger T-squared values next to it. This strategy, combined with the results from the Caliński–Harabasz results, indicates that the four-group solution is the most distinct from this hierarchical cluster analysis.

◁

❑ Technical note

There is a good reason that the word "pseudo" appears in "pseudo-F" and "pseudo-T-squared". Although these index values are based on well-known statistics, any p-values computed from these statistics would not be valid. Remember that cluster analysis searches for structure.

If you were to generate random observations, perform a cluster analysis, compute these stopping-rule statistics, and then follow that by computing what would normally be the p-values associated with the statistics, you would almost always end up with significant p-values.

Remember that you would expect, on average, five of every 100 groupings of your random data to show up as significant when you use .05 as your threshold for declaring significance. Cluster-analysis methods search for the best groupings, so there is no surprise that p-values show high significance, even when none exists.

Examining the stopping-rule index values relative to one another is useful, however, in finding relatively reasonable groupings that may exist in the data.

❑

❑ Technical note

As mentioned in *Methods and formulas*, ties in the hierarchical cluster structure cause some of the stopping-rule index values to be undefined. Discrete (as opposed to continuous) data tend to cause ties in a hierarchical clustering. The more discrete the data, the more likely it is that ties will occur (and the more of them you will encounter) within a hierarchy.

Even with so-called continuous data, ties in the hierarchical clustering can occur. We say "so-called" because most continuous data are truncated or rounded. For instance, miles per gallon, length, weight, etc., which may really be continuous, may be observed and recorded only to the tens, ones, tenths, or hundredths of a unit.

You can have data with no ties in the observations and still have many ties in the hierarchy. Ties in distances (or similarities) between observations and groups of observations cause the ties in the hierarchy.

Thus, do not be surprised when some (many) of the stopping-rule values that you request are not presented. Stata has decided not to break the ties arbitrarily, because the stopping-rule values may differ widely, depending on which split is made.

❑

❑ Technical note

The stopping rules also become less informative as the number of elements in the groups becomes small, that is, having many groups, each with few observations. We recommend that if you need to examine the stopping-rule values deep within your hierarchical cluster analysis, you do so skeptically.

❑

Saved results

`cluster stop` and `clustermat stop` with `rule(calinski)` saves the following in `r()`:

Scalars

`r(calinski_#)`	Caliński–Harabasz pseudo-F for # groups

Macros

`r(rule)`	`calinski`
`r(label)`	`C-H pseudo-F`
`r(longlabel)`	`Calinski & Harabasz pseudo-F`

`cluster stop` and `clustermat stop` with `rule(duda)` saves the following in `r()`:

Scalars

`r(duda_#)`	Duda–Hart Je(2)/Je(1) value for # groups
`r(dudat2_#)`	Duda–Hart pseudo-T-squared value for # groups

Macros

`r(rule)`	`duda`
`r(label)`	`D-H Je(2)/Je(1)`
`r(longlabel)`	`Duda & Hart Je(2)/Je(1)`
`r(label2)`	`D-H pseudo-T-squared`
`r(longlabel2)`	`Duda & Hart pseudo-T-squared`

Methods and formulas

`cluster stop` and `clustermat stop` are implemented as ado-files.

The Caliński–Harabasz pseudo-F stopping-rule index for g groups and N observations is

$$\frac{\text{trace}(\mathbf{B})/(g-1)}{\text{trace}(\mathbf{W})/(N-g)}$$

where $\mathbf{B}$ is the between-cluster sum of squares and cross-products matrix, and $\mathbf{W}$ is the within-cluster sum of squares and cross-products matrix.

Large values of the Caliński–Harabasz pseudo-F stopping-rule index indicate distinct cluster structure. Small values indicate less clearly defined cluster structure.

The Duda–Hart Je(2)/Je(1) stopping-rule index value is literally Je(2) divided by Je(1). Je(1) is the sum of squared errors within the group that is to be divided. Je(2) is the sum of squared errors in the two resulting subgroups.

Large values of the Duda–Hart pseudo-T-squared stopping-rule index indicate distinct cluster structure. Small values indicate less clearly defined cluster structure.

The Duda–Hart Je(2)/Je(1) index requires hierarchical clustering information. It needs to know at each level of the hierarchy which group is to be split and how. The Duda–Hart index is also local because the only information used comes from the group's being split. The information in the rest of the groups does not enter the computation.

In comparison, the Caliński–Harabasz rule does not require hierarchical information and is global because the information from each group is used in the computation.

A pseudo-T-squared value is also presented with the Duda and Hart Je(2)/Je(1) index. The relationship is

$$\frac{1}{\text{Je}(2)/\text{Je}(1)} = 1 + \frac{T^2}{N_1 + N_2 - 2}$$

where N_1 and N_2 are the numbers of observations in the two subgroups.

Je(2)/Je(1) will be zero when Je(2) is zero, that is, when the two subgroups each have no variability. An example of this is when the cluster being split has two distinct values that are being split into singleton subgroups. Je(1) will never be zero because we do not split groups that have no variability. When Je(2)/Je(1) is zero, the pseudo-T-squared value is undefined.

Ties in splitting a hierarchical cluster analysis create an ambiguity for the Je(2)/Je(1) measure. For example, to compute the measure for the case of going from five clusters to six, you need to identify the one cluster that will be split. With a tie in the hierarchy, you would instead go from five clusters directly to seven (just as an example). Stata refuses to produce an answer in this situation.

References

Caliński, T., and J. Harabasz. 1974. A dendrite method for cluster analysis. *Communications in Statistics* 3: 1–27.

Duda, R. O., P. E. Hart, and D. G. Stork. 2001. *Pattern Classification*. 2nd ed. New York: Wiley.

Everitt, B. S., S. Landau, M. Leese, and D. Stahl. 2011. *Cluster Analysis*. 5th ed. Chichester, UK: Wiley.

Gordon, A. D. 1999. *Classification*. 2nd ed. Boca Raton, FL: Chapman & Hall/CRC.

Milligan, G. W., and M. C. Cooper. 1985. An examination of procedures for determining the number of clusters in a dataset. *Psychometrika* 50: 159–179.

Also see

[MV] **cluster** — Introduction to cluster-analysis commands

[MV] **clustermat** — Introduction to clustermat commands

Title

cluster utility — List, rename, use, and drop cluster analyses

Syntax

Directory-style listing of currently defined clusters

`cluster dir`

Detailed listing of clusters

`cluster list` [*clnamelist*] [, *list_options*]

Drop cluster analyses

`cluster drop` { *clnamelist* | `_all` }

Mark a cluster analysis as the most recent one

`cluster use` *clname*

Rename a cluster

`cluster rename` *oldclname newclname*

Rename variables attached to a cluster

`cluster renamevar` *oldvarname newvar* [, `name(`*clname*`)`]

`cluster renamevar` *oldstub newstub* , `prefix` [`name(`*clname*`)`]

list_options	Description
Options	
`notes`	list cluster notes
`type`	list cluster analysis type
`method`	list cluster analysis method
`dissimilarity`	list cluster analysis dissimilarity measure
`similarity`	list cluster analysis similarity measure
`vars`	list variable names attached to the cluster analysis
`chars`	list any characteristics attached to the cluster analysis
`other`	list any "other" information
`all`	list all items and information attached to the cluster; the default

`all` does not appear in the dialog box.

Menu

cluster list

Statistics > Multivariate analysis > Cluster analysis > Postclustering > Detailed listing of clusters

cluster drop

Statistics > Multivariate analysis > Cluster analysis > Postclustering > Drop cluster analyses

cluster rename

Statistics > Multivariate analysis > Cluster analysis > Postclustering > Rename a cluster or cluster variables

Description

These `cluster` utility commands allow you to view and manipulate the cluster objects that you have created. See [MV] **cluster** for an overview of cluster analysis and for the available `cluster` commands. If you want even more control over your cluster objects, or if you are programming new cluster subprograms, more `cluster` programmer utilities are available; see [MV] **cluster programming utilities** for details.

The `cluster dir` command provides a directory-style listing of all the currently defined clusters. `cluster list` provides a detailed listing of the specified clusters or of all current clusters if no cluster names are specified. The default action is to list all the information attached to the clusters. You may limit the type of information listed by specifying particular options.

The `cluster drop` command removes the named clusters. The keyword `_all` specifies that all current cluster analyses be dropped.

Stata cluster analyses are referred to by name. Many `cluster` commands default to using the most recently defined cluster analysis if no cluster name is provided. The `cluster use` command sets the specified cluster analysis as the most recently executed cluster analysis, so that, by default, this cluster analysis will be used if the cluster name is omitted from many of the `cluster` commands. You may use the `*` and `?` name-matching characters to shorten the typing of cluster names; see [U] **11.2 Abbreviation rules**.

`cluster rename` allows you to rename a cluster analysis without changing any of the variable names attached to the cluster analysis. The `cluster renamevar` command, on the other hand, allows you to rename the variables attached to a cluster analysis and to update the cluster object with the new variable names. Do not use the `rename` command (see [D] **rename**) to rename variables attached to a cluster analysis because this would invalidate the cluster object. Use the `cluster renamevar` command instead.

Options for cluster list

Options

`notes` specifies that cluster notes be listed.

`type` specifies that the type of cluster analysis be listed.

`method` specifies that the cluster analysis method be listed.

`dissimilarity` specifies that the dissimilarity measure be listed.

`similarity` specifies that the similarity measure be listed.

`vars` specifies that the variables attached to the clusters be listed.

`chars` specifies that any Stata characteristics attached to the clusters be listed.

`other` specifies that information attached to the clusters under the heading "other" be listed.

The following option is available with `cluster list` but is not shown in the dialog box:

`all`, the default, specifies that all items and information attached to the cluster(s) be listed. You may instead pick among the `notes`, `type`, `method`, `dissimilarity`, `similarity`, `vars`, `chars`, and `other` options to limit what is presented.

Options for cluster renamevar

`name(`*clname*`)` indicates the cluster analysis within which the variable renaming is to take place. If `name()` is not specified, the most recently performed cluster analysis (or the one specified by `cluster use`) will be used.

`prefix` specifies that all variables attached to the cluster analysis that have *oldstub* as the beginning of their name be renamed, with *newstub* replacing *oldstub*.

Remarks

▷ Example 1

We demonstrate these `cluster` utility commands by beginning with four already-defined cluster analyses. The `dir` and `list` subcommands provide listings of the cluster analyses.

```
. cluster dir
bcx3kmed
ayz5kmeans
abc_clink
xyz_slink

. cluster list xyz_slink
xyz_slink  (type: hierarchical,  method: single,  dissimilarity: L2)
      vars: xyz_slink_id (id variable)
            xyz_slink_ord (order variable)
            xyz_slink_hgt (height variable)
     other: cmd: cluster singlelinkage x y z, name(xyz_slink)
            varlist: x y z
            range: 0 .

. cluster list
bcx3kmed  (type: partition,  method: kmedians,  dissimilarity: L2)
      vars: bcx3kmed (group variable)
     other: cmd: cluster kmedians b c x, k(3) name(bcx3kmed)
            varlist: b c x
            k: 3
            start: krandom
            range: 0 .
ayz5kmeans  (type: partition,  method: kmeans,  dissimilarity: L2)
      vars: ayz5kmeans (group variable)
     other: cmd: cluster kmeans a y z, k(5) name(ayz5kmeans)
            varlist: a y z
            k: 5
            start: krandom
            range: 0 .
```

```
abc_clink  (type: hierarchical,  method: complete,  dissimilarity: L2)
      vars: abc_clink_id (id variable)
            abc_clink_ord (order variable)
            abc_clink_hgt (height variable)
     other: cmd: cluster completelinkage a b c, name(abc_clink)
            varlist: a b c
            range: 0 .
xyz_slink  (type: hierarchical,  method: single,  dissimilarity: L2)
      vars: xyz_slink_id (id variable)
            xyz_slink_ord (order variable)
            xyz_slink_hgt (height variable)
     other: cmd: cluster singlelinkage x y z, name(xyz_slink)
            varlist: x y z
            range: 0 .
. cluster list a*, vars
ayz5kmeans
      vars: ayz5kmeans (group variable)
abc_clink
      vars: abc_clink_id (id variable)
            abc_clink_ord (order variable)
            abc_clink_hgt (height variable)
```

cluster dir listed the names of the four currently defined cluster analyses. cluster list followed by the name of one of the cluster analyses listed the information attached to that cluster analysis. The cluster list command, without an argument, listed the information for all currently defined cluster analyses. We demonstrated the vars option of cluster list to show that we can restrict the information that is listed. Notice also the use of a* as the cluster name. The * here indicates that any ending is allowed. For these four cluster analyses, Stata matches the names ayz5kmeans and abc_clink.

We now demonstrate the use of the renamevar subcommand.

```
. cluster renamevar ayz5kmeans g5km
variable ayz5kmeans not found in bcx3kmed
r(198);
. cluster renamevar ayz5kmeans g5km, name(ayz5kmeans)
. cluster list ayz5kmeans
ayz5kmeans  (type: partition,  method: kmeans,  dissimilarity: L2)
      vars: g5km (group variable)
     other: cmd: cluster kmeans a y z, k(5) name(ayz5kmeans)
            varlist: a y z
            k: 5
            start: krandom
            range: 0 .
```

The first use of cluster renamevar failed because we did not specify which cluster object to use (with the name() option), and the most recent cluster object, bcx3kmed, was not the appropriate one. After specifying the name() option with the appropriate cluster name, the renamevar subcommand changed the name as shown in the cluster list command that followed.

The cluster use command sets a particular cluster object as the default. We show this in conjunction with the prefix option of the renamevar subcommand.

```
. cluster use ayz5kmeans
. cluster renamevar g grp, prefix
. cluster renamevar xyz_slink_ wrk, prefix name(xyz*)
```

```
. cluster list ayz* xyz*
ayz5kmeans  (type: partition,  method: kmeans,  dissimilarity: L2)
      vars: grp5km (group variable)
     other: cmd: cluster kmeans a y z, k(5) name(ayz5kmeans)
            varlist: a y z
            k: 5
            start: krandom
            range: 0 .
xyz_slink  (type: hierarchical,  method: single,  dissimilarity: L2)
      vars: wrkid (id variable)
            wrkord (order variable)
            wrkhgt (height variable)
     other: cmd: cluster singlelinkage x y z, name(xyz_slink)
            varlist: x y z
            range: 0 .
```

The `cluster use` command placed `ayz5kmeans` as the current cluster object. The `cluster renamevar` command that followed capitalized on this placement by leaving off the `name()` option. The `prefix` option allowed us to change the variable names, as demonstrated in the `cluster list` of the two changed cluster objects.

`cluster rename` changes the name of cluster objects. `cluster drop` allows us to drop some or all of the cluster objects.

```
. cluster rename xyz_slink bob

. cluster rename ayz* sam

. cluster list, type method vars
sam  (type: partition,  method: kmeans)
      vars: grp5km (group variable)
bob  (type: hierarchical,  method: single)
      vars: wrkid (id variable)
            wrkord (order variable)
            wrkhgt (height variable)
bcx3kmed  (type: partition,  method: kmedians)
      vars: bcx3kmed (group variable)
abc_clink  (type: hierarchical,  method: complete)
      vars: abc_clink_id (id variable)
            abc_clink_ord (order variable)
            abc_clink_hgt (height variable)
. cluster drop bcx3kmed abc_clink

. cluster dir
sam
bob

. cluster drop _all

. cluster dir
```

We used options with `cluster list` to limit what was presented. The `_all` keyword with `cluster drop` removed all currently defined cluster objects.

◁

Methods and formulas

All `cluster` commands listed above are implemented as ado-files.

Also see

[MV] **cluster notes** — Place notes in cluster analysis

[MV] **cluster programming utilities** — Cluster-analysis programming utilities

[D] **notes** — Place notes in data

[P] **char** — Characteristics

[MV] **cluster** — Introduction to cluster-analysis commands

[MV] **clustermat** — Introduction to clustermat commands

Title

discrim — Discriminant analysis

Syntax

`discrim` *subcommand* ... [, ...]

subcommand	Description
`knn`	kth-nearest-neighbor discriminant analysis
`lda`	linear discriminant analysis
`logistic`	logistic discriminant analysis
`qda`	quadratic discriminant analysis

See [MV] **discrim knn**, [MV] **discrim lda**, [MV] **discrim logistic**, and [MV] **discrim qda** for details about the *subcommands*.

Description

`discrim` performs discriminant analysis, which is also known as classification. kth-nearest-neighbor (KNN) discriminant analysis, linear discriminant analysis (LDA), quadratic discriminant analysis (QDA), and logistic discriminant analysis are available.

Remarks

Remarks are presented under the following headings:

Introduction
A simple example
Prior probabilities, costs, and ties

Introduction

Discriminant analysis is used to describe the differences between groups and to exploit those differences in allocating (classifying) observations of unknown group membership to the groups. Discriminant analysis is also called classification in many references. However, several sources use the word classification to mean cluster analysis.

Some applications of discriminant analysis include medical diagnosis, market research, classification of specimens in anthropology, predicting company failure or success, placement of students (workers) based on comparing pretest results to those of past students (workers), discrimination of natural versus man-made seismic activity, fingerprint analysis, image pattern recognition, and signal pattern classification.

Most multivariate statistics texts have chapters on discriminant analysis, including (Rencher 1998, 2002), Johnson and Wichern (2007), Mardia, Kent, and Bibby (1979), Anderson (2003), Everitt and Dunn (2001), Tabachnick and Fidell (2007), and Albert and Harris (1987). Books dedicated to discriminant analysis include Lachenbruch (1975), Klecka (1980), Hand (1981), Huberty (1994), and McLachlan (2004). Of these, McLachlan (2004) gives the most extensive coverage, including 60 pages of references.

If you lack observations with known group membership, use cluster analysis to discover the natural groupings in the data; see [MV] **cluster**. If you have data with known group membership, possibly with other data of unknown membership to be classified, use discriminant analysis to examine the differences between the groups, based on data where membership is known, and to assign group membership for cases where membership is unknown.

Some researchers are not interested in classifying unknown observations and are interested only in the descriptive aspects of discriminant analysis. For others, the classification of unknown observations is the primary consideration. Huberty (1994), Rencher (1998 and 2002), and others split their discussion of discrimination into two parts. Huberty labels the two parts descriptive discriminant analysis and predictive discriminant analysis. Rencher reserves discriminant analysis for descriptive discriminant analysis and uses the label classification for predictive discriminant analysis.

There are many discrimination methods. `discrim` has both descriptive and predictive LDA; see [MV] **discrim lda**. If your interest is in descriptive LDA, `candisc` computes the same thing as `discrim lda`, but with output tailored for the descriptive aspects of the discrimination; see [MV] **candisc**.

The remaining `discrim` subcommands provide alternatives to LDA for predictive discrimination. [MV] **discrim qda** provides quadratic discriminant analysis (QDA). [MV] **discrim logistic** provides logistic discriminant analysis. [MV] **discrim knn** provides kth-nearest-neighbor (KNN) discrimination.

The discriminant analysis literature uses conflicting terminology for several features of discriminant analysis. For example, in descriptive LDA, what one source calls a classification function another source calls a discriminant function while calling something else a classification function. Check the *Methods and formulas* sections for the `discrim` subcommands for clarification.

A simple example

We demonstrate the predictive and descriptive aspects of discriminant analysis with a simple example.

▷ Example 1

Johnson and Wichern (2007, 578) introduce the concepts of discriminant analysis with a two-group dataset. A sample of 12 riding-lawnmower owners and 12 nonowners is sampled from a city and the income in thousands of dollars and lot size in thousands of square feet are recorded. A riding-mower manufacturer wants to see if these two variables adequately separate owners from nonowners, and if so to then direct their marketing on the basis of the separation of owners from nonowners.

```
. use http://www.stata-press.com/data/r12/lawnmower2
(Johnson and Wichern (2007) Table 11.1)
```

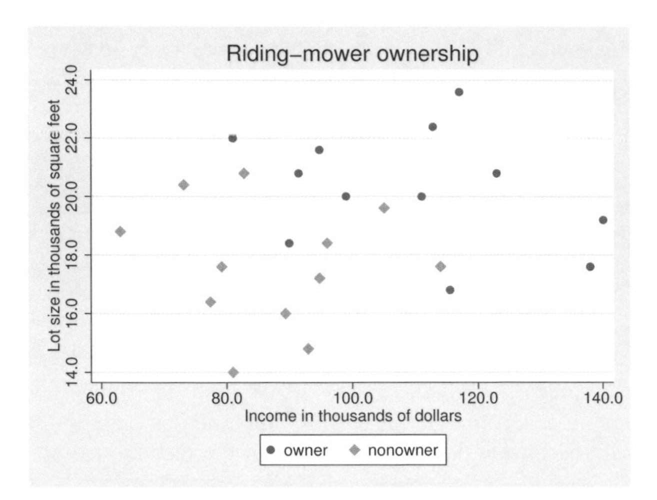

Do these two variables adequately separate riding-mower owners from nonowners so that the riding-mower manufacturer can base predictions of riding-mower ownership on income and lot size? The graph shows some separation of owners from nonowners, but with overlap. With predictive LDA we can quantify our ability to discriminate between riding-mower owners and nonowners.

```
. discrim lda lotsize income, group(owner)
Linear discriminant analysis
Resubstitution classification summary

  +---------+
  | Key     |
  |---------|
  | Number  |
  | Percent |
  +---------+

             | Classified
  True owner | nonowner     owner |    Total
-------------+--------------------+---------
    nonowner |       10         2 |       12
             |    83.33     16.67 |   100.00
             |                    |
       owner |        1        11 |       12
             |     8.33     91.67 |   100.00
-------------+--------------------+---------
       Total |       11        13 |       24
             |    45.83     54.17 |   100.00
             |                    |
      Priors |   0.5000    0.5000 |
```

The table presented by `discrim lda` (and the other `discrim` subcommands) is called a classification table or confusion matrix. It is labeled as a resubstitution classification table because the same observations used in estimating the discriminant model were classified using the model. The diagonal elements in the main body of the table show the number and percent correctly classified into each group. The off-diagonal elements show the misclassified number and percent. One owner and two nonowners were misclassified.

The resubstitution classification table provides an overly optimistic assessment of how well the linear discriminant function will predict the ownership status for observations that were not part of the training sample. A leave-one-out classification table provides a more realistic assessment for future prediction. The leave-one-out classification is produced by holding each observation out, one at a time; building an LDA model from the remaining training observations; and then classifying the held out observation using this model. The leave-one-out classification table is available at estimation time, at playback, or through the `estat classtable` postestimation command.

```
. estat classtable, loo nopriors
Leave-one-out classification table
```

Key
Number Percent

True owner	LOO Classified nonowner	owner	Total
nonowner	9 75.00	3 25.00	12 100.00
owner	2 16.67	10 83.33	12 100.00
Total	11 45.83	13 54.17	24 100.00

With leave-one-out classification we see that 5, instead of only 3, of the 24 observations are misclassified.

The `predict` and `estat` commands provide other predictive discriminant analysis tools. `predict` generates variables containing the posterior probabilities of group membership or generates a group membership classification variable. `estat` displays classification tables, displays error-rate tables, and lists classifications and probabilities for the observations.

We now use `estat list` to show the resubstitution and leave-one-out classifications and posterior probabilities for those observations that were misclassified by our LDA model.

```
. estat list, class(loo) probabilities(loo) misclassified
```

	Classification			Probabilities		LOO Probabilities	
Obs.	True	Class.	LOO Cl.	nonowner	owner	nonowner	owner
1	owner	nonown *	nonown *	0.7820	0.2180	0.8460	0.1540
2	owner	owner	nonown *	0.4945	0.5055	0.6177	0.3823
13	nonown	owner *	owner *	0.2372	0.7628	0.1761	0.8239
14	nonown	nonown	owner *	0.5287	0.4713	0.4313	0.5687
17	nonown	owner *	owner *	0.3776	0.6224	0.2791	0.7209

```
* indicates misclassified observations
```

◁

We have used `discrim lda` to illustrate predictive discriminant analysis. The other `discrim` subcommands could also be used for predictive discrimination of this data.

Postestimation commands after `discrim lda` provide descriptive discriminant analysis; see [MV] **discrim lda postestimation** and [MV] **candisc**.

▷ Example 2

The riding-mower manufacturer of the previous example wants to understand how income and lot size affect riding-mower ownership. Descriptive discriminant analysis provides tools for exploring how the groups are separated. Fisher's (1936) linear discriminant functions provide the basis for descriptive LDA; see [MV] **discrim lda** and [MV] **discrim lda postestimation**. The postestimation command `estat loadings` allows us to view the discriminant function coefficients, which are also called loadings.

```
. estat loadings, standardized unstandardized
Canonical discriminant function coefficients

             | function1
-------------+-----------
     lotsize |  .3795228
      income |  .0484468
       _cons | -11.96094

Standardized canonical discriminant function coefficients

             | function1
-------------+-----------
     lotsize |  .7845512
      income |  .8058419
```

We requested both the unstandardized and standardized coefficients. The unstandardized coefficients apply to unstandardized variables. The standardized coefficients apply to variables standardized using the pooled within-group covariance. Standardized coefficients are examined to assess the relative importance of the variables to the discriminant function.

The unstandardized coefficients determine the separating line between riding-mower owners and nonowners.

$$0 = 0.3795228\ \texttt{lotsize} + 0.0484468\ \texttt{income} - 11.96094$$

which can be reexpressed as

$$\texttt{lotsize} = -0.1276519\ \texttt{income} + 31.51574$$

We now display this line superimposed on the scatterplot of the data.

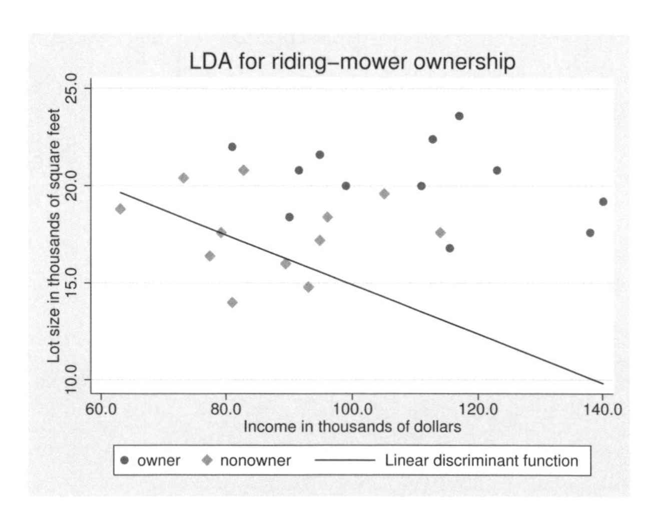

Other descriptive statistics and summaries are available; see [MV] **discrim lda postestimation**.

◁

Prior probabilities, costs, and ties

Classification is influenced by the selection of prior probabilities, assignment of costs to misclassification, and the method of handling ties in classification criteria.

Prior probabilities are the presumptive or a priori probabilities of group membership. Before you flip a balanced coin 10 times, you know the prior probability of getting heads is the same as getting tails—both are 0.5. Group prior probabilities, commonly called priors, must be taken into account in calculations of posterior probabilities; see *Methods and formulas* for details.

If the cost of misclassification is not equal over the groups, an optimal classification into groups must take misclassification cost into account. When there are two groups, members of the first group can be classified into the second, or members of the second group can be classified into the first. The relative undesirability of these two misclassifications may not be the same. Example 3 of [MV] **discrim knn** classifies poisonous and edible mushrooms. Misclassifying poisonous mushrooms as edible is a big deal at dinnertime.

The expected misclassification cost is the sum of the products of the cost for each misclassification multiplied by the probability of its occurrence. Let p_{ij} be the probability that an observation from group i is classified into group j, let c_{ij} be the cost of misclassifying an observation from group i into group j, and let q_i be the prior probability that the observation is from group i. The expected cost of misclassification is then

$$\text{cost} = \sum_{i,j \neq i}^{g} c_{ij} p_{ij} q_i$$

It is this expected cost that we wish to minimize. In the two-group case

$$\text{cost} = c_{12} p_{12} q_1 + c_{21} p_{21} q_2$$

and we can use cost-adjusted group prior probabilities, $\widehat{q}_i$, in the place of the prior probabilities to minimize the cost of misclassification.

$$\widehat{q}_1 = \frac{c_{12} q_1}{c_{12} q_1 + c_{21} q_2}$$

$$\widehat{q}_2 = \frac{c_{21} q_2}{c_{12} q_1 + c_{21} q_2}$$

With more than two groups, there is often not a simple rule to take costs into account. More discussion on this topic is provided by McLachlan (2004, 7–9), Huberty (1994, 68–69), Johnson and Wichern (2007, 606–609), and Anderson (2003, chap. 6).

See example 3 of [MV] **discrim knn** for an application of costs.

A tie in classification occurs when two or more group posterior probabilities are equal for an observation. Ties are most common with kth-nearest-neighbor discriminant analysis, though they can occur in other forms of discriminant analysis. There are several options for assigning tied observations. The default is to mark the observation as unclassified, that is, classified to a missing value. Ties can also be broken. For most forms of discriminant analysis ties can be broken in two ways—randomly or assigned to the first group that is tied. For kth-nearest-neighbor discriminant analysis, dissimilarities are calculated, and so ties may also be broken by choosing the group of the nearest of the tied observations. If this still results in a tie, the observation is unclassified.

Methods and formulas

discrim is implemented as an ado-file.

See [MV] **discrim lda** for the methods and formulas for descriptive discriminant analysis.

For predictive discriminant analysis, let g be the number of groups, n_i the number of observations for group i, and q_i the prior probability for group i. Let $\mathbf{x}$ denote an observation measured on p discriminating variables. For consistency with the discriminant analysis literature, $\mathbf{x}$ will be a column vector, though it corresponds to a row in your dataset. Let $f_i(\mathbf{x})$ represent the density function for group i, and let $P(\mathbf{x}|G_i)$ denote the probability of observing $\mathbf{x}$ conditional on belonging to group i. Denote the posterior probability of group i given observation $\mathbf{x}$ as $P(G_i|\mathbf{x})$. With Bayes' theorem, we have

$$P(G_i|\mathbf{x}) = \frac{q_i f_i(\mathbf{x})}{\sum_{j=1}^{g} q_j f_j(\mathbf{x})}$$

Substituting $P(\mathbf{x}|G_i)$ for $f_i(\mathbf{x})$, we have

$$P(G_i|\mathbf{x}) = \frac{q_i P(\mathbf{x}|G_i)}{\sum_{j=1}^{g} q_j P(\mathbf{x}|G_j)}$$

An observation is classified as belonging to the group with the highest posterior probability.

The difference between the discrim subcommands is in the choice of $f_i(\mathbf{x})$. LDA, discrim lda, assumes that the groups are multivariate normal with equal covariance matrices; see [MV] **discrim lda**. QDA, discrim qda, assumes that the groups are multivariate normal, allowing the groups to have unequal covariance matrices; see [MV] **discrim qda**. Logistic discriminant analysis, discrim logistic, uses the multinomial logistic model to obtain the posterior probabilities; see [MV] **discrim logistic**. kth-nearest neighbor, discrim knn, uses a simple nonparametric estimate of $f_i(\mathbf{x})$, based on examination of the k closest observations; see [MV] **discrim knn**.

References

Albert, A., and E. K. Harris. 1987. *Multivariate Interpretation of Clinical Laboratory Data.* New York: Marcel Dekker.

Anderson, T. W. 2003. *An Introduction to Multivariate Statistical Analysis.* 3rd ed. New York: Wiley.

Everitt, B. S., and G. Dunn. 2001. *Applied Multivariate Data Analysis.* 2nd ed. London: Arnold.

Fisher, R. A. 1936. The use of multiple measurements in taxonomic problems. *Annals of Eugenics* 7: 179–188.

Hand, D. J. 1981. *Discrimination and Classification.* New York: Wiley.

Huberty, C. J. 1994. *Applied Discriminant Analysis.* New York: Wiley.

Johnson, R. A., and D. W. Wichern. 2007. *Applied Multivariate Statistical Analysis.* 6th ed. Englewood Cliffs, NJ: Prentice Hall.

Klecka, W. R. 1980. *Discriminant Analysis.* Newbury Park, CA: Sage.

Lachenbruch, P. A. 1975. *Discriminant Analysis.* New York: Hafner Press.

Mardia, K. V., J. T. Kent, and J. M. Bibby. 1979. *Multivariate Analysis.* London: Academic Press.

McLachlan, G. J. 2004. *Discriminant Analysis and Statistical Pattern Recognition.* New York: Wiley.

Rencher, A. C. 1998. *Multivariate Statistical Inference and Applications.* New York: Wiley.

——. 2002. *Methods of Multivariate Analysis.* 2nd ed. New York: Wiley.

Tabachnick, B. G., and L. S. Fidell. 2007. *Using Multivariate Statistics.* 5th ed. Boston: Allyn and Bacon.

Also see

[MV] **discrim estat** — Postestimation tools for discrim

[MV] **candisc** — Canonical linear discriminant analysis

[MV] **cluster** — Introduction to cluster-analysis commands

[U] **20 Estimation and postestimation commands**

Title

discrim estat — Postestimation tools for discrim

Description

The following postestimation commands are of special interest after `candisc`, `discrim knn`, `discrim lda`, `discrim logistic`, and `discrim qda`:

Command	Description
`estat classtable`	classification table
`estat errorrate`	classification error-rate estimation
`estat grsummarize`	group summaries
`estat list`	classification listing
`estat summarize`	estimation sample summary

For information about these commands, see below.

There are more postestimation commands of special interest after `discrim lda` and `discrim qda`; see [MV] **discrim lda postestimation** and [MV] **discrim qda postestimation**.

Special-interest postestimation commands

`estat classtable` displays a cross-tabulation of the original groups with the classification groups. Classification percentages, average posterior probabilities, group prior probabilities, totals, and leave-one-out results are available.

`estat errorrate` displays error-rate estimates for the classification. Count-based estimates and both stratified and unstratified posterior-probability-based estimates of the error rate are available. These estimates can be resubstitution or leave-one-out estimates.

`estat grsummarize` presents estimation sample summary statistics for the discriminating variables for each group defined by the grouping variable. Means, medians, minimums, maximums, standard deviations, coefficients of variation, standard errors of the means, and group sizes may be displayed. Overall sample statistics are also available.

`estat list` lists group membership, classification, and probabilities for observations.

`estat summarize` summarizes the variables in the discriminant analysis over the estimation sample.

Syntax for estat classtable

`estat classtable` [*if*] [*in*] [*weight*] [, *options*]

options	Description
Main	
`class`	display the classification table; the default
`looclass`	display the leave-one-out classification table
Options	
`priors(`*priors*`)`	group prior probabilities; defaults to `e(grouppriors)`
`nopriors`	suppress display of prior probabilities
`ties(`*ties*`)`	how ties in classification are to be handled; defaults to `e(ties)`
`title(`*text*`)`	title for classification table
`probabilities`	display the average posterior probability of being classified into each group
`nopercents`	suppress display of percentages
`nototals`	suppress display of row and column totals
`norowtotals`	suppress display of row totals
`nocoltotals`	suppress display of column totals

priors	Description
`equal`	equal prior probabilities
`proportional`	group-size-proportional prior probabilities
matname	row or column vector containing the group prior probabilities
matrix_exp	matrix expression providing a row or column vector of the group prior probabilities

ties	Description
`missing`	ties in group classification produce missing values
`random`	ties in group classification are broken randomly
`first`	ties in group classification are set to the first tied group
`nearest`	ties in group classification are assigned based on the closest observation, or missing if this still results in a tie; after `discrim knn` only

`fweight`s are allowed; see [U] **11.1.6 weight**.

Menu

Statistics > Postestimation > Reports and statistics

Options for estat classtable

Main

`class`, the default, displays the classification table. With in-sample observations, this is called the resubstitution classification table.

`looclass` displays a leave-one-out classification table, instead of the default classification table. Leave-one-out classification applies only to the estimation sample, and so, in addition to restricting the observations to those chosen with `if` and `in` qualifiers, the observations are further restricted to those included in `e(sample)`.

Options

priors(*priors*) specifies the prior probabilities for group membership. If priors() is not specified, e(grouppriors) is used. If nopriors is specified with priors(), prior probabilities are used for calculation of the classification variable but not displayed. The following *priors* are allowed:

priors(equal) specifies equal prior probabilities.

priors(proportional) specifies group-size-proportional prior probabilities.

priors(*matname*) specifies a row or column vector containing the group prior probabilities.

priors(*matrix_exp*) specifies a matrix expression providing a row or column vector of the group prior probabilities.

nopriors suppresses display of the prior probabilities. This option does not change the computations that rely on the prior probabilities specified in priors() or as found by default in e(grouppriors).

ties(*ties*) specifies how ties in group classification will be handled. If ties() is not specified, e(ties) determines how ties are handled. The following *ties* are allowed:

ties(missing) specifies that ties in group classification produce missing values.

ties(random) specifies that ties in group classification are broken randomly.

ties(first) specifies that ties in group classification are set to the first tied group.

ties(nearest) specifies that ties in group classification are assigned based on the closest observation, or missing if this still results in a tie. ties(nearest) is available after discrim knn only.

title(*text*) customizes the title for the classification table.

probabilities specifies that the classification table show the average posterior probability of being classified into each group. probabilities implies norowtotals and nopercents.

nopercents specifies that percentages are to be omitted from the classification table.

nototals specifies that row and column totals are to be omitted from the classification table.

norowtotals specifies that row totals are to be omitted from the classification table.

nocoltotals specifies that column totals are to be omitted from the classification table.

Syntax for estat errorrate

`estat errorrate [if] [in] [weight] [, options]`

options	Description
Main	
`class`	display the classification-based error-rate estimates table; the default
`looclass`	display the leave-one-out classification-based error-rate estimates table
`count`	use a count-based error-rate estimate
`pp[(`*ppopts*`)]`	use a posterior-probability-based error-rate estimate
Options	
`priors(`*priors*`)`	group prior probabilities; defaults to `e(grouppriors)`
`nopriors`	suppress display of prior probabilities
`ties(`*ties*`)`	how ties in classification are to be handled; defaults to `e(ties)`
`title(`*text*`)`	title for error-rate estimate table
`nototal`	suppress display of total column

ppopts	Description
`stratified`	present stratified results
`unstratified`	present unstratified results

`fweight`s are allowed; see [U] **11.1.6 weight**.

Menu

Statistics > Postestimation > Reports and statistics

Options for estat errorrate

Main

`class`, the default, specifies that the classification-based error-rate estimates table be presented. The alternative to `class` is `looclass`.

`looclass` specifies that the leave-one-out classification error-rate estimates table be presented.

`count`, the default, specifies that the error-rate estimates be based on misclassification counts. The alternative to `count` is `pp()`.

`pp[(`*ppopts*`)]` specifies that the error-rate estimates be based on posterior probabilities. `pp` is equivalent to `pp(stratified unstratified)`. `stratified` indicates that stratified estimates be presented. `unstratified` indicates that unstratified estimates be presented. One or both may be specified.

Options

`priors(`*priors*`)` specifies the prior probabilities for group membership. If `priors()` is not specified, `e(grouppriors)` is used. If `nopriors` is specified with `priors()`, prior probabilities are used for calculation of the error-rate estimates but not displayed. The following *priors* are allowed:

priors(equal) specifies equal prior probabilities.

priors(proportional) specifies group-size-proportional prior probabilities.

priors(*matname*) specifies a row or column vector containing the group prior probabilities.

priors(*matrix_exp*) specifies a matrix expression providing a row or column vector of the group prior probabilities.

nopriors suppresses display of the prior probabilities. This option does not change the computations that rely on the prior probabilities specified in priors() or as found by default in e(grouppriors).

ties(*ties*) specifies how ties in group classification will be handled. If ties() is not specified, e(ties) determines how ties are handled. The following *ties* are allowed:

ties(missing) specifies that ties in group classification produce missing values.

ties(random) specifies that ties in group classification are broken randomly.

ties(first) specifies that ties in group classification are set to the first tied group.

ties(nearest) specifies that ties in group classification are assigned based on the closest observation, or missing if this still results in a tie. ties(nearest) is available after discrim knn only.

title(*text*) customizes the title for the error-rate estimates table.

nototal suppresses the total column containing overall sample error-rate estimates.

Syntax for estat grsummarize

estat grsummarize [, *options*]

options	Description
Main	
n[(%*fmt*)]	group sizes
mean[(%*fmt*)]	means
median[(%*fmt*)]	medians
sd[(%*fmt*)]	standard deviations
cv[(%*fmt*)]	coefficients of variation
semean[(%*fmt*)]	standard errors of the means
min[(%*fmt*)]	minimums
max[(%*fmt*)]	maximums
Options	
nototal	suppress overall statistics
transpose	display groups by row instead of column

Menu

Statistics > Postestimation > Reports and statistics

Options for estat grsummarize

Main

n[(%*fmt*)] specifies that group sizes be presented. The optional argument provides a display format. The default options are n and mean.

mean[(%*fmt*)] specifies that means be presented. The optional argument provides a display format. The default options are n and mean.

median[(%*fmt*)] specifies that medians be presented. The optional argument provides a display format.

sd[(%*fmt*)] specifies that standard deviations be presented. The optional argument provides a display format.

cv[(%*fmt*)] specifies that coefficients of variation be presented. The optional argument provides a display format.

semean[(%*fmt*)] specifies that standard errors of the means be presented. The optional argument provides a display format.

min[(%*fmt*)] specifies that minimums be presented. The optional argument provides a display format.

max[(%*fmt*)] specifies that maximums be presented. The optional argument provides a display format.

Options

nototal suppresses display of the total column containing overall sample statistics.

transpose specifies that the groups are to be displayed by row. By default, groups are displayed by column. If you have more variables than groups, you might prefer the output produced by transpose.

Syntax for estat list

estat list [*if*] [*in*] [, *options*]

options	Description
Main	
misclassified	list only misclassified and unclassified observations
classification(*clopts*)	control display of classification
probabilities(*propts*)	control display of probabilities
varlist[(*varopts*)]	display discriminating variables
[no]obs	display or suppress the observation number
id(*varname* [format(%*fmt*)])	display identification variable
Options	
weight[(*weightopts*)]	display frequency weights
priors(*priors*)	group prior probabilities; defaults to e(grouppriors)
ties(*ties*)	how ties in classification are to be handled; defaults to e(ties)
separator(#)	display a horizontal separator every # lines

clopts	Description
noclass	do not display the standard classification
looclass	display the leave-one-out classification
notrue	do not show the group variable
nostar	do not display stars indicating misclassified observations
nolabel	suppress display of value labels for the group and classification variables
format(%*fmt*)	format for group and classification variables; default is %5.0f for unlabeled numeric variables

propts	Description
nopr	suppress display of standard posterior probabilities
loopr	display leave-one-out posterior probabilities
format(%*fmt*)	format for probabilities; default %7.4f

varopts	Description
first	display input variables before classifications and probabilities
last	display input variables after classifications and probabilities
format(%*fmt*)	format for input variables; default is the input variable format

weightopts	Description
none	do not display the weights
format(%*fmt*)	format for the weight; default is %3.0f for weights < 1,000, %5.0f for 1,000 < weights < 100,000, and %8.0g otherwise

Menu

Statistics > Postestimation > Reports and statistics

Options for estat list

Main

misclassified lists only misclassified and unclassified observations.

classification(*clopts*) controls display of the group variable and classification. By default, the standard classification is calculated and displayed along with the group variable in e(groupvar), using labels from the group variable if they exist. *clopts* may be one or more of the following:

noclass suppresses display of the standard classification. If the observations are those used in the estimation, classification is called resubstitution classification.

looclass specifies that the leave-one-out classification be calculated and displayed. The default is that the leave-one-out classification is not calculated. looclass is not allowed after discrim logistic.

notrue suppresses the display of the group variable. By default, e(groupvar) is displayed. notrue implies nostar.

nostar suppresses the display of stars indicating misclassified observations. A star is displayed by default when the classification is not in agreement with the group variable. nostar is the default when notrue is specified.

nolabel specifies that value labels for the group variable, if they exist, not be displayed for the group or classification or used as labels for the probability column names.

format(%*fmt*) specifies the format for the group and classification variables. If value labels are used, string formats are permitted.

probabilities(*propts*) controls the display of group posterior probabilities. *propts* may be one or more of the following:

nopr suppresses display of the standard posterior probabilities. By default, the posterior probabilities are shown.

loopr specifies that leave-one-out posterior probabilities be displayed. loopr is not allowed after discrim logistic.

format(%*fmt*) specifies the format for displaying probabilities. %7.4f is the default.

varlist[(*varopts*)] specifies that the discriminating variables found in e(varlist) be displayed and specifies the display options for the variables.

first specifies variables be displayed before classifications and probabilities.

last specifies variables be displayed after classifications and probabilities.

format(%*fmt*) specifies the format for the input variables. By default, the variable's format is used.

[no]obs indicates that observation numbers be or not be displayed. Observation numbers are displayed by default unless id() is specified.

id(*varname* [format(%*fmt*)]) specifies the identification variable to display and, optionally, the format for that variable. By default, the format of *varname* is used.

Options

weight[(*weightopts*)] specifies options for displaying weights. By default, if e(wexp) exists, weights are displayed.

none specifies weights not be displayed. This is the default if weights were not used with discrim.

format(%*fmt*) specifies a display format for the weights. If the weights are < 1,000, %3.0f is the default, %5.0f is the default if 1,000 < weights < 100,000, else %8.0g is used.

priors(*priors*) specifies the prior probabilities for group membership. If priors() is not specified, e(grouppriors) is used. The following *priors* are allowed:

priors(equal) specifies equal prior probabilities.

priors(proportional) specifies group-size-proportional prior probabilities.

priors(*matname*) specifies a row or column vector containing the group prior probabilities.

priors(*matrix_exp*) specifies a matrix expression providing a row or column vector of the group prior probabilities.

ties(*ties*) specifies how ties in group classification will be handled. If ties() is not specified, e(ties) determines how ties are handled. The following *ties* are allowed:

ties(missing) specifies that ties in group classification produce missing values.

ties(random) specifies that ties in group classification are broken randomly.

ties(first) specifies that ties in group classification are set to the first tied group.

ties(nearest) specifies that ties in group classification are assigned based on the closest observation, or missing if this still results in a tie. ties(nearest) is available after discrim knn only.

separator(#) specifies a horizontal separator line be drawn every # observations. The default is separator(5).

Syntax for estat summarize

estat summarize [, labels noheader noweights]

Menu

Statistics > Postestimation > Reports and statistics

Options for estat summarize

labels, noheader, and noweights are the same as for the generic estat summarize; see [R] **estat**.

Remarks

Remarks are presented under the following headings:

Discriminating-variable summaries
Discrimination listings
Classification tables and error rates

There are several estat commands that apply after all the discrim subcommands. estat summarize and estat grsummarize summarize the discriminating variables over the estimation sample and by-group. estat list displays classifications, posterior probabilities, and more for selected observations. estat classtable and estat errorrate display the classification table, also known as a confusion matrix, and error-rate estimates based on the classification table.

Discriminating-variable summaries

estat summarize and estat grsummarize provide summaries of the variables involved in the preceding discriminant analysis model.

▷ Example 1

Example 3 of [MV] **discrim lda** introduces the famous iris data originally from Anderson (1935) and used by Fisher (1936) in the development of linear discriminant analysis. We continue our exploration of the linear discriminant analysis of the iris data and demonstrate the summary estat tools available after all discrim subcommands.

```
. use http://www.stata-press.com/data/r12/iris
(Iris data)
. discrim lda seplen sepwid petlen petwid, group(iris) notable
```

The `notable` option of `discrim` suppressed display of the classification table. We explore the use of `estat classtable` later.

What can we learn about the underlying discriminating variables? `estat summarize` gives a summary of the variables involved in the discriminant analysis, restricted to the estimation sample.

```
. estat summarize
Estimation sample discrim                    Number of obs =     150

    Variable         Mean    Std. Dev.         Min          Max

groupvar
        iris            2     .8192319           1            3

variables
      seplen     5.843333     .8280661         4.3          7.9
      sepwid     3.057333     .4358663           2          4.4
      petlen        3.758     1.765298           1          6.9
      petwid     1.199333     .7622377          .1          2.5
```

`estat summarize` displays the mean, standard deviation, minimum, and maximum for the group variable, `iris`, and the four discriminating variables, `seplen`, `sepwid`, `petlen`, and `petwid`. Also shown is the number of observations. If we had fit our discriminant model on a subset of the data, `estat summarize` would have restricted its summary to those observations.

More interesting than an overall summary of the discriminating variables is a summary by our group variable, `iris`.

```
. estat grsummarize
Estimation sample discrim lda
Summarized by iris

                   iris
Mean              Setosa  Versicolor   Virginica       Total

      seplen       5.006       5.936       6.588    5.843333
      sepwid       3.428        2.77       2.974    3.057333
      petlen       1.462        4.26       5.552       3.758
      petwid        .246       1.326       2.026    1.199333

           N          50          50          50         150
```

By default, `estat grsummarize` displays means of the discriminating variables for each group and overall (the total column), along with group sizes. The summary is restricted to the estimation sample.

The petal length and width of *Iris setosa* appear to be much smaller than those of the other two species. *Iris versicolor* has petal length and width between that of the other two species.

Other statistics may be requested. A look at the minimums and maximums might provide more insight into the separation of the three iris species.

```
. estat grsummarize, min max

Estimation sample discrim lda
Summarized by iris

                   iris
                       Setosa  Versicolor   Virginica       Total

seplen
            Min           4.3         4.9         4.9         4.3
            Max           5.8           7         7.9         7.9

sepwid
            Min           2.3           2         2.2           2
            Max           4.4         3.4         3.8         4.4

petlen
            Min             1           3         4.5           1
            Max           1.9         5.1         6.9         6.9

petwid
            Min            .1           1         1.4          .1
            Max            .6         1.8         2.5         2.5
```

Although this table is helpful, an altered view of it might make comparisons easier. `estat grsummarize` allows a format to be specified with each requested statistic. We can request a shorter format for the minimum and maximum and specify a fixed format so that the decimal point lines up. `estat grsummarize` also has a `transpose` option that places the variables and requested statistics as columns and the groups as rows. If you have fewer discriminating variables than groups, this might be the most natural way to view the statistics. Here we have more variables, but with a narrow display format, the transposed view still works well.

```
. estat grsummarize, min(%4.1f) max(%4.1f) transpose

Estimation sample discrim lda
Summarized by iris

                   seplen      sepwid      petlen      petwid
iris               Min  Max    Min  Max    Min  Max    Min  Max

        Setosa     4.3  5.8    2.3  4.4    1.0  1.9    0.1  0.6
    Versicolor     4.9  7.0    2.0  3.4    3.0  5.1    1.0  1.8
     Virginica     4.9  7.9    2.2  3.8    4.5  6.9    1.4  2.5

         Total     4.3  7.9    2.0  4.4    1.0  6.9    0.1  2.5
```

The maximum petal length and width for *Iris setosa* are much smaller than the minimum petal length and width for the other two species. The petal length and width clearly separate *Iris setosa* from the other two species.

You are not limited to one or two statistics with `estat grsummarize`, and each statistic may have different requested display formats. The total column, or row if the table is transposed, can also be suppressed.

Using Stata's `graph box` command is another way of seeing the differences among the three iris species for the four discriminating variables.

```
. graph box seplen, over(iris) name(sl)
. graph box sepwid, over(iris) name(sw)
. graph box petlen, over(iris) name(pl)
. graph box petwid, over(iris) name(pw)
. graph combine sl sw pl pw, title(Characteristics of three iris species)
```

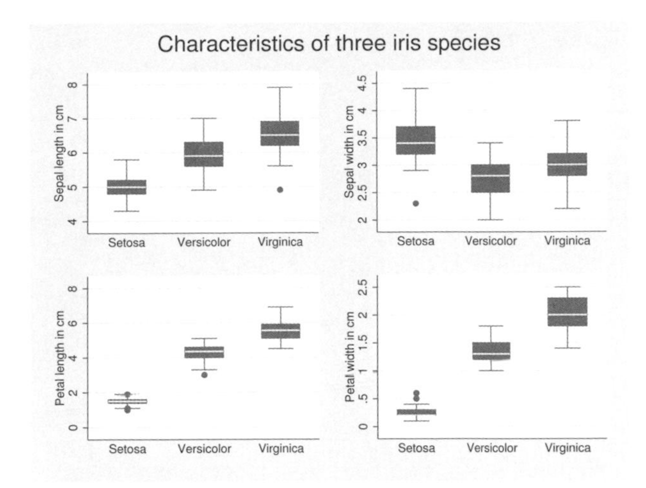

The box plots confirm the lack of overlap in the petal lengths and widths for *Iris setosa* compared with the other two iris species. Other differences between the species are also seen.

◁

More summary `estat` commands are available after `discrim lda`; see [MV] **discrim lda postestimation**.

Discrimination listings

Listing the true group, classified group, group posterior probabilities, and discriminating variables for observations of interest after `discrim` is easy with the `estat list` command.

▷ Example 2

Example 1 of [MV] **discrim** introduced the riding-mower data of Johnson and Wichern (2007) and presented a linear discriminant analysis that concluded with the use of `estat list` displaying the misclassified observations.

```
. use http://www.stata-press.com/data/r12/lawnmower2
(Johnson and Wichern (2007) Table 11.1)
. discrim lda income lotsize, group(owner) notable
. estat list, class(loo) pr(loo) misclassified
```

	Classification			Probabilities		LOO Probabilities	
Obs.	True	Class.	LOO Cl.	nonowner	owner	nonowner	owner
1	owner	nonown *	nonown *	0.7820	0.2180	0.8460	0.1540
2	owner	owner	nonown *	0.4945	0.5055	0.6177	0.3823
13	nonown	owner *	owner *	0.2372	0.7628	0.1761	0.8239
14	nonown	nonown	owner *	0.5287	0.4713	0.4313	0.5687
17	nonown	owner *	owner *	0.3776	0.6224	0.2791	0.7209

```
* indicates misclassified observations
```

The `misclassified` option limited the listing to those observations that were misclassified by the linear discriminant model. `class(loo)` and `pr(loo)` added leave-one-out (LOO) classifications and probabilities to the resubstitution classifications and probabilities.

We demonstrate a few other options available with `estat list`. We can limit which observations are displayed with `if` and `in` modifiers and can add the display of the discriminating variables with the `varlist` option. Here we limit the display to those observations that have `income` greater than $110,000.

```
. estat list if income > 110, varlist
```

	Data		Classification		Probabilities	
Obs.	income	lotsize	True	Class.	nonowner	owner
2	115.5	16.8	owner	owner	0.4945	0.5055
5	117.0	23.6	owner	owner	0.0040	0.9960
6	140.1	19.2	owner	owner	0.0125	0.9875
7	138.0	17.6	owner	owner	0.0519	0.9481
8	112.8	22.4	owner	owner	0.0155	0.9845
10	123.0	20.8	owner	owner	0.0196	0.9804
12	111.0	20.0	owner	owner	0.1107	0.8893
17	114.0	17.6	nonowner	owner *	0.3776	0.6224

```
* indicates misclassified observations
```

Starting with the command above, we specify `sep(0)` to suppress the separator line that, by default, displays after every 5 observations. We eliminate the observation numbers with the `noobs` option. With the `class()` option: the `looclass` suboption adds the LOO classification; the `noclass` suboption suppress the resubstitution classification; and the `nostar` suboption eliminates the marking of misclassified observations with asterisks. With `pr(loopr nopr)` we specify that LOO probabilities are to be displayed and resubstitution probabilities are to be suppressed.

```
. estat list if income > 110, sep(0) class(looclass noclass nostar) pr(loopr nopr)
> varlist noobs
```

Data		Classification		LOO Probabilities	
income	lotsize	True	LOO Cl	nonowner	owner
115.5	16.8	owner	nonowner	0.6177	0.3823
117.0	23.6	owner	owner	0.0029	0.9971
140.1	19.2	owner	owner	0.0124	0.9876
138.0	17.6	owner	owner	0.0737	0.9263
112.8	22.4	owner	owner	0.0168	0.9832
123.0	20.8	owner	owner	0.0217	0.9783
111.0	20.0	owner	owner	0.1206	0.8794
114.0	17.6	nonowner	owner	0.2791	0.7209

◁

Use the `if e(sample)` qualifier to restrict the listing from `estat list` to the estimation sample. Out-of-sample listings are obtained if your selected observations are not part of the estimation sample.

As an alternative to `estat list`, you can use `predict` after `discrim` to obtain the classifications, posterior probabilities, or whatever else is available for prediction from the `discrim` subcommand that you ran, and then use `list` to display your predictions; see [D] **list** and see example 2 of [MV] **discrim knn postestimation** for an example.

Classification tables and error rates

Classification tables (also known as confusion matrices) and error-rate estimate tables are available with the `estat classtable` and `estat errorrate` commands after `discrim`.

▷ Example 3

Example 2 of [MV] **discrim knn** introduces a head measurement dataset from Rencher (2002, 280–281) with six discriminating variables and three groups. We perform a quadratic discriminant analysis (QDA) on the dataset to illustrate classification tables and error-rate estimation.

```
. use http://www.stata-press.com/data/r12/head
(Table 8.3 Head measurements -- Rencher (2002))
. discrim qda wdim circum fbeye eyehd earhd jaw, group(group)
Quadratic discriminant analysis
Resubstitution classification summary
```

Key
Number Percent

True group	Classified high school	college	nonplayer	Total
high school	27 90.00	1 3.33	2 6.67	30 100.00
college	2 6.67	21 70.00	7 23.33	30 100.00
nonplayer	1 3.33	4 13.33	25 83.33	30 100.00
Total	30 33.33	26 28.89	34 37.78	90 100.00
Priors	0.3333	0.3333	0.3333	

By default, `discrim` displayed the resubstitution classification table. A resubstitution classification table is obtained by classifying the observations used in building the discriminant model. The resubstitution classification table is overly optimistic as an indicator of how well you might classify other observations.

This resubstitution classification table shows that from the high school group 27 observations were correctly classified, 1 observation was classified as belonging to the college group, and 2 observations were classified as belonging to the nonplayer group. The corresponding percentages were also presented: 90%, 3.33%, and 6.67%. The college and nonplayer rows are read in a similar manner. For instance, there were 7 observations from the college group that were misclassified as nonplayers. Row and column totals are presented along with the group prior probabilities. See table 9.4 of Rencher (2002, 309) for this same classification table.

There are various ways of estimating the error rate for a classification. `estat errorrate` presents the overall (total) error rate and the error rate for each group. By default, it uses a count-based estimate of the error rate.

```
. estat errorrate
Error rate estimated by error count
```

	group high school	college	nonplayer	Total
Error rate	.1	.3	.166666667	.188888889
Priors	.333333333	.333333333	.333333333	

This is a resubstitution count-based error-rate estimate corresponding to the classification table previously presented. Three of the 30 high school observations were misclassified—a proportion of 0.1; 9 of the 30 college observations were misclassified—a proportion of 0.3; and 5 of the 30 nonplayers were misclassified—a proportion of 0.1667. The total error rate is computed as the sum of the group error rates times their prior probabilities—here 0.1889.

An error-rate estimate based on the posterior probabilities is also available with `estat errorrate`.

```
. estat errorrate, pp
Error rate estimated from posterior probabilities
```

Error Rate	group high school	college	nonplayer	Total
Stratified	.08308968	.337824355	.2030882	.208000745
Unstratified	.08308968	.337824355	.2030882	.208000745
Priors	.333333333	.333333333	.333333333	

Because we did not specify otherwise, we obtained resubstitution error-rate estimates. By default both the stratified and unstratified estimates are shown. The stratified estimates give less weight to probabilities where the group sample size is large compared with the group prior probabilities; see *Methods and formulas* for details. Here the stratified and unstratified estimates are the same. This happens when the prior probabilities are proportional to the sample sizes—here we have equal prior probabilities and equal group sizes.

For this example, the count-based and posterior-probability-based estimates are similar to one another.

Leave-one-out (LOO) estimation provides a more realistic assessment of your potential classification success with observations that were not used in building the discriminant analysis model. The `loo` option of `estat classtable` and `estat errorrate` specify a LOO estimation.

```
. estat classtable, loo nopercents nopriors nototals
Leave-one-out classification table
```

Key
Number

True group	LOO Classified high school	college	nonplayer
high school	26	2	2
college	3	16	11
nonplayer	4	9	17

To demonstrate some of the available options, we specified the nopercents option to suppress the display of percentages; the nopriors option to suppress the display of the prior probabilities; and the nototals option to suppress row and column totals.

If you compare this LOO classification table with the resubstitution classification table, you will see that fewer observations appear on the diagonal (were correctly classified) in the LOO table. The LOO estimates are less biased than the resubstitution estimates.

We now examine the LOO error-rate estimates by using the loo option with the estat error command. We first produce the count-based estimates and then request the posterior-probability-based estimates. In the first case, we use the nopriors option to demonstrate that you can suppress the display of the prior probabilities. Suppressing the display does not remove their effect on the computations. In the second estat errorrate call, we specify that only the unstratified estimates be presented. (Because the prior probabilities and samples sizes match [are equal], the stratified results will be the same.)

```
. estat errorrate, loo nopriors
Error rate estimated by leave-one-out error count
```

	group high school	college	nonplayer	Total
Error rate	.133333333	.466666667	.433333333	.344444444

```
. estat errorrate, loo pp(unstratified)
Error rate estimated from leave-one-out posterior probabilities
```

Error Rate	group high school	college	nonplayer	Total
Unstratified	.049034154	.354290969	.294376541	.232567222
Priors	.333333333	.333333333	.333333333	

Instead of displaying percentages below the counts in the classification table, we can display average posterior probabilities. The probabilities option requests the display of average posterior probabilities. We add the nopriors option to demonstrate that the prior probabilities can be suppressed from the table. The classifications are still based on the prior probabilities; they are just not displayed.

```
. estat classtable, probabilities nopriors
Resubstitution average-posterior-probabilities classification table
```

Key
Number Average posterior probability

True group	Classified high school	college	nonplayer
high school	27 0.9517	1 0.6180	2 0.5921
college	2 0.6564	21 0.8108	7 0.5835
nonplayer	1 0.4973	4 0.5549	25 0.7456
Total	30 0.9169	26 0.7640	34 0.7032

Both `estat classtable` and `estat errorrate` allow `if` and `in` qualifiers so that you can select the observations to be included in the computations and displayed. If you want to limit the table to the estimation sample, use `if e(sample)`. You can also do out-of-sample classification tables and error-rate estimation by selecting observations that were not part of the estimation sample.

◁

❑ Technical note

As noted by Huberty (1994, 92), the posterior-probability-based error-rate estimates for the individual groups may be negative. This may happen when there is a discrepancy between group prior probabilities and relative sample size.

Continuing with our last example, if we use prior probabilities of 0.2, 0.1, and 0.7 for the high school, college, and nonplayer groups, the nonplayer stratified error-rate estimate and the high school group unstratified error-rate estimate are negative.

```
. estat errorrate, pp priors(.2, .1, .7)
Error rate estimated from posterior probabilities
```

Error Rate	group high school	college	nonplayer	Total
Stratified	.19121145	.737812235	-.001699715	.110833713
Unstratified	-.36619243	.126040785	.29616143	.146678593
Priors	.2	.1	.7	

❑

More examples of the use of `estat list`, `estat classtable`, and `estat errorrate` can be found in the other `discrim`-related manual entries.

Saved results

`estat classtable` saves the following in `r()`:

Matrices

`r(counts)`	group counts
`r(percents)`	percentages for each group (unless `nopercents` specified)
`r(avgpostprob)`	average posterior probabilities classified into each group (`probabilities` only)

`estat errorrate` saves the following in `r()`:

Matrices

`r(grouppriors)`	row vector of group prior probabilities used in the calculations
`r(erate_count)`	matrix of error rates estimated from error counts (`count` only)
`r(erate_strat)`	matrix of stratified error rates estimated from posterior probabilities (`pp` only)
`r(erate_unstrat)`	matrix of unstratified error rates estimated from posterior probabilities (`pp` only)

`estat grsummarize` saves the following in `r()`:

Matrices

`r(count)`	group counts
`r(mean)`	means (`mean` only)
`r(median)`	medians (`median` only)
`r(sd)`	standard deviations (`sd` only)
`r(cv)`	coefficients of variation (`cv` only)
`r(semean)`	standard errors of the means (`semean` only)
`r(min)`	minimums (`min` only)
`r(max)`	maximums (`max` only)

Methods and formulas

All postestimation commands listed above are implemented as ado-files.

Let **C** denote the classification table (also known as the confusion matrix), with rows corresponding to the true groups and columns corresponding to the assigned groups. Let C_{ij} denote the element from row i and column j of **C**. C_{ij} represents the number of observations from group i assigned to group j. n_i is the number of observations from group i and $N = \sum_{i=1}^{g} n_i$ is the total sample size. $\mathcal{N}_i = \sum_{j=1}^{g} C_{ij}$ is the number of observations from group i that were classified into one of the g groups. If some observations from group i are unclassified (because of ties), $\mathcal{N}_i \neq n_i$ and $\mathcal{N} \neq N$ (where $\mathcal{N} = \sum \mathcal{N}_i$). Let q_i be the prior probability of group i.

`estat classtable` displays **C**, with options controlling the display of cell percentages by row, average posterior probabilities, prior probabilities, row totals, and column totals.

McLachlan (2004, chap. 10) devotes a chapter to classification error-rate estimation. The `estat errorrate` command provides several popular error-rate estimates. Smith (1947) introduced the count-based apparent error-rate estimate. The count-based error-rate estimate for group i is

$$\widehat{E}_i^{(C)} = 1 - C_{ii}/\mathcal{N}_i$$

The overall (total) count-based error-rate estimate is

$$\widehat{E}^{(C)} = \sum_{i=1}^{g} q_i \widehat{E}_i^{(C)}$$

In general, $\widehat{E}^{(C)} \neq 1 - \sum_{i=1}^{g} C_{ii}/\mathcal{N}$, though some sources, Rencher (2002, 307), appear to report this latter quantity.

If **C** is based on the same data used in the estimation of the discriminant analysis model, the error rates are called apparent error rates. Leave-one-out (LOO) error rates are obtained if **C** is based on a leave-one-out analysis where each observation to be classified is classified based on the discriminant model built excluding that observation; see Lachenbruch and Mickey (1968) and McLachlan (2004, 342).

Error rates can also be estimated from the posterior probabilities. Huberty (1994, 90–91) discusses hit rates (one minus the error rates) based on posterior probabilities and shows two versions of the posterior-probability based estimate—stratified and unstratified.

Let $\mathcal{P}_{ji}$ be the sum of the posterior probabilities for all observations from group j assigned to group i. The posterior-probability-based unstratified error-rate estimate for group i is

$$\widehat{E}_i^{(Pu)} = 1 - \frac{1}{\mathcal{N} q_i} \sum_{j=1}^{g} \mathcal{P}_{ji}$$

The overall (total) posterior-probability-based unstratified error-rate estimate is

$$\widehat{E}^{(Pu)} = \sum_{i=1}^{g} q_i \widehat{E}_i^{(Pu)}$$

The posterior-probability-based stratified error-rate estimate for group i is

$$\widehat{E}_i^{(Ps)} = 1 - \frac{1}{q_i} \sum_{j=1}^{g} \frac{q_j}{\mathcal{N}_j} \mathcal{P}_{ji}$$

The overall (total) posterior-probability-based stratified error-rate estimate is

$$\widehat{E}^{(Ps)} = \sum_{i=1}^{g} q_i \widehat{E}_i^{(Ps)}$$

References

Anderson, E. 1935. The irises of the Gaspe Peninsula. *Bulletin of the American Iris Society* 59: 2–5.

Fisher, R. A. 1936. The use of multiple measurements in taxonomic problems. *Annals of Eugenics* 7: 179–188.

Huberty, C. J. 1994. *Applied Discriminant Analysis*. New York: Wiley.

Johnson, R. A., and D. W. Wichern. 2007. *Applied Multivariate Statistical Analysis*. 6th ed. Englewood Cliffs, NJ: Prentice Hall.

Lachenbruch, P. A., and M. R. Mickey. 1968. Estimation of error rates in discriminant analysis. *Technometrics* 10: 1–11.

McLachlan, G. J. 2004. *Discriminant Analysis and Statistical Pattern Recognition*. New York: Wiley.

Rencher, A. C. 2002. *Methods of Multivariate Analysis*. 2nd ed. New York: Wiley.

Smith, C. A. B. 1947. Some examples of discrimination. *Annals of Eugenics* 13: 272–282.

Also see

[MV] **discrim** — Discriminant analysis

[MV] **discrim knn postestimation** — Postestimation tools for discrim knn

[MV] **discrim lda postestimation** — Postestimation tools for discrim lda

[MV] **discrim logistic postestimation** — Postestimation tools for discrim logistic

[MV] **discrim qda postestimation** — Postestimation tools for discrim qda

[MV] **candisc** — Canonical linear discriminant analysis

[U] **20 Estimation and postestimation commands**

Title

discrim knn — kth-nearest-neighbor discriminant analysis

Syntax

`discrim knn` *varlist* [*if*] [*in*] [*weight*], `group(`*groupvar*`)` `k(`*#*`)` [*options*]

options	Description
Model	
*`group(`*groupvar*`)`	variable specifying the groups
`k(`#*`)`	number of nearest neighbors
`priors(`*priors*`)`	group prior probabilities
`ties(`*ties*`)`	how ties in classification are to be handled
Measure	
`measure(`*measure*`)`	similarity or dissimilarity measure; default is `measure(L2)`
`s2d(standard)`	convert similarity to dissimilarity: $d(ij) = \sqrt{s(ii) + s(jj) - 2s(ij)}$, the default
`s2d(oneminus)`	convert similarity to dissimilarity: $d(ij) = 1 - s(ij)$
`mahalanobis`	Mahalanobis transform continuous data before computing dissimilarities
Reporting	
`notable`	suppress resubstitution classification table
`lootable`	display leave-one-out classification table

priors	Description
`equal`	equal prior probabilities; the default
`proportional`	group-size-proportional prior probabilities
matname	row or column vector containing the group prior probabilities
matrix_exp	matrix expression providing a row or column vector of the group prior probabilities

ties	Description
`missing`	ties in group classification produce missing values; the default
`random`	ties in group classification are broken randomly
`first`	ties in group classification are set to the first tied group
`nearest`	ties in group classification are assigned based on the closest observation, or missing if this still results in a tie

*`group()` and `k()` are required.

`statsby` and `xi` are allowed; see [U] **11.1.10 Prefix commands**.

`fweight`s are allowed; see [U] **11.1.6 weight**.

See [U] **20 Estimation and postestimation commands** for more capabilities of estimation commands.

Menu

Statistics > Multivariate analysis > Discriminant analysis > Kth-nearest neighbor (KNN)

Description

`discrim knn` performs kth-nearest-neighbor discriminant analysis. A wide selection of similarity and dissimilarity measures is available; see the `measure()` option.

kth-nearest neighbor must retain the training data and search through the data for the k nearest observations each time a classification or prediction is performed. Consequently for large datasets, kth-nearest neighbor is slow and uses a lot of memory.

See [MV] **discrim** for other discrimination commands.

Options

Model

`group(`*groupvar*`)` is required and specifies the name of the grouping variable. *groupvar* must be a numeric variable.

`k(`*#*`)` is required and specifies the number of nearest neighbors on which to base computations. In the event of ties, the next largest value of `k()` is selected. Suppose that `k(3)` is selected. For a given observation, one must go out a distance d to find three nearest neighbors, but if, say, there are five data points all within distance d, then the computation will be based on all five nearest points.

`priors(`*priors*`)` specifies the prior probabilities for group membership. The following *priors* are allowed:

`priors(equal)` specifies equal prior probabilities. This is the default.

`priors(proportional)` specifies group-size-proportional prior probabilities.

`priors(`*matname*`)` specifies a row or column vector containing the group prior probabilities.

`priors(`*matrix_exp*`)` specifies a matrix expression providing a row or column vector of the group prior probabilities.

`ties(`*ties*`)` specifies how ties in group classification will be handled. The following *ties* are allowed:

`ties(missing)` specifies that ties in group classification produce missing values. This is the default.

`ties(random)` specifies that ties in group classification are broken randomly.

`ties(first)` specifies that ties in group classification are set to the first tied group.

`ties(nearest)` specifies that ties in group classification are assigned based on the closest observation, or missing if this still results in a tie.

Measure

`measure(`*measure*`)` specifies the similarity or dissimilarity measure. The default is `measure(L2)`; all measures in [MV] ***measure_option*** are supported except for `measure(Gower)`.

`s2d(standard | oneminus)` specifies how similarities are converted into dissimilarities.

The available `s2d()` options, `standard` and `oneminus`, are defined as

`standard`	$d(ij) = \sqrt{s(ii) + s(jj) - 2s(ij)} = \sqrt{2\{1 - s(ij)\}}$
`oneminus`	$d(ij) = 1 - s(ij)$

`s2d(standard)` is the default.

`mahalanobis` specifies performing a Mahalanobis transformation on continuous data before computing dissimilarities. The data is transformed via the Cholesky decomposition of the within-group covariance matrix, and then the selected dissimilarity measure is performed on the transformed data. If the `L2` (Euclidean) dissimilarity is chosen, this is the Mahalanobis distance. If the within-group covariance matrix does not have sufficient rank, an error is returned.

Reporting

`notable` suppresses the computation and display of the resubstitution classification table.

`lootable` displays the leave-one-out classification table.

Remarks

Remarks are presented under the following headings:

Introduction
A first example
Mahalanobis transformation
Binary data

Introduction

kth-nearest-neighbor (KNN) discriminant analysis dates at least as far back as Fix and Hodges (1951). An introductory treatment is available in Rencher (2002). More advanced treatments are in Hastie, Tibshirani, and Friedman (2009) and McLachlan (2004).

KNN is a nonparametric discrimination method based on the k nearest neighbors of each observation. KNN can deal with binary data via one of the binary measures; see [MV] ***measure_option***.

A first example

What distinguishes kth-nearest-neighbor analysis from other methods of discriminant analysis is its ability to distinguish irregular-shaped groups, including groups with multiple modes. We create a dataset with unusual boundaries that lends itself to KNN analysis and graphical interpretation.

▷ Example 1

We create a two-dimensional dataset on the plane with x and y values in $[-4, 4]$. In each quadrant we consider points within a circle of radius two, centered around the points $(2, 2)$, $(-2, 2)$, $(-2, -2)$, and $(2, -2)$. We set the group value to 1 to start and then replace it in the circles. In the first and third circles we set the group value to 2, and in the second and fourth circles we set the group value to 3. Outside the circles, the group value remains 1.

```
. set seed 98712321
. set obs 500
obs was 0, now 500
. gen x = 8*runiform() - 4
. gen y = 8*runiform() - 4
. gen group = 1
. replace group = 2 if (y+2)^2 + (x+2)^2 <= 2
(34 real changes made)
. replace group = 2 if (y-2)^2 + (x-2)^2 <= 2
(38 real changes made)
. replace group = 3 if (y+2)^2 + (x-2)^2 <= 2
(58 real changes made)
. replace group = 3 if (y-2)^2 + (x+2)^2 <= 2
(59 real changes made)
```

Next we define some local macros for function plots of the circles. This makes it easier to graph the boundary circles on top of the data. We set the graph option `aspectratio(1)` to force the aspect ratio to be 1; otherwise, the circles might appear to be ovals.

```
. local rp : di %12.10f 2+sqrt(2)
. local rm : di %12.10f 2-sqrt(2)
. local functionplot
> (function y =  sqrt(2-(x+2)^2) - 2, lpat(solid) range(-`rp' -`rm'))
> (function y = -sqrt(2-(x+2)^2) - 2, lpat(solid) range(-`rp' -`rm'))
> (function y =  sqrt(2-(x-2)^2) + 2, lpat(solid) range(-`rm'  `rp'))
> (function y = -sqrt(2-(x-2)^2) + 2, lpat(solid) range(-`rm'  `rp'))
> (function y =  sqrt(2-(x+2)^2) + 2, lpat(solid) range(-`rp' -`rm'))
> (function y = -sqrt(2-(x+2)^2) + 2, lpat(solid) range(-`rp' -`rm'))
> (function y =  sqrt(2-(x-2)^2) - 2, lpat(solid) range( `rm'  `rp'))
> (function y = -sqrt(2-(x-2)^2) - 2, lpat(solid) range( `rm'  `rp'))
. local graphopts
> aspectratio(1) legend(order(1 "group 1" 2 "group 2" 3 "group 3") rows(1))
. twoway (scatter y x if group==1)
>        (scatter y x if group==2)
>        (scatter y x if group==3)
>        `functionplot' , `graphopts' name(original, replace)
>        title("Training data")
```

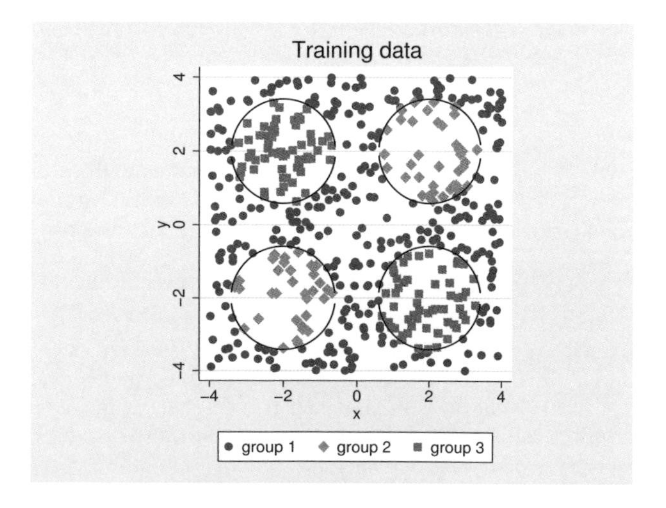

We perform three discriminant analyses on these data for comparison. We use linear discriminant analysis (LDA), quadratic discriminant analysis (QDA) and KNN. The results from logistic discriminant analysis are similar to those of LDA and are not included. With all three models we use proportional probabilities, `priors(proportional)`. The probability of landing in a given group is proportional to the geometric area of that group; they are certainly not equal. Rather than doing geometric calculations for the prior probabilities, we use `priors(proportional)` to approximate this. We suppress the standard classification table with `notable`. Instead we look at the `lootable`, that is, leave-one-out (LOO) table, where the observation in question is omitted and its result is predicted from the rest of the data. Likewise, we predict the LOO classification (`looclass`). With KNN we get to choose a `measure()`; here we want the straight line distance between the points. This is the default, Euclidean distance, so we do not have to specify `measure()`.

We choose $k = 7$ for this run with 500 observations. See *Methods and formulas* for more information on choosing k.

```
. discrim lda x y, group(group) notable lootable priors(proportional)
Linear discriminant analysis
Leave-one-out classification summary

  Key
  Number
  Percent
```

True group	Classified 1	2	3	Total
1	311 100.00	0 0.00	0 0.00	311 100.00
2	72 100.00	0 0.00	0 0.00	72 100.00
3	117 100.00	0 0.00	0 0.00	117 100.00
Total	500 100.00	0 0.00	0 0.00	500 100.00
Priors	0.6220	0.1440	0.2340	

LDA classifies all observations into group one, the group with the highest prior probability.

```
. discrim qda x y, group(group) notable lootable priors(proportional)
Quadratic discriminant analysis
Leave-one-out classification summary
```

Key
Number Percent

True group	Classified 1	2	3	Total
1	275 88.42	0 0.00	36 11.58	311 100.00
2	72 100.00	0 0.00	0 0.00	72 100.00
3	53 45.30	0 0.00	64 54.70	117 100.00
Total	400 80.00	0 0.00	100 20.00	500 100.00
Priors	0.6220	0.1440	0.2340	

QDA has 161 $(36 + 72 + 53)$ misclassified observations of 500, and it correctly classifies 64 of the 117 observations from group 3 but misclassifies all of group 2 into group 1.

```
. discrim knn x y, group(group) k(7) notable lootable priors(proportional)
Kth-nearest-neighbor discriminant analysis
Leave-one-out classification summary
```

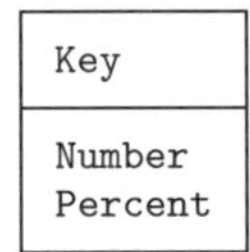

True group	Classified 1	2	3	Total
1	291 93.57	11 3.54	9 2.89	311 100.00
2	17 23.61	55 76.39	0 0.00	72 100.00
3	15 12.82	0 0.00	102 87.18	117 100.00
Total	323 64.60	66 13.20	111 22.20	500 100.00
Priors	0.6220	0.1440	0.2340	

In contrast to the other two models, KNN has only 52 $(11+9+17+15)$ misclassified observations.

We can see how points are classified by KNN by looking at the following graph.

```
. predict cknn, looclass
. twoway (scatter y x if cknn==1 )
>         (scatter y x if cknn ==2)
>         (scatter y x if cknn ==3)
>         `functionplot', `graphopts' name(knn, replace)
>         title("knn LOO classification")
```

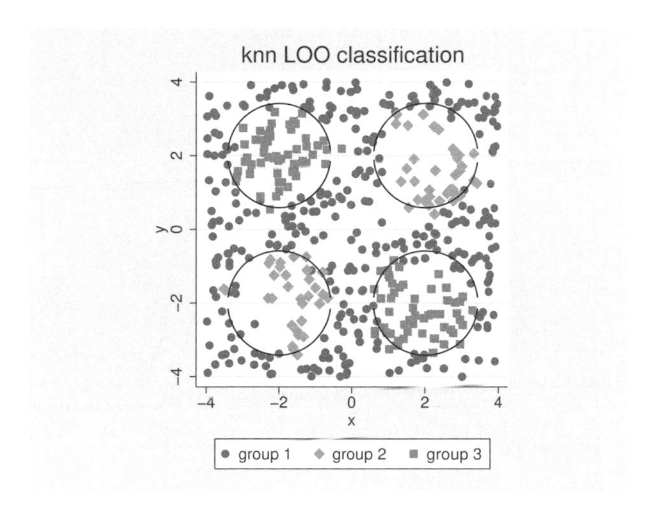

KNN has some resolution of the circles and correctly classifies most of the points. Its misclassified observations are near the boundaries of the circles, where nearest points fall on both sides of the boundary line.

◁

Mahalanobis transformation

The Mahalanobis transformation eliminates the correlation between the variables and standardizes the variance of each variable, as demonstrated in example 2 of [MV] **discrim lda**. When the Mahalanobis transformation is used in conjunction with Euclidean distance, it is called Mahalanobis distance. The Mahalanobis transformation may be applied when any continuous measure is chosen, not just `measure(Euclidean)`. See [MV] ***measure_option*** for a description of the available measures.

▷ Example 2

We will reproduce an example from Rencher (2002, 279–281) that uses the Mahalanobis distance. Rencher presents data collected by G. R. Bryce and R. M. Barker of Brigham Young University as part of a preliminary study of a possible link between football helmet design and neck injuries. Six head dimensions were measured for each subject. Thirty subjects were sampled in each of three groups: high school football players (group 1), college football players (group 2), and nonfootball players (group 3). The six variables are `wdim`, head width at its widest dimension; `circum`, head circumference; `fbeye`, front-to-back measurement at eye level; `eyehd`, eye to top of head measurement; `earhd`, ear to top of head measurement; and `jaw`, jaw width.

These measurements will not have the same ranges. For example, the head circumference should be much larger than eye to top of head measurement. Mahalanobis distance is used to standardize the measurements.

```
. use http://www.stata-press.com/data/r12/head, clear
(Table 8.3 Head measurements -- Rencher (2002))
. discrim knn wdim-jaw, k(5) group(group) mahalanobis
Kth-nearest neighbor discriminant analysis
Resubstitution classification summary
```

Key
Number Percent

True group	Classified high school	college	nonplayer	Unclassified	Total
high school	26 86.67	0 0.00	1 3.33	3 10.00	30 100.00
college	1 3.33	19 63.33	9 30.00	1 3.33	30 100.00
nonplayer	1 3.33	4 13.33	22 73.33	3 10.00	30 100.00
Total	28 31.11	23 25.56	32 35.56	7 7.78	90 100.00
Priors	0.3333	0.3333	0.3333		

A subset of this result is in Rencher (2002, 319). Of the 90 original observations, 16 were misclassified and 7 observations were unclassified. Rencher also states the error rate for this example is 0.193. We use `estat errorrate` to get the error rate.

```
. estat errorrate
Error rate estimated by error count
Note: 7 observations were not classified and are not included in the table
```

	group high school	college	nonplayer	Total
Error rate	.037037037	.344827586	.185185185	.189016603
Priors	.333333333	.333333333	.333333333	

Our error rate of 0.189 does not match that of Rencher. Why is this? Rencher calculates the error rate as the number of misclassified observations over the total number of observations classified. This is

$16/83 \approx 0.193$. We use the standard error-rate definition that takes into account the prior probabilities. From the high school group, there is one misclassified observation of 27 total observations classified from this group, so its error rate is $(1/27) \approx 0.037$, and its contribution to the total is $(1/27)(1/3)$. Likewise, the error rates for the college and nonplayer group are $(10/29) \approx 0.345$ and $(5/27) \approx 0.185$ respectively, with contributions of $(10/29)(1/3)$ and $(5/27)(1/3)$. Adding all contributions, we get the displayed error rate of 0.189. See *Methods and formulas* of [MV] **discrim estat** for details.

The unclassified observations are those that resulted in ties. We can force ties to be classified by changing the `ties()` option. The default is `ties(missing)`, which says that ties are to be classified as missing values. Here we choose `ties(nearest)`, which breaks the tie by classifying to the group of the nearest tied observation.

```
. discrim knn wdim-jaw, k(5) group(group) mahalanobis ties(nearest)
Kth-nearest-neighbor discriminant analysis
Resubstitution classification summary

  Key

  Number
  Percent
```

True group	Classified high school	college	nonplayer	Total
high school	28	0	2	30
	93.33	0.00	6.67	100.00
college	1	20	9	30
	3.33	66.67	30.00	100.00
nonplayer	1	4	25	30
	3.33	13.33	83.33	100.00
Total	30	24	36	90
	33.33	26.67	40.00	100.00
Priors	0.3333	0.3333	0.3333	

Compare this with example 1 in [MV] **candisc**, example 3 in [MV] **discrim estat**, and example 2 of [MV] **discrim logistic**.

◁

Binary data

In addition to the measures for continuous data, a variety of binary measures are available for KNN. Binary data can be created from any categorical dataset by using `xi`; see [R] **xi**.

▷ Example 3

You have invited some scientist friends over for dinner, including Mr. Mushroom (see vignette below), a real "fun guy". Mr. Mushroom is a researcher in mycology. He not only enjoys studying mushrooms; he is also an enthusiastic mushroom gourmand who likes nothing better than to combine his interests in classification and cookery. His current research is identification of poisonous mushrooms from photographs. From the photographs, he can identify the shape of a mushroom's cap, the cap's surface, the cap's color, the population of mushrooms, and, with some careful attention to detail in the surrounding area, the habitat.

William Alphonso Murrill (1867–1957) was a famous mycologist, taxonomist, and writer from the New York Botanical Gardens and was nicknamed "Mr. Mushroom". Although we borrowed his nickname, Mr. Mushroom and the events portrayed in this example are entirely fictitious. William Murrill's many scientific accomplishments include the 1916 book *Edible and Poisonous Mushrooms*.

Knowing your friend, you imagine that he will insist on bringing a mushroom dish to be unveiled and served at dinnertime—perhaps his experimental subjects. Although you know that he is a careful scientist and a gifted cook, you are stalked with worries about poisoning your other guests.

Late at night you cannot sleep for worrying about poisonous mushrooms, and you decide to do a little research into mushroom classification. You do a Google search online and find mushroom data at http://archive.ics.uci.edu/ml/datasets/Mushroom. For reference, these records are drawn from Lincoff (1981).

This is a large dataset of 8,124 observations on the *Agaricus* and *Lepiota*. Each species is identified as definitely edible, definitely poisonous, or of unknown edibility and not recommended. This last class was combined with the poisonous one. Lincoff (1981) clearly states that there is no simple rule for determining the edibility of a mushroom; no rule like "leaflets three, let it be" for Poison Oak and Ivy, a fact that does not seem comforting. Twenty-two attributes are collected, including those that Mr. Mushroom can identify from his photographs.

The mushroom data is a set of 23 variables that describe the cap of the mushroom, whether or not it has bruises, the gills, the veil, the stalk, the ring, the spores, the population, and the habitat. The variables that describe the cap, for example, are `capshape`, `capsurface`, and `capcolor`. The `capshape` variable, for example, has categories bell, conical, convex, flat, knobbed, and sunken. Other variables and categories are similar.

You read in this dataset by using `infile` and make some modifications, attaching notes to this dataset to describe what you did to the original mushroom data. Modifications include dropping categories of the variables of interest that completely determine whether a mushroom is poisonous. The full mushroom data are also available; `webuse mushroom_full` to obtain it.

```
. use http://www.stata-press.com/data/r12/mushroom
(Lincoff (1981) Audubon Guide; http://archive.ics.uci.edu/ml/datasets/Mushroom)
. tabulate habitat poison
```

habitat	poison edible	poisonous	Total
grasses	752	680	1,432
leaves	240	585	825
meadows	128	24	152
paths	136	1,008	1,144
urban	64	224	288
woods	1,848	1,268	3,116
Total	3,168	3,789	6,957

You can see by tabulating two of the variables, `habitat` and `poison`, that in each habitat you have some mushrooms that are poisonous as well as some that are edible. The other descriptive variables of interest produce similar results.

Each variable is a set of unordered categories. You can use xi to create indicator (binary) variables from the categorical data; see [R] **xi**. xi, by default, does not create collinear variables; it omits one of the categories. You want to take all categories into account, and so you use the noomit option.

With KNN you can choose a measure that is suited to this data. You expect data with many zeroes and few ones. A match of two ones is far more significant than two matching zeroes. Looking through the binary similarity measures in [MV] ***measure_option***, you see that the Jaccard binary similarity coefficient reports the proportion of matches of ones when at least one of the observations contains a one, and the Dice binary similarity measure weighs matches of ones twice as heavily as the Jaccard measure. Either suits the situation, and you choose the Dice measure. The conversion from a similarity to a dissimilarity measure will be s2d(standard) by default.

The poisonous and edible mushrooms are split about half and half in the original dataset, and in the current subset of these data the ratio is still approximately half and half, so you do not specify priors, obtaining priors(equal), the default.

Because of the size of the dataset and the number of indicator variables created by xi, KNN analysis is slow. You decide to discriminate based on 2,000 points selected at random, approximately a third of the data.

```
. set seed 12345678
. generate u = runiform()
. sort u
. xi, noomit: discrim knn i.population i.habitat i.bruises i.capshape
> i.capsurface i.capcolor in 1/2000, k(15) group(poison) measure(dice)
Kth-nearest-neighbor discriminant analysis
Resubstitution classification summary

  +---------+
  | Key     |
  |---------|
  | Number  |
  | Percent |
  +---------+
```

True poison	Classified edible	poisonous	Total
edible	839 94.06	53 5.94	892 100.00
poisonous	45 4.06	1,063 95.94	1,108 100.00
Total	884 44.20	1,116 55.80	2,000 100.00
Priors	0.5000	0.5000	

In some settings, these results would be considered good. Of the original 2,000 mushrooms, you see that only 45 poisonous mushrooms have been misclassified as edible. However, even sporadic classification of a poisonous mushroom as edible is a much bigger problem than classifying an edible mushroom as poisonous. This does not take the cost of misclassification into account. You decide that calling a poisonous mushroom edible is at least 10 times worse than calling an edible mushroom poisonous. In the two-group case, you can easily use the priors() option to factor in this cost; see [MV] **discrim** or McLachlan (2004, 9). We set the prior probability of poisonous mushrooms 10 times higher than that of the edible mushrooms.

```
. estat classtable in 1/2000, priors(.09, .91)
Resubstitution classification table

  +---------+
  | Key     |
  |---------|
  | Number  |
  | Percent |
  +---------+

             | Classified
 True poison |     edible  poisonous |     Total
-------------+-----------------------+----------
      edible |        674        218 |       892
             |      75.56      24.44 |    100.00
             |                       |
   poisonous |          0      1,108 |     1,108
             |       0.00     100.00 |    100.00
-------------+-----------------------+----------
       Total |        674      1,326 |     2,000
             |      33.70      66.30 |    100.00
             |                       |
      Priors |     0.0900     0.9100 |
```

These results are reassuring. There are no misclassified poisonous mushrooms, although 218 edible mushrooms of the total 2,000 mushrooms in our model are misclassified.

You now check to see how this subsample of the data performs in predicting the poison status of the rest of the data. This takes a few minutes of computer time, but unlike using `estat classtable` above, the variable predicted will stay with your dataset until you drop it. `tabulate` can be used instead of `estat classtable`.

```
. predict cpoison, classification priors(.09, .91)
. label values cpoison poison
. tabulate poison cpoison
           |    classification
    poison |    edible  poisonous |     Total
-----------+----------------------+----------
    edible |     2,439        729 |     3,168
 poisonous |         0      3,789 |     3,789
-----------+----------------------+----------
     Total |     2,439      4,518 |     6,957
```

This is altogether reassuring. Again, no poisonous mushrooms were misclassified. Perhaps there is no need to worry about dinnertime disasters, even with a fungus among us. You are so relieved that you plan on serving a Jello dessert to cap off the evening—your guests will enjoy a mold to behold. Under the circumstances, you think doing so might just be a “morel” imperative.

◁

Saved results

discrim knn saves the following in e():

Scalars

e(N)	number of observations
e(N_groups)	number of groups
e(k_nn)	number of nearest neighbors
e(k)	number of discriminating variables

Macros

e(cmd)	discrim
e(subcmd)	knn
e(cmdline)	command as typed
e(groupvar)	name of group variable
e(grouplabels)	labels for the groups
e(measure)	similarity or dissimilarity measure
e(measure_type)	dissimilarity or similarity
e(measure_binary)	binary, if binary measure specified
e(s2d)	standard or oneminus, if s2d() specified
e(varlist)	discriminating variables
e(wtype)	weight type
e(wexp)	weight expression
e(title)	title in estimation output
e(ties)	how ties are to be handled
e(mahalanobis)	mahalanobis, if Mahalanobis transform is performed
e(properties)	nob noV
e(estat_cmd)	program used to implement estat
e(predict)	program used to implement predict
e(marginsnotok)	predictions disallowed by margins

Matrices

e(groupcounts)	number of observations for each group
e(grouppriors)	prior probabilities for each group
e(groupvalues)	numeric value for each group
e(means)	group means on discriminating variables
e(SSCP_W)	pooled within-group SSCP matrix
e(W_eigvals)	eigenvalues of e(SSCP_W)
e(W_eigvecs)	eigenvectors of e(SSCP_W)
e(S)	pooled within group covariance matrix
e(Sinv)	inverse of e(S)
e(sqrtSinv)	Cholesky (square root) of e(Sinv)
e(community)	community of neighbors for prediction

Functions

e(sample)	marks estimation sample

Methods and formulas

discrim knn is implemented as an ado-file.

Let g be the number of groups, n_i the number of observations for group i, and q_i the prior probability for group i. Let $\mathbf{x}$ denote an observation measured on p discriminating variables. For consistency with the discriminant analysis literature, $\mathbf{x}$ will be a column vector, though it corresponds to a row in your dataset. Let $f_i(\mathbf{x})$ represent the density function for group i, and let $P(\mathbf{x}|G_i)$ denote the probability of observing $\mathbf{x}$ conditional on belonging to group i. Denote the posterior probability of group i given observation $\mathbf{x}$ as $P(G_i|\mathbf{x})$. With Bayes' theorem, we have

$$P(G_i|\mathbf{x}) = \frac{q_i f_i(\mathbf{x})}{\sum_{j=1}^{g} q_j f_j(\mathbf{x})}$$

Substituting $P(\mathbf{x}|G_i)$ for $f_i(\mathbf{x})$, we have

$$P(G_i|\mathbf{x}) = \frac{q_i P(\mathbf{x}|G_i)}{\sum_{j=1}^{g} q_j P(\mathbf{x}|G_j)}$$

For KNN discrimination, we let k_i be the number of the k nearest neighbors from group i, and the posterior-probability formula becomes

$$P(G_i|\mathbf{x}) = \frac{\dfrac{q_i k_i}{n_i}}{\displaystyle\sum_{j=1}^{g} \frac{q_j k_j}{n_j}}$$

In the event that there are ties among the nearest neighbors, k is increased to accommodate the ties. If five points are all nearest and equidistant from a given $\mathbf{x}$, then an attempt to calculate the three nearest neighbors of $\mathbf{x}$ will actually obtain five nearest neighbors.

Determining the nearest neighbors depends on a dissimilarity or distance calculation. The available dissimilarity measures are described in [MV] ***measure_option***. Continuous and binary measures are available. If a similarity measure is selected, it will be converted to a dissimilarity by either

`standard` $\quad d(ij) = \sqrt{s(ii) + s(jj) - 2s(ij)} = \sqrt{2\{1 - s(ij)\}}$

`oneminus` $\quad d(ij) = 1 - s(ij)$

With any of the continuous measures, a Mahalanobis transformation may be performed before computing the dissimilarities. For details on the Mahalanobis transformation, see *Methods and formulas* of [MV] **discrim lda**. The Mahalanobis transformation with Euclidean distance is called Mahalanobis distance.

Optimal choice of k for KNN is not an exact science. With two groups, k should be chosen as an odd integer to avoid ties. Rencher (2002, 319) cites the research of Loftsgaarden and Quesenberry (1965), which suggests that an optimal k is $\sqrt{n_i}$, where n_i is a typical group size. Rencher also suggests running with several different values of k and choosing the one that gives the best error rate. McLachlan (2004) cites Enas and Choi (1986), which suggests that when there are two groups of comparable size that k should be chosen approximately between $N^{3/8}$ or $N^{2/8}$, where N is the number of observations.

References

Enas, G. G., and S. C. Choi. 1986. Choice of the smoothing parameter and efficiency of k-nearest neighbor classification. *Computers and Mathematics with Applications* 12A: 235–244.

Fix, E., and J. L. Hodges. 1951. Discriminatory analysis: Nonparametric discrimination, consistency properties. In *Technical Report No. 4, Project No. 21-49-004*. Randolph Field, Texas: Brooks Air Force Base, USAF School of Aviation Medicine.

Hastie, T., R. J. Tibshirani, and J. Friedman. 2009. *The Elements of Statistical Learning: Data Mining, Inference, and Prediction*. 2nd ed. New York: Springer.

Kimbrough, J. W. 2003. The twilight years of William Alphonso Murrill. http://www.mushroomthejournal.com/bestof/Murrillpt2.pdf.

Lincoff, G. H. 1981. *National Audubon Society Field Guide to North American Mushrooms (National Audubon Society Field Guide Series)*. New York: Alfred A. Knopf.

Loftsgaarden, D. O., and C. P. Quesenberry. 1965. A nonparametric estimate of a multivariate density function. *Annals of Mathematical Statistics* 36: 1049–1051.

McLachlan, G. J. 2004. *Discriminant Analysis and Statistical Pattern Recognition.* New York: Wiley.

Murrill, W. A. 1916. *Edible and Poisonous Mushrooms.* Published by the author.

Rencher, A. C. 2002. *Methods of Multivariate Analysis.* 2nd ed. New York: Wiley.

Rose, D. W. 2002. William Alphonso Murrill: The legend of the naturalist. *Mushroom, The Journal of Wild Mushrooming.* http://www.mushroomthejournal.com/bestof/Murrillpt1.pdf.

Smith-Vikos, T. 2003. William Alphonso Murrill (1869–1957). *The New York Botanical Garden.* http://sciweb.nybg.org/science2/hcol/intern/murrill1.asp.

Also see

[MV] **discrim knn postestimation** — Postestimation tools for discrim knn

[MV] **discrim** — Discriminant analysis

[U] **20 Estimation and postestimation commands**

Title

discrim knn postestimation — Postestimation tools for discrim knn

Description

The following postestimation commands are of special interest after discrim knn:

Command	Description
estat classtable	classification table
estat errorrate	classification error-rate estimation
estat grsummarize	group summaries
estat list	classification listing
estat summarize	estimation sample summary

For information about these commands, see [MV] **discrim estat**.

The following standard postestimation commands are also available:

Command	Description
* estimates	cataloging estimation results
predict	group classification and posterior probabilities

*All estimates subcommands except table and stats are available; see [R] **estimates**.

Syntax for predict

predict [*type*] *newvar* [*if*] [*in*] [, *statistic options*]

predict [*type*] { *stub** | *newvarlist* } [*if*] [*in*] [, *statistic options*]

statistic	Description
Main	
classification	group membership classification; the default when one variable is specified and group() is not specified
pr	probability of group membership; the default when group() is specified or when multiple variables are specified
* looclass	leave-one-out group membership classification; may be used only when one new variable is specified
* loopr	leave-one-out probability of group membership

options	Description
Main	
group(*group*)	the group for which the statistic is to be calculated
Options	
priors(*priors*)	group prior probabilities; defaults to e(grouppriors)
ties(*ties*)	how ties in classification are to be handled; defaults to e(ties)
noupdate	do not update the within-group covariance matrix with leave-one-out predictions

priors	Description
equal	equal prior probabilities
proportional	group-size-proportional prior probabilities
matname	row or column vector containing the group prior probabilities
matrix_exp	matrix expression providing a row or column vector of the group prior probabilities

ties	Description
missing	ties in group classification produce missing values
random	ties in group classification are broken randomly
first	ties in group classification are set to the first tied group
nearest	ties in group classification are assigned based on the closest observation, or missing if this still results in a tie

You specify one new variable with classification or looclass and specify either one or e(N_groups) new variables with pr or loopr.

Unstarred statistics are available both in and out of sample; type predict ... if e(sample) ... if wanted only for the estimation sample. Starred statistics are calculated only for the estimation sample, even when if e(sample) is not specified.

group() is not allowed with classification or looclass.

noupdate is an advanced option and does not appear in the dialog box.

Menu

Statistics > Postestimation > Predictions, residuals, etc.

Options for predict

Main

classification, the default, calculates the group classification. Only one new variable may be specified.

pr calculates group membership posterior probabilities. If you specify the group() option, specify one new variable. Otherwise, you must specify e(N_groups) new variables.

looclass calculates the leave-one-out group classifications. Only one new variable may be specified. Leave-one-out calculations are restricted to e(sample) observations.

loopr calculates the leave-one-out group membership posterior probabilities. If you specify the group() option, specify one new variable. Otherwise, you must specify e(N_groups) new variables. Leave-one-out calculations are restricted to e(sample) observations.

group(*group*) specifies the group for which the statistic is to be calculated and can be specified using

#1, #2, ..., where #1 means the first category of the e(groupvar) variable, #2 the second category, etc.;

the values of the e(groupvar) variable; or

the value labels of the e(groupvar) variable if they exist.

group() is not allowed with classification or looclass.

Options

priors(*priors*) specifies the prior probabilities for group membership. If priors() is not specified, e(grouppriors) is used. The following *priors* are allowed:

priors(equal) specifies equal prior probabilities.

priors(proportional) specifies group-size-proportional prior probabilities.

priors(*matname*) specifies a row or column vector containing the group prior probabilities.

priors(*matrix_exp*) specifies a matrix expression providing a row or column vector of the group prior probabilities.

ties(*ties*) specifies how ties in group classification will be handled. If ties() is not specified, e(ties) is used. The following *ties* are allowed:

ties(missing) specifies that ties in group classification produce missing values.

ties(random) specifies that ties in group classification are broken randomly.

ties(first) specifies that ties in group classification are set to the first tied group.

ties(nearest) specifies that ties in group classification are assigned based on the closest observation, or missing if this still results in a tie.

The following option is available with predict after discrim knn but is not shown in the dialog box:

noupdate causes the within-group covariance matrix not to be updated with leave-one-out predictions. noupdate is an advanced, rarely used option that is valid only if a Mahalanobis transformation is specified.

Remarks

*k*th-nearest-neighbor (KNN) discriminant analysis and postestimation can be time consuming for large datasets. The training data must be retained and then searched to find the nearest neighbors each time a classification or prediction is performed.

You can find more examples of postestimation with KNN in [MV] **discrim knn**, and more examples of the common estat subcommands in [MV] **discrim estat**.

▷ Example 1

Recall example 1 of of [MV] **discrim knn**. We use a similar idea here, creating a two-dimensional dataset on the plane with x and y variables in $[-4, 4]$. Instead of random data, we choose data on a regular grid to make things easy to visualize, and once again, we assign groups on the basis of geometric calculations. To start, we assign all points a group value of one, then within four circles of radius 3, one in each quadrant, we change the group value to two in the circles in the first and third quadrants, and we change the group value to three in the circles in the second and fourth quadrants.

Instructions for creating this dataset and definitions for local macros associated with it are contained in its notes.

```
. use http://www.stata-press.com/data/r12/circlegrid
(Gridded circle data)
. local rp : di %12.10f 2+sqrt(3)
. local rm : di %12.10f 2-sqrt(3)
. local functionplot
> (function y =  sqrt(3-(x+2)^2) - 2, lpat(solid) range(-`rp' -`rm'))
> (function y = -sqrt(3-(x+2)^2) - 2, lpat(solid) range(-`rp' -`rm'))
> (function y =  sqrt(3-(x-2)^2) + 2, lpat(solid) range(-`rm'  `rp'))
> (function y = -sqrt(3-(x-2)^2) + 2, lpat(solid) range(-`rm'  `rp'))
> (function y =  sqrt(3-(x+2)^2) + 2, lpat(solid) range(-`rp' -`rm'))
> (function y = -sqrt(3-(x+2)^2) + 2, lpat(solid) range(-`rp' -`rm'))
> (function y =  sqrt(3-(x-2)^2) - 2, lpat(solid) range( `rm'  `rp'))
> (function y = -sqrt(3-(x-2)^2) - 2, lpat(solid) range( `rm'  `rp'))
. local graphopts
> aspectratio(1) legend(order(1 "group 1" 2 "group 2" 3 "group 3") rows(1))
. twoway (scatter y x if group==1)
>        (scatter y x if group==2)
>        (scatter y x if group==3)
>        `functionplot', `graphopts' name(original, replace)
>        title("Training data")
```

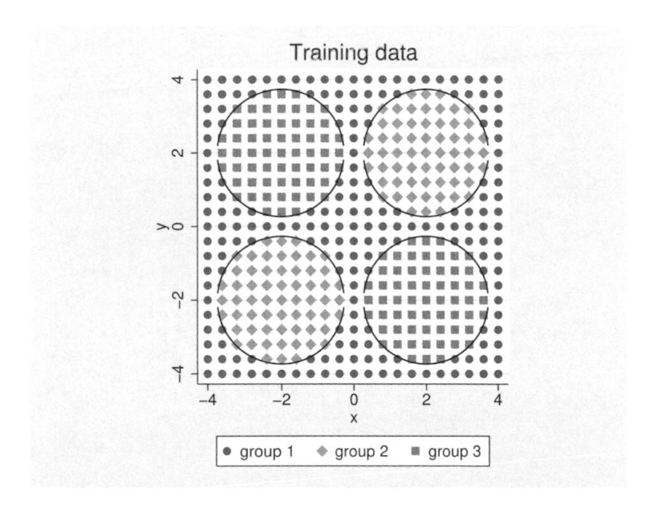

We do a KNN discriminant analysis, choosing `k(3)`. We elect to omit the standard classification table and instead take a look at the leave-one-out (LOO) classification table.

```
. discrim knn x y, group(group) k(3) priors(proportional) notable lootable
Kth-nearest-neighbor discriminant analysis
Leave-one-out classification summary
```

Key
Number Percent

True group	Classified 1	2	3	Total
1	173 87.82	12 6.09	12 6.09	197 100.00
2	8 6.56	114 93.44	0 0.00	122 100.00
3	8 6.56	0 0.00	114 93.44	122 100.00
Total	189 42.86	126 28.57	126 28.57	441 100.00
Priors	0.4467	0.2766	0.2766	

We will predict the LOO classification, changing to `priors(equal)`, and look at the plot.

```
. predict cknn, looclass priors(equal)
warning: 8 ties encountered
ties are assigned to missing values
(8 missing values generated)
. twoway (scatter y x if cknn==1)
>        (scatter y x if cknn==2)
>        (scatter y x if cknn ==3)
>        `functionplot', `graphopts' name(KNN, replace)
>        title("KNN classification")
```

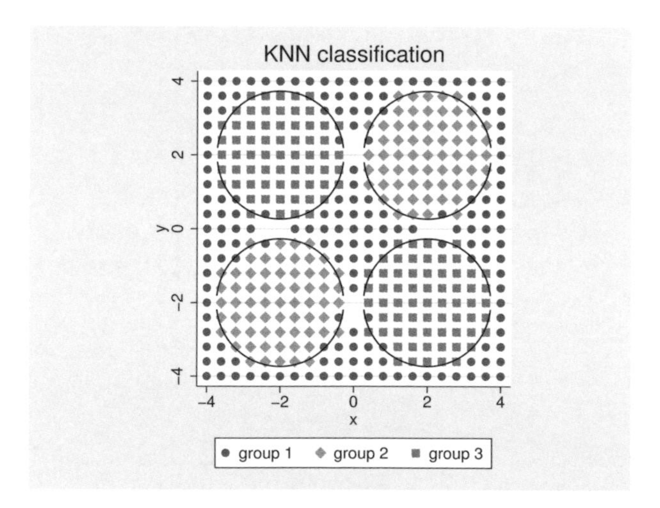

We see several empty spots on the grid. In our output, changing to `priors(equal)` created several ties that were assigned to missing values. Missing values are the blank places in our graph.

◁

▷ Example 2

Continuing where we left off, we use `estat list` to display LOO probabilities for the misclassified observations, but this produces a lot of output.

```
. estat list, misclass class(noclass looclass) pr(nopr loopr) priors(equal)
```

	Classification		LOO Probabilities		
Obs.	True	LOO Cl.	1	2	3
24	1	2 *	0.3836	0.6164	0.0000
28	1	2 *	0.2374	0.7626	0.0000
34	1	3 *	0.2374	0.0000	0.7626
38	1	3 *	0.3836	0.0000	0.6164
50	2	1 *	0.5513	0.4487	0.0000
54	3	1 *	0.5513	0.0000	0.4487

(output omitted)

```
* indicates misclassified observations
```

Instead, we predict the LOO probabilities and list only those where the LOO classification is missing.

```
. predict pr*, loopr priors(equal)
. list group cknn pr* if missing(cknn)
```

	group	cknn	pr1	pr2	pr3
94.	1	.	.2373541	.381323	.381323
115.	1	.	.2373541	.381323	.381323
214.	1	.	.2373541	.381323	.381323
215.	1	.	.2373541	.381323	.381323
225.	1	.	.2373541	.381323	.381323
226.	1	.	.2373541	.381323	.381323
325.	1	.	.2373541	.381323	.381323
346.	1	.	.2373541	.381323	.381323

The missing LOO classifications represent ties for the largest probability.

◁

▷ Example 3

LOO classification and LOO probabilities are available only in sample, but standard probabilities can be obtained out of sample. To demonstrate this, we continue where we left off, with the KNN model of example 2 still active. We drop our current data and generate some new data. We predict the standard classification with the new data and graph our results.

```
. clear
. set obs 500
obs was 0, now 500
. set seed 314159265
. gen x = 8*runiform() - 4
. gen y = 8*runiform() - 4
. predict cknn, class
```

```
. twoway (scatter y x if cknn==1)
>        (scatter y x if cknn==2)
>        (scatter y x if cknn ==3)
>        `functionplot', `graphopts' name(KNN2, replace)
>        title("Out-of-sample KNN classification", span)
```

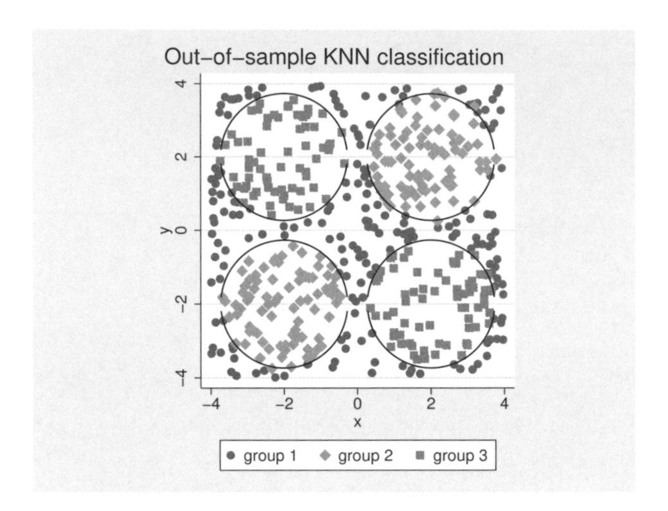

◁

Methods and formulas

See [MV] **discrim knn** for methods and formulas.

Also see

[MV] **discrim estat** — Postestimation tools for discrim

[MV] **discrim knn** — kth-nearest-neighbor discriminant analysis

[U] **20 Estimation and postestimation commands**

Title

discrim lda — Linear discriminant analysis

Syntax

`discrim lda` *varlist* [*if*] [*in*] [*weight*], `group(`*groupvar*`)` [*options*]

options	Description
Model	
* `group(`*groupvar*`)`	variable specifying the groups
`priors(`*priors*`)`	group prior probabilities
`ties(`*ties*`)`	how ties in classification are to be handled
Reporting	
`notable`	suppress resubstitution classification table
`lootable`	display leave-one-out classification table

priors	Description
`equal`	equal prior probabilities; the default
`proportional`	group-size-proportional prior probabilities
matname	row or column vector containing the group prior probabilities
matrix_exp	matrix expression providing a row or column vector of the group prior probabilities

ties	Description
`missing`	ties in group classification produce missing values; the default
`random`	ties in group classification are broken randomly
`first`	ties in group classification are set to the first tied group

* `group()` is required.

`statsby` and `xi` are allowed; see [U] **11.1.10 Prefix commands**.

`fweights` are allowed; see [U] **11.1.6 weight**.

See [U] **20 Estimation and postestimation commands** for more capabilities of estimation commands.

Menu

Statistics > Multivariate analysis > Discriminant analysis > Linear (LDA)

Description

`discrim lda` performs linear discriminant analysis. See [MV] **discrim** for other discrimination commands.

If by default you want canonical linear discriminant results displayed, see [MV] **candisc**. candisc and discrim lda compute the same things, but candisc displays more information. The same information can be displayed after discrim lda by using the estat suite of commands; see [MV] **discrim lda postestimation**.

Options

Main

group(*groupvar*) is required and specifies the name of the grouping variable. *groupvar* must be a numeric variable.

priors(*priors*) specifies the prior probabilities for group membership. The following *priors* are allowed:

priors(equal) specifies equal prior probabilities. This is the default.

priors(proportional) specifies group-size-proportional prior probabilities.

priors(*matname*) specifies a row or column vector containing the group prior probabilities.

priors(*matrix_exp*) specifies a matrix expression providing a row or column vector of the group prior probabilities.

ties(*ties*) specifies how ties in group classification will be handled. The following *ties* are allowed:

ties(missing) specifies that ties in group classification produce missing values. This is the default.

ties(random) specifies that ties in group classification are broken randomly.

ties(first) specifies that ties in group classification are set to the first tied group.

Reporting

notable suppresses the computation and display of the resubstitution classification table.

lootable displays the leave-one-out classification table.

Remarks

Remarks are presented under the following headings:

Introduction
Descriptive LDA
Predictive LDA
A classic example

Introduction

Linear discriminant analysis (LDA) was developed by different researchers, Fisher (1936) and Mahalanobis (1936), starting with different approaches to the problem of discriminating between groups. Kshirsagar and Arseven (1975), Green (1979), and Williams (1982) demonstrate the mathematical relationship between Fisher's linear discriminant functions and the classification functions from the Mahalanobis approach to LDA; see Rencher (1998, 239).

Fisher's approach to LDA forms the basis of descriptive LDA but can be used for predictive LDA. The Mahalanobis approach to LDA more naturally handles predictive LDA, allowing for prior probabilities and producing estimates of the posterior probabilities. The Mahalanobis approach to LDA also extends to quadratic discriminant analysis (QDA); see [MV] **discrim qda**.

Descriptive LDA

Fisher (1936) approached linear discriminant analysis by seeking the linear combination of the discriminating variables that provides maximal separation between the groups (originally two groups, but later extended to multiple groups). Maximal separation of groups is determined from an eigen analysis of $\mathbf{W}^{-1}\mathbf{B}$, where $\mathbf{B}$ is the between-group sum-of-squares and cross-products (SSCP) matrix, and $\mathbf{W}$ is the within-group SSCP matrix. The eigenvalues and eigenvectors of $\mathbf{W}^{-1}\mathbf{B}$ provide what are called Fisher's linear discriminant functions.

The first linear discriminant function is the eigenvector associated with the largest eigenvalue. This first discriminant function provides a linear transformation of the original discriminating variables into one dimension that has maximal separation between group means. The eigenvector associated with the second-largest eigenvalue is the second linear discriminant function and provides a dimension uncorrelated with (but usually not orthogonal to) the first discriminant function. The second discriminant function provides the maximal separation of groups in a second dimension. The third discriminant function provides the maximum separation of groups in a third dimension.

▷ Example 1

Two groups measured on two variables illustrate Fisher's approach to linear discriminant analysis.

```
. use http://www.stata-press.com/data/r12/twogroup
(Two Groups)
```

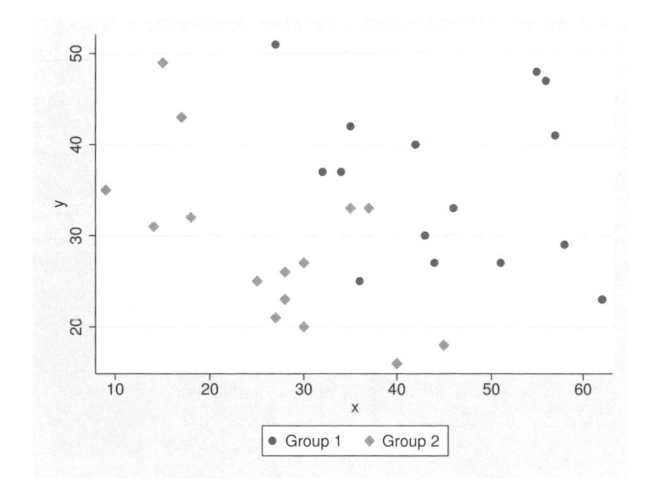

Splitting the two groups on the basis of y or x alone would leave a great deal of overlap between the groups. By eye it appears that a line with a slope of about -1 could be positioned between the two groups with only a few of the observations falling on the wrong side of the line.

Fisher's approach to LDA helps us find the best separating line.

```
. discrim lda y x, group(group) notable
```

`discrim lda` computes the information needed for both a predictive and descriptive linear discriminant analysis. We requested `notable`, which suppressed the display of the resubstitution classification table. We will examine this feature of `discrim lda` when we discuss predictive LDA. The descriptive features of LDA are available through postestimation commands.

```
. estat loadings, unstandardized
Canonical discriminant function coefficients

              function1

           y   .0862145
           x   .0994392
       _cons   -6.35128
```

Fisher's linear discriminant functions provide the basis for what are called the canonical discriminant functions; see *Methods and formulas*. The canonical discriminant function coefficients are also called unstandardized loadings because they apply to the unstandardized discriminating variables (`x` and `y`). Because we have only two groups, there is only one discriminant function. From the coefficients or loadings of this discriminant function, we obtain a one-dimensional projection of the data that gives maximal separation between the two groups relative to the spread within the groups. The `estat loadings` postestimation command displayed these loadings; see [MV] **discrim lda postestimation**. After `estat loadings`, the unstandardized loadings are available in matrix `r(L_unstd)`. We take these values and determine the equation of the separating line between the groups and a line perpendicular to the separating line.

The unstandardized canonical discriminant function coefficients indicate that

$$0 = 0.0862145\text{y} + 0.0994392\text{x} - 6.35128$$

which in standard $\text{y} = m\text{x} + b$ form is

$$\text{y} = -1.1534\text{x} + 73.6684$$

which is the dividing line for classifying observations into the two groups for this LDA. A line perpendicular to this dividing line has slope $-1/-1.153 = 0.867$. The following graph shows the data with this dividing line and a perpendicular projection line.

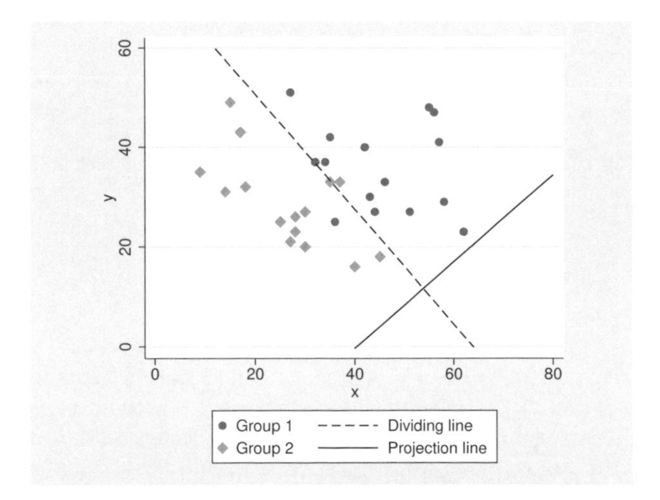

Another way of thinking about the discriminant function is that it projects the data from the original two dimensions down to one dimension—the line perpendicular to the dividing line. Classifications are based on which side of the separating line the observations fall.

Researchers often wish to know which of the discriminating variables is most important or helpful in discriminating between the groups. They want to examine the standardized loadings—the loadings that apply to standardized variables.

```
. estat loadings, standardized
Standardized canonical discriminant function coefficients

             | function1
-------------+-----------
           y |  .7798206
           x |  1.057076
```

These coefficients or loadings apply to x and y that have been standardized using the pooled within-group covariance matrix. The loading for x is larger than that for y, indicating that it contributes more to the discrimination between the groups. Look back at the scatterplot to see that there is more separation between the groups in the x dimension than the y dimension. See [MV] **discrim lda postestimation** for more details of the `estat loadings` command.

Some researchers prefer to examine what are called structure coefficients.

```
. estat structure
Canonical structure

             | function1
-------------+-----------
           y |  .3146309
           x |  .7138982
```

The `estat structure` command provides structure coefficients, which measure the correlation between each discriminating variable and the discriminant function; see [MV] **discrim lda postestimation**. Here the canonical structure coefficient for x is larger than that for y, leading to the same conclusion as with standardized loadings. There is disagreement in the literature concerning the use of canonical structure coefficients versus standardized loadings; see Rencher (2002, 291) and Huberty (1994, 262–264).

◁

In addition to loading and structure coefficients, there are other descriptive LDA features available after `discrim lda`. These include canonical correlations and tests of the canonical correlations, classification functions, scree plots, loading plots, score plots, and various group summaries; see [MV] **discrim lda postestimation**.

If your main interest is in descriptive LDA, you may find the `candisc` command of interest; see [MV] **candisc**. `discrim lda` and `candisc` differ only in their default output. `discrim lda` shows classification tables. `candisc` shows canonical correlations, standardized coefficients (loadings), structure coefficients, and more. All the features found in [MV] **discrim lda postestimation** are available for both commands.

Predictive LDA

Another approach to linear discriminant analysis starts with the assumption that the observations from each group are multivariate normal with the groups having equal covariance matrices but different means. Mahalanobis (1936) distance plays an important role in this approach. An observation with unknown group membership is classified as belonging to the group with smallest Mahalanobis distance between the observation and group mean. Classification functions for classifying observations of unknown group membership can also be derived from this approach to LDA and formulas for the posterior probability of group membership are available.

As shown in *Methods and formulas*, Mahalanobis distance can be viewed as a transformation followed by Euclidean distance. Group membership is assigned based on the Euclidean distance in this transformed space.

▷ Example 2

We illustrate the Mahalanobis transformation and show some of the features of predictive discriminant analysis with a simple three-group example dataset.

```
. use http://www.stata-press.com/data/r12/threegroup
(Three Groups)
```

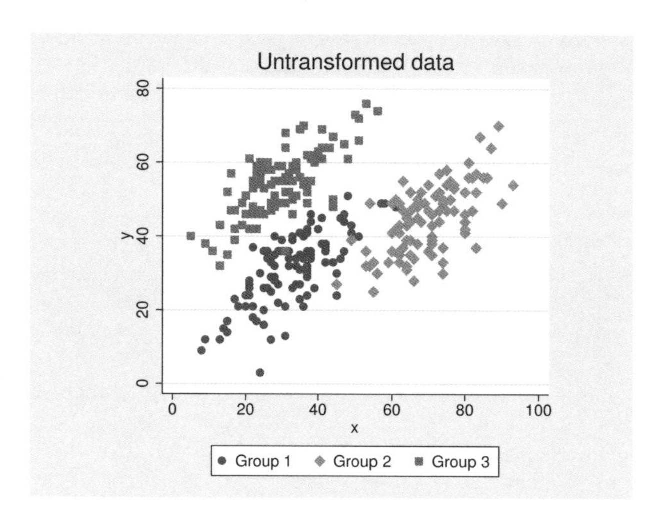

These three groups appear to have similar covariance structure—showing a positive correlation between x and y. There is some overlap of the three groups, but general identification of the groups does not appear difficult by human eye.

If we were to apply Euclidean distance for classifying this untransformed data, we would misclassify some observations that clearly should not be misclassified when judged by eye. For example, in the graph above, the observations from group 3 that have y values below 40 (found in the lower left of the group 3 cloud of points) are closer in Euclidean distance to the center of group 1.

The following graph shows the Mahalanobis-transformed data.

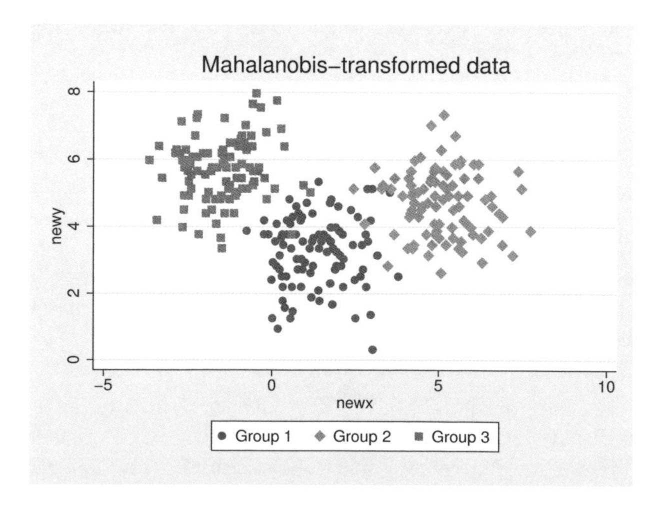

With the transformed data, using Euclidean distance between observations and group means works well.

Now let's see how well `discrim lda` can distinguish between the groups. By default, a resubstitution classification table is presented. The rows correspond to the known group and the columns to the grouping as assigned by the `discrim` model. The word resubstitution is used because the same observations that built the model are being classified by the model.

```
. discrim lda y x, group(group)
Linear discriminant analysis
Resubstitution classification summary

  +---------+
  | Key     |
  |---------|
  | Number  |
  | Percent |
  +---------+

              | Classified
   True group |        1         2         3 |    Total
 -------------+------------------------------+---------
            1 |       93         4         3 |      100
              |    93.00      4.00      3.00 |   100.00
              |                              |
            2 |        3        97         0 |      100
              |     3.00     97.00      0.00 |   100.00
              |                              |
            3 |        3         0        97 |      100
              |     3.00      0.00     97.00 |   100.00
 -------------+------------------------------+---------
        Total |       99       101       100 |      300
              |    33.00     33.67     33.33 |   100.00
              |                              |
       Priors |   0.3333    0.3333    0.3333 |
```

For these 300 observations, group 1 had 93 observations correctly classified, 4 observations misclassified into group 2, and 3 observations misclassified into group 3. Group 2 had 3 observations misclassified into group 1 and 97 observations correctly classified. Group 3 had 3 observations misclassified into group 1 and 97 observations correctly classified.

Generally, resubstitution classification tables give an overly optimistic view of how well you would classify an unknown observation. Leave-one-out (LOO) classification tables provide a more realistic assessment for classification success. With this 300-observation dataset, the LOO classification table gives the same results. We could see the LOO classification table by requesting it at estimation, by requesting it at replay, or by using the `estat classtable` command.

We now list the misclassified observations.

```
. estat list, varlist misclassified
```

Obs.	Data y	x	Classification True	Class.	Probabilities 1	2	3
19	49	37	1	3 *	0.2559	0.0000	0.7441
29	49	57	1	2 *	0.4245	0.5750	0.0005
47	49	37	1	3 *	0.2559	0.0000	0.7441
55	24	45	1	2 *	0.4428	0.5572	0.0000
70	48	61	1	2 *	0.0661	0.9339	0.0000
74	49	58	1	2 *	0.3041	0.6957	0.0003
92	37	22	1	3 *	0.3969	0.0000	0.6031
143	27	45	2	1 *	0.6262	0.3738	0.0000
161	39	49	2	1 *	0.8026	0.1973	0.0001
185	49	54	2	1 *	0.7782	0.2187	0.0030
238	48	44	3	1 *	0.8982	0.0017	0.1001
268	50	44	3	1 *	0.7523	0.0009	0.2469
278	36	31	3	1 *	0.9739	0.0000	0.0261

```
* indicates misclassified observations
```

The posterior probabilities for each displayed observation for each of the three groups is presented along with the true group and the classified group. The observation number is also shown. We added the discriminating variables `x` and `y` to the list with the `varlist` option. By default, `estat list` would list all the observations. The `misclassified` option restricts the list to those observations that were misclassified.

With `predict` we could generate classification variables, posterior probabilities, Mahalanobis squared distances from observations to group means, classification function scores (see *Methods and formulas*), and more. Fifteen `estat` commands provide more predictive and descriptive tools after `discrim lda`; see [MV] **discrim lda postestimation**.

◁

A classic example

We use the iris data from Fisher's (1936) pioneering LDA article to demonstrate the `discrim lda` command.

▷ Example 3

Fisher obtained the iris data from Anderson (1935). The data consist of four features measured on 50 samples from each of three iris species. The four features are the length and width of the sepal and petal. The three species are *Iris setosa*, *Iris versicolor*, and *Iris virginica*. Morrison (2005, app. B.2) is a modern source of the data.

```
. use http://www.stata-press.com/data/r12/iris, clear
(Iris data)
```

Running `discrim lda` produces the resubstitution classification table.

```
. discrim lda seplen sepwid petlen petwid, group(iris)
Linear discriminant analysis
Resubstitution classification summary
```

Key
Number Percent

True iris	Classified Setosa	Versicolor	Virginica	Total
Setosa	50 100.00	0 0.00	0 0.00	50 100.00
Versicolor	0 0.00	48 96.00	2 4.00	50 100.00
Virginica	0 0.00	1 2.00	49 98.00	50 100.00
Total	50 33.33	49 32.67	51 34.00	150 100.00
Priors	0.3333	0.3333	0.3333	

One *Iris virginica* observation was misclassified as a *versicolor*, two *Iris versicolor* observations were misclassified as *virginica*, and no *Iris setosa* observations were misclassified in our resubstitution classification.

Which observations were misclassified?

```
. estat list, misclassified
```

	Classification		Probabilities		
Obs.	True	Class.	Setosa	Versicolor	Virginica
71	Versicol	Virginic *	0.0000	0.2532	0.7468
84	Versicol	Virginic *	0.0000	0.1434	0.8566
134	Virginic	Versicol *	0.0000	0.7294	0.2706

```
* indicates misclassified observations
```

Postestimation command `estat list` shows that observations 71, 84, and 134 were misclassified and shows the estimated posterior probabilities for the three species for the misclassified observations.

We now examine the canonical discriminant functions for this LDA. The number of discriminant functions will be one fewer than the number of groups or will be the number of discriminating variables, whichever is less. With four discriminating variables and three species, we will have two discriminant functions. `estat loadings` displays the discriminant coefficients or loadings.

```
. estat loadings, unstandardized standardized
Canonical discriminant function coefficients

             | function1  function2
-------------+----------------------
      seplen | -.8293776  -.0241021
      sepwid | -1.534473  -2.164521
      petlen |  2.201212   .9319212
      petwid |   2.81046  -2.839188
       _cons | -2.105106   6.661473

Standardized canonical discriminant function coefficients

             | function1  function2
-------------+----------------------
      seplen | -.4269548  -.0124075
      sepwid | -.5212417  -.7352613
      petlen |  .9472572   .4010378
      petwid |  .5751608  -.5810399
```

We requested the display of both unstandardized and standardized loadings. The two unstandardized discriminant functions provide linear combinations of the `seplen`, `sepwid`, `petlen`, and `petwid` discriminating variables—producing two new dimensions. The standardized canonical discriminant function coefficients indicate the relative importance and relationship between the discriminating variables and the discriminant functions. The first discriminant function compares `seplen` and `sepwid`, which have negative standardized coefficients, to `petlen` and `petwid`, which have positive standardized coefficients. The second discriminant function appears to be contrasting the two width variables from the two length variables, though this is not as distinct of a difference as found in the first discriminant function because the `seplen` variable in the second standardized discriminant function is close to zero.

Understanding the composition of the discriminant functions is aided by plotting the coefficients. `loadingplot` graphs the discriminant function coefficients (loadings); see [MV] **discrim lda postestimation** and [MV] **scoreplot**.

```
. loadingplot
```

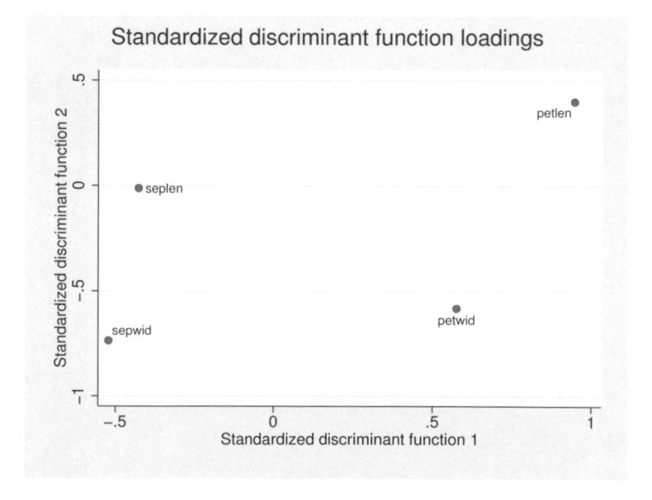

We now show a scatterplot of our three species projected onto the two dimensions of our discriminant solution. The `scoreplot` command takes care of most of the graphing details for us; see [MV] **discrim lda postestimation** and [MV] **scoreplot**. However, by default, `scoreplot` uses the full value labels for the three iris species and the resulting graph is busy. The `iris` dataset has two label languages predefined. The default label language has the full value labels. The other predefined label language is called `oneletter`, and it uses a one-letter code as value labels for the three iris species. The `label language` command will switch between these two label languages; see [D] **label language**. We also use the `msymbol(i)` graph option so that the points will not be displayed—only the one-letter value labels will be displayed for each observation.

```
. label language oneletter
. scoreplot, msymbol(i)
> note("S = Iris Setosa, C = Iris Versicolor, V = Iris Virginica")
```

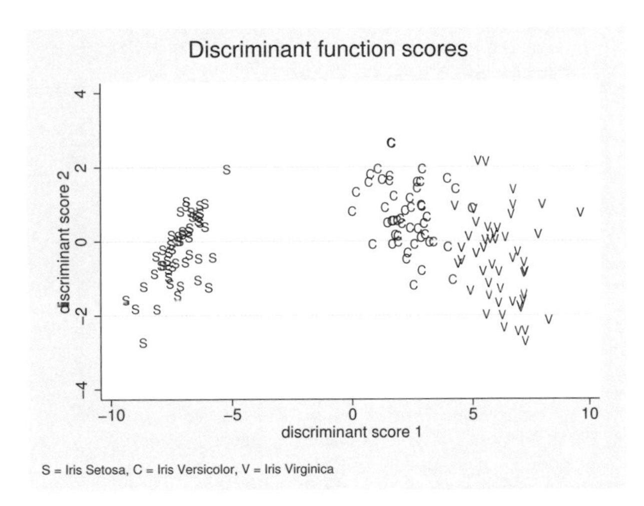

The *Iris setosa* are well separated from the other two species. *Iris versicolor* and *Iris virginica* show some overlap.

See example 1 of [MV] **discrim estat** and examples 6, 7, and 8, of [MV] **discrim lda postestimation** for more examples of what can be produced after `discrim lda` for this `iris` dataset.

◁

Saved results

`discrim lda` saves the following in `e()`:

Scalars

`e(N)`	number of observations
`e(N_groups)`	number of groups
`e(k)`	number of discriminating variables
`e(f)`	number of nonzero eigenvalues

Macros

`e(cmd)`	`discrim`
`e(subcmd)`	`lda`
`e(cmdline)`	command as typed
`e(groupvar)`	name of group variable
`e(grouplabels)`	labels for the groups
`e(varlist)`	discriminating variables
`e(wtype)`	weight type
`e(wexp)`	weight expression
`e(title)`	title in estimation output
`e(ties)`	how ties are to be handled
`e(properties)`	`nob noV eigen`
`e(estat_cmd)`	program used to implement `estat`
`e(predict)`	program used to implement `predict`
`e(marginsnotok)`	predictions disallowed by `margins`

Matrices

`e(groupcounts)`	number of observations for each group
`e(grouppriors)`	prior probabilities for each group
`e(groupvalues)`	numeric value for each group
`e(means)`	group means on discriminating variables
`e(SSCP_W)`	pooled within-group SSCP matrix
`e(SSCP_B)`	between-groups SSCP matrix
`e(SSCP_T)`	total SSCP matrix
`e(SSCP_W#)`	within-group SSCP matrix for group #
`e(W_eigvals)`	eigenvalues of `e(SSCP_W)`
`e(W_eigvecs)`	eigenvectors of `e(SSCP_W)`
`e(S)`	pooled within-group covariance matrix
`e(Sinv)`	inverse of `e(S)`
`e(sqrtSinv)`	Cholesky (square root) of `e(Sinv)`
`e(Ev)`	eigenvalues of $\mathbf{W}^{-1}\mathbf{B}$
`e(L_raw)`	eigenvectors of $\mathbf{W}^{-1}\mathbf{B}$
`e(L_unstd)`	unstandardized canonical discriminant function coefficients
`e(L_std)`	within-group standardized canonical discriminant function coefficients
`e(L_totalstd)`	total-sample standardized canonical discriminant function coefficients
`e(C)`	classification coefficients
`e(cmeans)`	unstandardized canonical discriminant functions evaluated at group means
`e(canstruct)`	canonical structure matrix
`e(candisc_stat)`	canonical discriminant analysis statistics

Functions

`e(sample)`	marks estimation sample

Methods and formulas

`discrim lda` is implemented as an ado-file.

Methods and formulas are presented under the following headings:

Predictive LDA
Descriptive LDA

Predictive LDA

Let g be the number of groups, n_i the number of observations for group i, and q_i the prior probability for group i. Let $\mathbf{x}$ denote an observation measured on p discriminating variables. For consistency with the discriminant analysis literature, $\mathbf{x}$ will be a column vector, though it corresponds to a row in your dataset. Let $f_i(\mathbf{x})$ represent the density function for group i, and let $P(\mathbf{x}|G_i)$ denote the probability of observing $\mathbf{x}$ conditional on belonging to group i. Denote the posterior probability of group i given observation $\mathbf{x}$ as $P(G_i|\mathbf{x})$. With Bayes' theorem, we have

$$P(G_i|\mathbf{x}) = \frac{q_i f_i(\mathbf{x})}{\sum_{j=1}^{g} q_j f_j(\mathbf{x})}$$

Substituting $P(\mathbf{x}|G_i)$ for $f_i(\mathbf{x})$, we have

$$P(G_i|\mathbf{x}) = \frac{q_i P(\mathbf{x}|G_i)}{\sum_{j=1}^{g} q_j P(\mathbf{x}|G_j)}$$

LDA assumes that the groups are multivariate normal with equal covariance matrices. Let $\mathbf{S}$ denote the pooled within-group sample covariance matrix and $\bar{\mathbf{x}}_i$ denote the sample mean of group i. The $\bar{\mathbf{x}}_i$ are returned as the columns of the `e(means)` matrix. The squared Mahalanobis distance between observation $\mathbf{x}$ and $\bar{\mathbf{x}}_i$ is

$$D_i^2 = (\mathbf{x} - \bar{\mathbf{x}}_i)'\mathbf{S}^{-1}(\mathbf{x} - \bar{\mathbf{x}}_i)$$

Plugging these sample estimates into the multivariate normal density gives

$$P(\mathbf{x}|G_i) = (2\pi)^{-p/2}|\mathbf{S}|^{-1/2}e^{-D_i^2/2}$$

Substituting this into the formula for $P(G_i|\mathbf{x})$ and simplifying gives

$$P(G_i|\mathbf{x}) = \frac{q_i e^{-D_i^2/2}}{\sum_{j=1}^{g} q_j e^{-D_j^2/2}}$$

as the LDA posterior probability of observation $\mathbf{x}$ belonging to group i.

Computation of Mahalanobis distance can be broken down into two steps. Step one: transform the data by using the Mahalanobis transformation. Step two: compute the Euclidean distance of the transformed data.

Let $\mathbf{L}$ be the Cholesky factorization of $\mathbf{S}^{-1}$ such that $\mathbf{S}^{-1} = \mathbf{L}'\mathbf{L}$ and $\mathbf{L}$ is lower triangular. $\mathbf{L}$ is returned in matrix `e(sqrtSinv)`. Squared Mahalanobis distance can be expressed in terms of $\mathbf{L}$.

$$\begin{aligned} D_i^2 &= (\mathbf{x} - \bar{\mathbf{x}}_i)'\mathbf{S}^{-1}(\mathbf{x} - \bar{\mathbf{x}}_i) \\ &= (\mathbf{x} - \bar{\mathbf{x}}_i)'\mathbf{L}'\mathbf{L}(\mathbf{x} - \bar{\mathbf{x}}_i) \\ &= (\mathbf{L}\mathbf{x} - \mathbf{L}\bar{\mathbf{x}}_i)'(\mathbf{L}\mathbf{x} - \mathbf{L}\bar{\mathbf{x}}_i) \\ &= (\mathbf{z} - \bar{\mathbf{z}}_i)'(\mathbf{z} - \bar{\mathbf{z}}_i) \end{aligned}$$

which is the squared Euclidean distance between $\mathbf{z}$ and $\bar{\mathbf{z}}_i$. We call $\mathbf{z} = \mathbf{Lx}$ the Mahalanobis transformation.

The squared Mahalanobis distance between group means is produced by `estat grdistances`; see [MV] **discrim lda postestimation**.

Classification functions can be derived from the Mahalanobis formulation for LDA; see Rencher (2002, 304–305) and Huberty (1994, 59). Let $L_i(\mathbf{x})$ denote the linear classification function for the ith group applied to observation $\mathbf{x}$. $L_i(\mathbf{x}) = \mathbf{c}_i'\mathbf{x} + c_{i0}$, where $\mathbf{c}_i = \bar{\mathbf{x}}_i'\mathbf{S}^{-1}$ and $c_{i0} = -(1/2)\bar{\mathbf{x}}_i'\mathbf{S}^{-1}\bar{\mathbf{x}}_i + \ln(q_i)$. The g linear classification functions are returned as the columns of matrix **e(C)** and through the `estat classfunction` command; see [MV] **discrim lda postestimation**. An observation can be classified based on largest posterior probability or based on largest classification function score.

Descriptive LDA

As with predictive LDA, let g be the number groups, n_i the number of training (sample) observations for group i, p the number of discriminating variables, and $N = \sum_{i=1}^{g} n_i$ the total number of observations. Also, let $\mathbf{W}$ be the within-group sums-of-squares and cross-products (SSCP) matrix and let $\mathbf{B}$ be the between-groups SSCP matrix. Fisher's (1936) linear discriminant functions are based on the eigenvalues and eigenvectors of $\mathbf{W}^{-1}\mathbf{B}$.

There are $s = \min(g-1, p)$ nonzero eigenvalues of $\mathbf{W}^{-1}\mathbf{B}$. Let $\lambda_1, \lambda_2, \ldots, \lambda_s$ denote the s eigenvalues in decreasing order. The eigenvalues are returned in `e(Ev)`. Let $\mathbf{v}_1, \mathbf{v}_2, \ldots, \mathbf{v}_s$ denote the corresponding eigenvectors. Rencher (2002, 279) outlines the approach for obtaining the eigenvalues and eigenvectors of the nonsymmetric $\mathbf{W}^{-1}\mathbf{B}$ matrix. Because $\mathbf{W}^{-1}\mathbf{B}$ is nonsymmetric, the resulting eigenvectors are not orthogonal but merely uncorrelated; see Rencher (2002, 278). A matrix with the $\mathbf{v}_i$ as columns is returned in `e(L_raw)`. The phrase raw coefficients is used by Klecka (1980, 22) to describe the $\mathbf{v}$ vectors.

Any constant multiple of the eigenvector $\mathbf{v}_i$ is still an eigenvector of $\mathbf{W}^{-1}\mathbf{B}$ associated with eigenvalue λ_i. Typically, $\mathbf{vu}_i = \mathbf{v}_i\sqrt{N-g}$ are used as the eigenvectors (discriminant functions) and are called unstandardized canonical discriminant functions because they correspond to the unstandardized discriminating variables. An extra element is added to the bottom of the $\mathbf{vu}$ vectors for the constant, so that if the $\mathbf{vu}$ vectors are applied as linear combinations of the discriminating variables, the resulting variables have mean zero; see Klecka (1980, 21–23). A matrix with the $\mathbf{vu}_i$ as columns is returned in `e(L_unstd)`.

The means of the discriminating variables for each group are returned as the columns of the matrix `e(means)`. These group means multiplied by the unstandardized discriminant-function coefficients, $\mathbf{vu}_i$, produce what are called group means on canonical variables and are returned in the matrix `e(cmeans)` and displayed with the command `estat grmeans, canonical`.

Standardized discriminant functions are obtained as $\mathbf{vs}_i = \mathbf{v}_i\sqrt{W_{ii}}$. The ith raw eigenvector is standardized by the square root of the ith diagonal element of the $\mathbf{W}$ matrix. These within-group standardized discriminant functions are used in assessing the importance and relationship of the original discriminating variables to the discriminant functions. A matrix with the $\mathbf{vs}_i$ as columns is returned in `e(L_std)`.

Let $\mathbf{T}$ denote the total sample SSCP matrix. Total-sample standardized discriminant functions are computed as $\mathbf{vt}_i = \mathbf{v}_i\sqrt{T_{ii}(N-g)/(N-1)}$. A matrix with the $\mathbf{vt}_i$ as columns is returned in `e(L_totalstd)`. There is debate as to which of $\mathbf{vs}$ and $\mathbf{vt}$ should be used for interpretation; see Mueller and Cozad (1988), Nordlund and Nagel (1991), and Mueller and Cozad (1993).

The `estat loadings` command displays `e(L_unstd)`, `e(L_std)`, and `e(L_totalstd)`; see [MV] **discrim lda postestimation**.

The canonical structure matrix measures the correlation between the discriminating variables and the discriminant function and is returned in matrix `e(canstruct)`. The canonical structure matrix is equal to $\mathbf{WV}$ with the ith row divided by $\sqrt{W_{ii}}$, where $\mathbf{V}$ contains the $\mathbf{v}_i$ eigenvectors as columns. Rencher (2002, 291) warns against the use of structure coefficients for interpretation, but Huberty (1994, 262–264) contends otherwise.

The returned matrix `e(candisc_stat)` contains columns for the information shown by `estat canontest`, including the eigenvalues, canonical correlations, proportion of variance, cumulative proportion of variance, likelihood-ratio test statistics, and the corresponding F tests, degrees of freedom, and p-values. See [MV] **canon**.

As noted in the *Introduction* section of *Remarks*, Kshirsagar and Arseven (1975), Green (1979), and Williams (1982) demonstrate the mathematical relationship between Fisher's linear discriminant functions (the basis for descriptive LDA) and the classification functions from the Mahalanobis approach to LDA (the basis for predictive LDA); see Rencher (1998, 239).

References

Anderson, E. 1935. The irises of the Gaspe Peninsula. *Bulletin of the American Iris Society* 59: 2–5.

Fisher, R. A. 1936. The use of multiple measurements in taxonomic problems. *Annals of Eugenics* 7: 179–188.

Green, B. F. 1979. The two kinds of linear discriminant functions and their relationship. *Journal of Educational Statistics* 4: 247–263.

Hilbe, J. M. 1992. smv3: Regression-based dichotomous discriminant analysis. *Stata Technical Bulletin* 5: 13–17. Reprinted in *Stata Technical Bulletin Reprints*, vol. 1, pp. 131–137. College Station, TX: Stata Press.

——. 1996. smv3.1: Discriminant analysis: An enhanced command. *Stata Technical Bulletin* 34: 34–36. Reprinted in *Stata Technical Bulletin Reprints*, vol. 6, pp. 190–192. College Station, TX: Stata Press.

——. 1997. smv3.2: Enhancements to discriminant analysis. *Stata Technical Bulletin* 38: 26–27. Reprinted in *Stata Technical Bulletin Reprints*, vol. 7, p. 229. College Station, TX: Stata Press.

Huberty, C. J. 1994. *Applied Discriminant Analysis*. New York: Wiley.

Johnson, R. A., and D. W. Wichern. 2007. *Applied Multivariate Statistical Analysis*. 6th ed. Englewood Cliffs, NJ: Prentice Hall.

Klecka, W. R. 1980. *Discriminant Analysis*. Newbury Park, CA: Sage.

Kshirsagar, A. M., and E. Arseven. 1975. A note on the equivalency of two discrimination procedures. *American Statistician* 29: 38–39.

Lachenbruch, P. A. 1975. *Discriminant Analysis*. New York: Hafner Press.

Mahalanobis, P. C. 1936. On the generalized distance in statistics. *National Institute of Science of India* 12: 49–55.

Mardia, K. V., J. T. Kent, and J. M. Bibby. 1979. *Multivariate Analysis*. London: Academic Press.

McLachlan, G. J. 2004. *Discriminant Analysis and Statistical Pattern Recognition*. New York: Wiley.

Morrison, D. F. 2005. *Multivariate Statistical Methods*. 4th ed. Belmont, CA: Duxbury.

Mueller, R. O., and J. B. Cozad. 1988. Standardized discriminant coefficients: Which variance estimate is appropriate? *Journal of Educational Statistics* 13: 313–318.

——. 1993. Standardized discriminant coefficients: A rejoinder. *Journal of Educational Statistics* 18: 108–114.

Nordlund, D. J., and R. Nagel. 1991. Standardized discriminant coefficients revisited. *Journal of Educational Statistics* 16: 101–108.

Rencher, A. C. 1998. *Multivariate Statistical Inference and Applications*. New York: Wiley.

——. 2002. *Methods of Multivariate Analysis*. 2nd ed. New York: Wiley.

Tabachnick, B. G., and L. S. Fidell. 2007. *Using Multivariate Statistics*. 5th ed. Boston: Allyn and Bacon.

Williams, B. K. 1982. A simple demonstration of the relationship between classification and canonical variates analysis. *American Statistician* 36: 363–365.

Also see

[MV] **discrim lda postestimation** — Postestimation tools for discrim lda

[MV] **discrim** — Discriminant analysis

[MV] **candisc** — Canonical linear discriminant analysis

[U] **20 Estimation and postestimation commands**

Title

discrim lda postestimation — Postestimation tools for discrim lda

Description

The following postestimation commands are of special interest after `discrim lda`:

Command	Description
`estat anova`	ANOVA summaries table
`estat canontest`	tests of the canonical discriminant functions
`estat classfunctions`	classification functions
`estat classtable`	classification table
`estat correlations`	correlation matrices and p-values
`estat covariance`	covariance matrices
`estat errorrate`	classification error-rate estimation
`estat grdistances`	Mahalanobis and generalized squared distances between the group means
`estat grmeans`	group means and variously standardized or transformed means
`estat grsummarize`	group summaries
`estat list`	classification listing
`estat loadings`	canonical discriminant-function coefficients (loadings)
`estat manova`	MANOVA table
`estat structure`	canonical structure matrix
`estat summarize`	estimation sample summary
`loadingplot`	plot standardized discriminant-function loadings
`scoreplot`	plot discriminant-function scores
`screeplot`	plot eigenvalues

For information about `estat anova`, `estat canontest`, `estat classfunctions`, `estat correlations`, `estat covariance`, `estat grdistances`, `estat grmeans`, `estat loadings`, `estat manova`, and `estat structure`, see below.

For information about `estat classtable`, `estat errorrate`, `estat grsummarize`, `estat list`, and `estat summarize`, see [MV] **discrim estat**.
For information about `loadingplot` and `scoreplot`, see [MV] **scoreplot**.
For information about `screeplot`, see [MV] **screeplot**.

The following standard postestimation commands are also available:

Command	Description
* `estimates`	cataloging estimation results
`predict`	group classification and posterior probabilities

*All `estimates` subcommands except `table` and `stats` are available; see [R] **estimates**.

Special-interest postestimation commands

estat anova presents a table summarizing the one-way ANOVAs for each variable in the discriminant analysis.

estat canontest presents tests of the canonical discriminant functions. Presented are the canonical correlations, eigenvalues, proportion and cumulative proportion of variance, and likelihood-ratio tests for the number of nonzero eigenvalues.

estat classfunctions displays the classification functions.

estat correlations displays the pooled within-group correlation matrix, between-groups correlation matrix, total-sample correlation matrix, and/or the individual group correlation matrices. Two-tailed p-values for the correlations may also be requested.

estat covariance displays the pooled within-group covariance matrix, between-groups covariance matrix, total-sample covariance matrix, and/or the individual group covariance matrices.

estat grdistances provides Mahalanobis squared distances between the group means along with the associated F statistics and significance levels. Also available are generalized squared distances.

estat grmeans provides group means, total-sample standardized group means, pooled within-group standardized means, and canonical functions evaluated at the group means.

estat loadings present the canonical discriminant-function coefficients (loadings). Unstandardized, pooled within-class standardized, and total-sample standardized coefficients are available.

estat manova presents the MANOVA table associated with the discriminant analysis.

estat structure presents the canonical structure matrix.

Syntax for predict

predict [*type*] *newvar* [*if*] [*in*] [, *statistic options*]

predict [*type*] {*stub** | *newvarlist*} [*if*] [*in*] [, *statistic options*]

statistic	Description
Main	
classification	group membership classification; the default when one variable is specified and group() is not specified
pr	probability of group membership; the default when group() is specified or when multiple variables are specified
mahalanobis	Mahalanobis squared distance between observations and groups
dscore	discriminant function score
clscore	group classification function score
* looclass	leave-one-out group membership classification; may be used only when one new variable is specified
* loopr	leave-one-out probability of group membership
* loomahal	leave-one-out Mahalanobis squared distance between observations and groups

options	Description
Main	
group(*group*)	the group for which the statistic is to be calculated
Options	
priors(*priors*)	group prior probabilities; defaults to e(grouppriors)
ties(*ties*)	how ties in classification are to be handled; defaults to e(ties)

priors	Description
equal	equal prior probabilities
proportional	group-size-proportional prior probabilities
matname	row or column vector containing the group prior probabilities
matrix_exp	matrix expression providing a row or column vector of the group prior probabilities

ties	Description
missing	ties in group classification produce missing values
random	ties in group classification are broken randomly
first	ties in group classification are set to the first tied group

You specify one new variable with classification or looclass; either one or e(N_groups) new variables with pr, loopr, mahalanobis, loomahal, or clscore; and one to e(f) new variables with dscore.

Unstarred statistics are available both in and out of sample; type predict ... if e(sample) ... if wanted only for the estimation sample. Starred statistics are calculated only for the estimation sample, even when if e(sample) is not specified.

group() is not allowed with classification, dscore, or looclass.

Menu

Statistics > Postestimation > Predictions, residuals, etc.

Options for predict

Main

classification, the default, calculates the group classification. Only one new variable may be specified.

pr calculates group membership posterior probabilities. If you specify the group() option, specify one new variable. Otherwise, you must specify e(N_groups) new variables.

mahalanobis calculates the squared Mahalanobis distance between the observations and group means. If you specify the group() option, specify one new variable. Otherwise, you must specify e(N_groups) new variables.

dscore produces the discriminant function score. Specify as many variables as leading discriminant functions that you wish to score. No more than e(f) variables may be specified.

clscore produces the group classification function score. If you specify the group() option, specify one new variable. Otherwise, you must specify e(N_groups) new variables.

looclass calculates the leave-one-out group classifications. Only one new variable may be specified. Leave-one-out calculations are restricted to e(sample) observations.

loopr calculates the leave-one-out group membership posterior probabilities. If you specify the group() option, specify one new variable. Otherwise, you must specify e(N_groups) new variables. Leave-one-out calculations are restricted to e(sample) observations.

loomahal calculates the leave-one-out squared Mahalanobis distance between the observations and group means. If you specify the group() option, specify one new variable. Otherwise, you must specify e(N_groups) new variables. Leave-one-out calculations are restricted to e(sample) observations.

group(*group*) specifies the group for which the statistic is to be calculated and can be specified using

#1, #2, ..., where #1 means the first category of the e(groupvar) variable, #2 the second category, etc.;

the values of the e(groupvar) variable; or

the value labels of the e(groupvar) variable if they exist.

group() is not allowed with classification, dscore, or looclass.

Options

priors(*priors*) specifies the prior probabilities for group membership. If priors() is not specified, e(grouppriors) is used. The following *priors* are allowed:

priors(equal) specifies equal prior probabilities.

priors(proportional) specifies group-size-proportional prior probabilities.

priors(*matname*) specifies a row or column vector containing the group prior probabilities.

priors(*matrix_exp*) specifies a matrix expression providing a row or column vector of the group prior probabilities.

ties(*ties*) specifies how ties in group classification will be handled. If ties() is not specified, e(ties) is used. The following *ties* are allowed:

ties(missing) specifies that ties in group classification produce missing values.

ties(random) specifies that ties in group classification are broken randomly.

ties(first) specifies that ties in group classification are set to the first tied group.

Syntax for estat anova

estat anova

Menu

Statistics > Postestimation > Reports and statistics

Syntax for estat canontest

`estat canontest`

Menu

Statistics > Postestimation > Reports and statistics

Syntax for estat classfunctions

`estat classfunctions [ , options ]`

options	Description
Main	
`adjustequal`	adjust the constant even when priors are equal
`format(%fmt)`	numeric display format; default is `%9.0g`
Options	
`priors(priors)`	group prior probabilities; defaults to `e(grouppriors)`
`nopriors`	suppress display of prior probabilities

Menu

Statistics > Postestimation > Reports and statistics

Options for estat classfunctions

Main

`adjustequal` specifies that the constant term in the classification function be adjusted for prior probabilities even though the priors are equal. By default, equal prior probabilities are not used in adjusting the constant term. `adjustequal` has no effect with unequal prior probabilities.

`format(%`*fmt*`)` specifies the matrix display format. The default is `format(%9.0g)`.

Options

`priors(`*priors*`)` specifies the group prior probabilities. The prior probabilities affect the constant term in the classification function. By default, *priors* is determined from `e(grouppriors)`. See *Options for predict* for the *priors* specification. By common convention, when there are equal prior probabilities the adjustment of the constant term is not performed. See `adjustequal` to override this convention.

`nopriors` specifies that the prior probabilities not be displayed. By default, the prior probabilities used in determining the constant in the classification functions are displayed as the last row in the classification functions table.

Syntax for estat correlations

estat correlations [, *options*]

options	Description
Main	
within	display pooled within-group correlation matrix; the default
between	display between-groups correlation matrix
total	display total-sample correlation matrix
groups	display the correlation matrix for each group
all	display all the above
p	display two-sided p-values for requested correlations
format(%*fmt*)	numeric display format; default is %9.0g
nohalf	display full matrix even if symmetric

Menu

Statistics > Postestimation > Reports and statistics

Options for estat correlations

Main

within specifies that the pooled within-group correlation matrix be displayed. This is the default.

between specifies that the between-groups correlation matrix be displayed.

total specifies that the total-sample correlation matrix be displayed.

groups specifies that the correlation matrix for each group be displayed.

all is the same as specifying within, between, total, and groups.

p specifies that two-sided p-values be computed and displayed for the requested correlations.

format(%*fmt*) specifies the matrix display format. The default is format(%8.5f).

nohalf specifies that, even though the matrix is symmetric, the full matrix be printed. The default is to print only the lower triangle.

Syntax for estat covariance

```
estat covariance [, options]
```

options	Description
Main	
within	display pooled within-group covariance matrix; the default
between	display between-groups covariance matrix
total	display total-sample covariance matrix
groups	display the covariance matrix for each group
all	display all the above
format(%*fmt*)	numeric display format; default is %9.0g
nohalf	display full matrix even if symmetric

Menu

Statistics > Postestimation > Reports and statistics

Options for estat covariance

Main

within specifies that the pooled within-group covariance matrix be displayed. This is the default.

between specifies that the between-groups covariance matrix be displayed.

total specifies that the total-sample covariance matrix be displayed.

groups specifies that the covariance matrix for each group be displayed.

all is the same as specifying within, between, total, and groups.

format(%*fmt*) specifies the matrix display format. The default is format(%9.0g).

nohalf specifies that, even though the matrix is symmetric, the full matrix be printed. The default is to print only the lower triangle.

Syntax for estat grdistances

```
estat grdistances [, options]
```

options	Description
Main	
mahalanobis[(f p)]	display Mahalanobis squared distances between group means; the default
generalized	display generalized Mahalanobis squared distances between group means
all	equivalent to mahalanobis(f p) generalized
format(%*fmt*)	numeric display format; default is %9.0g
Options	
priors(*priors*)	group prior probabilities; defaults to e(grouppriors)

Menu

Statistics > Postestimation > Reports and statistics

Options for estat grdistances

Main

`mahalanobis`[`(f p)`] specifies that a table of Mahalanobis squared distances between group means be presented. `mahalanobis(f)` adds F tests for each displayed distance and `mahalanobis(p)` adds the associated p-values. `mahalanobis(f p)` adds both. The default is `mahalanobis`.

`generalized` specifies that a table of generalized Mahalanobis squared distances between group means be presented. `generalized` starts with what is produced by the `mahalanobis` option and adds a term accounting for prior probabilities. Prior probabilities are provided with the `priors()` option, or if `priors()` is not specified, by the values in `e(grouppriors)`. By common convention, if prior probabilities are equal across the groups, the prior probability term is omitted and the results from `generalized` will equal those from `mahalanobis`.

`all` is equivalent to specifying `mahalanobis(f p)` and `generalized`.

`format(%`*fmt*`)` specifies the matrix display format. The default is `format(%9.0g)`.

Options

`priors(`*priors*`)` specifies the group prior probabilities and affects only the output of the `generalized` option. By default, *priors* is determined from `e(grouppriors)`. See *Options for predict* for the *priors* specification.

Syntax for estat grmeans

`estat grmeans` [, *options*]

options	Description
Main	
`raw`	display untransformed and unstandardized group means
`totalstd`	display total-sample standardized group means
`withinstd`	display pooled within-group standardized group means
`canonical`	display canonical functions evaluated at group means
`all`	display all the mean tables

Menu

Statistics > Postestimation > Reports and statistics

Options for estat grmeans

Main

raw, the default, displays a table of group means.

totalstd specifies that a table of total-sample standardized group means be presented.

withinstd specifies that a table of pooled within-group standardized group means be presented.

canonical specifies that a table of the unstandardized canonical discriminant functions evaluated at the group means be presented.

all is equivalent to specifying raw, totalstd, withinstd, and canonical.

Syntax for estat loadings

estat loadings [, *options*]

options	Description
Main	
standardized	display pooled within-group standardized canonical discriminant function coefficients; the default
totalstandardized	display the total-sample standardized canonical discriminant function coefficients
unstandardized	display unstandardized canonical discriminant function coefficients
all	display all the above
format(%*fmt*)	numeric display format; default is %9.0g

Menu

Statistics > Postestimation > Reports and statistics

Options for estat loadings

Main

standardized specifies that the pooled within-group standardized canonical discriminant function coefficients be presented. This is the default.

totalstandardized specifies that the total-sample standardized canonical discriminant function coefficients be presented.

unstandardized specifies that the unstandardized canonical discriminant function coefficients be presented.

all is equivalent to specifying standardized, totalstandardized, and unstandardized.

format(%*fmt*) specifies the matrix display format. The default is format(%9.0g).

Syntax for estat manova

```
estat manova
```

Menu

Statistics > Postestimation > Reports and statistics

Syntax for estat structure

```
estat structure [, format(%fmt) ]
```

Menu

Statistics > Postestimation > Reports and statistics

Option for estat structure

Main

`format(%fmt)` specifies the matrix display format. The default is `format(%9.0g)`.

Remarks

Remarks are presented under the following headings:

Classification tables, error rates, and listings
ANOVA, MANOVA, and canonical correlations
Discriminant and classification functions
Scree, loading, and score plots
Means and distances
Covariance and correlation matrices
Predictions

Classification tables, error rates, and listings

After `discrim`, including `discrim lda`, you can obtain classification tables, error-rate estimates, and listings; see [MV] **discrim estat**.

▷ Example 1

Example 1 of [MV] **manova** introduces the apple tree rootstock data from Andrews and Herzberg (1985, 357–360) and used in Rencher (2002, 171). Descriptive linear discriminant analysis is often used after a multivariate analysis of variance (MANOVA) to explore the differences between groups found to be significantly different in the MANOVA.

We first examine the predictive aspects of the linear discriminant model on these data by examining classification tables, error-rate estimate tables, and classification listings.

To illustrate the ability of `discrim lda` and the postestimation commands of handling unequal prior probabilities, we perform our LDA using prior probabilities of 0.2 for the first four rootstock groups and 0.1 for the last two rootstock groups.

```
. use http://www.stata-press.com/data/r12/rootstock
(Table 6.2 Rootstock Data -- Rencher (2002))
```

```
. discrim lda y1 y2 y3 y4, group(rootstock) priors(.2, .2, .2, .2, .1, .1)
Linear discriminant analysis
Resubstitution classification summary
```

Key
Number Percent

True rootstock	Classified 1	2	3	4	5	6	Total
1	7 87.50	0 0.00	0 0.00	1 12.50	0 0.00	0 0.00	8 100.00
2	0 0.00	4 50.00	2 25.00	1 12.50	1 12.50	0 0.00	8 100.00
3	0 0.00	1 12.50	6 75.00	1 12.50	0 0.00	0 0.00	8 100.00
4	3 37.50	0 0.00	1 12.50	4 50.00	0 0.00	0 0.00	8 100.00
5	0 0.00	3 37.50	2 25.00	0 0.00	2 25.00	1 12.50	8 100.00
6	3 37.50	0 0.00	0 0.00	0 0.00	2 25.00	3 37.50	8 100.00
Total	13 27.08	8 16.67	11 22.92	7 14.58	5 10.42	4 8.33	48 100.00
Priors	0.2000	0.2000	0.2000	0.2000	0.1000	0.1000	

The prior probabilities are reported at the bottom of the table. The classification results are based, in part, on the selection of prior probabilities.

With only 8 observations per rootstock and six rootstock groups, we have small cell counts in our table, with many zero cell counts. Because resubstitution classification tables give an overly optimistic view of classification ability, we use the `estat classtable` command to request a leave-one-out (LOO) classification table and request the reporting of average posterior probabilities in place of percentages.

```
. estat classtable, probabilities loo
Leave-one-out average-posterior-probabilities classification table
```

Key
Number Average posterior probability

True rootstock	LOO Classified 1	2	3	4	5	6
1	5 0.6055	0 .	0 .	2 0.6251	0 .	1 0.3857
2	0 .	4 0.6095	2 0.7638	1 0.3509	1 0.6607	0 .
3	0 .	1 0.5520	6 0.7695	1 0.4241	0 .	0 .
4	4 0.5032	0 .	1 0.7821	3 0.5461	0 .	0 .
5	0 .	3 0.7723	2 0.5606	0 .	2 0.4897	1 0.6799
6	3 0.6725	0 .	0 .	0 .	2 0.4296	3 0.5763
Total	12 0.5881	8 0.6634	11 0.7316	7 0.5234	5 0.4999	5 0.5589
Priors	0.2000	0.2000	0.2000	0.2000	0.1000	0.1000

Zero cell counts report a missing value for the average posterior probability. We did not specify the `priors()` option with `estat classtable`, so the prior probabilities used in our LDA model were used.

`estat errorrate` estimates the error rates for each group. We use the `pp` option to obtain estimates based on the posterior probabilities instead of the counts.

```
. estat errorrate, pp
Error rate estimated from posterior probabilities
```

Error Rate	rootstock 1	2	3	4	5
Stratified	.2022195	.431596	.0868444	.4899799	.627472
Unstratified	.2404022	.41446	.1889412	.5749832	.4953118
Priors	.2	.2	.2	.2	.1

Error Rate	rootstock 6	Total
Stratified	.6416429	.3690394
Unstratified	.4027382	.3735623
Priors	.1	

We did not specify the `priors()` option, and `estat errorrate` defaulted to using the prior probabilities from the LDA model. Both stratified and unstratified estimates are shown for each rootstock group and for the overall total. See [MV] **discrim estat** for an explanation of the error-rate estimation.

We can list the classification results and posterior probabilities from our discriminant analysis model by using the `estat list` command. `estat list` allows us to specify which observations we wish to examine and what classification and probability results to report.

We request the LOO classification and LOO probabilities for all misclassified observations from the fourth rootstock group. We also suppress the resubstitution classification and probabilities from being displayed.

```
. estat list if rootstock==4, misclassified class(loo noclass) pr(loo nopr)
```

	Classification		LOO Probabilities					
Obs.	True	LOO Cl.	1	2	3	4	5	6
25	4	1 *	0.5433	0.1279	0.0997	0.0258	0.0636	0.1397
26	4	3 *	0.0216	0.0199	0.7821	0.1458	0.0259	0.0048
27	4	1 *	0.3506	0.1860	0.0583	0.2342	0.0702	0.1008
29	4	1 *	0.6134	0.0001	0.0005	0.2655	0.0002	0.1202
32	4	1 *	0.5054	0.0011	0.0017	0.4856	0.0002	0.0059

```
* indicates misclassified observations
```

Four of the five misclassifications for rootstock group 4 were incorrectly classified as belonging to rootstock group 1.

◁

ANOVA, MANOVA, and canonical correlations

There is a mathematical relationship between Fisher's LDA and one-way MANOVA. They are both based on the eigenvalues and eigenvectors of the same matrix, $\mathbf{W}^{-1}\mathbf{B}$ (though in MANOVA the matrices are labeled $\mathbf{E}$ and $\mathbf{H}$ for error and hypothesis instead of $\mathbf{W}$ and $\mathbf{B}$ for within and between). See [MV] **manova** and [R] **anova** for more information on MANOVA and ANOVA. Researchers often wish to examine the MANOVA and univariate ANOVA results corresponding to their LDA model.

Canonical correlations are also mathematically related to Fisher's LDA. The canonical correlations between the discriminating variables and indicator variables constructed from the group variable are based on the same eigenvalues and eigenvectors as MANOVA and Fisher's LDA. The information from a canonical correlation analysis gives insight into the importance of each discriminant function in the discrimination. See [MV] **canon** for more information on canonical correlations.

The `estat manova`, `estat anova`, and `estat canontest` commands display MANOVA, ANOVA, and canonical correlation information after `discrim lda`.

▷ Example 2

Continuing with the apple tree rootstock example, we examine the MANOVA, ANOVA, and canonical correlation results corresponding to our LDA.

```
. estat manova
                          Number of obs =       48

                          W = Wilks' lambda      L = Lawley-Hotelling trace
                          P = Pillai's trace     R = Roy's largest root

                  Source | Statistic     df   F(df1,     df2) =    F   Prob>F
              -----------+--------------------------------------------------
               rootstock | W   0.1540      5    20.0    130.3      4.94 0.0000 a
                         | P   1.3055           20.0    168.0      4.07 0.0000 a
                         | L   2.9214           20.0    150.0      5.48 0.0000 a
                         | R   1.8757            5.0     42.0     15.76 0.0000 u
                         |--------------------------------------------------
                Residual |               42
              -----------+--------------------------------------------------
                   Total |               47
              --------------------------------------------------------------
                          e = exact, a = approximate, u = upper bound on F

. estat anova
Univariate ANOVA summaries

                                                         Adj.
       Variable |  Model MS   Resid MS   Total MS   R-sq   R-sq       F   Pr > F
   -------------+-------------------------------------------------------------
             y1 | .07356042  .31998754  .29377189 0.1869 0.0901   1.931 0.1094
             y2 | 4.1996621   12.14279  11.297777 0.2570 0.1685  2.9052 0.0243
             y3 | 6.1139358  4.2908128   4.484762 0.5876 0.5385  11.969 0.0000
             y4 | 2.4930912  1.7225248  1.8044999 0.5914 0.5428  12.158 0.0000
   ----------------------------------------------------------------------------
                  Number of obs = 48     Model df = 5      Residual df = 42
```

All four of the MANOVA tests reject the null hypothesis that the six rootstock groups have equal means. See example 1 of [MV] **manova** for an explanation of the MANOVA table.

`estat anova` presents a concise summary of univariate ANOVAs run on each of our four discriminating variables. Variables y3, trunk girth at 15 years, and y4, weight of tree above ground at 15 years, show the highest level of significance of the four variables.

`estat canontest` displays the canonical correlations and associated tests that correspond to our LDA model.

```
. estat canontest
Canonical linear discriminant analysis

       |                                     | Like-
       | Canon.   Eigen-      Variance       | lihood
   Fcn | Corr.    value    Prop.     Cumul.  | Ratio      F      df1     df2  Prob>F
   ----+-------------------------------------+------------------------------------
     1 | 0.8076  1.87567  0.6421  0.6421 | 0.1540  4.9369    20   130.3  0.0000 a
     2 | 0.6645  .790694  0.2707  0.9127 | 0.4429  3.1879    12   106.1  0.0006 a
     3 | 0.4317  .229049  0.0784  0.9911 | 0.7931  1.6799     6      82  0.1363 e
     4 | 0.1591  .025954  0.0089  1.0000 | 0.9747  .54503     2      42  0.5839 e
   --------------------------------------------------------------------------------
   Ho: this and smaller canon. corr. are zero;     e = exact F, a = approximate F
```

The number of nonzero eigenvalues in Fisher's LDA is $\min(g-1,p)$ With $g=6$ groups, and $p=4$ discriminating variables, there are four nonzero eigenvalues. The four eigenvalues and the corresponding canonical correlations of $\mathbf{W}^{-1}\mathbf{B}$, ordered from largest to smallest, are reported along with the proportion and cumulative proportion of variance accounted for by each of the discriminant functions. Using one discriminant dimension is insufficient for capturing the variability of our four-dimensional data. With two dimensions we account for 91% of the variance. Using three of the four dimensions accounts for 99% of the variance. Little is gained from the fourth discriminant dimension.

Also presented are the likelihood-ratio tests of the null hypothesis that each canonical correlation and all smaller canonical correlations from this model are zero. The letter a is placed beside the p-values of the approximate F tests, and the letter e is placed beside the p-values of the exact F tests. The first two tests are highly significant, indicating that the first two canonical correlations are likely not zero. The third test has a p-value of 0.1363, so that we fail to reject that the third and fourth canonical correlation are zero.

◁

Discriminant and classification functions

See [MV] **discrim lda** for a discussion of linear discriminant functions and linear classification functions for LDA.

Discriminant functions are produced from Fisher's LDA. The discriminant functions provide a set of transformations from the original p-dimensional (the number of discriminating variables) space to the minimum of p and $g - 1$ (the number of groups minus 1) dimensional space. The discriminant functions are ordered in importance.

Classification functions are by-products of the Mahalanobis approach to LDA. There are always g classification functions—one for each group. They are not ordered by importance, and you cannot use a subset of them for classification.

A table showing the discriminant function coefficients is available with `estat loadings` (see example 3), and a table showing the classification function coefficients is available with `estat classfunctions` (see example 4).

▷ Example 3

We continue with the apple tree rootstock example. The canonical discriminant function coefficients (loadings) are available through the `estat loadings` command. Unstandardized, pooled within-group standardized, and total-sample standardized coefficients are available. The `all` option requests all three, and the `format()` option provides control over the numeric display format used in the tables.

```
. estat loadings, all format(%6.2f)
Canonical discriminant function coefficients

             |  func~1  func~2  func~3  func~4
-------------+--------------------------------
          y1 |   -3.05   -1.14   -1.00   23.42
          y2 |    1.70   -1.22    1.67   -3.08
          y3 |   -4.23    7.17    3.05   -2.01
          y4 |    0.48  -11.52   -5.51    3.10
       _cons |   15.45  -12.20   -9.99  -12.47

Standardized canonical discriminant function coefficients

             |  func~1  func~2  func~3  func~4
-------------+--------------------------------
          y1 |   -0.27   -0.10   -0.09    2.04
          y2 |    0.92   -0.65    0.90   -1.65
          y3 |   -1.35    2.29    0.97   -0.64
          y4 |    0.10   -2.33   -1.12    0.63

Total-sample standardized canonical discriminant function coefficients

             |  func~1  func~2  func~3  func~4
-------------+--------------------------------
          y1 |   -0.28   -0.10   -0.09    2.14
          y2 |    1.00   -0.72    0.99   -1.81
          y3 |   -1.99    3.37    1.43   -0.95
          y4 |    0.14   -3.45   -1.65    0.93
```

The unstandardized canonical discriminant function coefficients shown in the first table are the function coefficients that apply to the unstandardized discriminating variables—y1 through y4 and a constant term. See example 5 for a graph, known as a score plot, that plots the observations transformed by these unstandardized canonical discriminant function coefficients.

The standardized canonical discriminant function coefficients are the coefficients that apply to the discriminating variables after they have been standardized by the pooled within-group covariance. These coefficients are appropriate for interpreting the importance and relationship of the discriminating variables within the discriminant functions. See example 5 for a graph, known as a loading plot, that plots these standardized coefficients.

The total-sample standardized canonical discriminant function coefficients are the coefficients that apply to the discriminating variables after they have been standardized by the total-sample covariance. See *Methods and formulas* of [MV] **discrim lda** for references discussing which of within-group and total-sample standardization is most appropriate.

For both styles of standardization, variable y1 has small (in absolute value) coefficients for the first three discriminant functions. This indicates that y1 does not play an important part in these discriminant functions. Because the fourth discriminant function accounts for such a small percentage of the variance, we ignore the coefficients from the fourth function when assessing the importance of the variables.

Some sources, see Huberty (1994), advocate the interpretation of structure coefficients, which measure the correlation between the discriminating variables and the discriminant functions, instead of standardized discriminant function coefficients; see the discussion in example 1 of [MV] **discrim lda** for references to this dispute. The `estat structure` command displays structure coefficients.

```
. estat structure, format(%9.6f)
Canonical structure
```

	function1	function2	function3	function4
y1	-0.089595	-0.261416	0.820783	0.499949
y2	-0.086765	-0.431180	0.898063	0.006158
y3	-0.836986	-0.281362	0.457902	-0.103031
y4	-0.793621	-0.572890	0.162901	-0.124206

Using structure coefficients for interpretation, we conclude that y1 is important for the second and third discriminant functions.

◁

▷ Example 4

Switching from Fisher's approach to LDA to Mahalanobis's approach to LDA, we examine what are called classification functions with the `estat classfunctions` command. Classification functions are applied to the unstandardized discriminating variables. The classification function that results in the largest value for an observation indicates the group to assign the observation.

Continuing with the rootstock LDA, we specify the `format()` option to control the display format of the classification coefficients.

```
. estat classfunctions, format(%8.3f)
Classification functions
```

	rootstock 1	2	3	4	5	6
y1	314.640	317.120	324.589	307.260	316.767	311.301
y2	-59.417	-63.981	-65.152	-59.373	-65.826	-63.060
y3	149.610	168.161	154.910	147.652	168.221	160.622
y4	-161.178	-172.644	-150.356	-153.387	-172.851	-175.477
_cons	-301.590	-354.769	-330.103	-293.427	-349.847	-318.099
Priors	0.200	0.200	0.200	0.200	0.100	0.100

The prior probabilities, used in constructing the coefficient for the constant term, are displayed as the last row in the table. We did not specify the `priors()` option, so the prior probabilities defaulted to those in our LDA model, which has rootstock group 5 and 6 with prior probabilities of 0.1, whereas the other groups have prior probabilities of 0.2.

See example 10 for applying the classification function to data by using the `predict` command.

◁

Scree, loading, and score plots

Examples of discriminant function loading plots and score plots (see [MV] **scoreplot**) can be found in example 3 of [MV] **discrim lda** and example 1 of [MV] **candisc**. Also available after `discrim lda` are scree plots; see [MV] **screeplot**.

▷ Example 5

Continuing with our rootstock example, the scree plot of the four nonzero eigenvalues we previously saw in the output of `estat canontest` in example 2 are graphed using the `screeplot` command.

```
. screeplot
```

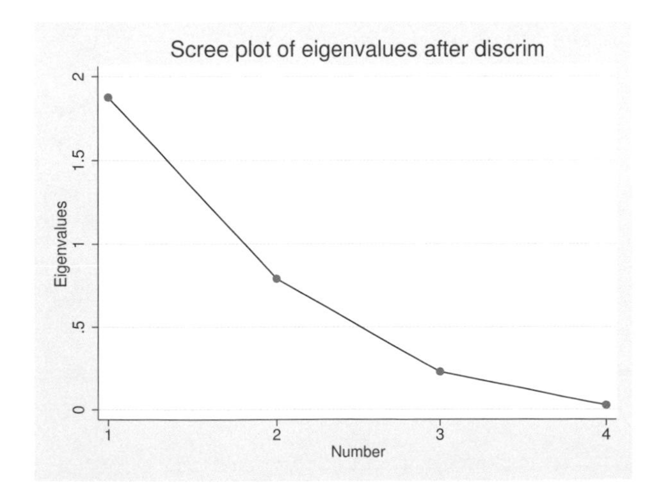

The *Remarks* in [MV] **screeplot** concerning the use of scree plots for selecting the number of components in the context of `pca` apply also for selecting the number of discriminant functions after `discrim lda`. With these four eigenvalues, it is not obvious whether to choose the top two or three eigenvalues. From the `estat canontest` output of example 2, the first two discriminant functions account for 91% of the variance, and three discriminant functions account for 99% of the variance.

The `loadingplot` command (see [MV] **scoreplot**) allows us to graph the pooled within-group standardized discriminant coefficients (loadings) that we saw in tabular form from the `estat loadings` command of example 3. By default only the loadings from the first two functions are graphed. We override this setting with the `components(3)` option, obtaining graphs of the first versus second, first versus third, and second versus third function loadings. The `combined` option switches from a matrix graph to a combined graph. The `msymbol(i)` option removes the plotting points, leaving the discriminating variable names in the graph, and the option `mlabpos(0)` places the discriminating variable names in the positions of the plotted points.

```
. loadingplot, components(3) combined msymbol(i) mlabpos(0)
```

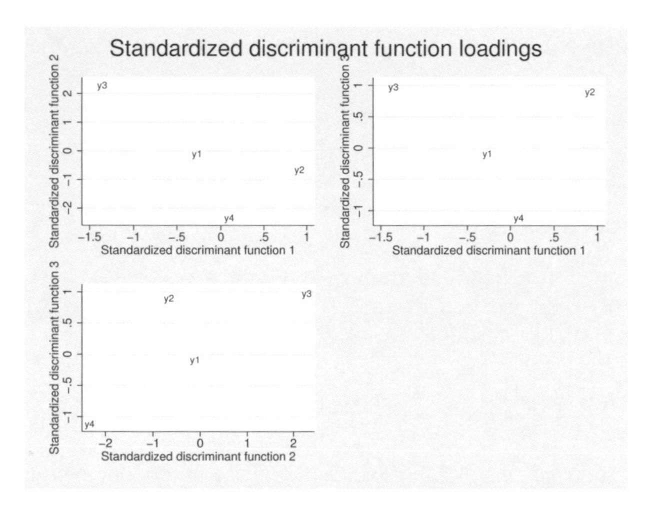

Variable `y1`, trunk girth at 4 years, is near the origin in all three graphs, indicating that it does not play a strong role in discriminating among our six rootstock groups. `y4`, weight of tree above ground at 15 years, does not play much of a role in the first discriminant function but does in the second and third discriminant functions.

The corresponding three score plots are easily produced with the `scoreplot` command; see [MV] **scoreplot**. Score plots graph the discriminant function–transformed observations (called scores).

```
. scoreplot, components(3) combined msymbol(i)
```

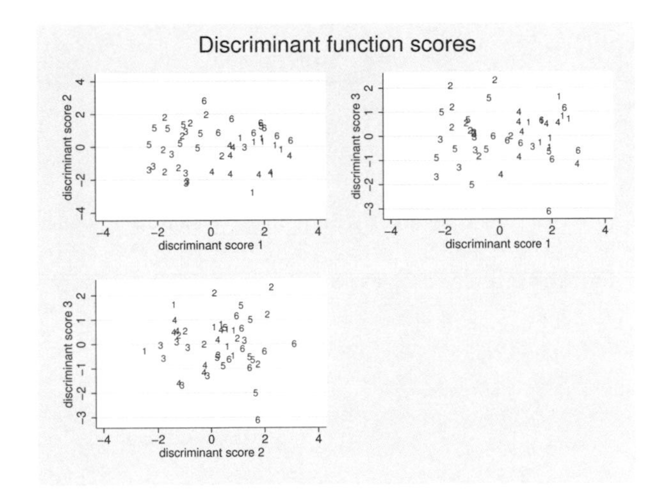

There is a lot of overlap, but some separation of the rootstock groups is apparent. One of the observations from group 6 seems to be sitting by itself in the bottom of the two graphs that have discriminant function 3 as the y axis. In example 11, we identify this point by using the `predict` command.

◁

Means and distances

The `estat grsummarize` command is available after all `discrim` commands and will display means, medians, minimums, maximums, standard deviations, group sizes, and more for the groups; see [MV] **discrim estat**. After `discrim lda`, the `estat grmeans` command will also display group means. It, however, has options for displaying the within-group standardized group means, the total-sample standardized group means, and the canonical discriminant functions evaluated at the group means.

▷ Example 6

We introduce the `estat grmeans` command with the iris data originally from Anderson (1935), introduced in example 3 of [MV] **discrim lda**.

```
. use http://www.stata-press.com/data/r12/iris
(Iris data)
. discrim lda seplen sepwid petlen petwid, group(iris) notable
```

The `notable` option of `discrim` suppressed the classification table.

By default, `estat grmeans` displays a table of the means of the discriminating variables for each group. You could obtain the same information along with other statistics with the `estat grsummarize` command; see [MV] **discrim estat**.

```
. estat grmeans
Group means

             |  iris
             |      Setosa  Versico~r  Virginica
-------------+----------------------------------
      seplen |       5.006      5.936      6.588
      sepwid |       3.428       2.77      2.974
      petlen |       1.462       4.26      5.552
      petwid |        .246      1.326      2.026
```

Differences in the iris species can be seen within these means. For instance, the petal lengths and widths of the *Iris setosa* are smaller than those of the other two species. See example 1 of [MV] **discrim estat** for further exploration of these differences.

The main purpose of `estat grmeans` is to present standardized or transformed means. The `totalstd` and `withinstd` options request the two available standardizations.

```
. estat grmeans, totalstd withinstd
Total-sample standardized group means

             |  iris
             |      Setosa  Versico~r  Virginica
-------------+----------------------------------
      seplen |   -1.011191   .1119073   .8992841
      sepwid |    .8504137  -.6592236  -.1911901
      petlen |    -1.30063   .2843712   1.016259
      petwid |   -1.250704   .1661774   1.084526
Pooled within-group standardized group means

             |  iris
             |      Setosa  Versico~r  Virginica
-------------+----------------------------------
      seplen |   -1.626555   .1800089   1.446546
      sepwid |    1.091198  -.8458749  -.2453234
      petlen |   -5.335385   1.166534    4.16885
      petwid |   -4.658359   .6189428   4.039416
```

The first table presents the total-sample standardized group means on the discriminating variables. These are the means for each group on the total-sample standardized discriminating variables.

The second table presents the pooled within-group standardized means on the discriminating variables. Instead of using the total-sample variance, the pooled within-group variance is used to standardize the variables. Of most interest in the context of an LDA is the within-group standardization.

The `canonical` option of `estat grmeans` displays the discriminant functions evaluated at the group means and gives insight into what the functions do to the groups.

```
. estat grmeans, canonical
Group means on canonical variables

        iris |  function1  function2
-------------+----------------------
      Setosa |    -7.6076   -.215133
  Versicolor |   1.825049   .7278996
   Virginica |    5.78255  -.5127666
```

The first function places *Iris setosa* strongly negative and *Iris virginica* strongly positive with *Iris versicolor* in between. The second function places *Iris virginica* and *Iris setosa* negative and *Iris versicolor* positive.

◁

The Mahalanobis distance between the groups in an LDA helps in assessing which groups are similar and which are different.

▷ Example 7

Continuing with the iris example, we use the `estat grdistances` command to view the squared Mahalanobis distances between the three iris species.

```
. estat grdistances
Mahalanobis squared distances between groups

             |  iris
        iris |     Setosa  Versicolor   Virginica
-------------+------------------------------------
      Setosa |          0
  Versicolor |  89.864186           0
   Virginica |  179.38471   17.201066           0
```

Iris setosa is farthest from *Iris virginica* with a squared Mahalanobis distance of 179. *Iris versicolor* and *Iris virginica* are closest with a squared Mahalanobis distance of 17.

Are these distances significant? Requesting F statistics and p-values associated with these Mahalanobis squared distances between means will help answer that question. The `mahalanobis()` option requests F tests, p-values, or both.

```
. estat grdistances, mahalanobis(f p)
Mahalanobis squared distances between groups

  +-----------------------------+
  | Key                         |
  |-----------------------------|
  | Mahalanobis squared distance|
  | F with 4 and 144 df         |
  | p-value                     |
  +-----------------------------+

             |  iris
        iris |     Setosa  Versicolor   Virginica
-------------+------------------------------------
      Setosa |          0
             |          0
             |          1
             |
  Versicolor |  89.864186           0
             |  550.18889           0
             |  3.902e-86           1
             |
   Virginica |  179.38471   17.201066           0
             |  1098.2738   105.31265           0
             | 9.20e-107   9.515e-42            1
```

All three of the means are statistically significantly different from one another.

The generalized squared distance between groups starts with the Mahalanobis squared distance between groups and adjusts for prior probabilities when they are not equal. With equal prior probabilities there will be no difference between the generalized squared distance and Mahalanobis squared distance. The `priors()` option specifies the prior probabilities for calculating the generalized squared distances.

To illustrate, we select prior probabilities of 0.2 for *I. setosa*, 0.3 for *I. versicolor*, and 0.5 for *I. virginica*.

```
. estat grdistances, generalized priors(.2, .3, .5)
Generalized squared distances between groups
```

iris	iris Setosa	Versicolor	Virginica
Setosa	3.2188758	92.272131	180.77101
Versicolor	93.083061	2.4079456	18.587361
Virginica	182.60359	19.609012	1.3862944

This matrix is not symmetric and does not have zeros on the diagonal.

◁

Covariance and correlation matrices

Equal group covariance matrices is an important assumption underlying LDA. The `estat covariance` command displays the group covariance matrices, the pooled within-group covariance matrix, the between-groups covariance matrix, and the total-sample covariance matrix. The `estat correlation` command provides the corresponding correlation matrices, with an option to present p-values with the correlations.

▷ Example 8

Continuing our examination of LDA on the iris data, we request to see the pooled within-group covariance matrix and the covariance matrices for the three iris species.

```
. estat covariance, within groups
Pooled within-group covariance matrix
```

	seplen	sepwid	petlen	petwid
seplen	.2650082			
sepwid	.0927211	.1153878		
petlen	.1675143	.0552435	.1851878	
petwid	.0384014	.0327102	.0426653	.0418816

Group covariance matrices

iris: Setosa

	seplen	sepwid	petlen	petwid
seplen	.124249			
sepwid	.0992163	.1436898		
petlen	.0163551	.011698	.0301592	
petwid	.0103306	.009298	.0060694	.0111061

iris: Versicolor

	seplen	sepwid	petlen	petwid
seplen	.2664327			
sepwid	.0851837	.0984694		
petlen	.182898	.0826531	.2208163	
petwid	.0557796	.0412041	.073102	.0391061

iris: Virginica

	seplen	sepwid	petlen	petwid
seplen	.4043429			
sepwid	.0937633	.1040041		
petlen	.3032898	.0713796	.3045878	
petwid	.0490939	.0476286	.0488245	.0754327

All variables have positive covariance—not surprising for physical measurements (length and width).

We could have requested the between-groups covariance matrix and the total-sample covariance matrix. Options of `estat covariance` control how the covariance matrices are displayed.

Correlation matrices are also easily displayed. With `estat correlations` we show the pooled within-group correlation matrix, and add the p option to request display of p-values with the correlations. The p-values help us evaluate whether the correlations are statistically significant.

```
. estat corr, p
Pooled within-group correlation matrix

  +-------------------+
  | Key               |
  |-------------------|
  | Correlation       |
  | Two-sided p-value |
  +-------------------+

             |    seplen     sepwid     petlen     petwid
-------------+--------------------------------------------
      seplen |   1.00000
             |
             |
      sepwid |   0.53024    1.00000
             |   0.00000
             |
      petlen |   0.75616    0.37792    1.00000
             |   0.00000    0.00000
             |
      petwid |   0.36451    0.47053    0.48446    1.00000
             |   0.00001    0.00000    0.00000
```

All correlations are statistically significant. The largest correlation is between the petal length and the sepal length.

◁

Predictions

The `predict` command after `discrim lda` has options for obtaining classifications, probabilities, Mahalanobis squared distances from observations to group means, and the leave-one-out (LOO) estimates of all of these. You can also obtain the discriminant scores and classification scores for observations. The predictions can be obtained in or out of sample.

▷ Example 9

We use the riding-mower data from Johnson and Wichern (2007) introduced in example 1 of [MV] **discrim** to illustrate out-of-sample prediction of classification and probabilities after an LDA.

```
. use http://www.stata-press.com/data/r12/lawnmower2
(Johnson and Wichern (2007) Table 11.1)
. discrim lda lotsize income, group(owner) notable
```

Now we see how the LDA model classifies observations with income of $90,000, $110,000, and $130,000, each with a lot size of 20,000 square feet. We add 3 observations to the bottom of our dataset containing these values and then use `predict` to obtain the classifications and probabilities.

```
. input

          owner   income   lots~e
 25. .   90 20
 26. .  110 20
 27. .  130 20
 28. end

. predict grp in 25/L, class
(24 missing values generated)

. predict pr* in 25/L, pr
(24 missing values generated)

. list in 25/L
```

	owner	income	lotsize	grp	pr1	pr2
25.	.	90.0	20.0	0	.5053121	.4946879
26.	.	110.0	20.0	1	.1209615	.8790385
27.	.	130.0	20.0	1	.0182001	.9818

The observation with income of $90,000 was classified as a nonowner, but it was a close decision with probabilities of 0.505 for nonowner and 0.495 for owner. The two other observations, with $110,000 and $130,000 income, were classified as owners, with higher probability of ownership for the higher income.

◁

The `estat list`, `estat classtable`, and `estat errorrate` commands (see [MV] **discrim estat**) obtain their information by calling `predict`. The LOO listings and tables from these commands are obtained by calling `predict` with the `looclass` and `loopr` options.

In addition to predictions and probabilities, we can obtain the classification scores for observations.

▷ Example 10

In example 4, we used the `estat classfunctions` command to view the classification functions for the LDA of the apple tree rootstock data. We can use `predict` to obtain the corresponding classification scores—the classification function applied to observations.

```
. use http://www.stata-press.com/data/r12/rootstock, clear
(Table 6.2 Rootstock Data -- Rencher (2002))

. discrim lda y1 y2 y3 y4, group(rootstock) priors(.2,.2,.2,.2,.1,.1) notable

. predict clscr*, clscore

. format clscr* %6.1f

. list rootstock clscr* in 1/3, noobs
```

rootst~k	clscr1	clscr2	clscr3	clscr4	clscr5	clscr6
1	308.1	303.7	303.1	307.1	303.5	307.1
1	327.6	324.1	322.9	326.1	323.3	326.0
1	309.5	308.2	306.3	309.3	307.5	309.0

We did not specify the `priors()` option, so `predict` used the prior probabilities that were specified with our LDA model in determining the constant term in the classification function; see example 4 for a table of the classification functions. Observations may be classified to the group with largest score. The first 3 observations belong to rootstock group 1 and are successfully classified as belonging to group 1 because the classification score in `clscr1` is larger than the classification scores for the other groups.

◁

Scoring the discriminating variables by using Fisher's canonical discriminant functions is accomplished with the `dscore` option of `predict`.

▷ Example 11

Using the rootstock data in example 5, we noticed 1 observation, from group 6, near the bottom of the score plot where the third discriminant function was the y axis. The observation has a score for the third discriminant function that appears to be below −3. We will use the `dscore` option of `predict` to find the observation.

```
. predict ds*, dscore
. format ds* %5.0g
. list rootstock y* ds* if ds3 < -3
```

	rootst~k	y1	y2	y3	y4	ds1	ds2	ds3	ds4
42.	6	0.75	0.840	3.14	0.606	1.59	1.44	-3.11	-1.93

Observation 42 is the one producing that third discriminant score.

◁

Saved results

`estat anova` saves the following in `r()`:

Scalars

`r(N)`	number of observations
`r(df_m)`	model degrees of freedom
`r(df_r)`	residual degrees of freedom

Matrices

`r(anova_stats)`	ANOVA statistics for the model

`estat canontest` saves the following in `r()`:

Scalars

`r(N)`	number of observations
`r(N_groups)`	number of groups
`r(k)`	number of variables
`r(f)`	number of canonical discriminant functions

Matrices

`r(stat)`	canonical discriminant statistics

`estat classfunction` saves the following in `r()`:

Matrices

`r(C)`	classification function matrix
`r(priors)`	group prior probabilities

`estat correlations` saves the following in `r()`:

Matrices

`r(Rho)`	pooled within-group correlation matrix (`within` only)
`r(P)`	two-sided p-values for pooled within-group correlations (`within` and `p` only)
`r(Rho_between)`	between-groups correlation matrix (`between` only)
`r(P_between)`	two-sided p-values for between-groups correlations (`between` and `p` only)
`r(Rho_total)`	total-sample correlation matrix (`total` only)
`r(P_total)`	two-sided p-values for total-sample correlations (`total` and `p` only)
`r(Rho_#)`	group # correlation matrix (`groups` only)
`r(P_#)`	two-sided p-values for group # correlations (`groups` and `p` only)

`estat covariance` saves the following in `r()`:

Matrices

`r(S)`	pooled within-group covariance matrix (`within` only)
`r(S_between)`	between-groups covariance matrix (`between` only)
`r(S_total)`	total-sample covariance matrix (`total` only)
`r(S_#)`	group # covariance matrix (`groups` only)

`estat grdistances` saves the following in `r()`:

Scalars

`r(df1)`	numerator degrees of freedom (`mahalanobis` only)
`r(df2)`	denominator degrees of freedom (`mahalanobis` only)

Matrices

`r(sqdist)`	Mahalanobis squared distances between group means (`mahalanobis` only)
`r(F_sqdist)`	F statistics for tests that the Mahalanobis squared distances between group means are zero (`mahalanobis` only)
`r(P_sqdist)`	p-value for tests that the Mahalanobis squared distances between group means are zero (`mahalanobis` only)
`r(gsqdist)`	generalized squared distances between group means (`generalized` only)

`estat grmeans` saves the following in `r()`:

Matrices

`r(means)`	group means (`raw` only)
`r(stdmeans)`	total-sample standardized group means (`totalstd` only)
`r(wstdmeans)`	pooled within-group standardized group means (`withinstd` only)
`r(cmeans)`	group means on canonical variables (`canonical` only)

`estat loadings` saves the following in `r()`:

Matrices

`r(L_std)`	Within-group standardized canonical discriminant function coefficients (`standardized` only)
`r(L_totalstd)`	total-sample standardized canonical discriminant function coefficients (`totalstandardized` only)
`r(L_unstd)`	unstandardized canonical discriminant function coefficients (`unstandardized` only)

`estat manova` saves the following in `r()`:

Scalars

`r(N)`	number of observations
`r(df_m)`	model degrees of freedom
`r(df_r)`	residual degrees of freedom

Matrices

`r(stat_m)`	multivariate statistics for the model

`estat structure` saves the following in `r()`:

Matrices
`r(canstruct)` canonical structure matrix

Methods and formulas

All postestimation commands listed above are implemented as ado-files.

See *Methods and formulas* of [MV] **discrim lda** for background on what is produced by `predict`, `estat classfunctions`, `estat grdistances`, `estat grmeans`, `estat loadings`, and `estat structure`. See [MV] **discrim estat** for more information on `estat classtable`, `estat errorrate`, `estat grsummarize`, and `estat list`. See [R] **anova** for background information on the ANOVAs summarized by `estat anova`; see [MV] **manova** for information on the MANOVA shown by `estat manova`; and see [MV] **canon** for background information on canonical correlations and related tests shown by `estat canontest`.

References

Anderson, E. 1935. The irises of the Gaspe Peninsula. *Bulletin of the American Iris Society* 59: 2–5.

Andrews, D. F., and A. M. Herzberg, ed. 1985. *Data: A Collection of Problems from Many Fields for the Student and Research Worker.* New York: Springer.

Huberty, C. J. 1994. *Applied Discriminant Analysis.* New York: Wiley.

Johnson, R. A., and D. W. Wichern. 2007. *Applied Multivariate Statistical Analysis.* 6th ed. Englewood Cliffs, NJ: Prentice Hall.

Rencher, A. C. 2002. *Methods of Multivariate Analysis.* 2nd ed. New York: Wiley.

Also see

[MV] **screeplot** — Scree plot

[MV] **scoreplot** — Score and loading plots

[MV] **discrim estat** — Postestimation tools for discrim

[MV] **discrim lda** — Linear discriminant analysis

[MV] **candisc** — Canonical linear discriminant analysis

[MV] **canon** — Canonical correlations

[MV] **manova** — Multivariate analysis of variance and covariance

[U] **20 Estimation and postestimation commands**

Title

discrim logistic — Logistic discriminant analysis

Syntax

discrim logistic *varlist* [*if*] [*in*] [*weight*], group(*groupvar*) [*options*]

options	Description
Model	
* group(*groupvar*)	variable specifying the groups
priors(*priors*)	group prior probabilities
ties(*ties*)	how ties in classification are to be handled
Reporting	
notable	suppress resubstitution classification table
nolog	suppress the mlogit log-likelihood iteration log

priors	Description
equal	equal prior probabilities; the default
proportional	group-size-proportional prior probabilities
matname	row or column vector containing the group prior probabilities
matrix_exp	matrix expression providing a row or column vector of the group prior probabilities

ties	Description
missing	ties in group classification produce missing values; the default
random	ties in group classification are broken randomly
first	ties in group classification are set to the first tied group

*group() is required.

statsby and xi are allowed; see [U] **11.1.10 Prefix commands**.

fweights are allowed; see [U] **11.1.6 weight**.

See [U] **20 Estimation and postestimation commands** for more capabilities of estimation commands.

Menu

Statistics > Multivariate analysis > Discriminant analysis > Logistic

Description

discrim logistic performs logistic discriminant analysis. See [MV] **discrim** for other discrimination commands.

Options

Main

`group(`*groupvar*`)` is required and specifies the name of the grouping variable. *groupvar* must be a numeric variable.

`priors(`*priors*`)` specifies the prior probabilities for group membership. The following *priors* are allowed:

`priors(equal)` specifies equal prior probabilities. This is the default.

`priors(proportional)` specifies group-size-proportional prior probabilities.

`priors(`*matname*`)` specifies a row or column vector containing the group prior probabilities.

`priors(`*matrix_exp*`)` specifies a matrix expression providing a row or column vector of the group prior probabilities.

`ties(`*ties*`)` specifies how ties in group classification will be handled. The following *ties* are allowed:

`ties(missing)` specifies that ties in group classification produce missing values. This is the default.

`ties(random)` specifies that ties in group classification are broken randomly.

`ties(first)` specifies that ties in group classification are set to the first tied group.

Reporting

`notable` suppresses the computation and display of the resubstitution classification table.

`nolog` suppress the `mlogit` log-likelihood iteration log.

Remarks

Albert and Lesaffre (1986) explain that logistic discriminant analysis is a partially parametric method falling between parametric discrimination methods such as LDA and QDA (see [MV] **discrim lda** and [MV] **discrim qda**) and nonparametric discrimination methods such as kth-nearest-neighbor (KNN) discrimination (see [MV] **discrim knn**). Albert and Harris (1987) provide a good explanation of logistic discriminant analysis. Instead of making assumptions about the distribution of the data within each group, logistic discriminant analysis is based on the assumption that the likelihood ratios of the groups have an exponential form; see *Methods and formulas*. Multinomial logistic regression provides the basis for logistic discriminant analysis; see [R] **mlogit**. Multinomial logistic regression can handle binary and continuous regressors, and hence logistic discriminant analysis is also appropriate for binary and continuous discriminating variables.

▷ Example 1

Morrison (2005, 443–445) provides data on 12 subjects with a senile-factor diagnosis and 37 subjects with a no-senile-factor diagnosis. The data consist of the Wechsler Adult Intelligence Scale (WAIS) subtest scores for information, similarities, arithmetic, and picture completion. Morrison (2005, 231) performs a logistic discriminant analysis on the two groups, using the similarities and picture completion scores as the discriminating variables.

```
. use http://www.stata-press.com/data/r12/senile
(Senility WAIS subtest scores)
```

```
. discrim logistic sim pc, group(sf) priors(proportional)
Iteration 0:   log likelihood = -27.276352
Iteration 1:   log likelihood = -19.531198
Iteration 2:   log likelihood = -19.036702
Iteration 3:   log likelihood = -19.018973
Iteration 4:   log likelihood = -19.018928
Logistic discriminant analysis
Resubstitution classification summary

  +---------+
  | Key     |
  |---------|
  | Number  |
  | Percent |
  +---------+

              | Classified
 True sf      |   No-SF        SF |   Total
--------------+-------------------+--------
        No-SF |      37         0 |      37
              |  100.00      0.00 |  100.00
              |                   |
           SF |       6         6 |      12
              |   50.00     50.00 |  100.00
--------------+-------------------+--------
        Total |      43         6 |      49
              |   87.76     12.24 |  100.00
              |                   |
       Priors |  0.7551    0.2449 |
```

We specified the `priors(proportional)` option to obtain proportional prior probabilities for our logistic classification. These results match those of Morrison (2005, 231), though he does not state that his results are based on proportional prior probabilities. If you change to equal prior probabilities you obtain different classification results.

Which observations were misclassified? `estat list` with the `misclassified` option shows the six misclassified observations and the estimated probabilities.

```
. estat list, misclassified varlist

  +----------------------------------------------------------+
  |      | Data      | Classification   | Probabilities      |
  |      |           |                  |                    |
  | Obs. |  sim   pc |   True   Class.  |   No-SF      SF    |
  |------+-----------+------------------+--------------------|
  |  38  |    5    8 |     SF   No-SF * |  0.7353  0.2647    |
  |  41  |    7    9 |     SF   No-SF * |  0.8677  0.1323    |
  |  44  |    9    8 |     SF   No-SF * |  0.8763  0.1237    |
  |  46  |    7    6 |     SF   No-SF * |  0.6697  0.3303    |
  |  48  |   10    3 |     SF   No-SF * |  0.5584  0.4416    |
  |------+-----------+------------------+--------------------|
  |  49  |   12   10 |     SF   No-SF * |  0.9690  0.0310    |
  +----------------------------------------------------------+

* indicates misclassified observations
```

See example 1 of [MV] **discrim logistic postestimation** for more postestimation analysis with this logistic discriminant analysis.

◁

▷ Example 2

Example 2 of [MV] **discrim knn** introduces a head measurement dataset with six discriminating variables and three groups; see Rencher (2002, 279–281). We now apply `discrim logistic` to see how well the logistic model can discriminate between the groups.

```
. use http://www.stata-press.com/data/r12/head
(Table 8.3 Head measurements -- Rencher (2002))
. discrim logistic wdim circum fbeye eyehd earhd jaw, group(group)
Iteration 0:   log likelihood = -98.875106
Iteration 1:   log likelihood = -60.790737
Iteration 2:   log likelihood = -53.746934
Iteration 3:   log likelihood = -51.114631
Iteration 4:   log likelihood = -50.249426
Iteration 5:   log likelihood = -50.081199
Iteration 6:   log likelihood = -50.072248
Iteration 7:   log likelihood = -50.072216
Logistic discriminant analysis
Resubstitution classification summary
```

Key
Number Percent

True group	Classified high school	college	nonplayer	Total
high school	27 90.00	2 6.67	1 3.33	30 100.00
college	1 3.33	20 66.67	9 30.00	30 100.00
nonplayer	2 6.67	8 26.67	20 66.67	30 100.00
Total	30 33.33	30 33.33	30 33.33	90 100.00
Priors	0.3333	0.3333	0.3333	

The counts on the diagonal of the resubstitution classification table are similar to those obtained by `discrim knn` (see example 2 of [MV] **discrim knn**) and `discrim lda` (see example 1 of [MV] **candisc**), whereas `discrim qda` seems to have classified the nonplayer group more accurately (see example 3 of [MV] **discrim estat**).

◁

Saved results

discrim logistic saves the following in e():

Scalars	
e(N)	number of observations
e(N_groups)	number of groups
e(k)	number of discriminating variables
e(ibaseout)	base outcome number
Macros	
e(cmd)	discrim
e(subcmd)	logistic
e(cmdline)	command as typed
e(groupvar)	name of group variable
e(grouplabels)	labels for the groups
e(varlist)	discriminating variables
e(dropped)	variables dropped because of collinearity
e(wtype)	weight type
e(wexp)	weight expression
e(title)	title in estimation output
e(ties)	how ties are to be handled
e(properties)	b noV
e(estat_cmd)	program used to implement estat
e(predict)	program used to implement predict
e(marginsnotok)	predictions disallowed by margins
Matrices	
e(b)	coefficient vector
e(groupcounts)	number of observations for each group
e(grouppriors)	prior probabilities for each group
e(groupvalues)	numeric value for each group
Functions	
e(sample)	marks estimation sample

Methods and formulas

discrim logistic is implemented as an ado-file.

Let g be the number of groups, n_i the number of observations for group i, and q_i the prior probability for group i. Let $\mathbf{x}$ denote an observation measured on p discriminating variables. For consistency with the discriminant analysis literature, $\mathbf{x}$ will be a column vector, though it corresponds to a row in your dataset. Let $f_i(\mathbf{x})$ represent the density function for group i, and let $P(\mathbf{x}|G_i)$ denote the probability of observing $\mathbf{x}$ conditional on belonging to group i. Denote the posterior probability of group i given observation $\mathbf{x}$ as $P(G_i|\mathbf{x})$. With Bayes' theorem, we have

$$P(G_i|\mathbf{x}) = \frac{q_i f_i(\mathbf{x})}{\sum_{j=1}^{g} q_j f_j(\mathbf{x})}$$

Substituting $P(\mathbf{x}|G_i)$ for $f_i(\mathbf{x})$, we have

$$P(G_i|\mathbf{x}) = \frac{q_i P(\mathbf{x}|G_i)}{\sum_{j=1}^{g} q_j P(\mathbf{x}|G_j)}$$

Dividing both the numerator and denominator by $P(\mathbf{x}|G_g)$, we can express this as

$$P(G_i|\mathbf{x}) = \frac{q_i L_{ig}(\mathbf{x})}{\sum_{j=1}^{g} q_j L_{jg}(\mathbf{x})}$$

where $L_{ig}(\mathbf{x}) = P(\mathbf{x}|G_i)/P(\mathbf{x}|G_g)$ is the likelihood ratio of $\mathbf{x}$ for groups i and g.

This formulation of the posterior probability allows easy insertion of the Multinomial logistic model into the discriminant analysis framework. The multinomial logistic model expresses $L_{ig}(\mathbf{x})$ in a simple exponential form

$$L_{ig}(\mathbf{x}) = \exp(a_{0i} + \mathbf{a}_i'\mathbf{x})$$

see Albert and Harris (1987, 117). Logistic discriminant analysis uses `mlogit` to compute the likelihood ratios, $L_{ig}(\mathbf{x})$, and hence the posterior probabilities $P(G_i|\mathbf{x})$; see [R] **mlogit**. However, `mlogit` and `predict` after `mlogit` assume proportional prior probabilities. `discrim logistic` assumes equal prior probabilities unless you specify the `priors(proportional)` option.

References

Albert, A., and E. K. Harris. 1987. *Multivariate Interpretation of Clinical Laboratory Data.* New York: Marcel Dekker.

Albert, A., and E. Lesaffre. 1986. Multiple group logistic discrimination. *Computers and Mathematics with Applications* 12A(2): 209–224.

Morrison, D. F. 2005. *Multivariate Statistical Methods.* 4th ed. Belmont, CA: Duxbury.

Rencher, A. C. 2002. *Methods of Multivariate Analysis.* 2nd ed. New York: Wiley.

Also see

[MV] **discrim logistic postestimation** — Postestimation tools for discrim logistic

[MV] **discrim** — Discriminant analysis

[R] **logistic** — Logistic regression, reporting odds ratios

[R] **mlogit** — Multinomial (polytomous) logistic regression

[U] **20 Estimation and postestimation commands**

Title

discrim logistic postestimation — Postestimation tools for discrim logistic

Description

The following postestimation commands are of special interest after `discrim logistic`:

Command	Description
`estat classtable`	classification table
`estat errorrate`	classification error-rate estimation
`estat grsummarize`	group summaries
`estat list`	classification listing
`estat summarize`	estimation sample summary

For information about these commands, see [MV] **discrim estat**.

The following standard postestimation commands are also available:

Command	Description
* `estimates`	cataloging estimation results
`predict`	group classification and posterior probabilities

* All `estimates` subcommands except `table` and `stats` are available; see [R] **estimates**.

Syntax for predict

`predict` [*type*] *newvar* [*if*] [*in*] [, *statistic options*]

`predict` [*type*] {*stub** | *newvarlist*} [*if*] [*in*] [, *statistic options*]

statistic	Description
Main	
`classification`	group membership classification; the default when one variable is specified and `group()` is not specified
`pr`	probability of group membership; the default when `group()` is specified or when multiple variables are specified

options	Description
Main	
`group(`*group*`)`	the group for which the statistic is to be calculated
Options	
`priors(`*priors*`)`	group prior probabilities; defaults to `e(grouppriors)`
`ties(`*ties*`)`	how ties in classification are to be handled; defaults to `e(ties)`

priors	Description
equal	equal prior probabilities
proportional	group-size-proportional prior probabilities
matname	row or column vector containing the group prior probabilities
matrix_exp	matrix expression providing a row or column vector of the group prior probabilities

ties	Description
missing	ties in group classification produce missing values
random	ties in group classification are broken randomly
first	ties in group classification are set to the first tied group

You specify one new variable with classification and specify either one or e(N_groups) new variables with pr.
group() is not allowed with classification.

Menu

Statistics > Postestimation > Predictions, residuals, etc.

Options for predict

Main

classification, the default, calculates the group classification. Only one new variable may be specified.

pr calculates group membership posterior probabilities. If you specify the group() option, specify one new variable. Otherwise, you must specify e(N_groups) new variables.

group(*group*) specifies the group for which the statistic is to be calculated and can be specified using

#1, #2, ..., where #1 means the first category of the e(groupvar) variable, #2 the second category, etc.;

the values of the e(groupvar) variable; or

the value labels of the e(groupvar) variable if they exist.

group() is not allowed with classification.

Options

priors(*priors*) specifies the prior probabilities for group membership. If priors() is not specified, e(grouppriors) is used. The following *priors* are allowed:

priors(equal) specifies equal prior probabilities.

priors(proportional) specifies group-size-proportional prior probabilities.

priors(*matname*) specifies a row or column vector containing the group prior probabilities.

priors(*matrix_exp*) specifies a matrix expression providing a row or column vector of the group prior probabilities.

`ties(ties)` specifies how ties in group classification will be handled. If `ties()` is not specified, `e(ties)` is used. The following *ties* are allowed:

`ties(missing)` specifies that ties in group classification produce missing values.

`ties(random)` specifies that ties in group classification are broken randomly.

`ties(first)` specifies that ties in group classification are set to the first tied group.

Remarks

Classifications and probabilities after `discrim logistic` are obtained with the `predict` command. The common `estat` subcommands after `discrim` are also available for producing classification tables, error-rate tables, classification listings, and group summaries; see [MV] **discrim estat**.

▷ Example 1

Continuing with our logistic discriminant analysis of the senility dataset of Morrison (2005), introduced in example 1 of [MV] **discrim logistic**, we illustrate the use of the `estat errorrate` postestimation command.

```
. use http://www.stata-press.com/data/r12/senile
(Senility WAIS subtest scores)
. discrim logistic sim pc, group(sf) priors(proportional) notable nolog
. estat errorrate, pp
Error rate estimated from posterior probabilities
```

Error Rate	sf No-SF	SF	Total
Stratified	.0305051	.5940575	.168518
Unstratified	.0305051	.5940575	.168518
Priors	.755102	.244898	

We specified the `pp` option to obtain the posterior probability–based error-rate estimates. The stratified and unstratified estimates are identical because proportional priors were used. The estimates were based on proportional priors because the logistic discriminant analysis model used proportional priors and we did not specify the `priors()` option in our call to `estat errorrate.`.

The error-rate estimate for the senile-factor group is much higher than for the no-senile-factor group.

What error-rate estimates would we obtain with equal group priors?

```
. estat errorrate, pp priors(equal)
Error rate estimated from posterior probabilities

                 | sf
      Error Rate |       No-SF          SF |      Total
-----------------+-------------------------+-----------
      Stratified |    .2508207    .2069481 |   .2288844
                 |                         |
    Unstratified |      .06308    .4289397 |   .2460098
-----------------+-------------------------+-----------
          Priors |          .5          .5 |
```

Stratified and unstratified estimates are now different. This happens when group sizes have a different proportion from that of the prior probabilities.

Morrison (2005, 231) shows a classification of the subjects where, if the estimated probability of belonging to the senile-factor group is less than 0.35, he classifies the subject to the no-senile-factor group; if the probability is more than 0.66, he classifies the subject to the senile-factor group; and if the probability is between those extremes, he classifies the subject to an uncertain group.

We can use `predict` to implement this same strategy. The `pr` option requests probabilities. Because the model was estimated with proportional prior probabilities, the prediction, by default, will also be based on proportional prior probabilities.

```
. predict prob0 prob1, pr
. gen newgrp = 1
. replace newgrp = 0 if prob1 <= 0.35
(38 real changes made)
. replace newgrp = 2 if prob1 >= 0.66
(5 real changes made)
. label define newgrp 0 "No-SF" 1 "Uncertain" 2 "SF"
. label values newgrp newgrp
. tabulate sf newgrp
Senile-fac |
       tor |               newgrp
 diagnosis |     No-SF  Uncertain         SF |     Total
-----------+---------------------------------+----------
     No-SF |        33          4          0 |        37
        SF |         5          2          5 |        12
-----------+---------------------------------+----------
     Total |        38          6          5 |        49
```

Six observations are placed in the uncertain group.

◁

Reference

Morrison, D. F. 2005. *Multivariate Statistical Methods*. 4th ed. Belmont, CA: Duxbury.

Also see

[MV] **discrim estat** — Postestimation tools for discrim

[MV] **discrim logistic** — Logistic discriminant analysis

[U] **20 Estimation and postestimation commands**

Title

discrim qda — Quadratic discriminant analysis

Syntax

`discrim qda` *varlist* [*if*] [*in*] [*weight*], `group(`*groupvar*`)` [*options*]

options	Description
Model	
* `group(`*groupvar*`)`	variable specifying the groups
`priors(`*priors*`)`	group prior probabilities
`ties(`*ties*`)`	how ties in classification are to be handled
Reporting	
`notable`	suppress resubstitution classification table
`lootable`	display leave-one-out classification table

priors	Description
`equal`	equal prior probabilities; the default
`proportional`	group-size-proportional prior probabilities
matname	row or column vector containing the group prior probabilities
matrix_exp	matrix expression providing a row or column vector of the group prior probabilities

ties	Description
`missing`	ties in group classification produce missing values; the default
`random`	ties in group classification are broken randomly
`first`	ties in group classification are set to the first tied group

* `group()` is required.

`statsby` and `xi` are allowed; see [U] **11.1.10 Prefix commands**.

`fweight`s are allowed; see [U] **11.1.6 weight**.

See [U] **20 Estimation and postestimation commands** for more capabilities of estimation commands.

Menu

Statistics > Multivariate analysis > Discriminant analysis > Quadratic (QDA)

Description

`discrim qda` performs quadratic discriminant analysis. See [MV] **discrim** for other discrimination commands.

Options

Main

`group(`*groupvar*`)` is required and specifies the name of the grouping variable. *groupvar* must be a numeric variable.

`priors(`*priors*`)` specifies the prior probabilities for group membership. The following *priors* are allowed:

`priors(equal)` specifies equal prior probabilities. This is the default.

`priors(proportional)` specifies group-size-proportional prior probabilities.

`priors(`*matname*`)` specifies a row or column vector containing the group prior probabilities.

`priors(`*matrix_exp*`)` specifies a matrix expression providing a row or column vector of the group prior probabilities.

`ties(`*ties*`)` specifies how ties in group classification will be handled. The following *ties* are allowed:

`ties(missing)` specifies that ties in group classification produce missing values. This is the default.

`ties(random)` specifies that ties in group classification are broken randomly.

`ties(first)` specifies that ties in group classification are set to the first tied group.

Reporting

`notable` suppresses the computation and display of the resubstitution classification table.

`lootable` displays the leave-one-out classification table.

Remarks

Quadratic discriminant analysis (QDA) was introduced by Smith (1947). It is a generalization of linear discriminant analysis (LDA). Both LDA and QDA assume that the observations come from a multivariate normal distribution. LDA assumes that the groups have equal covariance matrices. QDA removes this assumption, allowing the groups to have different covariance matrices.

One of the penalties associated with QDA's added flexibility is that if any groups have fewer observations, n_i, than discriminating variables, p, the covariance matrix for that group is singular and QDA cannot be performed. Even if there are enough observations to invert the covariance matrix, if the sample size is relatively small for a group, the estimation of the covariance matrix for that group may not do a good job of representing the group's population covariance, leading to inaccuracies in classification.

▷ Example 1

We illustrate QDA with a small dataset introduced in example 1 of [MV] **manova**. Andrews and Herzberg (1985, 357–360) present data on six apple tree rootstock groups with four measurements on eight trees from each group.

We request the display of the leave-one-out (LOO) classification table in addition to the standard resubstitution classification table produced by `discrim qda`.

```
. use http://www.stata-press.com/data/r12/rootstock
(Table 6.2 Rootstock Data -- Rencher (2002))

. discrim qda y1 y2 y3 y4, group(rootstock) lootable

Quadratic discriminant analysis
Resubstitution classification summary
```

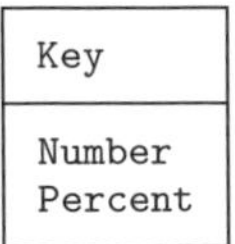

True rootstock	Classified 1	2	3	4	5	6	Total
1	8 100.00	0 0.00	0 0.00	0 0.00	0 0.00	0 0.00	8 100.00
2	0 0.00	7 87.50	0 0.00	1 12.50	0 0.00	0 0.00	8 100.00
3	1 12.50	0 0.00	6 75.00	0 0.00	1 12.50	0 0.00	8 100.00
4	0 0.00	0 0.00	1 12.50	7 87.50	0 0.00	0 0.00	8 100.00
5	0 0.00	3 37.50	0 0.00	0 0.00	4 50.00	1 12.50	8 100.00
6	2 25.00	0 0.00	0 0.00	0 0.00	1 12.50	5 62.50	8 100.00
Total	11 22.92	10 20.83	7 14.58	8 16.67	6 12.50	6 12.50	48 100.00
Priors	0.1667	0.1667	0.1667	0.1667	0.1667	0.1667	

```
Leave-one-out classification summary

  +---------+
  | Key     |
  |---------|
  | Number  |
  | Percent |
  +---------+

True         | Classified
rootstock    |       1       2       3       4       5       6 |   Total
-------------+-------------------------------------------------+--------
           1 |       2       0       0       3       1       2 |       8
             |   25.00    0.00    0.00   37.50   12.50   25.00 |  100.00
             |                                                 |
           2 |       0       3       0       2       2       1 |       8
             |    0.00   37.50    0.00   25.00   25.00   12.50 |  100.00
             |                                                 |
           3 |       1       2       4       0       1       0 |       8
             |   12.50   25.00   50.00    0.00   12.50    0.00 |  100.00
             |                                                 |
           4 |       1       1       3       2       0       1 |       8
             |   12.50   12.50   37.50   25.00    0.00   12.50 |  100.00
             |                                                 |
           5 |       0       4       1       0       2       1 |       8
             |    0.00   50.00   12.50    0.00   25.00   12.50 |  100.00
             |                                                 |
           6 |       3       1       0       0       2       2 |       8
             |   37.50   12.50    0.00    0.00   25.00   25.00 |  100.00
-------------+-------------------------------------------------+--------
       Total |       7      11       8       7       8       7 |      48
             |   14.58   22.92   16.67   14.58   16.67   14.58 |  100.00
             |                                                 |
      Priors |  0.1667  0.1667  0.1667  0.1667  0.1667  0.1667 |
```

Compare the counts on the diagonal of the resubstitution classification table with those on the LOO table. The LOO table has fewer of the observations with correct classifications. The resubstitution classification table is overly optimistic.

The `estat errorrate` postestimation command provides estimates of the error rates for the groups. We request the count-based estimates, first for the resubstitution classification and then for the LOO classification. We also suppress display of the prior probabilities, which will default to equal across the groups because that is how we estimated our QDA model. See [MV] **discrim estat** for details of the `estat errorrate` command.

```
. estat errorrate, nopriors
Error rate estimated by error count

             | rootstock
             |        1          2          3          4          5
-------------+-------------------------------------------------------
  Error rate |        0       .125        .25       .125         .5

             | rootstock
             |        6 |     Total
-------------+----------+----------
  Error rate |     .375 |  .2291667
```

```
. estat errorrate, nopriors looclass
Error rate estimated by leave-one-out error count

             | rootstock
             |       1        2        3        4        5
-------------+---------------------------------------------
  Error rate |     .75     .625       .5      .75      .75

             | rootstock
             |       6 |    Total
-------------+---------+---------
  Error rate |     .75 |    .6875
```

The estimated group error rates are much higher in the LOO table.

See example 2 of [MV] **discrim qda postestimation** for an examination of the squared Mahalanobis distances between the rootstock groups. We could also list the misclassified observations, produce group summaries, examine covariances and correlations, and generate classification and probability variables and more; see [MV] **discrim qda postestimation**.

◁

See example 3 of [MV] **discrim estat** and example 1 of [MV] **discrim qda postestimation** for other examples of the use of `discrim qda`.

Saved results

`discrim qda` saves the following in `e()`:

Scalars

`e(N)`	number of observations
`e(N_groups)`	number of groups
`e(k)`	number of discriminating variables

Macros

`e(cmd)`	`discrim`
`e(subcmd)`	`qda`
`e(cmdline)`	command as typed
`e(groupvar)`	name of group variable
`e(grouplabels)`	labels for the groups
`e(varlist)`	discriminating variables
`e(wtype)`	weight type
`e(wexp)`	weight expression
`e(title)`	title in estimation output
`e(ties)`	how ties are to be handled
`e(properties)`	`nob noV`
`e(estat_cmd)`	program used to implement `estat`
`e(predict)`	program used to implement `predict`
`e(marginsnotok)`	predictions disallowed by `margins`

Matrices

`e(groupcounts)`	number of observations for each group
`e(grouppriors)`	prior probabilities for each group
`e(groupvalues)`	numeric value for each group
`e(means)`	group means on discriminating variables
`e(SSCP_W#)`	within group SSCP matrix for group #
`e(W#_eigvals)`	eigenvalues of `e(SSCP_W#)`
`e(W#_eigvecs)`	eigenvectors of `e(SSCP_W#)`
`e(sqrtS#inv)`	Cholesky (square root) of the inverse covariance matrix for group #

Functions

`e(sample)`	marks estimation sample

Methods and formulas

`discrim qda` is implemented as an ado-file.

Let g be the number of groups, n_i the number of observations for group i, and q_i the prior probability for group i. Let $\mathbf{x}$ denote an observation measured on p discriminating variables. For consistency with the discriminant analysis literature, $\mathbf{x}$ will be a column vector, though it corresponds to a row in your dataset. Let $f_i(\mathbf{x})$ represent the density function for group i, and let $P(\mathbf{x}|G_i)$ denote the probability of observing $\mathbf{x}$ conditional on belonging to group i. Denote the posterior probability of group i given observation $\mathbf{x}$ as $P(G_i|\mathbf{x})$. With Bayes' theorem, we have

$$P(G_i|\mathbf{x}) = \frac{q_i f_i(\mathbf{x})}{\sum_{j=1}^{g} q_j f_j(\mathbf{x})}$$

Substituting $P(\mathbf{x}|G_i)$ for $f_i(\mathbf{x})$, we have

$$P(G_i|\mathbf{x}) = \frac{q_i P(\mathbf{x}|G_i)}{\sum_{j=1}^{g} q_j P(\mathbf{x}|G_j)}$$

QDA assumes that the groups are multivariate normal. Let $\mathbf{S}_i$ denote the within-group sample covariance matrix for group i and $\bar{\mathbf{x}}_i$ denote the sample mean of group i. The squared Mahalanobis distance between observation $\mathbf{x}$ and $\bar{\mathbf{x}}_i$ is

$$D_i^2 = (\mathbf{x} - \bar{\mathbf{x}}_i)'\mathbf{S}_i^{-1}(\mathbf{x} - \bar{\mathbf{x}}_i)$$

Plugging these sample estimates into the multivariate normal density gives

$$P(\mathbf{x}|G_i) = (2\pi)^{-p/2}|\mathbf{S}_i|^{-1/2}e^{-D_i^2/2}$$

Substituting this into the formula for $P(G_i|\mathbf{x})$ and simplifying gives

$$P(G_i|\mathbf{x}) = \frac{q_i|\mathbf{S}_i|^{-1/2}e^{-D_i^2/2}}{\sum_{j=1}^{g} q_j|\mathbf{S}_j|^{-1/2}e^{-D_j^2/2}}$$

as the QDA posterior probability of observation $\mathbf{x}$ belonging to group i.

The squared Mahalanobis distance between group means is produced by `estat grdistances`; see [MV] **discrim qda postestimation**.

Classification functions can be derived from the Mahalanobis QDA; see Huberty (1994, 58). Let $Q_i(\mathbf{x})$ denote the quadratic classification function for the ith group applied to observation $\mathbf{x}$.

$$Q_i(\mathbf{x}) = -D_i^2/2 - \ln|\mathbf{S}_i|/2 + \ln(q_i)$$

An observation can be classified based on largest posterior probability or based on largest quadratic classification function score.

References

Andrews, D. F., and A. M. Herzberg, ed. 1985. *Data: A Collection of Problems from Many Fields for the Student and Research Worker.* New York: Springer.

Huberty, C. J. 1994. *Applied Discriminant Analysis.* New York: Wiley.

Rencher, A. C. 1998. *Multivariate Statistical Inference and Applications.* New York: Wiley.

——. 2002. *Methods of Multivariate Analysis.* 2nd ed. New York: Wiley.

Smith, C. A. B. 1947. Some examples of discrimination. *Annals of Eugenics* 13: 272–282.

Also see

[MV] **discrim qda postestimation** — Postestimation tools for discrim qda

[MV] **discrim** — Discriminant analysis

[U] **20 Estimation and postestimation commands**

Title

discrim qda postestimation — Postestimation tools for discrim qda

Description

The following postestimation commands are of special interest after discrim qda:

Command	Description
estat classtable	classification table
estat correlations	group correlation matrices and p-values
estat covariance	group covariance matrices
estat errorrate	classification error-rate estimation
estat grdistances	Mahalanobis and generalized squared distances between the group means
estat grsummarize	group summaries
estat list	classification listing
estat summarize	estimation sample summary

For information about estat correlations, estat covariance, and estat grdistances, see below. For information about the rest of the commands, see [MV] **discrim estat**.

The following standard postestimation commands are also available:

Command	Description
* estimates	cataloging estimation results
predict	group classification and posterior probabilities

* All estimates subcommands except table and stats are available; see [R] **estimates**.

Special-interest postestimation commands

estat correlations displays group correlation matrices. Two-tailed p-values for the correlations are also available.

estat covariance displays group covariance matrices.

estat grdistances provides Mahalanobis squared distances and generalized squared distances between the group means.

Syntax for predict

predict [*type*] *newvar* [*if*] [*in*] [, *statistic options*]

predict [*type*] { *stub** | *newvarlist* } [*if*] [*in*] [, *statistic options*]

statistic	Description
Main	
`classification`	group membership classification; the default when one variable is specified and `group()` is not specified
`pr`	probability of group membership; the default when `group()` is specified or when multiple variables are specified
`mahalanobis`	Mahalanobis squared distance between observations and groups
`clscore`	group classification function score
* `looclass`	leave-one-out group membership classification; may be used only when one new variable is specified
* `loopr`	leave-one-out probability of group membership
* `loomahal`	leave-one-out Mahalanobis squared distance between observations and groups

options	Description
Main	
`group(`*group*`)`	the group for which the statistic is to be calculated
Options	
`priors(`*priors*`)`	group prior probabilities; defaults to `e(grouppriors)`
`ties(`*ties*`)`	how ties in classification are to be handled; defaults to `e(ties)`

priors	Description
`equal`	equal prior probabilities
`proportional`	group-size-proportional prior probabilities
matname	row or column vector containing the group prior probabilities
matrix_exp	matrix expression providing a row or column vector of the group prior probabilities

ties	Description
`missing`	ties in group classification produce missing values
`random`	ties in group classification are broken randomly
`first`	ties in group classification are set to the first tied group

You specify one new variable with `classification` or `looclass` and specify either one or `e(N_groups)` new variables with `pr`, `loopr`, `mahalanobis`, `loomahal`, or `clscore`.

Unstarred statistics are available both in and out of sample; type `predict ... if e(sample) ...` if wanted only for the estimation sample. Starred statistics are calculated only for the estimation sample, even when `if e(sample)` is not specified.

`group()` is not allowed with `classification` or `looclass`.

Menu

Statistics > Postestimation > Predictions, residuals, etc.

Options for predict

Main

`classification`, the default, calculates the group classification. Only one new variable may be specified.

`pr` calculates group membership posterior probabilities. If you specify the `group()` option, specify one new variable. Otherwise, you must specify `e(N_groups)` new variables.

`mahalanobis` calculates the squared Mahalanobis distance between the observations and group means. If you specify the `group()` option, specify one new variable. Otherwise, you must specify `e(N_groups)` new variables.

`clscore` produces the group classification function score. If you specify the `group()` option, specify one new variable. Otherwise, you must specify `e(N_groups)` new variables.

`looclass` calculates the leave-one-out group classifications. Only one new variable may be specified. Leave-one-out calculations are restricted to `e(sample)` observations.

`loopr` calculates the leave-one-out group membership posterior probabilities. If you specify the `group()` option, specify one new variable. Otherwise, you must specify `e(N_groups)` new variables. Leave-one-out calculations are restricted to `e(sample)` observations.

`loomahal` calculates the leave-one-out squared Mahalanobis distance between the observations and group means. If you specify the `group()` option, specify one new variable. Otherwise, you must specify `e(N_groups)` new variables. Leave-one-out calculations are restricted to `e(sample)` observations.

`group(`*group*`)` specifies the group for which the statistic is to be calculated and can be specified using

`#1`, `#2`, ..., where `#1` means the first category of the `e(groupvar)` variable, `#2` the second category, etc.;

the values of the `e(groupvar)` variable; or

the value labels of the `e(groupvar)` variable if they exist.

`group()` is not allowed with `classification` or `looclass`.

Options

`priors(`*priors*`)` specifies the prior probabilities for group membership. If `priors()` is not specified, `e(grouppriors)` is used. The following *priors* are allowed:

`priors(equal)` specifies equal prior probabilities.

`priors(proportional)` specifies group-size-proportional prior probabilities.

`priors(`*matname*`)` specifies a row or column vector containing the group prior probabilities.

`priors(`*matrix_exp*`)` specifies a matrix expression providing a row or column vector of the group prior probabilities.

`ties(`*ties*`)` specifies how ties in group classification will be handled. If `ties()` is not specified, `e(ties)` is used. The following *ties* are allowed:

`ties(missing)` specifies that ties in group classification produce missing values.

`ties(random)` specifies that ties in group classification are broken randomly.

`ties(first)` specifies that ties in group classification are set to the first tied group.

Syntax for estat correlations

`estat correlations` [, *options*]

options	Description
Main	
`p`	display two-sided *p*-values
`format(%`*fmt*`)`	numeric display format; default is `%9.0g`
`nohalf`	display full matrix even if symmetric

Menu

Statistics > Postestimation > Reports and statistics

Options for estat correlations

Main

`p` specifies that two-sided *p*-values be computed and displayed for the correlations.

`format(%`*fmt*`)` specifies the matrix display format. The default is `format(%8.5f)`.

`nohalf` specifies that, even though the matrix is symmetric, the full matrix be printed. The default is to print only the lower triangle.

Syntax for estat covariance

`estat covariance` [, *options*]

options	Description
Main	
`format(%`*fmt*`)`	numeric display format; default is `%9.0g`
`nohalf`	display full matrix even if symmetric

Menu

Statistics > Postestimation > Reports and statistics

Options for estat covariance

Main

`format(%fmt)` specifies the matrix display format. The default is `format(%9.0g)`.

`nohalf` specifies that, even though the matrix is symmetric, the full matrix be printed. The default is to print only the lower triangle.

Syntax for estat grdistances

`estat grdistances [, options]`

options	Description
Main	
`mahalanobis`	display Mahalanobis squared distances between group means; the default
`generalized`	display generalized Mahalanobis squared distances between group means
`all`	equivalent to `mahalanobis generalized`
`format(%fmt)`	numeric display format; default is `%9.0g`
Options	
`priors(priors)`	group prior probabilities; defaults to `e(grouppriors)`

Menu

Statistics > Postestimation > Reports and statistics

Options for estat grdistances

Main

`mahalanobis` specifies that a table of Mahalanobis squared distances between group means be presented.

`generalized` specifies that a table of generalized Mahalanobis squared distances between group means be presented. `generalized` starts with what is produced by the `mahalanobis` option and adds a term for the possibly unequal covariances and a term accounting for prior probabilities. Prior probabilities are provided with the `priors()` option, or if `priors()` is not specified, by the values in `e(grouppriors)`. By common convention, if prior probabilities are equal across the groups, the prior probability term is omitted.

`all` is equivalent to specifying `mahalanobis` and `generalized`.

`format(%fmt)` specifies the matrix display format. The default is `format(%9.0g)`.

Options

`priors(priors)` specifies the group prior probabilities and affects only the output of the `generalized` option. By default, *priors* is determined from `e(grouppriors)`. See *Options for predict* for the *priors* specification.

Remarks

The predict and estat commands after discrim qda help in exploring the QDA model. See [MV] **discrim estat** for details of the estat subcommands common to all discrim subcommands. Here we illustrate some of these common estat subcommands along with estat covariance, estat correlations, and estat grdistances that are specific to discrim qda.

▷ Example 1

Everitt and Dunn (2001, 269) show data for male Egyptian skulls from the early and late predynastic epochs. Ten observations from each epoch are provided. Four measurements were taken of each skull: x1, maximum breadth; x2, basibregmatic height; x3, basialveolar length; and x4, nasal height. All measurements were in millimeters. Everitt and Dunn obtained the data from Manly (2005).

We perform a quadratic discriminant analysis on this dataset and demonstrate the use of estat and predict.

```
. use http://www.stata-press.com/data/r12/skulls
(Egyptian Skulls)
. discrim qda x1 x2 x3 x4, group(predynastic)
Quadratic discriminant analysis
Resubstitution classification summary

    +---------+
    | Key     |
    |---------|
    | Number  |
    | Percent |
    +---------+

True         | Classified
predynastic  |  early    late  |  Total
-------------+-----------------+--------
       early |      9       1  |     10
             |  90.00   10.00  | 100.00
             |                 |
        late |      3       7  |     10
             |  30.00   70.00  | 100.00
-------------+-----------------+--------
       Total |     12       8  |     20
             |  60.00   40.00  | 100.00
             |                 |
      Priors | 0.5000  0.5000  |
```

What kind of covariance structure do the two groups have? If they are similar to one another, we might wish to switch to using LDA (see [MV] **discrim lda**) instead of QDA. estat covariance displays the covariance matrices for our two groups.

```
. estat covariance
Group covariance matrices
  predynastic: early

             |         x1          x2          x3          x4
-------------+------------------------------------------------
          x1 |   40.32222
          x2 |   7.188889    15.34444
          x3 |   13.18889   -7.322222        36.9
          x4 |       16.1    8.077778   -2.144444    11.43333
  predynastic: late

             |         x1          x2          x3          x4
-------------+------------------------------------------------
          x1 |   43.12222
          x2 |  -4.966667    38.98889
          x3 |   9.388889    6.611111    10.27778
          x4 |   5.211111    12.74444    4.388889    9.122222
```

There appear to be differences, including differences in sign between some of the elements of the covariance matrices of the two groups. How substantial are these differences? The `estat correlations` command displays the correlation matrices for the groups. The p option requests that p-values be presented with the correlations.

```
. estat correlations, p
Group correlation matrices
  predynastic: early

  +-------------------+
  | Key               |
  |-------------------|
  | Correlation       |
  | Two-sided p-value |
  +-------------------+

             |        x1         x2         x3         x4
-------------+--------------------------------------------
          x1 |   1.00000
             |
             |
          x2 |   0.28901    1.00000
             |   0.41800
             |
          x3 |   0.34192   -0.30772    1.00000
             |   0.33353    0.38707
             |
          x4 |   0.74984    0.60986   -0.10440    1.00000
             |   0.01251    0.06119    0.77409
```

```
predynastic: late

  +-------------------+
  | Key               |
  |-------------------|
  | Correlation       |
  | Two-sided p-value |
  +-------------------+

             |       x1        x2        x3        x4
-------------+----------------------------------------
          x1 |  1.00000
             |
             |
          x2 | -0.12113   1.00000
             |  0.73889
             |
          x3 |  0.44598   0.33026   1.00000
             |  0.19640   0.35133
             |
          x4 |  0.26274   0.67577   0.45327   1.00000
             |  0.46331   0.03196   0.18830
```

Few of the correlations in the two matrices are statistically significant. We are less sure of the apparent differences between the covariance structures for these two groups.

Let's press forward anyway. Everitt and Dunn (2001, 269) ask for the prediction for an unknown skull. We input the unknown observation and then use `predict` to obtain the classification and probabilities of group membership for the observation.

```
. input
      predyna~c     x1      x2      x3      x4
 21. . 127 129 95 51
 22. end
. predict grp
(option classification assumed; group classification)
. predict pr1 pr2, pr
. label values grp epoch
. list x* grp pr1 pr2 in 21

     +----------------------------------------------------+
     |  x1    x2   x3   x4    grp        pr1         pr2  |
     |----------------------------------------------------|
 21. | 127   129   95   51   late   .3654425    .6345575  |
     +----------------------------------------------------+
```

This skull is classified by our QDA model as belonging to the late predynastic epoch with probability 0.63.

`estat list` could also be used to obtain this same information; see [MV] **discrim estat**.

```
. estat list in 21, varlist

  +-------------------------------------------------------------------------+
  |      |  Data                     |  Classification  |  Probabilities    |
  |      |                           |                  |                   |
  | Obs. |   x1     x2     x3    x4  |    True   Class. |   early     late  |
  |------+---------------------------+------------------+-------------------|
  |   21 |  127    129     95    51  |             late |  0.3654   0.6346  |
  +-------------------------------------------------------------------------+
```

We could use `predict` and `estat` to explore other aspects of this QDA model, including leave-one-out (LOO) classifications, probabilities, classification tables, and error-rate estimates.

◁

▷ Example 2

Example 1 of [MV] **discrim qda** performs a QDA on the apple tree rootstock data found in Andrews and Herzberg (1985, 357–360). We now demonstrate the use of the `estat grdistances` command for examining the squared Mahalanobis distances and the squared generalized distances between the rootstock groups.

```
. use http://www.stata-press.com/data/r12/rootstock, clear
(Table 6.2 Rootstock Data -- Rencher (2002))
. discrim qda y1 y2 y3 y4, group(rootstock) notable
. estat grdistances, all
```

Mahalanobis squared distances between groups

rootstock	rootstock 1	2	3	4	5
1	0	18.37241	7.89988	1.622808	14.78843
2	42.19008	0	5.489408	14.08784	1.502462
3	36.81811	1.908369	0	6.406024	15.48121
4	2.281963	14.77928	6.742393	0	25.72128
5	33.70858	1.855704	4.617755	16.34139	0
6	3.860684	17.32868	12.5828	11.24491	3.49512

rootstock	rootstock 6
1	9.152132
2	30.45472
3	72.60112
4	29.01146
5	20.50925
6	0

Generalized squared distances between groups

rootstock	rootstock 1	2	3	4	5
1	-17.89946	2.47128	-9.577605	-14.60611	-1.796629
2	24.29063	-15.90113	-11.98808	-2.141072	-15.0826
3	18.91866	-13.99276	-17.47749	-9.822891	-1.103849
4	-15.61749	-1.121858	-10.73509	-16.22892	9.136221
5	15.80913	-14.04543	-12.85973	.1124762	-16.58506
6	-14.03877	1.427543	-4.894681	-4.984005	-13.08994

rootstock	rootstock 6
1	-7.241371
2	14.06121
3	56.20761
4	12.61796
5	4.115752
6	-16.3935

Both tables are nonsymmetric. For QDA the Mahalanobis distance depends on the covariance of the reference group. The Mahalanobis distance for group i (the rows in the tables above) to group j (the columns in the tables above) will use the covariance matrix of group j in determining the distance. The generalized distance also factors in the prior probabilities for the groups, and so the diagonal elements are not zero and the entries can be negative. In either matrix, the smaller the number, the closer the groups.

◁

Saved results

`estat correlations` saves the following in `r()`:

Matrices

`r(Rho_#)`	group # correlation matrix
`r(P_#)`	two-sided p-values for group # correlations

`estat covariance` saves the following in `r()`:

Matrices

`r(S_#)`	group # covariance matrix

`estat grdistances` saves the following in `r()`:

Matrices

`r(sqdist)`	Mahalanobis squared distances between group means (`mahalanobis` only)
`r(gsqdist)`	generalized squared distances between group means (`generalized` only)

Methods and formulas

All postestimation commands listed above are implemented as ado-files.

See *Methods and formulas* of [MV] **discrim qda** for background on what is produced by `predict` and `estat grdistances`. See [MV] **discrim estat** for more information on `estat classtable`, `estat errorrate`, `estat grsummarize`, and `estat list`.

References

Andrews, D. F., and A. M. Herzberg, ed. 1985. *Data: A Collection of Problems from Many Fields for the Student and Research Worker.* New York: Springer.

Everitt, B. S., and G. Dunn. 2001. *Applied Multivariate Data Analysis.* 2nd ed. London: Arnold.

Manly, B. F. J. 2005. *Multivariate Statistical Methods: A Primer.* 3rd ed. London: Chapman & Hall.

Rencher, A. C. 2002. *Methods of Multivariate Analysis.* 2nd ed. New York: Wiley.

Also see

[MV] **discrim estat** — Postestimation tools for discrim

[MV] **discrim qda** — Quadratic discriminant analysis

[U] **20 Estimation and postestimation commands**

Title

factor — Factor analysis

Syntax

Factor analysis of data

`factor` *varlist* [*if*] [*in*] [*weight*] [, *method options*]

Factor analysis of a correlation matrix

`factormat` *matname*`, n(`*#*`)` [*method options factormat_options*]

method	Description
Model 2	
`pf`	principal factor; the default
`pcf`	principal-component factor
`ipf`	iterated principal factor
`ml`	maximum likelihood factor

options	Description
Model 2	
`factors(`*#*`)`	maximum number of factors to be retained
`mineigen(`*#*`)`	minimum value of eigenvalues to be retained
`citerate(`*#*`)`	communality reestimation iterations (`ipf` only)
Reporting	
`blanks(`*#*`)`	display loadings as blank when $\lvert\text{loadings}\rvert < \#$
`altdivisor`	use trace of correlation matrix as the divisor for reported proportions
Maximization	
`protect(`*#*`)`	perform # optimizations and report the best solution (`ml` only)
`random`	use random starting values (`ml` only); seldom used
`seed(`*seed*`)`	random-number seed (`ml` with `protect()` or `random` only)
maximize_options	control the maximization process; seldom used (`ml` only)
`norotated`	display unrotated solution, even if rotated results are available (replay only)

`norotated` does not appear in the dialog box.

factormat_options	Description
Model	
shape(full)	*matname* is a square symmetric matrix; the default
shape(lower)	*matname* is a vector with the rowwise lower triangle (with diagonal)
shape(upper)	*matname* is a vector with the rowwise upper triangle (with diagonal)
names(*namelist*)	variable names; required if *matname* is triangular
forcepsd	modifies *matname* to be positive semidefinite
* n(#)	number of observations
sds(*matname*$_2$)	vector with standard deviations of variables
means(*matname*$_3$)	vector with means of variables

* n(#) is required for factormat.

bootstrap, by, jackknife, rolling, and statsby are allowed with factor; see [U] **11.1.10 Prefix commands**. However, bootstrap and jackknife results should be interpreted with caution; identification of the factor parameters involves data-dependent restrictions, possibly leading to badly biased and overdispersed estimates (Milan and Whittaker 1995).

Weights are not allowed with the bootstrap prefix; see [R] **bootstrap**.

aweights are not allowed with the jackknife prefix; see [R] **jackknife**.

aweights and fweights are allowed with factor; see [U] **11.1.6 weight**.

See [U] **20 Estimation and postestimation commands** for more capabilities of estimation commands.

Menu

factor

Statistics > Multivariate analysis > Factor and principal component analysis > Factor analysis

factormat

Statistics > Multivariate analysis > Factor and principal component analysis > Factor analysis of a correlation matrix

Description

factor and factormat perform a factor analysis of a correlation matrix. factor and factormat can produce principal factor, iterated principal factor, principal-component factor, and maximum-likelihood factor analyses. factor and factormat display the eigenvalues of the correlation matrix, the factor loadings, and the uniqueness (= 1 − communality) of the variables.

factor expects data in the form of variables, allows weights, and can be run for subgroups (see [D] **by**). factormat is for use with a correlation or covariance matrix in the form of a square Stata matrix or a vector containing the rowwise upper or lower triangle of the correlation or covariance matrix. This concept is explained in more detail below; see option shape(). If a covariance matrix is provided to factormat, it is transformed into a correlation matrix for the factor analysis. To replay estimation results, you may type either factor or factormat.

Options for factor and factormat

Model 2

pf, pcf, ipf, and ml indicate the type of estimation to be performed. The default is pf.

pf specifies that the principal-factor method be used to analyze the correlation matrix. The factor loadings, sometimes called the factor patterns, are computed using the squared multiple correlations as estimates of the communality. pf is the default.

pcf specifies that the principal-component factor method be used to analyze the correlation matrix. The communalities are assumed to be 1.

ipf specifies that the iterated principal-factor method be used to analyze the correlation matrix. This reestimates the communalities iteratively.

ml specifies the maximum-likelihood factor method, assuming multivariate normal observations. This estimation method is equivalent to Rao's canonical-factor method and maximizes the determinant of the partial correlation matrix. Hence, this solution is also meaningful as a descriptive method for nonnormal data. ml is not available for singular correlation matrices. At least three variables must be specified with method ml.

factors(*#*) and mineigen(*#*) specify the maximum number of factors to be retained. factors() specifies the number directly, and mineigen() specifies it indirectly, keeping all factors with eigenvalues greater than the indicated value. The options can be specified individually, together, or not at all.

factors(*#*) sets the maximum number of factors to be retained for later use by the postestimation commands. factor always prints the full set of eigenvalues but prints the corresponding eigenvectors only for retained factors. Specifying a number larger than the number of variables in the *varlist* is equivalent to specifying the number of variables in the *varlist* and is the default.

mineigen(*#*) sets the minimum value of eigenvalues to be retained. The default for all methods except pcf is 5×10^{-6} (effectively zero), meaning that factors associated with negative eigenvalues will not be printed or retained. The default for pcf is 1. Many sources recommend mineigen(1), although the justification is complex and uncertain. If # is less than 5×10^{-6}, it is reset to 5×10^{-6}.

citerate(*#*) is used only with ipf and sets the number of iterations for reestimating the communalities. If citerate() is not specified, iterations continue until the change in the communalities is small. ipf with citerate(0) produces the same results that pf does.

Reporting

blanks(*#*) specifies that factor loadings smaller than # (in absolute value) be displayed as blanks.

altdivisor specifies that reported proportions and cumulative proportions be computed using the trace of the correlation matrix, trace(e(C)), as the divisor. The default is to use the sum of all eigenvalues (even those that are negative) as the divisor.

Maximization

protect(*#*) is used only with ml and requests that # optimizations with random starting values be performed along with squared multiple correlation coefficient starting values and that the best of the solutions be reported. The output also indicates whether all starting values converged to the same solution. When specified with a large number, such as protect(50), this provides reasonable assurance that the solution found is global and not just a local maximum. If trace is also specified (see [R] **maximize**), the parameters and likelihoods of each maximization will be printed.

random is used only with ml and requests that random starting values be used. This option is rarely used and should be used only after protect() has shown the presence of multiple maximums.

seed(*seed*) is used only with ml when the random or protect() options are also specified. seed() specifies the random-number seed; see [R] **set seed**. If seed() is not specified, the random-number generator starts in whatever state it was last in.

maximize_options: iterate(#), [no]log, trace, tolerance(#), and ltolerance(#); see [R] **maximize**. These options are seldom used.

The following option is available with factor but is not shown in the dialog box:

norotated specifies that the unrotated factor solution be displayed, even if a rotated factor solution is available. norotated is for use only with replaying results.

Options unique to factormat

Model

shape(*shape*) specifies the shape (storage method) for the covariance or correlation matrix *matname*. The following shapes are supported:

full specifies that the correlation or covariance structure of k variables is a symmetric $k \times k$ matrix. This is the default.

lower specifies that the correlation or covariance structure of k variables is a vector with $k(k+1)/2$ elements in rowwise lower-triangular order,

$$\mathrm{C}_{11}\ \mathrm{C}_{21}\ \mathrm{C}_{22}\ \mathrm{C}_{31}\ \mathrm{C}_{32}\ \mathrm{C}_{33}\ \ldots\ \mathrm{C}_{k1}\ \mathrm{C}_{k2}\ \ldots\ \mathrm{C}_{kk}$$

upper specifies that the correlation or covariance structure of k variables is a vector with $k(k+1)/2$ elements in rowwise upper-triangular order,

$$\mathrm{C}_{11}\ \mathrm{C}_{12}\ \mathrm{C}_{13}\ \ldots\ \mathrm{C}_{1k}\ \mathrm{C}_{22}\ \mathrm{C}_{23}\ \ldots \mathrm{C}_{2k}\ \ldots\ \mathrm{C}_{(k-1,k-1)}\ \mathrm{C}_{(k-1,k)}\ \mathrm{C}_{kk}$$

names(*namelist*) specifies a list of k different names to be used to document output and label estimation results and as variable names by predict. names() is required if the correlation or covariance matrix is in vectorized storage mode (that is, shape(lower) or shape(upper) is specified). By default, factormat verifies that the row and column names of *matname* and the column or row names of $matname_2$ and $matname_3$ from the sds() and means() options are in agreement. Using the names() option turns off this check.

forcepsd modifies the matrix *matname* to be positive semidefinite (psd) and so be a proper covariance matrix. If *matname* is not positive semidefinite, it will have negative eigenvalues. By setting negative eigenvalues to 0 and reconstructing, we obtain the least-squares positive-semidefinite approximation to *matname*. This approximation is a singular covariance matrix.

n(#), a required option, specifies the number of observations on which *matname* is based.

sds($matname_2$) specifies a $k \times 1$ or $1 \times k$ matrix with the standard deviations of the variables. The row or column names should match the variable names, unless the names() option is specified. sds() may be specified only if *matname* is a correlation matrix. Specify sds() if you have variables in your dataset and want to use predict after factormat. sds() does not affect the computations of factormat but provides information so that predict does not assume that the standard deviations are one.

`means(`*matname*$_3$`)` specifies a $k \times 1$ or $1 \times k$ matrix with the means of the variables. The row or column names should match the variable names, unless the `names()` option is specified. Specify `means()` if you have variables in your dataset and want to use `predict` after `factormat`. `means()` does not affect the computations of `factormat` but provides information so that `predict` does not assume the means are zero.

Remarks

Remarks are presented under the following headings:

Introduction

Factor analysis, in the sense of exploratory factor analysis, is a statistical technique for data reduction. It reduces the number of variables in an analysis by describing linear combinations of the variables that contain most of the information and that, we hope, admit meaningful interpretations.

Factor analysis originated with the work of Spearman (1904), and has since witnessed an explosive growth, especially in the social sciences and, interestingly, in chemometrics. For an introduction, we refer to Kim and Mueller (1978a, 1978b), van Belle, Fisher, Heagerty, and Lumley (2004, chap. 14), and Hamilton (2009, chap. 12). Intermediate-level treatments include Gorsuch (1983) and Harman (1976). For mathematically more advanced discussions, see Mulaik (2010), Mardia, Kent, and Bibby (1979, chap. 9), and Fuller (1987).

Structural equation modeling provides a more general framework for performing factor analysis, including confirmatory factor analysis; see the *Stata Structural Equation Modeling Reference Manual*.

Also see Kolenikov (2009) for another implementation of confirmatory factor analysis.

Factor analysis

Factor analysis finds a few common factors (say, q of them) that linearly reconstruct the p original variables

$$y_{ij} = z_{i1}b_{1j} + z_{i2}b_{2j} + \cdots + z_{iq}b_{qj} + e_{ij}$$

where y_{ij} is the value of the ith observation on the jth variable, z_{ik} is the ith observation on the kth common factor, b_{kj} is the set of linear coefficients called the factor loadings, and e_{ij} is similar to a residual but is known as the jth variable's unique factor. Everything except the left-hand-side variable is to be estimated, so the model has an infinite number of solutions. Various constraints are introduced to make the model determinate.

"Reconstruction" is typically defined in terms of prediction of the correlation matrix of the original variables, unlike principal components (see [MV] **pca**), where reconstruction means minimum residual variance summed across all equations (variables).

Once the factors and their loadings have been estimated, they are interpreted—an admittedly subjective process. Interpretation typically means examining the b_{kj}'s and assigning names to each factor. Because of the indeterminacy of the factor solution, we are not limited to examining solely the b_{kj}'s. The loadings could be rotated. Rotations come in two forms—orthogonal and oblique. If we restrict to orthogonal rotations, the rotated b_{kj}s, despite appearing different, are every bit as good as (and no better than) the original loadings. Oblique rotations are often desired but do not retain some important properties of the original solution; see example 3. Because there are an infinite number of potential rotations, different rotations could lead to different interpretations of the same data. These are not to be viewed as conflicting, but instead as two different ways of looking at the same thing. See [MV] **factor postestimation** and [MV] **rotate** for more information on rotation.

▷ Example 1

We wish to analyze physicians' attitudes toward cost. Six questions about cost were asked of 568 physicians in the Medical Outcomes Study from Tarlov et al. (1989). We do not have the original data, so we used corr2data to create a dataset with the same correlation matrix. Factor analysis is often used to validate a combination of questions that looks meaningful at first glance. Here we wish to create a variable that summarizes the information on each physician's attitude toward cost.

Each response is coded on a five-point scale, where 1 means "agree" and 5 means "disagree":

```
. use http://www.stata-press.com/data/r12/bg2
(Physician-cost data)

. describe

Contains data from http://www.stata-press.com/data/r12/bg2.dta
  obs:           568                          Physician-cost data
 vars:             7                          11 Feb 2011 21:54
 size:        14,768                          (_dta has notes)
-------------------------------------------------------------------------------
              storage  display     value
variable name   type   format      label      variable label
-------------------------------------------------------------------------------
clinid          int    %9.0g                  Physician identifier
bg2cost1        float  %9.0g                  Best health care is expensive
bg2cost2        float  %9.0g                  Cost is a major consideration
bg2cost3        float  %9.0g                  Determine cost of tests first
bg2cost4        float  %9.0g                  Monitor likely complications
                                                only
bg2cost5        float  %9.0g                  Use all means regardless of cost
bg2cost6        float  %9.0g                  Prefer unnecessary tests to
                                                missing tests
-------------------------------------------------------------------------------
Sorted by:  clinid
```

We perform the factorization on bg2cost1, bg2cost2, ..., bg2cost6.

```
. factor bg2cost1-bg2cost6
(obs=568)

Factor analysis/correlation                        Number of obs    =      568
    Method: principal factors                      Retained factors =        3
    Rotation: (unrotated)                          Number of params =       15

    --------------------------------------------------------------------------
         Factor  |   Eigenvalue   Difference        Proportion   Cumulative
    -------------+------------------------------------------------------------
        Factor1  |      0.85389      0.31282            1.0310       1.0310
        Factor2  |      0.54107      0.51786            0.6533       1.6844
        Factor3  |      0.02321      0.17288            0.0280       1.7124
        Factor4  |     -0.14967      0.03951           -0.1807       1.5317
        Factor5  |     -0.18918      0.06197           -0.2284       1.3033
        Factor6  |     -0.25115            .           -0.3033       1.0000
    --------------------------------------------------------------------------
    LR test: independent vs. saturated:  chi2(15) =  269.07 Prob>chi2 = 0.0000
```

Factor loadings (pattern matrix) and unique variances

Variable	Factor1	Factor2	Factor3	Uniqueness
bg2cost1	0.2470	0.3670	-0.0446	0.8023
bg2cost2	-0.3374	0.3321	-0.0772	0.7699
bg2cost3	-0.3764	0.3756	0.0204	0.7169
bg2cost4	-0.3221	0.1942	0.1034	0.8479
bg2cost5	0.4550	0.2479	0.0641	0.7274
bg2cost6	0.4760	0.2364	-0.0068	0.7175

`factor` retained only the first three factors because the eigenvalues associated with the remaining factors are negative. According to the default `mineigen(0)` criterion, a factor must have an eigenvalue greater than zero to be retained. You can set this threshold higher by specifying `mineigen(#)`. Although `factor` elected to retain three factors, only the first two appear to be meaningful.

The first factor seems to describe the physician's average position on cost because it affects the responses to all the questions "positively", as shown by the signs in the first column of the factor-loading table. We say "positively" because, obviously, the signs on three of the loadings are negative. When we look back at the results of `describe`, however, we find that the direction of the responses on `bg2cost2`, `bg2cost3`, and `bg2cost4` are reversed. If the physician feels that cost should not be a major influence on medical treatment, he or she is likely to disagree with these three items and to agree with the other three.

The second factor loads positively (absolutely, not logically) on all six items and could be interpreted as describing the physician's tendency to agree with any good-sounding idea put forth. Psychologists refer to this as the "positive response set". On statistical grounds, we would probably keep this second factor, although on substantive grounds, we would be tempted to drop it.

We finally point to the column with the header "uniqueness". *Uniqueness* is the percentage of variance for the variable that is not explained by the common factors. The quantity "1 − *uniqueness*" is called *communality*. Uniqueness could be pure measurement error, or it could represent something that is measured reliably in that particular variable, but not by any of the others. The greater the uniqueness, the more likely that it is more than just measurement error. Values more than 0.6 are usually considered high; all the variables in this problem are even higher—more than 0.71. If the uniqueness is high, then the variable is not well explained by the factors.

◁

▷ Example 2

The cumulative proportions of the eigenvalues exceeded 1.0 in our factor analysis because of the negative eigenvalues. By default, the proportion and cumulative proportion columns are computed using the sum of all eigenvalues as the divisor. The `altdivisor` option allows you to display the proportions and cumulative proportions by using the trace of the correlation matrix as the divisor. This option is allowed at estimation time or when replaying results. We demonstrate by replaying the results with this option.

```
. factor, altdivisor
Factor analysis/correlation                      Number of obs    =      568
    Method: principal factors                    Retained factors =        3
    Rotation: (unrotated)                        Number of params =       15
```

Factor	Eigenvalue	Difference	Proportion	Cumulative
Factor1	0.85389	0.31282	0.1423	0.1423
Factor2	0.54107	0.51786	0.0902	0.2325
Factor3	0.02321	0.17288	0.0039	0.2364
Factor4	-0.14967	0.03951	-0.0249	0.2114
Factor5	-0.18918	0.06197	-0.0315	0.1799
Factor6	-0.25115	.	-0.0419	0.1380

```
    LR test: independent vs. saturated:  chi2(15) =  269.07 Prob>chi2 = 0.0000
Factor loadings (pattern matrix) and unique variances
```

Variable	Factor1	Factor2	Factor3	Uniqueness
bg2cost1	0.2470	0.3670	-0.0446	0.8023
bg2cost2	-0.3374	0.3321	-0.0772	0.7699
bg2cost3	-0.3764	0.3756	0.0204	0.7169
bg2cost4	-0.3221	0.1942	0.1034	0.8479
bg2cost5	0.4550	0.2479	0.0641	0.7274
bg2cost6	0.4760	0.2364	-0.0068	0.7175

Among the sources we examined, there was not a consensus on which divisor is most appropriate. Therefore, both are available.

◁

▷ Example 3

`factor` provides several alternative estimation strategies for the factor model. We specified no options on the `factor` command when we fit our first model, so we obtained the principal-factor solution. The *communalities* (defined as 1 − *uniqueness*) were estimated using the squared multiple correlation coefficients.

We could have instead obtained the estimates from "principal-component factors", treating the communalities as all 1—meaning that there are no unique factors—by specifying the `pcf` option:

```
. factor bg2cost1-bg2cost6, pcf
(obs=568)
Factor analysis/correlation                      Number of obs    =      568
    Method: principal-component factors         Retained factors =        2
    Rotation: (unrotated)                        Number of params =       11
```

Factor	Eigenvalue	Difference	Proportion	Cumulative
Factor1	1.70622	0.30334	0.2844	0.2844
Factor2	1.40288	0.49422	0.2338	0.5182
Factor3	0.90865	0.18567	0.1514	0.6696
Factor4	0.72298	0.05606	0.1205	0.7901
Factor5	0.66692	0.07456	0.1112	0.9013
Factor6	0.59236	.	0.0987	1.0000

```
    LR test: independent vs. saturated:  chi2(15) =  269.07 Prob>chi2 = 0.0000
```

```
Factor loadings (pattern matrix) and unique variances
```

Variable	Factor1	Factor2	Uniqueness
bg2cost1	0.3581	0.6279	0.4775
bg2cost2	-0.4850	0.5244	0.4898
bg2cost3	-0.5326	0.5725	0.3886
bg2cost4	-0.4919	0.3254	0.6521
bg2cost5	0.6238	0.3962	0.4539
bg2cost6	0.6543	0.3780	0.4290

Here we find that the principal-component factor model is inappropriate. It is based on the assumption that the uniquenesses are 0, but we find that there is considerable uniqueness—there is considerable variability left over after our two factors. We should use some other method.

◁

▷ Example 4

We could have fit our model using iterated principal factors by specifying the `ipf` option. Here the initial estimates of the communalities would be the squared multiple correlation coefficients, but the solution would then be iterated to obtain different (better) estimates:

```
. factor bg2cost1-bg2cost6, ipf
(obs=568)

Factor analysis/correlation                   Number of obs    =       568
    Method: iterated principal factors        Retained factors =         5
    Rotation: (unrotated)                     Number of params =        15
```

Factor	Eigenvalue	Difference	Proportion	Cumulative
Factor1	1.08361	0.31752	0.5104	0.5104
Factor2	0.76609	0.53816	0.3608	0.8712
Factor3	0.22793	0.19469	0.1074	0.9786
Factor4	0.03324	0.02085	0.0157	0.9942
Factor5	0.01239	0.01256	0.0058	1.0001
Factor6	-0.00017	.	-0.0001	1.0000

```
    LR test: independent vs. saturated:  chi2(15) =  269.07 Prob>chi2 = 0.0000

Factor loadings (pattern matrix) and unique variances
```

Variable	Factor1	Factor2	Factor3	Factor4	Factor5	Uniqueness
bg2cost1	0.2471	0.4059	-0.1349	-0.1303	0.0288	0.7381
bg2cost2	-0.4040	0.3959	-0.2636	0.0349	0.0040	0.6093
bg2cost3	-0.4479	0.4570	0.1290	0.0137	-0.0564	0.5705
bg2cost4	-0.3327	0.1943	0.2655	0.0091	0.0810	0.7744
bg2cost5	0.5294	0.3338	0.2161	-0.0134	-0.0331	0.5604
bg2cost6	0.5174	0.2943	-0.0801	0.1208	0.0265	0.6240

Here we retained too many factors. Unlike in principal factors or principal-component factors, we cannot simply ignore the unnecessary factors because the uniquenesses are reestimated from the data and therefore depend on the number of retained factors. We need to reestimate. We use the opportunity to demonstrate the option `blanks(#)` for displaying "small loadings" as blanks for easier reading:

```
. factor bg2cost1-bg2cost6, ipf factors(2) blanks(.30)
(obs=568)
Factor analysis/correlation                        Number of obs    =       568
    Method: iterated principal factors             Retained factors =         2
    Rotation: (unrotated)                          Number of params =        11

    --------------------------------------------------------------------------
         Factor  |   Eigenvalue   Difference        Proportion   Cumulative
    -------------+------------------------------------------------------------
        Factor1  |      1.03954      0.30810            0.5870       0.5870
        Factor2  |      0.73144      0.60785            0.4130       1.0000
        Factor3  |      0.12359      0.11571            0.0698       1.0698
        Factor4  |      0.00788      0.03656            0.0045       1.0743
        Factor5  |     -0.02867      0.07418           -0.0162       1.0581
        Factor6  |     -0.10285            .           -0.0581       1.0000
    --------------------------------------------------------------------------
    LR test: independent vs. saturated:  chi2(15) =  269.07 Prob>chi2 = 0.0000
Factor loadings (pattern matrix) and unique variances

    -------------------------------------------------
        Variable |  Factor1   Factor2 |   Uniqueness
    -------------+--------------------+--------------
        bg2cost1 |             0.3941 |      0.7937
        bg2cost2 |  -0.3590           |      0.7827
        bg2cost3 |  -0.5189    0.4935 |      0.4872
        bg2cost4 |  -0.3230           |      0.8699
        bg2cost5 |   0.4667    0.3286 |      0.6742
        bg2cost6 |   0.5179    0.3325 |      0.6212
    -------------------------------------------------
    (blanks represent abs(loading)<.3)
```

It is instructive to compare the reported uniquenesses for this model and the previous one, where five factors were retained. Also, compared with the results we obtained from principal factors, these results do not differ much.

◁

▷ Example 5

Finally, we could have fit our model using the maximum likelihood method by specifying the `ml` option. The maximum likelihood method assumes that the data are multivariate normal distributed. If the factor model provides an adequate approximation to the data, maximum likelihood estimates have favorable properties compared with the other estimation methods. Rao (1955) has shown that his canonical factor method is equivalent to the maximum likelihood method. This method seeks to maximize canonical correlations between the manifest variables and the common factors. Thus `ml` may be used descriptively, even if we are unwilling to assume multivariate normality.

As with `ipf`, if we do not specify the number of factors, Stata retains more than two factors (it retained three), and, as with `ipf`, we will need to reestimate with the number of factors that we really want. To save paper, we will start by retaining two factors:

```
. factor bg2cost1-bg2cost6, ml factors(2)
(obs=568)
Iteration 0:   log likelihood = -28.702162
Iteration 1:   log likelihood = -7.0065234
Iteration 2:   log likelihood = -6.8513798
Iteration 3:   log likelihood = -6.8429502
Iteration 4:   log likelihood = -6.8424747
Iteration 5:   log likelihood = -6.8424491
Iteration 6:   log likelihood = -6.8424477
```

```
Factor analysis/correlation                        Number of obs    =        568
    Method: maximum likelihood                     Retained factors =          2
    Rotation: (unrotated)                          Number of params =         11
                                                    Schwarz's BIC   =    83.4482
    Log likelihood = -6.842448                     (Akaike's) AIC   =    35.6849

    --------------------------------------------------------------------------
         Factor  |   Eigenvalue   Difference        Proportion   Cumulative
    -------------+------------------------------------------------------------
        Factor1  |      1.02766      0.28115            0.5792       0.5792
        Factor2  |      0.74651            .            0.4208       1.0000
    --------------------------------------------------------------------------
    LR test: independent vs. saturated:  chi2(15) =  269.07 Prob>chi2 = 0.0000
    LR test:    2 factors vs. saturated:  chi2(4)  =   13.58 Prob>chi2 = 0.0087
Factor loadings (pattern matrix) and unique variances

    -------------------------------------------------
        Variable |  Factor1   Factor2 |   Uniqueness 
    -------------+--------------------+--------------
        bg2cost1 |  -0.1371    0.4235 |      0.8018  
        bg2cost2 |   0.4140    0.1994 |      0.7888  
        bg2cost3 |   0.6199    0.3692 |      0.4794  
        bg2cost4 |   0.3577    0.0909 |      0.8638  
        bg2cost5 |  -0.3752    0.4355 |      0.6695  
        bg2cost6 |  -0.4295    0.4395 |      0.6224  
    -------------------------------------------------
```

`factor` displays a likelihood-ratio test of independence against the saturated model with each estimation method. Because we are factor analyzing a correlation matrix, independence implies sphericity. Passing this test is necessary for a factor analysis to be meaningful.

In addition to the "standard" output, when you use the `ml` option, Stata reports a likelihood-ratio test of the number of factors in the model against the saturated model. This test is only approximately chi-squared, and we have used the correction recommended by Bartlett (1951). There are many variations on this test in use by different statistical packages.

The following comments were made by the analyst looking at these results: "There is, in my opinion, weak evidence of more than two factors. The χ^2 test for more than two factors is really a test of how well you are fitting the correlation matrix. It is not surprising that the model does not fit it perfectly. The significance of 1%, however, suggests to me that there might be a third factor. As for the loadings, they yield a similar interpretation to other factor models we fit, although there are some noteworthy differences." When we challenged the analyst on this last statement, he added that he would want to rotate the resulting factors before committing himself further.

◁

❑ Technical note

Stata will sometimes comment, "Note: test formally not valid because a Heywood case was encountered". The approximations used in computing the χ^2 value and degrees of freedom are mathematically justified on the assumption that an *interior* solution to the factor maximum likelihood was found. This is the case in our example above, but that will not always be so.

Boundary solutions, called Heywood solutions, often produce uniquenesses of 0, and then at least at a formal level, the test cannot be justified. Nevertheless, we believe that the reported tests are useful, even in such circumstances, provided that they are interpreted cautiously. The maximum likelihood method seems to be particularly prone to producing Heywood solutions.

This message is also printed when, in principle, there are enough free parameters to completely fit the correlation matrix, another sort of boundary solution. We say "in principle" because the correlation matrix often cannot be fit perfectly, so you will see a positive χ^2 with zero degrees of freedom. This warning note is printed because the geometric assumptions underlying the likelihood-ratio test are violated.

❑

❑ Technical note

In a factor analysis with factors estimated with the maximum likelihood method, there may possibly be more than one local maximum, and you may want assurances that the maximum reported is the global maximum. Multiple maximums are especially likely when there is more than one group of variables, the groups are reasonably uncorrelated, and you attempt to fit a model with too few factors.

When you specify the `protect(#)` option, Stata performs # optimizations of the likelihood function, beginning each with random starting values, before continuing with the squared multiple correlations–initialized solution. Stata then selects the maximum of the maximums and reports it, along with a note informing you if other local maximums were found. `protect(50)` provides considerable assurance.

If you then wish to explore any of the nonglobal maximums, include the `random` option. This option, which is never specified with `protect()`, uses random starting values and reports the solution to which those random values converge. For multiple maximums, giving the command repeatedly will eventually report all local maximums. You are advised to set the random-number seed to ensure that your results are reproducible; see [R] **set seed**.

❑

Factor analysis from a correlation matrix

You may want to perform a factor analysis directly from a correlation matrix rather than from variables in a dataset. You may not have access to the dataset, or you may have used another method of estimating a correlation matrix—for example, as a matrix of tetrachoric correlations; see [R] **tetrachoric**. You can provide either a correlation or a covariance matrix—`factormat` will translate a covariance matrix into a correlation matrix.

▷ Example 6

We illustrate with a small example with three variables on respondent's senses (visual, hearing, and taste), with a correlation matrix.

```
. matrix C = ( 1.000, 0.943,  0.771 \
               0.943, 1.000,  0.605 \
               0.771, 0.605,  1.000 )
```

Elements within a row are separated by a comma, whereas rows are separated by a backslash, \. We now use `factormat` to analyze `C`. There are two required options here. First, the option `n(979)` specifies that the sample size is 979. Second, `factormat` has to have labels for the variables. It is possible to define row and column names for `C`. We did not explicitly set the names of `C`, so Stata has generated default row and columns names—`r1 r2 r3` for the rows, and `c1 c2 c3` for the columns. This will confuse `factormat`: why does a symmetric correlation matrix have different names for the rows and for the columns? `factormat` would complain about the problem and stop. We could set the row and column names of `C` to be the same and invoke `factormat` again. We can also specify the `names()` option with the variable names to be used.

```
. factormat C, n(979) names(visual hearing taste) fac(1) ipf
(obs=979)

Factor analysis/correlation                    Number of obs    =       979
    Method: iterated principal factors         Retained factors =         1
    Rotation: (unrotated)                      Number of params =         3

    Beware: solution is a Heywood case
            (i.e., invalid or boundary values of uniqueness)

        Factor  |   Eigenvalue   Difference        Proportion   Cumulative
    ------------+----------------------------------------------------------
       Factor1  |      2.43622      2.43609            1.0000       1.0000
       Factor2  |      0.00013      0.00028            0.0001       1.0001
       Factor3  |     -0.00015            .           -0.0001       1.0000

    LR test: independent vs. saturated:  chi2(3)  = 3425.87 Prob>chi2 = 0.0000

Factor loadings (pattern matrix) and unique variances

      Variable |  Factor1 |   Uniqueness
    -----------+----------+--------------
        visual |   1.0961 |     -0.2014
       hearing |   0.8603 |      0.2599
         taste |   0.7034 |      0.5053
```

If we have the correlation matrix already in electronic form, this is a fine method. But if we have to enter a correlation matrix by hand, we may rather want to exploit its symmetry to enter just the upper triangle or lower triangle. This is not an issue with our small three-variable example, but what about a correlation matrix of 25 variables? However, there is an advantage to entering the correlation matrix in full symmetric form: redundancy offers some protection against making data-entry errors; `factormat` will complain if the matrix is not symmetric.

`factormat` allows us to enter just one of the triangles of the correlation matrix as a vector, that is, a matrix with one row or column. We enter the upper triangle, including the diagonal,

```
. matrix Cup = (1.000, 0.943, 0.771,
                       1.000, 0.605,
                              1.000)
```

All elements are separated by a comma; indentation and the use of three lines are done for readability. We could have typed, all the numbers "in a row".

```
. matrix Cup = (1.000, 0.943, 0.771, 1.000, 0.605, 1.000)
```

We have to specify the option `shape(upper)` to inform `factormat` that the elements in the vector `Cup` are the upper triangle in rowwise order.

```
. factormat Cup, n(979) shape(upper) fac(2) names(visual hearing taste)
  (output omitted)
```

If we had entered the lower triangle of `C`, a vector `Clow`, it would have been defined as

```
. matrix Clow = ( 1.000, 0.943, 1.000, 0.771, 0.605, 1.000 )
```

The features of `factormat` and `factor` are the same for estimation. Postestimation facilities are also the same—except that `predict` will not work after `factormat`, unless variables corresponding to the `names()` option exist in the dataset; see [MV] **factor postestimation**.

◁

Saved results

factor and factormat save the following in e():

Scalars

e(N)	number of observations
e(f)	number of retained factors
e(evsum)	sum of all eigenvalues
e(df_m)	model degrees of freedom
e(df_r)	residual degrees of freedom
e(chi2_i)	likelihood-ratio test of "independence vs. saturated"
e(df_i)	degrees of freedom of test of "independence vs. saturated"
e(p_i)	p-value of "independence vs. saturated"
e(ll_0)	log likelihood of null model (ml only)
e(ll)	log likelihood (ml only)
e(aic)	Akaike's AIC (ml only)
e(bic)	Schwarz's BIC (ml only)
e(chi2_1)	likelihood-ratio test of "# factors vs. saturated" (ml only)
e(df_1)	degrees of freedom of test of "# factors vs. saturated" (ml only)

Macros

e(cmd)	factor
e(cmdline)	command as typed
e(method)	pf, pcf, ipf, or ml
e(wtype)	weight type (factor only)
e(wexp)	weight expression (factor only)
e(title)	Factor analysis
e(mtitle)	description of method (e.g., principal factors)
e(heywood)	Heywood case (when encountered)
e(matrixname)	input matrix (factormat only)
e(mineigen)	specified mineigen() option
e(factors)	specified factors() option
e(seed)	starting random-number seed (seed() option only)
e(properties)	nob noV eigen
e(rotate_cmd)	factor_rotate
e(estat_cmd)	factor_estat
e(predict)	factor_p
e(marginsnotok)	predictions disallowed by margins

Matrices

e(sds)	standard deviations of analyzed variables
e(means)	means of analyzed variables
e(C)	analyzed correlation matrix
e(Phi)	variance matrix common factors
e(L)	factor loadings
e(Psi)	uniqueness (variance of specific factors)
e(Ev)	eigenvalues

Functions

e(sample)	marks estimation sample (factor only)

rotate after factor and factormat stores items in e() along with the estimation command. See *Saved results* of [MV] **factor postestimation** and [MV] **rotate** for details.

Before Stata version 9, factor returned results in r(). This behavior is retained under version control.

Methods and formulas

factor and factormat are implemented as ado-files.

This section describes the statistical factor model. Suppose that there are p variables and q factors. Let $\mathbf{\Psi}$ represent the $p \times p$ diagonal matrix of uniquenesses, and let $\mathbf{\Lambda}$ represent the $p \times q$ factor loading matrix. Let $\mathbf{f}$ be a $1 \times q$ matrix of factors. The standardized (mean 0, variance 1) vector of observed variables $\mathbf{x}$ $(1 \times p)$ is given by the system of regression equations

$$\mathbf{x} = \mathbf{f}\mathbf{\Lambda}' + \mathbf{e}$$

where $\mathbf{e}$ is a $1 \times p$ vector of errors with diagonal covariance equal to the uniqueness matrix $\mathbf{\Psi}$. The common factors $\mathbf{f}$ and the specific factors $\mathbf{e}$ are assumed to be uncorrelated.

Under the factor model, the correlation matrix of $\mathbf{x}$, called $\mathbf{\Sigma}$, is decomposed by factor analysis as

$$\mathbf{\Sigma} = \mathbf{\Lambda}\mathbf{\Phi}\mathbf{\Lambda}' + \mathbf{\Psi}$$

There is an obvious freedom in reexpressing a given decomposition of $\mathbf{\Sigma}$. The default and unrotated form assumes uncorrelated common factors, $\mathbf{\Phi} = \mathbf{I}$. Stata performs this decomposition by an eigenvector calculation. First, an estimate is found for the uniqueness $\mathbf{\Psi}$, and then the columns of $\mathbf{\Lambda}$ are computed as the q leading eigenvectors, scaled by the square root of the appropriate eigenvalue.

See Harman (1976); Mardia, Kent, and Bibby (1979); Rencher (1998, chap. 10); and Rencher (2002, chap. 13) for discussions of estimation methods in factor analysis. Basilevsky (1994) places factor analysis in a wider statistical context and details many interesting examples and links to other methods. For details about maximum likelihood estimation, see also Lawley and Maxwell (1971) and Clarke (1970).

References

Bartlett, M. S. 1951. The effect of standardization on a χ^2 approximation in factor analysis. *Biometrika* 38: 337–344.

Basilevsky, A. T. 1994. *Statistical Factor Analysis and Related Methods: Theory and Applications.* New York: Wiley.

Clarke, M. R. B. 1970. A rapidly convergent method for maximum-likelihood factor analysis. *British Journal of Mathematical and Statistical Psychology* 23: 43–52.

Dinno, A. 2009. Implementing Horn's parallel analysis for principal component analysis and factor analysis. *Stata Journal* 9: 291–298.

Fuller, W. A. 1987. *Measurement Error Models.* New York: Wiley.

Gorsuch, R. L. 1983. *Factor Analysis.* 2nd ed. Hillsdale, NJ: Lawrence Erlbaum.

Hamilton, L. C. 2009. *Statistics with Stata (Updated for Version 10).* Belmont, CA: Brooks/Cole.

Harman, H. H. 1976. *Modern Factor Analysis.* 3rd ed. Chicago: University of Chicago Press.

Kim, J. O., and C. W. Mueller. 1978a. Introduction to factor analysis. What it is and how to do it. In *Sage University Paper Series on Quantitative Applications the Social Sciences*, vol. 07–013. Thousand Oaks, CA: Sage.

——. 1978b. Factor analysis: Statistical methods and practical issues. In *Sage University Paper Series on Quantitative Applications the Social Sciences*, vol. 07–014. Thousand Oaks, CA: Sage.

Kolenikov, S. 2009. Confirmatory factor analysis using confa. *Stata Journal* 9: 329–373.

Lawley, D. N., and A. E. Maxwell. 1971. *Factor Analysis as a Statistical Method.* 2nd ed. London: Butterworths.

Mardia, K. V., J. T. Kent, and J. M. Bibby. 1979. *Multivariate Analysis.* London: Academic Press.

Milan, L., and J. Whittaker. 1995. Application of the parametric bootstrap to models that incorporate a singular value decomposition. *Applied Statistics* 44: 31–49.

Mulaik, S. A. 2010. *Foundations of Factor Analysis.* 2nd ed. Boca Raton, FL: Chapman & Hall/CRC.

Rao, C. R. 1955. Estimation and tests of significance in factor analysis. *Psychometrika* 20: 93–111.

Rencher, A. C. 1998. *Multivariate Statistical Inference and Applications.* New York: Wiley.

——. 2002. *Methods of Multivariate Analysis.* 2nd ed. New York: Wiley.

Spearman, C. 1904. The proof and measurement of association between two things. *American Journal of Psychology* 15: 72–101.

Tarlov, A. R., J. E. Ware, Jr., S. Greenfield, E. C. Nelson, E. Perrin, and M. Zubkoff. 1989. The medical outcomes study. An application of methods for monitoring the results of medical care. *Journal of the American Medical Association* 262: 925–930.

van Belle, G., L. D. Fisher, P. J. Heagerty, and T. S. Lumley. 2004. *Biostatistics: A Methodology for the Health Sciences*. 2nd ed. New York: Wiley.

Also see

[MV] **factor postestimation** — Postestimation tools for factor and factormat

[MV] **canon** — Canonical correlations

[MV] **pca** — Principal component analysis

[R] **alpha** — Compute interitem correlations (covariances) and Cronbach's alpha

Stata Structural Equation Modeling Reference Manual

[U] **20 Estimation and postestimation commands**

Title

factor postestimation — Postestimation tools for factor and factormat

Description

The following postestimation commands are of special interest after `factor` and `factormat`:

Command	Description
`estat anti`	anti-image correlation and covariance matrices
`estat common`	correlation matrix of the common factors
`estat factors`	AIC and BIC model-selection criteria for different numbers of factors
`estat kmo`	Kaiser–Meyer–Olkin measure of sampling adequacy
`estat residuals`	matrix of correlation residuals
`estat rotatecompare`	compare rotated and unrotated loadings
`estat smc`	squared multiple correlations between each variable and the rest
`estat structure`	correlations between variables and common factors
* `estat summarize`	estimation sample summary
`loadingplot`	plot factor loadings
`rotate`	rotate factor loadings
`scoreplot`	plot score variables
`screeplot`	plot eigenvalues

* `estat summarize` is not available after `factormat`.

For information about `loadingplot` and `scoreplot`, see [MV] **scoreplot**; for information about `rotate`, see [MV] **rotate**; for information about `screeplot`, see [MV] **screeplot**; and for all other commands, see below.

The following standard postestimation commands are also available:

Command	Description
* `estimates`	cataloging estimation results; see [R] **estimates**
† `predict`	predict regression or Bartlett scores

* `estimates table` is not allowed, and `estimates stats` is allowed only with the `ml` factor method.

† `predict` after `factormat` works only if you have variables in memory that match the names specified in `factormat`. `predict` assumes mean zero and standard deviation one unless the `means()` and `sds()` options of `factormat` were provided.

See the corresponding entries in the *Base Reference Manual* for details.

Special-interest postestimation commands

`estat anti` displays the anti-image correlation and anti-image covariance matrices. These are minus the partial covariance and minus the partial correlation matrices of all pairs of variables, holding all other variables constant.

`estat common` displays the correlation matrix of the common factors. For orthogonal factor loadings, the common factors are uncorrelated, and hence an identity matrix is shown. `estat common` is of more interest after oblique rotations.

`estat factors` displays model-selection criteria (AIC and BIC) for models with 1, 2, ..., # factors. Each model is estimated using maximum likelihood (that is, using the `ml` option of `factor`).

`estat kmo` specifies that the Kaiser–Meyer–Olkin (KMO) measure of sampling adequacy be displayed. KMO takes values between 0 and 1, with small values meaning that overall the variables have too little in common to warrant a factor analysis. Historically, the following labels are given to values of KMO (Kaiser 1974):

0.00 to 0.49	unacceptable
0.50 to 0.59	miserable
0.60 to 0.69	mediocre
0.70 to 0.79	middling
0.80 to 0.89	meritorious
0.90 to 1.00	marvelous

`estat residuals` displays the raw or standardized residuals of the observed correlations with respect to the fitted (reproduced) correlation matrix.

`estat rotatecompare` displays the unrotated factor loadings and the most recent rotated factor loadings.

`estat smc` displays the squared multiple correlations between each variable and all other variables. SMC is a theoretical lower bound for communality, so it is an upper bound for uniqueness. The `pf` factor method estimates the communalities by `smc`.

`estat structure` displays the factor structure, that is, the correlations between the variables and the common factors.

`estat summarize` displays summary statistics of the variables in the factor analysis over the estimation sample. This subcommand is, of course, not available after `factormat`.

`rotate` modifies the results of the last `factor` or `factormat` command to create a set of loadings that are more interpretable than those originally produced. A variety of orthogonal and oblique rotations are available, including varimax, orthomax, promax, and oblimin. See [MV] **rotate** for more details. `rotate` stores results along with the original estimation results so that replaying `factor` or `factormat` and other postestimation commands may refer to the unrotated as well as the rotated results.

Syntax for predict

`predict` [*type*] {*stub**|*newvarlist*} [*if*] [*in*] [, *statistic options*]

statistic	Description
Main	
`regression`	regression scoring method
`bartlett`	Bartlett scoring method

options	Description
Main	
norotated	use unrotated results, even when rotated results are available
notable	suppress table of scoring coefficients
format(%*fmt*)	format for displaying the scoring coefficients

Menu

Statistics > Postestimation > Predictions, residuals, etc.

Options for predict

Main

`regression` produces factors scored by the regression method.

`bartlett` produces factors scored by the method suggested by Bartlett (1937, 1938). This method produces unbiased factors, but they may be less accurate than those produced by the default regression method suggested by Thomson (1951). Regression-scored factors have the smallest mean squared error from the true factors but may be biased.

`norotated` specifies that unrotated factors be scored even when you have previously issued a `rotate` command. The default is to use rotated factors if they are available and unrotated factors otherwise.

`notable` suppresses the table of scoring coefficients.

`format(%`*fmt*`)` specifies the display format for scoring coefficients.

Syntax for estat

Anti-image correlation/covariance matrices

`estat anti [, nocorr nocov format(%fmt)]`

Correlation of common factors

`estat common [, norotated format(%fmt)]`

Model-selection criteria

`estat factors [, factors(#) detail]`

Sample adequacy measures

`estat kmo [, novar format(%fmt)]`

Residuals of correlation matrix

`estat residuals [, fitted obs sresiduals format(%fmt)]`

Comparison of rotated and unrotated loadings

`estat rotatecompare [, format(%fmt)]`

Squared multiple correlations

`estat smc [, format(%fmt)]`

Correlations between variables and common factors

`estat structure [, norotated format(%fmt)]`

Summarize variables for estimation sample

`estat summarize [, labels noheader noweights]`

Menu

Statistics > Postestimation > Reports and statistics

Options for estat

Main

`nocorr`, an option used with `estat anti`, suppresses the display of the anti-image correlation matrix.

`nocov`, an option used with `estat anti`, suppresses the display of the anti-image covariance matrix.

`format(%fmt)` specifies the display format. The defaults differ between the subcommands.

`norotated`, an option used with `estat common` and `estat structure`, requests that the displayed and returned results be based on the unrotated original factor solution rather than on the last rotation (orthogonal or oblique).

`factors(#)`, an option used with `estat factors`, specifies the maximum number of factors to include in the summary table.

`detail`, an option used with `estat factors`, presents the output from each run of `factor` (or `factormat`) used in the computations of the AIC and BIC values.

`novar`, an option used with `estat kmo`, suppresses the KMO measures of sampling adequacy for the variables in the factor analysis, displaying the overall KMO measure only.

`fitted`, an option used with `estat residuals`, displays the fitted (reconstructed) correlation matrix on the basis of the retained factors.

`obs`, an option used with `estat residuals`, displays the observed correlation matrix.

`sresiduals`, an option used with `estat residuals`, displays the matrix of standardized residuals of the correlations. Be careful when interpreting these residuals; see Jöreskog and Sörbom (1988).

`labels`, `noheader`, and `noweights` are the same as for the generic `estat summarize` command; see [R] **estat**.

Remarks

Remarks are presented under the following headings:

Postestimation statistics
Plots of eigenvalues, factor loadings, and scores
Rotating the factor loadings
Factor scores

Postestimation statistics

Many postestimation statistics are available after `factor` and `factormat`.

▷ Example 1

After `factor` and `factormat` there are several "classical" methods for assessing whether the variables have enough in common to have warranted the use of a factor model. One method is to examine the squared multiple correlations of each variable with all other variables—this is usually an upper bound to *communality* and thus a lower bound to $1 - uniqueness (= communality)$ of the variables.

```
. use http://www.stata-press.com/data/r12/bg2
(Physician-cost data)
. quietly factor bg2cost1-bg2cost6, factors(2) ml
. estat smc
Squared multiple correlations of variables with all other variables
```

Variable	smc
bg2cost1	0.1054
bg2cost2	0.1370
bg2cost3	0.1637
bg2cost4	0.0866
bg2cost5	0.1671
bg2cost6	0.1683

Other diagnostic tools, such as examining the anti-image correlation and anti-image covariance matrices (`estat anti`) and the Kaiser–Meyer–Olkin measure of sampling adequacy (`estat kmo`), are also available. See [MV] **pca postestimation** for an illustration of their use.

◁

▷ Example 2

Another set of postestimation tools help in determining the number of factors that should be retained. Later we will show the use of `screeplot` for producing a scree plot—a plot of the explained variance by the common factors. This is often used as a visual guide for selecting the number of factors to retain.

Some authors advocate the standard model information criteria AIC and BIC for determining the number of factors (Schwarz 1978; Akaike 1987). This presupposes that the factors are extracted by maximum likelihood. `estat factors` provides these measures.

```
. estat factors
Factor analysis with different numbers of factors (maximum likelihood)
```

#factors	loglik	df_m	df_r	AIC	BIC
1	-60.53727	6	9	133.0745	159.1273
2	-6.842448	11	4	35.6849	83.44823
3	-3.37e-12	15	0	30	95.13182

```
no Heywood cases encountered
```

The table shows the AIC and BIC statistics for the models with 1, 2, and 3 factors. The three-factor model is saturated, with 0 degrees of freedom. In this trivial case, and excluding the saturated case, both criteria select the two-factor model.

◁

▷ Example 3

Two `estat` subcommands display statistics that help in interpreting the model and the results—in particular after an oblique rotation. `estat structure` displays the *structure* matrix containing the correlations between the (manifest) variables and the common factors.

```
. estat structure
Structure matrix: correlations between variables and common factors
```

Variable	Factor1	Factor2
bg2cost1	-0.1371	0.4235
bg2cost2	0.4140	0.1994
bg2cost3	0.6199	0.3692
bg2cost4	0.3577	0.0909
bg2cost5	-0.3752	0.4355
bg2cost6	-0.4295	0.4395

This matrix of correlations coincides with the pattern matrix, that is, the matrix with factor loadings. This holds true for the unrotated factor solution as well as after an orthogonal rotation, such as a varimax rotation. It does not hold true after an oblique rotation. After an oblique rotation, the common factors are correlated. This correlation between the common factors also influences the correlation between the common factors and the manifest variables. The correlation matrix of the common factors is displayed by the `common` subcommand of `estat`. Because we have not yet rotated, we would see only an identity matrix. Later we show `estat common` output after an oblique rotation.

To assess the quality of a factor model, we may compare the observed correlation matrix $\mathbf{C}$ with the fitted ("reconstructed") matrix $\widehat{\boldsymbol{\Sigma}} = \widehat{\boldsymbol{\Lambda}}\widehat{\boldsymbol{\Phi}}\widehat{\boldsymbol{\Lambda}}' + \widehat{\boldsymbol{\Psi}}$ by examining the raw residuals $\mathbf{C} - \widehat{\boldsymbol{\Sigma}}$.

```
. estat residuals, obs fit
Observed correlations
```

Variable	bg2co~1	bg2co~2	bg2co~3	bg2co~4	bg2co~5	bg2co~6
bg2cost1	1.0000					
bg2cost2	0.0920	1.0000				
bg2cost3	0.0540	0.3282	1.0000			
bg2cost4	-0.0380	0.1420	0.2676	1.0000		
bg2cost5	0.2380	-0.1394	-0.0550	-0.0567	1.0000	
bg2cost6	0.2431	-0.0671	-0.1075	-0.1329	0.3524	1.0000

```
Fitted ("reconstructed") values for correlations
```

Variable	bg2co~1	bg2co~2	bg2co~3	bg2co~4	bg2co~5	bg2co~6
bg2cost1	1.0000					
bg2cost2	0.0277	1.0000				
bg2cost3	0.0714	0.3303	0.9999			
bg2cost4	-0.0106	0.1662	0.2553	1.0000		
bg2cost5	0.2359	-0.0685	-0.0718	-0.0946	1.0000	
bg2cost6	0.2450	-0.0902	-0.1040	-0.1137	0.3525	1.0000

```
Raw residuals of correlations (observed-fitted)
```

Variable	bg2co~1	bg2co~2	bg2co~3	bg2co~4	bg2co~5	bg2co~6
bg2cost1	-0.0000					
bg2cost2	0.0643	-0.0000				
bg2cost3	-0.0174	-0.0021	0.0001			
bg2cost4	-0.0274	-0.0242	0.0124	-0.0000		
bg2cost5	0.0021	-0.0709	0.0168	0.0379	0.0000	
bg2cost6	-0.0019	0.0231	-0.0035	-0.0193	-0.0002	-0.0000

To gauge the size of the residuals, `estat residuals` can also display the standardized residuals.

```
. estat residuals, sres
Standardized residuals of correlations
```

Variable	bg2co~1	bg2co~2	bg2co~3	bg2co~4	bg2co~5	bg2co~6
bg2cost1	-0.0001					
bg2cost2	1.5324	-0.0003				
bg2cost3	-0.4140	-0.0480	0.0011			
bg2cost4	-0.6538	-0.5693	0.2859	-0.0000		
bg2cost5	0.0484	-1.6848	0.3993	0.9003	0.0001	
bg2cost6	-0.0434	0.5480	-0.0836	-0.4560	-0.0037	-0.0000

Be careful when interpreting these standardized residuals, as they tend to be smaller than normalized residuals; that is, these residuals tend to have a smaller variance than 1 if the model is true (see Bollen [1989]).

◁

Plots of eigenvalues, factor loadings, and scores

Scree plots, factor loading plots, and score plots are easily obtained after `factor` and `factormat`.

▷ Example 4

The scree plot is a popular tool for determining the number of factors to be retained. A scree plot is a plot of the eigenvalues shown in decreasing order (Cattell 1966). We fit a factor model, extracting factors with the principal factor method.

```
. use http://www.stata-press.com/data/r12/sp2
. factor ghp31-ghp05, pcf
  (output omitted)
```

How many factors should we retain? We issue the `screeplot` command with the `mean` option, specifying that a horizontal line be plotted at the mean of the eigenvalues (a height of 1 because we are dealing with the eigenvalues of a correlation matrix).

```
. screeplot, mean
```

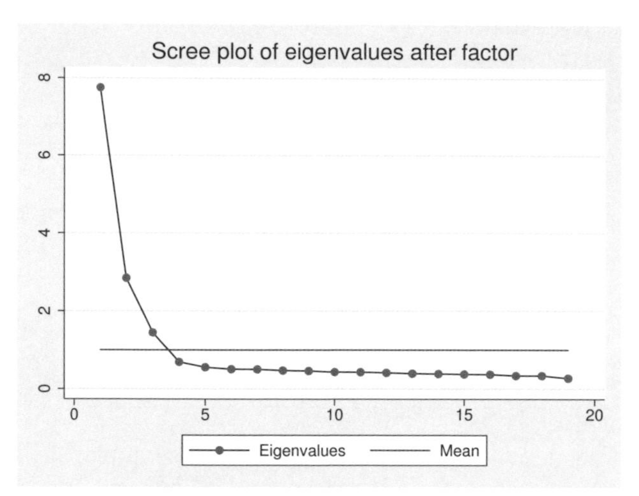

The plot suggests that we retain three factors, both because of the shape of the scree plot and because of Kaiser's well-known criterion suggesting that we retain factors with eigenvalue larger than 1. We may specify the option `mineigen(1)` during estimation to enforce this criterion. Here there is no need—`mineigen(1)` is the default with `pcf`.

◁

▷ Example 5

A second plot that is sometimes useful is the factor loadings plot. We display the plot with the loadings of the leading two factors.

```
. loadingplot, xline(0) yline(0) aspect(1) note(unrotated principal factors)
```

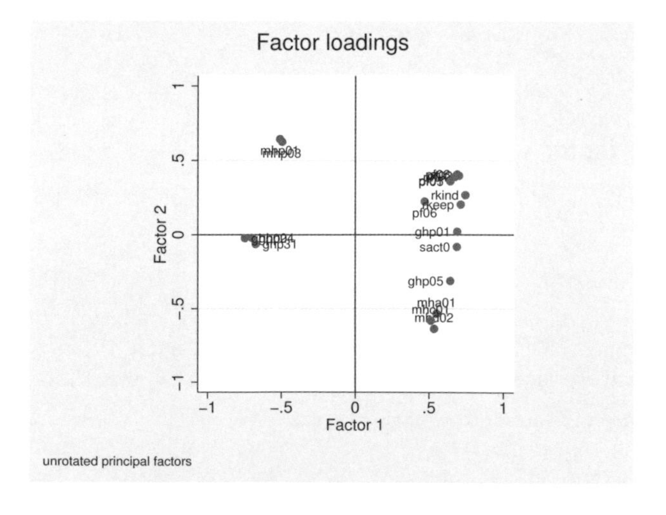

The plot makes it relatively easy to identify clusters of variables with similar loadings. With more than two factors, we can choose to see the multiple plots in a matrix style or a combined-graph style. The default is matrix style, but the combined style allows better control over various graph options—for instance, the addition of `xline(0)` and `yline(0)`. Here is a combined style graph.

```
. loadingplot, factors(3) combined xline(0) yline(0) aspect(1)
               xlabel(-0.8(0.4)0.8) ylabel(-0.8(0.4)0.8)
```

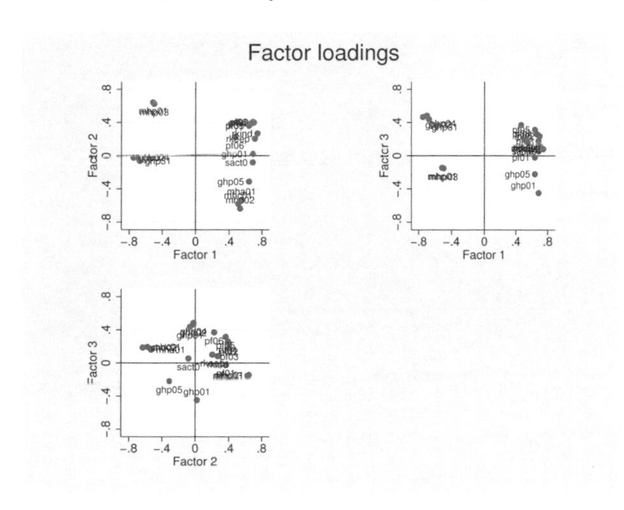

◁

▷ Example 6

Common factor scores can also be plotted for the observations by using the `scoreplot` command. (See the discussion of `predict` to see how you can produce score variables.)

```
. scoreplot, msymbol(smcircle) msize(tiny)
```

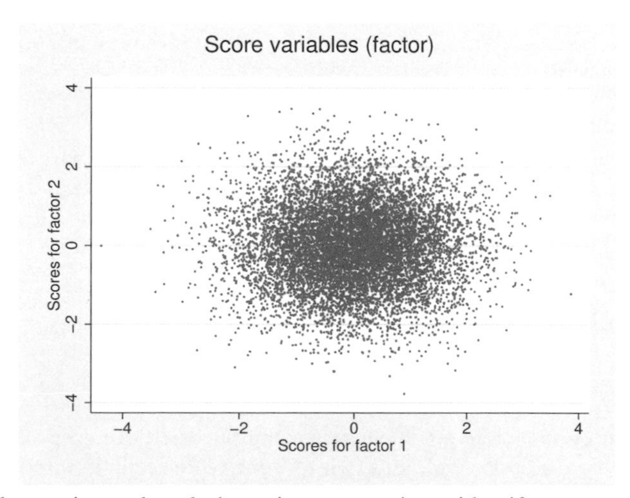

With so many observations, the plot's main purpose is to identify extreme cases. With smaller datasets with meaningful descriptions of the observations (for example, country names, brands), the score plot is good for visually clustering observations with similar loadings.

◁

See [MV] **scoreplot** for more examples of `loadingplot` and `scoreplot`.

❑ Technical note

The loading plots and score plots we have shown were for the original unrotated factor solution. After rotating (which we will discuss next), these plots display the most recent rotated solution. Specify option `norotated` to refer to the unrotated result. To display the plots of rotated and unrotated results at the same time, you may use either of the following two approaches. First, you may display them in different Graph windows.

```
. plotcmd, norotated name(name1)
. plotcmd, name(name2)
```

Alternatively, you may save the plots and create a combined graph

```
. plotcmd, norotated saving(name1)
. plotcmd, saving(name2)
. graph combine name1.gph name2.gph
```

See [G-2] **graph combine** for details.

❑

Rotating the factor loadings

Rotation is an attempt to describe the information in several factors by reexpressing them so that loadings on a few variables are as large as possible, and loadings on the rest of the variables are as small as possible. We have this freedom to reexpress because of the indeterminant nature of the factor model. For example, if you find that $\mathbf{z}_1$ and $\mathbf{z}_2$ are two factors, then $\mathbf{z}_1 + \mathbf{z}_2$ and $\mathbf{z}_1 - \mathbf{z}_2$ are equally valid solutions.

❑ Technical note

Said more technically: we are trying to find a set of f factor variables such that the observed variables can be best explained by regressing them on the f factor variables. Usually, f is a small number such as 1 or 2. If $f \geq 2$, there is an inherent indeterminacy in the construction of the factors because any linear combination of the calculated factors serves equally well as a set of regressors. Rotation capitalizes on this indeterminacy to create a set of variables that looks as much like the original variables as possible.

❑

The `rotate` command modifies the results of the last `factor` or `factormat` command to create a set of loadings that are more interpretable than those produced by `factor` or `factormat`. You may perform one factor analysis followed by several `rotate` commands, thus experimenting with different types of rotation. If you retain too few factors, the variables for several distinct concepts may be merged, as in our example below. If you retain too many factors, several factors may attempt to measure the same concept, causing the factors to get in each other's way, suggesting too many distinct concepts after rotation.

❑ Technical note

It is possible to restrict rotation to a number of leading factors. For instance, if you extracted three factors, you may specify the option `factors(2)` to `rotate` to exclude the third factor from being rotated. The new two leading factors are combinations of the initial two leading factors and are not affected by the fixed factor.

❑

▷ Example 7

We return to our physician-cost example and perform a factor analysis using the principal-component factor method, retaining two factors. We then tell `rotate` to apply the default orthogonal varimax rotation (Kaiser 1958).

```
. use http://www.stata-press.com/data/r12/bg2, clear
(Physician-cost data)
. quietly factor bg2cost1-bg2cost6, pcf factors(2)
. rotate
Factor analysis/correlation                        Number of obs    =      568
    Method: principal-component factors            Retained factors =        2
    Rotation: orthogonal varimax (Kaiser off)      Number of params =       11
```

Factor	Variance	Difference	Proportion	Cumulative
Factor1	1.57170	0.03430	0.2619	0.2619
Factor2	1.53740	.	0.2562	0.5182

```
    LR test: independent vs. saturated:  chi2(15) =  269.07 Prob>chi2 = 0.0000
Rotated factor loadings (pattern matrix) and unique variances
```

Variable	Factor1	Factor2	Uniqueness
bg2cost1	0.6853	0.2300	0.4775
bg2cost2	-0.0126	0.7142	0.4898
bg2cost3	-0.0161	0.7818	0.3886
bg2cost4	-0.1502	0.5703	0.6521
bg2cost5	0.7292	-0.1198	0.4539
bg2cost6	0.7398	-0.1537	0.4290

```
Factor rotation matrix
```

	Factor1	Factor2
Factor1	0.7460	-0.6659
Factor2	0.6659	0.7460

Here the factors are rotated so that the three "negative" items are grouped together and the three "positive" items are grouped.

Look at the uniqueness column. *Uniqueness* is the percentage of variance for the variable that is not explained by the common factors; we may also think of it as the variances of the specific factors for the variables. We stress that rotation involves the "common factors", so the *uniqueness* is not affected by the rotation. As we noted in [MV] **factor**, the uniqueness is relatively high in this example, placing doubt on the usefulness of the factor model here.

◁

▷ Example 8

Here we examine 19 variables describing various aspects of health. These variables were collected from a random selection of 9,999 visitors to doctors' offices by Tarlov et al. (1989). Factor analysis yields three clear factors. We then examine several rotations of these three factors.

```
. use http://www.stata-press.com/data/r12/sp2
. describe
Contains data from http://www.stata-press.com/data/r12/sp2.dta
  obs:         9,999
 vars:            20                          26 Jan 2011 09:26
 size:       779,922                           (_dta has notes)
--------------------------------------------------------------------------------
              storage  display     value
variable name   type   format      label      variable label
--------------------------------------------------------------------------------
patid           int    %9.0g                  Case ID
ghp31           float  %9.0g                  Health excellent, very good,
                                                good, fair, poor
pf01            float  %9.0g                  How long limit vigorous activity
pf02            float  %9.0g                  How long limit moderate activity
pf03            float  %9.0g                  How long limit walk/climb
pf04            float  %9.0g                  How long limit bend/stoop
pf05            float  %9.0g                  How long limit walk 1 block
pf06            float  %9.0g                  How long limit eat/dress/bath
rkeep           float  %9.0g                  Does health keep work-job-hse
rkind           float  %9.0g                  Can't do kind/amount of work
sact0           float  %9.0g                  Last month limit activities
mha01           float  %9.0g                  Last month very nervous
mhp03           float  %9.0g                  Last month calm/peaceful
mhd02           float  %9.0g                  Last month downhearted/blue
mhp01           float  %9.0g                  Last month a happy person
mhc01           float  %9.0g                  Last month down in the dumps
ghp01           float  %9.0g                  Somewhat ill
ghp04           float  %9.0g                  Healthy as anybody I know
ghp02           float  %9.0g                  Health is excellent
ghp05           float  %9.0g                  Feel bad lately
--------------------------------------------------------------------------------
Sorted by:  patid
```

We now perform our factorization, requesting that three factors be retained.

```
. factor ghp31-ghp05, factors(3)
(obs=9999)

Factor analysis/correlation                    Number of obs    =     9999
    Method: principal factors                  Retained factors =        3
    Rotation: (unrotated)                      Number of params =       54
```

Factor	Eigenvalue	Difference	Proportion	Cumulative
Factor1	7.27086	4.90563	0.7534	0.7534
Factor2	2.36523	1.38826	0.2451	0.9985
Factor3	0.97697	1.00351	0.1012	1.0997
Factor4	-0.02654	0.00538	-0.0027	1.0970
Factor5	-0.03191	0.00378	-0.0033	1.0937
Factor6	-0.03569	0.00353	-0.0037	1.0900
Factor7	-0.03922	0.00271	-0.0041	1.0859
Factor8	-0.04193	0.00662	-0.0043	1.0815
Factor9	-0.04855	0.01015	-0.0050	1.0765
Factor10	-0.05870	0.00250	-0.0061	1.0704
Factor11	-0.06120	0.00224	-0.0063	1.0641
Factor12	-0.06344	0.00376	-0.0066	1.0575
Factor13	-0.06720	0.00345	-0.0070	1.0506
Factor14	-0.07065	0.00185	-0.0073	1.0432
Factor15	-0.07250	0.00033	-0.0075	1.0357
Factor16	-0.07283	0.00772	-0.0075	1.0282
Factor17	-0.08055	0.01190	-0.0083	1.0198
Factor18	-0.09245	0.00649	-0.0096	1.0103
Factor19	-0.09894	.	-0.0103	1.0000

```
LR test: independent vs. saturated: chi2(171) = 1.0e+05 Prob>chi2 = 0.0000

Factor loadings (pattern matrix) and unique variances
```

Variable	Factor1	Factor2	Factor3	Uniqueness
ghp31	-0.6519	-0.0562	0.3440	0.4535
pf01	0.6150	0.3226	-0.0072	0.5177
pf02	0.6867	0.3737	0.2175	0.3415
pf03	0.6712	0.3774	0.1621	0.3807
pf04	0.6540	0.3588	0.2268	0.3921
pf05	0.6209	0.3258	0.2631	0.4392
pf06	0.4370	0.1803	0.2241	0.7263
rkeep	0.6868	0.1820	0.0870	0.4876
rkind	0.7244	0.2464	0.0780	0.4085
sact0	0.6556	-0.0719	0.0461	0.5628
mha01	0.5297	-0.4773	0.1268	0.4755
mhp03	-0.4810	0.5691	-0.1238	0.4294
mhd02	0.5208	-0.5949	0.1623	0.3485
mhp01	-0.4980	0.5955	-0.1225	0.3824
mhc01	0.4927	-0.5215	0.1531	0.4618
ghp01	0.6686	0.0194	-0.3621	0.4215
ghp04	-0.6833	-0.0195	0.4089	0.3656
ghp02	-0.7398	-0.0227	0.4212	0.2748
ghp05	0.6163	-0.2760	-0.1626	0.5175

The first factor is a general health factor. (To understand that claim, compare the factor loadings with the description of the variables as shown by `describe` above. Also, just as with the physician-cost data, the sense of some of the coded responses is reversed.) The second factor loads most highly on the five "mental health" items. The third factor loads most highly on "general health perception" items—those with names having the letters ghp in them. The other items describe "physical health".

These designations are based primarily on the wording of the questions, which is summarized in the variable labels.

```
. rotate, varimax
Factor analysis/correlation                        Number of obs    =     9999
    Method: principal factors                      Retained factors =        3
    Rotation: orthogonal varimax (Kaiser off)      Number of params =       54
```

Factor	Variance	Difference	Proportion	Cumulative
Factor1	4.20556	0.83302	0.4358	0.4358
Factor2	3.37253	0.33756	0.3495	0.7852
Factor3	3.03497	.	0.3145	1.0997

```
    LR test: independent vs. saturated: chi2(171) = 1.0e+05 Prob>chi2 = 0.0000
Rotated factor loadings (pattern matrix) and unique variances
```

Variable	Factor1	Factor2	Factor3	Uniqueness
ghp31	-0.2968	-0.1647	-0.6567	0.4535
pf01	0.5872	0.0263	0.3699	0.5177
pf02	0.7740	0.0848	0.2287	0.3415
pf03	0.7386	0.0580	0.2654	0.3807
pf04	0.7484	0.0842	0.2018	0.3921
pf05	0.7256	0.1063	0.1518	0.4392
pf06	0.5023	0.1268	0.0730	0.7263
rkeep	0.6023	0.2048	0.3282	0.4876
rkind	0.6590	0.1669	0.3597	0.4085
sact0	0.4187	0.3875	0.3342	0.5628
mha01	0.1467	0.6859	0.1803	0.4755
mhp03	-0.0613	-0.7375	-0.1514	0.4294
mhd02	0.0921	0.7893	0.1416	0.3485
mhp01	-0.0570	-0.7671	-0.1612	0.3824
mhc01	0.1102	0.7124	0.1359	0.4618
ghp01	0.2783	0.1977	0.6797	0.4215
ghp04	-0.2652	-0.1908	-0.7264	0.3656
ghp02	-0.2986	-0.2116	-0.7690	0.2748
ghp05	0.1755	0.4756	0.4748	0.5175

```
Factor rotation matrix
```

	Factor1	Factor2	Factor3
Factor1	0.6658	0.4796	0.5715
Factor2	0.5620	-0.8263	0.0387
Factor3	0.4908	0.2954	-0.8197

With rotation, the structure of the data becomes much clearer. The first rotated factor is physical health, the second is mental health, and the third is general health perception. The a priori designation of the items is confirmed.

After rotation, physical health is the first factor. `rotate` has ordered the factors by explained variance. Still, we warn that the importance of any factor must be gauged against the number of variables that purportedly measure it. Here we included nine variables that measured physical health, five that measured mental health, and five that measured general health perception. Had we started with only one mental health item, it would have had a high uniqueness, but we would not want to conclude that it was, therefore, largely noise.

◁

❑ Technical note

Some people prefer specifying the option `normalize` to apply a Kaiser normalization (Horst 1965), which places equal weight on all rows of the matrix to be rotated.

❑

▷ Example 9

The literature suggests that physical health and mental health are related. Also, general health perception may be largely a combination of the two. For these reasons, an oblique rotation of a two-factor solution is worth trying. We try the `oblique oblimin` rotation (Harman 1976).

```
. factor ghp31-ghp05, factors(2)
(obs=9999)

Factor analysis/correlation                  Number of obs    =      9999
    Method: principal factors                Retained factors =         2
    Rotation: (unrotated)                    Number of params =        37
```

Factor	Eigenvalue	Difference	Proportion	Cumulative
Factor1	7.27086	4.90563	0.7534	0.7534
Factor2	2.36523	1.38826	0.2451	0.9985
Factor3	0.97697	1.00351	0.1012	1.0997
Factor4	-0.02654	0.00538	-0.0027	1.0970
Factor5	-0.03191	0.00378	-0.0033	1.0937
Factor6	-0.03569	0.00353	-0.0037	1.0900
Factor7	-0.03922	0.00271	-0.0041	1.0859
Factor8	-0.04193	0.00662	-0.0043	1.0815
Factor9	-0.04855	0.01015	-0.0050	1.0765
Factor10	-0.05870	0.00250	-0.0061	1.0704
Factor11	-0.06120	0.00224	-0.0063	1.0641
Factor12	-0.06344	0.00376	-0.0066	1.0575
Factor13	-0.06720	0.00345	-0.0070	1.0506
Factor14	-0.07065	0.00185	-0.0073	1.0432
Factor15	-0.07250	0.00033	-0.0075	1.0357
Factor16	-0.07283	0.00772	-0.0075	1.0282
Factor17	-0.08055	0.01190	-0.0083	1.0198
Factor18	-0.09245	0.00649	-0.0096	1.0103
Factor19	-0.09894	.	-0.0103	1.0000

```
LR test: independent vs. saturated: chi2(171) = 1.0e+05 Prob>chi2 = 0.0000
```

```
Factor loadings (pattern matrix) and unique variances
```

Variable	Factor1	Factor2	Uniqueness
ghp31	-0.6519	-0.0562	0.5718
pf01	0.6150	0.3226	0.5178
pf02	0.6867	0.3737	0.3888
pf03	0.6712	0.3774	0.4070
pf04	0.6540	0.3588	0.4435
pf05	0.6209	0.3258	0.5084
pf06	0.4370	0.1803	0.7765
rkeep	0.6868	0.1820	0.4952
rkind	0.7244	0.2464	0.4145
sact0	0.6556	-0.0719	0.5650
mha01	0.5297	-0.4773	0.4916
mhp03	-0.4810	0.5691	0.4448
mhd02	0.5208	-0.5949	0.3748
mhp01	-0.4980	0.5955	0.3974
mhc01	0.4927	-0.5215	0.4853
ghp01	0.6686	0.0194	0.5526
ghp04	-0.6833	-0.0195	0.5327
ghp02	-0.7398	-0.0227	0.4522
ghp05	0.6163	-0.2760	0.5439

```
. rotate, oblimin oblique
Factor analysis/correlation                        Number of obs    =      9999
    Method: principal factors                      Retained factors =         2
    Rotation: oblique oblimin (Kaiser off)         Number of params =        37
```

Factor	Variance	Proportion	Rotated factors are correlated
Factor1	6.58719	0.6826	
Factor2	4.65444	0.4823	

```
    LR test: independent vs. saturated: chi2(171) = 1.0e+05 Prob>chi2 = 0.0000
Rotated factor loadings (pattern matrix) and unique variances
```

Variable	Factor1	Factor2	Uniqueness
ghp31	-0.5517	-0.2051	0.5718
pf01	0.7179	-0.0747	0.5178
pf02	0.8115	-0.0968	0.3888
pf03	0.8022	-0.1068	0.4070
pf04	0.7750	-0.0951	0.4435
pf05	0.7249	-0.0756	0.5084
pf06	0.4743	-0.0044	0.7765
rkeep	0.6712	0.0939	0.4952
rkind	0.7478	0.0449	0.4145
sact0	0.4608	0.3340	0.5650
mha01	0.0652	0.6869	0.4916
mhp03	0.0401	-0.7587	0.4448
mhd02	-0.0280	0.8003	0.3748
mhp01	0.0462	-0.7918	0.3974
mhc01	0.0039	0.7160	0.4853
ghp01	0.5378	0.2484	0.5526
ghp04	-0.5494	-0.2541	0.5327
ghp02	-0.5960	-0.2736	0.4522
ghp05	0.2805	0.5213	0.5439

Factor rotation matrix

	Factor1	Factor2
Factor1	0.9277	0.6831
Factor2	0.3733	-0.7303

The first factor is defined predominantly by physical health and the second by mental health. General health perception loads on both, but more on physical health than mental health. To compare the rotated and unrotated solution, looking at both in parallel form is often useful.

```
. estat rotatecompare
```

Rotation matrix — oblique oblimin (Kaiser off)

Variable	Factor1	Factor2
Factor1	0.9277	0.6831
Factor2	0.3733	-0.7303

Factor loadings

	Rotated		Unrotated	
Variable	Factor1	Factor2	Factor1	Factor2
ghp31	-0.5517	-0.2051	-0.6519	-0.0562
pf01	0.7179	-0.0747	0.6150	0.3226
pf02	0.8115	-0.0968	0.6867	0.3737
pf03	0.8022	-0.1068	0.6712	0.3774
pf04	0.7750	-0.0951	0.6540	0.3588
pf05	0.7249	-0.0756	0.6209	0.3258
pf06	0.4743	-0.0044	0.4370	0.1803
rkeep	0.6712	0.0939	0.6868	0.1820
rkind	0.7478	0.0449	0.7244	0.2464
sact0	0.4608	0.3340	0.6556	-0.0719
mha01	0.0652	0.6869	0.5297	-0.4773
mhp03	0.0401	-0.7587	-0.4810	0.5691
mhd02	-0.0280	0.8003	0.5208	-0.5949
mhp01	0.0462	-0.7918	-0.4980	0.5955
mhc01	0.0039	0.7160	0.4927	-0.5215
ghp01	0.5378	0.2484	0.6686	0.0194
ghp04	-0.5494	-0.2541	-0.6833	-0.0195
ghp02	-0.5960	-0.2736	-0.7398	-0.0227
ghp05	0.2805	0.5213	0.6163	-0.2760

Look again at the `factor` output. The variances of the first and second factor of the unrotated solution are 7.27 and 2.37, respectively. After an orthogonal rotation, the explained variance of $7.27 + 2.37$ is distributed differently over the two factors. For instance, after an orthogonal varimax rotation, the first factor has variance 5.75, and the second factor has 3.88—within rounding error $7.27 + 2.37 = 5.75 + 3.88$. The situation after an oblique rotation is different. The variances of the first and second factors are 6.59 and 4.65, which add up to more than in the orthogonal case. In the oblique case, the common factors are correlated and thus "partly explain the same variance". Therefore, the cumulative proportion of variance explained by the factors is not displayed here.

Most researchers would not be willing to accept a solution in which the common factors are highly correlated.

```
. estat common
Correlation matrix of the Oblimin(0) rotated common factors
```

Factors	Factor1	Factor2
Factor1	1	
Factor2	.3611	1

The correlation of .36 seems acceptable, so we think that the oblique rotation was a success here. ◁

Factor scores

The `predict` command creates a set of new variables that are estimates of the first k common factors produced by `factor`, `factormat`, or `rotate`. Two types of scoring are available: regression or Thomson scoring and Bartlett scoring.

The number of variables may be less than the number of factors. If so, the first such factors will be used. If the number of variables is greater than the number of factors created or rotated, the unused factors will be filled with missing values.

▷ Example 10

Using our automobile data, we wish to develop an index of roominess on the basis of a car's headroom, rear-seat leg room, and trunk space. We begin by extracting the factors of the three variables:

```
. use http://www.stata-press.com/data/r12/autofull
(Automobile Models)
. factor headroom rear_seat trunk
(obs=74)
Factor analysis/correlation                        Number of obs    =        74
    Method: principal factors                      Retained factors =         1
    Rotation: (unrotated)                          Number of params =         3
```

Factor	Eigenvalue	Difference	Proportion	Cumulative
Factor1	1.71426	1.79327	1.1799	1.1799
Factor2	-0.07901	0.10329	-0.0544	1.1255
Factor3	-0.18231	.	-0.1255	1.0000

```
    LR test: independent vs. saturated:  chi2(3)  =   82.93 Prob>chi2 = 0.0000
Factor loadings (pattern matrix) and unique variances
```

Variable	Factor1	Uniqueness
headroom	0.7280	0.4700
rear_seat	0.7144	0.4897
trunk	0.8209	0.3261

All the factor loadings are positive, so we have indeed obtained a "roominess" factor. The `predict` command will now create the one retained factor, which we will call `f1`:

```
. predict f1
(regression scoring assumed)
Scoring coefficients (method = regression)
```

Variable	Factor1
headroom	0.28323
rear_seat	0.26820
trunk	0.45964

The table with scoring coefficients informs us that the factor is obtained as a weighted sum of standardized versions of `headroom`, `rear_seat`, and `trunk` with weights 0.28, 0.26, and 0.46.

If `factor` had retained more than one factor, typing `predict f1` would still have added only the first factor to our data. Typing `predict f1 f2`, however, would have added the first two factors to our data. `f1` is now our "roominess" index, so we might compare the roominess of domestic and foreign cars:

```
. table foreign, c(mean f1 sd f1) row
```

Foreign	mean(f1)	sd(f1)
Domestic	.2022442	.9031404
Foreign	-.4780318	.6106609
Total	4.51e-09	.8804116

We find that domestic cars are, on average, roomier than foreign cars, at least in our data.

◁

❑ Technical note

Are common factors not supposed to be normalized to have mean 0 and standard deviation 1? In our example above, the mean is 4.5×10^{-9} and the standard deviation is 0.88. Why is that?

For the mean, the deviation from zero is due to numerical roundoff, which would diminish dramatically if we had typed `predict double f1` instead. The explanation for the standard deviation of 0.88, on the other hand, is not numerical roundoff. At a theoretical level, the factor is supposed to have standard deviation 1, but the estimation method almost never yields that result unless an exact solution to the factor model is found. This happens for the same reason that, when you regress y on x, you do not get the same equation as if you regress x on y, unless x and y are perfectly collinear.

By the way, if you had two factors, you would expect the correlation between the two factors to be zero because that is how they are theoretically defined. The matrix algebra, however, does not usually work out that way. It is somewhat analogous to the fact that if you regress y on x and the regression assumption that the errors are uncorrelated with the dependent variable is satisfied, then it automatically cannot be satisfied if you regress x on y.

The covariance matrix of the estimated factors is

$$E(\widehat{\mathbf{f}\mathbf{f}'}) = \mathbf{I} - (\mathbf{I} + \mathbf{\Gamma})^{-1}$$

where

$$\mathbf{\Gamma} = \mathbf{\Lambda}'\mathbf{\Psi}^{-1}\mathbf{\Lambda}$$

The columns of $\mathbf{\Lambda}$ are orthogonal to each other, but the inclusion of $\mathbf{\Psi}$ in the middle of the equation destroys that relationship unless all the elements of $\mathbf{\Psi}$ are equal.

❑

▷ Example 11

Let's pretend that we work for the K. E. Watt Company, a fictional industry group that generates statistics on automobiles. Our "roominess" index has mean 0 and standard deviation 0.88, but indexes we present to the public generally have mean 100 and standard deviation 10. First, we wish to rescale our index:

```
. generate roomidx = (f1/.88041161)*10 + 100
. table foreign, c(mean roomidx sd roomidx freq) row format(%9.2f)
```

Foreign	mean(roomidx)	sd(roomidx)	Freq.
Domestic	102.30	10.26	52
Foreign	94.57	6.94	22
Total	100.00	10.00	74

Now when we release our results, we can write, "The K. E. Watt index of roominess shows that domestic cars are, on average, roomier, with an index of 102 versus only 95 for foreign cars."

Now let's find the "roomiest" car in our data:

```
. sort roomidx
. list fullname roomidx in 1
```

	fullname	roomidx
74.	Merc. Marquis	116.7469

We can also write, "K. E. Watt finds that the Mercury Marquis is the roomiest automobile among those surveyed, with a roominess index of 117 versus an average of 100."

◁

❑ Technical note

`predict` provides two methods of scoring: the default regression scoring, which we have used above, and the optional Bartlett method. An artificial example will best illustrate the use and meaning of the methods. We begin by creating a known-to-be-correct factor model in which the true loadings are 0.4, 0.6, and 0.8. The variances of the unique factors are $1 - 0.4^2 = 0.84$, $1 - 0.6^2 = 0.64$, and $1 - 0.8^2 = 0.36$, respectively. We make the sample size large enough so that random fluctuations are not important.

```
. drop _all
. set seed 12345
. set obs 10000
obs was 0, now 10000
. generate ftrue = rnormal()
. generate x1 = .4*ftrue + sqrt(.84)*rnormal()
```

```
. generate x2 = .6*ftrue + sqrt(.64)*rnormal()
. generate x3 = .8*ftrue + sqrt(.36)*rnormal()
. summarize x1 x2 x3
    Variable |       Obs        Mean    Std. Dev.       Min        Max
-------------+--------------------------------------------------------
          x1 |     10000   -.0084217    1.001887  -3.804037   4.023879
          x2 |     10000   -.0190142     1.01522  -3.749022   3.870705
          x3 |     10000   -.0024562    1.002287  -3.606741   3.596839
```

Because we concocted our data, the iterated principal-factor method reproduces the true loadings most faithfully:

```
. factor x1 x2 x3, ipf factors(1)
(obs=10000)
Factor analysis/correlation                    Number of obs    =    10000
    Method: iterated principal factors         Retained factors =        1
    Rotation: (unrotated)                      Number of params =        3

    --------------------------------------------------------------------------
         Factor  |   Eigenvalue   Difference        Proportion   Cumulative
    -------------+------------------------------------------------------------
        Factor1  |      1.19856      1.19838            1.0000       1.0000
        Factor2  |      0.00018      0.00039            0.0001       1.0002
        Factor3  |     -0.00022            .           -0.0002       1.0000
    --------------------------------------------------------------------------
    LR test: independent vs. saturated:  chi2(3)  = 4029.71 Prob>chi2 = 0.0000
Factor loadings (pattern matrix) and unique variances

    -------------------------------------------
        Variable |  Factor1 |   Uniqueness 
    -------------+----------+--------------
              x1 |   0.4066 |      0.8346  
              x2 |   0.5946 |      0.6465  
              x3 |   0.8244 |      0.3203  
    -------------------------------------------
```

Let us now compare regression and Bartlett scoring:

```
. predict freg
(regression scoring assumed)
Scoring coefficients (method = regression)

    ----------------------------
        Variable |  Factor1 
    -------------+--------------
              x1 |  0.12596 
              x2 |  0.23782 
              x3 |  0.66561 
    ----------------------------

. predict fbar, bartlett
Scoring coefficients (method = Bartlett)

    ----------------------------
        Variable |  Factor1 
    -------------+--------------
              x1 |  0.16994 
              x2 |  0.32077 
              x3 |  0.89780 
    ----------------------------
```

Comparing the two scoring vectors, we see that Bartlett scoring yields larger coefficients. The regression scoring method is biased insofar as $E($freg$|$ftrue$)$ is not ftrue, something we can reveal by regressing freg on ftrue:

```
. regress freg ftrue

      Source |       SS       df       MS              Number of obs =   10000
-------------+------------------------------           F(  1,  9998) =26520.45
       Model |  5383.48877     1  5383.48877           Prob > F      =  0.0000
    Residual |  2029.53279  9998  .202993878           R-squared     =  0.7262
-------------+------------------------------           Adj R-squared =  0.7262
       Total |  7413.02156  9999  .741376294           Root MSE      =  .45055

------------------------------------------------------------------------------
        freg |      Coef.   Std. Err.      t    P>|t|     [95% Conf. Interval]
-------------+----------------------------------------------------------------
       ftrue |   .7309397   .0044884   162.85   0.000     .7221415    .7397379
       _cons |   .0047007   .0045056     1.04   0.297    -.0041312    .0135325
------------------------------------------------------------------------------
```

Note the coefficient on `ftrue` of $0.731 < 1$. The Bartlett scoring method, on the other hand, is unbiased:

```
. regress fbar ftrue

      Source |       SS       df       MS              Number of obs =   10000
-------------+------------------------------           F(  1,  9998) =26520.50
       Model |  9794.59885     1  9794.59885           Prob > F      =  0.0000
    Residual |  3692.47929  9998  .369321793           R-squared     =  0.7262
-------------+------------------------------           Adj R-squared =  0.7262
       Total |  13487.0781  9999   1.3488427           Root MSE      =  .60772

------------------------------------------------------------------------------
        fbar |      Coef.   Std. Err.      t    P>|t|     [95% Conf. Interval]
-------------+----------------------------------------------------------------
       ftrue |   .9859229   .0060541   162.85   0.000     .9740556    .9977903
       _cons |   .0063405   .0060773     1.04   0.297    -.0055723    .0182532
------------------------------------------------------------------------------
```

The zero bias of the Bartlett method comes at the costs of less accuracy, for example, in terms of the mean squared error.

```
. generate dbar = (fbar - ftrue)^2

. generate dreg = (freg - ftrue)^2

. summarize ftrue fbar freg dbar dreg

    Variable |       Obs        Mean    Std. Dev.       Min        Max
-------------+--------------------------------------------------------
       ftrue |     10000    -.006431    1.003858  -4.200537   3.712311
        fbar |     10000    1.00e-10    1.161397  -4.310825   4.389511
        freg |     10000   -7.55e-11    .8610321  -3.195944   3.254285
        dbar |     10000     .369489    .5175269   3.58e-09   4.625371
        dreg |     10000    .2759404     .387082   4.62e-11   4.098495
```

Neither estimator follows the assumption that the scaled factor has unit variance. The regression estimator has a variance less than 1, and the Bartlett estimator has a variance greater than 1.

The difference between the two scoring methods is not as important as it might seem because the bias in the regression method is only a matter of scaling and shifting.

```
. correlate freg fbar ftrue
(obs=10000)

             |     freg     fbar    ftrue
-------------+---------------------------
        freg |   1.0000
        fbar |   1.0000   1.0000
       ftrue |   0.8522   0.8522   1.0000
```

Therefore, the choice of which scoring method we apply is largely immaterial.

❑

Saved results

Let p be the number of variables and f, the number of factors.

predict, in addition to generating variables, also saves the following in r():

Macros

r(method)	regression or Bartlett

Matrices

r(scoef)	$p \times f$ matrix of scoring coefficients

estat anti saves the following in r():

Matrices

r(acov)	$p \times p$ anti-image covariance matrix
r(acorr)	$p \times p$ anti-image correlation matrix

estat common saves the following in r():

Matrices

r(Phi)	$f \times f$ correlation matrix of common factors

estat factors saves the following in r():

Matrices

r(stats)	$k \times 5$ matrix with log likelihood, degrees of freedom, AIC, and BIC for models with 1 to k factors estimated via maximum likelihood

estat kmo saves the following in r():

Scalars

r(kmo)	the Kaiser–Meyer–Olkin measure of sampling adequacy

Matrices

r(kmow)	column vector of KMO measures for each variable

estat residuals saves the following in r():

Matrices

r(fit)	fitted matrix for the correlations, $\widehat{\mathbf{C}}=\widehat{\mathbf{\Lambda}}\widehat{\mathbf{\Phi}}\widehat{\mathbf{\Lambda}}'+\widehat{\mathbf{\Psi}}$
r(res)	raw residual matrix $\mathbf{C}-\widehat{\mathbf{C}}$
r(SR)	standardized residuals (sresiduals option only)

estat smc saves the following in r():

Matrices

r(smc)	vector of squared multiple correlations of variables with all other variables

estat structure saves the following in r():

Matrices

r(st)	$p \times f$ matrix of correlations between variables and common factors

See [R] **estat** for the saved results of estat summarize.

`rotate` after `factor` and `factormat` add to the existing `e()`:

Scalars

`e(r_f)`	number of factors in rotated solution
`e(r_fmin)`	rotation criterion value

Macros

`e(r_class)`	`orthogonal` or `oblique`
`e(r_criterion)`	rotation criterion
`e(r_ctitle)`	title for rotation
`e(r_normalization)`	`kaiser` or `none`

Matrices

`e(r_L)`	rotated loadings
`e(r_T)`	rotation
`e(r_Phi)`	correlations between common factors
`e(r_Ev)`	explained variance by common factors

The factors in the rotated solution are in decreasing order of `e(r_Ev)`.

Methods and formulas

All postestimation commands listed above are implemented as ado-files.

Methods and formulas are presented under the following headings:

estat
rotate
predict

estat

See *Methods and formulas* of [MV] **pca postestimation** for the formulas for `estat anti`, `estat kmo`, and `estat smc`.

`estat residuals` computes the standardized residuals $\widetilde{r}_{ij}$ as

$$\widetilde{r}_{ij} = \frac{\sqrt{N}(r_{ij} - f_{ij})}{\sqrt{f_{ij}^2 + f_{ii}f_{jj}}}$$

suggested by Jöreskog and Sörbom (1986), where N is the number of observations, r_{ij} is the observed correlation of variables i and j, and f_{ij} is the fitted correlation of variables i and j. Also see Bollen (1989). Caution is warranted in interpretation of these residuals; see Jöreskog and Sörbom (1988).

`estat structure` computes the correlations of the variables and the common factors as $\mathbf{\Lambda\Phi}$.

rotate

See *Methods and formulas* of [MV] **rotatemat** for the details of rotation.

The correlation of common factors after rotation is $\mathbf{T}'\mathbf{T}$, where $\mathbf{T}$ is the factor rotation matrix, satisfying $\mathbf{L}_{\text{rotated}} = \mathbf{L}_{\text{unrotated}}(\mathbf{T}')^{-1}$

predict

The formula for regression scoring (Thomson 1951) in the orthogonal case is

$$\widehat{\mathbf{f}} = \mathbf{\Lambda}'\mathbf{\Sigma}^{-1}\mathbf{x}$$

where $\mathbf{\Lambda}$ is the unrotated or orthogonally rotated loading matrix. For oblique rotation, the regression scoring is defined as

$$\widehat{\mathbf{f}} = \mathbf{\Phi}\mathbf{\Lambda}'\mathbf{\Sigma}^{-1}\mathbf{x}$$

where $\mathbf{\Phi}$ is the correlation matrix of the common factors.

The formula for Bartlett scoring (Bartlett 1937, 1938) is

$$\mathbf{\Gamma}^{-1}\mathbf{\Lambda}'\mathbf{\Psi}^{-1}\mathbf{x}$$

where

$$\mathbf{\Gamma} = \mathbf{\Lambda}'\mathbf{\Psi}^{-1}\mathbf{\Lambda}$$

See Harman (1976) and Lawley and Maxwell (1971).

References

Akaike, H. 1987. Factor analysis and AIC. *Psychometrika* 52: 317–332.

Bartlett, M. S. 1937. The statistical conception of mental factors. *British Journal of Psychology* 28: 97–104.

——. 1938. Methods of estimating mental factors. *Nature, London* 141: 609–610.

Bollen, K. A. 1989. *Structural Equations with Latent Variables.* New York: Wiley.

Cattell, R. B. 1966. The scree test for the number of factors. *Multivariate Behavioral Research* 1: 245–276.

Harman, H. H. 1976. *Modern Factor Analysis.* 3rd ed. Chicago: University of Chicago Press.

Horst, P. 1965. *Factor Analysis of Data Matrices.* New York: Holt, Rinehart & Winston.

Jöreskog, K. G., and D. Sörbom. 1986. *Lisrel VI: Analysis of linear structural relationships by the method of maximum likelihood.* Mooresville, IN: Scientific Software.

——. 1988. *PRELIS: A program for multivariate data screening and data summarization. A preprocessor for LISREL.* 2nd ed. Mooresville, IN: Scientific Software.

Kaiser, H. F. 1958. The varimax criterion for analytic rotation in factor analysis. *Psychometrika* 23: 187–200.

——. 1974. An index of factor simplicity. *Psychometrika* 39: 31–36.

Lawley, D. N., and A. E. Maxwell. 1971. *Factor Analysis as a Statistical Method.* 2nd ed. London: Butterworths.

Schwarz, G. 1978. Estimating the dimension of a model. *Annals of Statistics* 6: 461–464.

Tarlov, A. R., J. E. Ware, Jr., S. Greenfield, E. C. Nelson, E. Perrin, and M. Zubkoff. 1989. The medical outcomes study. An application of methods for monitoring the results of medical care. *Journal of the American Medical Association* 262: 925–930.

Thomson, G. H. 1951. *The Factorial Analysis of Human Ability.* London: University of London Press.

Also see *References* in [MV] **factor**.

Also see

[MV] **factor** — Factor analysis

[MV] **rotate** — Orthogonal and oblique rotations after factor and pca

[MV] **scoreplot** — Score and loading plots

[MV] **screeplot** — Scree plot

Title

hotelling — Hotelling's T-squared generalized means test

Syntax

`hotelling` *varlist* [*if*] [*in*] [*weight*] [, `by(`*varname*`)` `notable`]

`aweight`s and `fweight`s are allowed; see [U] **11.1.6 weight**.

Note: `hotel` is a synonym for `hotelling`.

Menu

Statistics > Multivariate analysis > MANOVA, multivariate regression, and related > Hotelling's generalized means test

Description

`hotelling` performs Hotelling's T-squared test of whether a set of means is zero or, alternatively, equal between two groups.

See [MV] **mvtest means** for generalizations of Hotelling's one-sample test with more general hypotheses, two-sample tests that do not assume that the covariance matrices are the same in the two groups, and tests with more than two groups.

Options

Main

`by(`*varname*`)` specifies a variable identifying two groups; the test of equality of means between groups is performed. If `by()` is not specified, a test of means being jointly zero is performed.

`notable` suppresses printing a table of the means being compared.

Remarks

`hotelling` performs Hotelling's T-squared test of whether a set of means is zero or two sets of means are equal. It is a multivariate test that reduces to a standard t test if only one variable is specified.

▷ Example 1

You wish to test whether a new fuel additive improves gas mileage in both stop-and-go and highway situations. Taking 12 cars, you fill them with gas and run them on a highway-style track, recording their gas mileage. You then refill them and run them on a stop-and-go style track. Finally, you repeat the two runs, but this time you use fuel with the additive. Your dataset is

```
. use http://www.stata-press.com/data/r12/gasexp
. describe
Contains data from http://www.stata-press.com/data/r12/gasexp.dta
  obs:            12
 vars:             5                          15 Oct 2010 06:37
 size:           240
------------------------------------------------------------------------------
              storage  display     value
variable name   type   format      label      variable label
------------------------------------------------------------------------------
id              float  %9.0g                  car id
bmpg1           float  %9.0g                  track1 before additive
ampg1           float  %9.0g                  track1 after additive
bmpg2           float  %9.0g                  track 2 before additive
ampg2           float  %9.0g                  track 2 after additive
------------------------------------------------------------------------------
Sorted by:
```

To perform the statistical test, you jointly test whether the differences in before-and-after results are zero:

```
. gen diff1 = ampg1 - bmpg1
. gen diff2 = ampg2 - bmpg2
. hotelling diff1 diff2
    Variable |       Obs        Mean    Std. Dev.       Min        Max
-------------+--------------------------------------------------------
       diff1 |        12        1.75     2.70101         -3          5
       diff2 |        12    2.083333    2.906367       -3.5        5.5
1-group Hotelling's T-squared = 9.6980676
F test statistic: ((12-2)/(12-1)(2)) x 9.6980676 = 4.4082126
H0: Vector of means is equal to a vector of zeros
              F(2,10) =     4.4082
       Prob > F(2,10) =     0.0424
```

The means are different at the 4.24% significance level.

◁

❑ Technical note

We used Hotelling's T-squared test because we were testing two differences jointly. Had there been only one difference, we could have used a standard t test, which would have yielded the same results as Hotelling's test:

```
* We could have performed the test like this:
. ttest ampg1 = bmpg1
Paired t test
------------------------------------------------------------------------------
Variable |     Obs        Mean    Std. Err.   Std. Dev.   [95% Conf. Interval]
---------+--------------------------------------------------------------------
   ampg1 |      12       22.75    .9384465    3.250874    20.68449    24.81551
   bmpg1 |      12          21    .7881701    2.730301    19.26525    22.73475
---------+--------------------------------------------------------------------
    diff |      12        1.75    .7797144     2.70101    .0338602     3.46614
------------------------------------------------------------------------------
     mean(diff) = mean(ampg1 - bmpg1)                             t =   2.2444
 Ho: mean(diff) = 0                              degrees of freedom =       11
 Ha: mean(diff) < 0           Ha: mean(diff) != 0           Ha: mean(diff) > 0
 Pr(T < t) = 0.9768         Pr(|T| > |t|) = 0.0463          Pr(T > t) = 0.0232
```

```
* Or like this:
. ttest diff1 = 0

One-sample t test
------------------------------------------------------------------------------
Variable |     Obs        Mean    Std. Err.   Std. Dev.   [95% Conf. Interval]
---------+--------------------------------------------------------------------
   diff1 |      12        1.75    .7797144     2.70101    .0338602     3.46614
------------------------------------------------------------------------------
    mean = mean(diff1)                                            t =   2.2444
Ho: mean = 0                                     degrees of freedom =       11

    Ha: mean < 0                 Ha: mean != 0                 Ha: mean > 0
 Pr(T < t) = 0.9768         Pr(|T| > |t|) = 0.0463          Pr(T > t) = 0.0232

* Or like this:
. hotel diff1
    Variable |       Obs        Mean    Std. Dev.       Min        Max
-------------+--------------------------------------------------------
       diff1 |        12        1.75     2.70101         -3          5

1-group Hotelling's T-squared = 5.0373832
F test statistic: ((12-1)/(12-1)(1)) x 5.0373832 = 5.0373832

H0: Vector of means is equal to a vector of zeros
              F(1,11) =     5.0374
          Prob > F(1,11) =     0.0463
```

❑

▷ Example 2

Now consider a variation on the experiment: rather than using 12 cars and running each car with and without the fuel additive, you run 24 cars, 12 with the additive and 12 without. You have the following dataset:

```
. use http://www.stata-press.com/data/r12/gasexp2, clear

. describe

Contains data from http://www.stata-press.com/data/r12/gasexp2.dta
  obs:            24
 vars:             4                          17 Oct 2010 01:43
 size:           384
------------------------------------------------------------------------------
              storage  display     value
variable name   type   format      label      variable label
------------------------------------------------------------------------------
id              float  %9.0g                  car id
mpg1            float  %9.0g                  track 1
mpg2            float  %9.0g                  track 2
additive        float  %9.0g       yesno      additive?
------------------------------------------------------------------------------
Sorted by:

. tabulate additive

  additive? |      Freq.     Percent        Cum.
------------+-----------------------------------
         no |         12       50.00       50.00
        yes |         12       50.00      100.00
------------+-----------------------------------
      Total |         24      100.00
```

This is an unpaired experiment because there is no natural pairing of the cars; you want to test that the means of `mpg1` are equal for the two groups specified by `additive`, as are the means of `mpg2`:

```
. hotelling mpg1 mpg2, by(additive)

-> additive = no
    Variable |     Obs        Mean   Std. Dev.       Min        Max
-------------+-----------------------------------------------------
        mpg1 |      12          21   2.730301         17         25
        mpg2 |      12    19.91667   2.644319         16         24

-> additive = yes
    Variable |     Obs        Mean   Std. Dev.       Min        Max
-------------+-----------------------------------------------------
        mpg1 |      12       22.75   3.250874         17         28
        mpg2 |      12          22   3.316625       16.5       27.5

2-group Hotelling's T-squared = 7.1347584
F test statistic: ((24-2-1)/(24-2)(2)) x 7.1347584 = 3.4052256

H0: Vectors of means are equal for the two groups
              F(2,21) =     3.4052
       Prob > F(2,21) =     0.0524
```

◁

❑ Technical note

As in the paired experiment, had there been only one test track, the t test would have yielded the same results as Hotelling's test:

```
. hotel mpg1, by(additive)

-> additive = no
    Variable |     Obs        Mean   Std. Dev.       Min        Max
-------------+-----------------------------------------------------
        mpg1 |      12          21   2.730301         17         25

-> additive = yes
    Variable |     Obs        Mean   Std. Dev.       Min        Max
-------------+-----------------------------------------------------
        mpg1 |      12       22.75   3.250874         17         28

2-group Hotelling's T-squared = 2.0390921
F test statistic: ((24-1-1)/(24-2)(1)) x 2.0390921 = 2.0390921

H0: Vectors of means are equal for the two groups
              F(1,22) =     2.0391
       Prob > F(1,22) =     0.1673

. ttest mpg1, by(additive)

Two-sample t test with equal variances
------------------------------------------------------------------------------
   Group |     Obs        Mean    Std. Err.   Std. Dev.   [95% Conf. Interval]
---------+--------------------------------------------------------------------
      no |      12          21    .7881701    2.730301    19.26525    22.73475
     yes |      12       22.75    .9384465    3.250874    20.68449    24.81551
---------+--------------------------------------------------------------------
combined |      24      21.875    .6264476    3.068954    20.57909    23.17091
---------+--------------------------------------------------------------------
    diff |               -1.75    1.225518               -4.291568    .7915684
------------------------------------------------------------------------------
    diff = mean(no) - mean(yes)                                   t =  -1.4280
Ho: diff = 0                                     degrees of freedom =       22

    Ha: diff < 0                 Ha: diff != 0                 Ha: diff > 0
 Pr(T < t) = 0.0837         Pr(|T| > |t|) = 0.1673          Pr(T > t) = 0.9163
```

With more than one pair of means, however, there is no t test equivalent to Hotelling's test, although there are other logically (but not practically) equivalent solutions. One is the discriminant function: if the means of `mpg1` and `mpg2` are different, the discriminant function should separate the groups along that dimension.

```
. regress additive mpg1 mpg2
      Source |       SS       df       MS              Number of obs =      24
-------------+------------------------------           F(  2,    21) =    3.41
       Model |  1.46932917     2  .734664585           Prob > F      =  0.0524
    Residual |  4.53067083    21   .21574623           R-squared     =  0.2449
-------------+------------------------------           Adj R-squared =  0.1730
       Total |           6    23  .260869565           Root MSE      =  .46448

------------------------------------------------------------------------------
    additive |      Coef.   Std. Err.      t    P>|t|     [95% Conf. Interval]
-------------+----------------------------------------------------------------
        mpg1 |  -.4570407   .2416657    -1.89   0.072     -.959612    .0455306
        mpg2 |   .5014605   .2376762     2.11   0.047     .0071859    .9957352
       _cons |  -.0120115   .7437049    -0.02   0.987     -1.55863    1.534607
------------------------------------------------------------------------------
```

This test would declare the means different at the 5.24% level. You could also have fit this model by using logistic regression:

```
. logit additive mpg1 mpg2
Iteration 0:   log likelihood = -16.635532
Iteration 1:   log likelihood = -13.395178
Iteration 2:   log likelihood = -13.371971
Iteration 3:   log likelihood = -13.371143
Iteration 4:   log likelihood = -13.371143
Logistic regression                               Number of obs   =         24
                                                  LR chi2(2)      =       6.53
                                                  Prob > chi2     =     0.0382
Log likelihood = -13.371143                       Pseudo R2       =     0.1962

------------------------------------------------------------------------------
    additive |      Coef.   Std. Err.      z    P>|z|     [95% Conf. Interval]
-------------+----------------------------------------------------------------
        mpg1 |  -2.306844    1.36139    -1.69   0.090    -4.975119    .3614307
        mpg2 |   2.524477   1.367373     1.85   0.065    -.1555257     5.20448
       _cons |  -2.446527   3.689821    -0.66   0.507    -9.678443     4.78539
------------------------------------------------------------------------------
```

This test would have declared the means different at the 3.82% level.

Are the means different? Hotelling's T-squared and the discriminant function reject equality at the 5.24% level. The logistic regression rejects equality at the 3.82% level.

❑

Saved results

`hotelling` saves the following in `r()`:

Scalars

`r(N)`	number of observations	`r(T2)`	Hotelling's T-squared
`r(k)`	number of variables	`r(df)`	degrees of freedom

Methods and formulas

`hotelling` is implemented as an ado-file.

See Wilks (1962, 556–561) for a general discussion. The original formulation was by Hotelling (1931) and Mahalanobis (1930, 1936).

For the test that the means of k variables are 0, let $\overline{\mathbf{x}}$ be a $1 \times k$ matrix of the means and $\mathbf{S}$ be the estimated covariance matrix. Then $T^2 = \overline{\mathbf{x}}\mathbf{S}^{-1}\overline{\mathbf{x}}'$.

For two groups, the test of equality is $T^2 = (\overline{\mathbf{x}}_1 - \overline{\mathbf{x}}_2)\mathbf{S}^{-1}(\overline{\mathbf{x}}_1 - \overline{\mathbf{x}}_2)'$.

Harold Hotelling (1895–1973) was an American economist and statistician who made many important contributions to mathematical economics, multivariate analysis, and statistical inference. After obtaining degrees in journalism and mathematics, he taught and researched at Stanford, Columbia, and the University of North Carolina. His work generalizing Student's t ratio and on principal components, canonical correlation, multivariate analysis of variance, and correlation continues to be widely used.

Prasanta Chandra Mahalanobis (1893–1972) studied physics and mathematics at Calcutta and Cambridge. He became interested in statistics and on his return to India worked on applications in anthropology, meteorology, hydrology, and agriculture. Mahalanobis became the leader in Indian statistics, specializing in multivariate problems (including what is now called the Mahalanobis distance), the design of large-scale sample surveys, and the contribution of statistics to national planning.

References

Hotelling, H. 1931. The generalization of Student's ratio. *Annals of Mathematical Statistics* 2: 360–378.

Mahalanobis, P. C. 1930. On tests and measures of group divergence. *Journal of the Asiatic Society of Bengal* 26: 541–588.

——. 1936. On the generalized distance in statistics. *National Institute of Science of India* 12: 49–55.

Olkin, I., and A. R. Sampson. 2001. Harold Hotelling. In *Statisticians of the Centuries*, ed. C. C. Heyde and E. Seneta, 454–458. New York: Springer.

Rao, C. R. 1973. Prasantha Chandra Mahalanobis, 1893–1972. *Biographical Memoirs of Fellows of The Royal Society* 19: 455–492.

Wilks, S. S. 1962. *Mathematical Statistics*. New York: Wiley.

Also see

[MV] **manova** — Multivariate analysis of variance and covariance

[MV] **mvtest means** — Multivariate tests of means

[R] **regress** — Linear regression

[R] **ttest** — Mean-comparison tests

Title

manova — Multivariate analysis of variance and covariance

Syntax

`manova` *depvarlist* = *termlist* [*if*] [*in*] [*weight*] [, *options*]

where *termlist* is a factor-variable list (see [U] **11.4.3 Factor variables**) with the following additional features:

- Variables are assumed to be categorical; use the `c.` factor-variable operator to override this.
- The `|` symbol (indicating nesting) may be used in place of the `#` symbol (indicating interaction).
- The `/` symbol is allowed after a *term* and indicates that the following *term* is the error term for the preceding *terms*.

options	Description
Model	
`noconstant`	suppress constant term
`dropemptycells`	drop empty cells from the design matrix

`bootstrap`, `by`, `jackknife`, and `statsby` are allowed; see [U] **11.1.10 Prefix commands**.
Weights are not allowed with the `bootstrap` prefix; see [R] **bootstrap**.
`aweights` are not allowed with the `jackknife` prefix; see [R] **jackknife**.
`aweights` and `fweights` are allowed; see [U] **11.1.6 weight**.
See [U] **20 Estimation and postestimation commands** for more capabilities of estimation commands.

Menu

Statistics > Multivariate analysis > MANOVA, multivariate regression, and related > MANOVA

Description

The `manova` command fits multivariate analysis-of-variance (MANOVA) and multivariate analysis-of-covariance (MANCOVA) models for balanced and unbalanced designs, including designs with missing cells, and for factorial, nested, or mixed designs, or designs involving repeated measures.

The `mvreg` command (see [R] **mvreg**) will display the coefficients, standard errors, etc., of the multivariate regression model underlying the last run of `manova`.

See [R] **anova** for univariate ANOVA and ANCOVA models. See [MV] **mvtest covariances** for Box's test of MANOVA's assumption that the covariance matrices of the groups are the same, and see [MV] **mvtest means** for multivariate tests of means that do not make this assumption.

Options

Model

noconstant suppresses the constant term (intercept) from the model.

dropemptycells drops empty cells from the design matrix. If c(emptycells) is set to keep (see [R] **set emptycells**), this option temporarily resets it to drop before running the MANOVA model. If c(emptycells) is already set to drop, this option does nothing.

Remarks

Remarks are presented under the following headings:

Introduction
One-way MANOVA
Reporting coefficients
Two-way MANOVA
N-way MANOVA
MANCOVA
MANOVA for Latin-square designs
MANOVA for nested designs
MANOVA for mixed designs
MANOVA with repeated measures

Introduction

MANOVA is a generalization of ANOVA allowing multiple dependent variables. Several books discuss MANOVA, including Anderson (2003); Mardia, Kent, and Bibby (1979); Morrison (2005); Rencher (1998, 2002); Seber (1984); and Timm (1975). Introductory articles are provided by Pillai (1985) and Morrison (1998). Pioneering work is found in Wilks (1932), Pillai (1955), Lawley (1938), Hotelling (1951), and Roy (1939).

Four multivariate statistics are commonly computed in MANOVA: Wilks' lambda, Pillai's trace, Lawley–Hotelling trace, and Roy's largest root. See *Methods and formulas* for details.

Why four statistics? Arnold (1981), Rencher (1998, 2002), Morrison (1998), Pillai (1985), and Seber (1984) provide guidance. All four tests are admissible, unbiased, and invariant. Asymptotically, Wilks' lambda, Pillai's trace, and the Lawley–Hotelling trace are the same, but their behavior under various violations of the null hypothesis and with small samples is different. Roy's largest root is different from the other three, even asymptotically.

None of the four multivariate criteria appears to be most powerful against all alternative hypotheses. For instance, Roy's largest root is most powerful when the null hypothesis of equal mean vectors is violated in such a way that the mean vectors tend to lie in one line within p-dimensional space. For most other situations, Roy's largest root performs worse than the other three statistics. Pillai's trace tends to be more robust to nonnormality and heteroskedasticity than the other three statistics.

The # symbol indicates interaction. The | symbol indicates nesting (a|b is read "a is nested within b"). A / between *terms* indicates that the *term* to the right of the slash is the error term for the *terms* to the left of the slash.

One-way MANOVA

A one-way MANOVA is obtained by specifying the dependent variables followed by an equal sign, followed by the categorical variable defining the groups.

▷ Example 1: One-way MANOVA with balanced data

Rencher (2002) presents an example of a balanced one-way MANOVA by using data from Andrews and Herzberg (1985, 357–360). The data from eight trees from each of six apple tree rootstocks are from table 6.2 of Rencher (2002). Four dependent variables are recorded for each tree: trunk girth at 4 years (mm × 100), extension growth at 4 years (m), trunk girth at 15 years (mm × 100), and weight of tree above ground at 15 years (lb × 1000). The grouping variable is `rootstock`, and the four dependent variables are `y1`, `y2`, `y3`, and `y4`.

```
. use http://www.stata-press.com/data/r12/rootstock
(Table 6.2 Rootstock Data -- Rencher (2002))

. describe

Contains data from http://www.stata-press.com/data/r12/rootstock.dta
  obs:            48                          Table 6.2 Rootstock Data --
                                                Rencher (2002)
 vars:             5                          20 Apr 2011 20:03
 size:           816                          (_dta has notes)
------------------------------------------------------------------------------
              storage  display     value
variable name   type   format      label      variable label
------------------------------------------------------------------------------
rootstock       byte   %9.0g
y1              float  %4.2f                  trunk girth at 4 years (mm x 100)
y2              float  %5.3f                  extension growth at 4 years (m)
y3              float  %4.2f                  trunk girth at 15 years (mm x
                                                100)
y4              float  %5.3f                  weight of tree above ground at 15
                                                years (lb x 1000)
------------------------------------------------------------------------------
Sorted by:

. list in 7/10

     +------------------------------------------+
     | rootst~k     y1      y2     y3       y4  |
     |------------------------------------------|
  7. |        1   1.11   3.211   3.98    1.209  |
  8. |        1   1.16   3.037   3.62    0.750  |
  9. |        2   1.05   2.074   4.09    1.036  |
 10. |        2   1.17   2.885   4.06    1.094  |
     +------------------------------------------+
```

There are six rootstocks and four dependent variables. We test to see if the four-dimensional mean vectors of the six rootstocks are different. The null hypothesis is that the mean vectors are the same for the six rootstocks. To obtain one-way MANOVA results, we type

```
. manova y1 y2 y3 y4 = rootstock

                           Number of obs =      48

                           W = Wilks' lambda      L = Lawley-Hotelling trace
                           P = Pillai's trace     R = Roy's largest root

              Source |  Statistic      df   F(df1,    df2) =   F   Prob>F
          -----------+--------------------------------------------------
           rootstock | W   0.1540       5    20.0    130.3     4.94 0.0000 a
                     | P   1.3055            20.0    168.0     4.07 0.0000 a
                     | L   2.9214            20.0    150.0     5.48 0.0000 a
                     | R   1.8757             5.0     42.0    15.76 0.0000 u
                     |--------------------------------------------------
            Residual |                 42
          -----------+--------------------------------------------------
               Total |                 47
          --------------------------------------------------------------
                       e = exact, a = approximate, u = upper bound on F
```

All four multivariate tests reject the null hypothesis, indicating some kind of difference between the four-dimensional mean vectors of the six rootstocks.

Let's examine the output of manova. Above the table, it lists the number of observations used in the estimation. It also gives a key indicating that W stands for Wilks' lambda, P stands for Pillai's trace, L stands for Lawley–Hotelling trace, and R indicates Roy's largest root.

The first column of the table gives the source. Here we are testing the rootstock term (the only term in the model), and we are using residual error for the denominator of the test. Four lines of output are presented for rootstock, one line for each of the four multivariate tests, as indicated by the W, P, L, and R in the second column of the table.

The next column gives the multivariate statistics. Here Wilks' lambda is 0.1540, Pillai's trace is 1.3055, the Lawley–Hotelling trace is 2.9214, and Roy's largest root is 1.8757. Some authors report λ_1 and others (including Rencher) report $\theta = \lambda_1/(1+\lambda_1)$ for Roy's largest root. Stata reports λ_1.

The column labeled "df" gives the hypothesis degrees of freedom, the residual degrees of freedom, and the total degrees of freedom. These are just as they would be for an ANOVA. Because there are six rootstocks, we have 5 degrees of freedom for the hypothesis. There are 42 residual degrees of freedom and 47 total degrees of freedom.

The next three columns are labeled "F(df1, df2) = F", and for each of the four multivariate tests, the degrees of freedom and F statistic are listed. The following column gives the associated p-values for the F statistics. Wilks' lambda has an F statistic of 4.94 with 20 and 130.3 degrees of freedom, which produces a p-value small enough that 0.0000 is reported. The F statistics and p-values for the other three multivariate tests follow on the three lines after Wilks' lambda.

The final column indicates whether the F statistic is exactly F distributed, is approximately F distributed, or is an upper bound. The letters e, a, and u indicate these three possibilities, as described in the footer at the bottom of the table. For this example, the F statistics (and corresponding p-values) for Wilks' lambda, Pillai's trace, and the Lawley–Hotelling trace are approximate. The F statistic for Roy's largest root is an upper bound, which means that the p-value is a lower bound.

Examining some of the underlying matrices and values used in the calculation of the four multivariate statistics is easy. For example, you can list the sum of squares and cross products (SSCP) matrices for error and the hypothesis that are found in the e(E) and e(H_m) returned matrices, the eigenvalues of $\mathbf{E}^{-1}\mathbf{H}$ obtained from the e(eigvals_m) returned matrix, and the three auxiliary values (s, m, and n) that are returned in the e(aux_m) matrix.

```
. mat list e(E)

symmetric e(E)[4,4]
            y1          y2          y3          y4
y1   .31998754
y2   1.6965639    12.14279
y3   .55408744   4.3636123   4.2908128
y4   .21713994   2.1102135   2.4816563   1.7225248

. mat list e(H_m)

symmetric e(H_m)[4,4]
            y1          y2          y3          y4
y1   .07356042
y2   .53738525   4.1996621
y3   .33226448   2.3553887   6.1139358
y4   .20846994   1.6371084   3.7810439   2.4930912

. mat list e(eigvals_m)

e(eigvals_m)[1,4]
            c1          c2          c3          c4
r1   1.8756709   .79069412   .22904906   .02595358

. mat list e(aux_m)

e(aux_m)[3,1]
    value
s       4
m       0
n    18.5
```

The values `s`, `m`, and `n` are helpful when you do not want to rely on the approximate F tests but instead want to look up critical values for the multivariate tests. Tables of critical values can be found in many multivariate texts, including Rencher (1998, 2002).

See [MV] **manova postestimation** example 1 for an illustration of using `test` for Wald tests on expressions involving the underlying coefficients of the model and `lincom` for displaying linear combinations along with standard errors and confidence intervals from this MANOVA example.

See [MV] **discrim lda postestimation** examples 1–5 for a descriptive linear discriminant analysis of the `rootstock` data. Many researchers use linear discriminant analysis as a method of exploring the differences between groups after a MANOVA model.

◁

▷ Example 2: One-way MANOVA with unbalanced data

Table 4.5 of Rencher (1998) presents data reported by Allison, Zappasodi, and Lurie (1962). The dependent variables `y1`, recording the number of bacilli inhaled per tubercle formed, and `y2`, recording tubercle size (in millimeters), were measured for four groups of rabbits. Group one (unvaccinated control) and group two (infected during metabolic depression) have seven observations each, whereas group three (infected during heightened metabolic activity) has 5 observations, and group four (infected during normal activity) has only 2 observations.

```
. use http://www.stata-press.com/data/r12/metabolic
(Table 4.5 Metabolic Comparisons of Rabbits -- Rencher (1998))
. list
```

	group	y1	y2
1.	1	24	3.5
2.	1	13.3	3.5
3.	1	12.2	4
4.	1	14	4
5.	1	22.2	3.6
6.	1	16.1	4.3
7.	1	27.9	5.2
8.	2	7.4	3.5
9.	2	13.2	3
10.	2	8.5	3
11.	2	10.1	3
12.	2	9.3	2
13.	2	8.5	2.5
14.	2	4.3	1.5
15.	3	16.4	3.2
16.	3	24	2.5
17.	3	53	1.5
18.	3	32.7	2.6
19.	3	42.8	2
20.	4	25.1	2.7
21.	4	5.9	2.3

The one-way MANOVA for testing the null hypothesis that the two-dimensional mean vectors for the four groups of rabbits are equal is

```
. manova y1 y2 = group
                           Number of obs =      21
                           W = Wilks' lambda      L = Lawley-Hotelling trace
                           P = Pillai's trace     R = Roy's largest root

                  Source | Statistic     df   F(df1,    df2) =    F   Prob>F
              -----------+--------------------------------------------------
                   group | W   0.1596     3     6.0     32.0     8.02 0.0000 e
                         | P   1.2004           6.0     34.0     8.51 0.0000 a
                         | L   3.0096           6.0     30.0     7.52 0.0001 a
                         | R   1.5986           3.0     17.0     9.06 0.0008 u
                         |--------------------------------------------------
                Residual |                17
              -----------+--------------------------------------------------
                   Total |                20
              --------------------------------------------------------------
                           e = exact, a = approximate, u = upper bound on F
```

All four multivariate tests indicate rejection of the null hypothesis. This indicates that there are one or more differences among the two-dimensional mean vectors for the four groups. For this example, the F test for Wilks' lambda is exact because there are only two dependent variables in the model.

`manovatest` tests terms or linear combinations of the model's underlying design matrix. Example 2 of [MV] **manova postestimation** continues this example and illustrates `manovatest`.

◁

Reporting coefficients

The `mvreg` command (see [R] **mvreg**) is used as a coefficient displayer after `manova`. Simply type `mvreg` to view the coefficients, standard errors, t statistics, p-values, and confidence intervals of the multivariate regression model underlying the previous `manova`.

▷ Example 3: Reporting coefficients by using mvreg

Continuing with example 2, we now use `mvreg` to display the coefficients underlying our MANOVA.

```
. mvreg

Equation          Obs  Parms        RMSE    "R-sq"          F        P
------------------------------------------------------------------------
y1                 21      4    8.753754    0.5867   8.045716   0.0015
y2                 21      4    .6314183    0.6108   8.891362   0.0009

------------------------------------------------------------------------------
             |      Coef.   Std. Err.      t    P>|t|     [95% Conf. Interval]
-------------+----------------------------------------------------------------
y1           |
       group |
          2  |  -9.771429   4.679078    -2.09   0.052    -19.64342    .1005633
          3  |   15.25143   5.125673     2.98   0.008     4.437203    26.06565
          4  |  -3.028571   7.018617    -0.43   0.672    -17.83656    11.77942
             |
       _cons |   18.52857   3.308608     5.60   0.000     11.54802    25.50912
-------------+----------------------------------------------------------------
y2           |
       group |
          2  |  -1.371429   .3375073    -4.06   0.001    -2.083507   -.6593504
          3  |  -1.654286   .3697207    -4.47   0.000    -2.434328   -.8742432
          4  |  -1.514286   .5062609    -2.99   0.008    -2.582403   -.4461685
             |
       _cons |   4.014286   .2386537    16.82   0.000      3.51077    4.517801
------------------------------------------------------------------------------
```

◁

`mvreg` options allowed on replay, such as `level()`, `vsquish`, and `base`, may also be specified to alter what is displayed.

Two-way MANOVA

You can include multiple explanatory variables with the `manova` command, and you can specify interactions by placing ‘`#`’ between the variable names.

▷ Example 4: Two-way MANOVA with unbalanced data

Table 4.6 of Rencher (1998) presents unbalanced data from Woodard (1931) for a two-way MANOVA with three dependent variables (`y1`, `y2`, and `y3`) measured on patients with fractures of the jaw. `y1` is age of patient, `y2` is blood lymphocytes, and `y3` is blood polymorphonuclears. The two design factors are `gender` (1 = male, 2 = female) and `fracture` (indicating the type of fracture: 1 = one compound fracture, 2 = two compound fractures, and 3 = one simple fracture). `gender` and `fracture` are numeric variables with value labels.

```
. use http://www.stata-press.com/data/r12/jaw
(Table 4.6 Two-Way Unbalanced Data for Fractures of the Jaw -- Rencher (1998))

. describe

Contains data from http://www.stata-press.com/data/r12/jaw.dta
  obs:            27                          Table 4.6 Two-Way Unbalanced
                                                Data for Fractures of the Jaw
                                                -- Rencher (1998)
 vars:             5                          20 Apr 2011 14:53
 size:           135                          (_dta has notes)
-------------------------------------------------------------------------------
              storage  display     value
variable name   type   format      label      variable label
-------------------------------------------------------------------------------
gender          byte   %9.0g       gender
fracture        byte   %22.0g      fractype
y1              byte   %9.0g                  age
y2              byte   %9.0g                  blood lymphocytes
y3              byte   %9.0g                  blood polymorphonuclears
-------------------------------------------------------------------------------
Sorted by:

. list in 19/22

     +-----------------------------------------------------+
     | gender                 fracture   y1   y2   y3 |
     |-----------------------------------------------------|
 19. |   male       one simple fracture   55   32   60 |
 20. |   male       one simple fracture   30   34   62 |
 21. | female     one compound fracture   22   56   43 |
 22. | female   two compound fractures   22   29   68 |
     +-----------------------------------------------------+
```

The two-way factorial MANOVA for these data is

```
. manova y1 y2 y3 = gender fracture gender#fracture
                          Number of obs =         27

                          W = Wilks' lambda        L = Lawley-Hotelling trace
                          P = Pillai's trace       R = Roy's largest root

                Source |  Statistic      df   F(df1,     df2) =   F   Prob>F
        ---------------+--------------------------------------------------
                 Model | W   0.2419       5    15.0     52.9     2.37 0.0109 a
                       | P   1.1018            15.0     63.0     2.44 0.0072 a
                       | L   1.8853            15.0     53.0     2.22 0.0170 a
                       | R   0.9248             5.0     21.0     3.88 0.0119 u
                       |--------------------------------------------------
              Residual |                 21
        ---------------+--------------------------------------------------
                gender | W   0.7151       1     3.0     19.0     2.52 0.0885 e
                       | P   0.2849             3.0     19.0     2.52 0.0885 e
                       | L   0.3983             3.0     19.0     2.52 0.0885 e
                       | R   0.3983             3.0     19.0     2.52 0.0885 e
                       |--------------------------------------------------
              fracture | W   0.4492       2     6.0     38.0     3.12 0.0139 e
                       | P   0.6406             6.0     40.0     3.14 0.0128 a
                       | L   1.0260             6.0     36.0     3.08 0.0155 a
                       | R   0.7642             3.0     20.0     5.09 0.0088 u
                       |--------------------------------------------------
       gender#fracture | W   0.5126       2     6.0     38.0     2.51 0.0380 e
                       | P   0.5245             6.0     40.0     2.37 0.0472 a
                       | L   0.8784             6.0     36.0     2.64 0.0319 a
                       | R   0.7864             3.0     20.0     5.24 0.0078 u
                       |--------------------------------------------------
              Residual |                 21
        ---------------+--------------------------------------------------
                 Total |                 26
        ------------------------------------------------------------------
                          e = exact, a = approximate, u = upper bound on F
```

For MANOVA models with more than one term, the output of `manova` shows test results for the overall model, followed by results for each term in the MANOVA.

The interaction term, `gender#fracture`, is significant at the 0.05 level. Wilks' lambda for the interaction has an exact F that produces a p-value of 0.0380.

Example 3 of [MV] **manova postestimation** illustrates how the `margins` postestimation command can be used to examine details of this significant interaction. It also illustrates how to obtain residuals by using `predict`.

◁

N-way MANOVA

Higher-order MANOVA models are easily constructed using `#` to indicate the interaction terms.

▷ Example 5: MANOVA with interaction terms

Data on the wear of coated fabrics is provided by Box (1950) and is presented in table 6.20 of Rencher (2002). Variables `y1`, `y2`, and `y3` are the wear after successive 1,000 revolutions of an abrasive wheel. Three factors are also recorded. `treatment` is the surface treatment and has two levels. `filler` is the filler type, also with two levels. `proportion` is the proportion of filler and has three levels (25%, 50%, and 75%).

```
. use http://www.stata-press.com/data/r12/fabric
(Table 6.20 Wear of coated fabrics -- Rencher (2002))

. describe

Contains data from http://www.stata-press.com/data/r12/fabric.dta
  obs:            24                          Table 6.20 Wear of coated
                                                fabrics -- Rencher (2002)
 vars:             6                          21 Apr 2011 02:01
 size:           216                           (_dta has notes)
-------------------------------------------------------------------------------
              storage  display     value
variable name   type   format      label      variable label
-------------------------------------------------------------------------------
treatment       byte   %9.0g                  Surface treatment
filler          byte   %9.0g                  Filler type
proportion      byte   %9.0g       prop       Proportion of filler
y1              int    %9.0g                  First 1000 revolutions
y2              int    %9.0g                  Second 1000 revolutions
y3              int    %9.0g                  Third 1000 revolutions
-------------------------------------------------------------------------------
Sorted by:

. label list prop
prop:
           1 25%
           2 50%
           3 75%

. list

     +-------------------------------------------------+
     | treatm~t   filler   propor~n    y1    y2    y3 |
     |-------------------------------------------------|
  1. |        0        1        25%   194   192   141 |
  2. |        0        1        50%   233   217   171 |
  3. |        0        1        75%   265   252   207 |
  4. |        0        1        25%   208   188   165 |
  5. |        0        1        50%   241   222   201 |
     |-------------------------------------------------|
  6. |        0        1        75%   269   283   191 |
  7. |        0        2        25%   239   127    90 |
  8. |        0        2        50%   224   123    79 |
  9. |        0        2        75%   243   117   100 |
 10. |        0        2        25%   187   105    85 |
     |-------------------------------------------------|
 11. |        0        2        50%   243   123   110 |
 12. |        0        2        75%   226   125    75 |
 13. |        1        1        25%   155   169   151 |
 14. |        1        1        50%   198   187   176 |
 15. |        1        1        75%   235   225   166 |
     |-------------------------------------------------|
 16. |        1        1        25%   173   152   141 |
 17. |        1        1        50%   177   196   167 |
 18. |        1        1        75%   229   270   183 |
 19. |        1        2        25%   137    82    77 |
 20. |        1        2        50%   129    94    78 |
     |-------------------------------------------------|
 21. |        1        2        75%   155    76    92 |
 22. |        1        2        25%   160    82    83 |
 23. |        1        2        50%    98    89    48 |
 24. |        1        2        75%   132   105    67 |
     +-------------------------------------------------+
```

proportion is a numeric variable taking on values 1, 2, and 3, and is value-labeled with labels 25%, 50%, and 75%. treatment takes on values of 0 and 1, whereas filler is either 1 or 2.

First, we examine these data, ignoring the repeated-measures aspects of y1, y2, and y3. In example 12, we will take it into account.

```
. manova y1 y2 y3 = proportion##treatment##filler

                           Number of obs =      24

                           W = Wilks' lambda      L = Lawley-Hotelling trace
                           P = Pillai's trace     R = Roy's largest root

                    Source | Statistic     df   F(df1,    df2) =   F   Prob>F
      -------------------+--------------------------------------------------
                     Model | W   0.0007    11    33.0     30.2    10.10 0.0000 a
                           | P   2.3030          33.0     36.0     3.60 0.0001 a
                           | L  74.4794          33.0     26.0    19.56 0.0000 a
                           | R  59.1959          11.0     12.0    64.58 0.0000 u
                           |--------------------------------------------------
                  Residual |               12
      -------------------+--------------------------------------------------
                proportion | W   0.1375     2     6.0     20.0     5.65 0.0014 e
                           | P   0.9766           6.0     22.0     3.50 0.0139 a
                           | L   5.4405           6.0     18.0     8.16 0.0002 a
                           | R   5.2834           3.0     11.0    19.37 0.0001 u
                           |--------------------------------------------------
                 treatment | W   0.0800     1     3.0     10.0    38.34 0.0000 e
                           | P   0.9200           3.0     10.0    38.34 0.0000 e
                           | L  11.5032           3.0     10.0    38.34 0.0000 e
                           | R  11.5032           3.0     10.0    38.34 0.0000 e
                           |--------------------------------------------------
      proportion#treatment | W   0.7115     2     6.0     20.0     0.62 0.7134 e
                           | P   0.2951           6.0     22.0     0.63 0.7013 a
                           | L   0.3962           6.0     18.0     0.59 0.7310 a
                           | R   0.3712           3.0     11.0     1.36 0.3055 u
                           |--------------------------------------------------
                    filler | W   0.0192     1     3.0     10.0   170.60 0.0000 e
                           | P   0.9808           3.0     10.0   170.60 0.0000 e
                           | L  51.1803           3.0     10.0   170.60 0.0000 e
                           | R  51.1803           3.0     10.0   170.60 0.0000 e
                           |--------------------------------------------------
         proportion#filler | W   0.1785     2     6.0     20.0     4.56 0.0046 e
                           | P   0.9583           6.0     22.0     3.37 0.0164 a
                           | L   3.8350           6.0     18.0     5.75 0.0017 a
                           | R   3.6235           3.0     11.0    13.29 0.0006 u
                           |--------------------------------------------------
          treatment#filler | W   0.3552     1     3.0     10.0     6.05 0.0128 e
                           | P   0.6448           3.0     10.0     6.05 0.0128 e
                           | L   1.8150           3.0     10.0     6.05 0.0128 e
                           | R   1.8150           3.0     10.0     6.05 0.0128 e
                           |--------------------------------------------------
     proportion#treatment# | W   0.7518     2     6.0     20.0     0.51 0.7928 e
                    filler | P   0.2640           6.0     22.0     0.56 0.7589 a
                           | L   0.3092           6.0     18.0     0.46 0.8260 a
                           | R   0.2080           3.0     11.0     0.76 0.5381 u
                           |--------------------------------------------------
                  Residual |               12
      -------------------+--------------------------------------------------
                     Total |               23
      -------------------+--------------------------------------------------
                             e = exact, a = approximate, u = upper bound on F
```

The MANOVA table indicates that all the terms are significant, except for `proportion#treatment` and `proportion#treatment#filler`.

◁

❑ Technical note

MANOVA uses the same design matrix as ANOVA. manova saves the full variance–covariance matrix and coefficient vector. These need a dimension equal to the dimension of the design matrix times the number of variables in the *depvarlist*.

For large problems, you may need to increase matsize. With the fabric-wear data of example 5, a matsize of at least 108 ($108 = 36 \times 3$) is needed because there are three dependent variables and the design matrix has 36 columns. The 36 columns comprise 1 column for the overall mean, 3 columns for proportion, 2 columns for treatment, 6 columns for proportion#treatment, 2 columns for filler, 6 columns for proportion#filler, 4 columns for treatment#filler, and 12 columns for proportion#treatment#filler.

❑

MANCOVA

MANCOVA models are specified by including the covariates as *terms* in the manova preceded by the c. operator to indicate that they are to be treated as continuous instead of categorical variables.

▷ Example 6: MANCOVA

Table 4.9 of Rencher (1998) provides biochemical measurements on four weight groups. Rencher extracted the data from Brown and Beerstecher (1951) and Smith, Gnanadesikan, and Hughes (1962). Three dependent variables and two covariates are recorded for eight subjects within each of the four groups. The first two groups are underweight, and the last two groups are overweight. The dependent variables are modified creatinine coefficient (y1), pigment creatinine (y2), and phosphate in mg/mL (y3). The two covariates are volume in ml (x1) and specific gravity (x2).

```
. use http://www.stata-press.com/data/r12/biochemical
(Table 4.9, Rencher (1998))

. describe

Contains data from http://www.stata-press.com/data/r12/biochemical.dta
  obs:            32                          Table 4.9, Rencher (1998)
 vars:             6                          22 Apr 2011 21:48
 size:           512                          (_dta has notes)
-------------------------------------------------------------------------------
              storage  display     value
variable name   type   format      label      variable label
-------------------------------------------------------------------------------
group           byte   %9.0g
y1              float  %9.0g                  modified creatinine coefficient
y2              float  %9.0g                  pigment creatinine
y3              float  %9.0g                  phosphate (mg/ml)
x1              int    %9.0g                  volume (ml)
x2              byte   %9.0g                  specific gravity
-------------------------------------------------------------------------------
Sorted by:
```

Rencher performs three tests on these data. The first is a test of equality of group effects adjusted for the covariates. The second is a test that the coefficients for the covariates are jointly equal to zero. The third is a test that the coefficients for the covariates are equal across groups.

```
. manova y1 y2 y3 = group c.x1 c.x2
                           Number of obs =      32

                           W = Wilks' lambda      L = Lawley-Hotelling trace
                           P = Pillai's trace     R = Roy's largest root

                  Source | Statistic     df   F(df1,    df2) =   F   Prob>F
             ------------+----------------------------------------------------
                   Model | W   0.0619     5   15.0     66.7     7.73 0.0000 a
                         | P   1.4836         15.0     78.0     5.09 0.0000 a
                         | L   6.7860         15.0     68.0    10.25 0.0000 a
                         | R   5.3042          5.0     26.0    27.58 0.0000 u
                         |----------------------------------------------------
                Residual |               26
             ------------+----------------------------------------------------
                   group | W   0.1491     3    9.0     58.6     7.72 0.0000 a
                         | P   0.9041          9.0     78.0     3.74 0.0006 a
                         | L   5.3532          9.0     68.0    13.48 0.0000 a
                         | R   5.2872          3.0     26.0    45.82 0.0000 u
                         |----------------------------------------------------
                      x1 | W   0.6841     1    3.0     24.0     3.69 0.0257 e
                         | P   0.3159          3.0     24.0     3.69 0.0257 e
                         | L   0.4617          3.0     24.0     3.69 0.0257 e
                         | R   0.4617          3.0     24.0     3.69 0.0257 e
                         |----------------------------------------------------
                      x2 | W   0.9692     1    3.0     24.0     0.25 0.8576 e
                         | P   0.0308          3.0     24.0     0.25 0.8576 e
                         | L   0.0318          3.0     24.0     0.25 0.8576 e
                         | R   0.0318          3.0     24.0     0.25 0.8576 e
                         |----------------------------------------------------
                Residual |               26
             ------------+----------------------------------------------------
                   Total |               31
             -----------------------------------------------------------------
                           e = exact, a = approximate, u = upper bound on F
```

The test of equality of group effects adjusted for the covariates is shown in the MANCOVA table above. Rencher reports a Wilks' lambda value of 0.1491, which agrees with the value shown for the `group` term above. `group` is found to be significant.

The test that the coefficients for the covariates are jointly equal to zero is obtained using `manovatest`.

```
. manovatest c.x1 c.x2

                           W = Wilks' lambda      L = Lawley-Hotelling trace
                           P = Pillai's trace     R = Roy's largest root

                  Source | Statistic     df   F(df1,    df2) =   F   Prob>F
             ------------+----------------------------------------------------
                   x1 x2 | W   0.4470     2    6.0     48.0     3.97 0.0027 e
                         | P   0.5621          6.0     50.0     3.26 0.0088 a
                         | L   1.2166          6.0     46.0     4.66 0.0009 a
                         | R   1.1995          3.0     25.0    10.00 0.0002 u
                         |----------------------------------------------------
                Residual |               26
             -----------------------------------------------------------------
                           e = exact, a = approximate, u = upper bound on F
```

Wilks' lambda of 0.4470 agrees with the value reported by Rencher. With a p-value of 0.0027, we reject the null hypothesis that the coefficients for the covariates are jointly zero.

To test that the coefficients for the covariates are equal across groups, we perform a MANCOVA that includes our covariates (x1 and x2) interacted with group. We then use manovatest to obtain the combined test of equal coefficients for x1 and x2 across groups.

```
. manova y1 y2 y3 = group c.x1 c.x2 group#c.x1 group#c.x2
                           Number of obs =      32

                           W = Wilks' lambda       L = Lawley-Hotelling trace
                           P = Pillai's trace      R = Roy's largest root

                  Source | Statistic      df   F(df1,      df2) =    F   Prob>F
              -----------+--------------------------------------------------
                   Model | W    0.0205    11    33.0       53.7      4.47 0.0000 a
                         | P    1.9571          33.0       60.0      3.41 0.0000 a
                         | L   10.6273          33.0       50.0      5.37 0.0000 a
                         | R    7.0602          11.0       20.0     12.84 0.0000 u
                         |--------------------------------------------------
                Residual |                20
              -----------+--------------------------------------------------
                   group | W    0.4930     3     9.0       44.0      1.65 0.1317 a
                         | P    0.5942           9.0       60.0      1.65 0.1226 a
                         | L    0.8554           9.0       50.0      1.58 0.1458 a
                         | R    0.5746           3.0       20.0      3.83 0.0256 u
                         |--------------------------------------------------
                      x1 | W    0.7752     1     3.0       18.0      1.74 0.1947 e
                         | P    0.2248           3.0       18.0      1.74 0.1947 e
                         | L    0.2900           3.0       18.0      1.74 0.1947 e
                         | R    0.2900           3.0       18.0      1.74 0.1947 e
                         |--------------------------------------------------
                      x2 | W    0.8841     1     3.0       18.0      0.79 0.5169 e
                         | P    0.1159           3.0       18.0      0.79 0.5169 e
                         | L    0.1311           3.0       18.0      0.79 0.5169 e
                         | R    0.1311           3.0       18.0      0.79 0.5169 e
                         |--------------------------------------------------
                group#x1 | W    0.4590     3     9.0       44.0      1.84 0.0873 a
                         | P    0.6378           9.0       60.0      1.80 0.0869 a
                         | L    0.9702           9.0       50.0      1.80 0.0923 a
                         | R    0.6647           3.0       20.0      4.43 0.0152 u
                         |--------------------------------------------------
                group#x2 | W    0.5275     3     9.0       44.0      1.47 0.1899 a
                         | P    0.5462           9.0       60.0      1.48 0.1747 a
                         | L    0.7567           9.0       50.0      1.40 0.2130 a
                         | R    0.4564           3.0       20.0      3.04 0.0527 u
                         |--------------------------------------------------
                Residual |                20
              -----------+--------------------------------------------------
                   Total |                31
              -----------+--------------------------------------------------
                           e = exact, a = approximate, u = upper bound on F

. manovatest group#c.x1 group#c.x2

                           W = Wilks' lambda       L = Lawley-Hotelling trace
                           P = Pillai's trace      R = Roy's largest root

                  Source | Statistic      df   F(df1,      df2) =    F   Prob>F
     --------------------+--------------------------------------------------
     group#x1 group#x2   | W    0.3310     6    18.0       51.4      1.37 0.1896 a
                         | P    0.8600          18.0       60.0      1.34 0.1973 a
                         | L    1.4629          18.0       50.0      1.35 0.1968 a
                         | R    0.8665           6.0       20.0      2.89 0.0341 u
                         |--------------------------------------------------
                Residual |                20
     --------------------+--------------------------------------------------
                           e = exact, a = approximate, u = upper bound on F
```

Rencher reports 0.3310 for Wilks' lambda, which agrees with the results of `manovatest` above. Here we fail to reject the null hypothesis.

◁

MANOVA for Latin-square designs

▷ Example 7: MANOVA with Latin-square data

Exercise 5.11 from Timm (1975) presents data from a multivariate Latin-square design. Two dependent variables are measured in a 4 × 4 Latin square. `W` is the student's score on determining distances within the solar system. `B` is the student's score on determining distances beyond the solar system. The three variables comprising the square are `machine`, `ability`, and `treatment`, each at four levels.

```
. use http://www.stata-press.com/data/r12/solardistance
(Multivariate Latin Square, Timm (1975), Exercise 5.11 #1)

. describe

Contains data from http://www.stata-press.com/data/r12/solardistance.dta
  obs:            16                          Multivariate Latin Square, Timm
                                                (1975), Exercise 5.11 #1
 vars:             5                          23 Apr 2011 03:27
 size:            80                          (_dta has notes)

              storage  display     value
variable name   type   format      label      variable label

machine         byte   %9.0g                  teaching machine
ability         byte   %9.0g                  ability tracks
treatment       byte   %9.0g                  method of measuring astronomical
                                                distances
W               byte   %9.0g                  Solar system distances (within)
B               byte   %9.0g                  Solar system distances (beyond)

Sorted by:

. list

     +------------------------------------------+
     | machine   ability   treatm~t    W     B  |
     |------------------------------------------|
  1. |       1         1          2   33    15  |
  2. |       1         2          1   40     4  |
  3. |       1         3          3   31    16  |
  4. |       1         4          4   37    10  |
  5. |       2         1          4   25    20  |
     |------------------------------------------|
  6. |       2         2          3   30    18  |
  7. |       2         3          1   22     6  |
  8. |       2         4          2   25    18  |
  9. |       3         1          1   10     5  |
 10. |       3         2          4   20    16  |
     |------------------------------------------|
 11. |       3         3          2   17    16  |
 12. |       3         4          3   12     4  |
 13. |       4         1          3   24    15  |
 14. |       4         2          2   20    13  |
 15. |       4         3          4   19    14  |
     |------------------------------------------|
 16. |       4         4          1   29    20  |
     +------------------------------------------+
```

```
. manova W B = machine ability treatment
                          Number of obs =      16
                          W = Wilks' lambda       L = Lawley-Hotelling trace
                          P = Pillai's trace      R = Roy's largest root
                Source |  Statistic     df   F(df1,     df2) =   F    Prob>F
             ----------+-----------------------------------------------------
                 Model | W    0.0378     9    18.0      10.0      2.30 0.0898 e
                       | P    1.3658          18.0      12.0      1.44 0.2645 a
                       | L   14.7756          18.0       8.0      3.28 0.0455 a
                       | R   14.0137           9.0       6.0      9.34 0.0066 u
                       |-----------------------------------------------------
              Residual |                 6
             ----------+-----------------------------------------------------
               machine | W    0.0561     3     6.0      10.0      5.37 0.0101 e
                       | P    1.1853           6.0      12.0      2.91 0.0545 a
                       | L   12.5352           6.0       8.0      8.36 0.0043 a
                       | R   12.1818           3.0       6.0     24.36 0.0009 u
                       |-----------------------------------------------------
               ability | W    0.4657     3     6.0      10.0      0.78 0.6070 e
                       | P    0.5368           6.0      12.0      0.73 0.6322 a
                       | L    1.1416           6.0       8.0      0.76 0.6199 a
                       | R    1.1367           3.0       6.0      2.27 0.1802 u
                       |-----------------------------------------------------
             treatment | W    0.4697     3     6.0      10.0      0.77 0.6137 e
                       | P    0.5444           6.0      12.0      0.75 0.6226 a
                       | L    1.0988           6.0       8.0      0.73 0.6378 a
                       | R    1.0706           3.0       6.0      2.14 0.1963 u
                       |-----------------------------------------------------
              Residual |                 6
             ----------+-----------------------------------------------------
                 Total |                15
             ----------------------------------------------------------------
                        e = exact, a = approximate, u = upper bound on F
```

We find that machine is a significant factor in the model, whereas ability and treatment are not.

◁

MANOVA for nested designs

Nested terms are specified using a vertical bar. A|B is read as A nested within B. A|B|C is read as A nested within B, which is nested within C. A|B#C is read as A nested within the interaction of B and C. A#B|C is read as the interaction of A and B, which is nested within C.

Different error terms can be specified for different parts of the model. The forward slash is used to indicate that the next term in the model is the error term for what precedes it. For instance, manova y1 y2 = A / B|A indicates that the multivariate tests for A are to be tested using the SSCP matrix from B|A in the denominator. Error terms (terms following the slash) are generally not tested unless they are themselves followed by a slash. The residual-error SSCP matrix is the default error-term matrix.

For example, consider T_1 / T_2 / T_3, where T_1, T_2, and T_3 may be arbitrarily complex terms. manova will report T_1 tested by T_2 and T_2 tested by T_3. If we add one more slash on the end to form T_1 / T_2 / T_3 /, then manova will also report T_3 tested by the residual error.

When you have nested terms in your model, we recommend using the dropemptycells option of manova or setting c(emptycells) to drop; see [R] **set emptycells**. See the technical note at the end of the *Nested designs* section of [R] **anova** for details.

▷ Example 8: MANOVA with nested data

A chain of retail stores produced two training videos for sales associates. The videos teach how to increase sales of the store's primary product. The videos also teach how to follow up a primary sale with secondary sales of the accessories that consumers often use with the primary product. The company trainers are not sure which video will provide the best training. To decide which video to distribute to all their stores to train sales associates, they selected three stores to use one of the training videos and three other stores to use the other training video. From each store, two employees (sales associates) were selected to receive the training. The baseline weekly sales for each of these employees was recorded and then the increase in sales over their baseline was recorded for 3 or 4 different weeks. The `videotrainer` data are described below.

```
. use http://www.stata-press.com/data/r12/videotrainer
(video training)

. describe

Contains data from http://www.stata-press.com/data/r12/videotrainer.dta
  obs:            42                              video training
 vars:             5                              9 May 2011 12:50
 size:           462
-------------------------------------------------------------------------------
              storage  display     value
variable name   type   format      label      variable label
-------------------------------------------------------------------------------
video           byte   %9.0g                  training video
store           byte   %9.0g                  store (nested in video)
associate       byte   %9.0g                  sales associate (nested in store)
primary         float  %9.0g                  primary sales increase
extra           float  %9.0g                  secondary sales increase
-------------------------------------------------------------------------------
Sorted by:  video  store  associate
```

In this fully nested design, `video` is a fixed factor, whereas the remaining terms are random factors.

```
. manova primary extra =  video / store|video / associate|store|video /,
> dropemptycells

                           Number of obs =      42

                           W = Wilks' lambda      L = Lawley-Hotelling trace
                           P = Pillai's trace     R = Roy's largest root

                  Source | Statistic      df   F(df1,     df2) =   F   Prob>F
  -----------------------+--------------------------------------------------
                   Model | W   0.2455     11    22.0     58.0     2.68 0.0014 e
                         | P   0.9320           22.0     60.0     2.38 0.0042 a
                         | L   2.3507           22.0     56.0     2.99 0.0005 a
                         | R   1.9867           11.0     30.0     5.42 0.0001 u
                         |--------------------------------------------------
                Residual |                30
  -----------------------+--------------------------------------------------
                   video | W   0.1610      1     2.0      3.0     7.82 0.0646 e
                         | P   0.8390            2.0      3.0     7.82 0.0646 e
                         | L   5.2119            2.0      3.0     7.82 0.0646 e
                         | R   5.2119            2.0      3.0     7.82 0.0646 e
                         |--------------------------------------------------
             store|video |                 4
  -----------------------+--------------------------------------------------
             store|video | W   0.3515      4     8.0     10.0     0.86 0.5775 e
                         | P   0.7853            8.0     12.0     0.97 0.5011 a
                         | L   1.4558            8.0      8.0     0.73 0.6680 a
                         | R   1.1029            4.0      6.0     1.65 0.2767 u
                         |--------------------------------------------------
   associate|store|video |                 6
  -----------------------+--------------------------------------------------
   associate|store|video | W   0.5164      6    12.0     58.0     1.89 0.0543 e
                         | P   0.5316           12.0     60.0     1.81 0.0668 a
                         | L   0.8433           12.0     56.0     1.97 0.0451 a
                         | R   0.7129            6.0     30.0     3.56 0.0087 u
                         |--------------------------------------------------
                Residual |                30
  -----------------------+--------------------------------------------------
                   Total |                41
  --------------------------------------------------------------------------
                           e = exact, a = approximate, u = upper bound on F
```

There appears to be a difference in the videos (with significance levels just a bit above the standard 5% level). There also appears to be a sales associate effect but not a store effect.

See example 4 of [MV] **manova postestimation** for a continuation of this example. It illustrates how to test pooled terms against nonresidual error terms by using the `manovatest` postestimation command. In that example, `store` is pooled with `associate` from the original fully specified MANOVA. Another way of pooling is to refit the model, discarding the higher-level terms. Be careful in doing this to ensure that the remaining lower-level terms have a numbering scheme that will not mistakenly consider different subjects as being the same. The `videotrainer` dataset has `associate` numbered uniquely, so we can simply type

```
. manova primary extra = video / associate|video /, dropemptycells
                          Number of obs =      42
                          W = Wilks' lambda      L = Lawley-Hotelling trace
                          P = Pillai's trace     R = Roy's largest root
                  Source | Statistic      df   F(df1,    df2) =   F   Prob>F
          ---------------+-----------------------------------------------------
                   Model | W   0.2455     11    22.0     58.0     2.68 0.0014 e
                         | P   0.9320           22.0     60.0     2.38 0.0042 a
                         | L   2.3507           22.0     56.0     2.99 0.0005 a
                         | R   1.9867           11.0     30.0     5.42 0.0001 u
                         |-----------------------------------------------------
                Residual |                30
          ---------------+-----------------------------------------------------
                   video | W   0.4079      1     2.0      9.0     6.53 0.0177 e
                         | P   0.5921            2.0      9.0     6.53 0.0177 e
                         | L   1.4516            2.0      9.0     6.53 0.0177 e
                         | R   1.4516            2.0      9.0     6.53 0.0177 e
                         |-----------------------------------------------------
         associate|video |                10
          ---------------+-----------------------------------------------------
         associate|video | W   0.3925     10    20.0     58.0     1.73 0.0546 e
                         | P   0.7160           20.0     60.0     1.67 0.0647 a
                         | L   1.2711           20.0     56.0     1.78 0.0469 a
                         | R   0.9924           10.0     30.0     2.98 0.0100 u
                         |-----------------------------------------------------
                Residual |                30
          ---------------+-----------------------------------------------------
                   Total |                41
          ---------------------------------------------------------------------
                             e = exact, a = approximate, u = upper bound on F
```

and get the same results that we obtained using `manovatest` to get a pooled test after the full MANOVA; see example 4 of [MV] **manova postestimation**.

With `store` omitted from the model, `video` now has a significance level below 5%. The increase from 4 to 10 denominator degrees of freedom for the test of `video` provides a more powerful test.

The `margins` command provides a predictive marginal mean increase in sales based on the two videos. We could request the marginal means for primary sales increase or for extra sales increase, or we can use the `expression()` option to obtain the marginal means for combined primary and secondary sales increase. By default, the predicted means are constructed taking into account the number of observations in each cell.

```
. margins, within(video) expression(predict(eq(primary))+predict(eq(extra)))
Predictive margins                                Number of obs   =         42
Expression   : predict(eq(primary))+predict(eq(extra))
within       : video
Empty cells  : reweight

------------------------------------------------------------------------------
             |            Delta-method
             |     Margin   Std. Err.      z    P>|z|     [95% Conf. Interval]
-------------+----------------------------------------------------------------
       video |
          1  |   883.1395   30.01873    29.42   0.000     824.3039    941.9752
          2  |   698.0791   30.01873    23.25   0.000     639.2434    756.9147
------------------------------------------------------------------------------
```

Alternatively, we can examine the adjusted marginal mean increase in sales letting each cell have equal weight (regardless of its sample size) by using the `asbalanced` option of the `margins` command.

```
. margins, within(video) expression(predict(eq(primary))+predict(eq(extra)))
> asbalanced
Adjusted predictions                              Number of obs   =         42
Expression   : predict(eq(primary))+predict(eq(extra))
within       : video
Empty cells  : reweight
at           : 1.video
                   associate        (asbalanced)
               2.video
                   associate        (asbalanced)

------------------------------------------------------------------------------
             |            Delta-method
             |     Margin   Std. Err.      z    P>|z|     [95% Conf. Interval]
-------------+----------------------------------------------------------------
       video |
          1  |   1041.075   59.67154    17.45   0.000     924.1213    1158.029
          2  |   849.7187   68.23348    12.45   0.000     715.9836    983.4539
------------------------------------------------------------------------------
```

Though the values are different between the two tables, the conclusion is the same. Using training video 1 leads to increased primary and secondary sales.

◁

MANOVA for mixed designs

▷ Example 9: Split-plot MANOVA

`reading2.dta` has data from an experiment involving two reading programs and three skill-enhancement techniques. Ten classes of first-grade students were randomly assigned so that five classes were taught with one reading program and another five classes were taught with the other. The 30 students in each class were divided into six groups with 5 students each. Within each class, the six groups were divided randomly so that each of the three skill-enhancement techniques was taught to two of the groups within each class. At the end of the school year, a reading assessment test was administered to all the students. Two scores were recorded. The first was a reading score (`score`), and the second was a comprehension score (`comprehension`).

Example 13 of [R] **anova** uses `reading.dta` to illustrate mixed designs for ANOVA. `reading2.dta` is the same as `reading.dta`, except that the `comprehension` variable is added.

```
. use http://www.stata-press.com/data/r12/reading2
(Reading experiment data)
. describe
Contains data from http://www.stata-press.com/data/r12/reading2.dta
  obs:           300                          Reading experiment data
 vars:             6                          24 Apr 2011 08:31
 size:         1,800                          (_dta has notes)
-------------------------------------------------------------------------------
              storage  display     value
variable name   type   format      label      variable label
-------------------------------------------------------------------------------
score           byte   %9.0g                  reading score
comprehension   byte   %9.0g                  comprehension score
program         byte   %9.0g                  reading program
class           byte   %9.0g                  class nested in program
skill           byte   %9.0g                  skill enhancement technique
group           byte   %9.0g                  group nested in class and skill
-------------------------------------------------------------------------------
Sorted by:
```

In this split-plot MANOVA, the whole-plot treatment is the two reading programs, and the split-plot treatment is the three skill-enhancement techniques.

For this split-plot MANOVA, the error term for `program` is `class` nested within `program`. The error term for `skill` and the `program` by `skill` interaction is the `class` by `skill` interaction nested within `program`. Other terms are also involved in the model and can be seen below.

```
. manova score comp = pr / cl|pr sk pr#sk / cl#sk|pr / gr|cl#sk|pr /,
> dropemptycells

                            Number of obs =       300

                            W = Wilks' lambda      L = Lawley-Hotelling trace
                            P = Pillai's trace     R = Roy's largest root

                  Source |  Statistic      df   F(df1,     df2) =   F   Prob>F
      -------------------+----------------------------------------------------
                   Model | W   0.5234      59   118.0    478.0     1.55 0.0008 e
                         | P   0.5249           118.0    480.0     1.45 0.0039 a
                         | L   0.8181           118.0    476.0     1.65 0.0001 a
                         | R   0.6830            59.0    240.0     2.78 0.0000 u
                         |----------------------------------------------------
                Residual |                240
      -------------------+----------------------------------------------------
                 program | W   0.4543       1     2.0      7.0     4.20 0.0632 e
                         | P   0.5457             2.0      7.0     4.20 0.0632 e
                         | L   1.2010             2.0      7.0     4.20 0.0632 e
                         | R   1.2010             2.0      7.0     4.20 0.0632 e
                         |----------------------------------------------------
           class|program |                  8
      -------------------+----------------------------------------------------
                   skill | W   0.6754       2     4.0     30.0     1.63 0.1935 e
                         | P   0.3317             4.0     32.0     1.59 0.2008 a
                         | L   0.4701             4.0     28.0     1.65 0.1908 a
                         | R   0.4466             2.0     16.0     3.57 0.0522 u
                         |
           program#skill | W   0.3955       2     4.0     30.0     4.43 0.0063 e
                         | P   0.6117             4.0     32.0     3.53 0.0171 a
                         | L   1.5100             4.0     28.0     5.29 0.0027 a
                         | R   1.4978             2.0     16.0    11.98 0.0007 u
                         |----------------------------------------------------
     class#skill|program |                 16
      -------------------+----------------------------------------------------
     class#skill|program | W   0.4010      16    32.0     58.0     1.05 0.4265 e
                         | P   0.7324            32.0     60.0     1.08 0.3860 a
                         | L   1.1609            32.0     56.0     1.02 0.4688 a
                         | R   0.6453            16.0     30.0     1.21 0.3160 u
                         |----------------------------------------------------
       group|class#skill||                 30
                 program |
      -------------------+----------------------------------------------------
       group|class#skill|| W   0.7713      30    60.0    478.0     1.10 0.2844 e
                 program | P   0.2363            60.0    480.0     1.07 0.3405 a
                         | L   0.2867            60.0    476.0     1.14 0.2344 a
                         | R   0.2469            30.0    240.0     1.98 0.0028 u
                         |----------------------------------------------------
                Residual |                240
      -------------------+----------------------------------------------------
                   Total |                299
      -------------------+----------------------------------------------------
                            e = exact, a = approximate, u = upper bound on F
```

The `program#skill` interaction is significant.

◁

MANOVA with repeated measures

One approach to analyzing repeated measures in an ANOVA setting is to use correction factors for terms in an ANOVA that involve the repeated measures. These correction factors attempt to correct for the violated assumption of independence of observations; see [R] **anova**. In this approach, the data are in long form; see [D] **reshape**.

Another approach to repeated measures is to use MANOVA with the repeated measures appearing as dependent variables, followed by tests involving linear combinations of these repeated measures. This approach involves fewer assumptions than the repeated-measures ANOVA approach.

The simplest possible repeated-measures design has no between-subject factors and only one within-subject factor (the repeated measures).

▷ Example 10: MANOVA with repeated-measures data

Here are data on five subjects, each of whom took three tests.

```
. use http://www.stata-press.com/data/r12/nobetween
. list

     +-------------------------------------+
     | subject   test1   test2   test3     |
     |-------------------------------------|
  1. |       1      68      69      95     |
  2. |       2      50      74      69     |
  3. |       3      72      89      71     |
  4. |       4      61      64      61     |
  5. |       5      60      71      90     |
     +-------------------------------------+
```

manova must be tricked into fitting a constant-only model. To do this, you generate a variable equal to one, use that variable as the single *term* in your manova, and then specify the noconstant option. From the resulting MANOVA, you then test the repeated measures with the ytransform() option of manovatest; see [MV] **manova postestimation** for syntax details.

```
. generate mycons = 1
. manova test1 test2 test3 = mycons, noconstant
                           Number of obs =       5
                           W = Wilks' lambda      L = Lawley-Hotelling trace
                           P = Pillai's trace     R = Roy's largest root

                  Source |  Statistic     df   F(df1,     df2) =   F   Prob>F
              -----------+--------------------------------------------------
                  mycons | W    0.0076     1      3.0       2.0   86.91 0.0114 e
                         | P    0.9924            3.0       2.0   86.91 0.0114 e
                         | L  130.3722            3.0       2.0   86.91 0.0114 e
                         | R  130.3722            3.0       2.0   86.91 0.0114 e
                         |--------------------------------------------------
                Residual |                 4
              -----------+--------------------------------------------------
                   Total |                 5
              --------------------------------------------------------------
                           e = exact, a = approximate, u = upper bound on F
```

```
. mat c = (1,0,-1\0,1,-1)
. manovatest mycons, ytransform(c)
Transformations of the dependent variables
(1)     test1 - test3
(2)     test2 - test3

                   W = Wilks' lambda      L = Lawley-Hotelling trace
                   P = Pillai's trace     R = Roy's largest root

          Source | Statistic     df   F(df1,    df2) =    F   Prob>F
     ------------+--------------------------------------------------
          mycons | W   0.2352     1     2.0      3.0     4.88 0.1141 e
                 | P   0.7648           2.0      3.0     4.88 0.1141 e
                 | L   3.2509           2.0      3.0     4.88 0.1141 e
                 | R   3.2509           2.0      3.0     4.88 0.1141 e
                 |--------------------------------------------------
        Residual |                4
     ------------+--------------------------------------------------
                   e = exact, a = approximate, u = upper bound on F
```

The test produced directly with `manova` is not interesting. It is testing the hypothesis that the three test score means are zero. The test produced by `manovatest` is of interest. From the contrasts in the matrix c, you produce a test that there is a difference between the `test1`, `test2`, and `test3` scores. Here the test produces a p-value of 0.1141, and you fail to reject the null hypothesis of equality between the test scores.

You can compare this finding with the results obtained from a repeated-measures ANOVA,

```
. reshape long test, i(subject) j(testnum)
. anova test subject testnum, repeated(testnum)
```

which produced an uncorrected p-value of 0.1160 and corrected p-values of 0.1181, 0.1435, and 0.1665 by using the Huynh–Feldt, Greenhouse–Geisser, and Box's conservative correction, respectively.

◁

▷ Example 11: Randomized block design with repeated measures

Milliken and Johnson (2009) demonstrate using `manova` to analyze repeated measures from a randomized block design used in studying the differences among varieties of sorghum. Table 27.1 of Milliken and Johnson (2009) provides the data. Four sorghum varieties were each planted in five blocks. A leaf-area index measurement was recorded for each of 5 weeks, starting 2 weeks after emergence.

The tests of interest include a test for equal variety marginal means, equal time marginal means, and a test for the interaction of variety and time. The MANOVA below does not directly provide these tests. `manovatest` after the `manova` gives the three tests of interest.

```
. use http://www.stata-press.com/data/r12/sorghum, clear
(Leaf area index on 4 sorghum varieties, Milliken & Johnson (2009))
. manova time1 time2 time3 time4 time5 = variety block
                           Number of obs =       20
                           W = Wilks' lambda      L = Lawley-Hotelling trace
                           P = Pillai's trace     R = Roy's largest root

                  Source |  Statistic     df   F(df1,    df2) =   F   Prob>F
              -----------+--------------------------------------------------
                   Model | W    0.0001     7    35.0     36.1     9.50 0.0000 a
                         | P    3.3890          35.0     60.0     3.61 0.0000 a
                         | L  126.2712          35.0     32.0    23.09 0.0000 a
                         | R  109.7360           7.0     12.0   188.12 0.0000 u
                         |--------------------------------------------------
                Residual |                12
              -----------+--------------------------------------------------
                 variety | W    0.0011     3    15.0     22.5    16.11 0.0000 a
                         | P    2.5031          15.0     30.0    10.08 0.0000 a
                         | L   48.3550          15.0     20.0    21.49 0.0000 a
                         | R   40.0068           5.0     10.0    80.01 0.0000 u
                         |--------------------------------------------------
                   block | W    0.0047     4    20.0     27.5     5.55 0.0000 a
                         | P    1.7518          20.0     44.0     1.71 0.0681 a
                         | L   77.9162          20.0     26.0    25.32 0.0000 a
                         | R   76.4899           5.0     11.0   168.28 0.0000 u
                         |--------------------------------------------------
                Residual |                12
              -----------+--------------------------------------------------
                   Total |                19
              --------------------------------------------------------------
                           e = exact, a = approximate, u = upper bound on F
```

Two matrices are needed for transformations of the `time#` variables. `m1` is a row vector containing five ones. `m2` provides contrasts for `time#`. The `manovatest, showorder` command lists the underlying ordering of columns for constructing two more matrices used to obtain linear combinations from the design matrix. Matrix `c1` provides contrasts on `variety`. Matrix `c2` is used to collapse to the overall margin of the design matrix to obtain time marginal means.

```
. matrix m1 = J(1,5,1)
. matrix m2 = (1,-1,0,0,0 \ 1,0,-1,0,0 \ 1,0,0,-1,0 \ 1,0,0,0,-1)
. manovatest, showorder
 Order of columns in the design matrix
      1: (variety==1)
      2: (variety==2)
      3: (variety==3)
      4: (variety==4)
      5: (block==1)
      6: (block==2)
      7: (block==3)
      8: (block==4)
      9: (block==5)
     10: _cons
. matrix c1 = (1,-1,0,0,0,0,0,0,0,0\1,0,-1,0,0,0,0,0,0,0\1,0,0,-1,0,0,0,0,0,0)
. matrix c2 = (.25,.25,.25,.25,.2,.2,.2,.2,.2,1)
```

The test for equal variety marginal means uses matrix `m1` to obtain the sum of the `time#` variables and matrix `c1` to provide the contrasts on `variety`. The second test uses `m2` to provide contrasts on `time#` and matrix `c2` to collapse to the appropriate margin for the test of time marginal means. The final test uses `m2` for contrasts on `time#` and `c1` for contrasts on `variety` to test the `variety`-by-`time` interaction.

```
. manovatest, test(c1) ytransform(m1)
 Transformation of the dependent variables
 (1)    time1 + time2 + time3 + time4 + time5
 Test constraints
 (1)    1.variety - 2.variety = 0
 (2)    1.variety - 3.variety = 0
 (3)    1.variety - 4.variety = 0
                       W = Wilks' lambda      L = Lawley-Hotelling trace
                       P = Pillai's trace     R = Roy's largest root

              Source | Statistic     df   F(df1,    df2) =   F    Prob>F
          -----------+--------------------------------------------------
          manovatest | W   0.0435     3     3.0     12.0    88.05 0.0000 e
                     | P   0.9565           3.0     12.0    88.05 0.0000 e
                     | L  22.0133           3.0     12.0    88.05 0.0000 e
                     | R  22.0133           3.0     12.0    88.05 0.0000 e
                     |--------------------------------------------------
            Residual |               12
          -----------+--------------------------------------------------
                       e = exact, a = approximate, u = upper bound on F
. manovatest, test(c2) ytransform(m2)
 Transformations of the dependent variables
 (1)    time1 - time2
 (2)    time1 - time3
 (3)    time1 - time4
 (4)    time1 - time5
 Test constraint
 (1)    .25*1.variety + .25*2.variety + .25*3.variety + .25*4.variety +
        .2*1.block + .2*2.block + .2*3.block + .2*4.block + .2*5.block + _cons
        = 0
                       W = Wilks' lambda      L = Lawley-Hotelling trace
                       P = Pillai's trace     R = Roy's largest root

              Source | Statistic     df   F(df1,    df2) =   F    Prob>F
          -----------+--------------------------------------------------
          manovatest | W   0.0050     1     4.0      9.0   445.62 0.0000 e
                     | P   0.9950           4.0      9.0   445.62 0.0000 e
                     | L 198.0544           4.0      9.0   445.62 0.0000 e
                     | R 198.0544           4.0      9.0   445.62 0.0000 e
                     |--------------------------------------------------
            Residual |               12
          -----------+--------------------------------------------------
                       e = exact, a = approximate, u = upper bound on F
```

```
. manovatest, test(c1) ytransform(m2)
Transformations of the dependent variables
(1)    time1 - time2
(2)    time1 - time3
(3)    time1 - time4
(4)    time1 - time5
Test constraints
(1)    1.variety - 2.variety = 0
(2)    1.variety - 3.variety = 0
(3)    1.variety - 4.variety = 0
                        W = Wilks' lambda      L = Lawley-Hotelling trace
                        P = Pillai's trace     R = Roy's largest root

              Source |  Statistic      df   F(df1,    df2) =   F    Prob>F
          -----------+--------------------------------------------------
          manovatest | W   0.0143       3    12.0    24.1      8.00 0.0000 a
                     | P   2.1463            12.0    33.0      6.91 0.0000 a
                     | L  12.1760            12.0    23.0      7.78 0.0000 a
                     | R   8.7953             4.0    11.0     24.19 0.0000 u
                     |--------------------------------------------------
            Residual |                 12
          --------------------------------------------------------------
                        e = exact, a = approximate, u = upper bound on F
```

All three tests are significant, indicating differences in `variety`, in `time`, and in the `variety`-by-`time` interaction.

◁

▷ Example 12: MANOVA and dependent-variable effects

Recall the fabric-data example from Rencher (2002) that we used in example 6 to illustrate a three-way MANOVA. Rencher has as an additional exercise to test the period effect (the `y1`, `y2`, and `y3` repeated-measures variables) and the interaction of period with the other factors in the model. The `ytransform()` option of `manovatest` provides a method to do this; see [MV] **manova postestimation**. Here are the tests of the period effect interacted with each term in the model. We create the matrix `c` with rows corresponding to the linear and quadratic contrasts for the three dependent variables.

```
. quietly manova y1 y2 y3 = proportion##treatment##filler
. matrix c = (-1,0,1 \ -1,2,-1)
. manovatest proportion, ytransform(c)
Transformations of the dependent variables
(1)    - y1 + y3
(2)    - y1 + 2*y2 - y3
                        W = Wilks' lambda      L = Lawley-Hotelling trace
                        P = Pillai's trace     R = Roy's largest root

              Source |  Statistic      df   F(df1,    df2) =   F    Prob>F
          -----------+--------------------------------------------------
          proportion | W   0.4749       2     4.0    22.0      2.48 0.0736 e
                     | P   0.5454             4.0    24.0      2.25 0.0936 a
                     | L   1.0631             4.0    20.0      2.66 0.0630 a
                     | R   1.0213             2.0    12.0      6.13 0.0147 u
                     |--------------------------------------------------
            Residual |                 12
          --------------------------------------------------------------
                        e = exact, a = approximate, u = upper bound on F
```

```
. manovatest treatment, ytransform(c)
 Transformations of the dependent variables
 (1)     - y1 + y3
 (2)     - y1 + 2*y2 - y3

                         W = Wilks' lambda      L = Lawley-Hotelling trace
                         P = Pillai's trace     R = Roy's largest root

                  Source | Statistic     df   F(df1,    df2) =   F   Prob>F
             ------------+-------------------------------------------------
               treatment | W   0.1419     1     2.0     11.0    33.27 0.0000 e
                         | P   0.8581           2.0     11.0    33.27 0.0000 e
                         | L   6.0487           2.0     11.0    33.27 0.0000 e
                         | R   6.0487           2.0     11.0    33.27 0.0000 e
                         |-------------------------------------------------
                Residual |               12
             ------------+-------------------------------------------------
                           e = exact, a = approximate, u = upper bound on F
. manovatest proportion#treatment, ytransform(c)
 Transformations of the dependent variables
 (1)     - y1 + y3
 (2)     - y1 + 2*y2 - y3

                         W = Wilks' lambda      L = Lawley-Hotelling trace
                         P = Pillai's trace     R = Roy's largest root

                  Source | Statistic     df   F(df1,    df2) =   F   Prob>F
    ---------------------+-------------------------------------------------
    proportion#treatment | W   0.7766     2     4.0     22.0     0.74 0.5740 e
                         | P   0.2276           4.0     24.0     0.77 0.5550 a
                         | L   0.2824           4.0     20.0     0.71 0.5972 a
                         | R   0.2620           2.0     12.0     1.57 0.2476 u
                         |-------------------------------------------------
                Residual |               12
    ---------------------+-------------------------------------------------
                           e = exact, a = approximate, u = upper bound on F
. manovatest filler, ytransform(c)
 Transformations of the dependent variables
 (1)     - y1 + y3
 (2)     - y1 + 2*y2 - y3

                         W = Wilks' lambda      L = Lawley-Hotelling trace
                         P = Pillai's trace     R = Roy's largest root

                  Source | Statistic     df   F(df1,    df2) =   F   Prob>F
             ------------+-------------------------------------------------
                  filler | W   0.0954     1     2.0     11.0    52.17 0.0000 e
                         | P   0.9046           2.0     11.0    52.17 0.0000 e
                         | L   9.4863           2.0     11.0    52.17 0.0000 e
                         | R   9.4863           2.0     11.0    52.17 0.0000 e
                         |-------------------------------------------------
                Residual |               12
             ------------+-------------------------------------------------
                           e = exact, a = approximate, u = upper bound on F
```

```
. manovatest proportion#filler, ytransform(c)
 Transformations of the dependent variables
 (1)    - y1 + y3
 (2)    - y1 + 2*y2 - y3

                           W = Wilks' lambda      L = Lawley-Hotelling trace
                           P = Pillai's trace     R = Roy's largest root

                    Source | Statistic     df   F(df1,    df2) =   F   Prob>F
          -----------------+--------------------------------------------------
         proportion#filler | W   0.6217     2    4.0     22.0     1.48 0.2436 e
                           | P   0.3870          4.0     24.0     1.44 0.2515 a
                           | L   0.5944          4.0     20.0     1.49 0.2439 a
                           | R   0.5698          2.0     12.0     3.42 0.0668 u
                           |--------------------------------------------------
                  Residual |               12
          -----------------+--------------------------------------------------
                             e = exact, a = approximate, u = upper bound on F
. manovatest treatment#filler, ytransform(c)
 Transformations of the dependent variables
 (1)    - y1 + y3
 (2)    - y1 + 2*y2 - y3

                           W = Wilks' lambda      L = Lawley-Hotelling trace
                           P = Pillai's trace     R = Roy's largest root

                    Source | Statistic     df   F(df1,    df2) =   F   Prob>F
           ----------------+--------------------------------------------------
          treatment#filler | W   0.3867     1    2.0     11.0     8.72 0.0054 e
                           | P   0.6133          2.0     11.0     8.72 0.0054 e
                           | L   1.5857          2.0     11.0     8.72 0.0054 e
                           | R   1.5857          2.0     11.0     8.72 0.0054 e
                           |--------------------------------------------------
                  Residual |               12
           ----------------+--------------------------------------------------
                             e = exact, a = approximate, u = upper bound on F
. manovatest proportion#treatment#filler, ytransform(c)
 Transformations of the dependent variables
 (1)    - y1 + y3
 (2)    - y1 + 2*y2 - y3

                           W = Wilks' lambda      L = Lawley-Hotelling trace
                           P = Pillai's trace     R = Roy's largest root

                    Source | Statistic     df   F(df1,    df2) =   F   Prob>F
      ---------------------+--------------------------------------------------
     proportion#treatment# | W   0.7812     2    4.0     22.0     0.72 0.5857 e
                    filler | P   0.2290          4.0     24.0     0.78 0.5518 a
                           | L   0.2671          4.0     20.0     0.67 0.6219 a
                           | R   0.2028          2.0     12.0     1.22 0.3303 u
                           |--------------------------------------------------
                  Residual |               12
      ---------------------+--------------------------------------------------
                             e = exact, a = approximate, u = upper bound on F
```

The first test, `manovatest proportion, ytransform(c)`, provides the test of `proportion` interacted with the period effect. The F tests for Wilks' lambda, Pillai's trace, and the Lawley–Hotelling trace do not reject the null hypothesis with a significance level of 0.05 (p-values of 0.0736, 0.0936, and 0.0630). The F test for Roy's largest root is an upper bound, so the p-value of 0.0147 is a lower bound.

The tests of `treatment` interacted with the period effect, `filler` interacted with the period effect, and `treatment#filler` interacted with the period effect are significant. The remaining tests are not.

To test the period effect, we call manovatest with both the ytransform() and test() options. The showorder option guides us in constructing the matrix for the test() option.

```
. manovatest, showorder
 Order of columns in the design matrix
      1: (proportion==1)
      2: (proportion==2)
      3: (proportion==3)
      4: (treatment==0)
      5: (treatment==1)
      6: (proportion==1)*(treatment==0)
      7: (proportion==1)*(treatment==1)
      8: (proportion==2)*(treatment==0)
      9: (proportion==2)*(treatment==1)
     10: (proportion==3)*(treatment==0)
     11: (proportion==3)*(treatment==1)
     12: (filler==1)
     13: (filler==2)
     14: (proportion==1)*(filler==1)
     15: (proportion==1)*(filler==2)
     16: (proportion==2)*(filler==1)
     17: (proportion==2)*(filler==2)
     18: (proportion==3)*(filler==1)
     19: (proportion==3)*(filler==2)
     20: (treatment==0)*(filler==1)
     21: (treatment==0)*(filler==2)
     22: (treatment==1)*(filler==1)
     23: (treatment==1)*(filler==2)
     24: (proportion==1)*(treatment==0)*(filler==1)
     25: (proportion==1)*(treatment==0)*(filler==2)
     26: (proportion==1)*(treatment==1)*(filler==1)
     27: (proportion==1)*(treatment==1)*(filler==2)
     28: (proportion==2)*(treatment==0)*(filler==1)
     29: (proportion==2)*(treatment==0)*(filler==2)
     30: (proportion==2)*(treatment==1)*(filler==1)
     31: (proportion==2)*(treatment==1)*(filler==2)
     32: (proportion==3)*(treatment==0)*(filler==1)
     33: (proportion==3)*(treatment==0)*(filler==2)
     34: (proportion==3)*(treatment==1)*(filler==1)
     35: (proportion==3)*(treatment==1)*(filler==2)
     36: _cons
```

We create a row vector, m, starting with 1/3 for three columns (corresponding to proportion), followed by 1/2 for two columns (corresponding to treatment), followed by 1/6 for six columns (for proportion#treatment), followed by 1/2 for two columns (for filler), followed by 1/6 for six columns (for proportion#filler), followed by four columns of 1/4 (for treatment#filler), followed by 1/12 for 12 columns (corresponding to the proportion#treatment#filler term), and finally, a 1 for the last column (corresponding to the constant in the model). The test of period effect then uses this m matrix and the c matrix previously defined as the basis of the test for the period effect.

```
. matrix m = J(1,3,1/3), J(1,2,1/2), J(1,6,1/6), J(1,2,1/2), J(1,6,1/6),
> J(1,4,1/4), J(1,12,1/12), (1)
. manovatest, test(m) ytrans(c)
 Transformations of the dependent variables
 (1)    - y1 + y3
 (2)    - y1 + 2*y2 - y3
 Test constraint
 (1)    .3333333*1.proportion + .3333333*2.proportion + .3333333*3.proportion +
        .5*0.treatment + .5*1.treatment + .1666667*1.proportion#0.treatment +
        .1666667*1.proportion#1.treatment + .1666667*2.proportion#0.treatment +
        .1666667*2.proportion#1.treatment + .1666667*3.proportion#0.treatment +
        .1666667*3.proportion#1.treatment + .5*1.filler + .5*2.filler +
        .1666667*1.proportion#1.filler + .1666667*1.proportion#2.filler +
        .1666667*2.proportion#1.filler + .1666667*2.proportion#2.filler +
        .1666667*3.proportion#1.filler + .1666667*3.proportion#2.filler +
        .25*0.treatment#1.filler + .25*0.treatment#2.filler +
        .25*1.treatment#1.filler + .25*1.treatment#2.filler +
        .0833333*1.proportion#0.treatment#1.filler +
        .0833333*1.proportion#0.treatment#2.filler +
        .0833333*1.proportion#1.treatment#1.filler +
        .0833333*1.proportion#1.treatment#2.filler +
        .0833333*2.proportion#0.treatment#1.filler +
        .0833333*2.proportion#0.treatment#2.filler +
        .0833333*2.proportion#1.treatment#1.filler +
        .0833333*2.proportion#1.treatment#2.filler +
        .0833333*3.proportion#0.treatment#1.filler +
        .0833333*3.proportion#0.treatment#2.filler +
        .0833333*3.proportion#1.treatment#1.filler +
        .0833333*3.proportion#1.treatment#2.filler + _cons = 0
                          W = Wilks' lambda      L = Lawley-Hotelling trace
                          P = Pillai's trace     R = Roy's largest root

              Source |  Statistic     df   F(df1,     df2) =   F    Prob>F
          -----------+--------------------------------------------------
          manovatest | W    0.0208     1     2.0     11.0   259.04 0.0000 e
                     | P    0.9792           2.0     11.0   259.04 0.0000 e
                     | L   47.0988           2.0     11.0   259.04 0.0000 e
                     | R   47.0988           2.0     11.0   259.04 0.0000 e
                     |--------------------------------------------------
            Residual |                12
          -----------+--------------------------------------------------
                          e = exact, a = approximate, u = upper bound on F
```

This result agrees with the answers provided by Rencher (2002).

◁

In the previous three examples, one factor has been encoded within the dependent variables. We have seen that the `ytransform()` option of `manovatest` provides the method for testing this factor and its interactions with the factors that appear on the right-hand side of the MANOVA.

More than one factor could be encoded within the dependent variables. Again the `ytransform()` option of `manovatest` allows us to perform multivariate tests of interest.

▷ Example 13: MANOVA and multiple dependent-variable effects

Table 6.14 of Rencher (2002) provides an example with two within-subject factors represented in the dependent variables and one between-subject factor.

```
. use http://www.stata-press.com/data/r12/table614
(Table 6.14, Rencher (2002))
. list in 9/12, noobs compress
```

c	sub~t	ab11	ab12	ab13	ab21	ab22	ab23	ab31	ab32	ab33
1	9	41	32	23	37	51	39	27	28	30
1	10	39	32	24	30	35	31	26	29	32
2	1	47	36	25	31	36	29	21	24	27
2	2	53	43	32	40	48	47	46	50	54

There are 20 observations. Factors a and b are encoded in the names of the nine dependent variables. Variable name ab23, for instance, indicates factor a at level 2 and factor b at level 3. Factor c is the between-subject factor.

We first compute a MANOVA by using the dependent variables and our one between-subject term.

```
. manova ab11 ab12 ab13 ab21 ab22 ab23 ab31 ab32 ab33 = c
                           Number of obs =      20
                           W = Wilks' lambda      L = Lawley-Hotelling trace
                           P = Pillai's trace     R = Roy's largest root

                  Source | Statistic     df   F(df1,    df2) =   F   Prob>F
              -----------+--------------------------------------------------
                       c | W   0.5330      1     9.0     10.0     0.97 0.5114 e
                         | P   0.4670            9.0     10.0     0.97 0.5114 e
                         | L   0.8762            9.0     10.0     0.97 0.5114 e
                         | R   0.8762            9.0     10.0     0.97 0.5114 e
                         |--------------------------------------------------
                Residual |                18
              -----------+--------------------------------------------------
                   Total |                19
              --------------------------------------------------------------
                           e = exact, a = approximate, u = upper bound on F
```

This approach provides the basis for computing tests on all terms of interest. We use the ytransform() and test() options of manovatest with the following matrices to obtain the tests of interest.

```
. mat a = (2,2,2,-1,-1,-1,-1,-1,-1 \ 0,0,0,1,1,1,-1,-1,-1)
. mat b = (2,-1,-1,2,-1,-1,2,-1,-1 \ 0,1,-1,0,1,-1,0,1,-1)
. forvalues i = 1/2 {
  2.          forvalues j = 1/2 {
  3.                  mat g = nullmat(g) \ vecdiag(a[`i',1...]'*b[`j',1...])
  4.          }
  5. }
. mat list g
g[4,9]
    c1  c2  c3  c4  c5  c6  c7  c8  c9
r1   4  -2  -2  -2   1   1  -2   1   1
r1   0   2  -2   0  -1   1   0  -1   1
r1   0   0   0   2  -1  -1  -2   1   1
r1   0   0   0   0   1  -1   0  -1   1
. mat j = J(1,9,1/9)
. mat xall = (.5,.5,1)
```

Matrices a and b correspond to factors a and b. Matrix g is the elementwise multiplication of each row of a with each row of b and corresponds to the a#b interaction. Matrix j is used to average the dependent variables, whereas matrix xall collapses over factor c.

Here are the tests for a, b, and a#b.

```
. manovatest, test(xall) ytrans(a)
 Transformations of the dependent variables
 (1)    2*ab11 + 2*ab12 + 2*ab13 - ab21 - ab22 - ab23 - ab31 - ab32 - ab33
 (2)    ab21 + ab22 + ab23 - ab31 - ab32 - ab33

 Test constraint
 (1)    .5*1.c + .5*2.c + _cons = 0

                           W = Wilks' lambda      L = Lawley-Hotelling trace
                           P = Pillai's trace     R = Roy's largest root

                  Source | Statistic     df   F(df1,    df2) =   F   Prob>F
              -----------+--------------------------------------------------
              manovatest | W   0.6755     1     2.0     17.0     4.08 0.0356 e
                         | P   0.3245           2.0     17.0     4.08 0.0356 e
                         | L   0.4803           2.0     17.0     4.08 0.0356 e
                         | R   0.4803           2.0     17.0     4.08 0.0356 e
                         |--------------------------------------------------
                Residual |               18
              -----------+--------------------------------------------------
                           e = exact, a = approximate, u = upper bound on F

. manovatest, test(xall) ytrans(b)
 Transformations of the dependent variables
 (1)    2*ab11 - ab12 - ab13 + 2*ab21 - ab22 - ab23 + 2*ab31 - ab32 - ab33
 (2)    ab12 - ab13 + ab22 - ab23 + ab32 - ab33

 Test constraint
 (1)    .5*1.c + .5*2.c + _cons = 0

                           W = Wilks' lambda      L = Lawley-Hotelling trace
                           P = Pillai's trace     R = Roy's largest root

                  Source | Statistic     df   F(df1,    df2) =   F   Prob>F
              -----------+--------------------------------------------------
              manovatest | W   0.3247     1     2.0     17.0    17.68 0.0001 e
                         | P   0.6753           2.0     17.0    17.68 0.0001 e
                         | L   2.0799           2.0     17.0    17.68 0.0001 e
                         | R   2.0799           2.0     17.0    17.68 0.0001 e
                         |--------------------------------------------------
                Residual |               18
              -----------+--------------------------------------------------
                           e = exact, a = approximate, u = upper bound on F

. manovatest, test(xall) ytrans(g)
 Transformations of the dependent variables
 (1)    4*ab11 - 2*ab12 - 2*ab13 - 2*ab21 + ab22 + ab23 - 2*ab31 + ab32 + ab33
 (2)    2*ab12 - 2*ab13 - ab22 + ab23 - ab32 + ab33
 (3)    2*ab21 - ab22 - ab23 - 2*ab31 + ab32 + ab33
 (4)    ab22 - ab23 - ab32 + ab33

 Test constraint
 (1)    .5*1.c + .5*2.c + _cons = 0

                           W = Wilks' lambda      L = Lawley-Hotelling trace
                           P = Pillai's trace     R = Roy's largest root

                  Source | Statistic     df   F(df1,    df2) =   F   Prob>F
              -----------+--------------------------------------------------
              manovatest | W   0.2255     1     4.0     15.0    12.88 0.0001 e
                         | P   0.7745           4.0     15.0    12.88 0.0001 e
                         | L   3.4347           4.0     15.0    12.88 0.0001 e
                         | R   3.4347           4.0     15.0    12.88 0.0001 e
                         |--------------------------------------------------
                Residual |               18
              -----------+--------------------------------------------------
                           e = exact, a = approximate, u = upper bound on F
```

Factors a, b, and a#b are significant with p-values of 0.0356, 0.0001, and 0.0001, respectively. The multivariate statistics are equivalent to the T^2 values Rencher reports using the relationship $T^2 = (n_1 + n_2 - 2) \times (1 - \Lambda)/\Lambda$ that applies in this situation. For instance, Wilks' lambda for factor a is reported as 0.6755 (and the actual value recorded in r(stat) is 0.67554286) so that $T^2 = (10 + 10 - 2) \times (1 - 0.67554286)/0.67554286 = 8.645$, as reported by Rencher.

We now compute the tests for c and the interactions of c with the other terms in the model.

```
. manovatest c, ytrans(j)

Transformation of the dependent variables
(1)     .1111111*ab11 + .1111111*ab12 + .1111111*ab13 + .1111111*ab21 +
        .1111111*ab22 + .1111111*ab23 + .1111111*ab31 + .1111111*ab32 +
        .1111111*ab33

                      W = Wilks' lambda         L = Lawley-Hotelling trace
                      P = Pillai's trace        R = Roy's largest root

                Source | Statistic     df   F(df1,    df2) =   F   Prob>F
           ------------+----------------------------------------------------
                     c | W   0.6781      1     1.0     18.0     8.54 0.0091 e
                       | P   0.3219            1.0     18.0     8.54 0.0091 e
                       | L   0.4747            1.0     18.0     8.54 0.0091 e
                       | R   0.4747            1.0     18.0     8.54 0.0091 e
                       |----------------------------------------------------
              Residual |                18
           ------------+----------------------------------------------------
                         e = exact, a = approximate, u = upper bound on F

. manovatest c, ytrans(a)

Transformations of the dependent variables
(1)     2*ab11 + 2*ab12 + 2*ab13 - ab21 - ab22 - ab23 - ab31 - ab32 - ab33
(2)     ab21 + ab22 + ab23 - ab31 - ab32 - ab33

                      W = Wilks' lambda         L = Lawley-Hotelling trace
                      P = Pillai's trace        R = Roy's largest root

                Source | Statistic     df   F(df1,    df2) =   F   Prob>F
           ------------+----------------------------------------------------
                     c | W   0.9889      1     2.0     17.0     0.10 0.9097 e
                       | P   0.0111            2.0     17.0     0.10 0.9097 e
                       | L   0.0112            2.0     17.0     0.10 0.9097 e
                       | R   0.0112            2.0     17.0     0.10 0.9097 e
                       |----------------------------------------------------
              Residual |                18
           ------------+----------------------------------------------------
                         e = exact, a = approximate, u = upper bound on F

. manovatest c, ytrans(b)

Transformations of the dependent variables
(1)     2*ab11 - ab12 - ab13 + 2*ab21 - ab22 - ab23 + 2*ab31 - ab32 - ab33
(2)     ab12 - ab13 + ab22 - ab23 + ab32 - ab33

                      W = Wilks' lambda         L = Lawley-Hotelling trace
                      P = Pillai's trace        R = Roy's largest root

                Source | Statistic     df   F(df1,    df2) =   F   Prob>F
           ------------+----------------------------------------------------
                     c | W   0.9718      1     2.0     17.0     0.25 0.7845 e
                       | P   0.0282            2.0     17.0     0.25 0.7845 e
                       | L   0.0290            2.0     17.0     0.25 0.7845 e
                       | R   0.0290            2.0     17.0     0.25 0.7845 e
                       |----------------------------------------------------
              Residual |                18
           ------------+----------------------------------------------------
                         e = exact, a = approximate, u = upper bound on F
```

```
. manovatest c, ytrans(g)
Transformations of the dependent variables
(1)    4*ab11 - 2*ab12 - 2*ab13 - 2*ab21 + ab22 + ab23 - 2*ab31 + ab32 + ab33
(2)    2*ab12 - 2*ab13 - ab22 + ab23 - ab32 + ab33
(3)    2*ab21 - ab22 - ab23 - 2*ab31 + ab32 + ab33
(4)    ab22 - ab23 - ab32 + ab33

                          W = Wilks' lambda      L = Lawley-Hotelling trace
                          P = Pillai's trace     R = Roy's largest root

               Source |  Statistic     df   F(df1,    df2) =   F   Prob>F
          ------------+--------------------------------------------------
                    c | W   0.9029      1     4.0     15.0     0.40 0.8035 e
                      | P   0.0971            4.0     15.0     0.40 0.8035 e
                      | L   0.1075            4.0     15.0     0.40 0.8035 e
                      | R   0.1075            4.0     15.0     0.40 0.8035 e
                      |--------------------------------------------------
             Residual |                18
          ------------+--------------------------------------------------
                           e = exact, a = approximate, u = upper bound on F
```

The test of c is equivalent to an ANOVA using the sum or average of the dependent variables as the dependent variable. The test of c produces an F of 8.54 with a p-value of 0.0091, which agrees with the results of Rencher (2002).

The tests of a#c, b#c, and a#b#c produce p-values of 0.9097, 0.7845, and 0.8035, respectively.

In summary, the factors that are significant are a, b, a#b, and c.

◁

Saved results

manova saves the following in e():

Scalars

e(N)	number of observations
e(k)	number of parameters
e(k_eq)	number of equations in e(b)
e(df_m)	model degrees of freedom
e(df_r)	residual degrees of freedom
e(df_#)	degrees of freedom for term #
e(rank)	rank of e(V)

Macros

e(cmd)	manova
e(cmdline)	command as typed
e(depvar)	names of dependent variables
e(indepvars)	names of the right-hand-side variables
e(term_#)	term #
e(errorterm_#)	error term for term # (defined for terms using nonresidual error)
e(wtype)	weight type
e(wexp)	weight expression
e(r2)	R^2 for each equation
e(rmse)	RMSE for each equation
e(F)	F statistic for each equation
e(p_F)	significance of F for each equation
e(properties)	b V
e(estat_cmd)	program used to implement estat
e(predict)	program used to implement predict
e(marginsnotok)	predictions disallowed by margins
e(asbalanced)	factor variables fvset as asbalanced
e(asobserved)	factor variables fvset as asobserved

Matrices

e(b)	coefficient vector (a stacked version of e(B))
e(B)	coefficient matrix
e(E)	residual-error SSCP matrix
e(xpxinv)	generalized inverse of X′X
e(H_m)	hypothesis SSCP matrix for the overall model
e(stat_m)	multivariate statistics for the overall model
e(eigvals_m)	eigenvalues of $E^{-1}H$ for the overall model
e(aux_m)	s, m, and n values for the overall model
e(H_#)	hypothesis SSCP matrix for term #
e(stat_#)	multivariate statistics for term # (if computed)
e(eigvals_#)	eigenvalues of $E^{-1}H$ for term # (if computed)
e(aux_#)	s, m, and n values for term # (if computed)
e(V)	variance–covariance matrix of the estimators

Functions

e(sample)	marks estimation sample

Methods and formulas

manova is implemented as an ado-file.

Let $\mathbf{Y}$ denote the matrix of observations on the left-hand-side variables. Let $\mathbf{X}$ denote the design matrix based on the right-hand-side variables. The last column of $\mathbf{X}$ is equal to all ones (unless the noconstant option was specified). Categorical right-hand-side variables are placed in $\mathbf{X}$ as a set of indicator (sometimes called dummy) variables, whereas continuous variables enter as is. Columns of $\mathbf{X}$ corresponding to interactions are formed by multiplying the various combinations of columns for the variables involved in the interaction.

The multivariate model

$$\mathbf{Y} = \mathbf{X}\boldsymbol{\beta} + \boldsymbol{\epsilon}$$

leads to multivariate hypotheses of the form

$$\mathbf{C}\boldsymbol{\beta}\mathbf{A}' = \mathbf{0}$$

where $\boldsymbol{\beta}$ is a matrix of parameters, $\mathbf{C}$ specifies constraints on the design matrix $\mathbf{X}$ for a particular hypothesis, and $\mathbf{A}$ provides a transformation of $\mathbf{Y}$. $\mathbf{A}$ is often the identity matrix.

An estimate of $\boldsymbol{\beta}$ is provided by

$$\mathbf{B} = (\mathbf{X}'\mathbf{X})^{-}\mathbf{X}'\mathbf{Y}$$

The error sum of squares and cross products (SSCP) matrix is

$$\mathbf{E} = \mathbf{A}(\mathbf{Y}'\mathbf{Y} - \mathbf{B}'\mathbf{X}'\mathbf{X}\mathbf{B})\mathbf{A}'$$

and the SSCP matrix for the hypothesis is

$$\mathbf{H} = \mathbf{A}(\mathbf{C}\mathbf{B})'\{\mathbf{C}(\mathbf{X}'\mathbf{X})^{-}\mathbf{C}'\}^{-1}(\mathbf{C}\mathbf{B})\mathbf{A}'$$

The inclusion of weights, if specified, enters the formulas in a manner similar to that shown in [R] **regress**.

Let $\lambda_1 > \lambda_2 > \cdots > \lambda_s$ represent the nonzero eigenvalues of $\mathbf{E}^{-1}\mathbf{H}$. $s = \min(p, \nu_h)$, where p is the number of columns of $\mathbf{YA}'$ (that is, the number of y variables or number of resultant transformed left-hand-side variables), and ν_h is the hypothesis degrees of freedom.

Wilks' (1932) lambda statistic is

$$\Lambda = \prod_{i=1}^{s} \frac{1}{1+\lambda_i} = \frac{|\mathbf{E}|}{|\mathbf{H}+\mathbf{E}|}$$

and is a likelihood-ratio test. This statistic is distributed as the Wilks' Λ distribution if $\mathbf{E}$ has the Wishart distribution, $\mathbf{H}$ has the Wishart distribution under the null hypothesis, and $\mathbf{E}$ and $\mathbf{H}$ are independent. The null hypothesis is rejected for small values of Λ.

Pillai's (1955) trace is

$$V = \sum_{i=1}^{s} \frac{\lambda_i}{1+\lambda_i} = \text{trace}\left\{(\mathbf{E}+\mathbf{H})^{-1}\mathbf{H}\right\}$$

and the Lawley–Hotelling trace (Lawley 1938; Hotelling 1951) is

$$U = \sum_{i=1}^{s} \lambda_i = \text{trace}(\mathbf{E}^{-1}\mathbf{H})$$

and is also known as Hotelling's generalized T^2 statistic.

Roy's largest root is taken as λ_1, though some report $\theta = \lambda_1/(1+\lambda_1)$, which is bounded between zero and one. Roy's largest root provides a test based on the union-intersection approach to test construction introduced by Roy (1939).

Tables providing critical values for these four multivariate statistics are found in many of the books that discuss MANOVA, including Rencher (1998, 2002).

Let p be the number of columns of $\mathbf{YA}'$ (that is, the number of y variables or the number of resultant transformed y variables), ν_h be the hypothesis degrees of freedom, ν_e be the error degrees of freedom, $s = \min(\nu_h, p)$, $m = (|\nu_h - p| - 1)/2$, and $n = (\nu_e - p - 1)/2$. Transformations of these four multivariate statistics to F statistics are as follows.

For Wilks' lambda, an approximate F statistic (Rao 1951) with df_1 and df_2 degrees of freedom is

$$F = \frac{(1-\Lambda^{1/t})df_1}{(\Lambda^{1/t})df_2}$$

where

$$df_1 = p\nu_h \qquad df_2 = wt - 1 - p\nu_h/2$$

$$w = \nu_e + \nu_h - (p + \nu_h + 1)/2$$

$$t = \left(\frac{p^2\nu_h^2 - 4}{p^2 + \nu_h^2 - 5}\right)^{1/2}$$

t is set to one if either the numerator or the denominator equals zero. This F statistic is exact when p equals 1 or 2 or when ν_h equals 1 or 2.

An approximate F statistic for Pillai's trace (Pillai 1954, 1956b) with $s(2m+s+1)$ and $s(2n+s+1)$ degrees of freedom is

$$F = \frac{(2n+s+1)V}{(2m+s+1)(s-V)}$$

An approximate F statistic for the Lawley–Hotelling trace (Pillai 1954, 1956a) with $s(2m+s+1)$ and $2sn+2$ degrees of freedom is

$$F = \frac{2(sn+1)U}{s^2(2m+s+1)}$$

When p or ν_h are 1, an exact F statistic for Roy's largest root is

$$F = \lambda_1 \frac{\nu_e - p + 1}{p}$$

with $|\nu_h - p| + 1$ and $\nu_e - p + 1$ degrees of freedom. In other cases, an upper bound F statistic (providing a lower bound on the p-value) for Roy's largest root is

$$F = \lambda_1 \frac{\nu_e - d + \nu_h}{d}$$

with d and $\nu_e - d + \nu_h$ degrees of freedom, where $d = \max(p, \nu_h)$.

Samuel Stanley Wilks (1906–1964) was born in Texas. He gained degrees in architecture, mathematics, and statistics from North Texas Teachers' College and the universities of Texas and Iowa. After periods in Columbia and England, he moved to Princeton in 1933. Wilks published various widely used texts, was founding editor of the *Annals of Mathematical Statistics*, and made many key contributions to multivariate statistics. Wilks' lambda is named for him.

References

Allison, M. J., P. Zappasodi, and M. B. Lurie. 1962. The correlation of biphasic metabolic response with a biphasic response in resistance to tuberculosis in rabbits. *Journal of Experimental Medicine* 115: 881–890.

Anderson, T. W. 1965. Samuel Stanley Wilks, 1906–1964. *Annals of Mathematical Statistics* 36: 1–23.

——. 2003. *An Introduction to Multivariate Statistical Analysis*. 3rd ed. New York: Wiley.

Andrews, D. F., and A. M. Herzberg, ed. 1985. *Data: A Collection of Problems from Many Fields for the Student and Research Worker*. New York: Springer.

Arnold, S. F. 1981. *The Theory of Linear Models and Multivariate Analysis*. New York: Wiley.

Box, G. E. P. 1950. Problems in the analysis of growth and wear curves. *Biometrics* 6: 362–389.

Brown, J. D., and E. Beerstecher. 1951. Metabolic patterns of underweight and overweight individuals. In *Biochemical Institute Studies IV, No. 5109*. Austin, TX: University of Texas Press.

Hilbe, J. M. 1992. smv4: One-way multivariate analysis of variance (MANOVA). *Stata Technical Bulletin* 6: 5–7. Reprinted in *Stata Technical Bulletin Reprints*, vol. 1, pp. 138–139. College Station, TX: Stata Press.

Hotelling, H. 1951. A generalized t^2 test and measurement of multivariate dispersion. *Proceedings of the Second Berkeley Symposium on Mathematical Statistics and Probability* 1: 23–41.

Lawley, D. N. 1938. A generalization of Fisher's z-test. *Biometrika* 30: 180–187.

Mardia, K. V., J. T. Kent, and J. M. Bibby. 1979. *Multivariate Analysis*. London: Academic Press.

Milliken, G. A., and D. E. Johnson. 2009. *Analysis of Messy Data, Volume 1: Designed Experiments*. 2nd ed. Boca Raton, FL: CRC Press.

Morrison, D. F. 1998. Multivariate analysis of variance. In Vol. 4 of *Encyclopedia of Biostatistics*, ed. P. Armitage and T. Colton, 2820–2825. New York: Wiley.

——. 2005. *Multivariate Statistical Methods*. 4th ed. Belmont, CA: Duxbury.

Pillai, K. C. S. 1954. On some distribution problems in multivariate analysis. In *Mimeograph Series No. 88*. Institute of Statistics, University of North Carolina, Chapel Hill.

——. 1955. Some new test criteria in multivariate analysis. *Annals of Mathematical Statistics* 26: 117–121.

——. 1956a. Some results useful in multivariate analysis. *Annals of Mathematical Statistics* 27: 1106–1114.

——. 1956b. On the distribution of the largest or the smallest root of a matrix in multivariate analysis. *Biometrika* 43: 122–127.

——. 1985. Multivariate analysis of variance (MANOVA). In Vol. 6 of *Encyclopedia of Statistical Sciences*, ed. S. Kotz, N. L. Johnson, and C. B. Read, 20–29. New York: Wiley.

Rao, C. R. 1951. An asymptotic expansion of the distribution of Wilks' criterion. *Bulletin of the International Statistical Institute* 33: 177–180.

Rencher, A. C. 1998. *Multivariate Statistical Inference and Applications*. New York: Wiley.

——. 2002. *Methods of Multivariate Analysis*. 2nd ed. New York: Wiley.

Roy, S. N. 1939. p-statistics or some generalizations in analysis of variance appropriate to multivariate problems. *Sankhyā* 4: 381–396.

Seber, G. A. F. 1984. *Multivariate Observations*. New York: Wiley.

Smith, H., R. Gnanadesikan, and J. B. Hughes. 1962. Multivariate analysis of variance (MANOVA). *Biometrics* 18: 22–41.

Timm, N. H. 1975. *Multivariate Analysis with Applications in Education and Psychology*. Pacific Grove, CA: Brooks/Cole.

Wilks, S. S. 1932. Certain generalizations in the analysis of variance. *Biometrika* 24: 471–494.

Woodard, D. E. 1931. Healing time of fractures of the jaw in relation to delay before reduction, infection, syphilis and blood calcium and phosphorus content. *Journal of the American Dental Association* 18: 419–442.

Also see

[MV] **manova postestimation** — Postestimation tools for manova

[D] **encode** — Encode string into numeric and vice versa

[D] **reshape** — Convert data from wide to long form and vice versa

[MV] **mvtest** — Multivariate tests

[R] **anova** — Analysis of variance and covariance

[R] **mvreg** — Multivariate regression

Stata Structural Equation Modeling Reference Manual

[U] **13.5 Accessing coefficients and standard errors**

[U] **20 Estimation and postestimation commands**

Title

manova postestimation — Postestimation tools for manova

Description

The following postestimation commands are of special interest after `manova`:

Command	Description
`manovatest`	multivariate tests after `manova`
`screeplot`	plot eigenvalues

For information about `manovatest`, see below. For information about `screeplot`, see [MV] **screeplot**.

The following standard postestimation commands are also available:

Command	Description
`contrast`	contrasts and ANOVA-style joint tests of estimates
* `estat`	VCE and estimation sample summary
`estimates`	cataloging estimation results
`lincom`	point estimates, standard errors, testing, and inference for linear combinations of coefficients
`margins`	marginal means, predictive margins, marginal effects, and average marginal effects
`marginsplot`	graph the results from margins (profile plots, interaction plots, etc.)
`nlcom`	point estimates, standard errors, testing, and inference for nonlinear combinations of coefficients
`predict`	predictions, residuals, and standard errors
`predictnl`	point estimates, standard errors, testing, and inference for generalized predictions
`pwcompare`	pairwise comparisons of estimates
`test`	Wald tests of simple and composite linear hypotheses
`testnl`	Wald tests of nonlinear hypotheses

* `estat ic` is not available after `manova`.

For all except `predict` and `test`, see the corresponding entries in the *Base Reference Manual* for details. For `predict` and `test`, see below.

Special-interest postestimation commands

`manovatest` provides multivariate tests involving *terms* or linear combinations of the underlying design matrix from the most recently fit `manova`. The four multivariate test statistics are Wilks' lambda, Pillai's trace, Lawley–Hotelling trace, and Roy's largest root. The format of the output is similar to that shown by `manova`; see [MV] **manova**.

In addition to the standard syntax of `test` (see [R] **test**), `test` after `manova` has two additionally allowed syntaxes; see below. `test` performs Wald tests of expressions involving the coefficients of the underlying regression model. Simple and composite linear hypotheses are possible.

Syntax for predict

predict [*type*] *newvar* [*if*] [*in*] [, equation(*eqno*[, *eqno*]) *statistic*]

statistic	Description
Main	
xb	$\mathbf{x}_j\mathbf{b}$, fitted values; the default
stdp	standard error of the fitted value
residuals	residuals
difference	difference between the linear predictions of two equations
stddp	standard error of the fitted values for differences

These statistics are available both in and out of sample; type predict ... if e(sample) ... if wanted only for the estimation sample.

Menu

Statistics > Postestimation > Predictions, residuals, etc.

Options for predict

Main

equation(*eqno*[, *eqno*]) specifies the equation to which you are referring.

equation() is filled in with one *eqno* for the xb, stdp, and residuals options. equation(#1) would mean that the calculation is to be made for the first equation (that is, for the first dependent variable), equation(#2) would mean the second, and so on. You could also refer to the equations by their names. equation(income) would refer to the equation named income and equation(hours), to the equation named hours.

If you do not specify equation(), results are the same as if you had specified equation(#1).

difference and stddp refer to between-equations concepts. To use these options, you must specify two equations, for example, equation(#1,#2) or equation(income,hours). When two equations must be specified, equation() is required. With equation(#1,#2), difference computes the prediction of equation(#1) minus the prediction of equation(#2).

xb, the default, calculates the fitted values—the prediction of $\mathbf{x}_j\mathbf{b}$ for the specified equation.

stdp calculates the standard error of the prediction for the specified equation (the standard error of the estimated expected value or mean for the observation's covariate pattern). The standard error of the prediction is also referred to as the standard error of the fitted value.

residuals calculates the residuals.

difference calculates the difference between the linear predictions of two equations in the system.

stddp calculates the standard error of the difference in linear predictions ($\mathbf{x}_{1j}\mathbf{b} - \mathbf{x}_{2j}\mathbf{b}$) between equations 1 and 2.

For more information on using predict after multiple-equation estimation commands, see [R] **predict**.

Syntax for manovatest

manovatest *term* [*term* ...] [/ *term* [*term* ...]] [, ytransform(*matname*)]

manovatest , test(*matname*) [ytransform(*matname*)]

manovatest , showorder

where *term* is a term from the *termlist* in the previously run manova.

Menu

Statistics > Multivariate analysis > MANOVA, multivariate regression, and related > Multivariate tests after MANOVA

Options for manovatest

ytransform(*matname*) specifies a matrix for transforming the y variables (the *depvarlist* from manova) as part of the test. The multivariate tests are based on $(\mathbf{AEA}')^{-1}(\mathbf{AHA}')$. By default, $\mathbf{A}$ is the identity matrix. ytransform() is how you specify an $\mathbf{A}$ matrix to be used in the multivariate tests. Specifying ytransform() provides the same results as first transforming the y variables with $\mathbf{YA}'$, where $\mathbf{Y}$ is the matrix formed by binding the y variables by column and $\mathbf{A}$ is the matrix stored in *matname*; then performing manova on the transformed y's; and finally running manovatest without ytransform().

The number of columns of *matname* must equal the number of variables in the *depvarlist* from manova. The number of rows must be less than or equal to the number of variables in the *depvarlist* from manova. *matname* should have columns in the same order as the *depvarlist* from manova. The column and row names of *matname* are ignored.

When ytransform() is specified, a listing of the transformations is presented before the table containing the multivariate tests. You should examine this table to verify that you have applied the transformation you desired.

test(*matname*) is required with the second syntax of manovatest. The rows of *matname* specify linear combinations of the underlying design matrix of the MANOVA that are to be jointly tested. The columns correspond to the underlying design matrix (including the constant if it has not been suppressed). The column and row names of *matname* are ignored.

A listing of the constraints imposed by the test() option is presented before the table containing the multivariate tests. You should examine this table to verify that you have applied the linear combinations you desired. Typing manovatest, showorder allows you to examine the ordering of the columns for the design matrix from the MANOVA.

showorder causes manovatest to list the definition of each column in the design matrix. showorder is not allowed with any other option or when *terms* are specified.

Syntax for test following manova

In addition to the standard syntax of `test` (see [R] **test**), `test` after `manova` also allows the following.

`test , test(`*matname*`)` [`mtest`[`(`*opt*`)`] `matvlc(`*matname*`)`]	syntax A
`test , showorder`	syntax B

syntax A test expression involving the coefficients of the underlying multivariate regression model; you provide information as a matrix

syntax B show underlying order of design matrix, which is useful when constructing the *matname* argument of the `test()` option

Menu

Statistics > Multivariate analysis > MANOVA, multivariate regression, and related > Wald test after MANOVA

Options for test after manova

`test(`*matname*`)` is required with syntax A of `test`. The rows of *matname* specify linear combinations of the underlying design matrix of the MANOVA that are to be jointly tested. The columns correspond to the underlying design matrix (including the constant if it has not been suppressed). The column and row names of *matname* are ignored.

A listing of the constraints imposed by the `test()` option is presented before the table containing the tests. You should examine this table to verify that you have applied the linear combinations you desired. Typing `test, showorder` allows you to examine the ordering of the columns for the design matrix from the MANOVA.

matname should have as many columns as the number of dependent variables times the number of columns in the basic design matrix. The design matrix is repeated for each dependent variable.

`mtest`[`(`*opt*`)`] specifies that tests be performed for each condition separately. *opt* specifies the method for adjusting p-values for multiple testing. Valid values for *opt* are

`bonferroni`	Bonferroni's method
`holm`	Holm's method
`sidak`	Šidák's method
`noadjust`	no adjustment is to be made

Specifying `mtest` without an argument is equivalent to specifying `mtest(noadjust)`.

`matvlc(`*matname*`)`, a programmer's option, saves the variance–covariance matrix of the linear combinations involved in the suite of tests. For the test of H_0: $\mathbf{Lb} = \mathbf{c}$, what is returned in *matname* is $\mathbf{LVL}'$, where $\mathbf{V}$ is the estimated variance–covariance matrix of $\mathbf{b}$.

`showorder` causes `test` to list the definition of each column in the design matrix. `showorder` is not allowed with any other option.

Remarks

Several postestimation tools are available after manova. We demonstrate these tools by extending examples 1, 2, 4, and 8 of [MV] **manova**.

▷ Example 1

Example 1 of [MV] **manova** presented a balanced one-way MANOVA on the rootstock data.

```
. use http://www.stata-press.com/data/r12/rootstock
(Table 6.2 Rootstock Data -- Rencher (2002))
. manova y1 y2 y3 y4 = rootstock
  (output omitted )
```

test provides Wald tests on expressions involving the underlying coefficients of the model, and lincom provides linear combinations along with standard errors and confidence intervals.

```
. test [y3]3.rootstock = ([y3]1.rootstock + [y3]2.rootstock)/2
 ( 1)  - .5*[y3]1b.rootstock - .5*[y3]2.rootstock + [y3]3.rootstock = 0
       F(  1,    42) =    5.62
            Prob > F =    0.0224
. lincom [y3]4.rootstock - [y1]4.rootstock
 ( 1)  - [y1]4.rootstock + [y3]4.rootstock = 0
```

	Coef.	Std. Err.	t	P>|t|	[95% Conf.	Interval]
(1)	.2075001	.1443917	1.44	0.158	-.0838941	.4988943

If the equation portion of the expression is omitted, the first equation (first dependent variable) is assumed.

◁

The manovatest postestimation command provides multivariate tests of *terms* or linear combinations of the underlying design matrix from the most recent MANOVA model.

▷ Example 2

In example 2 of [MV] **manova**, a one-way MANOVA on the metabolic dataset was shown.

```
. use http://www.stata-press.com/data/r12/metabolic
(Table 4.5 Metabolic Comparisons of Rabbits -- Rencher (1998))
. manova y1 y2 = group
  (output omitted )
```

manovatest can test *terms* from the preceding manova. Here we test the group term from our one-way MANOVA:

```
. manovatest group

                          W = Wilks' lambda      L = Lawley-Hotelling trace
                          P = Pillai's trace     R = Roy's largest root

                 Source |  Statistic      df   F(df1,     df2) =   F   Prob>F
            ------------+-----------------------------------------------------
                  group | W   0.1596       3     6.0     32.0      8.02 0.0000 e
                        | P   1.2004             6.0     34.0      8.51 0.0000 a
                        | L   3.0096             6.0     30.0      7.52 0.0001 a
                        | R   1.5986             3.0     17.0      9.06 0.0008 u
                        |-----------------------------------------------------
               Residual |                 17
            ------------+-----------------------------------------------------
                          e = exact, a = approximate, u = upper bound on F
```

Using `manovatest` to test model *terms* is not interesting here. It merely repeats information already presented by `manova`. Later we will see useful applications of *term* testing via `manovatest`.

`manovatest` can also be used to test linear combinations of the underlying design matrix of the MANOVA model. Whereas the MANOVA indicates that there are differences in the groups, it does not indicate the nature of those differences. Rencher discusses three linear contrasts of interest for this example: group one (the control) versus the rest, group four versus groups two and three, and group two versus group three. The `test()` option of `manovatest` allows us to test these hypotheses.

Because we did not use the `noconstant` option with our `manova`, the underlying parameterization of the design matrix has the last column corresponding to the constant in the model, whereas the first four columns correspond to the four groups of rabbits. The `showorder` option of `manovatest` illustrates this point. The tests on the three contrasts of interest follow.

```
. manovatest, showorder

 Order of columns in the design matrix
      1: (group==1)
      2: (group==2)
      3: (group==3)
      4: (group==4)
      5: _cons

. matrix c1 = (3,-1,-1,-1,0)

. manovatest, test(c1)

 Test constraint
 (1)    3*1.group - 2.group - 3.group - 4.group = 0

                          W = Wilks' lambda      L = Lawley-Hotelling trace
                          P = Pillai's trace     R = Roy's largest root

                 Source |  Statistic      df   F(df1,     df2) =   F   Prob>F
            ------------+-----------------------------------------------------
             manovatest | W   0.4063       1     2.0     16.0     11.69 0.0007 e
                        | P   0.5937             2.0     16.0     11.69 0.0007 e
                        | L   1.4615             2.0     16.0     11.69 0.0007 e
                        | R   1.4615             2.0     16.0     11.69 0.0007 e
                        |-----------------------------------------------------
               Residual |                 17
            ------------+-----------------------------------------------------
                          e = exact, a = approximate, u = upper bound on F
```

```
. matrix c2 = (0,-1,-1,2,0)
. manovatest, test(c2)
Test constraint
(1)    - 2.group - 3.group + 2*4.group = 0
                          W = Wilks' lambda        L = Lawley-Hotelling trace
                          P = Pillai's trace       R = Roy's largest root

                  Source | Statistic      df   F(df1,    df2) =    F   Prob>F
              -----------+----------------------------------------------------
              manovatest | W   0.9567      1      2.0     16.0      0.36 0.7018 e
                         | P   0.0433             2.0     16.0      0.36 0.7018 e
                         | L   0.0453             2.0     16.0      0.36 0.7018 e
                         | R   0.0453             2.0     16.0      0.36 0.7018 e
                         |----------------------------------------------------
                Residual |                17
              -----------+----------------------------------------------------
                           e = exact, a = approximate, u = upper bound on F
. matrix c3 = (0,1,-1,0,0)
. manovatest, test(c3)
Test constraint
(1)     2.group - 3.group = 0
                          W = Wilks' lambda        L = Lawley-Hotelling trace
                          P = Pillai's trace       R = Roy's largest root

                  Source | Statistic      df   F(df1,    df2) =    F   Prob>F
              -----------+----------------------------------------------------
              manovatest | W   0.4161      1      2.0     16.0     11.23 0.0009 e
                         | P   0.5839             2.0     16.0     11.23 0.0009 e
                         | L   1.4033             2.0     16.0     11.23 0.0009 e
                         | R   1.4033             2.0     16.0     11.23 0.0009 e
                         |----------------------------------------------------
                Residual |                17
              -----------+----------------------------------------------------
                           e = exact, a = approximate, u = upper bound on F
```

Because there is only 1 degree of freedom for each of the hypotheses, the F tests are exact (and identical for the four multivariate methods). The first test indicates that the mean vector for the control group is significantly different from the mean vectors for the other three groups. The second test, with a p-value of 0.7018, fails to reject the null hypothesis that group four equals groups two and three. The third test, with a p-value of 0.0009, indicates differences between the mean vectors of groups two and three.

Rencher also tests using weighted orthogonal contrasts. `manovatest` can do these tests as well.

```
. matrix c1w = (14,-7,-5,-2,0)
. manovatest, test(c1w)
Test constraint
(1)     14*1.group - 7*2.group - 5*3.group - 2*4.group = 0
                          W = Wilks' lambda        L = Lawley-Hotelling trace
                          P = Pillai's trace       R = Roy's largest root

                  Source | Statistic      df   F(df1,    df2) =    F   Prob>F
              -----------+----------------------------------------------------
              manovatest | W   0.3866      1      2.0     16.0     12.70 0.0005 e
                         | P   0.6134             2.0     16.0     12.70 0.0005 e
                         | L   1.5869             2.0     16.0     12.70 0.0005 e
                         | R   1.5869             2.0     16.0     12.70 0.0005 e
                         |----------------------------------------------------
                Residual |                17
              -----------+----------------------------------------------------
                           e = exact, a = approximate, u = upper bound on F
```

```
. matrix c2w = (0,-7,-5,12,0)
. manovatest, test(c2w)
 Test constraint
 (1)    -7*2.group - 5*3.group + 12*4.group = 0
                           W = Wilks' lambda      L = Lawley-Hotelling trace
                           P = Pillai's trace     R = Roy's largest root

                  Source │  Statistic      df   F(df1,     df2) =   F   Prob>F
             ────────────┼──────────────────────────────────────────────────────
             manovatest  │ W   0.9810       1      2.0     16.0     0.15 0.8580 e
                         │ P   0.0190              2.0     16.0     0.15 0.8580 e
                         │ L   0.0193              2.0     16.0     0.15 0.8580 e
                         │ R   0.0193              2.0     16.0     0.15 0.8580 e
                         ├──────────────────────────────────────────────────────
               Residual  │                17
             ────────────┴──────────────────────────────────────────────────────
                           e = exact, a = approximate, u = upper bound on F
```

These two weighted contrasts do not lead to different conclusions compared with their unweighted counterparts.

◁

❑ Technical note

`manovatest, test(`*matname*`)` displays the linear combination (labeled "Test constraint") indicated by *matname*. You should examine this listing to make sure that the matrix you specify in `test()` provides the test you want.

❑

The `margins` postestimation command provides, among other things, tables of predicted means and confidence intervals that are based on the most recently fit model.

▷ Example 3

Example 4 of [MV] **manova** presented a two-way MANOVA model on the jaw data.

```
. use http://www.stata-press.com/data/r12/jaw
(Table 4.6 Two-Way Unbalanced Data for Fractures of the Jaw -- Rencher (1998))
. manova y1 y2 y3 = gender fracture gender#fracture
  (output omitted)
```

The interaction term, `gender#fracture`, was significant. `margins` may be used to examine the interaction; see [R] **margins**.

```
. margins gender#fracture, predict(equation(y1))
Predictive margins                                Number of obs   =         27
Expression   : Linear prediction:  y1, predict(equation(y1))

------------------------------------------------------------------------------
             |            Delta-method
             |     Margin   Std. Err.      z    P>|z|     [95% Conf. Interval]
-------------+----------------------------------------------------------------
     gender#
    fracture |
        1 1  |       39.5   4.171386     9.47   0.000     31.32423    47.67577
        1 2  |     26.875   3.612526     7.44   0.000     19.79458    33.95542
        1 3  |   45.16667   4.171386    10.83   0.000      36.9909    53.34243
        2 1  |         22   10.21777     2.15   0.031     1.973543    42.02646
        2 2  |      30.75   5.108884     6.02   0.000     20.73677    40.76323
        2 3  |       36.5   7.225053     5.05   0.000     22.33916    50.66084
------------------------------------------------------------------------------

. margins gender#fracture, predict(equation(y2))
Predictive margins                                Number of obs   =         27
Expression   : Linear prediction:  y2, predict(equation(y2))

------------------------------------------------------------------------------
             |            Delta-method
             |     Margin   Std. Err.      z    P>|z|     [95% Conf. Interval]
-------------+----------------------------------------------------------------
     gender#
    fracture |
        1 1  |       35.5   2.150966    16.50   0.000     31.28418    39.71582
        1 2  |     32.375   1.862791    17.38   0.000       28.724      36.026
        1 3  |   36.16667   2.150966    16.81   0.000     31.95085    40.38248
        2 1  |         56   5.268768    10.63   0.000      45.6734     66.3266
        2 2  |      33.25   2.634384    12.62   0.000      28.0867     38.4133
        2 3  |         33   3.725582     8.86   0.000     25.69799    40.30201
------------------------------------------------------------------------------

. margins gender#fracture, predict(equation(y3))
Predictive margins                                Number of obs   =         27
Expression   : Linear prediction:  y3, predict(equation(y3))

------------------------------------------------------------------------------
             |            Delta-method
             |     Margin   Std. Err.      z    P>|z|     [95% Conf. Interval]
-------------+----------------------------------------------------------------
     gender#
    fracture |
        1 1  |   61.16667   2.038648    30.00   0.000     57.17099    65.16234
        1 2  |      62.25   1.765521    35.26   0.000     58.78964    65.71036
        1 3  |   58.16667   2.038648    28.53   0.000     54.17099    62.16234
        2 1  |         43   4.993647     8.61   0.000     33.21263    52.78737
        2 2  |         64   2.496823    25.63   0.000     59.10632    68.89368
        2 3  |       63.5   3.531041    17.98   0.000     56.57929    70.42071
------------------------------------------------------------------------------
```

The first `margins` table shows the predicted mean (marginal mean), standard error, z statistic, p-value, and confidence interval of `y1` (age of patient) for each combination of `fracture` and `gender`. The second and third `margins` tables provide this information for `y2` (blood lymphocytes) and `y3` (blood polymorphonuclears). These three tables of predictions are the same as those you would obtain from `margins` after running `anova` for each of the three dependent variables separately.

The predicted `y2` value is larger than the predicted `y3` value for females with one compound fracture. For the other five combinations of `gender` and `fracture`, the relationship is reversed. There is only 1 observation for the combination of female and one compound fracture.

There are nine possible contrasts if we contrast women with men for every fracture type and every dependent variable. We will use `contrast` to estimate all nine contrasts and apply Scheffé's adjustment for multiple comparisons.

```
. contrast gender@fracture#_eqns, mcompare(scheffe)
Contrasts of marginal linear predictions
Margins      : asbalanced
```

	df	F	P>F	Scheffe P>F
gender@fracture#_eqns				
1 1	1	2.51	0.1278	0.9733
1 2	1	12.98	0.0017	0.2333
1 3	1	11.34	0.0029	0.3137
2 1	1	0.38	0.5424	1.0000
2 2	1	0.07	0.7889	1.0000
2 3	1	0.33	0.5732	1.0000
3 1	1	1.08	0.3107	0.9987
3 2	1	0.54	0.4698	0.9999
3 3	1	1.71	0.2050	0.9929
Joint	9	2.57	0.0361	
Residual	21			

Note: Scheffe-adjusted p-values are reported for tests on individual contrasts only.

	Number of Comparisons
gender@fracture#_eqns	9

	Contrast	Std. Err.	Scheffe [95% Conf.	Interval]
gender@fracture#_eqns				
(2 vs base) 1 1	-17.5	11.03645	-68.42869	33.42869
(2 vs base) 1 2	20.5	5.69092	-5.76126	46.76126
(2 vs base) 1 3	-18.16667	5.393755	-43.05663	6.723297
(2 vs base) 2 1	3.875	6.257079	-24.99885	32.74885
(2 vs base) 2 2	.875	3.226449	-14.01373	15.76373
(2 vs base) 2 3	1.75	3.057972	-12.36128	15.86128
(2 vs base) 3 1	-8.666667	8.342772	-47.16513	29.8318
(2 vs base) 3 2	-3.166667	4.301931	-23.01831	16.68498
(2 vs base) 3 3	5.333333	4.077296	-13.48171	24.14838

Women do not differ significantly from men in any of the nine comparisons.

Let's examine the residuals with the `predict` command:

```
. predict y1res, residual equation(y1)
. predict y2res, residual equation(y2)
. predict y3res, residual equation(y3)
```

```
. list gender fracture y1res y2res y3res
```

	gender	fracture	y1res	y2res	y3res
1.	male	one compound fracture	2.5	-.5	-.1666667
2.	male	one compound fracture	2.5	7.5	-6.166667
3.	male	one compound fracture	8.5	-.5	2.833333
4.	male	one compound fracture	-4.5	-2.5	3.833333
5.	male	one compound fracture	-14.5	-4.5	2.833333
6.	male	one compound fracture	5.5	.5	-3.166667
7.	male	two compound fractures	-3.875	-5.375	1.75
8.	male	two compound fractures	-4.875	-.375	1.75
9.	male	two compound fractures	-1.875	-2.375	1.75
10.	male	two compound fractures	1.125	6.625	-6.25
11.	male	two compound fractures	-2.875	-1.375	6.75
12.	male	two compound fractures	25.125	-4.375	-2.25
13.	male	two compound fractures	-9.875	-2.375	1.75
14.	male	two compound fractures	-2.875	9.625	-5.25
15.	male	one simple fracture	-13.16667	.8333333	-4.166667
16.	male	one simple fracture	6.833333	-2.166667	3.833333
17.	male	one simple fracture	7.833333	8.833333	-7.166667
18.	male	one simple fracture	3.833333	-1.166667	1.833333
19.	male	one simple fracture	9.833333	-4.166667	1.833333
20.	male	one simple fracture	-15.16667	-2.166667	3.833333
21.	female	one compound fracture	7.11e-15	-1.42e-14	1.42e-14
22.	female	two compound fractures	-8.75	-4.25	4
23.	female	two compound fractures	7.25	-8.25	9
24.	female	two compound fractures	-9.75	3.75	-5
25.	female	two compound fractures	11.25	8.75	-8
26.	female	one simple fracture	6.5	-3	3.5
27.	female	one simple fracture	-6.5	3	-3.5

The single observation for a female with one compound fracture has residuals that are within roundoff of zero. With only 1 observation for that cell of the design, this MANOVA model is forced to fit to that point. The largest residual (in absolute value) appears for observation 12, which has an age 25.125 higher than the model prediction for a male with two compound fractures. ◁

▷ Example 4

Example 8 of [MV] **manova** presents a nested MANOVA on the `videotrainer` data.

```
. use http://www.stata-press.com/data/r12/videotrainer, clear
(video training)
. manova primary extra = video / store|video / associate|store|video /,
> dropemptycells
  (output omitted )
```

The MANOVA indicated that `store` was not significant.

You decide to follow the rule of thumb that says to pool terms whose p-values are larger than 0.25. Wilks' lambda reports a p-value of 0.5775 for the test of `store|video` (see example 8 of [MV] **manova**). You decide to pool the `store` and `associate` terms in the MANOVA to gain power for the test of `video`. The forward-slash notation of `manova` is also allowed with `manovatest` to indicate nonresidual error terms. Here is the multivariate test of `video` using the pooled `store` and `associate` terms and then the multivariate test of the pooled term:

```
. manovatest video / store|video associate|store|video
                           W = Wilks' lambda      L = Lawley-Hotelling trace
                           P = Pillai's trace     R = Roy's largest root

                  Source | Statistic     df   F(df1,     df2) =    F   Prob>F
    ---------------------+----------------------------------------------------
                   video | W   0.4079     1     2.0      9.0      6.53 0.0177 e
                         | P   0.5921           2.0      9.0      6.53 0.0177 e
                         | L   1.4516           2.0      9.0      6.53 0.0177 e
                         | R   1.4516           2.0      9.0      6.53 0.0177 e
                         |----------------------------------------------------
    store|video associate|                10
             store|video |
    ---------------------+----------------------------------------------------
                           e = exact, a = approximate, u = upper bound on F
. manovatest store|video associate|store|video
                           W = Wilks' lambda      L = Lawley-Hotelling trace
                           P = Pillai's trace     R = Roy's largest root

                  Source | Statistic     df   F(df1,     df2) =    F   Prob>F
    ---------------------+----------------------------------------------------
    store|video associate| W   0.3925    10    20.0     58.0      1.73 0.0546 e
             store|video | P   0.7160          20.0     60.0      1.67 0.0647 a
                         | L   1.2711          20.0     56.0      1.78 0.0469 a
                         | R   0.9924          10.0     30.0      2.98 0.0100 u
                         |----------------------------------------------------
                Residual |                30
    ---------------------+----------------------------------------------------
                           e = exact, a = approximate, u = upper bound on F
```

Pooling `store` with `associate` helps increase the power for the test of `video`.

You can show the univariate analysis for one of your dependent variables by using the `ytransform()` option of `manovatest`:

```
. mat primonly = (1,0)

. manovatest video / store|video associate|store|video, ytransform(primonly)

 Transformation of the dependent variables
 (1)    primary

                           W = Wilks' lambda      L = Lawley-Hotelling trace
                           P = Pillai's trace     R = Roy's largest root

                  Source | Statistic     df   F(df1,     df2) =    F   Prob>F
    ---------------------+----------------------------------------------------
                   video | W   0.8449     1     1.0     10.0      1.84 0.2053 e
                         | P   0.1551           1.0     10.0      1.84 0.2053 e
                         | L   0.1835           1.0     10.0      1.84 0.2053 e
                         | R   0.1835           1.0     10.0      1.84 0.2053 e
                         |----------------------------------------------------
    store|video associate|                10
             store|video |
    ---------------------+----------------------------------------------------
                           e = exact, a = approximate, u = upper bound on F
```

```
. manovatest store|video associate|store|video, ytransform(primonly)
 Transformation of the dependent variables
 (1)    primary
                           W = Wilks' lambda      L = Lawley-Hotelling trace
                           P = Pillai's trace     R = Roy's largest root

                  Source | Statistic     df   F(df1,    df2) =   F   Prob>F
  -----------------------+--------------------------------------------------
  store|video associate| | W   0.6119    10    10.0     30.0     1.90 0.0846 e
             store|video | P   0.3881          10.0     30.0     1.90 0.0846 e
                         | L   0.6344          10.0     30.0     1.90 0.0846 e
                         | R   0.6344          10.0     30.0     1.90 0.0846 e
                         |--------------------------------------------------
                Residual |               30
  -----------------------+--------------------------------------------------
                           e = exact, a = approximate, u = upper bound on F
```

See the second `manova` run from example 8 of [MV] **manova** for an alternate way of pooling the terms by refitting the MANOVA model.

◁

See examples 6, 10, 11, 12, and 13 of [MV] **manova** for more examples of `manovatest`, including examples involving both the `test()` and the `ytransform()` options.

Saved results

`manovatest` saves the following in `r()`:

Scalars

`r(df)`	hypothesis degrees of freedom
`r(df_r)`	residual degrees of freedom

Matrices

`r(H)`	hypothesis SSCP matrix
`r(E)`	residual-error SSCP matrix
`r(stat)`	multivariate statistics
`r(eigvals)`	eigenvalues of $E^{-1}H$
`r(aux)`	s, m, and n values

`test` after `manova` saves the following in `r()`:

Scalars

`r(p)`	two-sided p-value
`r(F)`	F statistic
`r(df)`	hypothesis degrees of freedom
`r(df_r)`	residual degrees of freedom
`r(drop)`	0 if no constraints dropped, 1 otherwise
`r(dropped_#)`	index of #th constraint dropped

Macros

`r(mtmethod)`	method of adjustment for multiple testing

Matrices

`r(mtest)`	multiple test results

Methods and formulas

All postestimation commands listed above are implemented as ado-files.

See [MV] **manova** for methods and formulas for the multivariate tests performed by `manovatest`.

Also see

[MV] **manova** — Multivariate analysis of variance and covariance

[MV] **screeplot** — Scree plot

Title

matrix dissimilarity — Compute similarity or dissimilarity measures

Syntax

`matrix dissimilarity` *matname* = [*varlist*] [*if*] [*in*] [, *options*]

options	Description
measure	similarity or dissimilarity measure; default is L2 (Euclidean)
`observations`	compute similarities or dissimilarities between observations; the default
`variables`	compute similarities or dissimilarities between variables
`names(`*varname*`)`	row/column names for *matname* (allowed with `observations`)
`allbinary`	check that all values are 0, 1, or missing
`proportions`	interpret values as proportions of binary values
`dissim(`*method*`)`	change similarity measure to dissimilarity measure

where *method* transforms similarities to dissimilarities by using

`oneminus`	$d_{ij} = 1 - s_{ij}$
`standard`	$d_{ij} = \sqrt{s_{ii} + s_{jj} - 2s_{ij}}$

Description

`matrix dissimilarity` computes a similarity, dissimilarity, or distance matrix.

Options

measure specifies one of the similarity or dissimilarity measures allowed by Stata. The default is L2, Euclidean distance. Many similarity and dissimilarity measures are provided for continuous data and for binary data; see [MV] ***measure_option***.

`observations` and `variables` specify whether similarities or dissimilarities are computed between observations or variables. The default is `observations`.

`names(`*varname*`)` provides row and column names for *matname*. *varname* must be a string variable with a length of 32 or less. You will want to pick a *varname* that yields unique values for the row and column names. Uniqueness of values is not checked by `matrix dissimilarity`. `names()` is not allowed with the `variables` option. The default row and column names when the similarities or dissimilarities are computed between observations is `obs#`, where # is the observation number corresponding to that row or column.

`allbinary` checks that all values are 0, 1, or missing. Stata treats nonzero values as one (excluding missing values) when dealing with what are supposed to be binary data (including binary similarity *measures*). `allbinary` causes `matrix dissimilarity` to exit with an error message if the values are not truly binary. `allbinary` is not allowed with `proportions` or the `Gower` *measure*.

`proportions` is for use with binary similarity *measures*. It specifies that values be interpreted as proportions of binary values. The default action treats all nonzero values as one (excluding missing values). With `proportions`, the values are confirmed to be between zero and one, inclusive. See [MV] ***measure_option*** for a discussion of the use of proportions with binary *measures*. `proportions` is not allowed with `allbinary` or the `Gower` *measure*.

`dissim(`*method*`)` specifies that similarity measures be transformed into dissimilarity measures. *method* may be `oneminus` or `standard`. `oneminus` transforms similarities to dissimilarities by using $d_{ij} = 1 - s_{ij}$ (Kaufman and Rousseeuw 1990, 21). `standard` uses $d_{ij} = \sqrt{s_{ii} + s_{jj} - 2s_{ij}}$ (Mardia, Kent, and Bibby 1979, 402). `dissim()` does nothing when the *measure* is already a dissimilarity or distance. See [MV] ***measure_option*** to see which *measures* are similarities.

Remarks

Commands such as `cluster singlelinkage`, `cluster completelinkage`, and `mds` (see [MV] **cluster** and [MV] **mds**) have options allowing the user to select the similarity or dissimilarity measure to use for its computation. If you are developing a command that requires a similarity or dissimilarity matrix, the `matrix dissimilarity` command provides a convenient way to obtain it.

The similarity or dissimilarity between each observation (or variable if the `variables` option is specified) and the others is placed in *matname*. The element in the *i*th row and *j*th column gives either the similarity or dissimilarity between the *i*th and *j*th observation (or variable). Whether you get a similarity or a dissimilarity depends upon the requested *measure*; see [MV] ***measure_option***.

If there are many observations (variables when the `variables` option is specified), you may need to increase the maximum matrix size; see [R] **matsize**. If the number of observations (or variables) is so large that storing the results in a matrix is not practical, you may wish to consider using the `cluster measures` command, which stores similarities or dissimilarities in variables; see [MV] **cluster programming utilities**.

When computing similarities or dissimilarities between observations, the default row and column names of *matname* are set to `obs#`, where # is the observation number. The `names()` option allows you to override this default. For similarities or dissimilarities between variables, the row and column names of *matname* are set to the appropriate variable names.

The order of the rows and columns corresponds with the order of your observations when you are computing similarities or dissimilarities between observations. Warning: If you reorder your data (for example, using `sort` or `gsort`) after running `matrix dissimilarity`, the row and column ordering will no longer match your data.

Another use of `matrix dissimilarity` is in performing a cluster analysis on variables instead of observations. The `cluster` command performs a cluster analysis of the observations; see [MV] **cluster**. If you instead wish to cluster variables, you can use the `variables` option of `matrix dissimilarity` to obtain a dissimilarity matrix that can then be used with `clustermat`; see [MV] **clustermat** and example 2 below.

▷ Example 1

Example 1 of [MV] **cluster linkage** introduces data with four chemical laboratory measurements on 50 different samples of a particular plant. Let's find the Canberra distance between the measurements performed by lab technician Bill found among the first 25 observations of the labtech dataset.

```
. use http://www.stata-press.com/data/r12/labtech
. matrix dissim D = x1 x2 x3 x4 if labtech=="Bill" in 1/25, canberra
. matrix list D
symmetric D[6,6]
             obs7       obs18       obs20       obs22       obs23       obs25
 obs7           0
obs18   1.3100445           0
obs20   1.1134916   .87626565           0
obs22    1.452748   1.0363077   1.0621064           0
obs23   1.0380665   1.4952796   .81602718   1.6888123           0
obs25   1.4668898   1.5139834   1.4492336   1.0668425   1.1252514           0
```

By default, the row and column names of the matrix indicate the observations involved. The Canberra distance between the 23rd observation and the 18th observation is 1.4952796. See [MV] ***measure_option*** for a description of the Canberra distance.

◁

▷ Example 2

Example 2 of [MV] **cluster linkage** presents a dataset with 30 observations of 60 binary variables, `a1`, `a2`, ..., `a30`. In [MV] **cluster linkage**, the observations were clustered. Here we instead cluster the variables by computing the dissimilarity matrix by using `matrix dissimilarity` with the `variables` option followed by the `clustermat` command.

We use the `matching` option to obtain the simple matching similarity coefficient but then specify `dissim(oneminus)` to transform the similarities to dissimilarities by using the transformation $d_{ij} = 1 - s_{ij}$. The `allbinary` option checks that the variables really are binary (0/1) data.

```
. use http://www.stata-press.com/data/r12/homework
. matrix dissim Avars = a*, variables matching dissim(oneminus) allbinary
. matrix subA = Avars[1..5,1..5]
. matrix list subA
symmetric subA[5,5]
            a1          a2          a3          a4          a5
a1           0
a2          .4           0
a3          .4   .46666667           0
a4          .3          .3   .36666667           0
a5          .4          .4   .13333333          .3           0
```

We listed the first five rows and columns of the 60×60 matrix. The matrix row and column names correspond to the variable names.

To perform an average-linkage cluster analysis on the 60 variables, we supply the `Avars` matrix created by `matrix dissimilarity` to the `clustermat averagelinkage` command; see [MV] **cluster linkage**.

```
. clustermat averagelinkage Avars, clear
obs was 0, now 60
cluster name: _clus_1
. cluster generate g5 = groups(5)
```

```
. table g5
```

g5	Freq.
1	21
2	9
3	25
4	4
5	1

We generated a variable, g5, indicating the five-group cluster solution and then tabulated to show how many variables were clustered into each of the five groups. Group five has only one member.

```
. list g5 if g5==5
```

	g5
13.	5

The member corresponds to the 13th observation in the current dataset, which in turn corresponds to variable a13 from the original dataset. It appears that a13 is not like the other variables.

◁

▷ Example 3

matrix dissimilarity drops observations containing missing values, except when the Gower measure is specified. The computation of the Gower dissimilarity between 2 observations is based on the variables where the 2 observations both have nonmissing values.

We illustrate using a dataset with 6 observations and 4 variables where only 2 of the observations have complete data.

```
. use http://www.stata-press.com/data/r12/gower, clear
. list
```

	b1	b2	x1	x2
1.	0	1	.76	.75
2.	.	.	.	.
3.	1	0	.72	.88
4.	.	1	.4	.
5.	0	.	.	.14
6.	0	0	.55	.

```
. mat dissimilarity matL2 = b* x*, L2
. matlist matL2, format(%8.3f)
```

	obs1	obs3
obs1	0.000	
obs3	1.421	0.000

The resulting matrix is 2×2 and provides the dissimilarity between observations 1 and 3. All other observations contained at least one missing value.

However, with the gower measure we obtain a 6×6 matrix.

```
. matrix dissimilarity matgow = b1 b2 x1 x2, gower
. matlist matgow, format(%8.3f)
```

	obs1	obs2	obs3	obs4	obs5	obs6
obs1	0.000					
obs2	.	0.000				
obs3	0.572	.	0.000			
obs4	0.500	.	0.944	0.000		
obs5	0.412	.	1.000	.	0.000	
obs6	0.528	.	0.491	0.708	0.000	0.000

Because all the values for observation 2 are missing, the matrix contains missing values for the dissimilarity between observation 2 and the other observations. Notice the missing value in matgow for the dissimilarity between observations 4 and 5. There were no variables where observations 4 and 5 both had nonmissing values, and hence the Gower coefficient could not be computed.

◁

References

Kaufman, L., and P. J. Rousseeuw. 1990. *Finding Groups in Data: An Introduction to Cluster Analysis*. New York: Wiley.

Mardia, K. V., J. T. Kent, and J. M. Bibby. 1979. *Multivariate Analysis*. London: Academic Press.

Also see

[MV] **cluster** — Introduction to cluster-analysis commands

[MV] **clustermat** — Introduction to clustermat commands

[MV] **mdsmat** — Multidimensional scaling of proximity data in a matrix

[MV] **cluster programming utilities** — Cluster-analysis programming utilities

[MV] ***measure_option*** — Option for similarity and dissimilarity measures

[P] **matrix** — Introduction to matrix commands

Title

mca — Multiple and joint correspondence analysis

Syntax

Basic syntax for two or more categorical variables

mca *varlist* [*if*] [*in*] [*weight*] [, *options*]

Full syntax for use with two or more categorical or crossed (stacked) categorical variables

mca *speclist* [*if*] [*in*] [*weight*] [, *options*]

where

speclist = *spec* [*spec* ...]

spec = *varlist* | (*newvar* : *varlist*)

options	Description
Model	
`supplementary(`*speclist*`)`	supplementary (passive) variables
`method(burt)`	use the Burt matrix approach to MCA; the default
`method(indicator)`	use the indicator matrix approach to MCA
`method(joint)`	perform a joint correspondence analysis (JCA)
`dimensions(#)`	number of dimensions (factors, axes); default is `dim(2)`
`normalize(standard)`	display standard coordinates; the default
`normalize(principal)`	display principal coordinates
`iterate(#)`	maximum number of `method(joint)` iterations; default is `iterate(250)`
`tolerance(#)`	tolerance for `method(joint)` convergence criterion; default is `tolerance(1e-5)`
`missing`	treat missing values as ordinary values
`noadjust`	suppress the adjustment of eigenvalues (`method(burt)` only)
Codes	
`report(variables)`	report coding of crossing variables
`report(crossed)`	report coding of crossed variables
`report(all)`	report coding of crossing and crossed variables
`length(min)`	use minimal length unique codes of crossing variables
`length(#)`	use # as coding length of crossing variables

Reporting

ddimensions(#)	display # singular values; default is ddim(.) (all)
points(*varlist*)	display tables for listed variables; default is all variables
compact	display statistics table in a compact format
log	display the iteration log (method(joint) only)
plot	plot the coordinates (that is, mcaplot)
maxlength(#)	maximum number of characters for labels in plot; default is maxlength(12)

bootstrap, by, jackknife, rolling, and statsby may be used with mca; see [U] **11.1.10 Prefix commands**. However, bootstrap and jackknife results should be interpreted with caution; identification of the mca parameters involves data-dependent restrictions, possibly leading to badly biased and overdispersed estimates (Milan and Whittaker 1995).

Weights are not allowed with the bootstrap prefix; see [R] **bootstrap**.

fweights are allowed; see [U] **11.1.6 weight**.

See [U] **20 Estimation and postestimation commands** for more capabilities of estimation commands.

Menu

Statistics > Multivariate analysis > Correspondence analysis > Multiple correspondence analysis (MCA)

Description

mca performs multiple correspondence analysis (MCA) or joint correspondence analysis (JCA) on a series of categorical variables. MCA and JCA are two generalizations of correspondence analysis (CA) of a cross-tabulation of two variables (see [MV] **ca**) to the cross-tabulation of multiple variables.

mca performs an analysis of two or more integer-valued variables. Crossing (also called stacking) of integer-valued variables is also allowed.

Options

Model

supplementary(*speclist*) specifies that *speclist* are supplementary variables. Such variables do not affect the MCA solution, but their categories are mapped into the solution space. For method(indicator), this mapping uses the first method described by Greenacre (2006). For method(burt) and method(joint), the second and recommended method described by Greenacre (2006) is used, in which supplementary column principal coordinates are derived as a weighted average of the standard row coordinates, weighted by the supplementary profile. See the syntax diagram for the syntax of *speclist*.

method(*method*) specifies the method of MCA/JCA.

method(burt), the default, specifies MCA, a categorical variables analogue to principal component analysis (see [MV] **pca**). The Burt method performs a CA of the Burt matrix, a matrix of the two-way cross-tabulations of all pairs of variables.

method(indicator) specifies MCA via a CA on the indicator matrix formed from the variables.

method(joint) specifies JCA, a categorical variables analogue to factor analysis (see [MV] **factor**). This method analyzes a variant of the Burt matrix, in which the diagonal blocks are iteratively adjusted for the poor diagonal fit of MCA.

`dimensions(#)` specifies the number of dimensions (= factors = axes) to be extracted. The default is `dimensions(2)`. If you specify `dimensions(1)`, the categories are placed on one dimension. The number of dimensions is no larger than the number of categories in the active variables (regular and crossed) minus the number of active variables, and it can be less. This excludes supplementary variables. Specifying a larger number than dimensions available results in extracting all dimensions.

MCA is a hierarchical method so that extracting more dimensions does not affect the coordinates and decomposition of inertia of dimensions already included. The percentages of inertia accounting for the dimensions are in decreasing order as indicated by the singular values. The first dimension accounts for the most inertia, followed by the second dimension, and then the third dimension, etc.

`normalize(`*normalization*`)` specifies the normalization method, that is, how the row and column coordinates are obtained from the singular vectors and singular values of the matrix of standardized residuals.

`normalize(standard)` specifies that coordinates are returned in standard normalization (singular values divided by the square root of the mass). This is the default.

`normalize(principal)` specifies that coordinates are returned in principal normalization. Principal coordinates are standard coordinates multiplied by the square root of the corresponding principal inertia.

`iterate(#)` is a technical and rarely used option specifying the maximum number of iterations. `iterate()` is permitted only with `method(joint)`. The default is `iterate(250)`.

`tolerance(#)` is a technical and rarely used option specifying the tolerance for subsequent modification of the diagonal blocks of the Burt matrix. `tolerance()` is permitted only with `method(joint)`. The default is `tolerance(1e-5)`.

`missing` treats missing values as ordinary values to be included in the analysis. Observations with missing values are omitted from the analysis by default.

`noadjust` suppresses principal inertia adjustment and is allowed with `method(burt)` only. By default, the principal inertias (eigenvalues of the Burt matrix) are adjusted. The unmodified principal inertias present a pessimistic measure of fit because MCA fits the diagonal of $\mathbf{P}$ poorly (see Greenacre [1984]).

Codes

`report(`*opt*`)` displays coding information for the crossing variables, crossed variables, or both. `report()` is ignored if you do not specify at least one crossed variable.

`report(variables)` displays the coding schemes of the crossing variables, that is, the variables used to define the crossed variables.

`report(crossed)` displays a table explaining the value labels of the crossed variables.

`report(all)` displays the codings of the crossing and crossed variables.

`length(`*opt*`)` specifies the coding length of crossing variables.

`length(min)` specifies that the minimal-length unique codes of crossing variables be used.

`length(#)` specifies that the coding length # of crossing variables be used, where # must be between 4 and 32.

Reporting

`ddimensions(#)` specifies the number of singular values to be displayed. If `ddimensions()` is greater than the number of singular values, all the singular values will be displayed. The default is `ddimensions(.)`, meaning all singular values.

`points(`*varlist*`)` indicates the variables to be included in the tables. By default, tables are displayed for all variables. Regular categorical variables, crossed variables, and supplementary variables may be specified in `points()`.

`compact` specifies that point statistics tables be displayed multiplied by 1,000, enabling the display of more columns without wrapping output. The compact tables can be displayed without wrapping for models with two dimensions at line size 79 and with three dimensions at line size 99.

`log` displays an iteration log. This option is permitted with `method(joint)` only.

`plot` displays a plot of the row and column coordinates in two dimensions. Use `mcaplot` directly to select different plotting points or for other graph refinements; see [MV] **mca postestimation**.

`maxlength(#)` specifies the maximum number of characters for labels in plots. The default is `maxlength(12)`. # must be less than 32.

Note: the reporting options may be specified during estimation or replay.

Remarks

Remarks are presented under the following headings:

Introduction
Compare MCA on two variables and CA
MCA on four variables
CA of the indicator matrix
CA of the Burt matrix
Joint correspondence analysis

Introduction

Multiple correspondence analysis (MCA) and joint correspondence analysis (JCA) are methods for analyzing observations on categorical variables. MCA is usually viewed as an extension of simple correspondence analysis (CA) to more than two variables. CA analyzes a two-way contingency table; MCA and JCA analyze a multiway table.

MCA can be viewed as a generalization of principal component analysis where the variables to be analyzed are categorical, not continuous. MCA performs a CA on a Burt or indicator matrix; it explores the relationships within a set of variables, whereas CA has more focus on exploring relationships between two sets of variables. JCA is an extension of MCA that attempts to remedy discrepancies between CA and MCA.

For an introduction to MCA via CA, see Rencher (2002) or Everitt and Dunn (2001). For an advanced introduction to MCA without previous discussion of CA, see Gower and Hand (1996). Greenacre (2006) approaches MCA from CA, then JCA from MCA, and gives a more advanced treatment. [MV] **ca** also introduces MCA concepts. Gower (1990) explores MCA history.

Three methods are implemented here. We will start with the simplest and most intuitive representation of unordered categorical data: the indicator matrix, usually denoted $\mathbf{Z}$, a matrix of zeros and ones with columns for all categories of all variables and rows corresponding to observations. A value of one indicates that a category is observed; a zero indicates that it is not. MCA can be performed as a CA on the indicator matrix; an equivalent to this method is `method(indicator)` with `mca`.

Here is a manufactured indicator matrix $\mathbf{Z}$, with 10 observations on three categorical variables, w, x, and y, each with two categories indicated by $w_1, w_2, x_1, x_2, y_1, y_2$:

$$\mathbf{Z} = \begin{array}{r} \text{obs} \\ 1. \\ 2. \\ 3. \\ 4. \\ 5. \\ 6. \\ 7. \\ 8. \\ 9. \\ 10. \end{array} \begin{array}{c} \begin{array}{cccccc} w_1 & w_2 & x_1 & x_2 & y_1 & y_2 \end{array} \\ \begin{pmatrix} 1 & 0 & 1 & 0 & 1 & 0 \\ 0 & 1 & 1 & 0 & 1 & 0 \\ 0 & 1 & 0 & 1 & 1 & 0 \\ 1 & 0 & 0 & 1 & 0 & 1 \\ 1 & 0 & 0 & 1 & 0 & 1 \\ 1 & 0 & 1 & 0 & 0 & 1 \\ 0 & 1 & 0 & 1 & 1 & 0 \\ 1 & 0 & 0 & 1 & 1 & 0 \\ 1 & 0 & 1 & 0 & 0 & 1 \\ 1 & 0 & 1 & 0 & 1 & 0 \end{pmatrix} \end{array}$$

For large datasets with many variables and observations, the indicator matrix is burdensome and can exceed memory limitations. The Burt matrix, usually denoted as $\mathbf{B}$, is a cross-tabulation of all categories of all variables. $\mathbf{B} = \mathbf{Z}'\mathbf{Z}$. The Burt matrix is smaller than the indicator matrix $\mathbf{Z}$. Performing CA on the Burt matrix and performing CA on the indicator matrix are related but not the same: they give equivalent standard coordinates, but the total principal inertias from the indicator matrix approach and Burt matrix approach are different. Performing a CA of the Burt matrix without further calculation is `method(burt)` along with the `noadjust` option.

The Burt matrix, $\mathbf{B}$, corresponding to matrix $\mathbf{Z}$, above:

$$\mathbf{B} = \begin{array}{r} \\ w_1 \\ w_2 \\ x_1 \\ x_2 \\ y_1 \\ y_2 \end{array} \begin{array}{c} \begin{array}{cccccc} w_1 & w_2 & x_1 & x_2 & y_1 & y_2 \end{array} \\ \begin{pmatrix} 7 & 0 & 4 & 3 & 3 & 4 \\ 0 & 3 & 1 & 2 & 3 & 0 \\ 4 & 1 & 5 & 0 & 3 & 2 \\ 3 & 2 & 0 & 5 & 3 & 2 \\ 3 & 3 & 3 & 3 & 6 & 0 \\ 4 & 0 & 2 & 2 & 0 & 4 \end{pmatrix} \end{array}$$

Applying CA to the Burt matrix or indicator matrix artificially inflates the chi-squared distances between profiles and the total inertia; see Gower (2006). This can be partially remedied after CA of the Burt matrix by scale readjustments of the MCA solution. Performing adjustments after a CA of the Burt matrix is the default, that is, if no method is specified `method(burt)` is assumed; unadjusted estimates may be obtained by using the option `noadjust`.

The third method, JCA, generalizes CA more naturally to more than two categorical variables. JCA attempts to account for the variation in the off-diagonal submatrices of the Burt matrix. It corrects the inflation of the total inertia. The solution can no longer be obtained by one application of the singular value decomposition (as in CA). JCA is obtained by specifying the option `method(joint)` to `mca`.

Compare MCA on two variables and CA

We illustrate MCA with a dataset from the International Social Survey Program on environment (ISSP 1993). This example is used in the MCA literature; see Greenacre (2006). We will look at the questions on attitudes toward science. We use data from the West German sample only and remove all observations containing missing data; 871 observations remain. The survey questions are

How much do you agree or disagree with each of these statements?

A. We believe too often in science, and not enough in feelings and faith.

B. Overall, modern science does more harm than good.

C. Any change humans cause in nature—no matter how scientific—is likely to make things worse.

D. Modern science will solve our environmental problems with little change to our way of life.

Each question has five possible response categories:

1. Agree strongly
2. Agree
3. Neither agree nor disagree
4. Disagree
5. Disagree strongly

Supplementary demographic information is also available as coded categorical variables:

Sex: male, female

Age: (six groups) 16–24, 25–34, 35–44, 45–54, 55–64, 65 and older

Education: (six groups) primary incomplete, primary completed, secondary incomplete, secondary completed, tertiary incomplete, tertiary completed.

▷ Example 1

Here is a summary of these data.

```
. use http://www.stata-press.com/data/r12/issp93
(Selection from ISSP (1993))
. describe
Contains data from http://www.stata-press.com/data/r12/issp93.dta
  obs:           871                          Selection from ISSP (1993)
 vars:             8                          17 May 2011 09:36
 size:         7,839                          (_dta has notes)
------------------------------------------------------------------------------
              storage  display     value
variable name   type   format      label      variable label
------------------------------------------------------------------------------
id              int    %9.0g                  respondent identifier
A               byte   %26.0g      agree5     too much science, not enough
                                                feelings&faith
B               byte   %26.0g      agree5     science does more harm than good
C               byte   %26.0g      agree5     any change makes nature worse
D               byte   %26.0g      agree5     science will solve
                                                environmental problems
sex             byte   %9.0g       sex        sex
age             byte   %9.0g       age        age (6 categories)
edu             byte   %20.0g      edu        education (6 categories)
------------------------------------------------------------------------------
Sorted by:
```

We begin by comparing MCA with two variables to CA with two variables. The default MCA analysis is a CA of the Burt matrix, performed with adjustment to the principal inertias. The unadjusted results from MCA will give different principal inertias from the CA, although the standard coordinates would be the same. With adjustment, results are nearly identical. For a detailed discussion of the output of ca and mca, see [MV] **ca**.

```
. mca A B

Multiple/Joint correspondence analysis          Number of obs    =        871
                                                Total inertia    =   .2377535
    Method: Burt/adjusted inertias              Number of axes   =          2

                 |   principal                 cumul
       Dimension |    inertia    percent    percent
    -------------+----------------------------------
           dim 1 |   .1686131     70.92      70.92
           dim 2 |   .0586785     24.68      95.60
           dim 3 |    .010444      4.39      99.99
           dim 4 |   .0000178      0.01     100.00
    -------------+----------------------------------
           Total |   .2377535    100.00
Statistics for column categories in standard normalization

                 |            overall            |          dimension_1
      Categories |     mass  quality    %inert   |    coord   sqcorr  contrib
    -------------+-------------------------------+---------------------------
    A            |                               |
    agree stro~y |    0.068    1.000     0.100   |    1.017    0.502    0.071
           agree |    0.185    0.997     0.042   |    0.560    0.982    0.058
    neither ag~e |    0.117    0.930     0.062   |    0.248    0.083    0.007
        disagree |    0.102    0.922     0.123   |   -1.239    0.907    0.157
    disagree s~y |    0.028    0.954     0.174   |   -2.741    0.845    0.207
    -------------+-------------------------------+---------------------------
    B            |                               |
    agree stro~y |    0.041    0.982     0.146   |    1.571    0.490    0.101
           agree |    0.100    0.962     0.034   |    0.667    0.932    0.044
    neither ag~e |    0.118    0.840     0.043   |    0.606    0.716    0.043
        disagree |    0.161    0.769     0.043   |   -0.293    0.228    0.014
    disagree s~y |    0.080    0.994     0.235   |   -1.926    0.900    0.298
    ---------------------------------------------------------------------------

                 |          dimension_2
      Categories |    coord   sqcorr  contrib
    -------------+---------------------------
    A            |
    agree stro~y |    1.718    0.498    0.202
           agree |    0.116    0.015    0.002
    neither ag~e |   -1.344    0.847    0.212
        disagree |   -0.268    0.015    0.007
    disagree s~y |    1.672    0.109    0.077
    -------------+---------------------------
    B            |
    agree stro~y |    2.671    0.493    0.291
           agree |   -0.201    0.029    0.004
    neither ag~e |   -0.427    0.124    0.022
        disagree |   -0.764    0.541    0.094
    disagree s~y |    1.055    0.094    0.089
    -----------------------------------------
```

With ca we use the norm(standard) option to obtain results in the same normalization as the preceding mca. [MV] **ca** discusses the normalize() option; mca has some of the normalize() options of ca.

The top table in the output for both ca and mca reports the principal inertias. The principal inertias reported by ca are simply the squares of the singular values. Two pieces of information reported by ca that are not reported by mca are the chi-squared value in the table and the explained inertia in the header. The chi-squared value is taken as a measure of dissimilarity between the row and column profiles of the two-way table. Chi-squared distance makes sense for a two-way table but has less justification when applied to the rows and columns of either the indicator matrix or the Burt matrix (Greenacre 2006). The explained inertia is the value from the cumulative percent column in the top table corresponding to the number of dimensions selected; it is simply not reported in the mca header.

The coordinates reported in the second table are the same. The mass, inertia, and contributions are half as large in mca as they are in ca; in ca the row and column each sum to the principal inertia, but in mca there are only columns, which sum to the principal inertia.

```
. ca A B, norm(standard)
Correspondence analysis                        Number of obs     =       871
                                               Pearson chi2(16)  =    207.08
                                               Prob > chi2       =    0.0000
                                               Total inertia     =    0.2378
    5 active rows                              Number of dim.    =         2
    5 active columns                           Expl. inertia (%) =     95.60

                 |   singular    principal                              cumul
       Dimension |    values      inertia        chi2     percent     percent
    -------------+---------------------------------------------------------
           dim 1 |   .4106252    .1686131       146.86      70.92       70.92
           dim 2 |   .2422364    .0586785        51.11      24.68       95.60
           dim 3 |   .1021961     .010444         9.10       4.39       99.99
           dim 4 |   .0042238    .0000178         0.02       0.01      100.00
    -------------+---------------------------------------------------------
           total |               .2377535       207.08        100
Statistics for row and column categories in standard normalization

                 |         overall          |       dimension_1
      Categories |  mass  quality   %inert  |  coord   sqcorr  contrib
    -------------+--------------------------+---------------------------
    A            |                          |
    agree stro~y |  0.137    1.000    0.200 |  1.017    0.502    0.141
           agree |  0.370    0.997    0.084 |  0.560    0.982    0.116
    neither ag~e |  0.234    0.930    0.123 |  0.248    0.083    0.014
        disagree |  0.204    0.922    0.245 | -1.239    0.907    0.314
    disagree s~y |  0.055    0.954    0.348 | -2.741    0.845    0.414
    -------------+--------------------------+---------------------------
    B            |                          |
    agree stro~y |  0.082    0.982    0.291 |  1.571    0.490    0.201
           agree |  0.200    0.962    0.068 |  0.667    0.932    0.089
    neither ag~e |  0.235    0.840    0.086 |  0.606    0.716    0.087
        disagree |  0.323    0.769    0.086 | -0.293    0.228    0.028
    disagree s~y |  0.161    0.994    0.470 | -1.926    0.900    0.596
    ---------------------------------------------------------------------

                 |       dimension_2
      Categories |  coord   sqcorr  contrib
    -------------+---------------------------
    A            |
    agree stro~y |  1.718    0.498    0.403
           agree |  0.116    0.015    0.005
    neither ag~e | -1.344    0.847    0.423
        disagree | -0.268    0.015    0.015
    disagree s~y |  1.672    0.109    0.154
    -------------+---------------------------
    B            |
    agree stro~y |  2.671    0.493    0.582
           agree | -0.201    0.029    0.008
    neither ag~e | -0.427    0.124    0.043
        disagree | -0.764    0.541    0.188
    disagree s~y |  1.055    0.094    0.179
    -----------------------------------------
```

◁

MCA on four variables

Now we will take a look at MCA with more than two variables and at the different methods of performing MCA or JCA.

▷ Example 2

We continue to use the ISSP (1993) dataset, looking at all four questions on attitudes toward science. We use the default method of MCA, which is a CA of the Burt matrix for the data, followed by simple scale adjustments. We choose the principal normalization `normalize(principal)`, which scales the coordinates by the principal inertias. MCA with the Burt matrix and adjustments explains at least 79.1% of the total inertia in the first two dimensions.

```
. mca A-D, normalize(principal)
Multiple/Joint correspondence analysis          Number of obs    =        871
                                                Total inertia    =   .1702455
    Method: Burt/adjusted inertias              Number of axes   =          2
```

Dimension	principal inertia	percent	cumul percent
dim 1	.0764553	44.91	44.91
dim 2	.0582198	34.20	79.11
dim 3	.009197	5.40	84.51
dim 4	.0056697	3.33	87.84
dim 5	.0011719	0.69	88.53
dim 6	6.61e-06	0.00	88.53
Total	.1702455	100.00	

Statistics for column categories in principal normalization

Categories	overall mass	overall quality	overall %inert	dimension_1 coord	dimension_1 sqcorr	dimension_1 contrib
A						
agree stro~y	0.034	0.963	0.060	0.508	0.860	0.115
agree	0.092	0.659	0.023	0.151	0.546	0.028
neither ag~e	0.059	0.929	0.037	-0.124	0.143	0.012
disagree	0.051	0.798	0.051	-0.322	0.612	0.069
disagree s~y	0.014	0.799	0.067	-0.552	0.369	0.055
B						
agree stro~y	0.020	0.911	0.100	0.809	0.781	0.174
agree	0.050	0.631	0.027	0.177	0.346	0.021
neither ag~e	0.059	0.806	0.027	0.096	0.117	0.007
disagree	0.081	0.620	0.033	-0.197	0.555	0.041
disagree s~y	0.040	0.810	0.116	-0.374	0.285	0.074
C						
agree stro~y	0.044	0.847	0.122	0.597	0.746	0.203
agree	0.091	0.545	0.024	0.068	0.101	0.006
neither ag~e	0.057	0.691	0.045	-0.171	0.218	0.022
disagree	0.044	0.788	0.054	-0.373	0.674	0.080
disagree s~y	0.015	0.852	0.071	-0.406	0.202	0.032
D						
agree stro~y	0.017	0.782	0.039	0.333	0.285	0.025
agree	0.067	0.126	0.012	-0.061	0.126	0.003
neither ag~e	0.058	0.688	0.044	-0.106	0.087	0.009
disagree	0.065	0.174	0.014	-0.061	0.103	0.003
disagree s~y	0.043	0.869	0.034	0.196	0.288	0.022

Categories	dimension_2 coord	sqcorr	contrib
A			
agree stro~y	0.176	0.103	0.018
agree	-0.069	0.113	0.007
neither ag~e	-0.289	0.786	0.084
disagree	0.178	0.186	0.028
disagree s~y	0.596	0.430	0.084
B			
agree stro~y	0.331	0.131	0.038
agree	-0.161	0.285	0.022
neither ag~e	-0.233	0.690	0.055
disagree	-0.068	0.065	0.006
disagree s~y	0.509	0.526	0.179
C			
agree stro~y	0.219	0.101	0.036
agree	-0.143	0.444	0.032
neither ag~e	-0.252	0.473	0.062
disagree	0.153	0.114	0.018
disagree s~y	0.728	0.650	0.136
D			
agree stro~y	0.440	0.497	0.057
agree	-0.002	0.000	0.000
neither ag~e	-0.280	0.601	0.078
disagree	-0.051	0.071	0.003
disagree s~y	0.278	0.581	0.057

```
. mcaplot, overlay origin
```

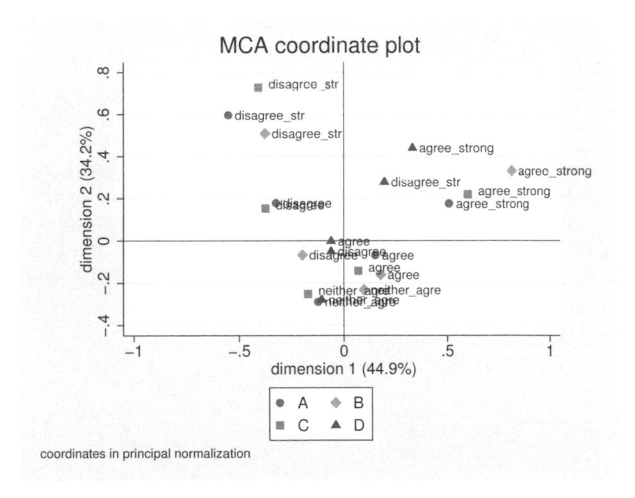

We displayed the origin axes in the plot with option `origin`. This plot makes it easier to see data associations. For more information on `mcaplot`, see [MV] **mca postestimation**.

A clear pattern is seen in the plot. Results from questions A, B, and C are clustered together, whereas question D forms a pattern of its own. Question D is formulated differently from A, B, and C, and the plot shows its incompatibility with the others.

Greenacre (2006, 70) produces this same plot. To obtain equivalent results, we reflect over the y axis with `xnegate` and adjust the scale of the graph with `xlabel()` and `ylabel()`.

```
. mcaplot, overlay xnegate origin ylabel(-1(.5)1.5) xlabel(-1.5(.5)1)
```

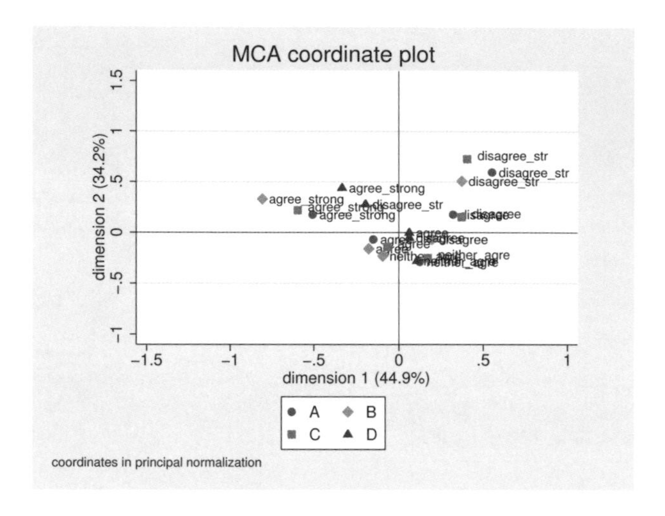

◁

❑ Technical note

The percentage of inertia in the top table of the `mca` output does not add up to 100%, although all singular values are displayed. Why? The percentages are lower-bound estimates in the Burt method with adjustments.

❑

❑ Technical note

Readers who are familiar with the discussion of the `normalize()` option in [MV] **ca** might be wondering why we are doing analysis of coordinates in the principal normalization. Principal normalization in `ca` does not allow good associations between the row and column categories, and so the symmetric normalization is the default used for `ca`. Principal normalization does allow studying the row categories or column categories separately from each other in `ca`. In `mca` there are only column categories. Consequently, the principal normalization is often preferred.

❑

CA of the indicator matrix

▷ Example 3

We compare the previous result with that obtained using the `method(indicator)` option to perform an equivalent analysis to CA on the indicator matrix for these data. The first two dimensions explain only 22.2% of the total principal inertia.

```
. mca A-D, method(indicator)

Multiple/Joint correspondence analysis          Number of obs     =        871
                                                Total inertia     =          4
    Method: Indicator matrix                    Number of axes    =          2

                 principal               cumul
    Dimension     inertia    percent    percent

        dim 1    .4573792      11.43      11.43
        dim 2    .4309658      10.77      22.21
        dim 3    .3219257       8.05      30.26
        dim 4    .3064732       7.66      37.92
        dim 5    .2756747       6.89      44.81
        dim 6     .251928       6.30      51.11
        dim 7    .2425591       6.06      57.17
        dim 8    .2349506       5.87      63.05
        dim 9     .225468       5.64      68.68
       dim 10    .2206291       5.52      74.20
       dim 11    .2098376       5.25      79.44
       dim 12    .1971485       4.93      84.37
       dim 13    .1778833       4.45      88.82
       dim 14    .1691119       4.23      93.05
       dim 15    .1528191       3.82      96.87
       dim 16    .1252462       3.13     100.00

        Total           4     100.00

Statistics for column categories in standard normalization

                           overall                 dimension_1
   Categories      mass  quality   %inert     coord   sqcorr  contrib

A
agree stro~y      0.034    0.280    0.054     1.837    0.244    0.078
       agree      0.092    0.100    0.039     0.546    0.080    0.019
neither ag~e      0.059    0.218    0.048    -0.447    0.028    0.008
    disagree      0.051    0.220    0.050    -1.166    0.160    0.047
disagree s~y      0.014    0.260    0.059    -1.995    0.106    0.037

B
agree stro~y      0.020    0.419    0.057     2.924    0.347    0.118
       agree      0.050    0.095    0.050     0.642    0.047    0.014
neither ag~e      0.059    0.140    0.048     0.346    0.017    0.005
    disagree      0.081    0.127    0.042    -0.714    0.111    0.028
disagree s~y      0.040    0.527    0.052    -1.354    0.161    0.050

C
agree stro~y      0.044    0.525    0.052     2.158    0.450    0.137
       agree      0.091    0.102    0.040     0.247    0.016    0.004
neither ag~e      0.057    0.189    0.048    -0.619    0.051    0.015
    disagree      0.044    0.216    0.051    -1.349    0.179    0.054
disagree s~y      0.015    0.312    0.059    -1.468    0.063    0.022

D
agree stro~y      0.017    0.155    0.058     1.204    0.049    0.017
       agree      0.067    0.008    0.046    -0.221    0.008    0.002
neither ag~e      0.058    0.195    0.048    -0.385    0.020    0.006
    disagree      0.065    0.015    0.046    -0.222    0.008    0.002
disagree s~y      0.043    0.168    0.052     0.708    0.048    0.015
```

Categories	dimension_2		
	coord	sqcorr	contrib
A			
agree stro~y	0.727	0.036	0.012
agree	-0.284	0.020	0.005
neither ag~e	-1.199	0.190	0.055
disagree	0.737	0.060	0.018
disagree s~y	2.470	0.153	0.055
B			
agree stro~y	1.370	0.072	0.025
agree	-0.667	0.048	0.015
neither ag~e	-0.964	0.123	0.036
disagree	-0.280	0.016	0.004
disagree s~y	2.108	0.367	0.117
C			
agree stro~y	0.909	0.075	0.024
agree	-0.592	0.086	0.021
neither ag~e	-1.044	0.137	0.040
disagree	0.635	0.037	0.012
disagree s~y	3.017	0.249	0.089
D			
agree stro~y	1.822	0.106	0.038
agree	-0.007	0.000	0.000
neither ag~e	-1.159	0.175	0.051
disagree	-0.211	0.007	0.002
disagree s~y	1.152	0.120	0.038

```
. mcaplot, overlay origin
```

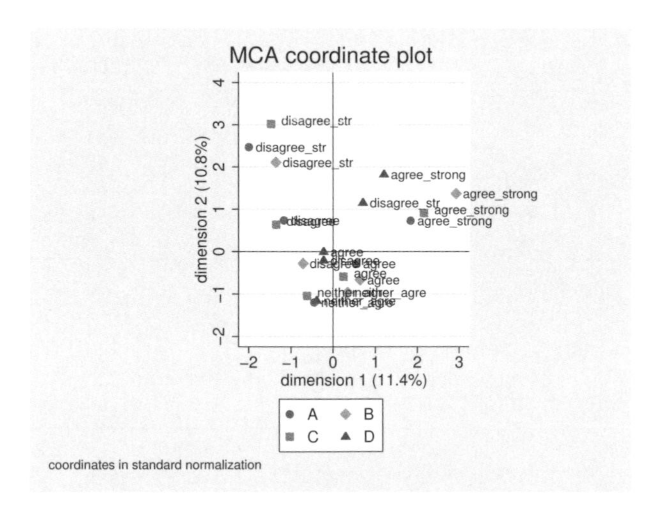

Even though the first two dimensions explain only 22.2% of the total inertia, and difficulties exist in justifying the full chi-squared geometry for the CA of the indicator matrix, the same clear pattern we saw in the previous plot is seen here. Variables A, B, and C are related, but variable D does not belong with the rest.

◁

CA of the Burt matrix

▷ Example 4

Now we will look at results with `method(burt)` and the `noadjust` option. This performs a CA on the Burt matrix without adjusting the principal inertias. This model does slightly better than the indicator matrix approach, explaining 35.0% of the principal inertia in the first two dimensions. We display column points only for variable A to reduce the output.

```
. mca A-D, method(burt) noadjust points(A)
Multiple/Joint correspondence analysis          Number of obs     =        871
                                                Total inertia     =   1.127684
    Method: Burt/unadjusted inertias            Number of axes    =          2

                 principal              cumul
    Dimension     inertia    percent   percent

        dim 1    .2091957     18.55     18.55
        dim 2    .1857315     16.47     35.02
        dim 3    .1036362      9.19     44.21
        dim 4    .0939258      8.33     52.54
        dim 5    .0759966      6.74     59.28
        dim 6    .0634677      5.63     64.91
        dim 7    .0588349      5.22     70.12
        dim 8    .0552018      4.90     75.02
        dim 9    .0508358      4.51     79.53
       dim 10    .0486772      4.32     83.84
       dim 11    .0440318      3.90     87.75
       dim 12    .0388675      3.45     91.20
       dim 13    .0316425      2.81     94.00
       dim 14    .0285988      2.54     96.54
       dim 15    .0233537      2.07     98.61
       dim 16    .0156866      1.39    100.00

        Total    1.127684    100.00
Statistics for column categories in standard normalization

                          overall                   dimension_1
    Categories      mass  quality  %inert     coord   sqcorr  contrib

    A
    agree stro~y   0.034    0.445   0.055     1.837    0.391    0.115
           agree   0.092    0.169   0.038     0.546    0.136    0.028
    neither ag~e   0.059    0.344   0.047    -0.447    0.047    0.012
        disagree   0.051    0.350   0.050    -1.166    0.258    0.069
    disagree s~y   0.014    0.401   0.060    -1.995    0.170    0.055

                      dimension_2
    Categories     coord   sqcorr  contrib

    A
    agree stro~y   0.727    0.054    0.018
           agree  -0.284    0.033    0.007
    neither ag~e  -1.199    0.298    0.084
        disagree   0.737    0.092    0.028
    disagree s~y   2.470    0.231    0.084
```

We do not provide a plot for this `mca` example; it would be the same as the previous one. MCA via the indicator matrix or the Burt matrix produces the same standard coordinates, although they produce different principal inertias. If `normalize(principal)` is used, different coordinates will be produced. Principal normalization relies on the principal inertias for normalization.

For more information on normalization, see either *Methods and formulas* in mca or the discussion of normalization in [MV] **ca**.

◁

Joint correspondence analysis

▷ Example 5

JCA attempts to remedy inflation of the total inertia by the block diagonal submatrices of the Burt matrix and is implemented as `method(joint)`. Results using JCA explain 90% of the principal inertia in the first two dimensions. With JCA, we must specify the dimensions we want in advance. The output contains principal inertias only for the first two dimensions (the default). For other methods, principal inertias for all dimensions are displayed. More dimensions can be requested with the `dim()` option. JCA is an iterative method, with repeated CAs and adjustments taking place until convergence is achieved.

We ask for coordinates in the principal normalization, `norm(principal)`, and ask for coordinates of only variable `A` to be displayed.

```
. mca A-D, method(joint) norm(principal) points(A)
Multiple/Joint correspondence analysis          Number of obs    =       871
                                                Total inertia    =  .1824248
    Method: Joint (JCA)                         Number of axes   =         2
```

Dimension	principal inertia	percent	cumul percent
dim 1	.099091	54.32	54.32
dim 2	.0650329	35.65	89.97
Total	.1824248	100.00	

Statistics for column categories in principal normalization

Categories	mass	overall quality	%inert	dimension_1 coord	sqcorr	contrib
A						
agree stro~y	0.034	0.964	0.052	-0.458	0.759	0.072
agree	0.092	0.774	0.020	-0.169	0.733	0.027
neither ag~e	0.059	0.884	0.030	0.048	0.025	0.001
disagree	0.051	0.887	0.046	0.364	0.810	0.068
disagree s~y	0.014	0.899	0.060	0.711	0.636	0.070

Categories	dimension_2 coord	sqcorr	contrib
A			
agree stro~y	-0.238	0.205	0.030
agree	0.040	0.042	0.002
neither ag~e	0.281	0.859	0.071
disagree	-0.112	0.077	0.010
disagree s~y	-0.458	0.264	0.044

```
. mcaplot, overlay origin
```

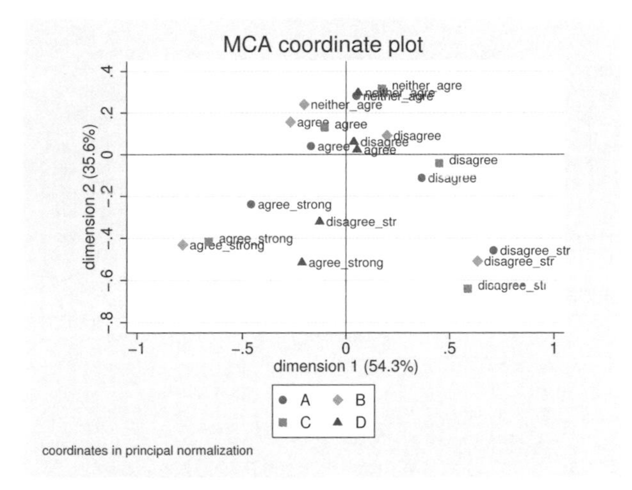

The plot shows the same relationships among the variables that we saw in the one after the indicator matrix approach to MCA. The main difference between the plots is the change in scale, partially because of the normalization and largely because of the change of method.

These same data are analyzed and plotted in Greenacre (2006, 66). To obtain an equivalent plot, we perform a reflection on the data with `ynegate`, add in the origin axes with `origin`, and use the same scale with options `xlabel()` and `ylabel()`,

```
. mcaplot, overlay ynegate origin ylabel(-1(.5)1.5) xlabel(-1.5(.5)1)
```

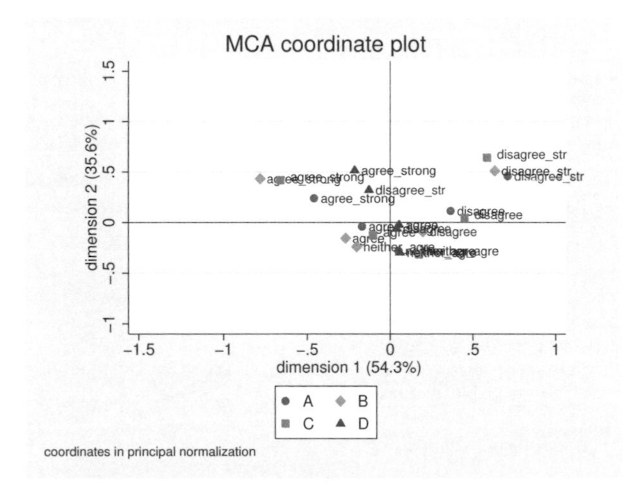

Note the similarities between this plot and the one obtained through the default approach to MCA via the Burt matrix and adjustments.

◁

Saved results

mca saves the following in e():

Scalars

e(N)	number of observations
e(f)	number of dimensions
e(inertia)	total inertia
e(ev_unique)	1 if all eigenvalues are distinct, 0 otherwise
e(adjust)	1 if eigenvalues are adjusted, 0 otherwise (method(burt) only)
e(converged)	1 if successful convergence, 0 otherwise (method(joint) only)
e(iter)	number of iterations (method(joint) only)
e(inertia_od)	proportion of off-diagonal inertia explained by extracted dimensions (method(joint) only)

Macros

e(cmd)	mca
e(cmdline)	command as typed
e(names)	names of MCA variables (crossed or actual)
e(supp)	names of supplementary variables
e(defs)	per crossed variable: crossing variables separated by "\"
e(missing)	missing if missing values are treated as ordinary values
e(crossed)	1 if there are crossed variables, 0 otherwise
e(wtype)	weight type
e(wexp)	weight expression
e(title)	title in output
e(method)	burt, indicator, or joint
e(norm)	standard or principal
e(properties)	nob noV eigen
e(estat_cmd)	program used to implement estat
e(predict)	program used to implement predict
e(marginsnotok)	predictions disallowed by margins

Matrices

e(Coding#)	row vector with coding of variable #
e(A)	standard coordinates for column categories
e(F)	principal coordinates for column categories
e(cMass)	column mass
e(cDist)	distance column to centroid
e(cInertia)	column inertia
e(cGS)	general statistics of column categories
	[.,1] column mass
	[.,2] overall quality
	[.,3] inertia/sum(inertia)
	[.,3*f+1] dim f: coordinate in e(norm) normalization
	[.,3*f+2] dim f: contribution of the profiles to principal axes
	[.,3*f+3] dim f: contribution of principal axes to profiles (= squared correlation of profile and axes)
e(rSCW)	weight matrix for row standard coordinates
e(Ev)	principal inertias/eigenvalues
e(inertia_e)	explained inertia (percent)
e(Bmod)	modified Burt matrix of active variables (method(joint) only)
e(inertia_sub)	variable-by-variable inertias (method(joint) only)

Functions

e(sample)	marks estimation sample

Methods and formulas

mca is implemented as an ado-file.

Methods and formulas are presented under the following headings:

Notation

We use notation that is fairly standard in the literature on correspondence analysis (for example, Greenacre [2006]). Let $x_1, \dots, x_q$ be categorical variables on N observations that are active in the analysis. To simplify notation, but without loss of generality, we assume that x_j is coded with consecutive integers $1, \dots, n_j$. Let $\mathbf{Z}^{(j)}$ be the $N \times n_j$ binary indicator matrix associated with x_j, $\mathbf{Z}^{(j)}_{ih} = 1$ if and only if $x_{ij} = h$. Let

$$\mathbf{Z} = \left(\mathbf{Z}^{(1)}, \mathbf{Z}^{(2)}, \dots, \mathbf{Z}^{(q)}\right)$$

be the $N \times J$ indicator matrix of the set of active x-variables, where $J = n_1 + \cdots + n_q$.

We will be consistent in letting i index observations $1, \dots, N$, j index variables $1, \dots, q$, and h index categories $1, \dots, n_j$, or $1, \dots, J$.

The $J \times J$ Burt matrix is defined as $\mathbf{B} = \mathbf{Z}'\mathbf{Z}$, or $\mathbf{B} = \mathbf{Z}'\mathbf{D}(w)\mathbf{Z}$, where w is the weight for the analysis and $\mathbf{D}(w)$ is a $J \times J$ square matrix with the weights on the diagonal and 0 off diagonal. The diagonal block of $\mathbf{B}$ associated with variable x_j is a diagonal matrix with the frequencies of x_j on the diagonal. The off-diagonal block of $\mathbf{B}$ associated with variables x_j and x_k is the two-way cross-tabulation of x_j and x_k.

In an analogous way, we define $\mathbf{B}^*$, the Burt matrix with more rows containing cross-tabulation from the supplementary variables. $\mathbf{B}^* = \mathbf{Z}^{*\prime}\mathbf{Z}$, where $\mathbf{Z}^*$ is the indicator matrix with more columns for the supplementary variables.

$\mathbf{D}(v)$, in general, represents a diagonal matrix with the elements of vector v on the diagonal and 0 off diagonal; $\mathbf{1}$ is a column vector of ones where length is defined by the context.

Using ca to compute MCA

The indicator approach to MCA involves an equivalent technique to performing a standard CA on the indicator matrix $\mathbf{Z}$; see Greenacre (2006). We refer to *Methods and formulas* in [MV] **ca** for details. The indicator approach could be handled by forming $\mathbf{Z}$ and invoking `camat` on $\mathbf{Z}$. If you had k categorical variables named `v1, ..., vk`, you could perform CA on the indicator matrix by taking the steps in the following code fragment:

```
tab v1, gen(V1_)
...
tab vk, gen(Vk_)
mkmat V1_* ... Vk_*, matrix(VALL)
camat VALL
```

CA of an indicator or Burt matrix

$\mathbf{Z}$ may be a rather large matrix: the number of rows is equal to the number of observations, and memory limitations would pose a limit for applications. Moreover, the interest in MCA is mostly on the column space, and so much computation would be wasted. Therefore, the implementation in `mca` exploits the close analogy between the indicator and Burt matrix approach for efficiency (see Greenacre [1984, chap. 5; 2006, app. A.3]). The consequence is that the CA results for the rows of $\mathbf{Z}$ are not computed. You may use the above scheme if you need row information.

Much literature (for example, Everitt and Dunn [2001]; Rencher [2002]) identifies MCA with the application of CA on the Burt matrix $\mathbf{B} = \mathbf{Z}'\mathbf{Z}$.

This discussion is related to that of *Methods and formulas* in [MV] **ca**. We define

$$\begin{aligned}
\mathbf{B}_{++} &= \sum_{k=1}^{J}\sum_{h=1}^{J} \mathbf{B}_{kh} \\
\mathbf{P} &= \mathbf{B}/\mathbf{B}_{++} \\
\mathbf{c} &= \sum_{k=1}^{J} \mathbf{P}_{k*} = \mathbf{P}_{+*} = \mathbf{P}'\mathbf{1} \qquad \text{here } * \text{ represents all possible values} \\
\mathbf{S} &= \mathbf{P} - \mathbf{c}\mathbf{c}'
\end{aligned}$$

$\mathbf{c}$ is called the column mass. $\mathbf{B}$ and $\mathbf{S}$ are symmetric. Thus the singular-value decomposition commonly associated with CA is equivalent to the spectral or eigen decomposition of $\mathbf{S}$.

$$\mathbf{S} = \mathbf{V}\boldsymbol{\Phi}\mathbf{V}', \qquad \phi_1 \geq \phi_2 \geq \dots$$

For identification purposes and to facilitate resampling methods, the eigenvectors of $\mathbf{V}$ are directed so that for h, $\mathbf{V}_{kh} > 0$ for the first k for which $\mathbf{V}_{kh} \neq 0$. The standard column coordinates $\mathbf{A}$ are the same for the indicator and the Burt approach and are defined as

$$\mathbf{A} = \mathbf{D}(\mathbf{c})^{-1}\mathbf{V}$$

$\mathbf{D}(\mathbf{c})$ is the diagonal matrix with diagonal $\mathbf{c}$; $\mathbf{D}(\mathbf{c})^{-1}$ is therefore the diagonal matrix with elements $1/c_t$, where c_t is an element of $\mathbf{c}$.

In the indicator approach to MCA, the tth principal inertia is defined as $\lambda_t = \phi_t$, the total inertia as $\sum_t \lambda_t$. The inertia of column j or variable j, $\mathbf{In}^{(j)}$, is computed elementwise as

$$\mathrm{In}_h^{(j)} = \sum_{i=1}^{N} w_i \frac{(Z_{ih}^{(j)} - qc_h^{(j)})^2}{q^2 c_h^{(j)} w_+}$$

where $Z_{ih}^{(j)}$ is the i,hth element of the indicator matrix for variable j, w_i is the weight for observation i, q is the number of active variables, $c_h^{(j)}$ is the column mass of variable j for category h, and w_+ is the sum of the weights over the observations.

In the Burt approach to MCA, the unadjusted principal inertia is defined as $\lambda_t^{\mathrm{unadj}} = \phi_t^2$, the total unadjusted inertia as $\sum_t \lambda_t^{\mathrm{unadj}}$, and the unadjusted column inertias as $\mathbf{1}'\mathbf{S} \odot \mathbf{S}$, with $\odot$ the Hadamard or elementwise matrix product. The adjusted principal inertia, λ_t^{adj}, is defined as

$$\lambda_t^{\mathrm{adj}} = \left(\frac{q}{q-1}\right)^2 \left(\phi_t^2 - \frac{1}{q}\right)^2 \qquad \text{provided } q\phi_t > 1$$

The total inertia is defined as

$$\text{total inertia} = \left(\frac{q}{q-1}\right) \sum \phi_t^2 - \frac{(J-q)}{q^2}$$

The standard coordinates are independent of the principal inertia; with or without adjustment, these are defined as before

$$\mathbf{A} = \mathbf{D}(\mathbf{c})^{-1}\mathbf{V}$$

The principal coordinates $\mathbf{F}$ are defined as

$$\mathbf{F} = \mathbf{A}\mathbf{D}(\mathbf{\Lambda})^{1/2}$$

where $\mathbf{\Lambda}$ is a vector of adjusted or unadjusted principal inertias and $\mathbf{D}(\mathbf{\Lambda})^{1/2}$ is the diagonal matrix with elements $\lambda_t^{1/2}$ on the diagonals.

JCA

The implementation of JCA uses the alternating least-squares method proposed by Greenacre (1988, 2006). This algorithm forms a modification of the Burt matrix, changing the diagonal blocks associated with the variables, keeping the off-diagonal blocks unchanged. In the first iteration, $\mathbf{B}_0 = \mathbf{B}$. In iteration m, the blocks are replaced by the f-dimensional MCA approximation of the adjusted Burt matrix $\mathbf{B}_{m-1}$. Iteration continues until the change in the elements of $\mathbf{B}_m$ and $\mathbf{B}_{m-1}$ falls below a convergence tolerance. The JCA coordinates and inertias are computed from the converged solution, $\mathbf{B}_\infty$, analogous to the (adjusted) Burt method. The total inertia of the modified Burt matrix is defined as the sum of the inertias of the off-diagonal blocks.

To compute the f-dimensional MCA approximation of the adjusted Burt matrix $\mathbf{B}_{m-1}$, we perform MCA on $\mathbf{B}_{m-1}$ and then reconstruct the approximation of the data from the solution

$$\widehat{\mathbf{B}}_{hk} = \mathbf{B}_{++} c_h c_k \left(1 + \sum_{t=1}^{f} \phi_t^2 A_{ht} A_{kt}\right)$$

where A_{ht} is an element of the standard coordinate matrix $\mathbf{A}$, c_h and c_k are the column masses, and ϕ_t are the eigenvalues as in the computation of the CA of the Burt matrix. We then update the main diagonal submatrices of $\mathbf{B}_{m-1}$ with the corresponding entries of $\widehat{\mathbf{B}}$ to obtain $\mathbf{B}_m$.

Supplementary variables

The coordinates of supplementary variables are computed as weighted averages of the column coordinates by using the so-called CA transition formulas. As outlined by Greenacre (2006), standard coordinates may be used for averaging, with the profiles of the indicator representation of supplementary columns as weights. Supplementary principal column coordinates are computed as weighted averages of the standard active column coordinates, and then supplementary standard coordinates are computed by division by the principal inertias.

To compute, we add the supplementary variables to the Burt matrix as more rows; if $\mathbf{B}$ is the Burt matrix of the active variables then $\mathbf{B}^*$ is the Burt matrix with the additional cross-tabulation from the supplementary variables. Define $\mathbf{P}$ as above and $\mathbf{P}^*$ analogously with the supplementary variables added. MCA is performed on $\mathbf{B}$ as before, and information from this solution is then applied to $\mathbf{B}^*$. Let p^* represent the elements of $\mathbf{P}^*$. Let k index categories of supplementary variables, h index categories of active variables, and t index dimensions of the solution. Let $\mathbf{A}$ be the standard coordinates of the active variables, as computed previously. Then the principal coordinate for category k and dimension s is computed as:

$$g_{kt} = \sum_{h=1}^{J} \frac{p^*_{hk}}{p^*_{+k}} A_{ht}$$

Coordinates in standard coordinates are obtained by division by the square root of the corresponding principal inertia as described above.

predict

`predict` after `mca` produces variables with the MCA coordinates as displayed by `mca` for both active and supplementary variables. Formulas are shown above. `predict` can also compute row coordinates also known as row scores. Row coordinates computed are always based on the indicator method. The standard row coordinate for the tth dimension for the ith observation with indicator matrix elements Z_{ih} is computed as

$$R_{it} = \sum_{h=1}^{J} \frac{Z_{ih} A_{ht}}{q\sqrt{\phi_t}}$$

where $\mathbf{A}$ is the matrix of standard coordinates, q is the number of active variables in the analysis, and ϕ_t is an eigenvalue of the CA on the Burt matrix. To get the row coordinate in principal normalization, one multiplies by the square root of the corresponding principal inertia.

References

Everitt, B. S., and G. Dunn. 2001. *Applied Multivariate Data Analysis*. 2nd ed. London: Arnold.

Gower, J. C. 1990. Fisher's optimal scores and multiple correspondence analysis. *Biometrics* 46: 947–961.

——. 2006. Divided by a common language—analyzing and visualizing two-way arrays. In *Multiple Correspondence Analysis and Related Methods*, ed. M. Greenacre and J. Blasius. Boca Raton, FL: Chapman & Hall/CRC.

Gower, J. C., and D. J. Hand. 1996. *Biplots*. London: Chapman & Hall.

Greenacre, M. J. 1984. *Theory and Applications of Correspondence Analysis*. London: Academic Press.

——. 1988. Correspondence analysis of multivariate categorical data by weighted least-squares. *Biometrika* 75: 457–467.

——. 2006. From simple to multiple correspondence analysis. In *Multiple Correspondence Analysis and Related Methods*, ed. M. Greenacre and J. Blasius. Boca Raton, FL: Chapman & Hall.

Greenacre, M. J., and J. Blasius, ed. 1994. *Correspondence Analysis in the Social Sciences*. London: Academic Press.

——. 2006. *Multiple Correspondence Analysis and Related Methods*. Boca Raton, FL: Chapman & Hall.

ISSP. 1993. International Social Survey Programme: Environment. http://www.issp.org.

Milan, L., and J. Whittaker. 1995. Application of the parametric bootstrap to models that incorporate a singular value decomposition. *Applied Statistics* 44: 31–49.

Rencher, A. C. 2002. *Methods of Multivariate Analysis*. 2nd ed. New York: Wiley.

Also see

[MV] **mca postestimation** — Postestimation tools for mca

[MV] **ca** — Simple correspondence analysis

[MV] **canon** — Canonical correlations

[MV] **factor** — Factor analysis

[MV] **pca** — Principal component analysis

Title

mca postestimation — Postestimation tools for mca

Description

The following postestimation commands are of special interest after `mca`:

Command	Description
`mcaplot`	plot of category coordinates
`mcaprojection`	MCA dimension projection plot
`estat coordinates`	display of category coordinates
`estat subinertia`	matrix of inertias of the active variables (after JCA only)
`estat summarize`	estimation sample summary
`screeplot`	plot principal inertias (eigenvalues)

For information about `screeplot`, see [MV] **screeplot**; for all other commands, see below.

The following standard postestimation commands are also available:

Command	Description
*`estimates`	cataloging estimation results
`predict`	row and category coordinates

*All `estimates` subcommands except `table` and `stats` are available; [R] **estimates**.

Special-interest postestimation commands

`mcaplot` produces a scatterplot of category points of the MCA variables in two dimensions.

`mcaprojection` produces a projection plot of the coordinates of the categories of the MCA variables.

`estat coordinates` displays the category coordinates, optionally with column statistics.

`estat subinertia` displays the matrix of inertias of the active variables (after JCA only).

`estat summarize` displays summary information of MCA variables over the estimation sample.

Syntax for predict

predict [*type*] *newvar* [*if*] [*in*] [, *statistic* normalize(*norm*) dimensions(*#*)]

predict [*type*] {*stub** | *newvarlist*} [*if*] [*in*] [, *statistic* normalize(*norm*) dimensions(*numlist*)]

statistic	Description
Main	
rowscores	row scores (coordinates); the default
score(*varname*)	scores (coordinates) for MCA variable *varname*

norm	Description
standard	use standard normalization
principal	use principal normalization

Menu

Statistics > Postestimation > Predictions, residuals, etc.

Options for predict

Main

rowscores specifies that row scores (row coordinates) be computed. The row scores returned are based on the indicator matrix approach to multiple correspondence analysis, even if another method was specified in the original mca estimation. The sample for which row scores are computed may exceed the estimation sample; for example, it may include supplementary rows (variables). score() and rowscores are mutually exclusive. rowscores is the default.

score(*varname*) specifies the name of a variable from the preceding MCA for which scores should be computed. The variable may be a regular categorical variable, a crossed variable, or a supplementary variable. score() and rowscores are mutually exclusive.

Options

normalize(*norm*) specifies the normalization of the scores (coordinates). normalize(standard) returns coordinates in standard normalization. normalize(principal) returns principal scores. The default is the normalization method specified with mca during estimation, or normalize(standard) if no method was specified.

dimensions(*#*) or dimensions(*numlist*) specifies the dimensions for which scores (coordinates) are computed. The number of dimensions specified should equal the number of variables in *newvarlist*. If dimensions() is not specified, scores for dimensions 1, ..., k are returned, where k is the number of variables in *newvarlist*. The number of variables in *newvarlist* should not exceed the number of dimensions extracted during estimation.

Syntax for estat coordinates

```
estat coordinates [varlist] [, normalize(norm) stats format(%fmt) ]
```

Note: variables in *varlist* must be from the preceding mca and may refer to either a regular categorical variable or a crossed variable. The variables in *varlist* may also be chosen from the supplementary variables.

options	Description
normalize(standard)	standard coordinates
normalize(principal)	principal coordinates
stats	include mass, distance, and inertia
format(%*fmt*)	display format; default is format(%9.4f)

Menu

Statistics > Postestimation > Reports and statistics

Options for estat coordinates

normalize(*norm*) specifies the normalization of the scores (coordinates). normalize(standard) returns coordinates in standard normalization. normalize(principal) returns principal scores. The default is the normalization method specified with mca during estimation, or normalize(standard) if no method was specified.

stats includes the column mass, the distance of the columns to the centroid, and the column inertias in the table.

format(%*fmt*) specifies the display format for the matrix, for example, format(%8.3f). The default is format(%9.4f).

Syntax for estat subinertia

```
estat subinertia
```

Menu

Statistics > Postestimation > Reports and statistics

Syntax for estat summarize

```
estat summarize [, crossed labels noheader noweights]
```

options	Description
Main	
crossed	summarize crossed and uncrossed variables as used
labels	display variable labels
noheader	suppress the header
noweights	ignore weights

Menu

Statistics > Postestimation > Reports and statistics

Options for estat summarize

Main

`crossed` specifies summarizing the crossed variables if crossed variables are used in the MCA, rather than the crossing variables from which they are formed. The default is to summarize the crossing variables and single categorical variables used in the MCA.

`labels` displays variable labels.

`noheader` suppresses the header.

`noweights` ignores the weights, if any. The default when weights are present is to perform a weighted summarize on all variables except the weight variable itself. An unweighted summarize is performed on the weight variable.

Syntax for mcaplot

`mcaplot` [*speclist*] [, *options*]

where

speclist = *spec* [*spec* ...]

spec = *varlist* | (*varname* [, *plot_options*])

and variables in *varlist* or *varname* must be from the preceding `mca` and may refer to either a regular categorical variable or a crossed variable. The variables may also be supplementary.

options	Description
Options	
combine_options	affect the rendition of the combined graphs
`overlay`	overlay the plots of the variables; default is to produce separate plots
`dimensions(`$\#_1$ $\#_2$`)`	display dimensions $\#_1$ and $\#_2$; default is `dimensions(2 1)`
`normalize(standard)`	display standard coordinates
`normalize(principal)`	display principal coordinates
`maxlength(#)`	use # as maximum number of characters for labels; default is `maxlength(12)`
`xnegate`	negate the coordinates relative to the x axis
`ynegate`	negate the coordinates relative to the y axis
`origin`	mark the origin and draw origin axes
`originlopts(`*line_options*`)`	affect the rendition of the origin axes
Y axis, X axis, Titles, Legend, Overall	
twoway_options	any options other than `by()` documented in [G-3] ***twoway_options***

plot_options	Description
marker_options	change look of markers (color, size, etc.)
marker_label_options	add marker labels; change look or position
twoway_options	titles, legends, axes, added lines and text, regions, etc.

Menu

Statistics > Multivariate analysis > Correspondence analysis > Postestimation after MCA or JCA > Plot of category coordinates

Options for mcaplot

Plots

plot_options affect the rendition of markers, including their shape, size, color, and outline (see [G-3] ***marker_options***) and specify if and how the markers are to be labeled (see [G-3] ***marker_label_options***). These options may be specified for each variable. If the overlay option is not specified, then for each variable you may also specify many of the *twoway_options* excluding by(), name(), and aspectratio(); see [G-3] ***twoway_options***. See *twoway_options* below for a warning against using options such as xlabel(), xscale(), ylabel(), and yscale().

Options

combine_options affect the rendition of the combined plot; see [G-2] **graph combine**. *combine_options* may not be specified with overlay.

overlay overlays the biplot graphs for the variables. The default is to produce a combined graph of the biplot graphs.

dimensions($\#_1$ $\#_2$) identifies the dimensions to be displayed. For instance, dimensions(3 2) plots the third dimension (vertically) versus the second dimension (horizontally). The dimension number cannot exceed the number of extracted dimensions. The default is dimensions(2 1).

normalize(*norm*) specifies the normalization of the coordinates. normalize(standard) returns coordinates in standard normalization. normalize(principal) returns principal coordinates. The default is the normalization method specified with mca during estimation, or normalize(standard) if no method was specified.

maxlength(*#*) specifies the maximum number of characters for row and column labels; the default is maxlength(12).

xnegate specifies that the x-axis coordinates be negated (multiplied by -1).

ynegate specifies that the y-axis coordinates be negated (multiplied by -1).

origin marks the origin and draws the origin axes.

originlopts(*line_options*) affect the rendition of the origin axes. See [G-3] ***line_options***.

Y axis, X axis, Titles, Legend, Overall

twoway_options are any of the options documented in [G-3] ***twoway_options*** excluding by().

mcaplot automatically adjusts the aspect ratio on the basis of the range of the data and ensures that the axes are balanced. As an alternative, the *twoway_option* aspectratio() can be used to override the default aspect ratio. mcaplot accepts the aspectratio() option as a suggestion only and will override it when necessary to produce plots with balanced axes; that is, distance on the x axis equals distance on the y axis.

twoway_options such as xlabel(), xscale(), ylabel(), and yscale() should be used with caution. These options *axis_options* are accepted but may have unintended side effects on the aspect ratio. See [G-3] ***twoway_options***.

Syntax for mcaprojection

mcaprojection [*speclist*] [, *options*]

where

speclist = *spec* [*spec* ...]

spec = *varlist* | (*varname* [, *plot_options*])

and variables in *varlist* or *varname* must be from the preceding mca and may refer to either a regular categorical variable or a crossed variable. The variables may also be supplementary.

options	Description
Options	
dimensions(*numlist*)	display *numlist* dimensions; default is all
normalize(principal)	scores (coordinates) should be in principal normalization
normalize(standard)	scores (coordinates) should be in standard normalization
alternate	alternate labels
maxlength(#)	use # as maximum number of characters for labels; default is maxlength(12)
combine_options	affect the rendition of the combined graphs
Y axis, X axis, Titles, Legend, Overall	
twoway_options	any options other than by() documented in [G-3] ***twoway_options***

plot_options	Description
marker_options	change look of markers (color, size, etc.)
marker_label_options	add marker labels; change look or position
twoway_options	titles, legends, axes, added lines and text, regions, etc.

Menu

Statistics > Multivariate analysis > Correspondence analysis > Postestimation after MCA or JCA > Dimension projection plot

Options for mcaprojection

Plots

plot_options affect the rendition of markers, including their shape, size, color, and outline (see [G-3] ***marker_options***) and specify if and how the markers are to be labeled (see [G-3] ***marker_label_options***). These options may be specified for each variable. If the overlay option is not specified then for each variable you may also specify *twoway_options* excluding by() and name(); see [G-3] ***twoway_options***.

Options

dimensions(*numlist*) identifies the dimensions to be displayed. By default all dimensions are displayed.

normalize(*norm*) specifies the normalization of the coordinates. normalize(standard) returns coordinates in standard normalization. normalize(principal) returns principal coordinates. The default is the normalization method specified with mca during estimation, or normalize(standard) if no method was specified.

alternate causes adjacent labels to alternate sides.

maxlength(#) specifies the maximum number of characters for row and column labels; the default is maxlength(12).

combine_options affect the rendition of the combined plot; see [G-2] **graph combine**. These options may not be used if only one variable is specified.

Y axis, X axis, Titles, Legend, Overall

twoway_options are any of the options documented in [G-3] ***twoway_options***, excluding by().

Remarks

Remarks are presented under the following headings:

Postestimation statistics
Postestimation graphs
Predicting new variables

Postestimation statistics

We continue to examine the ISSP (1993) dataset on the environment. We perform joint correspondence analysis.

▷ Example 1

```
. use http://www.stata-press.com/data/r12/issp93
(Selection from ISSP (1993))
. mca A-D, method(joint)
Multiple/Joint correspondence analysis        Number of obs     =       871
                                              Total inertia     =  .1824248
    Method: Joint (JCA)                       Number of axes    =         2

                    principal                cumul
    Dimension        inertia    percent    percent

        dim 1        .099091     54.32      54.32
        dim 2       .0650329     35.65      89.97

        Total       .1824248    100.00
Statistics for column categories in standard normalization
```

Categories	mass	overall quality	%inert	dimension_1 coord	sqcorr	contrib
A						
agree stro~y	0.034	0.964	0.052	-1.456	0.759	0.072
agree	0.092	0.774	0.020	-0.536	0.733	0.027
neither ag~e	0.059	0.884	0.030	0.154	0.025	0.001
disagree	0.051	0.887	0.046	1.157	0.810	0.068
disagree s~y	0.014	0.899	0.060	2.258	0.636	0.070
B						
agree stro~y	0.020	0.957	0.093	-2.490	0.735	0.126
agree	0.050	0.851	0.031	-0.850	0.636	0.036
neither ag~e	0.059	0.953	0.033	-0.639	0.393	0.024
disagree	0.081	0.705	0.029	0.618	0.579	0.031
disagree s~y	0.040	0.977	0.149	2.014	0.594	0.163
C						
agree stro~y	0.044	0.983	0.149	-2.104	0.704	0.193
agree	0.091	0.665	0.020	-0.327	0.257	0.010
neither ag~e	0.057	0.839	0.047	0.539	0.188	0.016
disagree	0.044	0.907	0.054	1.429	0.899	0.090
disagree s~y	0.015	0.944	0.065	1.864	0.431	0.052
D						
agree stro~y	0.017	0.850	0.034	-0.674	0.124	0.008
agree	0.067	0.145	0.008	0.165	0.120	0.002
neither ag~e	0.058	0.769	0.038	0.181	0.027	0.002
disagree	0.065	0.178	0.010	0.116	0.047	0.001
disagree s~y	0.043	0.929	0.030	-0.402	0.125	0.007

Categories	dimension_2 coord	sqcorr	contrib
A			
agree stro~y	-0.934	0.205	0.030
agree	0.158	0.042	0.002
neither ag~e	1.103	0.859	0.071
disagree	-0.440	0.077	0.010
disagree s~y	-1.796	0.264	0.044
B			
agree stro~y	-1.690	0.222	0.058
agree	0.611	0.216	0.019
neither ag~e	0.942	0.560	0.052
disagree	0.356	0.126	0.010
disagree s~y	-1.995	0.383	0.160
C			
agree stro~y	-1.634	0.279	0.116
agree	0.508	0.407	0.023
neither ag~e	1.237	0.651	0.087
disagree	-0.166	0.008	0.001
disagree s~y	-2.509	0.513	0.094
D			
agree stro~y	-2.018	0.727	0.070
agree	0.092	0.024	0.001
neither ag~e	1.166	0.741	0.079
disagree	0.239	0.131	0.004
disagree s~y	-1.256	0.804	0.068

If we wish to see the coordinates in the principal normalization, we do not need to rerun our analysis. We can use `estat coordinates` to display them.

```
. estat coordinates, norm(principal)
Column principal coordinates
```

Categories	dim1	dim2
A		
agree_stro~y	-.4582629	-.2381115
agree	-.1686314	.0402091
neither_ag~e	.0484366	.2811716
disagree	.3642677	-.1123168
disagree_s~y	.7106647	-.4578886
B		
agree_stro~y	-.783911	-.4310436
agree	-.2674646	.1558017
neither_ag~e	-.2010783	.2402487
disagree	.1944504	.0906804
disagree_s~y	.6341215	-.5088398
C		
agree_stro~y	-.6623101	-.4166016
agree	-.1029922	.1295649
neither_ag~e	.169804	.3155628
disagree	.4496893	-.0423339
disagree_s~y	.5867913	-.6397215
D		
agree_stro~y	-.2123187	-.5145647
agree	.05208	.0233723
neither_ag~e	.0569168	.297262
disagree	.0365233	.0609881
disagree_s~y	-.1264563	-.3203889

We may also be interested in the contributions of each submatrix of the Burt matrix to the total inertia. This can be obtained by `estat subinertia`. Compare with Greenacre (2006, table A.12).

```
. estat subinertia
Subinertias: decomposition of total inertia
```

Variable	A	B	C	D
A	.0074502			
B	.0148596	.022442		
C	.012149	.0185838	.0210336	
D	.0032898	.0053016	.0096583	.0038148

◁

Postestimation graphs

Several examples of mcaplot were displayed in [MV] **mca**, so we will not give more examples here. The discussion in [MV] **ca postestimation** is also relevant.

We will focus on mcaprojection, which is the mca analogue of caprojection, just as mcaplot is the analogue of cabiplot.

▷ Example 2

mcaprojection produces a projection plot of the column coordinates after mca. We continue where we left off with our previous example. Say that we want to examine the projections in the principal normalization.

```
. mcaprojection, norm(principal)
```

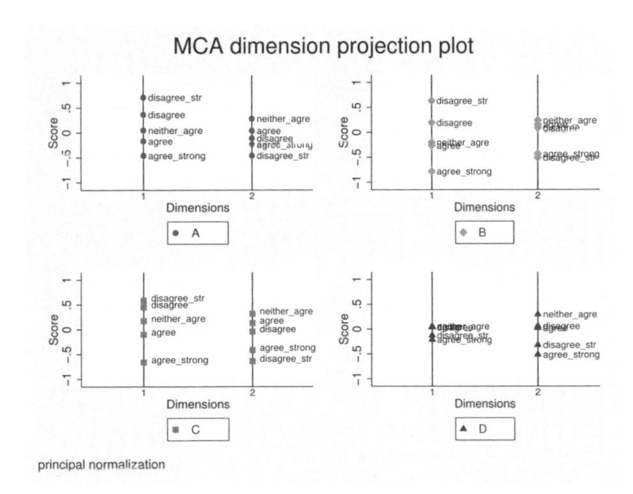

We would see the same alignment of points in the standard normalization, but the scale would be changed. We have noted previously that item D does not behave like the other variables in the MCA. Each of the first three variables, A, B, and C, has its responses arrayed in order from strong disagreement to strong agreement on the first dimension. Here again, variable D is the only one of the four that shows a different ordering in its projection. We do not see a nice projection from strong disagreement to strong agreement on the second dimension, but an inspection indicates that the first three variables are all in the same order, with the last, D, opposed to the remainder.

◁

Predicting new variables

Coordinates (scores) can be predicted after mca. You can predict either the column coordinates or the row coordinates.

▷ Example 3

We will calculate row scores by hand and via `predict, rowscore`. Row scores are calculated off the indicator method approach. This calculation would not work with JCA, because the standard coordinates and the principal inertias are different for JCA. The principal inertias are also a problem with the Burt method, with and without adjustment. We use the `points(A)` option to restrict the output from `mca`. Four variables were used in the original `mca`. `predict` without the `dimensions()` option predicts the first dimension.

```
. mca A-D, method(indicator) points(A)
Multiple/Joint correspondence analysis          Number of obs      =       871
                                                Total inertia      =         4
    Method: Indicator matrix                    Number of axes     =         2

                 principal               cumul
    Dimension      inertia    percent   percent

        dim 1     .4573792     11.43      11.43
        dim 2     .4309658     10.77      22.21
        dim 3     .3219257      8.05      30.26
        dim 4     .3064732      7.66      37.92
        dim 5     .2756747      6.89      44.81
        dim 6      .251928      6.30      51.11
        dim 7     .2425591      6.06      57.17
        dim 8     .2349506      5.87      63.05
        dim 9      .225468      5.64      68.68
       dim 10     .2206291      5.52      74.20
       dim 11     .2098376      5.25      79.44
       dim 12     .1971485      4.93      84.37
       dim 13     .1778833      4.45      88.82
       dim 14     .1691119      4.23      93.05
       dim 15     .1528191      3.82      96.87
       dim 16     .1252462      3.13     100.00

        Total            4    100.00
Statistics for column categories in standard normalization

                             overall                   dimension_1
    Categories       mass  quality   %inert     coord   sqcorr  contrib

    A
    agree stro~y    0.034    0.280    0.054     1.837    0.244    0.078
           agree    0.092    0.100    0.039     0.546    0.080    0.019
    neither ag~e    0.059    0.218    0.048    -0.447    0.028    0.008
        disagree    0.051    0.220    0.050    -1.166    0.160    0.047
    disagree s~y    0.014    0.260    0.059    -1.995    0.106    0.037

                          dimension_2
    Categories      coord   sqcorr  contrib

    A
    agree stro~y    0.727    0.036    0.012
           agree   -0.284    0.020    0.005
    neither ag~e   -1.199    0.190    0.055
        disagree    0.737    0.060    0.018
    disagree s~y    2.470    0.153    0.055

. predict double a1, score(A)
. predict double b1, score(B)
. predict double c1, score(C)
```

```
. predict double d1, score(D)
. predict double r1, rowscore
. mat Ev = e(Ev)
. scalar phi1 = Ev[1,1]
. gen double rc = (a1+b1+c1+d1)/(4*sqrt(phi1))
. assert reldif(rc, r1) < 1e-14
```

In the indicator method approach, we can also find Cronbach's alpha either via the `alpha` command (see [R] **alpha**) or by hand.

```
. alpha a1 b1 c1 d1
Test scale = mean(unstandardized items)
Average interitem covariance:     .2768234
Number of items in the scale:            4
Scale reliability coefficient:      0.6045
. scalar alpha = (4/(4-1))*(1-1/(4*phi1))
. di alpha
.60454338
```

◁

Saved results

`estat summarize` saves the following in `r()`:

Matrices

`r(stats)`	$k \times 4$ matrix of means, standard deviations, minimums, and maximums

`estat coordinates` saves the following in `r()`:

Macros

`r(norm)`	normalization method of the coordinates

Matrices

`r(Coord)`	column coordinates
`r(Stats)`	column statistics: mass, distance, and inertia (option `stats` only)

`estat subinertia` saves the following in `r()`:

Matrices

`r(inertia_sub)`	variable-by-variable inertias

Methods and formulas

All postestimation commands listed above are implemented as ado-files. See [MV] **mca** for methods and formulas.

References

Greenacre, M. J. 2006. From simple to multiple correspondence analysis. In *Multiple Correspondence Analysis and Related Methods*, ed. M. Greenacre and J. Blasius. Boca Raton, FL: Chapman & Hall.

ISSP. 1993. International Social Survey Programme: Environment. http://www.issp.org.

Also see

[MV] **mca** — Multiple and joint correspondence analysis

[MV] **ca** — Simple correspondence analysis,

[MV] **ca postestimation** — Postestimation tools for ca and camat

Title

mds — Multidimensional scaling for two-way data

Syntax

`mds` *varlist* [*if*] [*in*], `id(`*varname*`)` [*options*]

options	Description
Model	
* `id(`*varname*`)`	identify observations
`method(`*method*`)`	method for performing MDS
`loss(`*loss*`)`	loss function
`transform(`*tfunction*`)`	permitted transformations of dissimilarities
`normalize(`*norm*`)`	normalization method; default is `normalize(principal)`
`dimension(`*#*`)`	configuration dimensions; default is `dimension(2)`
`addconstant`	make distance matrix positive semidefinite
Model 2	
`unit`[`(`*varlist*$_2$`)`]	scale variables to min = 0 and max = 1
`std`[`(`*varlist*$_3$`)`]	scale variables to mean = 0 and sd = 1
`measure(`*measure*`)`	similarity or dissimilarity measure; default is `L2` (Euclidean)
`s2d(standard)`	convert similarity to dissimilarity: $\text{dissim}_{ij} = \sqrt{\text{sim}_{ii} + \text{sim}_{jj} - 2\text{sim}_{ij}}$; the default
`s2d(oneminus)`	convert similarity to dissimilarity: $\text{dissim}_{ij} = 1 - \text{sim}_{ij}$
Reporting	
`neigen(`*#*`)`	maximum number of eigenvalues to display; default is `neigen(10)`
`config`	display table with configuration coordinates
`noplot`	suppress configuration plot
Minimization	
`initialize(`*initopt*`)`	start with configuration given in *initopt*
`tolerance(`*#*`)`	tolerance for configuration matrix; default is `tolerance(1e-4)`
`ltolerance(`*#*`)`	tolerance for loss criterion; default is `ltolerance(1e-8)`
`iterate(`*#*`)`	perform maximum # of iterations; default is `iterate(1000)`
`protect(`*#*`)`	perform # optimizations and report best solution; default is `protect(1)`
`nolog`	suppress the iteration log
`trace`	display current configuration in iteration log
`gradient`	display current gradient matrix in iteration log
`sdprotect(`*#*`)`	advanced; see *Options* below

* `id(`*varname*`)` is required.

`bootstrap`, `by`, `jackknife`, `rolling`, `statsby`, and `xi` are allowed; see [U] **11.1.10 Prefix commands**.

The maximum number of observations allowed in `mds` is the maximum matrix size; see [R] **matsize**.

`sdprotect(`*#*`)` does not appear in the dialog box.

See [U] **20 Estimation and postestimation commands** for more capabilities of estimation commands.

method	Description
classical	classical MDS; default if neither loss() nor transform() is specified
modern	modern MDS; default if loss() or transform() is specified; except when loss(stress) and transform(monotonic) are specified
nonmetric	nonmetric (modern) MDS; default when loss(stress) and transform(monotonic) are specified

loss	Description
stress	stress criterion, normalized by distances; the default
nstress	stress criterion, normalized by disparities
sstress	squared stress criterion, normalized by distances
nsstress	squared stress criterion, normalized by disparities
strain	strain criterion (with transform(identity) is equivalent to classical MDS)
sammon	Sammon mapping

tfunction	Description
identity	no transformation; disparity = dissimilarity; the default
power	power α: disparity = dissimilarity$^{\alpha}$
monotonic	weakly monotonic increasing functions (nonmetric scaling); only with loss(stress)

norm	Description
principal	principal orientation; location = 0; the default
classical	Procrustes rotation toward classical solution
target(*matname*) [, copy]	Procrustes rotation toward *matname*; ignore naming conflicts if copy is specified

initopt	Description
classical	start with classical solution; the default
random[(#)]	start at random configuration, setting seed to #
from(*matname*) [, copy]	start from *matname*; ignore naming conflicts if copy is specified

Menu

Statistics > Multivariate analysis > Multidimensional scaling (MDS) > MDS of data

Description

`mds` performs multidimensional scaling (MDS) for dissimilarities between observations with respect to the variables in *varlist*. A wide selection of similarity and dissimilarity measures is available; see the `measure()` option. `mds` performs classical metric MDS (Torgerson 1952) as well as modern metric and nonmetric MDS; see the `loss()` and `transform()` options.

`mds` computes dissimilarities from the observations; `mdslong` and `mdsmat` are for use when you already have proximity information. `mdslong` and `mdsmat` offer the same statistical features but require different data organizations. `mdslong` expects the proximity information (and, optionally, weights) in a "long format" (pairwise or dyadic form), whereas `mdsmat` performs MDS on symmetric proximity and weight matrices; see [MV] **mdslong** and [MV] **mdsmat**.

Computing the classical solution is straightforward, but with modern MDS the minimization of the loss criteria over configurations is a high dimensional problem that is easily beset by convergence to local minimums. `mds`, `mdsmat`, and `mdslong` provide options to control the minimization process (1) by allowing the user to select the starting configuration and (2) by selecting the best solution among multiple minimization runs from random starting configurations.

Options

Model

`id(`*varname*`)` is required and specifies a variable that identifies observations. A warning message is displayed if *varname* has duplicate values.

`method(`*method*`)` specifies the method for MDS.

`method(classical)` specifies classical metric scaling, also known as "principal coordinates analysis" when used with Euclidean proximities. Classical MDS obtains equivalent results to modern MDS with `loss(strain)` and `transform(identity)` without weights. The calculations for classical MDS are fast; consequently, classical MDS is generally used to obtain starting values for modern MDS. If the options `loss()` and `transform()` are not specified, `mds` computes the classical solution, likewise if `method(classical)` is specified `loss()` and `transform()` are not allowed.

`method(modern)` specifies modern scaling. If `method(modern)` is specified but not `loss()` or `transform()`, then `loss(stress)` and `transform(identity)` are assumed. All values of `loss()` and `transform()` are valid with `method(modern)`.

`method(nonmetric)` specifies nonmetric scaling, which is a type of modern scaling. If `method(nonmetric)` is specified, `loss(stress)` and `transform(monotonic)` are assumed. Other values of `loss()` and `transform()` are not allowed.

`loss(`*loss*`)` specifies the loss criterion.

`loss(stress)` specifies that the stress loss function be used, normalized by the squared Euclidean distances. This criterion is often called Kruskal's stress-1. Optimal configurations for `loss(stress)` and for `loss(nstress)` are equivalent up to a scale factor, but the iteration paths may differ. `loss(stress)` is the default.

loss(nstress) specifies that the stress loss function be used, normalized by the squared disparities, that is, transformed dissimilarities. Optimal configurations for loss(stress) and for loss(nstress) are equivalent up to a scale factor, but the iteration paths may differ.

loss(sstress) specifies that the squared stress loss function be used, normalized by the fourth power of the Euclidean distances.

loss(nsstress) specifies that the squared stress criterion, normalized by the fourth power of the disparities (transformed dissimilarities) be used.

loss(strain) specifies the strain loss criterion. Classical scaling is equivalent to loss(strain) and transform(identity) but is computed by a faster noniterative algorithm. Specifying loss(strain) still allows transformations.

loss(sammon) specifies the Sammon (1969) loss criterion.

transform(*tfunction*) specifies the class of allowed transformations of the dissimilarities; transformed dissimilarities are called disparities.

transform(identity) specifies that the only allowed transformation is the identity; that is, disparities are equal to dissimilarities. transform(identity) is the default.

transform(power) specifies that disparities are related to the dissimilarities by a power function,

$$\text{disparity} = \text{dissimilarity}^{\alpha}, \qquad \alpha > 0$$

transform(monotonic) specifies that the disparities are a weakly monotonic function of the dissimilarities. This is also known as nonmetric MDS. Tied dissimilarities are handled by the primary method; that is, ties may be broken but are not necessarily broken. transform(monotonic) is valid only with loss(stress).

normalize(*norm*) specifies a normalization method for the configuration. Recall that the location and orientation of an MDS configuration is not defined ("identified"); an isometric transformation (that is, translation, reflection, or orthonormal rotation) of a configuration preserves interpoint Euclidean distances.

normalize(principal) performs a principal normalization, in which the configuration columns have zero mean and correspond to the principal components, with positive coefficient for the observation with lowest value of id(). normalize(principal) is the default.

normalize(classical) normalizes by a distance-preserving Procrustean transformation of the configuration toward the classical configuration in principal normalization; see [MV] **procrustes**. normalize(classical) is not valid if method(classical) is specified.

normalize(target(*matname*) [, copy]) normalizes by a distance-preserving Procrustean transformation toward *matname*; see [MV] **procrustes**. *matname* should be an $n \times p$ matrix, where n is the number of observations and p is the number of dimensions, and the rows of *matname* should be ordered with respect to id(). The rownames of *matname* should be set correctly but will be ignored if copy is also specified.

Note on normalize(classical) and normalize(target()): the Procrustes transformation comprises any combination of translation, reflection, and orthonormal rotation—these transformations preserve distance. Dilation (uniform scaling) would stretch distances and is not applied. However, the output reports the dilation factor, and the reported Procrustes statistic is for the dilated configuration.

dimension(#) specifies the dimension of the approximating configuration. The default # is 2 and should not exceed the number of observations; typically, # would be much smaller. With method(classical), it should not exceed the number of positive eigenvalues of the centered distance matrix.

`addconstant` specifies that if the double-centered distance matrix is not positive semidefinite (psd), a constant should be added to the squared distances to make it psd and, hence, Euclidean. `addconstant` is allowed with classical MDS only.

Model 2

`unit`[(*varlist*$_2$)] specifies variables that are transformed to min = 0 and max = 1 before entering in the computation of similarities or dissimilarities. `unit` by itself, without an argument, is a shorthand for `unit(_all)`. Variables in `unit()` should not be included in `std()`.

`std`[(*varlist*$_3$)] specifies variables that are transformed to mean = 0 and sd = 1 before entering in the computation of similarities or dissimilarities. `std` by itself, without an argument, is a shorthand for `std(_all)`. Variables in `std()` should not be included in `unit()`.

`measure(`*measure*`)` specifies the similarity or dissimilarity measure. The default is `measure(L2)`, Euclidean distance. This option is not case sensitive. See [MV] ***measure_option*** for detailed descriptions of the supported measures.

If a similarity measure is selected, the computed similarities will first be transformed into dissimilarities, before proceeding with the scaling; see the `s2d()` option below.

Classical metric MDS with Euclidean distance is equivalent to principal component analysis (see [MV] **pca**); the MDS configuration coordinates are the principal components.

`s2d(standard | oneminus)` specifies how similarities are converted into dissimilarities. By default, the command dissimilarity data. Specifying `s2d()` indicates that your proximity data are similarities.

Dissimilarity data should have zeros on the diagonal (that is, an object is identical to itself) and nonnegative off-diagonal values. Dissimilarities need not satisfy the triangular inequality, $D(i,j)^2 \leq D(i,h)^2 + D(h,j)^2$. Similarity data should have ones on the diagonal (that is, an object is identical to itself) and have off-diagonal values between zero and one. In either case, proximities should be symmetric.

The available `s2d()` options, `standard` and `oneminus`, are defined as follows:

`standard`	$\text{dissim}_{ij} = \sqrt{\text{sim}_{ii} + \text{sim}_{jj} - 2\text{sim}_{ij}} = \sqrt{2(1 - \text{sim}_{ij})}$
`oneminus`	$\text{dissim}_{ij} = 1 - \text{sim}_{ij}$

`s2d(standard)` is the default.

`s2d()` should be specified only with measures in similarity form.

Reporting

`neigen(`*#*`)` specifies the number of eigenvalues to be included in the table. The default is `neigen(10)`. Specifying `neigen(0)` suppresses the table. This option is allowed with classical MDS only.

`config` displays the table with the coordinates of the approximating configuration. This table may also be displayed using the postestimation command `estat config`; see [MV] **mds postestimation**.

`noplot` suppresses the graph of the approximating configuration. The graph can still be produced later via `mdsconfig`, which also allows the standard graphics options for fine-tuning the plot; see [MV] **mds postestimation**.

Minimization

These options are available only with `method(modern)` or `method(nonmetric)`:

`initialize(`*initopt*`)` specifies the initial values of the criterion minimization process.

`initialize(classical)`, the default, uses the solution from classical metric scaling as initial values. With `protect()`, all but the first run start from random perturbations from the classical solution. These random perturbations are independent and normally distributed with standard error equal to the product of `sdprotect(`*#*`)` and the standard deviation of the dissimilarities. `initialize(classical)` is the default.

`initialize(random)` starts an optimization process from a random starting configuration. These random configurations are generated from independent normal distributions with standard error equal to the product of `sdprotect(`*#*`)` and the standard deviation of the dissimilarities. The means of the configuration are irrelevant in MDS.

`initialize(from(`*matname*`)` [`, copy`]`)` sets the initial value to *matname*. *matname* should be an $n \times p$ matrix, where n is the number of observations and p is the number of dimensions, and the rows of *matname* should be ordered with respect to `id()`. The rownames of *matname* should be set correctly but will be ignored if `copy` is specified. With `protect()`, the second-to-last runs start from random perturbations from *matname*. These random perturbations are independent normal distributed with standard error equal to the product of `sdprotect(`*#*`)` and the standard deviation of the dissimilarities.

`tolerance(`*#*`)` specifies the tolerance for the configuration matrix. When the relative change in the configuration from one iteration to the next is less than or equal to `tolerance()`, the `tolerance()` convergence criterion is satisfied. The default is `tolerance(1e-4)`.

`ltolerance(`*#*`)` specifies the tolerance for the fit criterion. When the relative change in the fit criterion from one iteration to the next is less than or equal to `ltolerance()`, the `ltolerance()` convergence is satisfied. The default is `ltolerance(1e-8)`.

Both the `tolerance()` and `ltolerance()` criteria must be satisfied for convergence.

`iterate(`*#*`)` specifies the maximum number of iterations. The default is `iterate(1000)`.

`protect(`*#*`)` requests that # optimizations be performed and that the best of the solutions be reported. The default is `protect(1)`. See option `initialize()` on starting values of the runs. The output contains a table of the return code, the criterion value reached, and the seed of the random number used to generate the starting value. Specifying a large number, such as `protect(50)`, provides reasonable insight whether the solution found is a global minimum and not just a local minimum.

If any of the options `log`, `trace`, or `gradient` is also specified, iteration reports will be printed for each optimization run. Beware: this option will produce a lot of output.

`nolog` suppresses the iteration log, showing the progress of the minimization process.

`trace` displays the configuration matrices in the iteration report. Beware: this option may produce a lot of output.

`gradient` displays the gradient matrices of the fit criterion in the iteration report. Beware: this option may produce a lot of output.

The following option is available with `mds` but is not shown in the dialog box:

`sdprotect(`*#*`)` sets a proportionality constant for the standard deviations of random configurations (`init(random)`) or random perturbations of given starting configurations (`init(classical)` or `init(from())`). The default is `sdprotect(1)`.

Remarks

Remarks are presented under the following headings:

Introduction

Multidimensional scaling (MDS) is a dimension-reduction and visualization technique. Dissimilarities (for instance, Euclidean distances) between observations in a high-dimensional space are represented in a lower-dimensional space (typically two dimensions) so that the Euclidean distance in the lower-dimensional space approximates the dissimilarities in the higher-dimensional space. See Kruskal and Wish (1978) for a brief nontechnical introduction to MDS. Young and Hamer (1987) and Borg and Groenen (2005) offer more advanced textbook-sized treatments.

If you already have the similarities or dissimilarities of the n objects, you should continue by reading [MV] **mdsmat**.

In many applications of MDS, however, the similarity or dissimilarity of objects is not measured but rather *defined* by the researcher in terms of variables ("attributes") $x_1, \ldots, x_k$ that are measured on the objects. The pairwise dissimilarity of objects can be expressed using a variety of similarity or dissimilarity measures in the attributes (for example, Mardia, Kent, and Bibby [1979, sec. 13.4]; Cox and Cox [2001, sec. 1.3]). A common measure is the Euclidean distance L2 between the attributes of the objects i and j:

$$\texttt{L2}_{ij} = \left\{(x_{i1} - x_{j1})^2 + (x_{i2} - x_{j2})^2 + \cdots + (x_{ik} - x_{jk})^2\right\}^{1/2}$$

A popular alternative is the `L1` distance, also known as the `cityblock` or `Manhattan` distance. In comparison to `L2`, `L1` gives less influence to larger differences in attributes:

$$\texttt{L1}_{ij} = |x_{i1} - x_{j1}| + |x_{i2} - x_{j2}| + \cdots + |x_{ik} - x_{jk}|$$

In contrast, we may also define the extent of dissimilarity between 2 observations as the maximum absolute difference in the attributes and thus give a larger influence to larger differences:

$$\texttt{Linfinity}_{ij} = \max(|x_{i1} - x_{j1}|, |x_{i2} - x_{j2}|, \ldots, |x_{ik} - x_{jk}|)$$

These three measures are special cases of the Minkowski distance $L(q)$, for $q = 2$ (`L2`), $q = 1$ (`L1`), and $q = \infty$ (`Linfinity`), respectively. Minkowski distances with other values of q may be used as well. Stata supports a wide variety of other similarity and dissimilarity measures, both for continuous variables and for binary variables. See [MV] ***measure_option*** for details.

Multidimensional scaling constructs approximations for dissimilarities, not for similarities. Thus, if a similarity measure is specified, `mds` first transforms the similarities into dissimilarities. Two methods to do this are available. The default `standard` method,

$$\text{dissim}_{ij} = \sqrt{\text{sim}_{ii} - 2\text{sim}_{ij} + \text{sim}_{jj}}$$

has a useful property: if the similarity matrix is positive semidefinite, a property satisfied by most similarity measures, the standard dissimilarities are Euclidean.

Usually, the number of observations exceeds the number of variables on which the observations are compared, but this is not a requirement for MDS. MDS creates an $n \times n$ dissimilarity matrix $\mathbf{D}$ from the n observations on k variables. It then constructs an approximation of $\mathbf{D}$ by the Euclidean distances in a matching configuration $\mathbf{Y}$ of n points in p-dimensional space:

$$\text{dissimilarity}(x_i, x_j) \approx L2(y_i, y_j) \quad \text{for all } i, j$$

Typically, of course, $p << k$, and most often $p = 1$, 2, or 3.

A wide variety of MDS methods have been proposed. `mds` performs classical and modern scaling. Classical scaling has its roots in Young and Householder (1938) and Torgerson (1952). MDS requires complete and symmetric dissimilarity interval-level data. To explore modern scaling, see Borg and Groenen (2005). Classical scaling results in an eigen decomposition, whereas modern scaling is accomplished by the minimization of a loss function. Consequently, eigenvalues are not available after modern MDS.

Euclidean distances

▷ Example 1

The most popular dissimilarity measure is Euclidean distance. We illustrate with data from table 7.1 of Yang and Trewn (2004, 182). This dataset consists of eight variables with nutrition data on 25 breakfast cereals.

```
. use http://www.stata-press.com/data/r12/cerealnut
(Cereal Nutrition)

. describe

Contains data from http://www.stata-press.com/data/r12/cerealnut.dta
  obs:            25                          Cereal Nutrition
 vars:             9                          24 Feb 2011 17:19
 size:         1,050                          (_dta has notes)
-------------------------------------------------------------------------------
              storage  display     value
variable name   type   format      label      variable label
-------------------------------------------------------------------------------
brand           str25  %25s                   Cereal Brand
calories        int    %9.0g                  Calories (Cal/oz)
protein         byte   %9.0g                  Protein (g)
fat             byte   %9.0g                  Fat (g)
Na              int    %9.0g                  Na (mg)
fiber           float  %9.0g                  Fiber (g)
carbs           float  %9.0g                  Carbs (g)
sugar           byte   %9.0g                  Sugar (g)
K               int    %9.0g                  K (mg)
-------------------------------------------------------------------------------
Sorted by:
```

```
. summarize calories-K, sep(4)
```

Variable	Obs	Mean	Std. Dev.	Min	Max
calories	25	109.6	21.30728	50	160
protein	25	2.68	1.314027	1	6
fat	25	.92	.7593857	0	2
Na	25	195.8	71.32204	0	320
fiber	25	1.7	2.056494	0	9
carbs	25	15.3	4.028544	7	22
sugar	25	7.4	4.609772	0	14
K	25	90.6	77.5043	15	320

```
. replace brand = subinstr(brand," ","_",.)
(20 real changes made)
```

We replaced spaces in the cereal brand names with underscores to avoid confusing which words in the brand names are associated with which points in the graphs we are about to produce. Removing spaces is not required.

The default dissimilarity measure used by `mds` is the Euclidean distance `L2` computed on the raw data (unstandardized). The summary of the eight nutrition variables shows that `K`, `Na`, and `calories`—having much larger standard deviations—will largely determine the Euclidean distances.

```
. mds calories-K, id(brand)

Classical metric multidimensional scaling
    dissimilarity: L2, computed on 8 variables

                                            Number of obs        =        25
    Eigenvalues > 0       =         8       Mardia fit measure 1 =    0.9603
    Retained dimensions   =         2       Mardia fit measure 2 =    0.9970
```

Dimension	Eigenvalue	abs(eigenvalue) Percent	Cumul.	(eigenvalue)^2 Percent	Cumul.
1	158437.92	56.95	56.95	67.78	67.78
2	108728.77	39.08	96.03	31.92	99.70
3	10562.645	3.80	99.83	0.30	100.00
4	382.67849	0.14	99.97	0.00	100.00
5	69.761715	0.03	99.99	0.00	100.00
6	12.520822	0.00	100.00	0.00	100.00
7	5.7559984	0.00	100.00	0.00	100.00
8	2.2243244	0.00	100.00	0.00	100.00

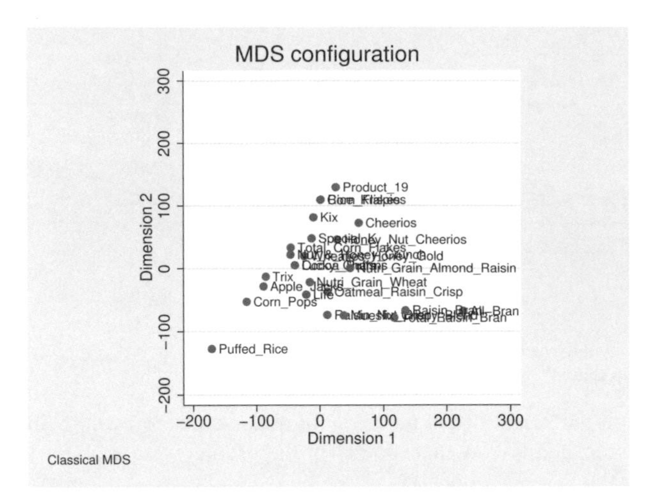

The default MDS configuration graph can be improved upon by using the `mdsconfig` postestimation command. We will demonstrate this in a moment. But first, we explain the output of `mds`.

`mds` has performed classical metric scaling and extracted two dimensions, which is the default action. To assess goodness of fit, the two statistics proposed by Mardia are reported (see Mardia, Kent, and Bibby [1979, sec. 14.4]). The statistics are defined in terms of the eigenvalues of the double-centered distance matrix. If the dissimilarities are truly Euclidean, all eigenvalues are nonnegative. Look at the eigenvalues. We may interpret these as the extent to which the dimensions account for dissimilarity between the cereals. Depending on whether you look at the eigenvalues or squared eigenvalues, it takes two or three dimensions to account for more than 99% of the dissimilarity.

We can produce a prettier configuration plot with the `mdsconfig` command; see [MV] **mds postestimation** for details.

```
. generate place = 3
. replace place = 9 if inlist(brand,"Rice_Krispies","Nut_&_Honey_Crunch",
> "Special_K","Raisin_Nut_Bran","Lucky_Charms")
(5 real changes made)
. replace place = 12 if inlist(brand,"Mueslix_Crispy_Blend")
(1 real change made)
. mdsconfig, autoaspect mlabvposition(place)
```

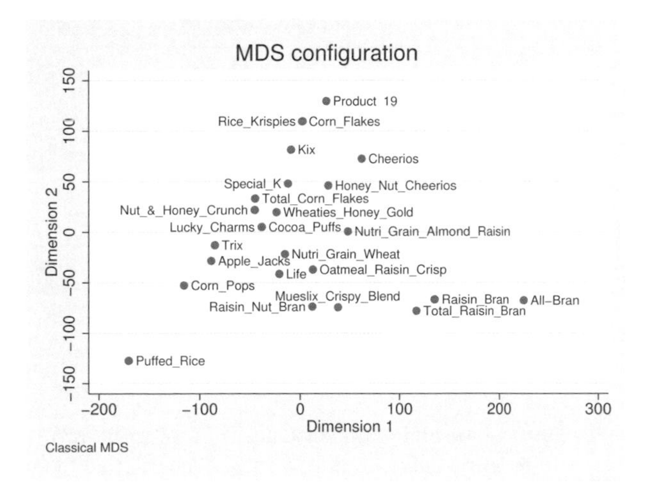

The *marker_label_option* `mlabvposition()` allowed fine control over the placement of the cereal brand names. We created a variable called `place` giving clock positions where the cereal names were to appear in relation to the plotted point. We set these to minimize overlap of the names. We also requested the `autoaspect` option to obtain better use of the graphing region while preserving the scale of the x and y axes.

MDS has placed the cereals so that all the brands fall within a triangle defined by Product 19, All-Bran, and Puffed Rice. You can examine the graph to see how close your favorite cereal is to the other cereals.

But, as we saw from the variable summary, three of the eight variables are controlling the distances. If we want to provide for a more equal footing for the eight variables, we can request that `mds` compute the Euclidean distances on standardized variables. Euclidean distance based on standardized variables is also known as the *Karl Pearson distance* (Pearson 1900). We obtain standardized measures with the option `std`.

```
. mds calories-K, id(brand) std noplot
Classical metric multidimensional scaling
    dissimilarity: L2, computed on 8 variables
                                          Number of obs        =         25
    Eigenvalues > 0       =          8    Mardia fit measure 1 =     0.5987
    Retained dimensions   =          2    Mardia fit measure 2 =     0.7697
```

Dimension	Eigenvalue	abs(eigenvalue) Percent	abs(eigenvalue) Cumul.	(eigenvalue)^2 Percent	(eigenvalue)^2 Cumul.
1	65.645395	34.19	34.19	49.21	49.21
2	49.311416	25.68	59.87	27.77	76.97
3	38.826608	20.22	80.10	17.21	94.19
4	17.727805	9.23	89.33	3.59	97.78
5	11.230087	5.85	95.18	1.44	99.22
6	8.2386231	4.29	99.47	0.78	99.99
7	.77953426	0.41	99.87	0.01	100.00
8	.24053137	0.13	100.00	0.00	100.00

In this and the previous example, we did not specify a `method()` for `mds` and got classical metric scaling. Classical scaling is the default when `method()` is omitted and neither the `loss()` nor `transform()` option is specified.

Accounting for more than 99% of the underlying distances now takes more MDS-retained dimensions. For this example, we have still retained only two dimensions. We specified the `noplot` option because we wanted to exercise control over the configuration plot by using the `mdsconfig` command. We generate a variable named pos that will help minimize cereal brand name overlap.

```
. generate pos = 3
. replace pos = 5 if inlist(brand,"Honey_Nut_Cheerios","Raisin_Nut_Bran",
> "Nutri_Grain_Almond_Raisin")
(3 real changes made)
. replace pos = 8 if inlist(brand,"Oatmeal_Raisin_Crisp")
(1 real change made)
. replace pos = 9 if inlist(brand,"Corn_Pops","Trix","Nut_&_Honey_Crunch",
> "Rice_Krispies","Wheaties_Honey_Gold")
(5 real changes made)
. replace pos = 12 if inlist(brand,"Life")
(1 real change made)
```

```
. mdsconfig, autoaspect mlabvpos(pos)
```

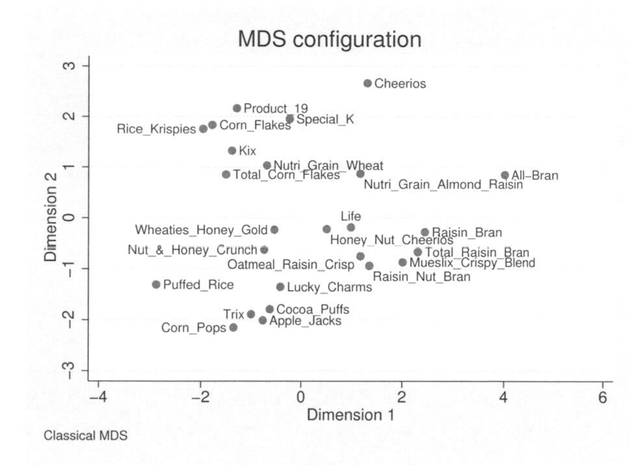

This configuration plot, based on the standardized variables, better incorporates all the nutrition data. If you are familiar with these cereal brands, spotting groups of similar cereals appearing near each other is easy. The bottom-left corner has several of the most sweetened cereals. The brands containing the word "Bran" all appear to the right of center. Rice Krispies and Puffed Rice are the farthest to the left.

Classical multidimensional scaling based on standardized Euclidean distances is actually equivalent to a principal component analysis of the correlation matrix of the variables. See Mardia, Kent, and Bibby (1979, sec. 14.3) for details.

We now demonstrate this property by doing a principal component analysis extracting the leading two principal components. See [MV] **pca** for details.

```
. pca calories-K, comp(2)
Principal components/correlation              Number of obs    =        25
                                              Number of comp.  =         2
                                              Trace            =         8
    Rotation: (unrotated = principal)         Rho              =    0.5987
```

Component	Eigenvalue	Difference	Proportion	Cumulative
Comp1	2.73522	.680583	0.3419	0.3419
Comp2	2.05464	.436867	0.2568	0.5987
Comp3	1.61778	.879117	0.2022	0.8010
Comp4	.738659	.270738	0.0923	0.8933
Comp5	.46792	.124644	0.0585	0.9518
Comp6	.343276	.310795	0.0429	0.9947
Comp7	.0324806	.0224585	0.0041	0.9987
Comp8	.0100221	.	0.0013	1.0000

```
Principal components (eigenvectors)
```

Variable	Comp1	Comp2	Unexplained
calories	0.1992	-0.0632	.8832
protein	0.3376	0.4203	.3253
fat	0.3811	-0.0667	.5936
Na	0.0962	0.5554	.3408
fiber	0.5146	0.0913	.2586
carbs	-0.2574	0.4492	.4043
sugar	0.2081	-0.5426	.2765
K	0.5635	0.0430	.1278

The proportion and cumulative proportion of the eigenvalues in the PCA match the percentages from MDS. We will ignore the interpretation of the principal components but move directly to the principal coordinates, also known as the scores of the PCA. We make a plot of the first and second scores, using the `scoreplot` command; see [MV] **scoreplot**. We specify the `mlabel()` option to label the cereals and the `mlabvpos()` option for fine control over placement of the brand names.

```
. replace pos = 11 if inlist(brand,"All-Bran")
(1 real change made)
. scoreplot, mlabel(brand) mlabvpos(pos)
```

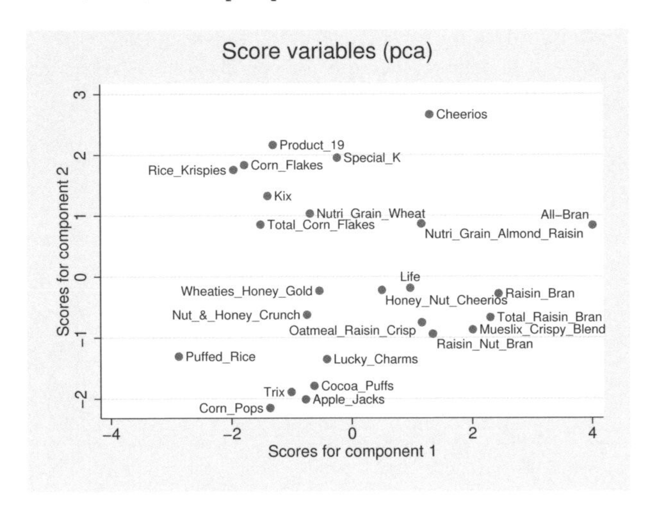

Compare this PCA score plot with the MDS configuration plot. Apart from some differences in how the graphs were rendered, they are the same.

◁

Non-Euclidean dissimilarity measures

With non-Euclidean dissimilarity measures, the parallel between PCA and MDS no longer holds.

▷ Example 2

To illustrate MDS with non-Euclidean distance measures, we will analyze books on multivariate statistics. Gifi (1990) reports on the number of pages devoted to six topics in 20 textbooks on multivariate statistics. We added similar data on five more recent books.

```
. use http://www.stata-press.com/data/r12/mvstatsbooks, clear
. describe
Contains data from http://www.stata-press.com/data/r12/mvstatsbooks.dta
  obs:            25
 vars:             8                          15 Mar 2011 16:27
 size:           725                          (_dta has notes)

              storage  display     value
variable name   type   format      label      variable label

author          str17  %17s
math            int    %9.0g                  math other than statistics
                                                (e.g., linear algebra)
corr            int    %9.0g                  correlation and regression,
                                                including linear structural and
                                                functional equations
fact            byte   %9.0g                  factor analysis and principal
                                                component analysis
cano            byte   %9.0g                  canonical correlation analysis
disc            int    %9.0g                  discriminant analysis,
                                                classification, and cluster
                                                analysis
stat            int    %9.0g                  statistics, incl. dist. theory,
                                                hypothesis testing & est.;
                                                categorical data
mano            int    %9.0g                  manova and the general linear
                                                model

Sorted by:
```

A brief description of the topics is given in the variable labels. For more details, we refer to Gifi (1990, 15). Here are the data:

```
. list, noobs
```

author	math	corr	fact	cano	disc	stat	mano
Roy57	31	0	0	0	0	164	11
Kendall57	0	16	54	18	27	13	14
Kendall75	0	40	32	10	42	60	0
Anderson58	19	0	35	19	28	163	52
CooleyLohnes62	14	7	35	22	17	0	56
(output omitted)							
GreenCaroll76	290	10	6	0	8	0	2
CailliezPages76	184	48	82	42	134	0	0
Giri77	29	0	0	0	41	211	32
Gnanadesikan77	0	19	56	0	39	75	0
Kshirsagar78	0	22	45	42	60	230	59
Thorndike78	30	128	90	28	48	0	0
MardiaKentBibby79	34	28	68	19	67	131	55
Seber84	16	0	59	13	116	129	101
Stevens96	23	87	67	21	30	43	249
EverittDunn01	0	54	65	0	56	20	30
Rencher02	38	0	71	19	105	135	131

For instance, the 1979 book by Mardia, Kent, and Bibby has 34 pages on mathematics (mostly linear algebra); 28 pages on correlation, regression, and related topics (in this particular case, simultaneous equations); etc. In most of these books, some pages are not classified. Anyway, the number of pages and the amount of information per page vary widely among the books. A Euclidean distance measure is not appropriate here. Standardization does not help us here—the problem is not differences in the scales of the variables but those in the observations. One possibility is to transform the data into *compositional data* by dividing the variables by the total number of classified pages. See Mardia, Kent, and Bibby (1979, 377–380) for a discussion of specialized dissimilarity measures for compositional data. However, we can also use the correlation between observations (not between variables) as the similarity measure. The higher the correlation between the attention given to the various topics, the more similar two textbooks are. We do a classical MDS, suppressing the plot to first assess the quality of a two-dimensional representation.

```
. mds math-mano, id(author) measure(corr) noplot
Classical metric multidimensional scaling
      similarity: correlation, computed on 7 variables
   dissimilarity: sqrt(2(1-similarity))
                                          Number of obs          =        25
   Eigenvalues > 0       =          6     Mardia fit measure 1 =     0.6680
   Retained dimensions   =          2     Mardia fit measure 2 =     0.8496
```

Dimension	Eigenvalue	abs(eigenvalue) Percent	Cumul.	(eigenvalue)^2 Percent	Cumul.
1	8.469821	38.92	38.92	56.15	56.15
2	6.0665813	27.88	66.80	28.81	84.96
3	3.8157101	17.53	84.33	11.40	96.35
4	1.6926956	7.78	92.11	2.24	98.60
5	1.2576053	5.78	97.89	1.24	99.83
6	.45929376	2.11	100.00	0.17	100.00

Again the quality of a two-dimensional approximation is somewhat unsatisfactory, with 67% and 85% of the variation accounted for according to the two Mardia criteria. Still, let's look at the plot, using a title that refers to the self-referential aspect of the analysis (Smullyan 1986). We reposition some of the author labels to enhance readability by using the `mlabvpos()` option.

```
. gen spot = 3
. replace spot = 5 if inlist(author,"Seber84","Kshirsagar78","Kendall75")
(3 real changes made)
. replace spot = 2 if author=="MardiaKentBibby79"
(1 real change made)
. replace spot = 9 if inlist(author, "Dagnelie75","Rencher02",
> "GreenCaroll76","EverittDunn01","CooleyLohnes62","Morrison67")
(6 real changes made)
. mdsconfig, mlabvpos(spot) title(This plot needs no title)
```

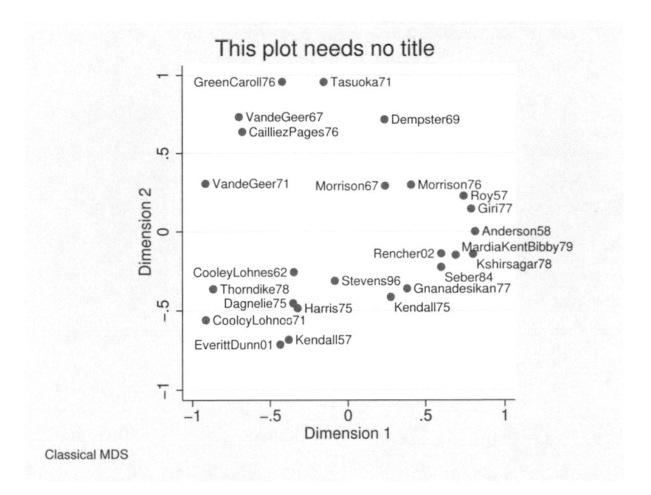

A striking characteristic of the plot is that the textbooks seem to be located on a circle. This is a phenomenon that is regularly encountered in multidimensional scaling and was labeled the "horseshoe effect" by Kendall (1971, 215–251). This phenomenon seems to occur especially in situations in which a one-dimensional representation of objects needs to be constructed, for example, in *seriation* applications, from data in which small dissimilarities were measured accurately but moderate and larger dissimilarities are "lumped together".

◁

❑ Technical note

These data could also be analyzed differently. A particularly interesting method is correspondence analysis (CA), which seeks a simultaneous geometric representation of the rows (textbooks) and columns (topics). We used `camat` to analyze these data. The results for the textbooks were not much different. Textbooks that were mapped as similar using MDS were also mapped this way by CA. The Green and Carroll book that appeared much different from the rest was also displayed away from the rest by CA. In the CA biplot, it was immediately clear that this book was so different because its pages were classified by Gifi (1990) as predominantly mathematical. But CA also located the topics in this space. The pattern was easy to interpret and was expected. The seven topics were mapped in three groups. `math` and `stat` appear as two groups by themselves, and the five applied topics were mapped close together. See [MV] **ca** for information on the `ca` command.

❑

Introduction to modern MDS

We return to the data on breakfast cereals explored above to introduce modern MDS. We repeat some steps taken previously and then perform estimation using options `loss(strain)` and `transform(identity)`, which we demonstrate are equivalent to classical MDS.

`mds` is an estimation or `eclass` command, see `program define` in [P] **program**. You can display its saved results using `ereturn list`. The configuration is saved as `e(Y)` and we will compare the configuration obtained from classical MDS with the equivalent one from modern MDS.

▷ Example 3

```
. use http://www.stata-press.com/data/r12/cerealnut, clear
(Cereal Nutrition)
. replace brand = subinstr(brand," ","_",.)
(20 real changes made)
. quietly mds calories-K, id(brand) noplot
. mat Yclass = e(Y)
. mds calories-K, id(brand) meth(modern) loss(strain) trans(ident) noplot
Iteration 1:   strain =  594.12657
Iteration 2:   strain =  594.12657
Modern multidimensional scaling
    dissimilarity: L2, computed on 8 variables
    Loss criterion: strain = loss for classical MDS
    Transformation: identity (no transformation)
                                              Number of obs    =        25
                                              Dimensions       =         2
    Normalization: principal                  Loss criterion   =  594.1266
. mat Ymod = e(Y)
. assert mreldif(Yclass, Ymod) < 1e-6
```

Note the output differences between modern and classical MDS. In modern MDS we have an iteration log from the minimization of the loss function. The method, measure, observations, dimensions, and number of variables are reported as before, but we do not have or display eigenvalues. The normalization is always reported in modern MDS and with `normalize(target())` for classical MDS. The loss criterion is simply the value of the loss function at the minimum.

◁

Protecting from local minimums

Modern MDS can sometimes converge to a local rather than a global minimum. To protect against this, multiple runs can be made, giving the best of the runs as the final answer. The option for performing this is `protect(#)`, where # is the number of runs to be performed. The `nolog` option is of particular use with `protect()`, because the iteration logs from the runs will create a lot of output. Repeating the minimization can take some time, depending on the number of runs selected and the number of iterations it takes to converge.

▷ Example 4

We choose `loss(stress)`, and `transform(identity)` is assumed with modern MDS. We omit the iteration logs to avoid a large amount of output. The number of iterations is available after estimation in `e(ic)`. We first do a run without the `protect()` option, and then we use `protect(50)` and compare our results.

```
. mds calories-K, id(brand) method(modern) loss(stress) nolog noplot
(transform(identity) assumed)
Modern multidimensional scaling
    dissimilarity: L2, computed on 8 variables
    Loss criterion: stress = raw_stress/norm(distances)
    Transformation: identity (no transformation)
                                                    Number of obs    =        25
                                                    Dimensions       =         2
    Normalization: principal                        Loss criterion   =    0.0263
. di e(ic)
45
. mat Ystress = e(Y)
. set seed 123456789
. mds calories-K, id(brand) method(modern) loss(stress) nolog protect(50)
(transform(identity) assumed)
run  mrc  #iter     lossval   seed random configuration
--------------------------------------------------------------------
  1    0     74   .02626681   Xdb3578617ea24d1bbd210b2e6541937c4bf2
  2    0    101   .02626681   X630dd08906daab6562fdcb6e21f566d13abb
  3    0     78   .02626681   X73b43d6df67516aea87005b9d405c7020c2c
  4    0     75   .02626681   Xc9913bd7b3258bd7215929fe40ee51c701be
  5    0     75   .02626681   X5fba1fdfb5219b6fd11daa1e739f9a3e0851
  6    0     57   .02626681   X3ea2b7553d633a5418c29f0da0b5cbf433c6
  7    0     84   .02626681   Xeb0a27a9aa7351c262bca0dccbf5dc8e0693
  8    0     75   .02626681   Xcb99c7f17ce97d3e7fa9d27bafc16ab60f2a
  9    0     85   .02626681   X8551c69beb028bbd48c91c1a1b6e6f0e2b08
 10    0     60   .02626681   X05d89a191a939001044c3710631948c12c41
 11    0     63   .02626681   X2c6eaeaf4dcd2394c628f466db148b3419c0
 12    0     45   .02626681       <initial nonrandom value>
 13    0     55   .02626681   X167c0bd9c43f462544a474abacbdd93d03c8
 14    0     57   .02626682   X51b9b6c5e05aadd50ca4b924a252124048e1
 15    0     82   .02626682   Xff9280cf913270440939762f65c2b4d622da
 16    0     63   .02626682   X14f4e343d3e32b22014308b4d2407e8949e3
 17    0     63   .02626682   X1b06e2ef52b30203908f0d10327044174a08
 18    0     66   .02626682   X70ceaf639c2f78374fd6a1181468489e3288
 19    0     72   .02626682   X5fa530eb49912716c3b27b00020b158c16cc
 20    0     71   .02626682   Xf2f8a723276c4a1f3c7e5848c07cc438343e
 21    0     52   .02626682   Xd5f821b18557b512b3ebc04c992e06b4115d
 22    0     66   .02626683   Xd0de57fd1b0a448b3450528326fab45d1681
 23    0     61   .02626683   X6d2da541fcdb0d9024a0a92d5d0496231d51
 24    0     59   .02626683   Xb8aae22160bc2fd1beaf5a4b98f46a254a25
 25    0     84   .02626684   Xe2b46cd1199d7ef41b8cd31479b25b274bdb
 26    0    138     .026303   Xb21a3be5f19dad75f7708eb425730c6e45b0
 27    0    100     .026303   X22ef41a50a68221b276cd98ee7dfeef51073
 28    0     74     .026303   X0efbcec71dbb1b3c7111cb0f622502830b54
 29    0     55     .026303   X01c889699f835483fd6182719be301f13e96
 30    0     56     .026303   X2f66c0cb554aca4ff44e5d6a6cd4b6273931
 31    0     67     .026303   X3a5b1f132e05f86d36e01eb46eff578b3d24
 32    0     67     .026303   Xac0226dd85d4f2c440745210acea6ceb12a4
 33    0     75     .026303   X9e59768de92f8d2ab8e9bc0fd0084c7d10ea
 34    0     58     .026303   X333e991d0bf2f21be1025e348a6825470b6c
 35    0     60     .026303   Xedef05bfdbdcddd5a2b8abeadcdd5ab74c9c
 36    0     59     .026303   X67e0caf9c38ba588e96cd01d5d908d7f022c
 37    0     53     .026303   X2af205b7aad416610a0dec141b66778a2eee
 38    0     52     .026303   X0b9944753b1c4b3bd3676f624643a915319c
 39    0     87     .026303   Xb175975333f6bdee5bc301e7d30556882042
 40    0     63   .02630301   X7e334ce7d25be1deb7b30539d716026639ef
 41    0     60   .02630301   Xf2e6bfadef621544c441e8363c853045203e
 42    0     60   .02630301   X45c7e0abd63ad668fa94cd4758d974eb2635
 43    0     58   .02630301   X60263a35772a812860431439cad14ad92943
```

```
   44    0     66    .02630301    X4bf3debb1c7e07f66b533ec5941e1e07433b
   45    0     63    .02630301    X01f186db4f0db540e749c79e59717c18247e
   46    0     56    .02630302    X66a301f734b575da6762a4edcf9ac6492715
   47    0     53    .02630302    X5c59c9ffd2e9f2e5bd45f3f9aa22b2f027b7
   48    0    131    .19899027    Xe2e15b07d97b0bcb086f194a133dd7b23f52
   49    0    138    .23020403    X065b9333ce65d69bf4d1596e8e8cc72904ef
   50    0    170    .23794378    X075bcd151f123bb5159a55e50022865746ad

Modern multidimensional scaling
    dissimilarity: L2, computed on 8 variables
    Loss criterion: stress = raw_stress/norm(distances)
    Transformation: identity (no transformation)
                                                Number of obs    =        25
                                                Dimensions       =         2
    Normalization: principal                    Loss criterion   =    0.0263
. mat YstressP = e(Y)
. assert mreldif(Ystress, YstressP) < 2e-3
```

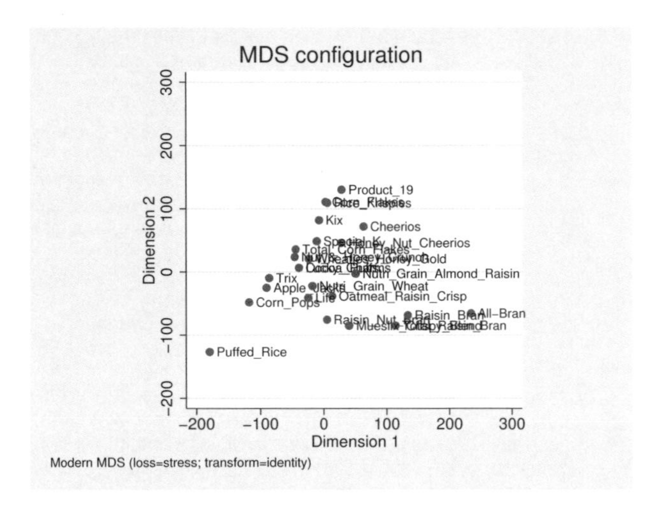

The output provided when `protect()` is specified includes a table with information on each run, sorted by the loss criterion value. The first column simply counts the runs. The second column gives the internal return code from modern MDS. This example only has values of 0, which indicate converged results. The column header `mrc` is clickable and opens a help file explaining the various MDS return codes. The number of iterations is in the third column. These runs converged in as few as 47 iterations to as many as 190. The loss criterion values are in the fourth column, and the final column contains the seeds used to calculate the starting values for the runs.

In this example, the results from our original run versus the protected run did not differ by much, approximately 1.3e–3. However, looking at runs 46–50 we see loss criterion values that are much higher than the rest. The loss criteria for runs 1–45 vary from .02627 to .02630, but these last runs' loss criteria are all more than .198. These runs clearly converged to local, not global, minimums.

The graph from this protected modern MDS run may be compared with the first one produced. There are obvious similarities, though inspection indicates that the two are not the same.

◁

Saved results

mds saves the following in e():

Scalars

e(N)	number of observations
e(p)	number of dimensions in the approximating configuration
e(np)	number of strictly positive eigenvalues
e(addcons)	constant added to squared dissimilarities to force positive semidefiniteness
e(mardia1)	Mardia measure 1
e(mardia2)	Mardia measure 2
e(critval)	loss criterion value
e(alpha)	parameter of transform(power)
e(ic)	iteration count
e(rc)	return code
e(converged)	1 if converged, 0 otherwise

Macros

e(cmd)	mds
e(cmdline)	command as typed
e(method)	classical or modern MDS method
e(method2)	nonmetric, if method(nonmetric)
e(loss)	loss criterion
e(losstitle)	description loss criterion
e(tfunction)	identity, power, or monotonic, transformation function
e(transftitle)	description of transformation
e(id)	ID variable name (mds)
e(idtype)	int or str; type of id() variable
e(duplicates)	1 if duplicates in id(), 0 otherwise
e(labels)	labels for ID categories
e(strfmt)	format for category labels
e(mxlen)	maximum length of category labels
e(varlist)	variables used in computing similarities or dissimilarities
e(dname)	similarity or dissimilarity measure name
e(dtype)	similarity or dissimilarity
e(s2d)	standard or oneminus (when e(dtype) is similarity)
e(unique)	1 if eigenvalues are distinct, 0 otherwise
e(init)	initialization method
e(iseed)	seed for init(random)
e(seed)	seed for solution
e(norm)	normalization method
e(targetmatrix)	name of target matrix for normalize(target)
e(properties)	nob noV for modern or nonmetric MDS; nob noV eigen for classical MDS
e(estat_cmd)	program used to implement estat
e(predict)	program used to implement predict
e(marginsnotok)	predictions disallowed by margins

Matrices	
e(D)	dissimilarity matrix
e(Disparities)	disparity matrix for nonmetric MDS
e(Y)	approximating configuration coordinates
e(Ev)	eigenvalues
e(idcoding)	coding for integer identifier variable
e(coding)	variable standardization values; first column has value to subtract and second column has divisor
e(norm_stats)	normalization statistics
e(linearf)	two element vector defining the linear transformation; distance equals first element plus second element times dissimilarity
Functions	
e(sample)	marks estimation sample

Methods and formulas

mds is implemented as an ado-file.

mds creates a dissimilarity matrix **D** according to the *measure* specified in option measure(). See [MV] ***measure_option*** for descriptions of these measures. Subsequently, mds uses the same subroutines as mdsmat to compute the MDS solution for **D**. See *Methods and formulas* in [MV] **mdsmat** for information.

References

Borg, I., and P. J. F. Groenen. 2005. *Modern Multidimensional Scaling: Theory and Applications.* 2nd ed. New York: Springer.

Corten, R. 2011. Visualization of social networks in Stata using multidimensional scaling. *Stata Journal* 11: 52–63.

Cox, T. F., and M. A. A. Cox. 2001. *Multidimensional Scaling.* 2nd ed. Boca Raton, FL: Chapman & Hall/CRC.

Gifi, A. 1990. *Nonlinear Multivariate Analysis.* New York: Wiley.

Kendall, D. G. 1971. Seriation from abundance matrices. In *Mathematics in the Archaeological and Historical Sciences.* Edinburgh: Edinburgh University Press.

Kruskal, J. B., and M. Wish. 1978. *Multidimensional Scaling.* Newbury Park, CA: Sage.

Lingoes, J. C. 1971. Some boundary conditions for a monotone analysis of symmetric matrices. *Psychometrika* 36: 195–203.

Mardia, K. V., J. T. Kent, and J. M. Bibby. 1979. *Multivariate Analysis.* London: Academic Press.

Pearson, K. 1900. On the criterion that a given system of deviations from the probable in the case of a correlated system of variables is such that it can be reasonably supposed to have arisen from random sampling. *Philosophical Magazine,* Series 5 50: 157–175.

Sammon, J. W., Jr. 1969. A nonlinear mapping for data structure analysis. *IEEE Transactions on Computers* 18: 401–409.

Smullyan, R. 1986. *This Book Needs No Title: A Budget of Living Paradoxes.* New York: Touchstone.

Torgerson, W. S. 1952. Multidimensional scaling: I. Theory and method. *Psychometrika* 17: 401–419.

Yang, K., and J. Trewn. 2004. *Multivariate Statistical Methods in Quality Management.* New York: McGraw–Hill.

Young, F. W., and R. M. Hamer. 1987. *Multidimensional Scaling: History, Theory, and Applications.* Hillsdale, NJ: Erlbaum Associates.

Young, G., and A. S. Householder. 1938. Discussion of a set of points in terms of their mutual distances. *Psychometrika* 3: 19–22.

Also see *References* in [MV] **mdsmat**.

Joseph Bernard Kruskal (1928–2010) was born in New York. His brothers were statistician William Henry Kruskal (1919–2005) and mathematician and physicist Martin David Kruskal (1925–2006). He earned degrees in mathematics from Chicago and Princeton and worked at Bell Labs until his retirement in 1993. In statistics, Kruskal made major contributions to multidimensional scaling. In computer science, he devised an algorithm for computing the minimal spanning tree of a weighted graph. His other interests include clustering and statistical linguistics.

Also see

[MV] **mds postestimation** — Postestimation tools for mds, mdsmat, and mdslong

[MV] **biplot** — Biplots

[MV] **ca** — Simple correspondence analysis

[MV] **factor** — Factor analysis

[MV] **mdslong** — Multidimensional scaling of proximity data in long format

[MV] **mdsmat** — Multidimensional scaling of proximity data in a matrix

[MV] **pca** — Principal component analysis

[U] **20 Estimation and postestimation commands**

Title

mds postestimation — Postestimation tools for mds, mdsmat, and mdslong

Description

The following postestimation commands are of special interest after mds, mdslong, and mdsmat:

Command	Description
estat config	coordinates of the approximating configuration
estat correlations	correlations between dissimilarities and approximating distances
estat pairwise	pairwise dissimilarities, approximating distances, and raw residuals
estat quantiles	quantiles of the residuals per object
estat stress	Kruskal stress (loss) measure (only after classical MDS)
*estat summarize	estimation sample summary
mdsconfig	plot of approximating configuration
mdsshepard	Shepard diagram
screeplot	plot eigenvalues (only after classical MDS)

* estat summarize is not available after mdsmat.

For more information on these commands, except screeplot, see below. For information on screeplot, see [MV] **screeplot**.

The following standard postestimation commands are also available:

Command	Description
*estimates	cataloging estimation results
predict	approximating configuration, disparities, dissimilarities, distances, and residuals

* All estimates subcommands except table and stats are available.

See the corresponding entries in the *Base Reference Manual* for more information.

Special-interest postestimation commands

estat config lists the coordinates of the approximating configuration.

estat correlations lists the Pearson and Spearman correlations between the disparities or dissimilarities and the Euclidean distances for each object.

estat pairwise lists the pairwise statistics: the disparities, the distances, and the residuals.

estat quantiles lists the quantiles of the residuals per object.

estat stress displays the Kruskal stress (loss) measure between the (transformed) dissimilarities and fitted distances per object (only after classical MDS).

estat summarize summarizes the variables in the MDS over the estimation sample. After mds, estat summarize also reports whether and how variables were transformed before computing similarities or dissimilarities.

mdsconfig produces a plot of the approximating Euclidean configuration. By default, dimensions 1 and 2 are plotted.

mdsshepard produces a Shepard diagram of the disparities against the Euclidean distances. Ideally, the points in the plot should be close to the $y = x$ line. Optionally, separate plots are generated for each "row" (value of id()).

> Roger Newland Shepard (1929–) was born in Palo Alto, California, earned degrees at Stanford and Yale, and worked at Bell Labs and Harvard before returning to Stanford in 1968. One of the world's leading psychologists and cognitive scientists, he has worked on perception, mental imagery, representation, learning, and generalization. Shepard is noted within statistical science primarily for his work on nonmetric multidimensional scaling. He is a member of the U.S. National Academy of Sciences and a recipient of the National Medal of Science.

Syntax for predict

predict [*type*] {*stub** | *newvarlist*} [*if*] [*in*] [, *statistic options*]

statistic	Description
Main	
config	approximating configuration; specify dimension() or fewer variables
pairwise(*pstats*)	selected pairwise statistics; specify same number of variables

pstats	Description
disparities	disparities = transformed(dissimilarities)
dissimilarities	dissimilarities
distances	Euclidean distances between configuration points
rresiduals	raw residual = dissimilarity − distance
tresiduals	transformed residual = disparity − distance
weights	weights

options	Description
Main	
* saving(*filename*, replace)	save results to *filename*; use replace to overwrite existing *filename*
full	create predictions for all pairs of objects; pairwise() only

* saving() is required after mdsmat, after mds if pairwise() is selected, and after mdslong if config is selected.

Menu

Statistics > Postestimation > Predictions, residuals, etc.

Options for predict

Main

`config` generates variables containing the approximating configuration in Euclidean space. Specify as many new variables as approximating dimensions (as determined by the `dimension()` option of `mds`, `mdsmat`, or `mdslong`), though you may specify fewer. `estat config` displays the same information but does not store the information in variables. After `mdsmat` and `mdslong`, you must also specify the `saving()` option.

`pairwise(`*pstats*`)` generates new variables containing pairwise statistics. The number of new variables should be the same as the number of specified statistics. The following statistics are allowed:

`disparities` generates the disparities, that is, the transformed dissimilarities. If no transformation is applied (modern MDS with `transform(identity)`), disparities are the same as dissimilarities.

`dissimilarities` generates the dissimilarities used in MDS. If `mds`, `mdslong`, or `mdsmat` was invoked on similarity data, the associated dissimilarities are returned.

`distances` generates the (unsquared) Euclidean distances between the fitted configuration points.

`rresiduals` generates the raw residuals: dissimilarities − distances.

`tresiduals` generates the transformed residuals: disparities − distances.

`weights` generates the weights. Missing proximities are represented by zero weights.

`estat pairwise` displays some of the same information but does not store the information in variables.

After `mds` and `mdsmat`, you must also specify the `saving()` option. With n objects, the pairwise dataset has $n(n-1)/2$ observations. In addition to the three requested variables, `predict` produces variables *id*`1` and *id*`2`, which identify pairs of objects. With `mds`, *id* is the name of the identification variable (`id()` option), and with `mdsmat` it is "`Category`".

`saving(`*filename* [`, replace`]`)` is required after `mdsmat`, after `mds` if `pairwise` is selected, and after `mdslong` if `config` is selected. `saving()` indicates that the generated variables are to be created in a new Stata dataset and saved in the file named *filename*. Unless `saving()` is specified, the variables are generated in the current dataset.

`replace` indicates that *filename* specified in `saving()` may be overwritten.

`full` creates predictions for all pairs of objects (j_1, j_2). The default is to generate predictions only for pairs (j_1, j_2) where $j_1 > j_2$. `full` may be specified only with `pairwise`.

Syntax for estat

List the coordinates of the approximating configuration

`estat config` [`, maxlength(`*#*`) format(%`*fmt*`)`]

List the Pearson and Spearman correlations

`estat correlations` [`, maxlength(`*#*`) format(%`*fmt*`) notransform nototal`]

List the pairwise statistics: disparities, distances, and residuals

```
estat pairwise [, maxlength(#) notransform full separator]
```

List the quantiles of the residuals

```
estat quantiles [, maxlength(#) format(%fmt) nototal notransform]
```

Display the Kruskal stress (loss) measure per point (only after classical MDS)

```
estat stress [, maxlength(#) format(%fmt) nototal notransform]
```

Summarize the variables in MDS

```
estat summarize [, labels]
```

options	Description
`maxlength(#)`	maximum number of characters for displaying object names; default is 12
`format(%fmt)`	display format
`nototal`	suppress display of overall summary statistics
`notransform`	use dissimilarities instead of disparities
`full`	display all pairs (j_1, j_2); default is $(j_1 > j_2)$ only
`separator`	draw separating lines
`labels`	display variable labels

Menu

Statistics > Postestimation > Reports and statistics

Options for estat

`maxlength(#)`, an option used with all but `estat summarize`, specifies the maximum number of characters of the object names to be displayed; the default is `maxlength(12)`.

`format(%fmt)`, an option used with `estat config`, `estat correlations`, `estat quantiles`, and `estat stress`, specifies the display format; the default differs between the subcommands.

`nototal`, an option used with `estat correlations`, `estat quantiles`, and `estat stress`, suppresses the overall summary statistics.

`notransform`, an option used with `estat correlations`, `estat pairwise`, `estat quantiles`, and `estat stress`, specifies that the untransformed dissimilarities be used instead of the transformed dissimilarities (disparities).

`full`, an option used with `estat pairwise`, displays a row for all pairs (j_1, j_2). The default is to display rows only for pairs where $j_1 > j_2$.

`separator`, an option used with `estat pairwise`, draws separating lines between blocks of rows corresponding to changes in the first of the pair of objects.

`labels`, an option used with `estat summarize`, displays variable labels.

Syntax for mdsconfig

mdsconfig [, *options*]

options	Description
Main	
dimensions(# #)	two dimensions to be displayed; default is dimensions(2 1)
xnegate	negate data relative to the x axis
ynegate	negate data relative to the y axis
autoaspect	adjust aspect ratio on the basis of the data; default aspect ratio is 1
maxlength(#)	maximum number of characters used in marker labels
cline_options	affect rendition of the lines connecting points
marker_options	change look of markers (color, size, etc.)
marker_label_options	change look or position of marker labels
Y axis, X axis, Titles, Legend, Overall	
twoway_options	any options other than by() documented in [G-3] ***twoway_options***

Menu

Statistics > Multivariate analysis > Multidimensional scaling (MDS) > Postestimation > Approximating configuration plot

Options for mdsconfig

Main

dimensions(# #) identifies the dimensions to be displayed. For instance, dimensions(3 2) plots the third dimension (vertically) versus the second dimension (horizontally). The dimension number cannot exceed the number of extracted dimensions. The default is dimensions(2 1).

xnegate specifies that the data be negated relative to the x axis.

ynegate specifies that the data be negated relative to the y axis.

autoaspect specifies that the aspect ratio be automatically adjusted based on the range of the data to be plotted. This option can make some plots more readable. By default, mdsconfig uses an aspect ratio of one, producing a square plot. Some plots will have little variation in the y-axis direction, and use of the autoaspect option will better fill the available graph space while preserving the equivalence of distance in the x and y axes.

As an alternative to autoaspect, the *twoway_option* aspectratio() can be used to override the default aspect ratio. mdsconfig accepts the aspectratio() option as a suggestion only and will override it when necessary to produce plots with balanced axes; that is, distance on the x axis equals distance on the y axis.

twoway_options, such as xlabel(), xscale(), ylabel(), and yscale(), should be used with caution. These *axis_options* are accepted but may have unintended side effects on the aspect ratio. See [G-3] ***twoway_options***.

maxlength(#) specifies the maximum number of characters for object names used to mark the points; the default is maxlength(12).

cline_options affect the rendition of the lines connecting the plotted points; see [G-3] ***cline_options***. If you are drawing connected lines, the appearance of the plot depends on the sort order of the data.

marker_options affect the rendition of the markers drawn at the plotted points, including their shape, size, color, and outline; see [G-3] ***marker_options***.

marker_label_options specify if and how the markers are to be labeled; see [G-3] ***marker_label_options***.

Y axis, X axis, Titles, Legend, Overall

twoway_options are any of the options documented in [G-3] ***twoway_options***, excluding `by()`. These include options for titling the graph (see [G-3] ***title_options***) and for saving the graph to disk (see [G-3] ***saving_option***). See `autoaspect` above for a warning against using options such as `xlabel()`, `xscale()`, `ylabel()`, and `yscale()`.

Syntax for mdsshepard

`mdsshepard` [, *options*]

options	Description
Main	
`notransform`	use dissimilarities instead of disparities
`autoaspect`	adjust aspect ratio on the basis of the data; default aspect ratio is 1
`separate`	draw separate Shepard diagrams for each object
marker_options	change look of markers (color, size, etc.)
Y axis, X axis, Titles, Legend, Overall	
twoway_options	any options other than `by()` documented in [G-3] ***twoway_options***
`byopts(`*by_option*`)`	affect the rendition of combined graphs; `separate` only

Menu

Statistics > Multivariate analysis > Multidimensional scaling (MDS) > Postestimation > Shepard diagram

Options for mdsshepard

Main

`notransform` uses dissimilarities instead of disparities, that is, suppresses the transformation of the dissimilarities.

`autoaspect` specifies that the aspect ratio be automatically adjusted based on the range of the data to be plotted. By default, `mdsshepard` uses an aspect ratio of one, producing a square plot.

See the description of the `autoaspect` option of `mdsconfig` for more details.

`separate` displays separate plots of each value of the ID variable. This may be time consuming if the number of distinct ID values is not small.

marker_options affect the rendition of the markers drawn at the plotted points, including their shape, size, color, and outline; see [G-3] ***marker_options***.

Y axis, X axis, Titles, Legend, Overall

twoway_options are any of the options documented in [G-3] ***twoway_options***, excluding by(). These include options for titling the graph (see [G-3] ***title_options***) and for saving the graph to disk (see [G-3] ***saving_option***). See the autoaspect option of mdsconfig for a warning against using options such as xlabel(), xscale(), ylabel(), and yscale().

byopts(*by_option*) is documented in [G-3] ***by_option***. This option affects the appearance of the combined graph and is allowed only with the separate option.

Remarks

Remarks are presented under the following headings:

Postestimation statistics
Matching configuration plot and the Shepard diagram
Predictions

Postestimation statistics

After an MDS analysis, several facilities can help you better understand the analysis and, in particular, to assess the quality of the lower-dimensional Euclidean representation. We display results after classical MDS. All are available after modern MDS except for estat stress.

▷ Example 1

We illustrate the MDS postestimation facilities with the Morse code digit-similarity dataset; see example 1 in [MV] **mdslong**.

```
. use http://www.stata-press.com/data/r12/morse_long
(Morse data (Rothkopf 1957))
. gen sim = freqsame/100
. mdslong sim, id(digit1 digit2) s2d(standard) noplot
  (output omitted)
```

MDS has produced a two-dimensional configuration with Euclidean distances approximating the dissimilarities between the Morse codes for digits. This configuration may be examined using the estat config command; see mdsconfig if you want to plot the configuration.

```
. estat config
Approximating configuration in 2-dimensional Euclidean space
```

digit1	dim1	dim2
0	0.5690	-0.0162
1	0.4561	0.3384
2	0.0372	0.5854
3	-0.3878	0.4516
4	-0.5800	0.0770
5	-0.5458	0.0196
6	-0.3960	-0.4187
7	-0.0963	-0.5901
8	0.3124	-0.3862
9	0.6312	-0.0608

This configuration is not unique. A translation, a reflection, and an orthonormal rotation of the configuration do not affect the interpoint Euclidean distances. All such transformations are equally reasonable MDS solutions. Thus you should not interpret aspects of these numbers (or of the configuration plot) that are not invariant to these transformations.

We now turn to the three `estat` subcommands that analyze the MDS residuals, that is, the differences between the disparities or dissimilarities and the matching Euclidean distances. There is a catch here. The raw residuals of MDS are not well behaved. For instance, the sum of the raw residuals is not zero—often it is not even close. The MDS solution does *not* minimize the sum of squares of the raw residuals Mardia, Kent, and Bibby (1979, 406–408). To create reasonable residuals with MDS, the dissimilarities can be transformed to disparities approximating the Euclidean distances. In classical MDS we use a linear transform f, fit by least squares. This is equivalent to Kruskal's Stress1 loss function. The modified residuals are defined as the differences between the linearly transformed dissimilarities and the matching Euclidean distances.

In modern MDS we have three types of transformations from dissimilarities to disparities to choose from: the identity (which does not transform the dissimilarities), a power transformation, and a monotonic transformation.

The three `estat` subcommands summarize the residuals in different ways. After classical MDS, `estat stress` displays the Kruskal loss or stress measures for each object and the overall total.

```
. estat stress

Stress between disparities and Euclidean distances
```

digit1	Kruskal
0	0.1339
1	0.1255
2	0.1972
3	0.2028
4	0.2040
5	0.2733
6	0.1926
7	0.1921
8	0.1715
9	0.1049
Total	0.1848

Second, after classical or modern MDS, the quantiles of the residuals are available, both overall and for the subgroup of object pairs in which an object is involved.

```
. estat quantiles

Quantiles of adjusted residuals
```

digit1	N	min	p25	q50	q75	max
0	9	-.111732	-.088079	-.028917	.11202	.220399
1	9	-.170063	-.137246	-.041244	.000571	.11202
2	9	-.332717	-.159472	-.072359	.074999	.234866
3	9	-.136251	-.120398	-.072359	.105572	.365833
4	9	-.160797	-.014099	.03845	.208215	.355053
5	9	-.09971	-.035357	.176337	.325043	.365833
6	0	.137240	.113504	-.075008	.177448	.325043
7	9	-.332717	-.170063	-.124129	.03845	.176337
8	9	-.186452	-.134831	-.041244	.075766	.220399
9	9	-.160797	-.104403	-.088079	-.064316	-.030032
Total	90	-.332717	-.113564	-.041244	.105572	.365833

The dissimilarities for the Morse code of digit 5 are fitted considerably worse than for all other Morse codes. Digit 5 has the largest Kruskal stress (0.273) and median residual (0.176).

Finally, after classical or modern MDS, `estat correlations` displays the Pearson and Spearman correlations between the (transformed or untransformed) dissimilarities and the Euclidean distances.

```
. estat correlations
Correlations of disparities and Euclidean distances
```

digit1	N	Pearson	Spearman
0	9	0.9510	0.9540
1	9	0.9397	0.7782
2	9	0.7674	0.4017
3	9	0.7922	0.7815
4	9	0.9899	0.9289
5	9	0.9412	0.9121
6	9	0.8226	0.8667
7	9	0.8444	0.4268
8	9	0.8505	0.7000
9	9	0.9954	0.9333
Total	90	0.8602	0.8301

◁

Matching configuration plot and the Shepard diagram

The matching configuration plot and Shepard diagram are easily obtained after an MDS analysis.

▷ Example 2

By default, `mds`, `mdsmat`, and `mdslong` display the MDS matching configuration plot. If you want to exercise control over the graph, you can specify the `noplot` option of `mds`, `mdsmat`, or `mdslong` and then use the `mdsconfig` postestimation graph command.

Continuing with the Morse code digit example: we produce a configuration plot with an added title and subtitle.

```
. mdsconfig, title(Morse code digit dissimilarity) subtitle(data: Rothkopf 1957)
```

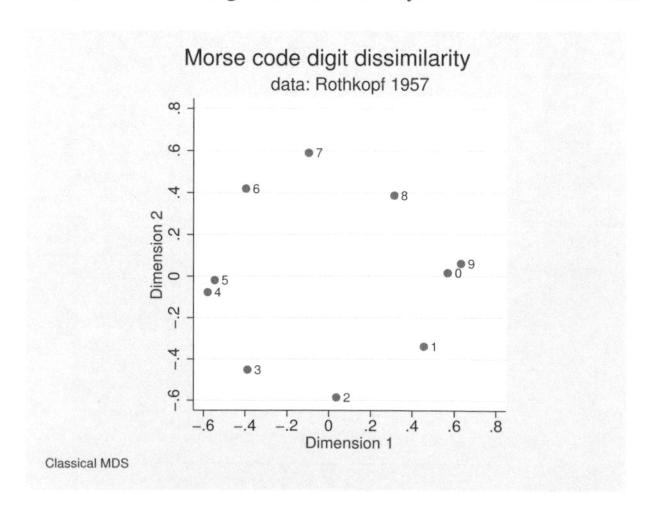

The plot has an aspect ratio of one so that 1 unit on the horizontal dimension equals 1 unit on the vertical dimension. Thus the "straight-line" distance in the plot is really (proportional to) the Euclidean distance between the points in the configuration and hence approximates the dissimilarities between the objects—here the Morse codes for digits.

◁

▷ Example 3

A second popular plot for MDS is the Shepard diagram. This is a plot of the Euclidean distances in the matching configuration against the "observed" dissimilarities. As we explained before, in classical MDS a linear transformation is applied to the dissimilarities to fit the Euclidean distances as close as possible (in the least-squares sense). In modern MDS the transformation may be the identity (no transformation), a power function, or a monotonic function. A Shepard diagram is a plot of the $n(n-1)/2$ transformed dissimilarities, called disparities, against the Euclidean distances.

```
. mdsshepard
```

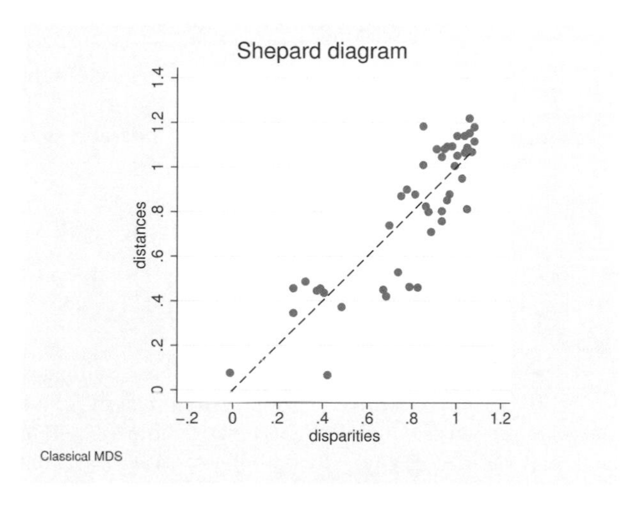

If the Euclidean configuration is close to the disparities between the objects, all points would be close to the $y = x$ line. Deviations indicate lack of fit. To simplify the diagnosis of whether there are specific objects that are poorly represented, Shepard diagrams can be produced for each object separately. Such plots consist of n small plots with $n - 1$ points each, namely, the disparities and Euclidean distances to all other objects.

```
. mdsshepard, separate
(mdsshepard is producing a separate plot for each obs; this may take a while)
```

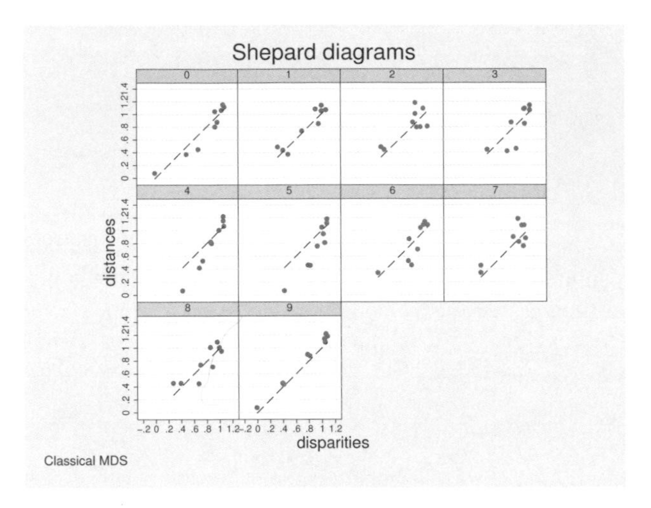

◁

Other examples of mdsconfig are found in [MV] **mds**, [MV] **mdslong**, and [MV] **mdsmat**.

Predictions

It is possible to generate variables containing the results from the MDS analysis. MDS operates at two levels: first at the level of the objects and second at the level of relations between the objects or pairs of objects. You can generate variables at both of these levels.

The `config` option of `predict` after an MDS requests that the coordinates of the objects in the matching configuration be stored in variables. You may specify as many variables as there are retained dimensions. You may also specify fewer variables. The first variable will store the coordinates from the first dimension. The second variable, if any, will store the coordinates from the second dimension, and so on.

The `pairwise()` option specifies that a given selection of the pairwise statistics are stored in variables. The statistics available are the disparities, dissimilarities, fitted distances, raw residuals, transformed residuals, and weights. The raw residuals are the difference between dissimilarities and the fitted distances, and the transformed residuals are the difference between the disparities and the fitted distances.

There is a complicating issue. With n objects, there are $n(n-1)/2$ pairs of objects. So, to store properties of objects, you need n observations, but to store properties of pairs of objects, you need $n(n-1)/2$ observations. So, where do you store the variables? `predict` after MDS commands can save the predicted variables in a new dataset. Specify the option `saving(`*filename*`)`. Such a dataset will automatically have the appropriate object identification variable or variables.

Sometimes it is also possible to store the variables in the dataset you have in memory: object-level variables in an object-level dataset and pairwise-level variables in a pairwise-level dataset.

After `mds` you have a dataset in memory in which the observations are associated with the MDS objects. Here you can store object-level variables directly in the dataset in memory. To do so, you just omit the `saving()` option. Here it is not possible to store the pairwise statistics in the dataset in memory. The pairwise statistics have to be stored in a new dataset.

After `mdslong`, the dataset in memory is in a pairwise form, so the variables predicted with the option `pairwise()` can be stored in the dataset in memory. It is, of course, also possible to store the pairwise variables in a new dataset; the choice is yours. With pairwise data in memory, you cannot store the object-level predicted variables into the data in memory; you need to specify the name of a new dataset.

After `mdsmat`, you always need to save the predicted variables in a new dataset.

▷ Example 4

Continuing with our Morse code example that used `mdslong`: the dataset in memory is in long form. Thus we can store the pairwise statistics with the dataset in memory.

```
. predict tdissim eudist resid, pairwise
. list in 1/10
```

	digit1	digit2	freqsame	sim	tdissim	eudist	resid
1.	2	1	62	.62	.3227682	.4862905	-.1635224
2.	3	1	16	.16	.957076	.851504	.105572
3.	3	2	59	.59	.3732277	.4455871	-.0723594
4.	4	1	6	.06	1.069154	1.068583	.0005709
5.	4	2	23	.23	.8745967	.7995979	.749988
6.	4	3	38	.38	.6841489	.4209922	.2631567
7.	5	1	12	.12	1.002667	1.051398	-.048731
8.	5	2	8	.08	1.047234	.8123672	.2348665
9.	5	3	27	.27	.8257753	.4599419	.3658335
10.	5	4	56	.56	.4218725	.0668193	.3550532

Because we used `mdslong`, the object-level statistics must be saved in a file.

```
. predict d1 d2, config saving(digitdata)
. describe digit* d1 d2 using digitdata
```

variable name	storage type	display format	value label	variable label
digit1	str1	%9s		
d1	float	%9.0g		MDS dimension 1
d2	float	%9.0g		MDS dimension 2

```
Sorted by:  digit1
```

The information in these variables was already shown with `estat config`. The dataset created has variables `d1` and `d2` with the coordinates of the Morse digits on the two retained dimensions and an identification variable `digit1`. Use `merge` to add these variables to the data in memory; see [D] **merge**.

◁

Saved results

`estat correlations` saves the following in `r()`:

Matrices

`r(R)`	statistics per object; columns with # of obs., Pearson corr., and Spearman corr.
`r(T)`	overall statistics; # of obs., Pearson corr., and Spearman corr.

`estat quantiles` saves the following in `r()`:

Macros

`r(dtype)`	`adjusted` or `raw`; dissimilarity transformation

Matrices

`r(Q)`	statistics per object; columns with # of obs., min., p25, p50, p75, and max.
`r(T)`	overall statistics; # of obs., min., p25, p50, p75, and max.

`estat stress` saves the following in `r()`:

Macros

`r(dtype)`	`adjusted` or `raw`; dissimilarity transformation

Matrices

`r(S)`	Kruskal's stress/loss measure per object
`r(T)`	1×1 matrix with the overall Kruskal stress/loss measure

Methods and formulas

All postestimation commands listed above are implemented as ado-files.

See [MV] **mdsmat** for information on the methods and formulas for multidimensional scaling.

For classical MDS, let D_{ij} be the dissimilarity between objects i and j, $1 \leq i, j \leq n$. We assume $D_{ii} = 0$ and $D_{ij} = D_{ji}$. Let E_{ij} be the Euclidean distance between rows i and j of the matching configuration $\mathbf{Y}$. In classical MDS, $\mathbf{D} - \mathbf{E}$ is not a well-behaved residual matrix. We follow the approach used in metric and nonmetric MDS to transform D_{ij} to "optimally match" E_{ij}, with $\widehat{D}_{ij} = a + bD_{ij}$, where a and b are chosen to minimize the residual sum of squares. This is a simple regression problem and is equivalent to minimizing Kruskal's stress measure (Kruskal 1964; Cox and Cox 2001, 63)

$$\text{Kruskal}(\widehat{\mathbf{D}}, \mathbf{E}) = \left\{ \frac{\sum (E_{ij} - \widehat{D}_{ij})^2}{\sum E_{ij}^2} \right\}^{1/2}$$

with summation over all pairs (i, j). We call the $\widehat{D}_{ij}$ the adjusted or transformed dissimilarities. If the transformation step is skipped by specifying the option `notransform`, we set $\widehat{D}_{ij} = D_{ij}$.

In `estat stress`, the decomposition of Kruskal's stress measure over the objects is displayed. $\text{Kruskal}(\widehat{\mathbf{D}}, \mathbf{E})_i$ is defined analogously with summation over all $j \neq i$.

For modern MDS, the optimal transformation to disparities, $f(\mathbf{D}) \rightarrow \widehat{\mathbf{D}}$, is calculated during the estimation. See [MV] **mdsmat** for details. For `transform(power)` the power is saved in `e(alpha)`. For `transform(monotonic)` the disparities themselves are saved as `e(Disparities)`.

References

Borg, I., and P. J. F. Groenen. 2005. *Modern Multidimensional Scaling: Theory and Applications.* 2nd ed. New York: Springer.

Cox, T. F., and M. A. A. Cox. 2001. *Multidimensional Scaling.* 2nd ed. Boca Raton, FL: Chapman & Hall/CRC.

Kruskal, J. B. 1964. Multidimensional scaling by optimizing goodness of fit to a nonmetric hypothesis. *Psychometrika* 29: 1–27.

Mardia, K. V., J. T. Kent, and J. M. Bibby. 1979. *Multivariate Analysis.* London: Academic Press.

Also see *References* in [MV] **mdsmat**.

Also see

[MV] **mds** — Multidimensional scaling for two-way data

[MV] **mdslong** — Multidimensional scaling of proximity data in long format

[MV] **mdsmat** — Multidimensional scaling of proximity data in a matrix

[MV] **screeplot** — Scree plot

Title

mdslong — Multidimensional scaling of proximity data in long format

Syntax

`mdslong` *depvar* [*if*] [*in*] [*weight*], `id(`var_1 var_2`)` [*options*]

options	Description
Model	
* `id(`var_1 var_2`)`	identify comparison pairs (object_1, object_2)
`method(`*method*`)`	method for performing MDS
`loss(`*loss*`)`	loss function
`transform(`*tfunction*`)`	permitted transformations of dissimilarities
`normalize(`*norm*`)`	normalization method; default is `normalize(principal)`
`s2d(standard)`	convert similarity to dissimilarity: $\text{dissim}_{ij} = \sqrt{\text{sim}_{ii} + \text{sim}_{jj} - 2\text{sim}_{ij}}$; the default
`s2d(oneminus)`	convert similarity to dissimilarity: $\text{dissim}_{ij} = 1 - \text{sim}_{ij}$
`force`	correct problems in proximity information
`dimension(`*#*`)`	configuration dimensions; default is `dimension(2)`
`addconstant`	make distance matrix positive semidefinite (classical MDS only)
Reporting	
`neigen(`*#*`)`	maximum number of eigenvalues to display; default is `neigen(10)` (classical MDS only)
`config`	display table with configuration coordinates
`noplot`	suppress configuration plot
Minimization	
`initialize(`*initopt*`)`	start with configuration given in *initopt*
`tolerance(`*#*`)`	tolerance for configuration matrix; default is `tolerance(1e-4)`
`ltolerance(`*#*`)`	tolerance for loss criterion; default is `ltolerance(1e-8)`
`iterate(`*#*`)`	perform maximum # of iterations; default is `iterate(1000)`
`protect(`*#*`)`	perform # optimizations and report best solution; default is `protect(1)`
`nolog`	suppress the iteration log
`trace`	display current configuration in iteration log
`gradient`	display current gradient matrix in iteration log
`sdprotect(`*#*`)`	advanced; see description below

* `id(`var_1 var_2`)` is required.

`by` and `statsby` are allowed; see [U] **11.1.10 Prefix commands**.

`aweight`s and `fweight`s are allowed; see [U] **11.1.6 weight**

The maximum number of compared objects allowed is the maximum matrix size; see [R] **matsize**.

`sdprotect(`*#*`)` does not appear in the dialog box.

See [U] **20 Estimation and postestimation commands** for more capabilities of estimation commands.

method	Description
classical	classical MDS; default if neither loss() nor transform() is specified
modern	modern MDS; default if loss() or transform() is specified; except when loss(stress) and transform(monotonic) are specified
nonmetric	nonmetric (modern) MDS; default when loss(stress) and transform(monotonic) are specified

loss	Description
stress	stress criterion, normalized by distances; the default
nstress	stress criterion, normalized by disparities
sstress	squared stress criterion, normalized by distances
nsstress	squared stress criterion, normalized by disparities
strain	strain criterion (with transform(identity) is equivalent to classical MDS)
sammon	Sammon mapping

tfunction	Description
identity	no transformation; disparity = dissimilarity; the default
power	power α: disparity = dissimilarity$^{\alpha}$
monotonic	weakly monotonic increasing functions (nonmetric scaling); only with loss(stress)

norm	Description
principal	principal orientation; location = 0; the default
classical	Procrustes rotation toward classical solution
target(*matname*) [, copy]	Procrustes rotation toward *matname*; ignore naming conflicts if copy is specified

initopt	Description
classical	start with classical solution; the default
random [(#)]	start at random configuration, setting seed to #
from(*matname*) [, copy]	start from *matname*; ignore naming conflicts if copy is specified

Menu

Statistics > Multivariate analysis > Multidimensional scaling (MDS) > MDS of proximity-pair data

Description

`mdslong` performs multidimensional scaling (MDS) for two-way proximity data in long format with an explicit measure of similarity or dissimilarity between objects. `mdslong` performs classical metric MDS (Torgerson 1952) as well as modern metric and nonmetric MDS; see the `method()`, `loss()`, and `transform()` options.

For MDS with two-way proximity data in a matrix, see [MV] **mdsmat**. If you are looking for MDS on a dataset, based on dissimilarities between observations over variables, see [MV] **mds**.

Computing the classical solution is straightforward, but with modern MDS the minimization of the loss criteria over configurations is a high-dimensional problem that is easily beset by convergence to local minimums. `mds`, `mdsmat`, and `mdslong` provide options to control the minimization process (1) by allowing the user to select the starting configuration and (2) by selecting the best solution among multiple minimization runs from random starting configurations.

Options

Model

`id(`var_1 var_2`)` is required. The pair of variables var_1 and var_2 should uniquely identify comparisons. var_1 and var_2 are string or numeric variables that identify the objects to be compared. var_1 and var_2 should be of the same data type; if they are value labeled, they should be labeled with the same value label. Using value-labeled variables or string variables is generally helpful in identifying the points in plots and tables.

Example data layout for `mdslong proxim, id(i1 i2)`.

proxim	i1	i2
7	1	2
10	1	3
12	1	4
4	2	3
6	2	4
3	3	4

If you have multiple measurements per pair, we suggest that you specify the mean of the measures as the proximity and the inverse of the variance as the weight.

`method(`*method*`)` specifies the method for MDS.

`method(classical)` specifies classical metric scaling, also known as "principal coordinates analysis" when used with Euclidean proximities. Classical MDS obtains equivalent results to modern MDS with `loss(strain)` and `transform(identity)` without weights. The calculations for classical MDS are fast; consequently, classical MDS is generally used to obtain starting values for modern MDS. If the options `loss()` and `transform()` are not specified, `mds` computes the classical solution, likewise if `method(classical)` is specified `loss()` and `transform()` are not allowed.

method(modern) specifies modern scaling. If method(modern) is specified but not loss() or transform(), then loss(stress) and transform(identity) are assumed. All values of loss() and transform() are valid with method(modern).

method(nonmetric) specifies nonmetric scaling, which is a type of modern scaling. If method(nonmetric) is specified, loss(stress) and transform(monotonic) are assumed. Other values of loss() and transform() are not allowed.

loss(*loss*) specifies the loss criterion.

loss(stress) specifies that the stress loss function be used, normalized by the squared Euclidean distances. This criterion is often called Kruskal's stress-1. Optimal configurations for loss(stress) and for loss(nstress) are equivalent up to a scale factor, but the iteration paths may differ. loss(stress) is the default.

loss(nstress) specifies that the stress loss function be used, normalized by the squared disparities, that is, transformed dissimilarities. Optimal configurations for loss(stress) and for loss(nstress) are equivalent up to a scale factor, but the iteration paths may differ.

loss(sstress) specifies that the squared stress loss function be used, normalized by the fourth power of the Euclidean distances.

loss(nsstress) specifies that the squared stress criterion, normalized by the fourth power of the disparities (transformed dissimilarities) be used.

loss(strain) specifies the strain loss criterion. Classical scaling is equivalent to loss(strain) and transform(identity) but is computed by a faster noniterative algorithm. Specifying loss(strain) still allows transformations.

loss(sammon) specifies the Sammon (1969) loss criterion.

transform(*tfunction*) specifies the class of allowed transformations of the dissimilarities; transformed dissimilarities are called disparities.

transform(identity) specifies that the only allowed transformation is the identity; that is, disparities are equal to dissimilarities. transform(identity) is the default.

transform(power) specifies that disparities are related to the dissimilarities by a power function,

$$\text{disparity} = \text{dissimilarity}^{\alpha}, \qquad \alpha > 0$$

transform(monotonic) specifies that the disparities are a weakly monotonic function of the dissimilarities. This is also known as nonmetric MDS. Tied dissimilarities are handled by the primary method; that is, ties may be broken but are not necessarily broken. transform(monotonic) is valid only with loss(stress).

normalize(*norm*) specifies a normalization method for the configuration. Recall that the location and orientation of an MDS configuration is not defined ("identified"); an isometric transformation (that is, translation, reflection, or orthonormal rotation) of a configuration preserves interpoint Euclidean distances.

normalize(principal) performs a principal normalization, in which the configuration columns have zero mean and correspond to the principal components, with positive coefficient for the observation with lowest value of id(). normalize(principal) is the default.

normalize(classical) normalizes by a distance-preserving Procrustean transformation of the configuration toward the classical configuration in principal normalization; see [MV] **procrustes**. normalize(classical) is not valid if method(classical) is specified.

normalize(target(*matname*) [, copy]) normalizes by a distance-preserving Procrustean transformation toward *matname*; see [MV] **procrustes**. *matname* should be an $n \times p$ matrix, where n is the number of observations and p is the number of dimensions, and the rows of *matname* should be ordered with respect to id(). The rownames of *matname* should be set correctly but will be ignored if copy is also specified.

Note on normalize(classical) and normalize(target()): the Procrustes transformation comprises any combination of translation, reflection, and orthonormal rotation—these transformations preserve distance. Dilation (uniform scaling) would stretch distances and is not applied. However, the output reports the dilation factor, and the reported Procrustes statistic is for the dilated configuration.

s2d(standard | oneminus) specifies how similarities are converted into dissimilarities. By default, the command assumes dissimilarity data. Specifying s2d() indicates that your proximity data are similarities.

Dissimilarity data should have zeros on the diagonal (that is, an object is identical to itself) and nonnegative off-diagonal values. Dissimilarities need not satisfy the triangular inequality, $D(i,j)^2 \leq D(i,h)^2 + D(h,j)^2$. Similarity data should have ones on the diagonal (that is, an object is identical to itself) and have off-diagonal values between zero and one. In either case, proximities should be symmetric. See option force if your data violate these assumptions.

The available s2d() options, standard and oneminus, are defined as follows:

standard	$\text{dissim}_{ij} = \sqrt{\text{sim}_{ii} + \text{sim}_{jj} - 2\text{sim}_{ij}} = \sqrt{2(1 - \text{sim}_{ij})}$
oneminus	$\text{dissim}_{ij} = 1 - \text{sim}_{ij}$

s2d(standard) is the default.

s2d() should be specified only with measures in similarity form.

force corrects problems with the supplied proximity information. In the long format used by mdslong, multiple measurements on (i,j) may be available. Including both (i,j) and (j,i) would be treated as multiple measurements. This is an error, even if the measures are identical. Option force uses the mean of the measurements. force also resolves problems on the diagonal, that is, comparisons of objects with themselves; these should have zero dissimilarity or unit similarity. force does not resolve incomplete data, that is, pairs (i,j) for which no measurement is available. Out-of-range values are also not fixed.

dimension(#) specifies the dimension of the approximating configuration. The default # is 2 and should not exceed the number of positive eigenvalues of the centered distance matrix.

addconstant specifies that if the double-centered distance matrix is not positive semidefinite (psd), a constant should be added to the squared distances to make it psd and, hence, Euclidean. This option is allowed with classical MDS only.

Reporting

neigen(#) specifies the number of eigenvalues to be included in the table. The default is neigen(10). Specifying neigen(0) suppresses the table. This option is allowed with classical MDS only.

config displays the table with the coordinates of the approximating configuration. This table may also be displayed using the postestimation command estat config; see [MV] **mds postestimation**.

noplot suppresses the graph of the approximating configuration. The graph can still be produced later via mdsconfig, which also allows the standard graphics options for fine-tuning the plot; see [MV] **mds postestimation**.

Minimization

These options are available only with method(modern) or method(nonmetric):

initialize(*initopt*) specifies the initial values of the criterion minimization process.

initialize(classical), the default, uses the solution from classical metric scaling as initial values. With protect(), all but the first run start from random perturbations from the classical solution. These random perturbations are independent and normally distributed with standard error equal to the product of sdprotect(*#*) and the standard deviation of the dissimilarities. initialize(classical) is the default.

initialize(random) starts an optimization process from a random starting configuration. These random configurations are generated from independent normal distributions with standard error equal to the product of sdprotect(*#*) and the standard deviation of the dissimilarities. The means of the configuration are irrelevant in MDS.

initialize(from(*matname*) [, copy]) sets the initial value to *matname*. *matname* should be an $n \times p$ matrix, where n is the number of observations and p is the number of dimensions, and the rows of *matname* should be ordered with respect to id(). The rownames of *matname* should be set correctly but will be ignored if copy is specified. With protect(), the second-to-last runs start from random perturbations from *matname*. These random perturbations are independent normal distributed with standard error equal to the product of sdprotect(*#*) and the standard deviation of the dissimilarities.

tolerance(*#*) specifies the tolerance for the configuration matrix. When the relative change in the configuration from one iteration to the next is less than or equal to tolerance(), the tolerance() convergence criterion is satisfied. The default is tolerance(1e-4).

ltolerance(*#*) specifies the tolerance for the fit criterion. When the relative change in the fit criterion from one iteration to the next is less than or equal to ltolerance(), the ltolerance() convergence is satisfied. The default is ltolerance(1e-8).

Both the tolerance() and ltolerance() criteria must be satisfied for convergence.

iterate(*#*) specifies the maximum number of iterations. The default is iterate(1000).

protect(*#*) requests that # optimizations be performed and that the best of the solutions be reported. The default is protect(1). See option initialize() on starting values of the runs. The output contains a table of the return code, the criterion value reached, and the seed of the random number used to generate the starting value. Specifying a large number, such as protect(50), provides reasonable insight whether the solution found is a global minimum and not just a local minimum.

If any of the options log, trace, or gradient is also specified, iteration reports will be printed for each optimization run. Beware: this option will produce a lot of output.

nolog suppresses the iteration log, showing the progress of the minimization process.

trace displays the configuration matrices in the iteration report. Beware: this option may produce a lot of output.

gradient displays the gradient matrices of the fit criterion in the iteration report. Beware: this option may produce a lot of output.

The following option is available with mdslong but is not shown in the dialog box:

sdprotect(*#*) sets a proportionality constant for the standard deviations of random configurations (init(random)) or random perturbations of given starting configurations (init(classical) or init(from())). The default is sdprotect(1).

Remarks

Remarks are presented under the following headings:

Introduction

Multidimensional scaling (MDS) is a dimension-reduction and visualization technique. Dissimilarities (for instance, Euclidean distances) between observations in a high-dimensional space are represented in a lower-dimensional space (typically two dimensions) so that the Euclidean distance in the lower-dimensional space approximates the dissimilarities in the higher-dimensional space. See Kruskal and Wish (1978) for a brief nontechnical introduction to MDS. Young and Hamer (1987) and Borg and Groenen (2005) are more advanced textbook-sized treatments.

`mdslong` performs MDS on data in long format. *depvar* specifies proximity data in either dissimilarity or similarity form. The comparison pairs are identified by two variables specified in the required option `id()`. Exactly 1 observation with a nonmissing *depvar* should be included for each pair (i,j). Pairs are unordered; you do not include observations for both (i,j) and (j,i). Observations for comparisons of objects with themselves (i,i) are optional. See option `force` if your data violate these assumptions.

When you have multiple independent measurements of the dissimilarities, you may specify the mean of these dissimilarity measurements as the combined measurement and specify $1/(\#\text{of measurements})$ or $1/\text{variance}(\text{measurements})$ as weights. For more discussion of weights in MDS, we refer to Borg and Groenen (2005, sec. 11.3). Weights should be irreducible; that is, it is not possible to split the objects into disjointed groups with all intergroup weights 0.

In some applications, the similarity or dissimilarity of objects is defined by the researcher in terms of variables (attributes) measured on the objects. If you need MDS of this form, you should continue by reading [MV] **mds**.

Often, however, proximities—that is, similarities or dissimilarities—are measured directly. For instance, psychologists studying the similarities or dissimilarities in a set of stimuli—smells, sounds, faces, concepts, etc.—may have subjects rate the dissimilarity of pairs of stimuli. Linguists have subjects rate the similarity or dissimilarity of pairs of dialects. Political scientists have subjects rate the similarity or dissimilarity of political parties or candidates for political office. In other fields, relational data are studied that may be interpreted as proximities in a more abstract sense. For instance, sociologists study interpersonal contact frequencies in groups ("social networks"); these measures are sometimes interpreted in terms of similarities.

A wide variety of MDS methods have been proposed. `mdslong` performs classical and modern scaling. Classical scaling has its roots in Young and Householder (1938) and Torgerson (1952). MDS requires complete and symmetric dissimilarity interval-level data. To explore modern scaling, see Borg and Groenen (2005). Classical scaling results in an eigen decomposition, whereas modern scaling is accomplished by the minimization of a loss function. Consequently, eigenvalues are not available after modern MDS.

Proximity data in long format

One format for proximity data is called the "long format", with an observation recording the dissimilarity d_{ij} of the "objects" i and j. This requires three variables: one variable to record the dissimilarities and two variables to identify the comparison pair. The MDS command `mdslong` requires

- Complete data without duplicates: there is exactly 1 observation for each combination (i, j) or (j, i).
- Optional diagonal: you may, but need not, specify dissimilarities for the reflexive pairs (i, i). If you do, you need not supply values for all (i, i).

▷ Example 1

We illustrate the use of `mdslong` with a popular dataset from the MDS literature. Rothkopf (1957) had 598 subjects listen to pairs of Morse codes for the 10 digits and for the 26 letters, recording for each pair of codes the percentage of subjects who declared the codes to be the same. The data on the 10 digits are reproduced in Mardia, Kent, and Bibby (1979, 395).

```
. use http://www.stata-press.com/data/r12/morse_long
(Morse data (Rothkopf 1957))
. list in 1/10
```

	digit1	digit2	freqsame
1.	2	1	62
2.	3	1	16
3.	3	2	59
4.	4	1	6
5.	4	2	23
6.	4	3	38
7.	5	1	12
8.	5	2	8
9.	5	3	27
10.	5	4	56

Sixty-two percent of the subjects declare that the Morse codes for 1 and 2 are the same, 16% declare that 1 and 3 are the same, 59% declare 2 and 3 to be the same, etc. We may think that these percentages are similarity measures between the Morse codes: the more similar two Morse codes, the higher the percentage is of subjects who do not distinguish them. The reported percentages suggest, for example, that 1 and 2 are similar to approximately the same extent as 2 and 3, whereas 1 and 3 are much less similar. This is the kind of relationship you would expect with data that can be adequately represented with MDS.

We transform our data to a zero-to-one scale.

```
. gen sim = freqsame/100
```

and invoke `mdslong` on `sim`, specifying that the proximity variable `sim` be interpreted as similarities, and we use option `s2d(standard)` to convert to dissimilarities by using the standard conversion.

```
. mdslong sim, id(digit1 digit2) s2d(standard)
Classical metric multidimensional scaling
    similarity variable: sim in long format
          dissimilarity: sqrt(2(1-similarity))
                                              Number of obs         =         10
    Eigenvalues > 0     =          9          Mardia fit measure 1 =     0.5086
    Retained dimensions =          2          Mardia fit measure 2 =     0.7227
```

Dimension	Eigenvalue	abs(eigenvalue) Percent	Cumul.	(eigenvalue)^2 Percent	Cumul.
1	1.9800226	30.29	30.29	49.47	49.47
2	1.344165	20.57	50.86	22.80	72.27
3	1.063133	16.27	67.13	14.26	86.54
4	.66893922	10.23	77.36	5.65	92.18
5	.60159396	9.20	86.56	4.57	96.75
6	.42722301	6.54	93.10	2.30	99.06
7	.21220785	3.25	96.35	0.57	99.62
8	.1452025	2.22	98.57	0.27	99.89
9	.09351288	1.43	100.00	0.11	100.00

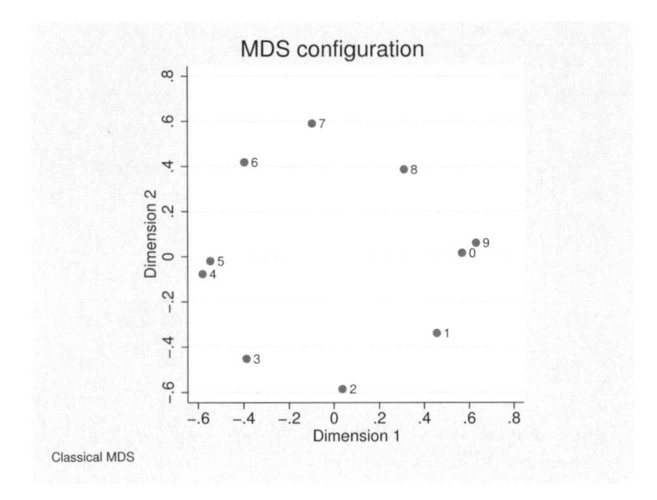

The two-dimensional representation provides a reasonable, but certainly not impressive, fit to the data. The plot itself is interesting, though, with the digits being roughly 45 degrees apart, except for the pairs (0,9) and (4,5), which are mapped almost at the same locations. Interpretation is certainly helped if you see the circular structure in the Morse codes.

digit	morse code
1	. - - - -
2	. . - - -
3	. . . - -
4	. . . . -
5	. . . . .
6	- . . . .
7	- - . . .
8	- - - . .
9	- - - - .
0	- - - - -

◁

▷ Example 2

You might have your data in wide instead of long format. The Morse code dataset in wide format has 10 observations, 10 data variables d1, ..., d9, d0, and one case identifier.

```
. use http://www.stata-press.com/data/r12/morse_wide, clear
(Morse data (Rothkopf 1957))

. describe

Contains data from http://www.stata-press.com/data/r12/morse_wide.dta
  obs:            10                          Morse data (Rothkopf 1957)
 vars:            11                          14 Feb 2011 20:28
 size:           110                          (_dta has notes)
-----------------------------------------------------------------------------
              storage  display     value
variable name   type   format      label      variable label
-----------------------------------------------------------------------------
digit           byte   %9.0g
d1              byte   %9.0g
d2              byte   %9.0g
d3              byte   %9.0g
d4              byte   %9.0g
d5              byte   %9.0g
d6              byte   %9.0g
d7              byte   %9.0g
d8              byte   %9.0g
d9              byte   %9.0g
d0              byte   %9.0g
-----------------------------------------------------------------------------
Sorted by:
```

```
. list
```

	digit	d1	d2	d3	d4	d5	d6	d7	d8	d9	d0
1.	1	84	62	16	6	12	12	20	37	57	52
2.	2	62	89	59	23	8	14	25	25	28	18
3.	3	16	59	86	38	27	33	17	16	9	9
4.	4	6	23	38	89	56	34	24	13	7	7
5.	5	12	8	27	56	90	30	18	10	5	5
6.	6	12	14	33	34	30	86	65	22	8	18
7.	7	20	25	17	24	18	65	85	65	31	15
8.	8	37	25	16	13	10	22	65	88	58	39
9.	9	57	28	9	7	5	8	31	58	91	79
10.	0	52	18	9	7	5	18	15	39	79	94

Stata does not provide an MDS command to deal directly with the wide format because it is easy to convert the wide format into the long format with the `reshape` command; see [D] **reshape**.

```
. reshape long d, i(digit) j(other)
(note: j = 0 1 2 3 4 5 6 7 8 9)
Data                                wide   ->   long
-----------------------------------------------------------------------------
Number of obs.                        10   ->    100
Number of variables                   11   ->      3
j variable (10 values)                     ->   other
xij variables:
                               d0 d1 ... d9   ->   d
-----------------------------------------------------------------------------
```

Now our data are in long format, and we can use `mdslong` to obtain a MDS analysis.

```
. gen sim = d/100
. mdslong sim, id(digit other) s2d(standard) noplot
objects should have unit similarity to themselves
r(198);
```

`mdslong` complains. The wide data—and hence also the long data that we now have—also contain the frequencies in which two identical Morse codes were recognized as the same. This is not 100%. Auditive memory is not perfect, posing a problem for the standard MDS model. We can solve this by ignoring the diagonal observations:

```
. mdslong ... if digit != other ...
```

We may also specify the `force` option. The `force` option will take care of a problem that has not yet surfaced, namely, that `mdslong` requires 1 observation for each pair (i, j). In the long data as now created, we have duplicates observations (i, j) and (j, i). `force` will symmetrize the proximity information, but it will not deal with multiple measurements in general; having 2 or more observations for (i, j) is an error. If you have multiple measurements, you may average the measurements and use weights.

```
. mdslong sim, id(digit other) s2d(standard) force noplot
Classical metric multidimensional scaling
    similarity variable: sim in long format
          dissimilarity: sqrt(2(1-similarity))

                                           Number of obs        =       10
    Eigenvalues > 0       =         9      Mardia fit measure 1 =   0.5086
    Retained dimensions   =         2      Mardia fit measure 2 =   0.7227

    ---------------------------------------------------------------------------
                |                 abs(eigenvalue)          (eigenvalue)^2
      Dimension |  Eigenvalue    Percent     Cumul.      Percent     Cumul.
    ------------+--------------------------------------------------------------
              1 |   1.9800226      30.29      30.29        49.47      49.47
              2 |    1.344165      20.57      50.86        22.80      72.27
    ------------+--------------------------------------------------------------
              3 |    1.063133      16.27      67.13        14.26      86.54
              4 |   .66893922      10.23      77.36         5.65      92.18
              5 |   .60159396       9.20      86.56         4.57      96.75
              6 |   .42722301       6.54      93.10         2.30      99.06
              7 |   .21220785       3.25      96.35         0.57      99.62
              8 |    .1452025       2.22      98.57         0.27      99.89
              9 |   .09351288       1.43     100.00         0.11     100.00
    ---------------------------------------------------------------------------
```

The output produced by `mdslong` here is identical to what we saw earlier.

◁

Modern nonmetric MDS

Unlike classical MDS, modern MDS is calculated via the minimization of the loss function. Eigenvalues are no longer calculated. We look at nonmetric MDS, which is a type of modern MDS in which the transformation from distances to disparities is not an identifiable function as in modern metric MDS but is instead a general monotonic function.

▷ Example 3

We return to the Rothkopf (1957) Morse codes in long format. When we specify `method(nonmetric)`, we assume `loss(stress)` and `transform(monotonic)`.

```
. use http://www.stata-press.com/data/r12/morse_long, clear
(Morse data (Rothkopf 1957))
. gen sim = freqsame/100
```

```
. mdslong sim, id(digit1 digit2) s2d(standard) meth(nonmetric)
(loss(stress) assumed)
(transform(monotonic) assumed)
Iteration 1t:   stress =  .14719847
Iteration 1c:   stress =  .11378737
  (output omitted)
Iteration 89t:   stress =  .07228281
Iteration 89c:   stress =  .07228281
Modern multidimensional scaling
    similarity variable: sim in long format
          dissimilarity: sqrt(2(1-similarity))
    Loss criterion: stress = raw_stress/norm(distances)
    Transformation: monotonic (nonmetric)
                                              Number of obs    =        10
                                              Dimensions       =         2
    Normalization: principal                  Loss criterion   =    0.0723
```

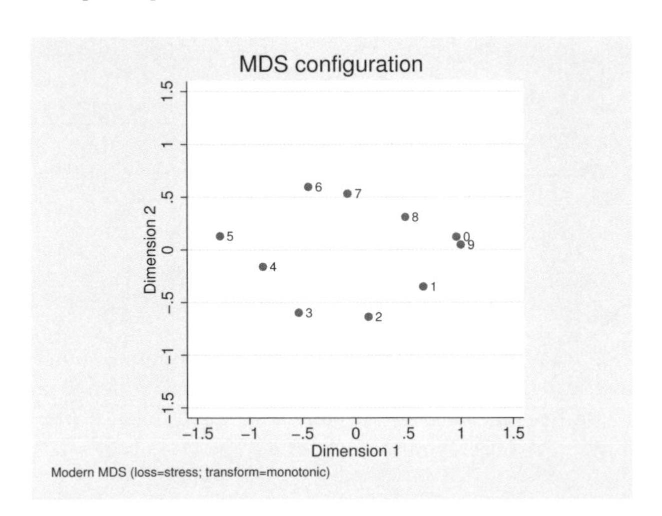

Each iteration has two steps associated with it. The two parts to each iteration consist of modifying the transformation (the T-step) and modifying the configuration (the C-step). If the `transform(identity)` option had been used, there would not be a T-step. In the iteration log, you see these as *Iteration 1t* and *Iteration 1c*. The rest of the output from modern MDS is explained in [MV] **mds**.

Although there is a resemblance between this graph and the first graph, the points are not as circular or as evenly spaced as they are in the first example, and a great deal more distance is seen between points 4 and 5. Nonmetric MDS depends only on the ordinal properties of the data and admits transformations that may radically change the appearance of the configuration.

◁

After `mdslong`, all MDS postestimation tools are available. For instance, you may analyze residuals with `estat quantile`, you may produce a Shepard diagram, etc.; see [MV] **mds postestimation**.

Saved results

mdslong saves the following in e():

Scalars	
e(N)	number of underlying observations
e(p)	number of dimensions in the approximating configuration
e(np)	number of strictly positive eigenvalues
e(addcons)	constant added to squared dissimilarities to force positive semidefiniteness
e(mardia1)	Mardia measure 1
e(mardia2)	Mardia measure 2
e(critval)	loss criterion value
e(npos)	number of pairs with positive weights
e(wsum)	sum of weights
e(alpha)	parameter of transform(power)
e(ic)	iteration count
e(rc)	return code
e(converged)	1 if converged, 0 otherwise

Macros	
e(cmd)	mdslong
e(cmdline)	command as typed
e(method)	classical or modern MDS method
e(method2)	nonmetric, if method(nonmetric)
e(loss)	loss criterion
e(losstitle)	description loss criterion
e(tfunction)	identity, power, or monotonic, transformation function
e(transftitle)	description of transformation
e(id)	two ID variable names identifying compared object pairs
e(idtype)	int or str; type of id() variable
e(duplicates)	1 if duplicates in id(), 0 otherwise
e(labels)	labels for ID categories
e(mxlen)	maximum length of category labels
e(depvar)	dependent variable containing dissimilarities
e(dtype)	similarity or dissimilarity; type of proximity data
e(s2d)	standard or oneminus (when e(dtype) is similarity)
e(wtype)	weight type
e(wexp)	weight expression
e(unique)	1 if eigenvalues are distinct, 0 otherwise
e(init)	initialization method
e(iseed)	seed for init(random)
e(seed)	seed for solution
e(norm)	normalization method
e(targetmatrix)	name of target matrix for normalize(target)
e(properties)	nob noV for modern or nonmetric MDS; nob noV eigen for classical MDS
e(estat_cmd)	program used to implement estat
e(predict)	program used to implement predict
e(marginsnotok)	predictions disallowed by margins

Matrices	
`e(D)`	dissimilarity matrix
`e(Disparities)`	disparity matrix for nonmetric MDS
`e(Y)`	approximating configuration coordinates
`e(Ev)`	eigenvalues
`e(W)`	weight matrix
`e(idcoding)`	coding for integer identifier variable
`e(norm_stats)`	normalization statistics
`e(linearf)`	two-element vector defining the linear transformation; distance equals first element plus second element times dissimilarity
Functions	
`e(sample)`	marks estimation sample

Methods and formulas

`mdslong` is implemented as an ado-file.

See *Methods and formulas* in [MV] **mdsmat** for information.

References

Borg, I., and P. J. F. Groenen. 2005. *Modern Multidimensional Scaling: Theory and Applications.* 2nd ed. New York: Springer.

Kruskal, J. B., and M. Wish. 1978. *Multidimensional Scaling.* Newbury Park, CA: Sage.

Lingoes, J. C. 1971. Some boundary conditions for a monotone analysis of symmetric matrices. *Psychometrika* 36: 195–203.

Mardia, K. V., J. T. Kent, and J. M. Bibby. 1979. *Multivariate Analysis.* London: Academic Press.

Rothkopf, E. Z. 1957. A measure of stimulus similarity and errors in some paired-associate learning tasks. *Journal of Experimental Psychology* 53: 94–101.

Sammon, J. W., Jr. 1969. A nonlinear mapping for data structure analysis. *IEEE Transactions on Computers* 18: 401–409.

Torgerson, W. S. 1952. Multidimensional scaling: I. Theory and method. *Psychometrika* 17: 401–419.

Young, F. W., and R. M. Hamer. 1987. *Multidimensional Scaling: History, Theory, and Applications.* Hillsdale, NJ: Erlbaum Associates.

Young, G., and A. S. Householder. 1938. Discussion of a set of points in terms of their mutual distances. *Psychometrika* 3: 19–22.

See [MV] **mdsmat** for more references.

Also see

[MV] **mds postestimation** — Postestimation tools for mds, mdsmat, and mdslong

[MV] **biplot** — Biplots

[MV] **ca** — Simple correspondence analysis

[MV] **factor** — Factor analysis

[MV] **mds** — Multidimensional scaling for two-way data

[MV] **mdsmat** — Multidimensional scaling of proximity data in a matrix

[MV] **pca** — Principal component analysis

[U] **20 Estimation and postestimation commands**

Title

mdsmat — Multidimensional scaling of proximity data in a matrix

Syntax

`mdsmat` *matname* [, *options*]

options	Description
Model	
`method(`*method*`)`	method for performing MDS
`loss(`*loss*`)`	loss function
`transform(`*tfunction*`)`	permitted transformations of dissimilarities
`normalize(`*norm*`)`	normalization method; default is `normalize(principal)`
`names(`*namelist*`)`	variable names corresponding to row and column names of the matrix; required with all but `shape(full)`
`shape(full)`	*matname* is a square symmetric matrix; the default
`shape(lower)`	*matname* is a vector with the rowwise lower triangle (with diagonal)
`shape(llower)`	*matname* is a vector with the rowwise strictly lower triangle (no diagonal)
`shape(upper)`	*matname* is a vector with the rowwise upper triangle (with diagonal)
`shape(uupper)`	*matname* is a vector with the rowwise strictly upper triangle (no diagonal)
`s2d(standard)`	convert similarity to dissimilarity: $d_{ij} = \sqrt{s_{ii} + s_{jj} - 2s_{ij}}$
`s2d(oneminus)`	convert similarity to dissimilarity: $d_{ij} = 1 - s_{ij}$
Model 2	
`dimension(`*#*`)`	configuration dimensions; default is `dimension(2)`
`force`	fix problems in proximity information
`addconstant`	make distance matrix positive semidefinite (classical MDS only)
`weight(`*matname*`)`	specifies a weight matrix with the same shape as the proximity matrix
Reporting	
`neigen(`*#*`)`	maximum number of eigenvalues to display; default is `neigen(10)` (classical MDS only)
`config`	display table with configuration coordinates
`noplot`	suppress configuration plot
Minimization	
`initialize(`*initopt*`)`	start with configuration given in *initopt*
`tolerance(`*#*`)`	tolerance for configuration matrix; default is `tolerance(1e-4)`
`ltolerance(`*#*`)`	tolerance for loss criterion; default is `ltolerance(1e-8)`
`iterate(`*#*`)`	perform maximum # of iterations; default is `iterate(1000)`
`protect(`*#*`)`	perform # optimizations and report best solution; default is `protect(1)`
`nolog`	suppress the iteration log
`trace`	display current configuration in iteration log
`gradient`	display current gradient matrix in iteration log
`sdprotect(`*#*`)`	advanced; see *Options* below

sdprotect(#) does not appear in the dialog box.

See [MV] **mds postestimation** for features available after estimation.

method	Description
classical	classical MDS; default if neither loss() nor transform() is specified
modern	modern MDS; default if loss() or transform() is specified; except when loss(stress) and transform(monotonic) are specified
nonmetric	nonmetric (modern) MDS; default when loss(stress) and transform(monotonic) are specified

loss	Description
stress	stress criterion, normalized by distances; the default
nstress	stress criterion, normalized by disparities
sstress	squared stress criterion, normalized by distances
nsstress	squared stress criterion, normalized by disparities
strain	strain criterion (with transform(identity) is equivalent to classical MDS)
sammon	Sammon mapping

tfunction	Description
identity	no transformation; disparity = dissimilarity; the default
power	power α: disparity = dissimilarity$^\alpha$
monotonic	weakly monotonic increasing functions (nonmetric scaling); only with loss(stress)

norm	Description
principal	principal orientation; location = 0; the default
classical	Procrustes rotation toward classical solution
target(*matname*) [, copy]	Procrustes rotation toward *matname*; ignore naming conflicts if copy is specified

initopt	Description
classical	start with classical solution; the default
random[(#)]	start at random configuration, setting seed to #
from(*matname*) [, copy]	start from *matname*; ignore naming conflicts if copy is specified

Menu

Statistics > Multivariate analysis > Multidimensional scaling (MDS) > MDS of proximity matrix

Description

`mdsmat` performs multidimensional scaling (MDS) for two-way proximity data with an explicit measure of similarity or dissimilarity between objects, where the proximities are found in matrix *matname*. `mdsmat` performs classical metric MDS (Torgerson 1952) as well as modern metric and nonmetric MDS; see the `loss()` and `transform()` options.

If your proximities are stored as variables in long format, see [MV] **mdslong**. If you are looking for MDS on a dataset on the basis of dissimilarities between observations over variables, see [MV] **mds**.

Computing the classical solution is straightforward, but with modern MDS the minimization of the loss criteria over configurations is a high-dimensional problem that is easily beset by convergence to local minimums. `mds`, `mdsmat`, and `mdslong` provide options to control the minimization process (1) by allowing the user to select the starting configuration and (2) by selecting the best solution among multiple minimization runs from random starting configurations.

Options

Model

`method(`*method*`)` specifies the method for MDS.

`method(classical)` specifies classical metric scaling, also known as "principal coordinates analysis" when used with Euclidean proximities. Classical MDS obtains equivalent results to modern MDS with `loss(strain)` and `transform(identity)` without weights. The calculations for classical MDS are fast; consequently, classical MDS is generally used to obtain starting values for modern MDS. If the options `loss()` and `transform()` are not specified, `mds` computes the classical solution, likewise if `method(classical)` is specified `loss()` and `transform()` are not allowed.

`method(modern)` specifies modern scaling. If `method(modern)` is specified but not `loss()` or `transform()`, then `loss(stress)` and `transform(identity)` are assumed. All values of `loss()` and `transform()` are valid with `method(modern)`.

`method(nonmetric)` specifies nonmetric scaling, which is a type of modern scaling. If `method(nonmetric)` is specified, `loss(stress)` and `transform(monotonic)` are assumed. Other values of `loss()` and `transform()` are not allowed.

`loss(`*loss*`)` specifies the loss criterion.

`loss(stress)` specifies that the stress loss function be used, normalized by the squared Euclidean distances. This criterion is often called Kruskal's stress-1. Optimal configurations for `loss(stress)` and for `loss(nstress)` are equivalent up to a scale factor, but the iteration paths may differ. `loss(stress)` is the default.

`loss(nstress)` specifies that the stress loss function be used, normalized by the squared disparities, that is, transformed dissimilarities. Optimal configurations for `loss(stress)` and for `loss(nstress)` are equivalent up to a scale factor, but the iteration paths may differ.

`loss(sstress)` specifies that the squared stress loss function be used, normalized by the fourth power of the Euclidean distances.

loss(nsstress) specifies that the squared stress criterion, normalized by the fourth power of the disparities (transformed dissimilarities) be used.

loss(strain) specifies the strain loss criterion. Classical scaling is equivalent to loss(strain) and transform(identity) but is computed by a faster noniterative algorithm. Specifying loss(strain) still allows transformations.

loss(sammon) specifies the Sammon (1969) loss criterion.

transform(*tfunction*) specifies the class of allowed transformations of the dissimilarities; transformed dissimilarities are called disparities.

transform(identity) specifies that the only allowed transformation is the identity; that is, disparities are equal to dissimilarities. transform(identity) is the default.

transform(power) specifies that disparities are related to the dissimilarities by a power function,

$$\text{disparity} = \text{dissimilarity}^{\alpha}, \qquad \alpha > 0$$

transform(monotonic) specifies that the disparities are a weakly monotonic function of the dissimilarities. This is also known as nonmetric MDS. Tied dissimilarities are handled by the primary method; that is, ties may be broken but are not necessarily broken. transform(monotonic) is valid only with loss(stress).

normalize(*norm*) specifies a normalization method for the configuration. Recall that the location and orientation of an MDS configuration is not defined ("identified"); an isometric transformation (that is, translation, reflection, or orthonormal rotation) of a configuration preserves interpoint Euclidean distances.

normalize(principal) performs a principal normalization, in which the configuration columns have zero mean and correspond to the principal components, with positive coefficient for the observation with lowest value of id(). normalize(principal) is the default.

normalize(classical) normalizes by a distance-preserving Procrustean transformation of the configuration toward the classical configuration in principal normalization; see [MV] **procrustes**. normalize(classical) is not valid if method(classical) is specified.

normalize(target(*matname*) [, copy]) normalizes by a distance-preserving Procrustean transformation toward *matname*; see [MV] **procrustes**. *matname* should be an $n \times p$ matrix, where n is the number of observations and p is the number of dimensions, and the rows of *matname* should be ordered with respect to id(). The rownames of *matname* should be set correctly but will be ignored if copy is also specified.

Note on normalize(classical) and normalize(target()): the Procrustes transformation comprises any combination of translation, reflection, and orthonormal rotation—these transformations preserve distance. Dilation (uniform scaling) would stretch distances and is not applied. However, the output reports the dilation factor, and the reported Procrustes statistic is for the dilated configuration.

names(*namelist*) is required with all but shape(full). The number of names should equal the number of rows (and columns) of the full similarity or dissimilarity matrix and should not contain duplicates.

shape(*shape*) specifies the storage mode of the existing similarity or dissimilarity matrix *matname*. The following storage modes are allowed:

full specifies that *matname* is a symmetric $n \times n$ matrix.

lower specifies that *matname* is a row or column vector of length $n(n+1)/2$, with the rowwise lower triangle of the similarity or dissimilarity matrix including the diagonal.

$$D_{11}\ D_{21}\ D_{22}\ D_{31}\ D_{32}\ D_{33}\ \ldots\ D_{n1}\ D_{n2}\ \ldots\ D_{nn}$$

`llower` specifies that *matname* is a row or column vector of length $n(n-1)/2$, with the rowwise lower triangle of the similarity or dissimilarity matrix excluding the diagonal.

$$D_{21}\ D_{31}\ D_{32}\ D_{41}\ D_{42}\ D_{43}\ \ldots\ D_{n1}\ D_{n2}\ \ldots\ D_{n,n-1}$$

`upper` specifies that *matname* is a row or column vector of length $n(n+1)/2$, with the rowwise upper triangle of the similarity or dissimilarity matrix including the diagonal.

$$D_{11}\ D_{12}\ \ldots\ D_{1n}\ D_{22}\ D_{23}\ \ldots\ D_{2n}\ D_{33}\ D_{34}\ \ldots\ D_{3n}\ \ldots\ D_{nn}$$

`uupper` specifies that *matname* is a row or column vector of length $n(n-1)/2$, with the rowwise upper triangle of the similarity or dissimilarity matrix excluding the diagonal.

$$D_{12}\ D_{13}\ \ldots\ D_{1n}\ D_{23}\ D_{24}\ \ldots\ D_{2n}\ D_{34}\ D_{35}\ \ldots\ D_{3n}\ \ldots\ D_{n-1,n}$$

`s2d(standard | oneminus)` specifies how similarities are converted into dissimilarities. By default, the command dissimilarity data. Specifying `s2d()` indicates that your proximity data are similarities.

Dissimilarity data should have zeros on the diagonal (that is, an object is identical to itself) and nonnegative off-diagonal values. Dissimilarities need not satisfy the triangular inequality, $D(i,j)^2 \le D(i,h)^2 + D(h,j)^2$. Similarity data should have ones on the diagonal (that is, an object is identical to itself) and have off-diagonal values between zero and one. In either case, proximities should be symmetric. See option `force` if your data violate these assumptions.

The available `s2d()` options, `standard` and `oneminus`, are defined as follows:

`standard`	$\text{dissim}_{ij} = \sqrt{\text{sim}_{ii} + \text{sim}_{jj} - 2\text{sim}_{ij}} = \sqrt{2(1-\text{sim}_{ij})}$
`oneminus`	$\text{dissim}_{ij} = 1 - \text{sim}_{ij}$

`s2d(standard)` is the default.

`s2d()` should be specified only with measures in similarity form.

Model 2

`dimension(#)` specifies the dimension of the approximating configuration. # defaults to 2 and should not exceed the number of positive eigenvalues of the centered distance matrix.

`force` corrects problems with the supplied proximity information. `force` specifies that the dissimilarity matrix be symmetrized; the mean of D_{ij} and D_{ji} is used. Also, problems on the diagonal (similarities: $D_{ii} \ne 1$; dissimilarities: $D_{ii} \ne 0$) are fixed. `force` does not fix missing values or out-of-range values (that is, $D_{ij} < 0$ or similarities with $D_{ij} > 1$). Analogously, `force` symmetrizes the weight matrix.

`addconstant` specifies that if the double-centered distance matrix is not positive semidefinite (psd), a constant should be added to the squared distances to make it psd and, hence, Euclidean.

`weight(`*matname*`)` specifies a symmetric weight matrix with the same shape as the proximity matrix; that is, if `shape(lower)` is specified, the weight matrix must have this shape. Weights should be nonnegative. Missing weights are assumed to be 0. Weights must also be irreducible; that is, it is not possible to split the objects into disjointed groups with all intergroup weights 0. `weight()` is not allowed with `method(classical)`, but see `loss(strain)`.

Reporting

`neigen(#)` specifies the number of eigenvalues to be included in the table. The default is `neigen(10)`. Specifying `neigen(0)` suppresses the table. This option is allowed with classical MDS only.

`config` displays the table with the coordinates of the approximating configuration. This table may also be displayed using the postestimation command `estat config`; see [MV] **mds postestimation**.

`noplot` suppresses the graph of the approximating configuration. The graph can still be produced later via `mdsconfig`, which also allows the standard graphics options for fine-tuning the plot; see [MV] **mds postestimation**.

Minimization

These options are available only with `method(modern)` or `method(nonmetric)`:

`initialize(`*initopt*`)` specifies the initial values of the criterion minimization process.

`initialize(classical)`, the default, uses the solution from classical metric scaling as initial values. With `protect()`, all but the first run start from random perturbations from the classical solution. These random perturbations are independent and normally distributed with standard error equal to the product of `sdprotect(`*#*`)` and the standard deviation of the dissimilarities. `initialize(classical)` is the default.

`initialize(random)` starts an optimization process from a random starting configuration. These random configurations are generated from independent normal distributions with standard error equal to the product of `sdprotect(`*#*`)` and the standard deviation of the dissimilarities. The means of the configuration are irrelevant in MDS.

`initialize(from(`*matname*`)` [`, copy`]`)` sets the initial value to *matname*. *matname* should be an $n \times p$ matrix, where n is the number of observations and p is the number of dimensions, and the rows of *matname* should be ordered with respect to `id()`. The rownames of *matname* should be set correctly but will be ignored if `copy` is specified. With `protect()`, the second-to-last runs start from random perturbations from *matname*. These random perturbations are independent normal distributed with standard error equal to the product of `sdprotect(`*#*`)` and the standard deviation of the dissimilarities.

`tolerance(`*#*`)` specifies the tolerance for the configuration matrix. When the relative change in the configuration from one iteration to the next is less than or equal to `tolerance()`, the `tolerance()` convergence criterion is satisfied. The default is `tolerance(1e-4)`.

`ltolerance(`*#*`)` specifies the tolerance for the fit criterion. When the relative change in the fit criterion from one iteration to the next is less than or equal to `ltolerance()`, the `ltolerance()` convergence is satisfied. The default is `ltolerance(1e-8)`.

Both the `tolerance()` and `ltolerance()` criteria must be satisfied for convergence.

`iterate(`*#*`)` specifies the maximum number of iterations. The default is `iterate(1000)`.

`protect(`*#*`)` requests that # optimizations be performed and that the best of the solutions be reported. The default is `protect(1)`. See option `initialize()` on starting values of the runs. The output contains a table of the return code, the criterion value reached, and the seed of the random number used to generate the starting value. Specifying a large number, such as `protect(50)`, provides reasonable insight whether the solution found is a global minimum and not just a local minimum.

If any of the options `log`, `trace`, or `gradient` is also specified, iteration reports will be printed for each optimization run. Beware: this option will produce a lot of output.

`nolog` suppresses the iteration log, showing the progress of the minimization process.

`trace` displays the configuration matrices in the iteration report. Beware: this option may produce a lot of output.

`gradient` displays the gradient matrices of the fit criterion in the iteration report. Beware: this option may produce a lot of output.

The following option is available with `mdsmat` but is not shown in the dialog box:

`sdprotect(#)` sets a proportionality constant for the standard deviations of random configurations (`init(random)`) or random perturbations of given starting configurations (`init(classical)` or `init(from())`). The default is `sdprotect(1)`.

Remarks

Remarks are presented under the following headings:

Introduction

Multidimensional scaling (MDS) is a dimension-reduction and visualization technique. Dissimilarities (for instance, Euclidean distances) between observations in a high-dimensional space are represented in a lower-dimensional space (typically two dimensions) so that the Euclidean distance in the lower-dimensional space approximates the dissimilarities in the higher-dimensional space. See Kruskal and Wish (1978) for a brief nontechnical introduction to MDS. Young and Hamer (1987) and Borg and Groenen (2005) are more advanced textbook-sized treatments.

`mdsmat` performs MDS on a similarity or dissimilarity matrix *matname*. You may enter the matrix as a symmetric square matrix or as a vector (matrix with one row or column) with only the upper or lower triangle; see option `shape()` for details. *matname* should not contain missing values. The diagonal elements should be 0 (dissimilarities) or 1 (similarities). If you provide a square matrix (that is, `shape(full)`), names of the objects are obtained from the matrix row and column names. The row names should all be distinct, and the column names should equal the row names. Equation names, if any, are ignored. In any of the vectorized shapes, names are specified with option `names()`, and the matrix row and column names are ignored.

See option `force` if your matrix violates these assumptions.

In some applications, the similarity or dissimilarity of objects is defined by the researcher in terms of variables (attributes) measured on the objects. If you need to do MDS of this form, you should continue by reading [MV] **mds**.

Often, however, proximities—that is, similarities or dissimilarities—are measured directly. For instance, psychologists studying the similarities or dissimilarities in a set of stimuli—smells, sounds, faces, concepts, etc.—may have subjects rate the dissimilarity of pairs of stimuli. Linguists have subjects rate the similarity or dissimilarity of pairs of dialects. Political scientists have subjects rate the similarity or dissimilarity of political parties or candidates for political office. In other fields, relational data are studied that may be interpreted as proximities in a more abstract sense. For instance, sociologists study interpersonal contact frequencies in groups ("social networks"); these measures are sometimes interpreted in terms of similarities.

A wide variety of MDS methods have been proposed. `mdsmat` performs classical and modern scaling. Classical scaling has its roots in Young and Householder (1938) and Torgerson (1952). MDS requires complete and symmetric dissimilarity interval-level data. To explore modern scaling, see Borg and Groenen (2005). Classical scaling results in an eigen decomposition, whereas modern scaling is accomplished by the minimization of a loss function. Consequently, eigenvalues are not available after modern MDS.

Proximity data in a Stata matrix

To perform MDS of relational data, you must enter the data in a suitable format. One convenient format is a Stata matrix. You may want to use this format for analyzing data that you obtain from a printed source.

▷ Example 1

Many texts on multidimensional scaling illustrate how locations can be inferred from a table of geographic distances. We will do this too, using an example of distances in miles between 14 locations in Texas, representing both manufactured and natural treasures:

Big Bend	0	523	551	243	322	412	263	596	181	313	553
Corpus Christi	523	0	396	280	705	232	619	226	342	234	30
Dallas	551	396	0	432	643	230	532	243	494	317	426
Del Rio	243	280	432	0	427	209	339	353	62	70	310
El Paso	322	705	643	427	0	528	110	763	365	525	735
Enchanted Rock	412	232	230	209	528	0	398	260	271	69	262
Guadalupe Mnt	263	619	532	339	110	398	0	674	277	280	646
Houston	596	226	243	353	763	260	674	0	415	292	256
Langtry	181	342	494	62	365	271	277	415	0	132	372
Lost Maples	313	234	317	70	525	69	280	292	132	0	264
Padre Island	553	30	426	310	735	262	646	256	372	264	0
Pedernales Falls	434	216	235	231	550	40	420	202	293	115	246
San Antonio	397	141	274	154	564	91	475	199	216	93	171
StataCorp	426	205	151	287	606	148	512	83	318	202	316

Big Bend	434	397	426
Corpus Christi	216	141	205
Dallas	235	274	151
Del Rio	231	154	287
El Paso	550	564	606
Enchanted Rock	40	91	148
Guadalupe Mnt	420	475	512
Houston	202	199	83
Langtry	293	216	318
Lost Maples	115	93	202
Padre Island	246	171	316
Pedernales Falls	0	75	116
San Antonio	75	0	154
StataCorp	116	154	0

Note the inclusion of StataCorp, which is located in the twin cities of Bryan/College Station (BCS). To get the data into Stata, we will enter only the strictly upper triangle as a Stata one-dimensional matrix and collect the names in a `global` macro for later use. We are using the strictly upper triangle (that is, omitting the diagonal) because the diagonal of a dissimilarity matrix contains all zeros—there is no need to enter them.

```
. matrix input D = (
>    523  551  243  322  412  263  596  181  313  553  434  397  426
>         396  280  705  232  619  226  342  234   30  216  141  205
>              432  643  230  532  243  494  317  426  235  274  151
>                   427  209  339  353   62   70  310  231  154  287
>                        528  110  763  365  525  735  550  564  606
>                             398  260  271   69  262   40   91  148
>                                  674  277  280  646  420  475  512
>                                       415  292  256  202  199   83
>                                            132  372  293  216  318
>                                                 264  115   93  202
>                                                      246  171  316
>                                                            75  116
>                                                                154 )
. global names
>    Big_Bend      Corpus_Christi  Dallas         Del_Rio
>    El_Paso       Enchanted_Rock  Guadalupe_Mnt  Houston
>    Langtry       Lost_Maples     Padre_Island   Pedernales_Falls
>    San_Antonio   StataCorp
```

The triangular data entry is just typographical and is useful for catching data-entry errors. As far as Stata is concerned, we could have typed all the numbers in one long row. We use `matrix input D =` rather than `matrix define D =` or just `matrix D =` so that we do not have to separate entries with commas.

With the data now in Stata, we may use `mdsmat` to infer the locations in Texas and produce a map:

```
. mdsmat D, names($names) shape(uupper)
Classical metric multidimensional scaling
    dissimilarity matrix: D
                                           Number of obs         =        14
    Eigenvalues > 0        =       8       Mardia fit measure 1 =    0.7828
    Retained dimensions    =       2       Mardia fit measure 2 =    0.9823
```

Dimension	Eigenvalue	abs(eigenvalue) Percent	Cumul.	(eigenvalue)^2 Percent	Cumul.
1	691969.62	62.63	62.63	92.45	92.45
2	172983.05	15.66	78.28	5.78	98.23
3	57771.995	5.23	83.51	0.64	98.87
4	38678.916	3.50	87.01	0.29	99.16
5	19262.579	1.74	88.76	0.07	99.23
6	9230.7695	0.84	89.59	0.02	99.25
7	839.70996	0.08	89.67	0.00	99.25
8	44.989372	0.00	89.67	0.00	99.25

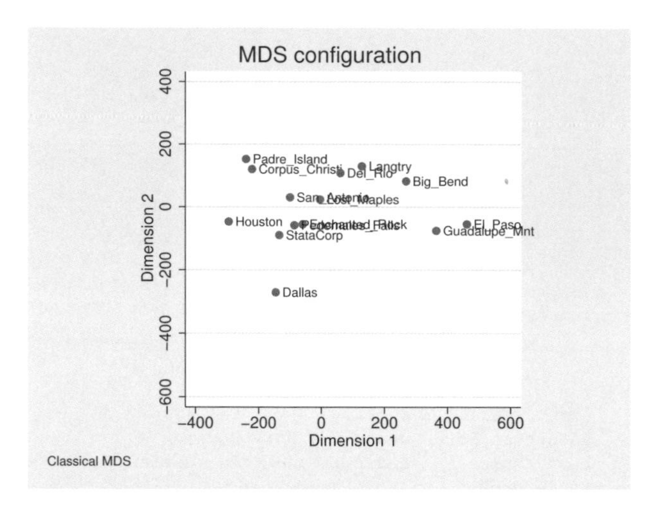

The representation of the distances in two dimensions provides a reasonable, but not great, fit; the percentage of eigenvalues accounted for is 78%.

By default, mdsmat produces a configuration plot. Enhancements to the configuration plot are possible using the mdsconfig postestimation graphics command; see [MV] **mds postestimation**. We present the configuration plot with the autoaspect option to obtain better use of the available space while preserving the equivalence of distance in the x and y axes. We negate the direction of the x axis with the xnegate option to flip the configuration horizontally and flip the direction of the y axis with the ynegate option. We also change the default title and control the placement of labels.

```
. set obs 14
obs was 0, now 14
. generate pos = 3
. replace pos = 4 in 6
(1 real change made)
. replace pos = 2 in 10
(1 real change made)
. mdsconfig, autoaspect xnegate ynegate mlabvpos(pos)
> title(MDS for 14 Texas locations)
```

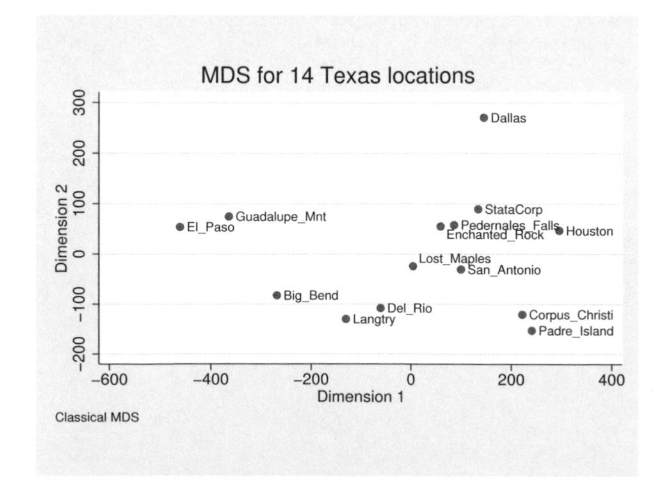

Look at the graph produced by `mdsconfig` after `mdsmat`. You will probably recognize a twisted (and slightly distorted) map of Texas. The vertical orientation of the map is not correctly north–south; you would probably want to turn the map some 20 degrees clockwise. Why didn't `mdsmat` get it right? It could not have concluded the correct rotation from the available distance information. Any orthogonal rotation of the map would produce the same distances. The orientation of the map is *not identified.* Finally, the "location" of the map cannot be inferred from the distances. Translating the coordinates does not change the distances. As far as `mdsmat` is concerned, Texas could be part of China.

◁

Modern MDS and local minimums

Modern MDS can converge to a local rather than a global minimum. We give an example where this happens and show how the `protect()` option can guard against this. `protect(#)` performs multiple minimizations and reports the best one. The output is explained in [MV] **mds**.

▷ Example 2

Continuing from where we left off, we perform modern MDS, using an initial random configuration with the `init(random(512308))` option. The number 512,308 sets the seed so that this run may be replicated.

```
. mdsmat D, names($names) shape(uupper) meth(modern) init(random(512308)) nolog
> noplot
(loss(stress) assumed)
(transform(identity) assumed)
Modern multidimensional scaling
    dissimilarity matrix: D
    Loss criterion: stress = raw_stress/norm(distances)
    Transformation: identity (no transformation)
                                                  Number of obs    =        14
                                                  Dimensions       =         2
    Normalization: principal                      Loss criterion   =    0.1639
. mdsconfig, autoaspect xnegate ynegate mlabvpos(pos) title(Padre Island Heads North?)
```

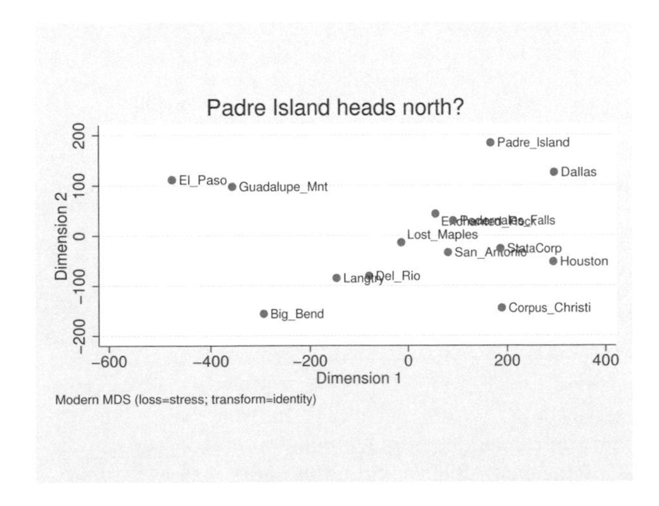

This graph has some resemblance to the one we saw before, but any Texan can assure you that Padre Island should not end up north of Dallas.

We check this result by rerunning with `protect(10)`. This will repeat the minimization and report the best result. Larger values of `protect()` give us more assurance that we have found the global minimum, but here `protect(10)` is sufficient to tell us that our original `mdsmat` found a local, not a global, minimum.

```
. mdsmat D, names($names) shape(uupper) meth(modern) init(random(512308)) nolog
> protect(10) noplot
(loss(stress) assumed)
(transform(identity) assumed)
run  mrc  #iter      lossval    seed random configuration

  1    0     54    .06180059    X4f2d0d0cc0f1b3343a20359ad2ef90f12c9e
  2    0     50    .06180059    X75445b1482f7cbeca4262f9391b6f5e1438e
  3    0     42     .0618006    X7fa3d0c0ff14ace95d0e18ed6b30fed811f1
  4    0     54     .0618006    Xbca058982f163fb5735186b9c7f2926234fd
  5    0     46     .0618006    X7ce217b44e3a967be8b4ef2ca63b10034c83
  6    0     47     .0618006    Xcaf38d70eba11618b2169652cd28c9a63803
  7    0     83    .08581202    Xa3aacf488b86c16fde811ec95b22fdf71d0a
  8    0    111    .08581202    Xaaa0cebc256ebbd6da55f007559421272862
  9    0     74    .08581202    X448b7f64f30e23ca382ef3eeeaec6da41116
 10    0     98     .1639279    X6c6495e8c43f462544a474abacbdd93d0fe1

Modern multidimensional scaling
    dissimilarity matrix: D

    Loss criterion: stress = raw_stress/norm(distances)
    Transformation: identity (no transformation)
                                                Number of obs    =        14
                                                Dimensions       =         2
    Normalization: principal                    Loss criterion   =    0.0618
. mdsconfig, autoaspect xnegate ynegate mlabvpos(pos)
> title(Padre Island is back where it belongs)
```

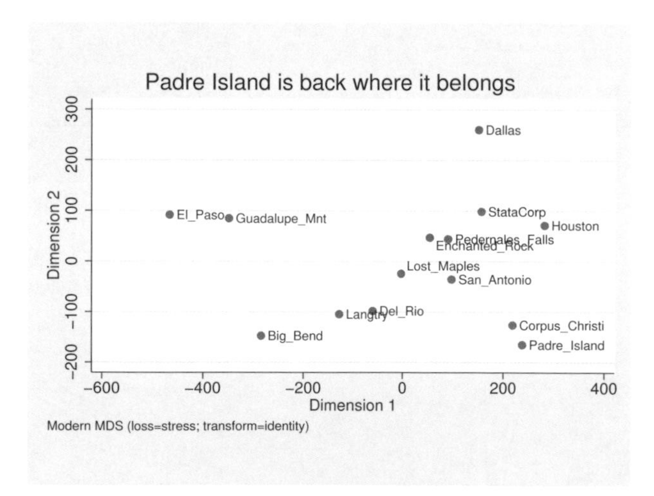

The original run had a loss criterion of 0.1639, but after using the `protect()` option the loss criterion was much lower—0.0618. We also see that Padre Island is back down south where it belongs. It is clear that the original run converged to a local minimum. You can see the original results appear as the final output line of the first table in the output after using `protect(10)`. The seed in the table is a hexadecimal representation of how the seed is stored internally. The number 512,308 in `init(random(512308))` is convenient shorthand for specifying the seed; the two are equivalent. If we wish, we could repeat the command with a larger value of `protect()` to assure ourselves that 0.0618 is indeed the global minimum.

◁

After `mdsmat`, all MDS postestimation tools are available. For instance, you may analyze residuals with `estat quantile`, you may produce a Shepard diagram, etc.; see [MV] **mds postestimation**.

Saved results

`mdsmat` saves the following in `e()`:

Scalars	
`e(N)`	number of rows or columns (i.e., number of observations)
`e(p)`	number of dimensions in the approximating configuration
`e(np)`	number of strictly positive eigenvalues
`e(addcons)`	constant added to squared dissimilarities to force positive semidefiniteness
`e(mardia1)`	Mardia measure 1
`e(mardia2)`	Mardia measure 2
`e(critval)`	loss criterion value
`e(wsum)`	sum of weights
`e(alpha)`	parameter of `transform(power)`
`e(ic)`	iteration count
`e(rc)`	return code
`e(converged)`	`1` if converged, `0` otherwise
Macros	
`e(cmd)`	`mdsmat`
`e(cmdline)`	command as typed
`e(method)`	`classical` or `modern` MDS method
`e(method2)`	`nonmetric`, if `method(nonmetric)`
`e(loss)`	loss criterion
`e(losstitle)`	description loss criterion
`e(dmatrix)`	name of analyzed matrix
`e(tfunction)`	`identity`, `power`, or `monotonic`, transformation function
`e(transftitle)`	description of transformation
`e(mxlen)`	maximum length of category labels
`e(dtype)`	`similarity` or `dissimilarity`; type of proximity data
`e(s2d)`	`standard` or `oneminus` (when `e(dtype)` is `similarity`)
`e(unique)`	`1` if eigenvalues are distinct, `0` otherwise
`e(init)`	initialization method
`e(iseed)`	seed for `init(random)`
`e(seed)`	seed for solution
`e(norm)`	normalization method
`e(targetmatrix)`	name of target matrix for `normalize(target)`
`e(properties)`	`nob noV` for modern or nonmetric MDS; `nob noV eigen` for classical MDS
`e(estat_cmd)`	program used to implement `estat`
`e(predict)`	program used to implement `predict`
`e(marginsnotok)`	predictions disallowed by `margins`
Matrices	
`e(D)`	dissimilarity matrix
`e(Disparities)`	disparity matrix for nonmetric MDS
`e(Y)`	approximating configuration coordinates
`e(Ev)`	eigenvalues
`e(W)`	weight matrix
`e(norm_stats)`	normalization statistics
`e(linearf)`	two element vector defining the linear transformation; distance equals first element plus second element times dissimilarity

Methods and formulas

`mdsmat` is implemented as an ado-file.

Methods and formulas are presented under the following headings:

Classical multidimensional scaling
Modern multidimensional scaling
Conversion of similarities to dissimilarities

Classical multidimensional scaling

Let $\mathbf{D}$ be an $n \times n$ dissimilarity matrix. The matrix $\mathbf{D}$ is said to be *Euclidean* if there are coordinates $\mathbf{Y}$ so that

$$D_{ij}^2 = (\mathbf{Y}_i - \mathbf{Y}_j)(\mathbf{Y}_i - \mathbf{Y}_j)'$$

Here $\mathbf{Y}_i$ and $\mathbf{Y}_j$ are the ith and jth column vectors extracted from $\mathbf{Y}$. Let $\mathbf{A} = -(1/2)\mathbf{D} \odot \mathbf{D}$, with $\odot$ being the Hadamard or elementwise matrix product, and define $\mathbf{B}$ as the double-centered distance matrix

$$\mathbf{B} = \mathbf{HAH} \qquad \text{with} \qquad \mathbf{H} = \mathbf{I} - \frac{1}{n}\mathbf{11}'$$

$\mathbf{D}$ is Euclidean if and only if $\mathbf{B}$ is positive semidefinite. Assume for now that $\mathbf{D}$ is indeed Euclidean. The spectral or eigen decomposition of $\mathbf{B}$ is written as $\mathbf{B} = \mathbf{U\Lambda U}'$, with $\mathbf{U}$ the orthonormal matrix of eigenvectors normed to 1, and $\mathbf{\Lambda}$ a diagonal matrix with nonnegative values (the eigenvalues of $\mathbf{B}$) in decreasing order. The coordinates $\mathbf{Y}$ are defined in terms of the spectral decomposition $\mathbf{Y} = \mathbf{U\Lambda}^{1/2}$. These coordinates are centered $\mathbf{Y}'\mathbf{1} = \mathbf{0}$.

The spectral decomposition can also be used to obtain a low-dimensional configuration $\widetilde{\mathbf{Y}}$, $n \times p$, so that the interrow distances of $\widetilde{\mathbf{Y}}$ approximate $\mathbf{D}$. Mardia, Kent, and Bibby (1979, sec. 14.4) discuss some characterizations under which the leading p columns of $\mathbf{Y}$ are an optimal choice of $\widetilde{\mathbf{Y}}$. These characterizations also apply to the case when $\mathbf{B}$ is not positive semidefinite, so some of the λ's are negative; we require that $\lambda_p > 0$.

Various other approaches have been proposed to deal with the case when the matrix $\mathbf{B}$ is not positive semidefinite, that is, when $\mathbf{B}$ has negative eigenvalues (see Cox and Cox 2001, 45–48). An easy solution is to add a constant to the off-diagonal elements of $\mathbf{D} \odot \mathbf{D}$ to make $\mathbf{B}$ positive semidefinite. The smallest such constant is $-2\lambda_n$, where λ_n is the smallest eigenvalue of $\mathbf{B}$ (Lingoes 1971). See Cailliez (1983) for a solution to the additive constant problem in terms of the dissimilarities instead of the squared dissimilarities.

Goodness-of-fit statistics for a configuration in p dimensions have been proposed by Mardia (1978) in characterizations of optimality properties of the classical solution

$$\text{Mardia}_1 = \frac{\sum_{i=1}^p |\lambda_i|}{\sum_{i=1}^n |\lambda_i|}$$

and

$$\text{Mardia}_2 = \frac{\sum_{i=1}^p \lambda_i^2}{\sum_{i=1}^n \lambda_i^2}$$

Modern multidimensional scaling

Let $\mathbf{D}$ be a symmetric $n \times n$ matrix of observed dissimilarities. We assume that proximity data in the form of similarities have already been transformed into dissimilarities. Let $\mathbf{W}$ be an $n \times n$ matrix of nonnegative weights. With unweighted MDS, we define $W_{ij} = 1$. For a configuration of n points in k-dimensional space represented by the $n \times k$ matrix $\mathbf{Y}$, let $\mathbf{B(Y)}$ be the $n \times n$ matrix of Euclidean distances between the rows of $\mathbf{Y}$. We consider $\mathcal{F}$ to be some class of permitted transformations from $n \times n$ real matrices to $n \times n$ real matrices.

Modern metric and nonmetric multidimensional scaling involves the minimization of a loss criterion

$$L\left\{f(\mathbf{D}), \mathbf{B}(\mathbf{Y}), \mathbf{W}\right\}$$

over the configurations $\mathbf{Y}$ and transformations $f \in \mathcal{F}$. Whether a scaling method is labeled metric or nonmetric depends on the class $\mathcal{F}$. In nonmetric scaling, $\mathcal{F}$ is taken to be the class of monotonic functions. If $\mathcal{F}$ is a regular parametrized set of functions, one commonly labels the scaling as metric.

$\mathbf{D}$ is the matrix of proximities or dissimilarities, $\mathbf{B(Y)}$ is the matrix of distances, and the result of $f(\mathbf{D}) = \widehat{\mathbf{D}}$ is the matrix of disparities.

The `mdsmat` command supports the following loss criteria:

(1) **stress** specifies Kruskal's stress-1 criterion: the Euclidean norm of the difference between the distances and the disparities, normalized by the Euclidean norm of the distances.

$$\textbf{stress}(\widehat{\mathbf{D}}, \mathbf{B}, \mathbf{W}) = \left\{ \frac{\sum_{ij} W_{ij}(B_{ij} - \widehat{D}_{ij})^2}{\sum_{ij} W_{ij} B_{ij}^2} \right\}^{1/2}$$

(2) **nstress** specifies the square root of the normalized stress criterion: the Euclidean norm of the difference between the distances and the disparities, normalized by the Euclidean norm of the disparities.

$$\textbf{nstress}(\widehat{\mathbf{D}}, \mathbf{B}, \mathbf{W}) = \left\{ \frac{\sum_{ij} W_{ij}(B_{ij} - \widehat{D}_{ij})^2}{\sum_{ij} W_{ij} \widehat{D}_{ij}^2} \right\}^{1/2}$$

nstress normalizes with the disparities, **stress** with the distances.

(3) **sammon** specifies the Sammon mapping criterion (Sammon 1969; Neimann and Weiss 1979): the sum of the scaled, squared differences between the distances and the disparities, normalized by the sum of the disparities.

$$\textbf{sammon}(\widehat{\mathbf{D}}, \mathbf{B}, \mathbf{W}) = \frac{\sum_{ij} W_{ij}(B_{ij} - \widehat{D}_{ij})^2 / \widehat{D}_{ij}}{\sum_{ij} W_{ij} \widehat{D}_{ij}}$$

(4) **sstress** specifies the squared stress criterion: the Euclidean norm of the difference between the squared distances and the squared disparities, normalized by the Euclidean norm of the squared distances.

$$\textbf{sstress}(\widehat{\mathbf{D}}, \mathbf{B}, \mathbf{W}) = \left\{ \frac{\sum_{ij} W_{ij}(B_{ij}^2 - \widehat{D}_{ij}^2)^2}{\sum_{ij} W_{ij} B_{ij}^4} \right\}^{1/2}$$

(5) **nsstress** specifies the normalized squared stress criterion: the Euclidean norm of the difference between the squared distances and the squared disparities, normalized by the Euclidean norm of the squared disparities.

$$\textbf{nsstress}(\widehat{\mathbf{D}}, \mathbf{B}, \mathbf{W}) = \left\{ \frac{\sum_{ij} W_{ij} (B_{ij}^2 - \widehat{D}_{ij}^2)^2}{\sum_{ij} W_{ij} \widehat{D}_{ij}^4} \right\}^{1/2}$$

nsstress normalizes with the disparities, **sstress** with the distances.

(6) **strain** specifies the strain criterion:

$$\textbf{strain}(\widehat{\mathbf{D}}, \mathbf{B}, \mathbf{W}) = \frac{\sqrt{\text{trace}(\mathbf{X}'\mathbf{X})}}{\sum_{ij} W_{ij}}$$

where

$$\mathbf{X} = \mathbf{W} \odot \left\{ \widehat{\mathbf{D}} - \mathbf{B}(\widetilde{\mathbf{Y}}) \right\}$$

where $\widetilde{\mathbf{Y}}$ is the centered configuration of $\mathbf{Y}$. Without weights, $W_{ij} = 1$, and without transformation, that is, $\widehat{\mathbf{D}} = \mathbf{D}$, minimization of the strain criterion is equivalent to classical metric scaling.

The `mdsmat` command supports three classes of permitted transformations, $f \in \mathcal{F}$: (1) the class of all weakly monotonic transformations, (2) the power class of functions where f is defined elementwise on $\mathbf{D}$ as $f(D_{ij}, \alpha) = D_{ij}^{\alpha}$ (Critchley 1978; Cox and Cox 2001), and (3) the trivial identity case of $f(\mathbf{D}) = \mathbf{D}$.

Minimization of a loss criterion with respect to the configuration $\mathbf{Y}$ and the permitted transformation $f \in \mathcal{F}$ is performed with an alternating algorithm in which the configuration $\mathbf{Y}$ is modified (the C-step) and the transformation f is adjusted (the T-step) to reduce loss. Obviously, no T-step is made with the identity transformation. The classical solution is the default starting configuration. Iteration continues until the C-step and T-step reduce loss by less than the tolerance for convergence or the maximum number of iterations is performed. The C-step is taken by steepest descent using analytical gradients and an optimal stepsize computed using Brent's bounded minimization (Brent 1973). The implementation of the T-step varies with the specified class of transformations. In the nonmetric case of monotonic transformations, we use isotonic regression (Kruskal 1964a, 1964b; Cox and Cox 2001), using the primary approach to ties (Borg and Groenen 2005, 40). For power transformations, we again apply Brent's minimization method.

Given enough iterations, convergence is usually not a problem. However, the alternating algorithm may not converge to a global minimum. `mdsmat` provides some protection by repeated runs from different initial configurations. However, as Euclidean distances $\mathbf{B}(\mathbf{Y})$ are invariant with respect to isometric transformations (rotations, translations) of $\mathbf{Y}$, some caution is required to compare different runs and, similarly, to compare the configurations obtained from different scaling methods. `mdsmat` normalizes the optimal configuration by centering and via the orthogonal Procrustean rotation without dilation toward the classical or a user-specified solution; see [MV] **procrustes**.

Conversion of similarities to dissimilarities

If a similarity measure was selected, it is turned into a dissimilarity measure by using one of two methods. The *standard* conversion method is

$$\text{dissim}_{ij} = \sqrt{\text{sim}_{ii} + \text{sim}_{jj} - 2\text{sim}_{ij}}$$

With the similarity of an object to itself being 1, this is equivalent to

$$\text{dissim}_{ij} = \sqrt{2(1 - \text{sim}_{ij})}$$

This conversion method has the attractive property that it transforms a positive-semidefinite similarity matrix into a Euclidean distance matrix (see Mardia, Kent, and Bibby 1979, 402).

We also offer the *one-minus* method

$$\text{dissim}_{ij} = 1 - \text{sim}_{ij}$$

References

Borg, I., and P. J. F. Groenen. 2005. *Modern Multidimensional Scaling: Theory and Applications.* 2nd ed. New York: Springer.

Brent, R. P. 1973. *Algorithms for Minimization without Derivatives.* Englewood Cliffs, NJ: Prentice Hall. (Reprinted in paperback by Dover Publications, Mineola, NY, January 2002).

Cailliez, F. 1983. The analytical solution of the additive constant problem. *Psychometrika* 48: 305–308.

Cox, T. F., and M. A. A. Cox. 2001. *Multidimensional Scaling.* 2nd ed. Boca Raton, FL: Chapman & Hall/CRC.

Critchley, F. 1978. Multidimensional scaling: A short critique and a new method. In *COMPSTAT 1978: Proceedings in Computational Statistics*, ed. L. C. A. Corsten and J. Hermans. Vienna: Physica.

Kruskal, J. B. 1964a. Multidimensional scaling by optimizing goodness of fit to a nonmetric hypothesis. *Psychometrika* 29: 1–27.

——. 1964b. Nonmetric multidimensional scaling: A numerical method. *Psychometrika* 29: 115–129.

Kruskal, J. B., and M. Wish. 1978. *Multidimensional Scaling.* Newbury Park, CA: Sage.

Lingoes, J. C. 1971. Some boundary conditions for a monotone analysis of symmetric matrices. *Psychometrika* 36: 195–203.

Mardia, K. V. 1978. Some properties of classical multidimensional scaling. *Communications in Statistics, Theory and Methods* 7: 1233–1241.

Mardia, K. V., J. T. Kent, and J. M. Bibby. 1979. *Multivariate Analysis.* London: Academic Press.

Neimann, H., and J. Weiss. 1979. A fast-converging algorithm for nonlinear mapping of high-dimensional data to a plane. *IEEE Transactions on Computers* 28: 142–147.

Sammon, J. W., Jr. 1969. A nonlinear mapping for data structure analysis. *IEEE Transactions on Computers* 18: 401–409.

Torgerson, W. S. 1952. Multidimensional scaling: I. Theory and method. *Psychometrika* 17: 401–419.

Young, F. W., and R. M. Hamer. 1987. *Multidimensional Scaling: History, Theory, and Applications.* Hillsdale, NJ: Erlbaum Associates.

Young, G., and A. S. Householder. 1938. Discussion of a set of points in terms of their mutual distances. *Psychometrika* 3: 19–22.

Also see

[MV] **mds postestimation** — Postestimation tools for mds, mdsmat, and mdslong

[MV] **biplot** — Biplots

[MV] **ca** — Simple correspondence analysis

[MV] **factor** — Factor analysis

[MV] **mds** — Multidimensional scaling for two-way data

[MV] **mdslong** — Multidimensional scaling of proximity data in long format

[MV] **pca** — Principal component analysis

[U] **20 Estimation and postestimation commands**

Title

measure_option — Option for similarity and dissimilarity measures

Syntax

command ..., ... measure(*measure*) ...

or

command ..., ... *measure* ...

measure	Description
cont_measure	similarity or dissimilarity measure for continuous data
binary_measure	similarity measure for binary data
mixed_measure	dissimilarity measure for a mix of binary and continuous data

cont_measure	Description
L2	Euclidean distance (Minkowski with argument 2)
Euclidean	alias for L2
L(2)	alias for L2
L2squared	squared Euclidean distance
Lpower(2)	alias for L2squared
L1	absolute-value distance (Minkowski with argument 1)
absolute	alias for L1
cityblock	alias for L1
manhattan	alias for L1
L(1)	alias for L1
Lpower(1)	alias for L1
Linfinity	maximum-value distance (Minkowski with infinite argument)
maximum	alias for Linfinity
L(#)	Minkowski distance with # arguments
Lpower(#)	Minkowski distance with # arguments raised to # power
Canberra	Canberra distance
correlation	correlation coefficient similarity measure
angular	angular separation similarity measure
angle	alias for angular

binary_measure	Description
matching	simple matching similarity coefficient
Jaccard	Jaccard binary similarity coefficient
Russell	Russell and Rao similarity coefficient
Hamann	Hamann similarity coefficient
Dice	Dice similarity coefficient
antiDice	anti-Dice similarity coefficient
Sneath	Sneath and Sokal similarity coefficient
Rogers	Rogers and Tanimoto similarity coefficient
Ochiai	Ochiai similarity coefficient
Yule	Yule similarity coefficient
Anderberg	Anderberg similarity coefficient
Kulczynski	Kulczynski similarity coefficient
Pearson	Pearson's ϕ similarity coefficient
Gower2	similarity coefficient with same denominator as Pearson

mixed_measure	Description
Gower	Gower's dissimilarity coefficient

Description

Several commands have options that allow you to specify a similarity or dissimilarity measure designated as *measure* in the syntax; see [MV] **cluster**, [MV] **mds**, [MV] **discrim knn**, and [MV] **matrix dissimilarity**. These options are documented here. Most analysis commands (for example, cluster and mds) transform similarity measures to dissimilarity measures as needed.

Options

Measures are divided into those for continuous data and binary data. *measure* is not case sensitive. Full definitions are presented in *Similarity and dissimilarity measures for continuous data*, *Similarity measures for binary data*, and *Dissimilarity measures for mixed data*.

The similarity or dissimilarity measure is most often used to determine the similarity or dissimilarity between observations. However, sometimes the similarity or dissimilarity between variables is of interest.

Similarity and dissimilarity measures for continuous data

Here are the similarity and dissimilarity measures for continuous data available in Stata. In the following formulas, p represents the number of variables, N is the number of observations, and x_{iv} denotes the value of observation i for variable v.

The formulas are presented in two forms. The first is the formula used when computing the similarity or dissimilarity between observations. The second is the formula used when computing the similarity or dissimilarity between variables.

`L2` (aliases `Euclidean` and `L(2)`)
requests the Minkowski distance metric with argument 2. For comparing observations i and j, the formula is

$$\left\{\sum_{a=1}^{p}(x_{ia}-x_{ja})^2\right\}^{1/2}$$

and for comparing variables u and v the formula is

$$\left\{\sum_{k=1}^{N}(x_{ku}-x_{kv})^2\right\}^{1/2}$$

`L2` is best known as Euclidean distance and is the default dissimilarity measure for `discrim knn`, `mds`, `matrix dissimilarity`, and all the `cluster` subcommands except for `centroidlinkage`, `medianlinkage`, and `wardslinkage`, which default to using `L2squared`; see [MV] **discrim knn**, [MV] **mds**, [MV] **matrix dissimilarity**, and [MV] **cluster**.

`L2squared` (alias `Lpower(2)`)
requests the square of the Minkowski distance metric with argument 2. For comparing observations i and j, the formula is

$$\sum_{a=1}^{p}(x_{ia}-x_{ja})^2$$

and for comparing variables u and v, the formula is

$$\sum_{k=1}^{N}(x_{ku}-x_{kv})^2$$

`L2squared` is best known as squared Euclidean distance and is the default dissimilarity measure for the `centroidlinkage`, `medianlinkage`, and `wardslinkage` subcommands of `cluster`; see [MV] **cluster**.

`L1` (aliases `absolute`, `cityblock`, `manhattan`, and `L(1)`)
requests the Minkowski distance metric with argument 1. For comparing observations i and j, the formula is

$$\sum_{a=1}^{p}|x_{ia}-x_{ja}|$$

and for comparing variables u and v, the formula is

$$\sum_{k=1}^{N}|x_{ku}-x_{kv}|$$

`L1` is best known as absolute-value distance.

`Linfinity` (alias `maximum`)
requests the Minkowski distance metric with infinite argument. For comparing observations i and j, the formula is

$$\max_{a=1,\ldots,p} |x_{ia} - x_{ja}|$$

and for comparing variables u and v, the formula is

$$\max_{k=1,\ldots,N} |x_{ku} - x_{kv}|$$

`Linfinity` is best known as maximum-value distance.

`L(#)`
requests the Minkowski distance metric with argument #. For comparing observations i and j, the formula is

$$\left(\sum_{a=1}^{p} |x_{ia} - x_{ja}|^{\#}\right)^{1/\#} \qquad \# \geq 1$$

and for comparing variables u and v, the formula is

$$\left(\sum_{k=1}^{N} |x_{ku} - x_{kv}|^{\#}\right)^{1/\#} \qquad \# \geq 1$$

We discourage using extremely large values for #. Because the absolute value of the difference is being raised to the value of #, depending on the nature of your data, you could experience numeric overflow or underflow. With a large value of #, the `L()` option will produce results similar to those of the `Linfinity` option. Use the numerically more stable `Linfinity` option instead of a large value for # in the `L()` option.

See Anderberg (1973) for a discussion of the Minkowski metric and its special cases.

`Lpower(#)`
requests the Minkowski distance metric with argument #, raised to the # power. For comparing observations i and j, the formula is

$$\sum_{a=1}^{p} |x_{ia} - x_{ja}|^{\#} \qquad \# \geq 1$$

and for comparing variables u and v, the formula is

$$\sum_{k=1}^{N} |x_{ku} - x_{kv}|^{\#} \qquad \# \geq 1$$

As with `L(#)`, we discourage using extremely large values for #; see the discussion above.

`Canberra`
requests the following distance metric when comparing observations i and j

$$\sum_{a=1}^{p} \frac{|x_{ia} - x_{ja}|}{|x_{ia}| + |x_{ja}|}$$

and the following distance metric when comparing variables u and v

$$\sum_{k=1}^{N} \frac{|x_{ku} - x_{kv}|}{|x_{ku}| + |x_{kv}|}$$

When comparing observations, the Canberra metric takes values between 0 and p, the number of variables. When comparing variables, the Canberra metric takes values between 0 and N, the number of observations; see Gordon (1999) and Gower (1985). Gordon (1999) explains that the Canberra distance is sensitive to small changes near zero.

`correlation`
requests the correlation coefficient similarity measure. For comparing observations i and j, the formula is

$$\frac{\sum_{a=1}^{p}(x_{ia} - \overline{x}_{i.})(x_{ja} - \overline{x}_{j.})}{\left\{\sum_{a=1}^{p}(x_{ia} - \overline{x}_{i.})^2 \sum_{b=1}^{p}(x_{jb} - \overline{x}_{j.})^2\right\}^{1/2}}$$

and for comparing variables u and v, the formula is

$$\frac{\sum_{k=1}^{N}(x_{ku} - \overline{x}_{.u})(x_{kv} - \overline{x}_{.v})}{\left\{\sum_{k=1}^{N}(x_{ku} - \overline{x}_{.u})^2 \sum_{l=1}^{N}(x_{lv} - \overline{x}_{.v})^2\right\}^{1/2}}$$

where $\overline{x}_{i.} = (\sum_{a=1}^{p} x_{ia})/p$ and $\overline{x}_{.u} = (\sum_{k=1}^{N} x_{ku})/N$.

The correlation similarity measure takes values between -1 and 1. With this measure, the relative direction of the two vectors is important. The correlation similarity measure is related to the angular separation similarity measure (described next). The correlation similarity measure gives the cosine of the angle between the two vectors measured from the mean; see Gordon (1999).

`angular` (alias `angle`)
requests the angular separation similarity measure. For comparing observations i and j, the formula is

$$\frac{\sum_{a=1}^{p} x_{ia}x_{ja}}{\left(\sum_{a=1}^{p} x_{ia}^2 \sum_{b=1}^{p} x_{jb}^2\right)^{1/2}}$$

and for comparing variables u and v, the formula is

$$\frac{\sum_{k=1}^{N} x_{ku}x_{kv}}{\left(\sum_{k=1}^{N} x_{ku}^2 \sum_{l=1}^{N} x_{lv}^2\right)^{1/2}}$$

The angular separation similarity measure is the cosine of the angle between the two vectors measured from zero and takes values from -1 to 1; see Gordon (1999).

Similarity measures for binary data

Similarity measures for binary data are based on the four values from the cross-tabulation of observation i and j (when comparing observations) or variables u and v (when comparing variables).

For comparing observation i and j, the cross-tabulation is

		obs. j	
		1	0
obs.	1	a	b
i	0	c	d

a is the number of variables where observations i and j both had ones, and d is the number of variables where observations i and j both had zeros. The number of variables where observation i is one and observation j is zero is b, and the number of variables where observation i is zero and observation j is one is c.

For comparing variables u and v, the cross-tabulation is

		var. v	
		1	0
var.	1	a	b
u	0	c	d

a is the number of observations where variables u and v both had ones, and d is the number of observations where variables u and v both had zeros. The number of observations where variable u is one and variable v is zero is b, and the number of observations where variable u is zero and variable v is one is c.

Stata treats nonzero values as one when a binary value is expected. Specifying one of the binary similarity measures imposes this behavior unless some other option overrides it (for instance, the `allbinary` option of `matrix dissimilarity`; see [MV] **matrix dissimilarity**).

Hubálek (1982) gives an extensive list of binary similarity measures. Gower (1985) lists 15 binary similarity measures, 14 of which are implemented in Stata. (The excluded measure has many cases where the quantity is undefined, so it was not implemented.) Anderberg (1973) gives an interesting table where many of these measures are compared based on whether the zero–zero matches are included in the numerator, whether these matches are included in the denominator, and how the weighting of matches and mismatches is handled. Hilbe (1992b, 1992a) implemented an early Stata command for computing some of these (as well as other) binary similarity measures.

The formulas for some of these binary similarity measures are undefined when either one or both of the vectors (observations or variables depending on which are being compared) are all zeros (or, sometimes, all ones). Gower (1985) says concerning these cases, "These coefficients are then conventionally assigned some appropriate value, usually zero."

The following binary similarity coefficients are available. Unless stated otherwise, the similarity measures range from 0 to 1.

`matching`
requests the simple matching (Zubin 1938, Sokal and Michener 1958) binary similarity coefficient

$$\frac{a+d}{a+b+c+d}$$

which is the proportion of matches between the 2 observations or variables.

`Jaccard`
requests the Jaccard (1901, 1908) binary similarity coefficient

$$\frac{a}{a+b+c}$$

which is the proportion of matches when at least one of the vectors had a one. If both vectors are all zeros, this measure is undefined. Stata then declares the answer to be one, meaning perfect agreement. This is a reasonable choice for most applications and will cause an all-zero vector to have similarity of one only with another all-zero vector. In all other cases, an all-zero vector will have Jaccard similarity of zero to the other vector.

The Jaccard coefficient was discovered earlier by Gilbert (1884).

`Russell`
requests the Russell and Rao (1940) binary similarity coefficient

$$\frac{a}{a+b+c+d}$$

`Hamann`
requests the Hamann (1961) binary similarity coefficient

$$\frac{(a+d)-(b+c)}{a+b+c+d}$$

which is the number of agreements minus disagreements divided by the total. The Hamann coefficient ranges from -1, perfect disagreement, to 1, perfect agreement. The Hamann coefficient is equal to twice the simple matching coefficient minus 1.

`Dice`
requests the Dice binary similarity coefficient

$$\frac{2a}{2a+b+c}$$

suggested by Czekanowski (1932), Dice (1945), and Sørensen (1948). The Dice coefficient is similar to the Jaccard similarity coefficient but gives twice the weight to agreements. Like the Jaccard coefficient, the Dice coefficient is declared by Stata to be one if both vectors are all zero, thus avoiding the case where the formula is undefined.

`antiDice`
requests the binary similarity coefficient

$$\frac{a}{a+2(b+c)}$$

which is credited to Anderberg (1973) but was shown earlier by Sokal and Sneath (1963, 129). We did not call this the Anderberg coefficient because there is another coefficient better known by that name; see the `Anderberg` option. The name anti-Dice is our creation. This coefficient takes the opposite view from the Dice coefficient and gives double weight to disagreements. As with the Jaccard and Dice coefficients, the anti-Dice coefficient is declared to be one if both vectors are all zeros.

`Sneath`
requests the Sneath and Sokal (1962) binary similarity coefficient

$$\frac{2(a+d)}{2(a+d)+(b+c)}$$

which is similar to the simple matching coefficient but gives double weight to matches. Also compare the Sneath and Sokal coefficient with the Dice coefficient, which differs only in whether it includes d.

`Rogers`
requests the Rogers and Tanimoto (1960) binary similarity coefficient

$$\frac{a+d}{(a+d)+2(b+c)}$$

which takes the opposite approach from the Sneath and Sokal coefficient and gives double weight to disagreements. Also compare the Rogers and Tanimoto coefficient with the anti-Dice coefficient, which differs only in whether it includes d.

`Ochiai`
requests the Ochiai (1957) binary similarity coefficient

$$\frac{a}{\left\{(a+b)(a+c)\right\}^{1/2}}$$

The formula for the Ochiai coefficient is undefined when one or both of the vectors being compared are all zeros. If both are all zeros, Stata declares the measure to be one, and if only one of the two vectors is all zeros, the measure is declared to be zero.

The Ochiai coefficient was presented earlier by Driver and Kroeber (1932).

`Yule`
requests the Yule (see Yule [1900] and Yule and Kendall [1950]) binary similarity coefficient

$$\frac{ad-bc}{ad+bc}$$

which ranges from -1 to 1. The formula for the Yule coefficient is undefined when one or both of the vectors are either all zeros or all ones. Stata declares the measure to be 1 when $b+c=0$, meaning that there is complete agreement. Stata declares the measure to be -1 when $a+d=0$, meaning that there is complete disagreement. Otherwise, if $ad-bc=0$, Stata declares the measure to be 0. These rules, applied before using the Yule formula, avoid the cases where the formula would produce an undefined result.

`Anderberg`

requests the Anderberg binary similarity coefficient

$$\left(\frac{a}{a+b}+\frac{a}{a+c}+\frac{d}{c+d}+\frac{d}{b+d}\right)\Big/4$$

The Anderberg coefficient is undefined when one or both vectors are either all zeros or all ones. This difficulty is overcome by first applying the rule that if both vectors are all ones (or both vectors are all zeros), the similarity measure is declared to be one. Otherwise, if any of the marginal totals ($a+b$, $a+c$, $c+d$, $b+d$) are zero, then the similarity measure is declared to be zero.

Though this similarity coefficient is best known as the Anderberg coefficient, it appeared earlier in Sokal and Sneath (1963, 130).

`Kulczynski`

requests the Kulczynski (1927) binary similarity coefficient

$$\left(\frac{a}{a+b}+\frac{a}{a+c}\right)\Big/2$$

The formula for this measure is undefined when one or both of the vectors are all zeros. If both vectors are all zeros, Stata declares the similarity measure to be one. If only one of the vectors is all zeros, the similarity measure is declared to be zero.

`Pearson`

requests Pearson's (1900) ϕ binary similarity coefficient

$$\frac{ad-bc}{\left\{(a+b)(a+c)(d+b)(d+c)\right\}^{1/2}}$$

which ranges from -1 to 1. The formula for this coefficient is undefined when one or both of the vectors are either all zeros or all ones. Stata declares the measure to be 1 when $b+c=0$, meaning that there is complete agreement. Stata declares the measure to be -1 when $a+d=0$, meaning that there is complete disagreement. Otherwise, if $ad-bc=0$, Stata declares the measure to be 0. These rules, applied before using Pearson's ϕ coefficient formula, avoid the cases where the formula would produce an undefined result.

`Gower2`

requests the binary similarity coefficient

$$\frac{ad}{\left\{(a+b)(a+c)(d+b)(d+c)\right\}^{1/2}}$$

which is presented by Gower (1985) but appeared earlier in Sokal and Sneath (1963, 130). Stata uses the name `Gower2` to avoid confusion with the better-known Gower coefficient, which is used with a mix of binary and continuous data.

The formula for this similarity measure is undefined when one or both of the vectors are all zeros or all ones. This is overcome by first applying the rule that if both vectors are all ones (or both vectors are all zeros) then the similarity measure is declared to be one. Otherwise, if $ad = 0$, the similarity measure is declared to be zero.

Dissimilarity measures for mixed data

Here is one measure that works with a mix of binary and continuous data. Binary variables are those containing only zeros, ones, and missing values; all other variables are treated as continuous.

`Gower`

requests the Gower (1971) dissimilarity coefficient for a mix of binary and continuous variables. For comparing observations i and j, the formula is

$$\frac{\sum_v \delta_{ijv} d_{ijv}}{\sum_v \delta_{ijv}}$$

where δ_{ijv} is a binary indicator equal to 1 whenever both observations i and j are nonmissing for variable v, and zero otherwise. Observations with missing values are not included when using `cluster` or `mds`, and so if an observation is included, $\delta_{ijv} = 1$ and $\sum_v \delta_{ijv}$ is the number of variables. However, using `matrix dissimilarity` with the `Gower` option does not exclude observations with missing values. See [MV] **cluster**, [MV] **mds**, and [MV] **matrix dissimilarity**.

For binary variables v,

$$d_{ijv} = \begin{cases} 0 & \text{if } x_{iv} = x_{jv} \\ 1 & \text{otherwise} \end{cases}$$

This is the same as the `matching` measure.

For continuous variables v,

$$d_{ijv} = \frac{|x_{iv} - x_{jv}|}{\left\{\max_k(x_{kv}) - \min_k(x_{kv})\right\}}$$

d_{ijv} is set to 0 if $\max_k(x_{kv}) - \min_k(x_{kv}) = 0$, that is, if the range of the variable is zero. This is the `L1` measure divided by the range of the variable.

For comparing variables u and v, the formula is

$$\frac{\sum_i \delta_{iuv} d_{iuv}}{\sum_i \delta_{iuv}}$$

where δ_{iuv} is a binary indicator equal to 1 whenever both variables u and v are nonmissing for observation i, and zero otherwise. If there are no missing values, $\sum_i \delta_{iuv}$ is the number of observations; otherwise, it is the number of observations for which neither variable u nor v has a missing value.

If all the variables are binary,

$$d_{iuv} = \begin{cases} 0 & \text{if } x_{iu} = x_{iv} \\ 1 & \text{otherwise} \end{cases}$$

If at least one variable is continuous,

$$d_{iuv} = \frac{|x_{iu} - x_{iv}|}{\left\{\max_v(x_{iv}) - \min_v(x_{iv})\right\}}$$

d_{iuv} is set to 0 if $\max_v(x_{iv}) - \min_v(x_{iv}) = 0$, that is, if the range of the observation is zero.

The Gower measure interprets binary variables as those with only 0, 1, or missing values. All other variables are treated as continuous.

In [MV] **matrix dissimilarity**, missing observations are included only in the calculation of the Gower dissimilarity, but the formula for this dissimilarity measure is undefined when all the values of δ_{ijv} or δ_{iuv} are zero. The dissimilarity is then set to missing.

❑ Technical note

Matrix dissimilarity and the Gower measure

Normally the commands

```
. matrix dissimilarity gm = x1 x2 y1, Gower
. clustermat waverage gm, add
```

and

```
. cluster waverage x1 x2 y1, measure(Gower)
```

will yield the same results, and likewise with `mdsmat` and `mds`. However, if any of the variables contain missing observations, this will not be the case. `cluster` and `mds` exclude all observations that have missing values for any of the variables of interest, whereas `matrix dissimilarity` with the `Gower` option does not. See [MV] **cluster**, [MV] **mds**, and [MV] **matrix dissimilarity** for more information.

Note: `matrix dissimilarity` without the `Gower` option does exclude all observations that have missing values for any of the variables of interest.

❑

❑ Technical note

Binary similarity measures applied to averages

Some cluster-analysis methods (such as Stata's kmeans and kmedians clustering) need to compute the similarity or dissimilarity between observations and group averages or group medians; see [MV] **cluster**. With binary data, a group average is interpreted as a proportion.

A group median for binary data will be zero or one, except when there are an equal number of zeros and ones. Here Stata calls the median 0.5, which can also be interpreted as a proportion.

In Stata's `cluster kmeans` and `cluster kmedians` commands for comparing a binary observation to a group proportion (see *Partition cluster-analysis methods* in [MV] **cluster**), the values of a, b, c, and d are obtained by assigning the appropriate fraction of the count to these values. In our earlier table showing the relationship of a, b, c, and d in the cross-tabulation of observation i and observation j, we replace observation j by the group-proportions vector. Then when observation i is 1, we add the corresponding proportion to a and add one minus that proportion to b. When observation i is 0, we add the corresponding proportion to c and add one minus that proportion to d. After the values of a, b, c, and d are computed in this way, the binary similarity measures are computed using the formulas as already described.

❑

Paul Jaccard (1868–1944) was a Swiss botanist who was born in Sainte-Croix (Vaud) and died in Zürich. He studied at Lausanne, Zürich, and Paris before being appointed to posts at Lausanne in 1894, where he specialized in plant geography, undertaking fieldwork in Egypt, Sweden, and Turkestan. In 1903, Jaccard returned to Zürich to a chair in general botany and plant physiology at ETH. His interests there centered on the microscopic analysis of wood, and anatomical and physiological studies of the growth of trees.

Robert Reuven Sokal (1926–) was born in Vienna to a Jewish family. He gained degrees from St. John's University in Shanghai and the University of Chicago. Sokal has worked at the University of Kansas–Lawrence and (from 1969) the State University of New York–Stony Brook. He was one of the leaders in the development of numerical taxonomy (Sokal and Sneath 1963; Sneath and Sokal 1973) and has been prominent in the application of statistical methods within biological systematics. With F. J. Rohlf, he has authored one of the leading biometrics texts (Sokal and Rohlf 1995). His current interests center on genetic variation in human populations, European ethnohistory, and spatial statistics. Sokal is a member of the U.S. National Academy of Sciences.

Peter Henry Andrews Sneath (1923–) studied medicine in Cambridge and London. After military service, he specialized in microbial systematics and worked for the Medical Research Council in the UK and the University of Leicester. With Robert Sokal, Sneath wrote the two initial texts on numerical taxonomy. He is a Fellow of the Royal Society.

References

Anderberg, M. R. 1973. *Cluster Analysis for Applications*. New York: Academic Press.

Czekanowski, J. 1932. "Coefficient of racial likeness" und "durchschnittliche Differenz". *Anthropologischer Anzeiger* 9: 227–249.

Dice, L. R. 1945. Measures of the amount of ecologic association between species. *Ecology* 26: 297–302.

Driver, H. E., and A. L. Kroeber. 1932. Quantitative expression of cultural relationships. *University of California Publications in American Archaeology and Ethnology* 31: 211–256.

Gilbert, G. K. 1884. Finley's tornado predictions. *American Meteorological Journal* 1: 166–172.

Gordon, A. D. 1999. *Classification*. 2nd ed. Boca Raton, FL: Chapman & Hall/CRC.

Gower, J. C. 1971. A general coefficient of similarity and some of its properties. *Biometrics* 27: 857–871.

——. 1985. Measures of similarity, dissimilarity, and distance. In Vol. 5 of *Encyclopedia of Statistical Sciences*, ed. S. Kotz, N. L. Johnson, and C. B. Read, 397–405. New York: Wiley.

Hamann, U. 1961. Merkmalsbestand und Verwandtschaftsbeziehungen der Farinosae. Ein Beitrag zum System der Monokotyledonen. *Willdenowia* 2: 639–768.

Hilbe, J. M. 1992a. sg9.1: Additional statistics to similari output. *Stata Technical Bulletin* 10: 22. Reprinted in *Stata Technical Bulletin Reprints*, vol. 2, p. 132. College Station, TX: Stata Press.

——. 1992b. sg9: Similarity coefficients for 2 × 2 binary data. *Stata Technical Bulletin* 9: 14–15. Reprinted in *Stata Technical Bulletin Reprints*, vol. 2, pp. 130–131. College Station, TX: Stata Press.

Hubálek, Z. 1982. Coefficients of association and similarity, based on binary (presence-absence) data: An evaluation. *Biological Reviews* 57: 669–689.

Jaccard, P. 1901. Distribution de la flore alpine dans le Bassin des Dranses et dans quelques régions voisines. *Bulletin de la Société Vaudoise des Sciences Naturelles* 37: 241–272.

——. 1908. Nouvelles recherches sur la distribution florale. *Bulletin de la Société Vaudoise des Sciences Naturelles* 44: 223–270.

Kaufman, L., and P. J. Rousseeuw. 1990. *Finding Groups in Data: An Introduction to Cluster Analysis*. New York: Wiley.

Kulczynski, S. 1927. Die Pflanzenassoziationen der Pieninen [In Polish, German summary]. *Bulletin International de l'Academie Polonaise des Sciences et des Lettres, Classe des Sciences Mathematiques et Naturelles, B (Sciences Naturelles)* Suppl. II: 57–203.

Ochiai, A. 1957. Zoogeographic studies on the soleoid fishes found in Japan and its neighbouring regions [in Japanese, English summary]. *Bulletin of the Japanese Society of Scientific Fisheries* 22: 526–530.

Pearson, K. 1900. Mathematical contributions to the theory of evolution—VII. On the correlation of characters not quantitatively measureable. *Philosophical Transactions of the Royal Society of London, Series A* 195: 1–47.

Rogers, D. J., and T. T. Tanimoto. 1960. A computer program for classifying plants. *Science* 132: 1115–1118.

Russell, P. F., and T. R. Rao. 1940. On habitat and association of species of anopheline larvae in south-eastern Madras. *Journal of the Malaria Institute of India* 3: 153–178.

Sneath, P. H. A., and R. R. Sokal. 1962. Numerical taxonomy. *Nature* 193: 855–860.

——. 1973. *Numerical Taxonomy: The Principles and Practice of Numerical Classification*. San Francisco: Freeman.

Sokal, R. R., and C. D. Michener. 1958. A statistical method for evaluating systematic relationships. *University of Kansas Science Bulletin* 28: 1409–1438.

Sokal, R. R., and F. J. Rohlf. 1995. *Biometry*. 3rd ed. San Francisco: Freeman.

Sokal, R. R., and P. H. A. Sneath. 1963. *Principles of Numerical Taxonomy*. San Francisco: Freeman.

Sørensen, T. 1948. A method of establishing groups of equal amplitude in plant sociology based on similarity of species content and its application to analyses of the vegetation on Danish commons. *Royal Danish Academy of Sciences and Letters, Biological Series* 5: 1–34.

Yule, G. U. 1900. On the association of attributes in statistics: With illustrations from the material of the Childhood Society, etc. *Philosophical Transactions of the Royal Society, Series A* 194: 257–319.

Yule, G. U., and M. G. Kendall. 1950. *An Introduction to the Theory of Statistics*. 14th ed. New York: Hafner.

Zubin, J. 1938. A technique for measuring like-mindedness. *Journal of Abnormal and Social Psychology* 33: 508–516.

Also see

[MV] **matrix dissimilarity** — Compute similarity or dissimilarity measures

[MV] **cluster** — Introduction to cluster-analysis commands

[MV] **clustermat** — Introduction to clustermat commands

Title

mvtest — Multivariate tests

Syntax

`mvtest` *subcommand* ... [, ...]

subcommand	Description	See
`means`	test means	[MV] **mvtest means**
`covariances`	test covariances	[MV] **mvtest covariances**
`correlations`	test correlations	[MV] **mvtest correlations**
`normality`	test multivariate normality	[MV] **mvtest normality**

Description

`mvtest` performs multivariate tests on means, covariances, and correlations and tests of univariate, bivariate, and multivariate normality. The tests of means, covariances, and correlations assume multivariate normality (Mardia, Kent, and Bibby 1979). Both one-sample and multiple-sample tests are provided. All multiple-sample tests provided by `mvtest` assume independent samples.

Structural equation modeling provides a more general framework for testing multivariate normality; see the *Stata Structural Equation Modeling Reference Manual.*

References

Achenback, T. M. 1991. *Manual for the Youth Self-Report and 1991 Profile.* Burlington, VT: University of Vermont.

Anderson, E. 1935. The irises of the Gaspe Peninsula. *Bulletin of the American Iris Society* 59: 2–5.

Baum, C. F., and N. J. Cox. 2007. omninorm: Stata module to calculate omnibus test for univariate/multivariate normality. Boston College Department of Economics, Statistical Software Components S417501. http://ideas.repec.org/c/boc/bocode/s417501.html.

Beall, G. 1945. Approximate methods in calculating discriminant functions. *Psychometrika* 10: 205–217.

Doornik, J. A., and H. Hansen. 2008. An omnibus test for univariate and multivariate normality. *Oxford Bulletin of Economics and Statistics* 70: 927–939.

Henze, N. 1994. On Mardia's kurtosis test for multivariate normality. *Communications in Statistics, Theory and Methods* 23: 1031–1045.

——. 1997. Extreme smoothing and testing for multivariate normality. *Statistics and Probability Letters* 35: 203–213.

Henze, N., and T. Wagner. 1997. A new approach to the BHEP tests for multivariate normality. *Journal of Multivariate Analysis* 62: 1–23.

Henze, N., and B. Zirkler. 1990. A class of invariant consistent tests for multivariate normality. *Communications in Statistics, Theory and Methods* 19: 3595–3617.

James, G. S. 1954. Tests of linear hypotheses in univariate and multivariate analysis when the ratios of the population variances are unknown. *Biometrika* 41: 19–43.

Jennrich, R. I. 1970. An asymptotic χ^2 test for the equality of two correlation matrices. *Journal of the American Statistical Association* 65: 904–912.

Johnson, R. A., and D. W. Wichern. 2007. *Applied Multivariate Statistical Analysis.* 6th ed. Englewood Cliffs, NJ: Prentice Hall.

Korin, B. P., and E. H. Stevens. 1973. Some approximations for the distribution of a multivariate likelihood ratio criterion. *Journal of the Royal Statistical Society, Series B* 29: 24–27.

Kramer, C. Y., and D. R. Jensen. 1969. Fundamentals of multivariate analysis, part I. Inference about means. *Journal of Quality Technology* 1: 120–133.

Krishnaiah, P. R., and J. C. Lee. 1980. Likelihood ratio tests for mean vectors and covariance matrices. In *Handbook of Statistics, Volume 1: Analysis of Variance*, ed. P. R. Krishnaiah. Amsterdam: North-Holland.

Krishnamoorthy, K., and J. Yu. 2004. Modified Nel and Van der Merwe test for the multivariate Behrens–Fisher problem. *Statistics and Probability Letters* 66: 161–169.

Lawley, D. N. 1963. On testing a set of correlation coefficients for equality. *Annals of Mathematical Statistics* 34: 149–151.

Mardia, K. V. 1970. Measures of multivariate skewness and kurtosis with applications. *Biometrika* 57: 519–530.

——. 1974. Applications of some measures of multivariate skewness and kurtosis for testing normality and robustness studies. *Sankhyā, Series B* 36: 115–128.

——. 1980. Tests of univariate and multivariate normality. In *Handbook of Statistics, Volume 1: Analysis of Variance*, ed. P. R. Krishnaiah. Amsterdam: North-Holland.

Mardia, K. V., J. T. Kent, and J. M. Bibby. 1979. *Multivariate Analysis.* London: Academic Press.

Nel, D. G., and C. A. Van der Merwe. 1986. A solution to the multivariate Behrens–Fisher problem. *Communications in Statistics, Theory and Methods* 15: 3719–3735.

Rencher, A. C. 2002. *Methods of Multivariate Analysis.* 2nd ed. New York: Wiley.

Seber, G. A. F. 1984. *Multivariate Observations.* New York: Wiley.

Vollebergh, W. A. M., S. van Dorsselaer, K. Monshouwer, J. Verdurmen, J. van der Ende, and T. ter Bogt. 2006. Mental health problems in early adolescents in the Netherlands: Differences between school and household surveys. *Social Psychiatry and Psychiatric Epidemiology* 41: 156–163.

Also see

[MV] **canon** — Canonical correlations

[MV] **hotelling** — Hotelling's T-squared generalized means test

[MV] **manova** — Multivariate analysis of variance and covariance

[R] **correlate** — Correlations (covariances) of variables or coefficients

[R] **mean** — Estimate means

[R] **sdtest** — Variance-comparison tests

[R] **sktest** — Skewness and kurtosis test for normality

[R] **swilk** — Shapiro–Wilk and Shapiro–Francia tests for normality

[R] **ttest** — Mean-comparison tests

Stata Structural Equation Modeling Reference Manual

Title

mvtest correlations — Multivariate tests of correlations

Syntax

Multiple-sample tests

`mvtest correlations` *varlist* [*if*] [*in*] [*weight*], `by(`*groupvars*`)` [*multisample_options*]

One-sample tests

`mvtest correlations` *varlist* [*if*] [*in*] [*weight*], [*one-sample_options*]

multisample_options	Description
Model	
* `by(`*groupvars*`)`	compare subsamples with same values in *groupvars*
`missing`	treat missing values in *groupvars* as ordinary values

* `by(`*groupvars*`)` is required.

one-sample_options	Description
Options	
`compound`	test that correlation matrix is compound symmetric (equal correlations); the default
`equals(`*C*`)`	test that correlation matrix equals matrix *C*

`bootstrap`, `by`, `jackknife`, `rolling`, and `statsby` are allowed; see [U] **11.1.10 Prefix commands**.
Weights are not allowed with the `bootstrap` prefix; see [R] **bootstrap**.
`aweights` are not allowed with the `jackknife` prefix; see [R] **jackknife**.
`aweights` and `fweights` are allowed; see [U] **11.1.6 weight**.

Menu

Statistics > Multivariate analysis > MANOVA, multivariate regression, and related > Multivariate test of means, covariances, and normality

Description

`mvtest correlations` performs one-sample and multiple-sample tests on correlations. These tests assume multivariate normality.

See [MV] **mvtest** for more multivariate tests.

Options for multiple-sample tests

Model

by(*groupvars*) is required with the multiple-sample version of the test. Observations with the same values in *groupvars* form each sample. Observations with missing values in *groupvars* are ignored, unless the missing option is specified. A Wald test due to Jennrich (1970) is displayed.

missing specifies that missing values in *groupvars* are treated like ordinary values.

Options for one-sample tests

Options

compound, the default, tests the hypothesis that the correlation matrix of the variables is compound symmetric, that is, that the correlations of all variables in *varlist* are the same. Lawley's (1963) chi-squared test is displayed.

equals(C) tests the hypothesis that the correlation matrix of *varlist* is C. The matrix C should be $k \times k$, symmetric, and positive definite. C is converted to a correlation matrix if needed. The row and column names of C are immaterial. A Wald test due to Jennrich (1970) is displayed.

Remarks

Remarks are presented under the following headings:

One-sample tests for correlation matrices
A multiple-sample test for correlation matrices

One-sample tests for correlation matrices

Both one-sample and multiple-sample tests of correlation matrices are provided with the mvtest correlations command. The one-sample tests include Lawley's (1963) test that the correlation matrix is compound symmetric (that is, all correlations are equal), and the Wald test proposed by Jennrich (1970) that the correlation matrix equals a given correlation matrix.

▷ Example 1

The gasoline-powered milk-truck dataset introduced in example 1 of [MV] **mvtest means** has price per mile for fuel, repair, and capital. We test if the correlations between these three variables are equal (that is, the correlation matrix is compound symmetric) using the compound option of mvtest correlations.

```
. use http://www.stata-press.com/data/r12/milktruck
(Milk transportation costs for 25 gasoline trucks (Johnson and Wichern 2007))
. mvtest correlations fuel repair capital, compound
Test that correlation matrix is compound symmetric (all correlations equal)
        Lawley chi2(2) =      7.75
           Prob > chi2 =    0.0208
```

We reject the null hypothesis and conclude that there are probably differences in the correlations of the three cost variables.

◁

▷ Example 2

Using the `equals()` option of `mvtest correlations`, we test the hypothesis that fuel and repair costs have a correlation of 0.75, while the correlation between capital and these two variables is zero.

```
. matrix C = (1, 0.75, 0 \ 0.75, 1, 0 \ 0, 0, 1)
. matrix list C
symmetric C[3,3]
     c1   c2   c3
r1    1
r2  .75    1
r3    0    0    1
. mvtest correlations fuel repair capital, equals(C)
Test that correlation matrix equals specified pattern C
      Jennrich chi2(3) =      4.55
           Prob > chi2 =    0.2077
```

We fail to reject this null hypothesis.

◁

A multiple-sample test for correlation matrices

A multiple-sample test of equality of correlation matrices is provided by the `mvtest correlations` command with the `by()` option defining the multiple samples (groups).

▷ Example 3

Psychological test score data are introduced in example 2 of [MV] **mvtest covariances**. We test whether the correlation matrices for the four test scores are the same for males and females.

```
. use http://www.stata-press.com/data/r12/genderpsych
(Four Psychological Test Scores on 32 Males and 32 Females, Rencher (2002))
. mvtest correlations y1 y2 y3 y4, by(gender)
Test of equality of correlation matrices across samples
      Jennrich chi2(6) =      5.01
           Prob > chi2 =    0.5422
```

We fail to reject the null hypothesis of equal correlation matrices for males and females.

◁

Saved results

`mvtest correlations` saves the following in `r()`:

Scalars

`r(chi2)`	chi-squared
`r(df)`	degrees of freedom for chi-squared test
`r(p_chi2)`	significance

Macros

`r(chi2type)`	type of model chi-squared test

Methods and formulas

mvtest correlations is implemented as an ado-file.

Methods and formulas are presented under the following headings:

One-sample tests for correlation matrices

Let the sample consist of N i.i.d. observations from a k-variate multivariate normal distribution $\text{MVN}_k(\boldsymbol{\mu}, \boldsymbol{\Sigma})$, with sample correlation matrix $\mathbf{R}$.

To test that a correlation matrix equals a given matrix, $\mathbf{R}_0$, mvtest correlations computes a Wald test proposed by Jennrich (1970):

$$\chi^2_{ocf} = \frac{1}{2}\text{trace}(\mathbf{ZZ}) - \text{diagonal}(\mathbf{Z})' \left(\mathbf{I} + \mathbf{R}_0 \bullet \mathbf{R}_0^{-1}\right)^{-1} \text{diagonal}(\mathbf{Z})$$

where $\mathbf{Z} = \sqrt{N}\mathbf{R}_0^{-1}(\mathbf{R} - \mathbf{R}_0)$ and $\bullet$ denotes the Hadamard product. χ^2_{ocf} is asymptotically χ^2 distributed with $k(k-1)/2$ degrees of freedom.

To test that the correlation matrix is compound symmetric, that is, to test that all correlations are equal, the likelihood-ratio test is somewhat cumbersome. Lawley (1963) offers an asymptotically equivalent test that is computationally simple (Johnson and Wichern 2007, 457–458):

$$\chi^2_{occ} = \frac{N-1}{(1-\overline{R})^2} \left\{ \sum_{i=2}^{k}\sum_{j=1}^{i-1}(R_{ij} - \overline{R})^2 - u\sum_{h=1}^{k}(\overline{R}_h - \overline{R})^2 \right\}$$

where

$$\overline{R} = \frac{2}{k(k-1)}\sum_{i=2}^{k}\sum_{j=1}^{i-1} R_{ij}$$

$$\overline{R}_h = \frac{1}{k-1}\sum_{i=1;i\neq h}^{k} R_{ih}$$

$$u = \frac{(k-1)^2\left\{1-(1-\overline{R})^2\right\}}{k-(k-2)(1-\overline{R})^2}$$

and R_{ij} denotes element (i,j) of the $k \times k$ correlation matrix $\mathbf{R}$. χ^2_{occ} is asymptotically χ^2 distributed with $(k-2)(k+1)/2$ degrees of freedom. Aitkin, Nelson, and Reinfurt (1968) study the quality of this χ^2 approximation for k up to six and various correlations, and conclude that the approximation is adequate for N as small as 25.

A multiple-sample test for correlation matrices

Let there be $m \geq 2$ independent samples with the jth sample containing N_j i.i.d. observations from a k-variate multivariate normal distribution, $\text{MVN}_k(\boldsymbol{\mu}_j, \boldsymbol{\Sigma}_j)$, with sample correlation matrix $\mathbf{R}_j$, $j = 1, \ldots, m$. Let $N = \sum_{j=1}^{m} N_j$.

To test for the equality of correlation matrices across m independent samples, `mvtest correlations` computes a Wald test proposed by Jennrich (1970):

$$\chi^2_{mc} = \sum_{j=1}^{m} \left\{ \frac{1}{2}\text{trace}\left(\mathbf{Z}_j^2\right) - \text{diagonal}\left(\mathbf{Z}_j\right)' \left(\mathbf{I} + \overline{\mathbf{R}} \bullet \overline{\mathbf{R}}^{-1}\right)^{-1} \text{diagonal}\left(\mathbf{Z}_j\right) \right\}$$

where $\overline{\mathbf{R}} = 1/N \sum_{j=1}^{m} N_j \mathbf{R}_j$, $\mathbf{Z}_j = \sqrt{N_j}\, \overline{\mathbf{R}}^{-1} \left(\mathbf{R}_j - \overline{\mathbf{R}}\right)$, and $\bullet$ denotes the Hadamard product. χ^2_{mc} is asymptotically χ^2 distributed with $(m-1)k(k-1)/2$ degrees of freedom.

References

Aitkin, M. A., W. C. Nelson, and K. H. Reinfurt. 1968. Tests for correlation matrices. *Biometrika* 55: 327–334.

Jennrich, R. I. 1970. An asymptotic χ^2 test for the equality of two correlation matrices. *Journal of the American Statistical Association* 65: 904–912.

Johnson, R. A., and D. W. Wichern. 2007. *Applied Multivariate Statistical Analysis*. 6th ed. Englewood Cliffs, NJ: Prentice Hall.

Lawley, D. N. 1963. On testing a set of correlation coefficients for equality. *Annals of Mathematical Statistics* 34: 149–151.

Rencher, A. C. 2002. *Methods of Multivariate Analysis*. 2nd ed. New York: Wiley.

Also see

[MV] **canon** — Canonical correlations

[R] **correlate** — Correlations (covariances) of variables or coefficients

Title

mvtest covariances — Multivariate tests of covariances

Syntax

Multiple-sample tests

mvtest covariances *varlist* [*if*] [*in*] [*weight*], by(*groupvars*) [*multisample_options*]

One-sample tests

mvtest covariances *varlist* [*if*] [*in*] [*weight*], [*one-sample_options*]

multisample_options	Description
Model	
* by(*groupvars*)	compare subsamples with same values in *groupvars*
missing	treat missing values in *groupvars* as ordinary values

* by(*groupvars*) is required.

one-sample_options	Description
Options	
diagonal	test that covariance matrix is diagonal; the default
spherical	test that covariance matrix is spherical
compound	test that covariance matrix is compound symmetric
equals(*C*)	test that covariance matrix equals matrix *C*
* block(*varlist*$_1$ [|| ...])	test that covariance matrix is block diagonal with blocks corresponding to *varlist#*

* The full specification is block(*varlist*$_1$ [|| *varlist*$_2$ [|| ...]]).

bootstrap, by, jackknife, rolling, and statsby are allowed; see [U] **11.1.10 Prefix commands**.
Weights are not allowed with the bootstrap prefix; see [R] **bootstrap**.
aweights are not allowed with the jackknife prefix; see [R] **jackknife**.
aweights and fweights are allowed; see [U] **11.1.6 weight**.

Menu

Statistics > Multivariate analysis > MANOVA, multivariate regression, and related > Multivariate test of means, covariances, and normality

Description

mvtest covariances performs one-sample and multiple-sample multivariate tests on covariances. These tests assume multivariate normality.

See [MV] **mvtest** for other multivariate tests. See [R] **sdtest** for univariate tests of standard deviations.

Options for multiple-sample tests

Model

`by(`*groupvars*`)` is required with the multiple-sample version of the test. Observations with the same values in *groupvars* form a sample. Observations with missing values in *groupvars* are ignored, unless the `missing` option is specified.

A modified likelihood-ratio statistic testing the equality of covariance matrices for the multiple independent samples defined by `by()` is presented along with an F and chi-squared approximation due to Box (1949). This test is also known as Box's M test.

`missing` specifies that missing values in *groupvars* are treated like ordinary values.

Options for one-sample tests

Options

`diagonal`, the default, tests the hypothesis that the covariance matrix is diagonal, that is, that the variables in *varlist* are independent. A likelihood-ratio test with first-order Bartlett correction is displayed.

`spherical` tests the hypothesis that the covariance matrix is diagonal with constant diagonal values, that is, that the variables in *varlist* are homoskedastic and independent. A likelihood-ratio test with first-order Bartlett correction is displayed.

`compound` tests the hypothesis that the covariance matrix is compound symmetric, that is, that the variables in *varlist* are homoskedastic and that every pair of two variables has the same covariance. A likelihood-ratio test with first-order Bartlett correction is displayed.

`equals(`*C*`)` specifies that the hypothesized covariance matrix for the k variables in *varlist* is *C*. The matrix *C* must be $k \times k$, symmetric, and positive definite. The row and column names of *C* are ignored. A likelihood-ratio test with first-order Bartlett correction is displayed.

`block(`$varlist_1$ [|| $varlist_2$ [|| ...]]`)` tests the hypothesis that the covariance matrix is block diagonal with blocks $varlist_1$, $varlist_2$, etc. Variables in *varlist* not included in $varlist_1$, $varlist_2$, etc., are treated as an additional block. With this pattern, variables in different blocks are independent, but no assumptions are made on the within-block covariance structure. A likelihood-ratio test with first-order Bartlett correction is displayed.

Remarks

Remarks are presented under the following headings:

One-sample tests for covariance matrices
A multiple-sample test for covariance matrices

One-sample tests for covariance matrices

One-sample and multiple-sample tests for covariance matrices are provided by the `mvtest covariances` command. One-sample tests include the test that the covariance matrix of *varlist* is diagonal, spherical, compound symmetric, block diagonal, or equal to a given matrix.

▷ Example 1

The gasoline-powered milk-truck dataset introduced in example 1 of [MV] **mvtest means** has price per mile for fuel, repair, and capital. We test if the covariance matrix for these three variables has any special structure.

```
. use http://www.stata-press.com/data/r12/milktruck
(Milk transportation costs for 25 gasoline trucks (Johnson and Wichern 2007))
. mvtest covariances fuel repair capital, diagonal
Test that covariance matrix is diagonal
   Adjusted LR chi2(3) =     17.91
           Prob > chi2 =    0.0005
. mvtest covariances fuel repair capital, spherical
Test that covariance matrix is spherical
   Adjusted LR chi2(5) =     21.53
           Prob > chi2 =    0.0006
. mvtest covariances fuel repair capital, compound
Test that covariance matrix is compound symmetric
   Adjusted LR chi2(4) =     11.29
           Prob > chi2 =    0.0235
```

We reject the hypotheses that the covariance is diagonal, spherical, or compound symmetric.

We now test whether there is covariance between `fuel` and `repair`, with `capital` not covarying with these two variables. Thus we hypothesize a block diagonal structure of the form

$$\boldsymbol{\Sigma} = \begin{pmatrix} \sigma_{11}^2 & \sigma_{12} & 0 \\ \sigma_{21} & \sigma_{22}^2 & 0 \\ 0 & 0 & \sigma_{33}^2 \end{pmatrix}$$

for the covariance matrix. The `block()` option of `mvtest covariances` provides the test:

```
. mvtest covariances fuel repair capital, block(fuel repair || capital)
Test that covariance matrix is block diagonal
   Adjusted LR chi2(2) =      3.52
           Prob > chi2 =    0.1722
```

We fail to reject the null hypothesis. The covariance matrix might have the block diagonal structure we hypothesized.

The same p-value could have been obtained from Stata's canonical correlation command:

```
. canon (fuel repair) (capital)
  (output omitted)
```

See [MV] **canon**.

Now, in addition to hypothesizing that the covariance is block diagonal, we specifically hypothesize that the variance for `capital` is 10, the variance of `fuel` is three times that of `capital`, the variance of `repair` is two times that of `capital`, and that there is no covariance between `capital` and the other two variables, while there is a covariance of 15 between `fuel` and `repair`. We test that hypothesis by using the `equals()` option.

```
. mat B = (30, 15, 0 \ 15, 20, 0 \ 0, 0, 10)
. matrix list B
symmetric B[3,3]
    c1  c2  c3
r1  30
r2  15  20
r3   0   0  10
. mvtest covariances fuel repair capital, equals(B)
Test that covariance matrix equals matrix B
   Adjusted LR chi2(6) =      5.48
           Prob > chi2 =    0.4837
```

We fail to reject the null hypothesis; the covariance might follow the structure hypothesized.

◁

❑ Technical note

If each block comprises a single variable, the test of independent subvectors reduces to a test that the covariance matrix is diagonal. Thus the following two commands are equivalent:

```
mvtest covariances x1 x2 x3 x4 x5, block(x1 || x2 || x3 || x4 || x5)
```

and

```
mvtest covariances x1 x2 x3 x4 x5, diagonal
```

❑

A multiple-sample test for covariance matrices

The `by()` option of `mvtest covariances` provides a modified likelihood-ratio statistic testing the equality of covariance matrices for the multiple independent samples defined by `by()`. This test is also known as Box's M test. There are both F and chi-squared approximations for the null distribution of the test.

▷ Example 2

We illustrate the multiple-sample test of equality of covariance matrices by using four psychological test scores on 32 men and 32 women (Rencher 2002, 125; Beall 1945).

```
. use http://www.stata-press.com/data/r12/genderpsych
(Four Psychological Test Scores on 32 Males and 32 Females, Rencher (2002))
. mvtest covariances y1 y2 y3 y4, by(gender)
Test of equality of covariance matrices across 2 samples
      Modified LR chi2 =   14.5606
     Box F(10,18377.7) =      1.35       Prob > F =  0.1950
          Box chi2(10) =     13.55    Prob > chi2 =  0.1945
```

Both the F and the chi-squared approximations indicate that we cannot reject the null hypothesis that the covariance matrices for males and females are equal (Rencher 2002, 258–259).

◁

Equality of group covariance matrices is an assumption of multivariate analysis of variance (see [MV] **manova**) and linear discriminant analysis (see [MV] **discrim lda**). Box's M test, produced by `mvtest covariances` with the `by()` option, is often recommended for testing this assumption.

Saved results

`mvtest covariances` saves the following in `r()`:

Scalars

`r(chi2)`	chi-squared
`r(df)`	degrees of freedom for chi-squared test
`r(p_chi2)`	significance
`r(F_Box)`	F statistic for Box test (`by()` only)
`r(df_m_Box)`	model degrees of freedom for Box test (`by()` only)
`r(df_r_Box)`	residual degrees of freedom for Box test (`by()` only)
`r(p_F_Box)`	significance of Box F test (`by()` only)

Macros

`r(chi2type)`	type of model chi-squared test

Methods and formulas

`mvtest covariances` is implemented as an ado-file.

When comparing the formulas in this section with those found in some multivariate texts, be aware of whether they define the sample covariance matrix with a divisor of N or $N-1$. We use N. The formulas for several of the statistics are presented differently depending on your choice of divisor (but are still equivalent).

Methods and formulas are presented under the following headings:

One-sample tests for covariance matrices
A multiple-sample test for covariance matrices

One-sample tests for covariance matrices

Let the sample consist of N i.i.d. observations, $\mathbf{x}_i$, $i=1,\ldots,N$, from a k-variate multivariate normal distribution, $\text{MVN}_k(\boldsymbol{\mu},\boldsymbol{\Sigma})$, with sample mean $\overline{\mathbf{x}} = 1/N\sum_{i=1}^N \mathbf{x}_i$, sample covariance matrix $\mathbf{S} = 1/N\sum_{i=1}^N(\mathbf{x}_i-\overline{\mathbf{x}})(\mathbf{x}_i-\overline{\mathbf{x}})'$, and sample correlation matrix $\mathbf{R}$.

To test that a covariance matrix equals a given matrix, $H_0: \boldsymbol{\Sigma} = \boldsymbol{\Sigma}_0$, `mvtest covariances` computes a likelihood-ratio test with Bartlett correction (Rencher 2002, 248–249):

$$\chi^2_{ovf} = (N-1)\left\{1 - \frac{1}{6(N-1)-1}\left(2k+1-\frac{2}{k+1}\right)\right\}$$
$$\times\left\{\ln|\boldsymbol{\Sigma}_0| - \ln\left|\frac{N}{N-1}\mathbf{S}\right| + \text{trace}\left(\frac{N}{N-1}\mathbf{S}\boldsymbol{\Sigma}_0^{-1}\right) - k\right\}$$

which is approximately χ^2 distributed with $k(k+1)/2$ degrees of freedom.

To test for a spherical covariance matrix, $H_0\colon \mathbf{\Sigma} = \sigma^2\mathbf{I}$, `mvtest covariances` computes a likelihood-ratio test with Bartlett correction (Rencher 2002, 250–251):

$$\chi^2_{ovs} = \left\{(N-1) - \frac{2k^2+k+2}{6k}\right\}\left[k\ln\{\text{trace}\,(\mathbf{S})\} - \ln|\mathbf{S}| - k\ln(k)\right]$$

which is approximately χ^2 distributed with $k(k+1)/2 - 1$ degrees of freedom.

To test for a diagonal covariance matrix, $H_0\colon \mathbf{\Sigma}_{ij} = 0$ for $i \neq j$, `mvtest covariances` computes a likelihood-ratio test with first-order Bartlett correction (Rencher 2002, 265):

$$\chi^2_{ovd} = -\left(N - 1 - \frac{2k+5}{6}\right)\ln|\mathbf{R}|$$

which is approximately χ^2 distributed with $k(k-1)/2$ degrees of freedom.

To test for a compound-symmetric covariance matrix, $H_0\colon \mathbf{\Sigma} = \sigma^2\{(1-\rho)\mathbf{I} + \rho\mathbf{1}\mathbf{1}'\}$, that is, a covariance matrix with common variance σ^2 and common correlation ρ, `mvtest covariances` computes a likelihood-ratio test with first-order Bartlett correction (Rencher 2002, 252–253):

$$\chi^2_{ovc} = \left\{N - 1 - \frac{k(k+1)^2(2k-3)}{6(k-1)(k^2+k-4)}\right\}$$
$$\times\left[k\ln\left(s^2\right) + (k-1)\ln(1-r) + \ln\{1+(k-1)r\} - \ln|\mathbf{S}|\right]$$

where

$$s^2 = \frac{1}{k}\sum_{j=1}^{k} s_{jj} \qquad \text{and} \qquad r = \frac{1}{k(k-1)s^2}\sum_{j=1}^{k}\sum_{h=1,h\neq j}^{k} s_{jh}$$

where s_{jh} is the (j,h) element of $\mathbf{S}$. χ^2_{ovc} is approximately χ^2 distributed with $k(k+1)/2-2$ degrees of freedom.

To test that a covariance matrix is block diagonal with b diagonal blocks and with k_j variables in block j, `mvtest covariances` computes a likelihood-ratio test with first-order Bartlett correction (Rencher 2002, 261–262). Thus variables in different blocks are hypothesized to be independent.

$$\chi^2_{ovb} = \left(N - 1 - \frac{2a_3+3a_2}{6a_2}\right)\left(\sum_{j=1}^{b}\ln|\mathbf{S}_j| - \ln|\mathbf{S}|\right)$$

where $a_2 = k^2 - \sum_{j=1}^{b} k_j^2$, $a_3 = k^3 - \sum_{j=1}^{b} k_j^3$, and $\mathbf{S}_j$ is the covariance matrix for the jth block. χ^2_{ovb} is approximately χ^2 distributed with $a_2/2$ degrees of freedom.

A multiple-sample test for covariance matrices

Let there be $m \geq 2$ independent samples with the jth sample containing N_j i.i.d. observations, $\mathbf{x}_{ji}$, $i = 1,\dots,N_j$, from a k-variate multivariate normal distribution $\text{MVN}_k(\boldsymbol{\mu}_j, \mathbf{\Sigma}_j)$. The observed jth sample mean is $\overline{\mathbf{x}}_j = 1/N_j\sum_{i=1}^{N_j}\mathbf{x}_{ji}$ and covariance is $\mathbf{S}_j = 1/N_j\sum_{i=1}^{N_j}(\mathbf{x}_{ji}-\overline{\mathbf{x}}_j)(\mathbf{x}_{ji}-\overline{\mathbf{x}}_j)'$. Let $N = \sum_{j=1}^{m} N_j$.

To test the equality of covariance matrices in m independent samples, $H_0 : \mathbf{\Sigma}_1 = \mathbf{\Sigma}_2 = \cdots = \mathbf{\Sigma}_m$, `mvtest covariances` computes a modified likelihood-ratio statistic, which is an unbiased variant of the likelihood-ratio statistic (Rencher 2002, 255–258):

$$-2\ln(M) = (N-m)\ln\left|\mathbf{S}_{\text{pooled}}\right| - \sum_{j=1}^{m}\left\{(N_j-1)\ln\left|\frac{N_j}{N_j-1}\mathbf{S}_j\right|\right\}$$

where $\mathbf{S}_{\text{pooled}} = \sum_{j=1}^{m} N_j/(N_j-1)\mathbf{S}_j$. Asymptotically, $-2\ln(M)$ is χ^2 distributed. Box (1949, 1950) derived more accurate χ^2 and F approximations (Rencher 2002, 257–258).

Box's χ^2 approximation is given by

$$\chi^2_{\text{mv}} = -2(1-c_1)\ln(M)$$

which is approximately χ^2 distributed with $(m-1)k(k+1)/2$ degrees of freedom.

Box's F approximation is given by

$$F_{\text{mv}} = \begin{cases} -2b_1\ln(M) & \text{if } c_2 > c_1^2 \\[2ex] \dfrac{2a_2b_2\ln(M)}{a_1\left\{1+2b_2\ln(M)\right\}} & \text{otherwise} \end{cases}$$

which is approximately F distributed with a_1 and a_2 degrees of freedom.

In the χ^2 and F approximations, we have

$$c_1 = \left\{\sum_{j=1}^{m}(N_j-1)^{-1} - (N-m)^{-1}\right\}\frac{2k^2+3k-1}{6(k+1)(m-1)}$$

$$c_2 = \left\{\sum_{j=1}^{m}(N_j-1)^{-2} - (N-m)^{-2}\right\}\frac{(k-1)(k+2)}{6(m-1)}$$

$a_1 = (m-1)k(k+1)/2$, $a_2 = (a_1+2)/|c_2 - c_1^2|$, $b_1 = (1 - c_1 - a_1/a_2)/a_1$, and $b_2 = (1 - c_1 + 2/a_2)/a_2$.

References

Beall, G. 1945. Approximate methods in calculating discriminant functions. *Psychometrika* 10: 205–217.

Box, G. E. P. 1949. A general distribution theory for a class of likelihood criteria. *Biometrika* 36: 317–346.

——. 1950. Problems in the analysis of growth and wear curves. *Biometrics* 6: 362–389.

Johnson, R. A., and D. W. Wichern. 2007. *Applied Multivariate Statistical Analysis*. 6th ed. Englewood Cliffs, NJ: Prentice Hall.

Rencher, A. C. 2002. *Methods of Multivariate Analysis*. 2nd ed. New York: Wiley.

Also see

[MV] **candisc** — Canonical linear discriminant analysis

[MV] **canon** — Canonical correlations

[R] **correlate** — Correlations (covariances) of variables or coefficients

[MV] **discrim lda** — Linear discriminant analysis

[MV] **manova** — Multivariate analysis of variance and covariance

[R] **sdtest** — Variance-comparison tests

Title

mvtest means — Multivariate tests of means

Syntax

Multiple-sample tests

mvtest means *varlist* [*if*] [*in*] [*weight*], by(*groupvars*) [*multisample_options*]

One-sample tests

mvtest means *varlist* [*if*] [*in*] [*weight*], [*one-sample_options*]

multisample_options	Description
Model	
* by(*groupvars*)	compare subsamples with same values in *groupvars*
missing	treat missing values in *groupvars* as ordinary values
Options	
homogeneous	test for equal means with homogeneous covariance matrices across by-groups; the default
heterogeneous	James' test for equal means, allowing heterogeneous covariance matrices across by-groups
lr	likelihood-ratio test for equal means, allowing heterogeneous covariance matrices across by-groups
protect(*spec*)	run protection as a safeguard against local minimum with the group means as initial values; use only with lr option

* by(*groupvars*) is required.

one-sample_options	Description
Options	
equal	test that variables in *varlist* have equal means; the default
zero	test that means of *varlist* are all equal to 0
equals(*M*)	test that mean vector equals vector *M*
linear(*V*)	test that mean vector of *varlist* satisfies linear hypothesis described by matrix *V*

bootstrap, by, jackknife, rolling, and statsby are allowed; see [U] **11.1.10 Prefix commands**.
Weights are not allowed with the bootstrap prefix; see [R] **bootstrap**.
aweights are not allowed with the jackknife prefix; see [R] **jackknife**.
aweights and fweights are allowed; see [U] **11.1.6 weight**.

Menu

Statistics > Multivariate analysis > MANOVA, multivariate regression, and related > Multivariate test of means, covariances, and normality

Description

mvtest means performs one-sample and multiple-sample multivariate tests on means. These tests assume multivariate normality.

See [MV] **mvtest** for other multivariate tests.

Options for multiple-sample tests

Model

by(*groupvars*) is required with the multiple-sample version of the test. Observations with the same values in *groupvars* form a sample. Observations with missing values in *groupvars* are ignored, unless the missing option is specified.

missing specifies that missing values in *groupvars* are treated like ordinary values.

Options

homogeneous, the default, specifies the hypothesis that the mean vectors are the same across the by-groups, assuming homogeneous covariance matrices across the by-groups. homogeneous produces the four standard tests of multivariate means (Wilks' lambda, Pillai's trace, Lawley–Hotelling trace, and Roy's largest root).

heterogeneous removes the assumption that the covariance matrices are the same across the by-groups. This is the multivariate Behrens–Fisher problem. With two groups, the MNV test, an affine-invariant modification by Krishnamoorthy and Yu (2004) of the Nel–Van der Merwe (1986) test, is displayed. With more than two groups, the Wald test, with p-values adjusted as proposed by James (1954), is displayed.

lr removes the assumption that the covariance matrices are the same across the by-groups and specifies that a likelihood-ratio test be presented. The associated estimation problem may have multiple local optima, though this seems rare with two groups.

protect(*spec*) is a technical option accompanying lr, specifying that the "common means" model is fit from different starting values to ascertain with some confidence whether a global optimum to the underlying estimation problem was reached. The Mardia–Kent–Bibby (1979) proposal for initialization of the common means is always used as well. If the different trials do not converge to the same solution, the "best" one is used to obtain the test, and a warning message is displayed.

protect(groups) specifies to fit the common means model using each of the group means as starting values for the common means.

protect(randobs, reps(#)) specifies to fit the common means model using # random observations as starting values for the common means.

protect(#) is a convenient shorthand for protect(randobs, reps(#)).

Options with one-sample tests

Options

equal performs Hotelling's test of the hypothesis that the means of all variables in *varlist* are equal.

zero performs Hotelling's test of the hypothesis that the means of all variables in *varlist* are 0.

equals(*M*) performs Hotelling's test that the vector of means of the k variables in *varlist* equals *M*. The matrix *M* must be a $k \times 1$ or $1 \times k$ vector. The row and column names of *M* are ignored.

`linear(`*V*`)` performs Hotelling's test that the means satisfy a user-specified set of linear constraints, represented by V. V must be a matrix vector with k or $k+1$ columns, where k is the number of variables in *varlist*. Let A be a matrix of the first k columns of V. Let b be the last column of V if V has $k+1$ columns and a column of 0s otherwise. The linear hypothesis test is that A times a column vector of the means of *varlist* equals b. `mvtest` ignores matrix row and column names.

Remarks

Remarks are presented under the following headings:

One-sample tests for mean vectors
Multiple-sample tests for mean vectors

One-sample tests for mean vectors

One-sample and multiple-sample tests of means are available with the `mvtest means` command. One-sample tests include tests that the means of *varlist* are equal, the means of *varlist* equal a given vector, the means of *varlist* are zero, and linear combinations of the means of *varlist* equal a given vector.

We first explore the use of `mvtest means` for testing the one-sample hypothesis that the means of *varlist* are equal.

▷ Example 1

The cost on a per-mile basis of 25 gasoline trucks used for transporting milk are provided in three categories: fuel, repair, and capital (Johnson and Wichern 2007, 269).

```
. use http://www.stata-press.com/data/r12/milktruck
(Milk transportation costs for 25 gasoline trucks (Johnson and Wichern 2007))

. summarize
```

Variable	Obs	Mean	Std. Dev.	Min	Max
fuel	25	12.56	5.382	4.24	29.11
repair	25	8.1612	4.631723	1.35	17.44
capital	25	10.5444	3.687688	3.28	17.59

Are the means of the three costs equal? The `equal` option of `mvtest means` provides a way of testing this hypothesis.

```
. mvtest means fuel repair capital, equal

Test that all means are the same

          Hotelling T2 =     35.25
     Hotelling F(2,23) =     16.89
              Prob > F =    0.0000
```

We reject the null hypothesis of equal means for fuel, repair, and capital costs.

◁

Hotelling's T-squared statistic is a multivariate generalization of the univariate t statistic; see [R] **ttest**. A test of the bivariate hypothesis that the means of the repair and capital costs are equal could be obtained with

```
ttest repair == capital
```

or with

```
mvtest means repair capital, equal
```

The square of the t statistic from `ttest` equals the T-squared value from `mvtest means`. With `ttest`, you are limited to comparing the means of two variables; with `mvtest means`, you can simultaneously compare the means of two or more variables.

The `equals()` option of `mvtest means` provides Hotelling's T-squared statistic for the test that the mean vector for *varlist* equals a given vector. This provides a multivariate generalization of the univariate t statistic obtained using the `ttest` *varname* == # syntax of [R] **ttest**.

▷ Example 2

We compare the measurements of the available and exchangeable soil calcium (`y1` and `y2`) and turnip-green calcium (`y3`) at 10 locations in the South (Rencher 2002, 56; Kramer and Jensen 1969) to the values 15.0, 6.0, and 2.85 respectively (Rencher 2002, 120).

```
. use http://www.stata-press.com/data/r12/turnip
(Calcium in soil and turnip greens (Rencher 2002))
. summarize y*
    Variable |       Obs        Mean    Std. Dev.       Min        Max
-------------+--------------------------------------------------------
          y1 |        10        28.1    11.85514          6         40
          y2 |        10        7.18    8.499908        1.6         30
          y3 |        10       3.089    .5001211        2.7       4.38
. matrix Mstd = (15.0, 6.0, 2.85)
. mvtest means y* , equals(Mstd)
Test that means equal vector Mstd
         Hotelling T2 =     24.56
     Hotelling F(3,7) =      6.37
             Prob > F =    0.0207
```

The calcium measurements from these 10 locations in the South do not appear to match the hypothesized values.

◁

The `zero` option of `mvtest means` tests the hypothesis that the means of *varlist* are zero. The same result could be obtained by creating a column or row vector of the appropriate length filled with zeros and supplying that to the `equals()` option.

```
mvtest means y1 y2 y3, zero
```

would give the same test as

```
matrix Zero = 0,0,0
mvtest means y1 y2 y3, equals(Zero)
```

This same test against a zero-mean vector can be obtained with the `hotelling` command; see [MV] **hotelling**. For example,

```
hotelling y1 y2 y3
```

`mvtest means` also tests that linear combinations of the means of *varlist* equal a given vector.

▷ Example 3

The linear() option of mvtest means can be used to obtain the same result as in example 1, testing that the fuel, repair, and capital costs are equal. We do this by constructing two appropriate linear combinations of our three variables and testing that the means of these two linear combinations are zero.

```
. use http://www.stata-press.com/data/r12/milktruck
(Milk transportation costs for 25 gasoline trucks (Johnson and Wichern 2007))
. matrix C = 1, -1, 0 \ 0, 1, -1
. matrix list C
C[2,3]
    c1  c2  c3
r1   1  -1   0
r2   0   1  -1
. mvtest means fuel repair capital, linear(C)
Test that mean vector satisfies linear hypothesis C
            Hotelling T2 =     35.25
       Hotelling F(2,23) =     16.89
                Prob > F =    0.0000
```

We formed a matrix C that contrasted fuel to repair (the first row of C) and repair to capital (the second row of C). Note that we need not set the matrix row and column names of C. By default, the linear contrast was tested equal to a vector of zeros.

We could explicitly append an extra column of zeros in our matrix and obtain the same result.

```
. matrix Czero = C, (0 \ 0)
. matrix list Czero
Czero[2,4]
    c1  c2  c3  c4
r1   1  -1   0   0
r2   0   1  -1   0
. mvtest means fuel repair capital, linear(Czero)
Test that mean vector satisfies linear hypothesis Czero
            Hotelling T2 =     35.25
       Hotelling F(2,23) =     16.89
                Prob > F =    0.0000
```

Values other than zeros could have been appended to C to test if the linear combinations equal those other values.

◁

Rencher (2002, 139–141) discusses one-sample profile analysis. The linear() option of mvtest means allows you to do this and other one-sample comparisons of interest.

Multiple-sample tests for mean vectors

Multiple-sample tests of mean vectors are also supported by mvtest means. The groups defining the multiple samples are specified with the by() option. The test that is presented depends on whether homogeneity of variance is assumed and whether there are more than two groups. The homogeneity option, the default, provides four standard multivariate tests (Wilks' lambda, Pillai's trace, Lawley–Hotelling trace, and Roy's largest root) under the assumption that the group covariance matrices are equal. The remaining possibilities do not assume equal covariances for the groups. The heterogeneous option with two by-groups presents the affine-invariant modification by Krishnamoorthy and

Yu (2004) of the test proposed by Nel and Van der Merwe (1986). The heterogeneous option with more than two by-groups presents a Wald test along with James' approximation to the p-value of the test. The lr option also removes the assumption of equal covariance matrices for the groups and produces a likelihood-ratio test for the equality of the group means.

▷ Example 4

In example 2 of [MV] **manova**, we introduce two variables measured on four groups of rabbits. The groups have differing sample sizes. mvtest means with the by() option can test the hypothesis that the means of the two variables are the same for the four groups of rabbits under the assumption that the groups have equal covariance matrices.

```
. use http://www.stata-press.com/data/r12/metabolic
(Table 4.5 Metabolic Comparisons of Rabbits -- Rencher (1998))
. mvtest means y1 y2, by(group)
Test for equality of 4 group means, assuming homogeneity
```

	Statistic	F(df1,	df2)	= F	Prob>F	
Wilks' lambda	0.1596	6.0	32.0	8.02	0.0000	e
Pillai's trace	1.2004	6.0	34.0	8.51	0.0000	a
Lawley-Hotelling trace	3.0096	6.0	30.0	7.52	0.0001	a
Roy's largest root	1.5986	3.0	17.0	9.06	0.0008	u

```
e = exact, a = approximate, u = upper bound on F
```

We reject the null hypothesis and conclude that the means are likely different between the four groups of rabbits.

The statistics reported above are the same as reported by manova y1 y2 = group in example 2 of [MV] **manova**. mvtest means y1 y2, by(group) homogeneous would also have produced the same results because homogeneous is the default when by() is specified.

◁

▷ Example 5

Continuing with the rabbit data, restricting ourselves to examining only the first two groups of rabbits and continuing to assume equal covariance matrices, we obtain the following:

```
. mvtest means y1 y2 if group < 3, by(group)
Test for equality of 2 group means, assuming homogeneity
```

	Statistic	F(df1,	df2)	= F	Prob>F	
Wilks' lambda	0.3536	2.0	11.0	10.05	0.0033	e
Pillai's trace	0.6464	2.0	11.0	10.05	0.0033	e
Lawley-Hotelling trace	1.8279	2.0	11.0	10.05	0.0033	e
Roy's largest root	1.8279	2.0	11.0	10.05	0.0033	e

```
e = exact, a = approximate, u = upper bound on F
```

We reject the null hypothesis of equal means for the two groups of rabbits.

With only two groups, the four multivariate tests above are equivalent. Because there were only two groups, we could have also produced this same F test with hotelling y1 y2 if group < 3, by(group); see [MV] **hotelling**.

◁

▷ Example 6

We now remove the assumption of equal covariance matrices for these two groups of rabbits and see if our conclusions change.

```
. mvtest means y1 y2 if group < 3, by(group) heterogeneous
Test for equality of 2 group means, allowing for heterogeneity
          MNV F(2,9.5) =      9.92
              Prob > F =    0.0047
```

Removing the assumption of equal covariance matrices still leads to rejection of the null hypothesis that the means for the two groups of rabbits are equal.

Because there were only two groups, an F statistic based on an affine-invariant modification by Krishnamoorthy and Yu (2004) of the test proposed by Nel and Van der Merwe (1986) was presented.

◁

▷ Example 7

If we attempt to test all four groups of rabbits while removing the assumption of equal covariance matrices,

```
. mvtest means y1 y2, by(group) heterogeneous
```

we receive an error message indicating that we have a singular covariance matrix. This is because there are only two observations for the fourth group of rabbits.

If we omit the fourth group, we obtain

```
. mvtest means y1 y2 if group < 4, by(group) heterogeneous
Test for equality of 3 group means, allowing for heterogeneity
          Wald chi2(4) =     34.08
           Prob > chi2 =    0.0000  (chi-squared approximation)
           Prob > chi2 =    0.0017  (James' approximation)
```

With more than two groups, a Wald chi-squared statistic is presented along with two p-values. The first p-value is the approximate one based on the χ^2 distribution with four degrees of freedom. The second, more accurate, p-value is based on James' (1954) approximation. Both p-values lead to rejection of the null hypothesis; the three groups of rabbits are unlikely to have equal means.

◁

▷ Example 8

We can request a likelihood-ratio test instead of a Wald test by specifying the `lr` option. Like `heterogeneous`, the `lr` option does not assume that the group covariance matrices are equal.

```
. mvtest means y1 y2 if group < 4, by(group) lr
Test for equality of 3 group means, allowing for heterogeneity
            LR chi2(4) =     21.32
           Prob > chi2 =    0.0003
```

The likelihood-ratio test also leads us to reject the null hypothesis of equal means for the three groups of rabbits.

◁

The computation of the likelihood-ratio test requires fitting the multivariate normal distribution with common means while allowing for different covariance matrices. The iterative fitting process may converge to local solutions rather than to the global solution, invalidating the reported test (Buot, Hoşten, and Richards 2007). As a precaution, you may use the protect() option to request fitting from different starting values so that the test results are based on the best solution found, and you can check whether the likelihood surface has multiple optima.

Saved results

mvtest means without the by() option (that is, a one-sample means test) saves the following in r():

Scalars

r(T2)	Hotelling T-squared
r(F)	F statistic
r(df_m)	model degrees of freedom
r(df_r)	residual degrees of freedom
r(p_F)	significance

Macros

r(Ftype)	type of model F test

mvtest means with by() but without the lr or heterogeneous options (that is, a multiple-sample means test assuming homogeneity) saves the following in r():

Scalars

r(F)	F statistic
r(df_m)	model degrees of freedom
r(df_r)	residual degrees of freedom
r(p_F)	significance

Macros

r(Ftype)	type of model F test

Matrices

r(stat_m)	MANOVA model tests

mvtest means with by() defining two groups and with the heterogeneous option (that is, a two-sample test of means, allowing for heterogeneity) saves the following in r():

Scalars

r(F)	F statistic
r(df_m)	model degrees of freedom
r(df_r)	residual degrees of freedom
r(p_F)	significance

Macros

r(Ftype)	type of model F test

mvtest means with by() defining more than two groups and with the heterogeneous option (that is, a multiple-sample test of means, allowing for heterogeneity) saves the following in r():

Scalars

r(chi2)	chi-squared statistic
r(df)	degrees of freedom for chi-squared test
r(p_chi2)	significance
r(p_chi2_James)	significance via James' approximation

Macros

r(chi2type)	type of model chi-squared test

`mvtest means` with the `by()` and `lr` options (that is, a likelihood-ratio multiple-sample test of means, allowing for heterogeneity) saves the following in `r()`:

Scalars

`r(chi2)`	chi-squared statistic
`r(df)`	degrees of freedom for chi-squared test
`r(p_chi2)`	significance
`r(rc)`	return code
`r(uniq)`	1/0 if protection runs yielded/did not yield same solution (`protect()` only)
`r(nprotect)`	number of protection runs (`protect()` only)

Macros

`r(chi2type)`	type of model chi-squared test

Matrices

`r(M)`	maximum likelihood estimate of means

Methods and formulas

`mvtest means` is implemented as an ado-file.

When comparing the formulas in this section with those found in some multivariate texts, be aware of whether they define the sample covariance matrix with a divisor of N or $N-1$. We use N. The formulas for several of the statistics are presented differently depending on your choice of divisor (but are still equivalent).

Methods and formulas are presented under the following headings:

One-sample tests for mean vectors
Multiple-sample tests for mean vectors

One-sample tests for mean vectors

Let the sample consist of N i.i.d. observations, $\mathbf{x}_i$, $i = 1,\ldots,N$, from a k-variate multivariate normal distribution, $\text{MVN}_k(\boldsymbol{\mu}, \boldsymbol{\Sigma})$, with sample mean $\overline{\mathbf{x}} = 1/N \sum_{i=1}^N \mathbf{x}_i$ and sample covariance matrix $\mathbf{S} = 1/N \sum_{i=1}^N (\mathbf{x}_i - \overline{\mathbf{x}})(\mathbf{x}_i - \overline{\mathbf{x}})'$.

`mvtest means` with the `equals()` option tests that a mean vector equals a fixed vector, $H_0: \boldsymbol{\mu} = \boldsymbol{\mu}_0$, and produces a Hotelling T-squared statistic, which is equivalent to the likelihood-ratio test (Mardia, Kent, and Bibby 1979, 125–126)

$$T^2_{omf} = (N-1)(\overline{\mathbf{x}} - \boldsymbol{\mu}_0)'\mathbf{S}^{-1}(\overline{\mathbf{x}} - \boldsymbol{\mu}_0)$$

Under the null hypothesis,

$$F_{omf} = \frac{N-k}{(N-1)k} T^2_{omf}$$

is distributed $F(k, N-k)$.

`mvtest means` with the `zero` option tests that the mean vector equals $\mathbf{0}$ and is obtained from T^2_{omf} by setting $\boldsymbol{\mu}_0 = \mathbf{0}$. For this case, denote Hotelling's T-squared as T^2_{omz} and the corresponding F statistic as F_{omz}.

`mvtest means` with the `linear()` option tests that the mean vector $\boldsymbol{\mu}$ satisfies a linear hypothesis, $H_0: \mathbf{C}\boldsymbol{\mu} = \mathbf{b}$, and produces a Hotelling T^2 test, which is equivalent to the likelihood-ratio test (Mardia, Kent, and Bibby 1979, 132–133)

$$T^2_{omc} = (N-1)(\mathbf{C}\overline{\mathbf{x}} - \mathbf{b})'(\mathbf{CSC}')^{-1}(\mathbf{C}\overline{\mathbf{x}} - \mathbf{b})$$

Under the null hypothesis,

$$F_{omc} = \frac{N-q}{(N-1)q} T^2_{omc}$$

is distributed $F(q, N-q)$, where q is the rank of $\mathbf{CSC}'$, typically the number of rows of $\mathbf{C}$.

`mvtest means` with the `equal` option tests that all means are equal and is obtained from T^2_{omc} by setting $\mathbf{C} = (\mathbf{I}, -\mathbf{1})$ and $\mathbf{b} = \mathbf{0}$. For this case, denote Hotelling's T^2 as T^2_{ome} and the corresponding F statistic as F_{ome}.

Multiple-sample tests for mean vectors

Let there be $m \geq 2$ independent samples with the jth sample containing N_j i.i.d. observations, $\mathbf{x}_{ji}$, $i = 1, \ldots, N_j$, from a k-variate multivariate normal distribution, $\text{MVN}_k(\boldsymbol{\mu}_j, \boldsymbol{\Sigma}_j)$. The observed jth sample mean is $\overline{\mathbf{x}}_j = 1/N_j \sum_{i=1}^{N_j} \mathbf{x}_{ji}$ and covariance is $\mathbf{S}_j = 1/N_j \sum_{i=1}^{N_j} (\mathbf{x}_{ji} - \overline{\mathbf{x}}_j)(\mathbf{x}_{ji} - \overline{\mathbf{x}}_j)'$. Let $N = \sum_{j=1}^{m} N_j$.

The tests for the hypothesis that the mean vector is the same across m independent samples, obtained from `mvtest means` with the `by()` option, come in four different flavors, depending on whether the additional assumption is made that the covariance matrix $\mathbf{S}_j$ is the same over the m samples (the `homogeneous` and `heterogeneous` options) and on the number of samples (whether the `by()` option defines more than two groups). If equal covariance is not assumed, the problem is commonly referred to as the multivariate Behrens–Fisher problem.

When assuming equal covariance matrices for the `by()` groups (the `homogeneous` option, the default), `mvtest means` tests the equality of the group means by using `manova` to compute the classic quartet of test statistics: Wilks' lambda, Pillai's trace, Lawley–Hotelling trace, and Roy's largest root. See [MV] **manova** for details. For $m = 2$ samples, the four tests are equivalent.

`mvtest means` has the `heterogeneous` and `lr` options, which remove the assumption of equal covariance matrices for the `by()` groups. The statistic produced with the `heterogeneous` option depends on whether there are $m = 2$ groups or $m > 2$ groups.

With the `heterogeneous` option and $m = 2$ samples, the test for equal means is computed using the affine-invariant modification by Krishnamoorthy and Yu (2004) of the test proposed by Nel and Van der Merwe (1986):

$$T^2_{mm2} = (\overline{\mathbf{x}}_1 - \overline{\mathbf{x}}_2)' \, \widetilde{\mathbf{S}}^{-1} \, (\overline{\mathbf{x}}_1 - \overline{\mathbf{x}}_2)$$

where

$$F_{mm2} = \frac{v-k+1}{vk} T^2_{mm2}$$

is approximately $F(k, v-k+1)$ distributed, and where $\widetilde{\mathbf{S}}_j = \mathbf{S}_j/(N_j - 1)$, $\widetilde{\mathbf{S}} = \widetilde{\mathbf{S}}_1 + \widetilde{\mathbf{S}}_2$, $v = k(k+1)/(a_1 + a_2)$, and

$$a_j = \left[\text{trace}\left\{ \left(\widetilde{\mathbf{S}}_j \widetilde{\mathbf{S}}^{-1} \right)^2 \right\} + \text{trace}\left(\widetilde{\mathbf{S}}_j \widetilde{\mathbf{S}}^{-1} \right)^2 \right] / \, (N_j - 1)$$

With the `heterogeneous` option and $m > 2$ samples, `mvtest means` computes the Wald test for equal means (Seber 1984, 445–447)

$$T_{mmw} = \sum_{j=1}^{m} (\overline{\mathbf{x}}_j - \overline{\overline{\mathbf{x}}})' \mathbf{W}_j (\overline{\mathbf{x}}_j - \overline{\overline{\mathbf{x}}})$$

where $\mathbf{W}_j = \{\mathbf{S}_j/(N_j - 1)\}^{-1}$, $\mathbf{W} = \sum_{j=1}^{m} \mathbf{W}_j$, and $\overline{\overline{\mathbf{x}}} = \mathbf{W}^{-1} \sum_{j=1}^{m} \mathbf{W}_j \overline{\mathbf{x}}_j$.

James (1954) showed that the upper α quantile of T_{mmw} under the null hypothesis can, to order N_i^{-1}, be obtained as $\chi^2_{r,\alpha}(a + b\chi^2_{r,\alpha})$, where $\chi^2_{r,\alpha}$ is the upper α quantile of a χ^2 with $r = k(m-1)$ degrees of freedom, and

$$a = 1 + \frac{1}{2r}\sum_{j=1}^{m} \frac{\left\{\text{trace}\left(\mathbf{I} - \mathbf{W}^{-1}\mathbf{W}_j\right)\right\}^2}{N_j - 1}$$

$$b = \frac{1}{r(r+2)}\left[\sum_{j=1}^{m} \frac{\text{trace}\left\{\left(\mathbf{I} - \mathbf{W}^{-1}\mathbf{W}_j\right)^2\right\}}{N_j - 1} + \frac{\left\{\text{trace}\left(\mathbf{I} - \mathbf{W}^{-1}\mathbf{W}_j\right)\right\}^2}{2(N_j - 1)}\right]$$

`mvtest` computes the p-value associated with the observed T_{mmw} by inverting James' expansion with a scalar solver using Brent's (1973) algorithm.

With the `lr` option, `mvtest means` provides a likelihood-ratio test for constant means across $m \geq 2$ samples, without the assumption of constant covariances. To fit the null model with a common mean, $\boldsymbol{\mu}_c$, an iterative procedure proposed by Mardia, Kent, and Bibby (1979, 142–143) is used. Let h denote the iteration. Let $\widehat{\boldsymbol{\Sigma}}_j^{(0)} = \mathbf{S}_j$, and $\widehat{\boldsymbol{\mu}}_c^{(0)}$ be obtained using the second formula below, then iterate these two formulas starting with $h = 1$ until convergence is achieved.

$$\widehat{\boldsymbol{\Sigma}}_j^{(h)} = \mathbf{S}_j + \left(\overline{\mathbf{x}}_j - \widehat{\boldsymbol{\mu}}_c^{(h-1)}\right)\left(\overline{\mathbf{x}}_j - \widehat{\boldsymbol{\mu}}_c^{(h-1)}\right)'$$

$$\widehat{\boldsymbol{\mu}}_c^{(h)} = \left\{\sum_{j=1}^{m} N_j \left(\widehat{\boldsymbol{\Sigma}}_j^{(h)}\right)^{-1}\right\}^{-1} \left\{\sum_{j=1}^{m} N_j \left(\widehat{\boldsymbol{\Sigma}}_j^{(h)}\right)^{-1} \overline{\mathbf{x}}_j\right\}$$

The likelihood-ratio chi-squared statistic

$$\chi^2_{mml} = \sum_{j=1}^{m} N_j \ln\left\{1 + \left(\overline{\mathbf{x}}_j - \widehat{\boldsymbol{\mu}}_c^{(h)}\right)' \left(\widehat{\boldsymbol{\Sigma}}_j^{(h)}\right)^{-1} \left(\overline{\mathbf{x}}_j - \widehat{\boldsymbol{\mu}}_c^{(h)}\right)\right\}$$

is approximately χ^2 distributed with $k(m-1)$ degrees of freedom.

References

Brent, R. P. 1973. *Algorithms for Minimization without Derivatives*. Englewood Cliffs, NJ: Prentice Hall. (Reprinted in paperback by Dover Publications, Mineola, NY, January 2002).

Buot, M.-L. G., S. Hoşten, and D. S. P. Richards. 2007. Counting and locating the solutions of polynomial systems of maximum likelihood equations, II: The Behrens–Fisher problem. *Statistica Sinica* 17: 1343–1354.

James, G. S. 1954. Tests of linear hypotheses in univariate and multivariate analysis when the ratios of the population variances are unknown. *Biometrika* 41: 19–43.

Johnson, R. A., and D. W. Wichern. 2007. *Applied Multivariate Statistical Analysis*. 6th ed. Englewood Cliffs, NJ: Prentice Hall.

Kramer, C. Y., and D. R. Jensen. 1969. Fundamentals of multivariate analysis, part I. Inference about means. *Journal of Quality Technology* 1: 120–133.

Krishnamoorthy, K., and J. Yu. 2004. Modified Nel and Van der Merwe test for the multivariate Behrens–Fisher problem. *Statistics and Probability Letters* 66: 161–169.

Mardia, K. V., J. T. Kent, and J. M. Bibby. 1979. *Multivariate Analysis*. London: Academic Press.

Nel, D. G., and C. A. Van der Merwe. 1986. A solution to the multivariate Behrens–Fisher problem. *Communications in Statistics, Theory and Methods* 15: 3719–3735.

Rencher, A. C. 1998. *Multivariate Statistical Inference and Applications*. New York: Wiley.

——. 2002. *Methods of Multivariate Analysis*. 2nd ed. New York: Wiley.

Seber, G. A. F. 1984. *Multivariate Observations*. New York: Wiley.

Also see

[MV] **hotelling** — Hotelling's T-squared generalized means test

[MV] **manova** — Multivariate analysis of variance and covariance

[R] **mean** — Estimate means

[R] **ttest** — Mean-comparison tests

Title

mvtest normality — Multivariate normality tests

Syntax

`mvtest normality` *varlist* [*if*] [*in*] [*weight*] [, *options*]

options	Description
Options	
`univariate`	display tests for univariate normality (`sktest`)
`bivariate`	display tests for bivariate normality (Doornik–Hansen)
`stats(`*stats*`)`	statistics to be computed

stats	Description
`dhansen`	Doornik–Hansen omnibus test; the default
`hzirkler`	Henze–Zirkler's consistent test
`kurtosis`	Mardia's multivariate kurtosis test
`skewness`	Mardia's multivariate skewness test
`all`	all tests listed here

`bootstrap`, `by`, `jackknife`, `rolling`, and `statsby` are allowed; see [U] **11.1.10 Prefix commands**.
Weights are not allowed with the `bootstrap` prefix; see [R] **bootstrap**.
`aweight`s are not allowed with the `jackknife` prefix; see [R] **jackknife**.
`fweight`s are allowed; see [U] **11.1.6 weight**.

Menu

Statistics > Multivariate analysis > MANOVA, multivariate regression, and related > Multivariate test of means, covariances, and normality

Description

`mvtest normality` performs tests for univariate, bivariate, and multivariate normality.

See [MV] **mvtest** for more multivariate tests.

Options

Options

`univariate` specifies that tests for univariate normality be displayed, as obtained from `sktest`; see [R] **sktest**.

`bivariate` specifies that the Doornik–Hansen (2008) test for bivariate normality be displayed for each pair of variables.

`stats(`*stats*`)` specifies one or more test statistics for multivariate normality. Multiple *stats* are separated by white space. The following *stats* are available:

`dhansen` produces the Doornik–Hansen (2008) omnibus test.

`hzirkler` produces Henze–Zirkler's (1990) consistent test.

`kurtosis` produces the test based on Mardia's (1970) measure of multivariate kurtosis.

`skewness` produces the test based on Mardia's (1970) measure of multivariate skewness.

`all` is a convenient shorthand for `stats(dhansen hzirkler kurtosis skewness)`.

Remarks

Univariate and multivariate tests of normality are provided by the `mvtest normality` command.

▷ Example 1

The classic Fisher iris data from Anderson (1935) consists of four features measured on 50 samples from each of three iris species. The four features are the length and width of the sepal and petal. The three species are *Iris setosa*, *Iris versicolor*, and *Iris virginica*. We hypothesize that these features might be normally distributed within species, though they are likely not normally distributed across species. We will examine the *Iris setosa* data.

```
. use http://www.stata-press.com/data/r12/iris
(Iris data)
. kdensity petlen if iris==1, name(petlen, replace) title(Petal Length)
. kdensity petwid if iris==1, name(petwid, replace) title(Petal Width)
. kdensity sepwid if iris==1, name(sepwid, replace) title(Sepal Width)
. kdensity seplen if iris==1, name(seplen, replace) title(Sepal Length)
. graph combine petlen petwid seplen sepwid, title("Iris Setosa Data")
```

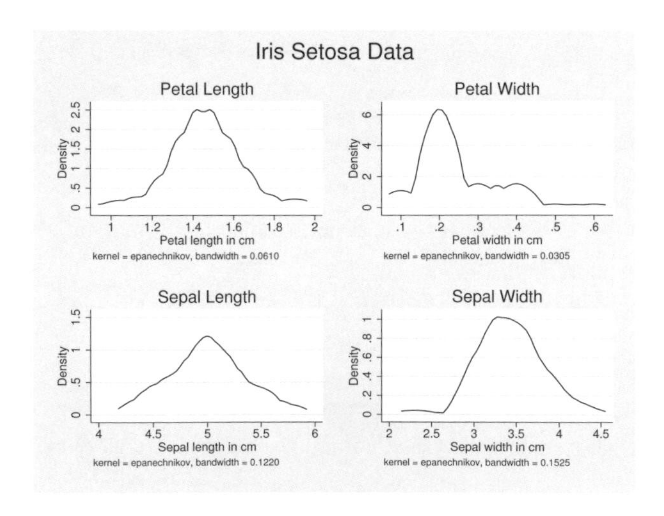

We perform all multivariate, univariate, and bivariate tests of normality.

```
. mvtest norm pet* sep* if iris==1, bivariate univariate stats(all)
Test for univariate normality
```

Variable	Pr(Skewness)	Pr(Kurtosis)	joint adj chi2(2)	joint Prob>chi2
petlen	0.7403	0.1447	2.36	0.3074
petwid	0.0010	0.0442	12.03	0.0024
seplen	0.7084	0.8157	0.19	0.9075
sepwid	0.8978	0.1627	2.07	0.3553

```
Doornik-Hansen test for bivariate normality
```

Pair of variables		chi2	df	Prob>chi2
petlen	petwid	17.47	4	0.0016
	seplen	5.76	4	0.2177
	sepwid	8.50	4	0.0748
petwid	seplen	14.97	4	0.0048
	sepwid	19.15	4	0.0007
seplen	sepwid	5.92	4	0.2049

```
Test for multivariate normality
    Mardia mSkewness =  3.079721    chi2(20) =   27.860    Prob>chi2 =  0.1128
    Mardia mKurtosis =  26.53766    chi2(1) =     1.677    Prob>chi2 =  0.1953
    Henze-Zirkler    =  .9488453    chi2(1) =     2.707    Prob>chi2 =  0.0999
    Doornik-Hansen                  chi2(8) =    24.414    Prob>chi2 =  0.0020
```

From the univariate tests of normality, petwid does not appear to be normally distributed: p-values of 0.0010 for skewness, 0.0442 for kurtosis, and 0.0024 for the joint univariate test. The univariate tests of the other three variables do not lead to a rejection of the null hypothesis of normality.

The bivariate tests of normality show a rejection (at the 5% level) of the null hypothesis of bivariate normality for all pairs of variables that include petwid. Other pairings fail to reject the null hypothesis of bivariate normality.

Of the four multivariate normality tests, only the Doornik–Hansen test rejects the null hypothesis of multivariate normality, p-value of 0.0020.

◁

The Doornik-Hansen (2008) test and Mardia's (1970) test for multivariate kurtosis take computing time roughly proportional to the number of observations. In contrast, the computing time of the test by Henze-Zirkler (1990) and Mardia's (1970) test for multivariate skewness are roughly proportional to the square of the number of observations.

Saved results

mvtest normality saves the following in r():

Scalars

r(p_dh)	significance of chi2_dh (stats(dhansen))
r(df_dh)	degrees of freedom of chi2_dh (stats(dhansen))
r(chi2_dh)	Doornik–Hansen statistic (stats(dhansen))
r(rank_hz)	rank of covariance matrix (stats(hzirkler))
r(p_hz)	two-sided significance of hz (stats(hzirkler))
r(z_hz)	normal variate associated with hz (stats(hzirkler))
r(V_hz)	expected variance of log(hz) (stats(hzirkler))
r(E_hz)	expected value of log(hz) (stats(hzirkler))
r(hz)	Henze–Zirkler discrepancy statistic (stats(hzirkler))
r(rank_mkurt)	rank of covariance matrix (stats(kurtosis))
r(p_mkurt)	significance of Mardia mKurtosis test (stats(kurtosis))
r(z_mkurt)	normal variate associated with Mardia mKurtosis (stats(kurtosis))
r(chi2_mkurt)	chi-squared of Mardia mKurtosis (stats(kurtosis))
r(mkurt)	Mardia mKurtosis test statistic (stats(kurtosis))
r(rank_mskew)	rank for Mardia mSkewness test (stats(skewness))
r(p_mskew)	significance of Mardia mSkewness test (stats(skewness))
r(df_mskew)	degrees of freedom of Mardia mSkewness test (stats(skewness))
r(chi2_mskew)	chi-squared of Mardia mSkewness test (stats(skewness))
r(mskew)	Mardia mSkewness test statistic (stats(skewness))

Matrices

r(U_dh)	matrix with the skewness and kurtosis of orthonormalized variables (used in the Doornik–Hansen test): b1, b2, z(b1), and z(b2) (stats(dhansen))
r(Btest)	bivariate test statistics (bivariate)
r(Utest)	univariate test statistics (univariate)

Methods and formulas

mvtest normality is implemented as an ado-file.

There are N independent k-variate observations, $\mathbf{x}_i$, $i = 1, \dots, N$. Let $\mathbf{X}$ denote the $N \times k$ matrix of observations. We wish to test whether these observations are multivariate normal distributed, $\text{MVN}_k(\boldsymbol{\mu}, \boldsymbol{\Sigma})$. The sample mean is $\overline{\mathbf{x}} = 1/N \sum_i \mathbf{x}_i$, and the sample covariance matrix is $\mathbf{S} = 1/N \sum (\mathbf{x}_i - \overline{\mathbf{x}})(\mathbf{x}_i - \overline{\mathbf{x}})'$.

Methods and formulas are presented under the following headings:

Mardia mSkewness and mKurtosis
Henze–Zirkler
Doornik–Hansen

Mardia mSkewness and mKurtosis

Mardia (1970) defined multivariate skewness, $b_{1,k}$, and kurtosis, $b_{2,k}$, as

$$b_{1,k} = \frac{1}{N^2} \sum_{i=1}^{N} \sum_{j=1}^{N} g_{ij}^3 \qquad \text{and} \qquad b_{2,k} = \frac{1}{N} \sum_{i=1}^{N} g_{ii}^2$$

where $g_{ij} = (\mathbf{x}_i - \overline{\mathbf{x}})'\mathbf{S}^{-1}(\mathbf{x}_j - \overline{\mathbf{x}})$. The test statistic

$$z_1 = \frac{(k+1)(N+1)(N+3)}{6\{(N+1)(k+1)-6\}} b_{1,k}$$

is approximately χ^2 distributed with $k(k+1)(k+2)/6$ degrees of freedom. The test statistic

$$z_2 = \frac{b_{2,k} - k(k+2)}{\sqrt{8k(k+2)/N}}$$

is approximately $N(0,1)$ distributed. Also see Rencher (2002, 99); Mardia, Kent, and Bibby (1979, 20–22); and Seber (1984, 148–149).

Henze–Zirkler

The Henze–Zirkler (1990) test, under the assumption that $\mathbf{S}$ is nonsingular, is

$$\begin{aligned} T = & \frac{1}{N}\sum_{i=1}^{N}\sum_{j=1}^{N} \exp\left\{-\frac{\beta^2}{2}(\mathbf{x}_i - \mathbf{x}_j)'\mathbf{S}^{-1}(\mathbf{x}_i - \mathbf{x}_j)\right\} \\ & - 2(1+\beta^2)^{-k/2}\sum_{i=1}^{N}\exp\left\{-\frac{\beta^2}{2(1+\beta^2)}(\mathbf{x}_i - \overline{\mathbf{x}})'\mathbf{S}^{-1}(\mathbf{x}_i - \overline{\mathbf{x}})\right\} \\ & + N(1+2\beta^2)^{-k/2} \end{aligned}$$

where

$$\beta = \frac{1}{\sqrt{2}}\left\{\frac{N(2k+1)}{4}\right\}^{1/(k+4)}$$

As $N \to \infty$, the first two moments of T are given by

$$\mathrm{E}(T) = 1 - (1+2\beta^2)^{-k/2}\left\{1 + \frac{k\beta^2}{1+2\beta^2} + \frac{k(k+2)\beta^4}{2(1+2\beta^2)^2}\right\}$$

$$\begin{aligned} \mathrm{Var}(T) = & \, 2(1+4\beta^2)^{-k/2} + 2(1+2\beta^2)^{-k}\left\{1 + \frac{2k\beta^4}{(1+2\beta^2)^2} + \frac{3k(k+2)\beta^8}{4(1+2\beta^2)^4}\right\} \\ & - 4w^{-k/2}\left\{1 + \frac{3k\beta^4}{2w} + \frac{k(k+2)\beta^8}{2w^2}\right\} \end{aligned}$$

where $w = (1+\beta^2)(1+3\beta^2)$.

Henze–Zirkler suggest obtaining a p-value from the assumption, supported by a series of simulations, that T is approximately lognormal distributed. Thus let $VZ = \ln\left\{1 + \mathrm{Var}(T)/E(T)^2\right\}$ and $EZ = \ln\{\mathrm{E}(T)\} - VZ/2$. The transformation $Z = \{\ln(T) - EZ\}/\sqrt{VZ}$. The p-value of Z is computed as $p = 2\Phi(-|Z|)$, where $\Phi()$ is the cumulative normal distribution.

Doornik–Hansen

For the Doornik–Hansen (2008) test, the multivariate observations are transformed, then the univariate skewness and kurtosis for each transformed variable is computed, and then these are combined into an approximate χ^2 statistic.

Let $\mathbf{V}$ be a matrix with ith diagonal element equal to $S_{ii}^{-1/2}$, where S_{ii} is the ith diagonal element of $\mathbf{S}$. $\mathbf{C} = \mathbf{VSV}$ is then the correlation matrix. Let $\mathbf{H}$ be a matrix with columns equal to the eigenvectors of $\mathbf{C}$, and let $\mathbf{\Lambda}$ be a diagonal matrix with the corresponding eigenvalues. Let $\breve{\mathbf{X}}$ be the centered version of $\mathbf{X}$, that is, $\overline{\mathbf{x}}$ subtracted from each row. The data are then transformed using $\dot{\mathbf{X}} = \breve{\mathbf{X}}\mathbf{VH}\mathbf{\Lambda}^{-1/2}\mathbf{H}'$.

The univariate skewness and kurtosis for each column of $\dot{\mathbf{X}}$ is then computed. The general formula for univariate skewness is $\sqrt{b_1} = m_3/m_2^{3/2}$ and kurtosis is $b_2 = m_4/m_2^2$, where $m_p = 1/N\sum_{i=1}^{N}(x_i - \overline{x})^p$. Let $\dot{x}_i$ denote the ith observation from the selected column of $\dot{\mathbf{X}}$. Because by construction the mean of $\dot{x}$ is zero and the variance m_2 is one, the formulas simplify to $\sqrt{b_1} = m_3$ and $b_2 = m_4$, where $m_p = 1/N\sum_{i=1}^{N}\dot{x}_i^p$.

The univariate skewness, $\sqrt{b_1}$, is transformed into an approximately normal variate, z_1, as in D'Agostino (1970):

$$z_1 = \delta \log\left(y + \sqrt{1+y^2}\right)$$

where

$$y = \left\{\frac{b_1(\omega^2-1)(N+1)(N+3)}{12(N-2)}\right\}^{1/2}$$

$$\delta = \left(\log\sqrt{\omega^2}\right)^{-1/2}$$

$$\omega^2 = -1 + \sqrt{2(\beta-1)}$$

$$\beta = \frac{3(N^2+27N-70)(N+1)(N+3)}{(N-2)(N+5)(N+7)(N+9)}$$

The univariate kurtosis, b_2, is transformed from a gamma variate into a χ^2-variate and then into a standard normal variable, z_2, using the Wilson–Hilferty (1931) transform:

$$z_2 = \sqrt{9\alpha}\left\{\left(\frac{\chi}{2\alpha}\right)^{1/3} - 1 + \frac{1}{9\alpha}\right\}$$

where

$$\chi = 2f(b_2 - 1 - b_1)$$

$$\alpha = a + b_1 c$$

$$f = \frac{(N+5)(N+7)(N^3+37N^2+11N-313)}{12\delta}$$

$$c = \frac{(N-7)(N+5)(N+7)(N^2+2N-5)}{6\delta}$$

$$a = \frac{(N-2)(N+5)(N+7)(N^2+27N-70)}{6\delta}$$

$$\delta = (N-3)(N+1)(N^2+15N-4)$$

The z_1 and z_2 associated with the columns of $\dot{\mathbf{X}}$ are collected into vectors $\mathbf{Z}_1$ and $\mathbf{Z}_2$. The statistic $\mathbf{Z}_1'\mathbf{Z}_1 + \mathbf{Z}_2'\mathbf{Z}_2$ is approximately χ^2 distributed with $2k$ degrees of freedom.

Acknowledgment

An earlier implementation of the Doornik and Hansen (2008) test is the `omninorm` package of Baum and Cox (2007).

References

Anderson, E. 1935. The irises of the Gaspe Peninsula. *Bulletin of the American Iris Society* 59: 2–5.

Baum, C. F., and N. J. Cox. 2007. omninorm: Stata module to calculate omnibus test for univariate/multivariate normality. Boston College Department of Economics, Statistical Software Components S417501. http://ideas.repec.org/c/boc/bocode/s417501.html.

D'Agostino, R. B. 1970. Transformation to normality of the null distribution of g_1. *Biometrika* 57: 679–681.

Doornik, J. A., and H. Hansen. 2008. An omnibus test for univariate and multivariate normality. *Oxford Bulletin of Economics and Statistics* 70: 927–939.

Henze, N., and B. Zirkler. 1990. A class of invariant consistent tests for multivariate normality. *Communications in Statistics, Theory and Methods* 19: 3595–3617.

Mardia, K. V. 1970. Measures of multivariate skewness and kurtosis with applications. *Biometrika* 57: 519–530.

Mardia, K. V., J. T. Kent, and J. M. Bibby. 1979. *Multivariate Analysis*. London: Academic Press.

Rencher, A. C. 2002. *Methods of Multivariate Analysis*. 2nd ed. New York: Wiley.

Seber, G. A. F. 1984. *Multivariate Observations*. New York: Wiley.

Wilson, E. B., and M. M. Hilferty. 1931. The distribution of chi-square. *Proceedings of the National Academy of Sciences* 17: 684–688.

Also see

[R] **sktest** — Skewness and kurtosis test for normality

[R] **swilk** — Shapiro–Wilk and Shapiro–Francia tests for normality

Title

pca — Principal component analysis

Syntax

Principal component analysis of data

pca *varlist* [*if*] [*in*] [*weight*] [, *options*]

Principal component analysis of a correlation or covariance matrix

pcamat *matname*, n(#) [*options pcamat_options*]

options	Description
Model 2	
components(#)	retain maximum of # principal components; factors() is a synonym
mineigen(#)	retain eigenvalues larger than #; default is 1e-5
correlation	perform PCA of the correlation matrix; the default
covariance	perform PCA of the covariance matrix
vce(none)	do not compute VCE of the eigenvalues and vectors; the default
vce(normal)	compute VCE of the eigenvalues and vectors assuming multivariate normality
Reporting	
level(#)	set confidence level; default is level(95)
blanks(#)	display loadings as blanks when $\lvert$loadings$\rvert < \#$
novce	suppress display of SEs even though calculated
*means	display summary statistics of variables
Advanced	
tol(#)	advanced option; see *Options* for details
ignore	advanced option; see *Options* for details
norotated	display unrotated results, even if rotated results are available (replay only)

* means is not allowed with pcamat.
norotated is not shown in the dialog box.

pcamat_options	Description
Model	
shape(full)	*matname* is a square symmetric matrix; the default
shape(lower)	*matname* is a vector with the rowwise lower triangle (with diagonal)
shape(upper)	*matname* is a vector with the rowwise upper triangle (with diagonal)
names(*namelist*)	variable names; required if *matname* is triangular
forcepsd	modifies *matname* to be positive semidefinite
*n(#)	number of observations
sds($matname_2$)	vector with standard deviations of variables
means($matname_3$)	vector with means of variables

* n() is required for pcamat.

bootstrap, by, jackknife, rolling, statsby, and xi are allowed with pca; see [U] **11.1.10 Prefix commands**. However, bootstrap and jackknife results should be interpreted with caution; identification of the pca parameters involves data-dependent restrictions, possibly leading to badly biased and overdispersed estimates (Milan and Whittaker 1995).

Weights are not allowed with the bootstrap prefix; see [R] **bootstrap**.

aweights are not allowed with the jackknife prefix; see [R] **jackknife**.

aweights and fweights are allowed with pca; see [U] **11.1.6 weight**.

See [MV] **pca postestimation** for features available after estimation.

Menu

pca

Statistics > Multivariate analysis > Factor and principal component analysis > Principal component analysis (PCA)

pcamat

Statistics > Multivariate analysis > Factor and principal component analysis > PCA of a correlation or covariance matrix

Description

Principal component analysis (PCA) is a statistical technique used for data reduction. The leading eigenvectors from the eigen decomposition of the correlation or covariance matrix of the variables describe a series of uncorrelated linear combinations of the variables that contain most of the variance. In addition to data reduction, the eigenvectors from a PCA are often inspected to learn more about the underlying structure of the data.

pca and pcamat display the eigenvalues and eigenvectors from the PCA eigen decomposition. The eigenvectors are returned in orthonormal form, that is, orthogonal (uncorrelated) and normalized (with unit length, $\mathbf{L}'\mathbf{L} = \mathbf{I}$). pcamat provides the correlation or covariance matrix directly. For pca, the correlation or covariance matrix is computed from the variables in *varlist*.

pcamat allows the correlation or covariance matrix $\mathbf{C}$ to be specified as a $k \times k$ symmetric matrix with row and column names set to the variable names or as a $k(k+1)/2$ long row or column vector containing the lower or upper triangle of $\mathbf{C}$ along with the names() option providing the variable names. See the shape() option for details.

The vce(normal) option of pca and pcamat provides standard errors of the eigenvalues and eigenvectors and aids in interpreting the eigenvectors. See the second technical note under *Remarks* for a discussion of the underlying assumptions.

Scores, residuals, rotations, scree plots, score plots, loading plots, and more are available after `pca` and `pcamat`, see [MV] **pca postestimation**.

Options

Model 2

`components(#)` and `mineigen(#)` specify the maximum number of components (eigenvectors or factors) to be retained. `components()` specifies the number directly, and `mineigen()` specifies it indirectly, keeping all components with eigenvalues greater than the indicated value. The options can be specified individually, together, or not at all. `factors()` is a synonym for `components()`.

`components(#)` sets the maximum number of components (factors) to be retained. `pca` and `pcamat` always display the full set of eigenvalues but display eigenvectors only for retained components. Specifying a number larger than the number of variables in *varlist* is equivalent to specifying the number of variables in *varlist* and is the default.

`mineigen(#)` sets the minimum value of eigenvalues to be retained. The default is `1e-5` or the value of `tol()` if specified.

Specifying `components()` and `mineigen()` affects only the number of components to be displayed and stored in `e()`; it does not enforce the assumption that the other eigenvalues are 0. In particular, the standard errors reported when `vce(normal)` is specified do not depend on the number of retained components.

`correlation` and `covariance` specify that principal components be calculated for the correlation matrix and covariance matrix, respectively. The default is `correlation`. Unlike factor analysis, PCA is not scale invariant; the eigenvalues and eigenvectors of a covariance matrix differ from those of the associated correlation matrix. Usually, a PCA of a covariance matrix is meaningful only if the variables are expressed in the same units.

For `pcamat`, do not confuse the type of the matrix to be analyzed with the type of *matname*. Obviously, if *matname* is a correlation matrix and the option `sds()` is not specified, it is not possible to perform a PCA of the covariance matrix.

`vce(none | normal)` specifies whether standard errors are to be computed for the eigenvalues, the eigenvectors, and the (cumulative) percentage of explained variance (confirmatory PCA). These standard errors are obtained assuming multivariate normality of the data and are valid only for a PCA of a covariance matrix. Be cautious if applying these to correlation matrices.

Reporting

`level(#)` specifies the confidence level, as a percentage, for confidence intervals. The default is `level(95)` or as set by `set level`; see [U] **20.7 Specifying the width of confidence intervals**. `level()` is allowed only with `vce(normal)`.

`blanks(#)` shows blanks for loadings with absolute value smaller than #. This option is ignored when specified with `vce(normal)`.

`novce` suppresses the display of standard errors, even though they are computed, and displays the PCA results in a matrix/table style. You can specify `novce` during estimation in combination with `vce(normal)`. More likely, you will want to use `novce` during replay.

`means` displays summary statistics of the variables over the estimation sample. This option is not available with `pcamat`.

Advanced

`tol(#)` is an advanced, rarely used option and is available only with `vce(normal)`. An eigenvalue, ev_i, is classified as being close to zero if $ev_i < \text{tol} \times \max(ev)$. Two eigenvalues, ev_1 and ev_2, are "close" if $\text{abs}(ev_1 - ev_2) < \text{tol} \times \max(ev)$. The default is `tol(1e-5)`. See option `ignore` below and the technical note later in this entry.

`ignore` is an advanced, rarely used option and is available only with `vce(normal)`. It continues the computation of standard errors and tests, even if some eigenvalues are suspiciously close to zero or suspiciously close to other eigenvalues, violating crucial assumptions of the asymptotic theory used to estimate standard errors and tests. See the technical note later in this entry.

The following option is available with `pca` and `pcamat` but is not shown in the dialog box:

`norotated` displays the unrotated principal components, even if rotated components are available. This option may be specified only when replaying results.

Options unique to pcamat

Model

`shape(`*shape_arg*`)` specifies the shape (storage mode) for the covariance or correlation matrix *matname*. The following modes are supported:

`full` specifies that the correlation or covariance structure of k variables is stored as a symmetric $k \times k$ matrix. Specifying `shape(full)` is optional in this case.

`lower` specifies that the correlation or covariance structure of k variables is stored as a vector with $k(k+1)/2$ elements in rowwise lower-triangular order:

$$C_{11}\ C_{21}\ C_{22}\ C_{31}\ C_{32}\ C_{33}\ \ldots\ C_{k1}\ C_{k2}\ \ldots\ C_{kk}$$

`upper` specifies that the correlation or covariance structure of k variables is stored as a vector with $k(k+1)/2$ elements in rowwise upper-triangular order:

$$C_{11}\ C_{12}\ C_{13}\ \ldots\ C_{1k}\ C_{22}\ C_{23}\ \ldots C_{2k}\ \ldots\ C_{(k-1k-1)}\ C_{(k-1k)}\ C_{kk}$$

`names(`*namelist*`)` specifies a list of k different names, which are used to document output and to label estimation results and are used as variable names by `predict`. By default, `pcamat` verifies that the row and column names of *matname* and the column or row names of $matname_2$ and $matname_3$ from the `sds()` and `means()` options are in agreement. Using the `names()` option turns off this check.

`forcepsd` modifies the matrix *matname* to be positive semidefinite (psd) and so to be a proper covariance matrix. If *matname* is not positive semidefinite, it will have negative eigenvalues. By setting negative eigenvalues to 0 and reconstructing, we obtain the least-squares positive-semidefinite approximation to *matname*. This approximation is a singular covariance matrix.

`n(#)` is required and specifies the number of observations.

`sds(`$matname_2$`)` specifies a $k \times 1$ or $1 \times k$ matrix with the standard deviations of the variables. The row or column names should match the variable names, unless the `names()` option is specified. `sds()` may be specified only if *matname* is a correlation matrix.

`means(`$matname_3$`)` specifies a $k \times 1$ or $1 \times k$ matrix with the means of the variables. The row or column names should match the variable names, unless the `names()` option is specified. Specify `means()` if you have variables in your dataset and want to use `predict` after `pcamat`.

Remarks

Principal component analysis (PCA) is commonly thought of as a statistical technique for data reduction. It helps you reduce the number of variables in an analysis by describing a series of uncorrelated linear combinations of the variables that contain most of the variance.

PCA originated with the work of Pearson (1901) and Hotelling (1933). For an introduction, see Rabe-Hesketh and Everitt (2007, chap. 14) or van Belle, Fisher, Heagerty, and Lumley (2004). More advanced treatments are Mardia, Kent, and Bibby (1979, chap. 8), and Rencher (2002, chap. 12). For monograph-sized treatments, including extensive discussions of the relationship between PCA and related approaches, see Jackson (2003) and Jolliffe (2002).

The objective of PCA is to find unit-length linear combinations of the variables with the greatest variance. The first principal component has maximal overall variance. The second principal component has maximal variance among all unit length linear combinations that are uncorrelated to the first principal component, etc. The last principal component has the smallest variance among all unit-length linear combinations of the variables. All principal components combined contain the same information as the original variables, but the important information is partitioned over the components in a particular way: the components are orthogonal, and earlier components contain more information than later components. PCA thus conceived is just a linear transformation of the data. It does not assume that the data satisfy a specific statistical model, though it does require that the data be interval-level data—otherwise taking linear combinations is meaningless.

PCA is scale dependent. The principal components of a covariance matrix and those of a correlation matrix are different. In applied research, PCA of a covariance matrix is useful only if the variables are expressed in commensurable units.

Structural equation modeling provides a more general framework for performing principal component analysis; see the *Stata Structural Equation Modeling Reference Manual.*

❑ Technical note

Principal components have several useful properties. Some of these are geometric. Both the principal components and the principal scores are uncorrelated (orthogonal) among each other. The f leading principal components have maximal generalized variance among all f unit-length linear combinations.

It is also possible to interpret PCA as a fixed effects factor analysis with homoskedastic residuals

$$y_{ij} = \mathbf{a}_i'\mathbf{b}_j + e_{ij} \qquad i = 1, \dots, n \qquad j = 1, \dots, p$$

where y_{ij} are the elements of the matrix $\mathbf{Y}$, $\mathbf{a}_i$ (scores) and $\mathbf{b}_j$ (loadings) are f-vectors of parameters, and e_{ij} are independent homoskedastic residuals. (In factor analysis, the scores $\mathbf{a}_i$ are random rather than fixed, and the residuals are allowed to be heteroskedastic in j.) It follows that $E(\mathbf{Y})$ is a matrix of rank f, with f typically substantially less than n or p. Thus we may think of PCA as a regression model with a restricted number but unknown independent variables. We may also say that the expected values of the rows (or columns) of $\mathbf{Y}$ are in some unknown f-dimensional space.

For more information on these properties and for other characterizations of PCA, see Jackson (2003) and Jolliffe (2002).

❑

▷ Example 1

We consider a dataset of audiometric measurements on 100 males, age 9. The measurements are minimal discernible intensities at four different frequencies with the left and right ear (see Jackson 2003, 106). The variable lft1000 refers to the left ear at 1,000 Hz.

```
. use http://www.stata-press.com/data/r12/audiometric
(Audiometric measures)

. correlate lft* rght*
(obs=100)
```

	lft500	lft1000	lft2000	lft4000	rght500	rght1000	rght2000
lft500	1.0000						
lft1000	0.7775	1.0000					
lft2000	0.4012	0.5366	1.0000				
lft4000	0.2554	0.2749	0.4250	1.0000			
rght500	0.6963	0.5515	0.2391	0.1790	1.0000		
rght1000	0.6416	0.7070	0.4460	0.2632	0.6634	1.0000	
rght2000	0.2372	0.3597	0.7011	0.3165	0.1589	0.4142	1.0000
rght4000	0.2041	0.2169	0.3262	0.7097	0.1321	0.2201	0.3746

	rght4000
rght4000	1.0000

As you may have expected, measurements on the same ear are more highly correlated than measurements on different ears. Also, measurements on different ears at the same frequency are more highly correlated than at different frequencies. Because the variables are in commensurable units, it would make theoretical sense to analyze the covariance matrix of these variables. However, the variances of the measures differ widely:

```
. summarize lft* rght*, sep(4)
```

Variable	Obs	Mean	Std. Dev.	Min	Max
lft500	100	-2.8	6.408643	-10	15
lft1000	100	-.5	7.571211	-10	20
lft2000	100	2	10.94061	-10	45
lft4000	100	21.35	19.61569	-10	70
rght500	100	-2.6	7.123726	-10	25
rght1000	100	-.7	6.396811	-10	20
rght2000	100	1.6	9.289942	-10	35
rght4000	100	21.35	19.33039	-10	75

In an analysis of the covariances, the higher frequency measures would dominate the results. There is no clinical reason for such an effect (see also Jackson [2003]). Therefore, we will analyze the correlation matrix.

```
. pca lft* rght*
Principal components/correlation                  Number of obs    =       100
                                                  Number of comp.  =         8
                                                  Trace            =         8
    Rotation: (unrotated = principal)             Rho              =    1.0000

    --------------------------------------------------------------------------
       Component |   Eigenvalue   Difference         Proportion   Cumulative
    -------------+------------------------------------------------------------
           Comp1 |      3.92901      2.31068             0.4911       0.4911
           Comp2 |      1.61832      .642997             0.2023       0.6934
           Comp3 |      .975325      .508543             0.1219       0.8153
           Comp4 |      .466782      .126692             0.0583       0.8737
           Comp5 |       .34009     .0241988             0.0425       0.9162
           Comp6 |      .315891       .11578             0.0395       0.9557
           Comp7 |      .200111     .0456375             0.0250       0.9807
           Comp8 |      .154474            .             0.0193       1.0000
    --------------------------------------------------------------------------

Principal components (eigenvectors)

    --------------------------------------------------------------------------
        Variable |    Comp1     Comp2     Comp3     Comp4     Comp5     Comp6
    -------------+------------------------------------------------------------
          lft500 |   0.4011   -0.3170    0.1582   -0.3278    0.0231    0.4459
         lft1000 |   0.4210   -0.2255   -0.0520   -0.4816   -0.3792   -0.0675
         lft2000 |   0.3664    0.2386   -0.4703   -0.2824    0.4392   -0.0638
         lft4000 |   0.2809    0.4742    0.4295   -0.1611    0.3503   -0.4169
         rght500 |   0.3433   -0.3860    0.2593    0.4876    0.4975    0.1948
        rght1000 |   0.4114   -0.2318   -0.0289    0.3723   -0.3513   -0.6136
        rght2000 |   0.3115    0.3171   -0.5629    0.3914   -0.1108    0.2650
        rght4000 |   0.2542    0.5135    0.4262    0.1591   -0.3960    0.3660
    --------------------------------------------------------------------------

    ----------------------------------------------
        Variable |    Comp7     Comp8 | Unexplained
    -------------+--------------------+-----------
          lft500 |   0.3293   -0.5463 |          0
         lft1000 |  -0.0331    0.6227 |          0
         lft2000 |  -0.5255   -0.1863 |          0
         lft4000 |   0.4269    0.0839 |          0
         rght500 |  -0.1594    0.3425 |          0
        rght1000 |  -0.0837   -0.3614 |          0
        rght2000 |   0.4778    0.1466 |          0
        rght4000 |  -0.4139   -0.0508 |          0
    ----------------------------------------------
```

pca shows two panels. The first panel lists the eigenvalues of the correlation matrix, ordered from largest to smallest. The corresponding eigenvectors are listed in the second panel. These are the principal components and have unit length; the columnwise sum of the squares of the loadings is 1 ($0.4011^2 + 0.4210^2 + \cdots + 0.2542^2 = 1$).

Remark: Literature and software that treat principal components in combination with factor analysis tend to display principal components normed to the associated eigenvalues rather than to 1. This normalization is available in the postestimation command `estat loadings`; see [MV] **pca postestimation**.

The eigenvalues add up to the sum of the variances of the variables in the analysis—the "total variance" of the variables. Because we are analyzing a correlation matrix, the variables are standardized to have unit variance, so the total variance is 8. The eigenvalues are the variances of the principal components. The first principal component has variance 3.93, explaining 49% (3.93/8) of the total variance. The second principal component has variance 1.61 or 20% (1.61/8) of the total variance. Principal components are uncorrelated. You may want to verify that; for instance,

$$0.4011(-0.3170) + 0.4210(-0.2255) + \cdots + 0.2542(0.5135) = 0$$

As a consequence, we may also say that the first two principal components explain the sum of the variances of the individual components, or $49 + 20 = 69\%$ of the total variance. Had the components been correlated, they would have partly represented the same information, so the information contained in the combination would not have been equal to the sum of the information of the components. All eight principal components combined explain all variance in all variables; therefore, the unexplained variances listed in the second panel are all zero, and `Rho` = 1.00 as shown above the first panel.

More than 85% of the variance is contained in the first four principal components. We can list just these components with the option `components(4)`.

```
. pca lft* rght*, components(4)
Principal components/correlation                  Number of obs    =       100
                                                  Number of comp.  =         4
                                                  Trace            =         8
    Rotation: (unrotated = principal)             Rho              =    0.8737

    --------------------------------------------------------------------------
       Component |   Eigenvalue   Difference         Proportion   Cumulative
    -------------+------------------------------------------------------------
           Comp1 |      3.92901      2.31068             0.4911       0.4911
           Comp2 |      1.61832      .642997             0.2023       0.6934
           Comp3 |      .975325      .508543             0.1219       0.8153
           Comp4 |      .466782      .126692             0.0583       0.8737
           Comp5 |       .34009     .0241988             0.0425       0.9162
           Comp6 |      .315891       .11578             0.0395       0.9557
           Comp7 |      .200111     .0456375             0.0250       0.9807
           Comp8 |      .154474            .             0.0193       1.0000
    --------------------------------------------------------------------------

Principal components (eigenvectors)

    ------------------------------------------------------------------
        Variable |    Comp1     Comp2     Comp3     Comp4 | Unexplained
    -------------+----------------------------------------+-------------
          lft500 |   0.4011   -0.3170    0.1582   -0.3278 |       .1308
         lft1000 |   0.4210   -0.2255   -0.0520   -0.4816 |       .1105
         lft2000 |   0.3664    0.2386   -0.4703   -0.2824 |       .1275
         lft4000 |   0.2809    0.4742    0.4295   -0.1611 |       .1342
         rght500 |   0.3433   -0.3860    0.2593    0.4876 |       .1194
        rght1000 |   0.4114   -0.2318   -0.0289    0.3723 |       .1825
        rght2000 |   0.3115    0.3171   -0.5629    0.3914 |      .07537
        rght4000 |   0.2542    0.5135    0.4262    0.1591 |       .1303
    ------------------------------------------------------------------
```

The first panel is not affected. The second panel now lists the first four principal components. These four components do not contain all information in the data, and therefore some of the variances in the variables are unaccounted for or unexplained. These equal the sums of squares of the loadings in the deleted components, weighted by the associated eigenvalues. The unexplained variances in all variables are of similar order. The average unexplained variance is equal to the overall unexplained variance of 13% $(1 - 0.87)$.

Look more closely at the principal components. The first component has positive loadings of roughly equal size on all variables. It can be interpreted as overall sensitivity of a person's ears. The second principal component has positive loadings on the higher frequencies with both ears and negative loadings for the lower frequencies. Thus the second principal component distinguishes sensitivity for higher frequencies versus lower frequencies. The third principal component similarly differentiates sensitivity at medium frequencies from sensitivity at other frequencies. Finally, the fourth principal component has negative loadings on the left ear and positive loadings on the right ear; it differentiates the left and right ear.

We stated earlier that the first principal component had similar loadings on all eight variables. This can be tested if we are willing to assume that the data are multivariate normal distributed. For this case, pca can estimate the standard errors and related statistics. To conserve paper, we request only the results of the first two principal components and specify the option vce(normal).

```
. pca l* r*, comp(2) vce(normal)
(with PCA/correlation, SEs and tests are approximate)
Principal components/correlation                 Number of obs    =        100
                                                 Number of comp.  =          2
                                                 Trace            =          8
                                                 Rho              =     0.6934
SEs assume multivariate normality                SE(Rho)          =     0.0273

------------------------------------------------------------------------------
             |      Coef.   Std. Err.      z    P>|z|     [95% Conf. Interval]
-------------+----------------------------------------------------------------
Eigenvalues  |
       Comp1 |   3.929005   .5556453     7.07   0.000     2.839961     5.01805
       Comp2 |   1.618322   .2288653     7.07   0.000     1.169754    2.066889
-------------+----------------------------------------------------------------
Comp1        |
      lft500 |   .4010948   .0429963     9.33   0.000     .3168236     .485366
     lft1000 |   .4209908   .0359372    11.71   0.000     .3505551    .4914264
     lft2000 |   .3663748   .0463297     7.91   0.000     .2755702    .4571794
     lft4000 |   .2808559   .0626577     4.48   0.000     .1580491    .4036628
     rght500 |    .343251   .0528285     6.50   0.000     .2397091     .446793
    rght1000 |   .4114209   .0374312    10.99   0.000     .3380571    .4847846
    rght2000 |   .3115483   .0551475     5.65   0.000     .2034612    .4196354
    rght4000 |   .2542212    .066068     3.85   0.000     .1247303    .3837121
-------------+----------------------------------------------------------------
Comp2        |
      lft500 |  -.3169638    .067871    -4.67   0.000    -.4499885   -.1839391
     lft1000 |   -.225464   .0669887    -3.37   0.001    -.3567595   -.0941686
     lft2000 |   .2385933   .1079073     2.21   0.027     .0270989    .4500877
     lft4000 |   .4741545   .0967918     4.90   0.000      .284446    .6638629
     rght500 |  -.3860197   .0803155    -4.81   0.000    -.5434352   -.2286042
    rght1000 |  -.2317725   .0674639    -3.44   0.001    -.3639994   -.0995456
    rght2000 |    .317059   .1215412     2.61   0.009     .0788427    .5552752
    rght4000 |   .5135121   .0951842     5.39   0.000     .3269544    .7000697
------------------------------------------------------------------------------
LR test for independence:     chi2(28)  =    448.21   Prob > chi2 =  0.0000
LR test for   sphericity:     chi2(35)  =    451.11   Prob > chi2 =  0.0000
Explained variance by components
  Components | Eigenvalue  Proportion  SE_Prop  Cumulative   SE_Cum       Bias
-------------+----------------------------------------------------------------
       Comp1 |   3.929005      0.4911   0.0394      0.4911   0.0394     .056663
       Comp2 |   1.618322      0.2023   0.0271      0.6934   0.0273     .015812
       Comp3 |   .9753248      0.1219   0.0178      0.8153   0.0175    -.014322
       Comp4 |   .4667822      0.0583   0.0090      0.8737   0.0127     .007304
       Comp5 |     .34009      0.0425   0.0066      0.9162   0.0092     .026307
       Comp6 |   .3158912      0.0395   0.0062      0.9557   0.0055    -.057717
       Comp7 |   .2001111      0.0250   0.0040      0.9807   0.0031    -.013961
       Comp8 |   .1544736      0.0193   0.0031      1.0000   0.0000    -.020087
------------------------------------------------------------------------------
```

Here pca acts like an estimation command. The output is organized in different equations. The first equation contains the eigenvalues. The second equation named, Comp1, is the first principal component, etc. pca reports, for instance, standard errors of the eigenvalues. Although testing the values of eigenvalues may, up to now, be rare in applied research, interpretation of results should take stability into consideration. It makes little sense to report the first eigenvalue as 3.929 if you see that the standard error is 0.55.

pca has also reported the standard errors of the principal components. It has also estimated the covariances.

```
. estat vce
  (output omitted )
```

Showing the large amount of information contained in the VCE matrix is not useful by itself. The fact that it has been estimated, however, enables us to test properties of the principal components. Does it make good sense to talk about the loadings of the first principal component being of the same size? We use `testparm` with two options; see [R] **test**. `eq(Comp1)` specifies that we are testing coefficients for equation `Comp1`, that is, the first principal component. `equal` specifies that instead of testing that the coefficients are zero, we want to test that the coefficients are equal to each other—a more sensible hypothesis because principal components are normalized to 1.

```
. testparm lft* rght*, equal eq(Comp1)
 ( 1)  - [Comp1]lft500 + [Comp1]lft1000 = 0
 ( 2)  - [Comp1]lft500 + [Comp1]lft2000 = 0
 ( 3)  - [Comp1]lft500 + [Comp1]lft4000 = 0
 ( 4)  - [Comp1]lft500 + [Comp1]rght500 = 0
 ( 5)  - [Comp1]lft500 + [Comp1]rght1000 = 0
 ( 6)  - [Comp1]lft500 + [Comp1]rght2000 = 0
 ( 7)  - [Comp1]lft500 + [Comp1]rght4000 = 0
           chi2(  7) =     7.56
         Prob > chi2 =    0.3729
```

We cannot reject the null hypothesis of equal loadings, so our interpretation of the first component does not seem to conflict with the data.

pca also displays standard errors of the proportions of variance explained by the leading principal components. Again this information is useful primarily to indicate the strength of formulations of results rather than to test hypotheses about these statistics. The information is also useful to compare studies: if in one study the leading two principal components explain 70% of variance, whereas in a replicating study they explain 80%, are these differences significant given the sampling variation?

Because `pca` is an estimation command just like `regress` or `xtlogit`, you may replay the output by typing just `pca`. If you have used `pca` with the `vce(normal)` option, you may use the option `novce` at estimation or during replay to display the standard PCA output.

```
. pca, novce
Principal components/correlation                 Number of obs    =       100
                                                 Number of comp.  =         2
                                                 Trace            =         8
    Rotation: (unrotated = principal)            Rho              =    0.6934
```

Component	Eigenvalue	Difference	Proportion	Cumulative
Comp1	3.92901	2.31068	0.4911	0.4911
Comp2	1.61832	.642997	0.2023	0.6934
Comp3	.975325	.508543	0.1219	0.8153
Comp4	.466782	.126692	0.0583	0.8737
Comp5	.34009	.0241988	0.0425	0.9162
Comp6	.315891	.11578	0.0395	0.9557
Comp7	.200111	.0456375	0.0250	0.9807
Comp8	.154474	.	0.0193	1.0000

```
Principal components (eigenvectors)

    ------------------------------------------------
        Variable |    Comp1     Comp2 | Unexplained
    -------------+--------------------+-------------
          lft500 |   0.4011   -0.3170 |       .2053
         lft1000 |   0.4210   -0.2255 |       .2214
         lft2000 |   0.3664    0.2386 |       .3805
         lft4000 |   0.2809    0.4742 |       .3262
         rght500 |   0.3433   -0.3860 |       .2959
        rght1000 |   0.4114   -0.2318 |        .248
        rght2000 |   0.3115    0.3171 |        .456
        rght4000 |   0.2542    0.5135 |       .3193
    ------------------------------------------------
```

◁

❑ Technical note

Inference on the eigenvalues and eigenvectors of a covariance matrix is based on a series of assumptions:

(A1) The variables are multivariate normal distributed.

(A2) The variance–covariance matrix of the observations has all distinct and strictly positive eigenvalues.

Under assumptions A1 and A2, the eigenvalues and eigenvectors of the sample covariance matrix can be seen as maximum likelihood estimates for the population analogues that are asymptotically (multivariate) normally distributed (Anderson 1963; Jackson 2003). See Tyler (1981) for related results for elliptic distributions. Be cautious when interpreting because the asymptotic variances are rather sensitive to violations of assumption A1 (and A2). Wald tests of hypotheses that are in conflict with assumption A2 (for example, testing that the first and second eigenvalues are the same) produce incorrect p-values.

Because the statistical theory for a PCA of a correlation matrix is much more complicated, `pca` and `pcamat` compute standard errors and tests of a correlation matrix as if it were a covariance matrix. This practice is in line with the application of asymptotic theory in Jackson (2003). This will usually lead to some underestimation of standard errors, but we believe that this problem is smaller than the consequences of deviations from normality.

You may conduct tests for multivariate normality using the `mvtest normality` command (see [MV] **mvtest normality**):

```
. mvtest normality lft* rght*, stats(all)
Test for multivariate normality
    Mardia mSkewness =  14.52785   chi2(120) =  251.052   Prob>chi2 =  0.0000
    Mardia mKurtosis =  94.53331     chi2(1) =   33.003   Prob>chi2 =  0.0000
    Henze-Zirkler    =  1.272529     chi2(1) =  118.563   Prob>chi2 =  0.0000
    Doornik-Hansen                  chi2(16) =   95.318   Prob>chi2 =  0.0000
```

These tests cast serious doubt on the multivariate normality of the variables. We advise caution in interpreting the inference results. Time permitting, you may want to turn to bootstrap methods for inference on the principal components and eigenvalues, but you should be aware of some serious identification problems in using the bootstrap here (Milan and Whittaker 1995).

❑

▷ Example 2

We remarked before that the principal components of a correlation matrix are generally different from the principal components of a covariance matrix. pca defaults to performing the PCA of the correlation matrix. To obtain a PCA of the covariance matrix, specify the covariance option.

```
. pca l* r*, comp(4) covariance
Principal components/covariance                  Number of obs    =        100
                                                 Number of comp.  =          4
                                                 Trace            =     1154.5
    Rotation: (unrotated = principal)            Rho              =     0.9396
```

Component	Eigenvalue	Difference	Proportion	Cumulative
Comp1	706.795	527.076	0.6122	0.6122
Comp2	179.719	68.3524	0.1557	0.7679
Comp3	111.366	24.5162	0.0965	0.8643
Comp4	86.8501	57.4842	0.0752	0.9396
Comp5	29.366	9.53428	0.0254	0.9650
Comp6	19.8317	6.67383	0.0172	0.9822
Comp7	13.1578	5.74352	0.0114	0.9936
Comp8	7.41432	.	0.0064	1.0000

Principal components (eigenvectors)

Variable	Comp1	Comp2	Comp3	Comp4	Unexplained
lft500	0.0835	0.2936	-0.0105	0.3837	7.85
lft1000	0.1091	0.3982	0.0111	0.3162	11.71
lft2000	0.2223	0.5578	0.0558	-0.4474	11.13
lft4000	0.6782	-0.1163	-0.7116	-0.0728	.4024
rght500	0.0662	0.2779	-0.0226	0.4951	12.42
rght1000	0.0891	0.3119	0.0268	0.2758	11.14
rght2000	0.1707	0.3745	0.2721	-0.4496	14.71
rght4000	0.6560	-0.3403	0.6441	0.1550	.4087

As expected, the results are less clear. The total variance to be analyzed is 1,154.5; this is the sum of the variances of the eight variables, that is, the trace of the covariance matrix. The leading principal components now account for a larger fraction of the variance; this is often the case with covariance matrices where the variables have widely different variances. The principal components are somewhat harder to interpret; mainly the loadings are no longer of roughly comparable size.

◁

▷ Example 3

Sometimes you do not have the original data but have only the correlation or covariance matrix. pcamat performs a PCA for such a matrix. To simplify presentation, we use the data on the left ear.

```
. correlate lft*, cov
(obs=100)
```

	lft500	lft1000	lft2000	lft4000
lft500	41.0707			
lft1000	37.7273	57.3232		
lft2000	28.1313	44.4444	119.697	
lft4000	32.101	40.8333	91.2121	384.775

Suppose that we have the covariances of the variables but not the original data. `correlate` stores the covariances in `r(C)`, so we can use that matrix and invoke `pcamat` with the options `n(100)`, specifying the number of observations, and `names()`, providing the variable names.

```
. matrix Cfull = r(C)

. pcamat Cfull, comp(2) n(100) names(lft500 lft1000 lft2000 lft4000)

Principal components/correlation                 Number of obs    =       100
                                                 Number of comp.  =         2
                                                 Trace            =         4
    Rotation: (unrotated = principal)            Rho              =    0.8169

    --------------------------------------------------------------------------
       Component |   Eigenvalue   Difference         Proportion   Cumulative
    -------------+------------------------------------------------------------
           Comp1 |      2.37181      1.47588             0.5930       0.5930
           Comp2 |      .895925      .366238             0.2240       0.8169
           Comp3 |      .529687      .327106             0.1324       0.9494
           Comp4 |      .202581            .             0.0506       1.0000
    --------------------------------------------------------------------------

Principal components (eigenvectors)

    ------------------------------------------------
        Variable |    Comp1     Comp2 | Unexplained
    -------------+--------------------+-------------
          lft500 |   0.5384   -0.4319 |      .1453
         lft1000 |   0.5730   -0.3499 |      .1116
         lft2000 |   0.4958    0.2955 |      .3387
         lft4000 |   0.3687    0.7770 |      .1367
    ------------------------------------------------
```

If we had to type in the covariance matrix, to avoid excess typing `pcamat` allows you to provide the covariance (or correlation) matrix with just the upper or lower triangular elements including the diagonal. (Thus, for correlations, you have to enter the 1s for the diagonal.) For example, we could enter the lower triangle of our covariance matrix row by row up to and including the diagonal as a one-row Stata matrix.

```
. matrix Clow = ( 41.0707, 37.7273, 57.3232, 28.1313, 44.4444,
                  119.697, 32.101,  40.8333, 91.2121, 384.775 )
```

The matrix `Clow` has one row and 10 columns. To make seeing the structure easier, we prefer to enter these numbers in the following way:

```
. matrix Clow = ( 41.0707,
                  37.7273, 57.3232,
                  28.1313, 44.4444, 119.697,
                  32.101,  40.8333, 91.2121, 384.775 )
```

When using the lower or upper triangle stored in a row or column vector, it is not possible to define the variable names as row or column names of the matrix; the option `names()` is required. Moreover, we have to specify the option `shape(lower)` to inform `pcamat` that the vector contains the lower triangle, not the upper triangle.

```
. pcamat Clow, comp(2) shape(lower) n(100) names(lft500 lft1000 lft2000 lft4000)
  (output omitted)
```

◁

Saved results

pca and pcamat without the vce(normal) option save the following in e():

Scalars

e(N)	number of observations
e(f)	number of retained components
e(rho)	fraction of explained variance
e(trace)	trace of e(C)
e(lndet)	ln of the determinant of e(C)
e(cond)	condition number of e(C)

Macros

e(cmd)	pca (even for pcamat)
e(cmdline)	command as typed
e(Ctype)	correlation or covariance
e(wtype)	weight type
e(wexp)	weight expression
e(title)	title in output
e(properties)	nob noV eigen
e(rotate_cmd)	program used to implement rotate
e(estat_cmd)	program used to implement estat
e(predict)	program used to implement predict
e(marginsnotok)	predictions disallowed by margins

Matrices

e(C)	$p \times p$ correlation or covariance matrix
e(means)	$1 \times p$ matrix of means
e(sds)	$1 \times p$ matrix of standard deviations
e(Ev)	$1 \times p$ matrix of eigenvalues (sorted)
e(L)	$p \times f$ matrix of eigenvectors = components
e(Psi)	$1 \times p$ matrix of unexplained variance

Functions

e(sample)	marks estimation sample

pca and pcamat with the vce(normal) option save the above, as well as the following:

Scalars

e(v_rho)	variance of e(rho)
e(chi2_i)	χ^2 statistic for test of independence
e(df_i)	degrees of freedom for test of independence
e(p_i)	significance of test of independence
e(chi2_s)	χ^2 statistic for test of sphericity
e(df_s)	degrees of freedom for test of sphericity
e(p_s)	significance of test of sphericity
e(rank)	rank of e(V)

Macros

e(vce)	multivariate normality
e(properties)	b V

Matrices

e(b)	$1 \times p + fp$ coefficient vector (all eigenvalues and retained eigenvectors)
e(Ev_bias)	$1 \times p$ matrix: bias of eigenvalues
e(Ev_stats)	$p \times 5$ matrix with statistics on explained variance
e(V)	variance–covariance matrix of the estimates e(b)

Methods and formulas

pca and pcamat are implemented as ado-files.

Methods and formulas are presented under the following headings:

Notation
Inference on eigenvalues and eigenvectors
More general tests for multivariate normal distributions

Notation

Let $\mathbf{C}$ be the $p \times p$ correlation or covariance matrix to be analyzed. The spectral or eigen decomposition of $\mathbf{C}$ is

$$\mathbf{C} = \mathbf{V}\boldsymbol{\Lambda}\mathbf{V}' = \sum_{i=1}^{p} \lambda_i \mathbf{v}_i \mathbf{v}_i'$$

$$\mathbf{v}_i'\mathbf{v}_j = \delta_{ij} \qquad \text{(that is, orthonormality)}$$

$$\lambda_1 \geq \lambda_2 \geq \ldots \geq \lambda_p \geq 0$$

The eigenvectors $\mathbf{v}_i$ are also known as the principal components. The direction (sign) of principal components is not defined. pca returns principal components signed so that $\mathbf{1}'\mathbf{v}_i > 0$. In PCA, "total variance" equals trace$(\mathbf{C}) = \sum \lambda_j$.

Inference on eigenvalues and eigenvectors

The asymptotic distribution of the eigenvectors $\widehat{\mathbf{v}}_i$ and eigenvalues $\widehat{\lambda}_i$ of a covariance matrix $\mathbf{S}$ for a sample from a multivariate normal distribution $N(\mu, \boldsymbol{\Sigma})$ was derived by Girshick (1939); for more results, see also Anderson (1963) and Jackson (2003). Higher-order expansions are discussed in Lawley (1956). See Tyler (1981) for related results for elliptic distributions. The theory of the exact distribution is rather complicated (Muirhead 1982, chap. 9) and hard to implement. If we assume that eigenvalues of $\boldsymbol{\Sigma}$ are distinct and strictly positive, the eigenvalues and eigenvectors of $\mathbf{S}$ are jointly asymptotically multivariate normal distributed with the following moments (up to order n^{-3}):

$$E(\widehat{\lambda}_i) = \lambda_i \left\{ 1 + \frac{1}{n} \sum_{j \neq i}^{k} \left(\frac{\lambda_j}{\lambda_i - \lambda_j} \right) \right\} + O(n^{-3})$$

$$\text{Var}(\widehat{\lambda}_i) = \frac{2\lambda_i^2}{n} \left\{ 1 - \frac{1}{n} \sum_{j \neq i}^{k} \left(\frac{\lambda_j}{\lambda_i - \lambda_j} \right)^2 \right\} + O(n^{-3})$$

$$\text{Cov}(\widehat{\lambda}_i, \widehat{\lambda}_j) = \frac{2}{n^2} \left(\frac{\lambda_i \lambda_j}{\lambda_i - \lambda_j} \right)^2 + O(n^{-3})$$

$$\text{Var}(\widehat{\mathbf{v}}_i) = \frac{1}{n} \sum_{j \neq i}^{k} \frac{\lambda_i \lambda_j}{(\lambda_i - \lambda_j)^2} \mathbf{v}_j \mathbf{v}_j'$$

$$\text{Cov}(\widehat{\mathbf{v}}_i, \widehat{\mathbf{v}}_j) = -\frac{1}{n} \frac{\lambda_i \lambda_j}{(\lambda_i - \lambda_j)^2} \mathbf{v}_i \mathbf{v}_j'$$

For the asymptotic theory of the cumulative proportion of variance explained, see Kshirsagar (1972, 454).

More general tests for multivariate normal distributions

The likelihood-ratio χ^2 test of independence (Basilevsky 1994, 187) is

$$\chi^2 = -\left(n - \frac{2p+5}{6}\right) \ln\{\det(\mathbf{C})\}$$

with $p(p-1)/2$ degrees of freedom.

The likelihood-ratio χ^2 test of sphericity (Basilevsky 1994, 192) is

$$\chi^2 = -\left(n - \frac{2p^2+p+2}{6p}\right) \left[\ln\{\det(\widetilde{\mathbf{\Lambda}})\} - p\ln\left\{\frac{\operatorname{trace}(\widetilde{\mathbf{\Lambda}})}{p}\right\}\right]$$

with $(p+2)(p-1)/2$ degrees of freedom and with $\widetilde{\mathbf{\Lambda}}$ the eigenvalues of the correlation matrix.

References

Anderson, T. W. 1963. Asymptotic theory for principal component analysis. *Annals of Mathematical Statistics* 34: 122–148.

Basilevsky, A. T. 1994. *Statistical Factor Analysis and Related Methods: Theory and Applications.* New York: Wiley.

Dinno, A. 2009. Implementing Horn's parallel analysis for principal component analysis and factor analysis. *Stata Journal* 9: 291–298.

Girshick, M. A. 1939. On the sampling theory of roots of determinantal equations. *Annals of Mathematical Statistics* 10: 203–224.

Hannachi, A., I. T. Jolliffe, and D. B. Stephenson. 2007. Empirical orthogonal functions and related techniques in atmospheric science: A review. *International Journal of Climatology* 27: 1119–1152.

Hotelling, H. 1933. Analysis of a complex of statistical variables into principal components. *Journal of Educational Psychology* 24: 417–441, 498–520.

Jackson, J. E. 2003. *A User's Guide to Principal Components.* New York: Wiley.

Jolliffe, I. T. 2002. *Principal Component Analysis.* 2nd ed. New York: Springer.

Kshirsagar, A. M. 1972. *Multivariate Analysis.* New York: Dekker.

Lawley, D. N. 1956. Tests of significance for the latent roots of covariance and correlation matrices. *Biometrika* 43: 128–136.

Mardia, K. V., J. T. Kent, and J. M. Bibby. 1979. *Multivariate Analysis.* London: Academic Press.

Milan, L., and J. Whittaker. 1995. Application of the parametric bootstrap to models that incorporate a singular value decomposition. *Applied Statistics* 44: 31–49.

Muirhead, R. J. 1982. *Aspects of Multivariate Statistical Theory.* New York: Wiley.

Pearson, K. 1901. On lines and planes of closest fit to systems of points in space. *Philosophical Magazine, Series 6* 2: 559–572.

Rabe-Hesketh, S., and B. S. Everitt. 2007. *A Handbook of Statistical Analyses Using Stata.* 4th ed. Boca Raton, FL: Chapman & Hall/CRC.

Rencher, A. C. 2002. *Methods of Multivariate Analysis.* 2nd ed. New York: Wiley.

Tyler, D. E. 1981. Asymptotic inference for eigenvectors. *Annals of Statistics* 9: 725–736.

van Belle, G., L. D. Fisher, P. J. Heagerty, and T. S. Lumley. 2004. *Biostatistics: A Methodology for the Health Sciences.* 2nd ed. New York: Wiley.

Weesie, J. 1997. smv7: Inference on principal components. *Stata Technical Bulletin* 37: 22–23. Reprinted in *Stata Technical Bulletin Reprints*, vol. 7, pp. 229–231. College Station, TX: Stata Press.

Also see

[MV] **pca postestimation** — Postestimation tools for pca and pcamat

[R] **tetrachoric** — Tetrachoric correlations for binary variables

[MV] **biplot** — Biplots

[MV] **canon** — Canonical correlations

[MV] **factor** — Factor analysis

[D] **corr2data** — Create dataset with specified correlation structure

[R] **alpha** — Compute interitem correlations (covariances) and Cronbach's alpha

Stata Structural Equation Modeling Reference Manual

[U] **20 Estimation and postestimation commands**

Title

pca postestimation — Postestimation tools for pca and pcamat

Description

The following postestimation commands are of special interest after `pca` and `pcamat`:

Command	Description
`estat anti`	anti-image correlation and covariance matrices
`estat kmo`	Kaiser–Meyer–Olkin measure of sampling adequacy
`estat loadings`	component-loading matrix in one of several normalizations
`estat residuals`	matrix of correlation or covariance residuals
`estat rotatecompare`	compare rotated and unrotated components
`estat smc`	squared multiple correlations between each variable and the rest
* `estat summarize`	display summary statistics over the estimation sample
`loadingplot`	plot component loadings
`rotate`	rotate component loadings
`scoreplot`	plot score variables
`screeplot`	plot eigenvalues

* `estat summarize` is not available after `pcamat`.

For information about `loadingplot` and `scoreplot`, see [MV] **scoreplot**; for information about `rotate`, see [MV] **rotate**; for information about `screeplot`, see [MV] **screeplot**; and for all other commands, see below.

The following standard postestimation commands are also available:

Command	Description
† `estat`	examine the VCE matrix
`estimates`	cataloging estimation results
* `lincom`	point estimates, standard errors, testing, and inference for linear combinations of coefficients
* `nlcom`	point estimates, standard errors, testing, and inference for nonlinear combinations of coefficients
`predict`	score variables, predictions, and residuals
* `predictnl`	point estimates, standard errors, testing, and inference for generalized predictions
* `test`	Wald tests of simple and composite linear hypotheses
* `testnl`	Wald tests of nonlinear hypotheses

† `estat` is available after `pca` and `pcamat` with the `vce(normal)` option.

* `lincom`, `nlcom`, `predictnl`, `test`, and `testnl` are available only after `pca` with the `vce(normal)` option.

See the corresponding entries in the *Base Reference Manual* for details.

Special-interest postestimation commands

`estat anti` displays the anti-image correlation and anti-image covariance matrices. These are minus the partial covariance and minus the partial correlation of all pairs of variables, holding all other variables constant.

`estat kmo` displays the Kaiser–Meyer–Olkin (KMO) measure of sampling adequacy. KMO takes values between 0 and 1, with small values indicating that overall the variables have too little in common to warrant a PCA. Historically, the following labels are given to values of KMO (Kaiser 1974):

0.00 to 0.49	unacceptable
0.50 to 0.59	miserable
0.60 to 0.69	mediocre
0.70 to 0.79	middling
0.80 to 0.89	meritorious
0.90 to 1.00	marvelous

`estat loadings` displays the component-loading matrix in one of several normalizations of the columns (eigenvectors).

`estat residuals` displays the difference between the observed correlation or covariance matrix and the fitted (reproduced) matrix using the retained factors.

`estat rotatecompare` displays the unrotated (principal) components next to the most recent rotated components.

`estat smc` displays the squared multiple correlations between each variable and all other variables. SMC is a theoretical lower bound for communality and thus an upper bound for the unexplained variance.

`estat summarize` displays summary statistics of the variables in the principal component analysis over the estimation sample. This subcommand is not available after `pcamat`.

Syntax for predict

`predict` [*type*] {*stub*∗ | *newvarlist*} [*if*] [*in*] [, *statistic options*]

statistic	# of vars.	Description (k = # of orig. vars.; f = # of components)
Main		
`score`	$1, \ldots, f$	scores based on the components; the default
`fit`	k	fitted values using the retained components
`residual`	k	raw residuals from the fit using the retained components
`q`	1	residual sum of squares

options	Description
Main	
`norotated`	use unrotated results, even when rotated results are available
`center`	base scores on centered variables
`notable`	suppress table of scoring coefficients
`format(%`*fmt*`)`	format for displaying the scoring coefficients

Menu

Statistics > Postestimation > Predictions, residuals, etc.

Options for predict

Note on `pcamat`: `predict` requires that variables with the correct names be available in memory. Apart from centered scores, `means()` should have been specified with `pcamat`. If you used `pcamat` because you have access only to the correlation or covariance matrix, you cannot use `predict`.

Main

`score` calculates the scores for components 1, ..., #, where # is the number of variables in *newvarlist*.

`fit` calculates the fitted values, using the retained components, for each variable. The number of variables in *newvarlist* should equal the number of variables in the *varlist* of `pca`; see [MV] **pca**.

`residual` calculates for each variable the raw residuals (residual = observed − fitted), with the fitted values computed using the retained components.

`q` calculates the Rao statistics (that is, the sum of squares of the omitted components) weighted by the respective eigenvalues. This equals the residual sum of squares between the original variables and the fitted values.

`norotated` uses unrotated results, even when rotated results are available.

`center` bases scores on centered variables. This option is relevant only for a PCA of a covariance matrix, in which the scores are based on uncentered variables by default. Scores for a PCA of a correlation matrix are always based on the standardized variables.

`notable` suppresses the table of scoring coefficients.

`format(%fmt)` specifies the display format for scoring coefficients. The default is `format(%8.4f)`.

Syntax for estat

Display the anti-image correlation and covariance matrices

```
estat anti [, nocorr nocov format(%fmt)]
```

Display the Kaiser–Meyer–Olkin measure of sampling adequacy

```
estat kmo [, novar format(%fmt)]
```

Display the component-loading matrix

```
estat loadings [, cnorm(unit|eigen|inveigen) format(%fmt)]
```

Display the differences in matrices

```
estat residuals [, obs fitted format(%fmt)]
```

Display the unrotated and rotated components

    estat rotatecompare [, format(%fmt)]

Display the squared multiple correlations

    estat smc [, format(%fmt)]

Display the summary statistics

    estat summarize [, labels noheader noweights]

Menu

Statistics > Postestimation > Reports and statistics

Options for estat

`nocorr`, an option used with `estat anti`, suppresses the display of the anti-image correlation matrix, that is, minus the partial correlation matrix of all pairs of variables, holding constant all other variables.

`nocov`, an option used with `estat anti`, suppresses the display of the anti-image covariance matrix, that is, minus the partial covariance matrix of all pairs of variables, holding constant all other variables.

`format(%`*fmt*`)` specifies the display format. The defaults differ between the subcommands.

`novar`, an option used with `estat kmo`, suppresses the Kaiser–Meyer–Olkin measures of sampling adequacy for the variables in the principal component analysis, displaying the overall KMO measure only.

`cnorm(unit | eigen | inveigen)`, an option used with `estat loadings`, selects the normalization of the eigenvectors, the columns of the principal-component loading matrix. The following normalizations are available

`unit`	ssq(column) = 1; the default
`eigen`	ssq(column) = eigenvalue
`inveigen`	ssq(column) = 1/eigenvalue

with ssq(column) being the sum of squares of the elements in a column and eigenvalue, the eigenvalue associated with the column (eigenvector).

`obs`, an option used with `estat residuals`, displays the observed correlation or covariance matrix for which the PCA was performed.

`fitted`, an option used with `estat residuals`, displays the fitted (reconstructed) correlation or covariance matrix based on the retained components.

`labels`, `noheader`, and `noweights` are the same as for the generic `estat summarize` command; see [R] **estat**.

Remarks

After computing the principal components and the associated eigenvalues, you have more issues to resolve. How many components do you want to retain? How well is the correlation or covariance matrix approximated by the retained components? How can you interpret the principal components? Is it possible to improve the interpretability by rotating the retained principal components? And, when these issues have been settled, the component scores are probably needed for later research.

The rest of this entry describes the specific tools available for these purposes.

Remarks are presented under the following headings:

Postestimation statistics
Plots of eigenvalues, component loadings, and scores
Rotating the components
How rotate interacts with pca
Predicting the component scores

In addition to these specific postestimation tools, general tools are available as well. `pca` is an estimation command, so it is possible to manage a series of PCA analyses with the `estimates` command; see [R] **estimates**. If you have specified the `vce(normal)` option, `pca` has saved the coefficients `e(b)` and the associated variance–covariance matrix `e(V)`, and you can use standard Stata commands to test hypotheses about the principal components and eigenvalues ("confirmatory principal component analysis"), for instance, with the `test`, `lincom`, and `testnl` commands. We caution you to test only hypotheses that do not violate the assumptions of the theory underlying the derivation of the covariance matrix. In particular, all eigenvalues are assumed to be different and strictly positive. Thus it makes no sense to use `test` to test the hypothesis that the smallest four eigenvalues are equal (let alone that they are equal to zero.)

Postestimation statistics

`pca` displays the principal components in unit normalization; the sum of squares of the principal loadings equals 1. This parallels the standard conventions in mathematics concerning eigenvectors. Some texts and some software use a different normalization. Some texts multiply the eigenvectors by the square root of the eigenvalues. In this normalization, the sum of the squared loadings equals the variance explained by that component. `estat loadings` can display the loadings in this normalization.

```
. use http://www.stata-press.com/data/r12/audiometric
(Audiometric measures)

. pca l* r*, comp(4)
  (output omitted )

. estat loadings, cnorm(eigen)

Principal component loadings (unrotated)
    component normalization: sum of squares(column) = eigenvalue
```

	Comp1	Comp2	Comp3	Comp4
lft500	.795	-.4032	.1562	-.2239
lft1000	.8345	-.2868	-.05132	-.3291
lft2000	.7262	.3035	-.4645	-.193
lft4000	.5567	.6032	.4242	-.1101
rght500	.0004	-.4911	.2561	.3331
rght1000	.8155	-.2948	-.0285	.2544
rght2000	.6175	.4033	-.5559	.2674
rght4000	.5039	.6533	.4209	.1087

How close the retained principal components approximate the correlation matrix can be seen from the fitted (reconstructed) correlation matrix and from the residuals, that is, the difference between the observed and fitted correlations.

```
. estat residual, fit format(%7.3f)
```

Fitted correlation matrix

Variable	lft500	lft1000	lft2000	lft4000	rght500	rg~1000
lft500	0.869					
lft1000	0.845	0.890				
lft2000	0.426	0.606	0.872			
lft4000	0.290	0.306	0.412	0.866		
rght500	0.704	0.586	0.162	0.155	0.881	
rght1000	0.706	0.683	0.467	0.236	0.777	0.818
rght2000	0.182	0.340	0.778	0.322	0.169	0.469
rght4000	0.179	0.176	0.348	0.841	0.166	0.234

Variable	rg~2000	rg~4000
rght2000	0.925	
rght4000	0.370	0.870

Residual correlation matrix

Variable	lft500	lft1000	lft2000	lft4000	rght500	rg~1000
lft500	0.131					
lft1000	-0.067	0.110				
lft2000	-0.024	-0.070	0.128			
lft4000	-0.035	-0.031	0.013	0.134		
rght500	-0.008	-0.034	0.077	0.024	0.119	
rght1000	-0.064	0.024	-0.021	0.027	-0.114	0.182
rght2000	0.056	0.020	-0.076	-0.005	-0.010	-0.054
rght4000	0.025	0.041	-0.022	-0.131	-0.034	-0.014

Variable	rg~2000	rg~4000
rght2000	0.075	
rght4000	0.005	0.130

All off diagonal residuals are small, except perhaps the two measurements at the highest frequency.

estat also provides some of the standard methods for studying correlation matrices to assess whether the variables have strong linear relations with each other. In a sense, these methods could be seen as preestimation rather than as postestimation methods. The first method is the inspection of the squared multiple correlation (the regression R^2) of each variable on all other variables.

```
. estat smc
Squared multiple correlations of variables with all other variables
```

Variable	smc
lft500	0.7113
lft1000	0.7167
lft2000	0.6229
lft4000	0.5597
rght500	0.5893
rght1000	0.6441
rght2000	0.5611
rght4000	0.5409

The SMC measures help identify variables that cannot be explained well from the other variables. For such variables, you should reevaluate whether they should be included in the analysis. In our examples, none of the SMCs are so small as to warrant exclusion. Two other statistics are offered. First, we can inspect the anti-image correlation and covariance matrices, that is, the negative of correlations (covariances) of the variables partialing out all other variables. If many of these correlations or covariances are "high", the relationships between some of the variables have little to do with the other variables, indicating that it will not be possible to obtain a low-dimensional reduction of the data.

```
. estat anti, nocov format(%7.3f)
Anti-image correlation coefficients —— partialing out all other variables
```

Variable	lft500	lft1000	lft2000	lft4000	rght500	rg~1000
lft500	1.000					
lft1000	-0.561	1.000				
lft2000	-0.051	-0.267	1.000			
lft4000	-0.014	0.026	-0.285	1.000		
rght500	-0.466	0.131	0.064	-0.017	1.000	
rght1000	0.023	-0.389	0.043	-0.042	-0.441	1.000
rght2000	0.085	0.068	-0.617	0.161	0.067	-0.248
rght4000	-0.047	-0.002	0.150	-0.675	0.019	0.023

Variable	rg~2000	rg~4000
rght2000	1.000	
rght4000	-0.266	1.000

The Kaiser–Meyer–Olkin measure of sampling adequacy compares the correlations and the partial correlations between variables. If the partial correlations are relatively high compared to the correlations, the KMO measure is small, and a low-dimensional representation of the data is not possible.

```
. estat kmo
Kaiser-Meyer-Olkin measure of sampling adequacy
```

Variable	kmo
lft500	0.7701
lft1000	0.7767
lft2000	0.7242
lft4000	0.6449
rght500	0.7562
rght1000	0.8168
rght2000	0.6673
rght4000	0.6214
Overall	0.7328

Using the Kaiser (1974) characterization of KMO values,

0.00 to 0.49	unacceptable
0.50 to 0.59	miserable
0.60 to 0.69	mediocre
0.70 to 0.79	middling
0.80 to 0.89	meritorious
0.90 to 1.00	marvelous

we declare our KMO value, 0.73, middling.

Plots of eigenvalues, component loadings, and scores

After computing the principal components, we probably wish to determine how many components to keep. In factor analysis the question of the "true" number of factors is a complicated one. With PCA, it is a little more straightforward. We may set a percentage of variance we wish to account for, say, 90%, and retain just enough components to account for at least that much of the variance. Usually you will want to weigh the costs associated with using more components in later analyses against the benefits of the extra variance they account for. The relative magnitudes of the eigenvalues indicate the amount of variance they account for. A useful tool for visualizing the eigenvalues relative to one another, so that you can decide the number of components to retain, is the scree plot proposed by Cattell (1966); see [MV] **screeplot**.

```
. screeplot, mean
```

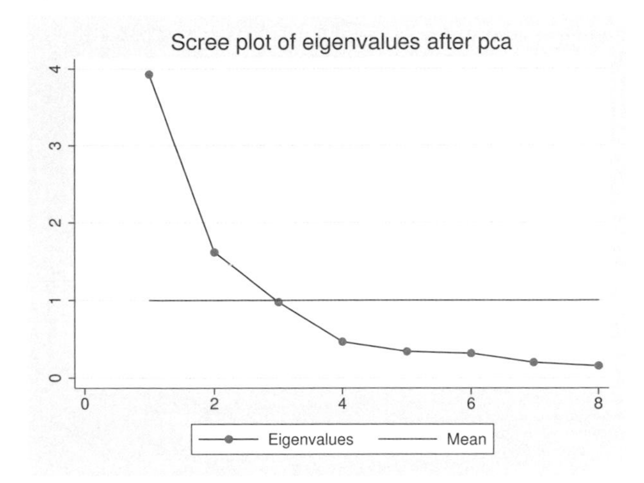

Because we are analyzing a correlation matrix, the mean eigenvalue is 1. We wish to retain the components associated with the high part of the scree plot and drop the components associated with the lower flat part of the scree plot. The boundary between high and low is not clear here, but we would choose two or three components, although the fourth component had the nice interpretation of the left versus the right ear; see [MV] **pca**.

A problem in interpreting the scree plot is that no guidance is given with respect to its stability under sampling. How different could the plot be with different samples? The approximate variance of an eigenvalue $\widehat{\lambda}$ of a covariance matrix for multivariate normal distributed data is $2\lambda^2/n$. From this we can derive confidence intervals for the eigenvalues. These scree plot confidence intervals aid in the selection of important components.

```
. screeplot, ci
(caution is advised in interpreting an asymptotic theory-based confidence
 interval of eigenvalues of a correlation matrix)
```

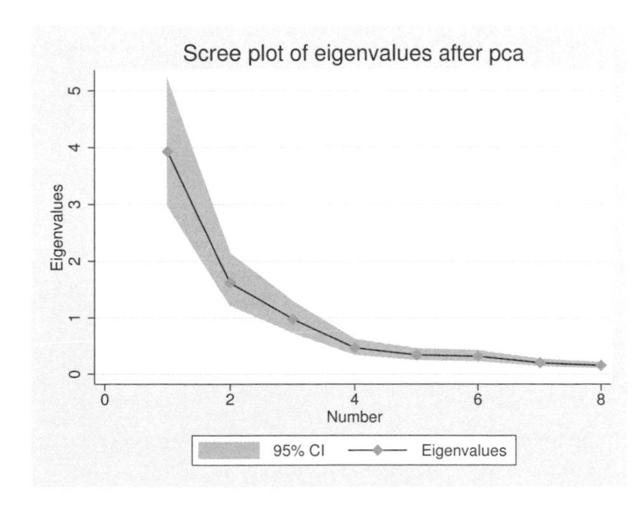

Despite our appreciation of the underlying interpretability of the fourth component, the evidence still points to retaining two or three principal components.

Plotting the components is sometimes useful in interpreting a PCA. We may look at the components from the perspective of the columns (variables) or the rows (observations). The associated plots are produced by the commands `loadingplot` (variables) and `scoreplot` (observations).

By default, the first two components are used to produce the loading plot.

```
. loadingplot
```

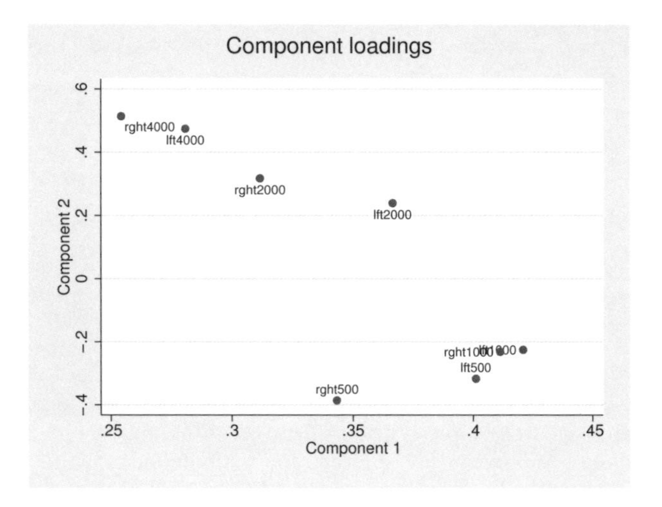

You may request more components, in which case each possible pair of requested components will be graphed. You can choose between a matrix or combined graph layout for the multiple graphs. Here we show the combined layout.

```
. loadingplot, comp(3) combined
```

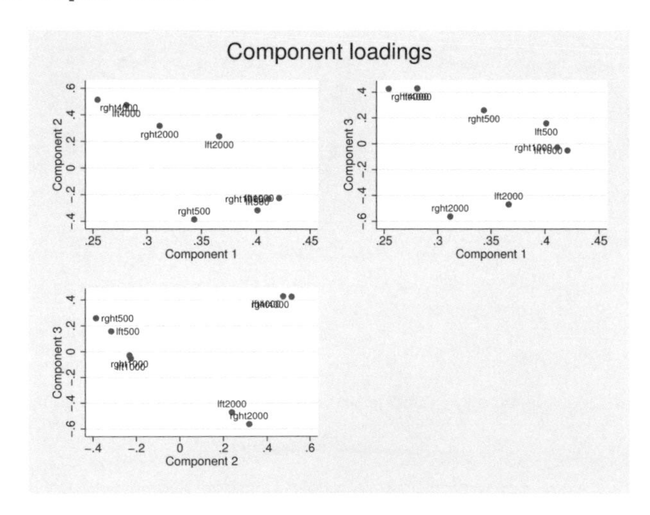

Score plots approach the display of principal components from the perspective of the observations. `scoreplot` and `loadingplot` have most of their options in common; see [MV] **scoreplot**. Unlike `loadingplot`, which automatically uses the variable names as marker labels, with `scoreplot` you use the `mlabel()` graph option to provide meaningful marker labels. Score plots are especially helpful if the observations are well-known objects, such as countries, firms, or brands. The score plot may help you visualize the principal components with your background knowledge of these objects. Score plots are sometimes useful for detecting outliers; see Jackson (2003).

❑ Technical note

In [MV] **pca** we noted that PCA may also be interpreted as fixed effects factor analysis; in that interpretation, the selection of the number of components to be retained is of comparable complexity as in factor analysis.

❑

Rotating the components

Rotating principal components is a disputed issue and one in which reasonable people may disagree. pca computes the principal components. Rotating the solution destroys some of the properties of principal components. In particular, the first rotated component no longer has maximal variance, the second rotated component no longer has maximal variance among those linear combinations uncorrelated to the first component, etc. If preserving the maximal variance property is very important to your interpretations, do not rotate.

On the other hand, when we rotate, say, the leading three principal components, the total variance explained by the three rotated components is equal to the variance explained by the three principal components. If you applied an orthogonal rotation, the rotated components are still uncorrelated. The only thing that has changed is that the explanation is distributed differently among the three rotated components. If the rotated components have a clearer interpretation, you may actually prefer to use them in your subsequent work.

After pca, a wide variety of rotations are available; see [MV] **rotate**. The default method of rotation is varimax, rotating the principal components to maximize the sum over the columns of the within-column variances.

```
. rotate
Principal components/correlation                  Number of obs    =       100
                                                  Number of comp.  =         4
                                                  Trace            =         8
    Rotation: orthogonal varimax (Kaiser off)     Rho              =    0.8737
```

Component	Variance	Difference	Proportion	Cumulative
Comp1	2.11361	.400444	0.2642	0.2642
Comp2	1.71316	.118053	0.2141	0.4783
Comp3	1.59511	.0275517	0.1994	0.6777
Comp4	1.56756	.	0.1959	0.8737

Rotated components

Variable	Comp1	Comp2	Comp3	Comp4	Unexplained
lft500	0.5756	0.0265	-0.1733	0.1781	.1308
lft1000	0.6789	-0.0289	-0.0227	-0.0223	.1105
lft2000	0.3933	0.0213	0.5119	-0.2737	.1275
lft4000	0.1231	0.6987	-0.0547	-0.0885	.1342
rght500	-0.0005	0.0158	-0.0380	0.7551	.1194
rght1000	0.0948	-0.0248	0.2289	0.5481	.1825
rght2000	-0.1173	-0.0021	0.8047	0.0795	.07537
rght4000	-0.1232	0.7134	0.0550	0.0899	.1303

```
Component rotation matrix
```

	Comp1	Comp2	Comp3	Comp4
Comp1	0.6663	0.3784	0.4390	0.4692
Comp2	-0.3055	0.6998	0.4012	-0.5059
Comp3	-0.0657	0.6059	-0.7365	0.2936
Comp4	-0.6770	-0.0022	0.3224	0.6616

`rotate` now labels one of the columns of the first table as "Variance" instead of "Eigenvalue"; the rotated components have been ordered in decreasing order of variance. The variance explained by the four rotated components equals 87.37%, which is identical to the explained variance by the four leading principal components. But whereas the principal components have rather dispersed eigenvalues, the four rotated components all explain about the same fraction of the variance.

You may also choose to rotate only a few of the retained principal components. In contrast to most methods of factor analysis, the principal components are not affected by the number of retained components. However, the first two rotated components are different if you are rotating all four components or only the leading two or three principal components.

```
. rotate, comp(3)
Principal components/correlation                  Number of obs    =       100
                                                  Number of comp.  =         4
                                                  Trace            =         8
    Rotation: orthogonal varimax (Kaiser off)     Rho              =    0.8737
```

Component	Variance	Difference	Proportion	Cumulative
Comp1	2.99422	1.16842	0.3743	0.3743
Comp2	1.8258	.123163	0.2282	0.6025
Comp3	1.70264	1.23585	0.2128	0.8153
Comp4	.466782	.	0.0583	0.8737

```
Rotated components
```

Variable	Comp1	Comp2	Comp3	Comp4	Unexplained
lft500	0.5326	-0.0457	0.0246	-0.3278	.1308
lft1000	0.4512	0.1618	-0.0320	-0.4816	.1105
lft2000	0.0484	0.6401	0.0174	-0.2824	.1275
lft4000	0.0247	0.0011	0.6983	-0.1611	.1342
rght500	0.5490	-0.1799	0.0163	0.4876	.1194
rght1000	0.4521	0.1368	-0.0259	0.3723	.1825
rght2000	-0.0596	0.7148	-0.0047	0.3914	.07537
rght4000	-0.0200	0.0059	0.7138	0.1591	.1303

```
Component rotation matrix
```

	Comp1	Comp2	Comp3	Comp4
Comp1	0.7790	0.5033	0.3738	0.0000
Comp2	-0.5932	0.3987	0.6994	0.0000
Comp3	0.2030	-0.7666	0.6092	-0.0000
Comp4	-0.0000	0.0000	0.0000	1.0000

The three-component varimax-rotated solution differs from the leading three components from the four component varimax-rotated solution. The fourth component is not affected by a rotation among the leading three component—it is still the fourth principal component.

So, how interpretable are rotated components? We believe that for this example the original components had a much clearer interpretation than the rotated components. Notice how the clear symmetry in the treatment of left and right ears has been broken.

To add further to an already controversial method, we may use oblique rotation methods. An example is the oblique oblimin method.

```
. rotate, oblimin oblique
Principal components/correlation                    Number of obs    =       100
                                                    Number of comp.  =         4
                                                    Trace            =         8
    Rotation: oblique oblimin (Kaiser off)          Rho              =    0.8737
```

Component	Variance	Proportion	Rotated comp. are correlated
Comp1	2.21066	0.2763	
Comp2	1.71164	0.2140	
Comp3	1.69708	0.2121	
Comp4	1.62592	0.2032	

Rotated components

Variable	Comp1	Comp2	Comp3	Comp4	Unexplained
lft500	0.5834	0.0259	0.1994	-0.1649	.1308
lft1000	0.6797	-0.0292	0.0055	-0.0157	.1105
lft2000	0.3840	0.0216	-0.2489	0.5127	.1275
lft4000	0.1199	0.6988	-0.0857	-0.0545	.1342
rght500	0.0261	0.0146	0.7561	-0.0283	.1194
rght1000	0.1140	-0.0257	0.5575	0.2370	.1825
rght2000	-0.1158	-0.0022	0.0892	0.8048	.07537
rght4000	-0.1209	0.7134	0.0848	0.0549	.1303

Component rotation matrix

	Comp1	Comp2	Comp3	Comp4
Comp1	0.6836	0.3773	0.5053	0.4523
Comp2	-0.3250	0.7008	-0.5137	0.3916
Comp3	-0.0550	0.6054	0.2774	-0.7337
Comp4	-0.6557	-0.0029	0.6408	0.3238

The oblique rotation methods do not change the variance that is unexplained by the components. But this time, the rotated components are no longer uncorrelated. This makes measuring the importance of the rotated components more ambiguous, a problem that is similar to ambiguities in interpreting importance of correlated independent variables. In this oblique case, the sum of the variances of the rotated components equals 90.5% $(0.2763 + 0.2140 + 0.2121 + 0.2032)$ of the total variance. This is larger than the 87.37% of variance explained by the four principal components. The oblique rotated components partly explain the same variance, and this shared variance is entering multiple times into the total.

How rotate interacts with pca

rotate stores the rotated component loadings and associated statistics in e(), the estimation storage area, along with the regular pca estimation results. Replaying pca will display the rotated results again.

Other postestimation statistics also use the rotated results whenever this is meaningful. For instance, loadingplot would display the rotated loadings. These postestimation commands have an option norotated that specifies that the unrotated results, that is, the principal components, be used. Thus

```
. pca, norotated
  (output omitted)
```

displays the standard pca output for the unrotated (principal) solution, and

```
. loadingplot, norotated
  (output omitted)
```

produces the loading plot for the unrotated (principal) solution.

If you execute rotate again, the new rotate results are stored with the pca estimation, replacing the previous rotate results. Thus pca knows about at most one rotation.

To compare rotated and unrotated results, it is of course possible to replay the rotated results (pca) and unrotated results (pca, norotate) consecutively. You would especially seek to compare the loadings. Such a comparison is easier if the loadings are displayed in parallel. This feature is provided with the estat command rotatecompare.

```
. estat rotatecompare
```

Rotation matrix — oblique oblimin (Kaiser off)

Variable	Comp1	Comp2	Comp3	Comp4
Comp1	0.6836	0.3773	0.5053	0.4523
Comp2	-0.3250	0.7008	-0.5137	0.3916
Comp3	-0.0550	0.6054	0.2774	-0.7337
Comp4	-0.6557	-0.0029	0.6408	0.3238

Rotated component loadings

Variable	Comp1	Comp2	Comp3	Comp4
lft500	0.5834	0.0259	0.1994	-0.1649
lft1000	0.6797	-0.0292	0.0055	-0.0157
lft2000	0.3840	0.0216	-0.2489	0.5127
lft4000	0.1199	0.6988	-0.0857	-0.0545
rght500	0.0261	0.0146	0.7561	-0.0283
rght1000	0.1140	-0.0257	0.5575	0.2370
rght2000	-0.1158	-0.0022	0.0892	0.8048
rght4000	-0.1209	0.7134	0.0848	0.0549

```
Unrotated component loadings
```

Variable	Comp1	Comp2	Comp3	Comp4
lft500	0.4011	-0.3170	0.1582	-0.3278
lft1000	0.4210	-0.2255	-0.0520	-0.4816
lft2000	0.3664	0.2386	-0.4703	-0.2824
lft4000	0.2809	0.4742	0.4295	-0.1611
rght500	0.3433	-0.3860	0.2593	0.4876
rght1000	0.4114	-0.2318	-0.0289	0.3723
rght2000	0.3115	0.3171	-0.5629	0.3914
rght4000	0.2542	0.5135	0.4262	0.1591

Finally, sometimes you may want to remove rotation results permanently; for example, you decide to continue with the unrotated (principal) solution. Because all postestimation commands operate on the rotated solution by default, you would have to add the option `norotated` over and over again. Instead, you can remove the rotated solution with the command

```
. rotate, clear
```

❑ Technical note

`pca` results may be stored and restored with `estimates`, just like other estimation results. If you have stored PCA estimation results without rotated results, and later `rotate` the solution, the rotated results are not automatically stored as well. The `pca` would need to be stored again.

❑

Predicting the component scores

After deciding on the number of components and, possibly, the rotation of the components, you may want to estimate the component scores for all respondents. To estimate only the first component scores, which here is called `pc1`:

```
. predict pc1
(score assumed)
(3 components skipped)
Scoring coefficients
    sum of squares(column-loading) = 1
```

Variable	Comp1	Comp2	Comp3	Comp4
lft500	0.4011	-0.3170	0.1582	-0.3278
lft1000	0.4210	-0.2255	-0.0520	-0.4816
lft2000	0.3664	0.2386	-0.4703	-0.2824
lft4000	0.2809	0.4742	0.4295	-0.1611
rght500	0.3433	-0.3860	0.2593	0.4876
rght1000	0.4114	-0.2318	-0.0289	0.3723
rght2000	0.3115	0.3171	-0.5629	0.3914
rght4000	0.2542	0.5135	0.4262	0.1591

The table is informing you that `pc1` could be obtained as a weighted sum of standardized variables,

```
. egen std_lft500   = std(lft500)
. egen std_lft1000  = std(lft1000)
. egen std_rght4000 = std(rght4000)
. gen pc1 = 0.4011*std_lft500 + 0.4210*std_lft500 + ... + 0.2542*std_rght4000
```

(egen's std() function converts a variable to its standardized form (mean 0, variance 1); see [D] **egen**.) The principal-component scores are in standardized units after a PCA of a correlation matrix and in the original units after a PCA of a covariance matrix.

It is possible to predict other statistics as well. For instance, the fitted values of the eight variables by the first four principal components are obtained as

```
. predict f_1-f_8, fit
```

The predicted values are in the units of the original variables, with the means substituted back in. If we had retained all eight components, the fitted values would have been identical to the observations.

❑ Technical note

The fitted values are meaningful in the interpretation of PCA as rank-restricted multivariate regression. The component scores are the "x variables"; the component loadings are the regression coefficients. If the PCA was computed for a correlation matrix, you could think of the regression as being in standardized units. The fitted values are transformed from the standardized units back to the original units.

❑

❑ Technical note

You may have observed that the scoring coefficients were equal to the component loadings. This holds true for the principal components in unit normalization and for the orthogonal rotations thereof; it does not hold for oblique rotations.

❑

Saved results

Let p be the number of variables and f, the number of factors.

predict, in addition to generating variables, also saves the following in r():

Matrices
- r(scoef) — $p \times f$ matrix of scoring coefficients

estat anti saves the following in r():

Matrices
- r(acov) — $p \times p$ anti-image covariance matrix
- r(acorr) — $p \times p$ anti-image correlation matrix

estat kmo saves the following in r():

Scalars
- r(kmo) — the Kaiser–Meyer–Olkin measure of sampling adequacy

Matrices
- r(kmow) — column vector of KMO measures for each variable

estat loadings saves the following in r():

Macros
- r(cnorm) — component normalization: eigen, inveigen, or unit

Matrices
- r(A) — $p \times f$ matrix of normalized component loadings

estat residuals saves the following in r():

Matrices

r(fit)	$p \times p$ matrix of fitted values
r(residual)	$p \times p$ matrix of residuals

estat smc saves the following in r():

Matrices

r(smc)	vector of squared multiple correlations of variables with all other variables

See [R] **estat** for the returned results of estat summarize and estat vce (available when vce(normal) is specified with pca or pcamat).

rotate after pca and pcamat add to the existing e():

Scalars

e(r_f)	number of components in rotated solution
e(r_fmin)	rotation criterion value

Macros

e(r_class)	orthogonal or oblique
e(r_criterion)	rotation criterion
e(r_ctitle)	title for rotation
e(r_normalization)	kaiser or none

Matrices

e(r_L)	rotated loadings
e(r_T)	rotation
e(r_Ev)	explained variance by rotated components

The components in the rotated solution are in decreasing order of e(r_Ev).

Methods and formulas

All postestimation commands listed above are implemented as ado-files.

estat anti computes and displays the anti-image covariance matrix $\mathbf{C}$ and the anti-image correlation matrix $\mathbf{A}$

$$\mathbf{C} = \{\text{diag}(\mathbf{R})\}^{-1/2}\, \mathbf{R}\, \{\text{diag}(\mathbf{R})\}^{-1/2}$$

$$\mathbf{A} = \{\text{diag}(\mathbf{R})\}^{-1}\, \mathbf{R}\, \{\text{diag}(\mathbf{R})\}^{-1}$$

where $\mathbf{R}$ is the inverse of the correlation matrix.

estat kmo computes the "Kaiser–Meyer–Olkin measure of sampling adequacy" (KMO) and is defined as

$$\text{KMO} = \frac{\sum_{\mathcal{S}} r_{ij}^2}{\sum_{\mathcal{S}} (a_{ij}^2 + r_{ij}^2)}$$

where $\mathcal{S} = (i, j; i \neq j)$; r_{ij} is the correlation of variables i and j; and a_{ij} is the anti-image correlation. The variable-wise measure KMO_i is defined analogously as

$$\text{KMO}_i = \frac{\sum_{\mathcal{P}} r_{ij}^2}{\sum_{\mathcal{P}} (a_{ij}^2 + r_{ij}^2)}$$

where $\mathcal{P} = (j; i \neq j)$.

`estat loadings` displays the component loadings in different normalizations (see Jackson [2003, 16–18]; he labels them as $\mathbf{U}$, $\mathbf{V}$, and $\mathbf{W}$ vectors). Let $\mathbf{C} = \mathbf{L}\mathbf{\Lambda}\mathbf{L}'$ be the spectral or eigen decomposition of the analyzed correlation or covariance matrix $\mathbf{C}$, with $\mathbf{L}$ the orthonormal eigenvectors of $\mathbf{C}$, and $\mathbf{\Lambda}$ a diagonal matrix of eigenvalues. The principal components $\mathbf{A}$, that is, the eigenvectors $\mathbf{L}$, are displayed in one of the following normalizations:

cnorm(unit)	$\mathbf{A} = \mathbf{L}$	and so $\mathbf{A}'\mathbf{A} = \mathbf{I}$
normal(eigen)	$\mathbf{A} = \mathbf{L}\mathbf{\Lambda}^{1/2}$	and so $\mathbf{A}'\mathbf{A} = \mathbf{\Lambda}$
normal(inveigen)	$\mathbf{A} = \mathbf{L}\mathbf{\Lambda}^{-1/2}$	and so $\mathbf{A}'\mathbf{A} = \mathbf{\Lambda}^{-1}$

Normalization of the component loadings affects the normalization of the component scores.

The standard errors of the components are available only in unit normalization, that is, as normalized eigenvectors.

`estat residuals` computes the fitted values $\mathbf{F}$ for the analyzed correlation or covariance matrix $\mathbf{C}$ as $\mathbf{F} = \mathbf{L}\mathbf{\Lambda}\mathbf{L}'$ over the retained components, with $\mathbf{L}$ being the retained components in unit normalization and $\mathbf{\Lambda}$ being the associated eigenvalues. The residuals are simply $\mathbf{C} - \mathbf{F}$.

`estat smc` displays the squared multiple correlation coefficients SMC_i of each variable on the other variables in the analysis. These are conveniently computed from the inverse $\mathbf{R}$ of the correlation matrix $\mathbf{C}$,

$$\text{SMC}_i = 1 - \mathbf{R}_{ii}^{-1}$$

See [MV] **rotate** and [MV] **rotatemat** for details concerning the rotation methods and algorithms used.

The variance of the rotated loadings $\mathbf{L}_\text{r}$ is computed as $\mathbf{L}_{\text{r}'}\mathbf{C}\mathbf{L}_\text{r}$.

To understand `predict` after `pca` and `pcamat`, think of PCA as a fixed-effect factor analysis with homoskedastic residuals

$$\mathbf{Z} = \mathbf{A}\mathbf{L}' + \mathbf{E}$$

$\mathbf{L}$ contains the loadings, and $\mathbf{A}$ contains the scores. $\mathbf{Z}$ is the centered variables for a PCA of a covariance matrix and standardized variables for a PCA of a correlation matrix. $\mathbf{A}$ is estimated by OLS regression of $\mathbf{Z}$ on $\mathbf{L}$

$$\widehat{\mathbf{A}} = \mathbf{Z}\mathbf{B} \qquad \mathbf{B} = \mathbf{L}(\mathbf{L}'\mathbf{L})^{-}$$

The columns of $\mathbf{A}$ are called the scores. The matrix $\mathbf{B}$ contains the scoring coefficients. The PCA-fitted values for $\mathbf{Z}$ are defined as the fitted values from this regression, or in matrix terms,

$$\widehat{\mathbf{Z}} = \mathbf{Z}\mathbf{P}_\mathbf{L} = \mathbf{Z}\mathbf{L}(\mathbf{L}'\mathbf{L})^{-}\mathbf{L}'$$

with $\mathbf{P}_\mathbf{L}$ the orthogonal projection on (the rowspace of) $\mathbf{L}$.

This formulation allows orthogonal as well as oblique loadings $\mathbf{L}$ as well as loadings in different normalizations.

The above formulation is in transformed units. `predict` transforms the fitted values back to the original units. The component scores are left in transformed units, with one exception. After a PCA of covariances, means are substituted back in unless the option `centered` is specified. The residuals are returned in the original units. The residual sum of squares (over the variables) and the normalized versions are in transformed units.

References

Cattell, R. B. 1966. The scree test for the number of factors. *Multivariate Behavioral Research* 1: 245–276.

Jackson, J. E. 2003. *A User's Guide to Principal Components*. New York: Wiley.

Kaiser, H. F. 1974. An index of factor simplicity. *Psychometrika* 39: 31–36.

Also see *References* in [MV] **pca**.

Also see

[MV] **pca** — Principal component analysis

[MV] **rotate** — Orthogonal and oblique rotations after factor and pca

[MV] **scoreplot** — Score and loading plots

[MV] **screeplot** — Scree plot

Title

procrustes — Procrustes transformation

Syntax

`procrustes` (*varlist*$_y$) (*varlist*$_x$) [*if*] [*in*] [*weight*] [, *options*]

options	Description
Model	
`transform(orthogonal)`	orthogonal rotation and reflection transformation; the default
`transform(oblique)`	oblique rotation transformation
`transform(unrestricted)`	unrestricted transformation
`noconstant`	suppress the constant
`norho`	suppress the dilation factor ρ (set $\rho = 1$)
`force`	allow overlap and duplicates in *varlist*$_y$ and *varlist*$_x$ (advanced)
Reporting	
`nofit`	suppress table of fit statistics by target variable

`bootstrap`, `by`, `jackknife`, and `statsby` are allowed; see [U] **11.1.10 Prefix commands**.
Weights are not allowed with the `bootstrap` prefix; see [R] **bootstrap**.
`aweights` are not allowed with the `jackknife` prefix; see [R] **jackknife**.
`aweights` and `fweights` are allowed; see [U] **11.1.6 weight**.
See [U] **20 Estimation and postestimation commands** for more capabilities of estimation commands.

Menu

Statistics > Multivariate analysis > Procrustes transformations

Description

`procrustes` performs the Procrustean analysis, one of the standard methods of multidimensional scaling. For given "target" variables *varlist*$_y$ and "source" variables *varlist*$_x$, the goal is to transform the source **X** to be as close as possible to the target **Y**. The permitted transformations are any combination of dilation (uniform scaling), rotation and reflection (that is, orthogonal or oblique transformations), and translation. Closeness is measured by the residual sum of squares. `procrustes` deals with complete cases only.

`procrustes` assumes equal weights or scaling for the dimensions. It would be inappropriate, for example, to have the first variable measured in grams ranging from 5,000 to 8,000, the second variable measured in dollars ranging from 3 to 12, and the third variable measured in meters ranging from 100 to 280. In such cases, you would want to operate on standardized variables.

Options

Model

`transform(`*transform*`)` specifies the transformation method. The following transformation methods are allowed:

`orthogonal` specifies that the linear transformation matrix $\mathbf{A}$ should be orthogonal, $\mathbf{A}'\mathbf{A} = \mathbf{A}\mathbf{A}' = \mathbf{I}$. This is the default.

`oblique` specifies that the linear transformation matrix $\mathbf{A}$ should be oblique, $\text{diag}(\mathbf{A}\mathbf{A}') = \mathbf{1}$.

`unrestricted` applies no restrictions to $\mathbf{A}$, making the `procrustes` transformation equivalent to multivariate regression with uncorrelated errors; see [R] **mvreg**.

`noconstant` specifies that the translation component $\mathbf{c}$ is fixed at $\mathbf{0}$ (the 0 vector).

`norho` specifies that the dilation (scaling) constant ρ is fixed at 1. This option is not relevant with `transform(unrestricted)`; here ρ is always fixed at 1.

`force`, an advanced option, allows overlap and duplicates in the target variables $varlist_y$ and source variables $varlist_x$.

Reporting

`nofit` suppresses the table of fit statistics per target variable. This option may be specified during estimation and upon replay.

Remarks

Remarks are presented under the following headings:

Introduction to Procrustes methods
Orthogonal Procrustes analysis
Is an orthogonal Procrustes analysis symmetric?
Other transformations

Introduction to Procrustes methods

The name *Procrustes analysis* was applied to optimal matching of configurations by Hurley and Cattell (1962) and refers to Greek mythology. The following account follows Cox and Cox (2001, 123). Travelers from Eleusis to Athens were kindly invited by Damastes to spend the night at his place. Damastes, however, practiced a queer kind of hospitality. If guests would not fit the bed, Damastes would either stretch them to make them fit, or chop off extremities if they were too long. Therefore, he was given the nickname Procrustes—ancient Greek for "the stretcher". Theseus, a warrior, finally gave Procrustes some of his own medicine.

Procrustes methods have been applied in many areas. Gower and Dijksterhuis (2004) mention applications in psychometrics (for example, the matching of factor loading matrices), image analysis, market research, molecular biology, biometric identification, and shape analysis.

Formally, `procrustes` solves the minimization problem

$$\text{Minimize } |\,\mathbf{Y} - (\mathbf{1}\mathbf{c}' + \rho\,\mathbf{X}\,\mathbf{A})\,|$$

where $\mathbf{c}$ is a row vector representing the translation, ρ is the scalar "dilation factor", $\mathbf{A}$ is the rotation and reflection matrix (orthogonal, oblique, or unrestricted), and $|.|$ denotes the L2 norm.

Some of the early work on Procrustes analysis was done by Mosier (1939), Green (1952), Hurley and Cattell (1962), and Browne (1967); see Gower and Dijksterhuis (2004).

Orthogonal Procrustes analysis

▷ Example 1

We illustrate `procrustes` with John Speed's historical 1610 map of the Worcestershire region in England, engraved and printed by Jodocus Hondius in Amsterdam in 1611–1612. Used with permission of Peen (2007).

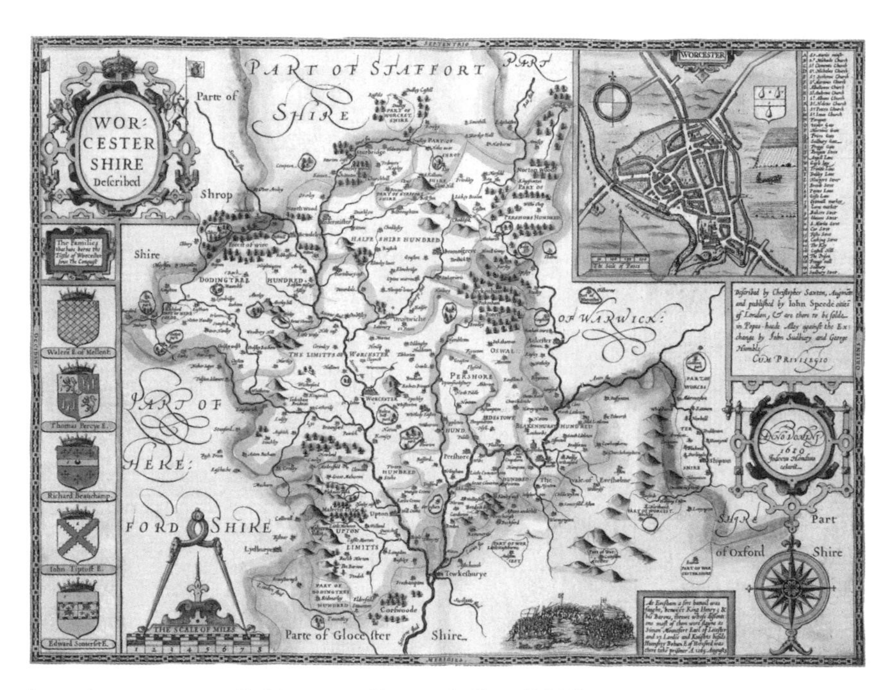

We analyze the accuracy of this map. Cox and Cox (2001) present data on the locations of 20 towns and villages on this old map, as well as the locations on a modern map from the Landranger Series of Ordnance Survey Maps. The locations were measured relative to the lower-left corner of the maps. We list this small dataset, specifying the `noobs` option to prevent wrapping and `sep(0)` to suppress internal horizontal lines.

```
. use http://www.stata-press.com/data/r12/speed_survey
(Data on Speed's Worcestershire map (1610))
. list name lname speed_x speed_y survey_x survey_y, sep(0) noobs
```

name	lname	speed_x	speed_y	survey_x	survey_y
Alvc	Alvechurch	192	211	1027	725
Arro	Arrow	217	155	1083	565
Astl	Astley	88	180	787	677
Beck	Beckford	193	66	976	358
Beng	Bengeworth	220	99	1045	435
Crad	Cradley	79	93	736	471
Droi	Droitwich	136	171	893	633
Ecki	Eckington	169	81	922	414
Eves	Evesham	211	105	1037	437
Hall	Hallow	113	142	828	579
Hanb	Hanbury	162	180	944	637
Inkb	Inkberrow	188	156	1016	573
Kemp	Kempsey	128	108	848	490
Kidd	Kidderminster	104	220	826	762
Mart	Martley	78	145	756	598
Stud	Studley	212	185	1074	632
Tewk	Tewkesbury	163	40	891	324
UpSn	UpperSnodsbury	163	138	943	544
Upto	Upton	138	71	852	403
Worc	Worcester	125	132	850	545

You will probably conclude immediately that the scales of the two maps differ and that the coordinates are expressed with respect to different reference points; the lower-left corners of the maps correspond to different physical locations. Another distinction will not be so obvious—at least not by looking at these numbers: the orientations of the maps may well differ. We display as scatterplots Speed's data (speed_x, speed_y) and the modern survey data (survey_x, survey_y).

```
. scatter speed_y speed_x, mlabel(name)
    ytitle("") xtitle("") yscale(off) xscale(off) ylabel(,nogrid)
    title(Historic map of 20 towns and villages in Worcestershire)
    subtitle((Speed 1610))
```

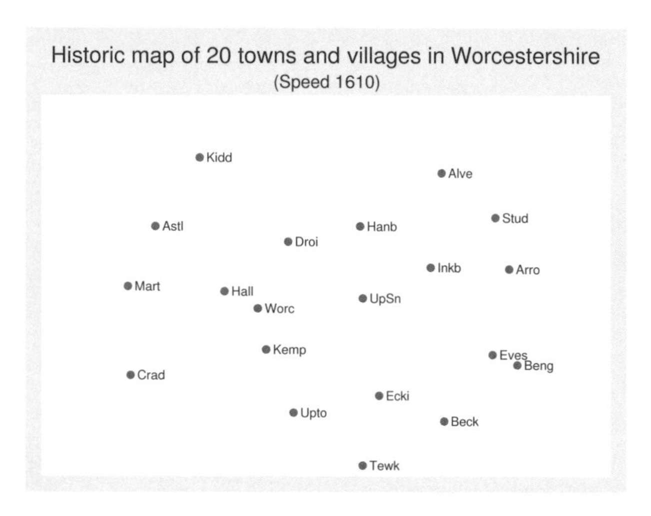


```
. scatter survey_y survey_x, mlabel(name)
    ytitle("") xtitle("") yscale(off) xscale(off) ylabel(,nogrid)
    title(Modern map of 20 towns and villages in Worcestershire)
    subtitle((Landranger series of Ordnance Survey Maps))
```

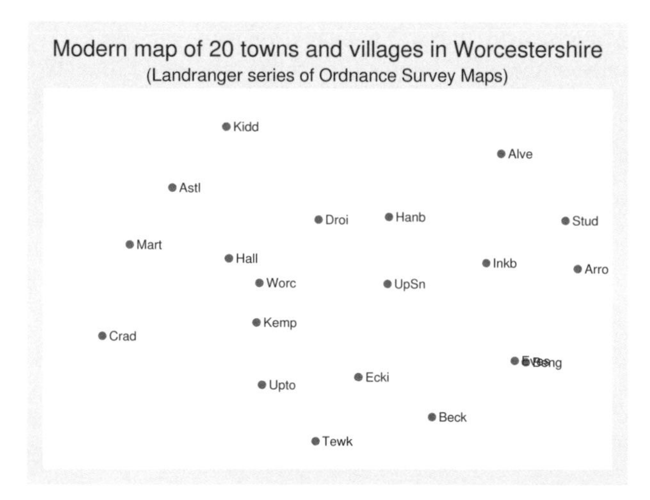

To gauge the accuracy of the historic map, we must account for differences in scale, location, and orientation. Because the area depicted on the map is relatively small, we think that it is justified to ignore the curvature of the earth and approximate distances along the globe with Euclidean distances, and apply a Procrustes analysis,

$$\text{survey\_map} = \text{transformation}(\text{speed\_map}) + \text{residual}$$

choosing the transformation (from among the allowed transformations) to minimize the residual in terms of the residual sum of squares. The transformation should allow for, in mathematical terms, translation, uniform scaling, and two-dimensional orthogonal rotation. The uniform scaling factor is often described as the dilation factor, a positive scalar. The transformation from source to target configuration can be written as

$$(\,\text{survey\_x} \quad \text{survey\_y}\,) = (\,c_x \quad c_y\,) + \rho\,(\,\text{speed\_x} \quad \text{speed\_y}\,) \begin{pmatrix} a_{11} & a_{12} \\ a_{21} & a_{22} \end{pmatrix} + (\,\text{res\_x} \quad \text{res\_y}\,)$$

or simply as

$$\text{survey\_map} = \text{translation} + \text{dilation} \times \text{speed\_map} \times \text{rotation} + \text{residual}$$

The matrix

$$\mathbf{A} = \begin{pmatrix} a_{11} & a_{12} \\ a_{21} & a_{22} \end{pmatrix}$$

should satisfy the constraint that it represents an orthogonal rotation—it should maintain the lengths of vectors and the angles between vectors. We estimate the translation $(\,c_x \quad c_y\,)$, dilation factor ρ, and the rotation matrix $\mathbf{A}$ with the `procrustes` command.

```
. procrustes (survey_x survey_y) (speed_x speed_y)
Procrustes analysis (orthogonal)              Number of observations =        20
                                              Model df (df_m)        =         4
                                              Residual df (df_r)     =        36
                                              SS(target)             =    495070
                                              RSS(target)            =  1973.384
                                              RMSE = root(RSS/df_r)  =  7.403797
                                              Procrustes = RSS/SS    =    0.0040

Translation c
```

	survey_x	survey_y
_cons	503.8667	293.9878

```
Rotation & reflection matrix A (orthogonal)
```

	survey_x	survey_y
speed_x	.9841521	-.1773266
speed_y	.1773266	.9841521

```
Dilation factor
             rho =    2.3556
Fit statistics by target variable
```

Statistics	survey_x	survey_y
SS	216310.2	278759.8
RSS	1081.36	892.0242
RMSE	7.750841	7.039666
Procrustes	.0049991	.0032
Corr_y_yhat	.9976669	.9985076

We can read the elements of the transformation from the output: the translation from the Speed map onto the survey map is (504, 294). The scale of the survey and Speed maps differ by a factor of 2.36. The orientations of the maps also differ somewhat; if the maps had been oriented the same, we would have expected the rotation to be an identity matrix. Note that $.984^2 + .177^2 = 1$, subject to rounding error—indeed the rotation is "norm preserving". A counterclockwise rotation in a plane over θ can be written as

$$\begin{pmatrix} \cos(\theta) & \sin(\theta) \\ -\sin(\theta) & \cos(\theta) \end{pmatrix}$$

See appendix B in Gower and Dijksterhuis (2004). Here $\cos(\theta) = 0.984$, so the difference in orientation of the two maps is $\theta = 10.2$ degrees.

The other output produced by `procrustes` may be more familiar. `procrustes` estimated four parameters: the angle of rotation, two translation parameters, and the dilation factor ρ. SS(target) is the centered sum of squares of the survey data, and is meaningful mostly in relation to the residual sum of squares RSS(target). The Procrustes statistic, defined as RSS/SS, measures the size of the residuals relative to the variation in the target variables; it is equivalent to $1 - R^2$ in a regression analysis context. The smaller the Procrustes statistic, the closer the correspondence of Speed's map to the survey map. The number in this case, 0.004, is small indeed. Another way of looking at fit is via the square root of the mean squared residual error, RMSE, a measure for the average size of residuals.

The last output table describes how well the transformed Speed coordinates match the survey coordinates, separately for the horizontal (x) and the vertical (y) coordinates. In this case, we do not see disturbing differences between the coordinates. By definition, the overall Procrustes statistic and the overall RMSE are averages of the coordinate statistics. Because Procrustes analysis treats (weights) both coordinates the same and independently, analogous to the sphericity assumption in multivariate regression or MANOVA analysis, the comparable statistics for the different coordinates is reassuring.

This example is continued in [MV] **procrustes postestimation**, demonstrating how to generate fitted values and residual sum of squares with `predict`, how to produce a graph showing the target overlaid with the transformed source values with `procoverlay`, and how to produce various summaries and comparisons with `estat`.

◁

A Procrustes analysis fits the transformation subject to the constraint that $\mathbf{A}$ is orthogonal; for other constraints, see below. In two dimensions, there are actually two types of orthogonal matrices: rotations and reflections. Think of left and right hands. A rotation transforms a left hand into a left hand, never into a right hand; rotation preserves orientation. A reflection changes a left hand into a right hand; reflections invert orientation. In algebraic terms, an orthogonal matrix $\mathbf{A}$ satisfies $\det(\mathbf{A}) = \pm 1$. $\mathbf{A}$ is a rotation if $\det(\mathbf{A}) = 1$, and $\mathbf{A}$ is a reflection if $\det(\mathbf{A}) = -1$. In more than two dimensions, the classification of orthogonal transformations is more complicated.

▷ Example 2

In example 1, we treated the location, dilation, and orientation as estimable aspects of the transformation. It is possible to omit the location and dilation aspects—though, admittedly, from a casual inspection as well as the substantial understanding of the data, these aspects are crucial. For instance, we may omit the dilation factor—that is, assume $\rho = 1$—with the `norho` option.

```
. procrustes (survey_x survey_y) (speed_x speed_y), norho
Procrustes analysis (orthogonal)              Number of observations =        20
                                              Model df (df_m)        =         3
                                              Residual df (df_r)     =        37
                                              SS(target)             =    495070
                                              RSS(target)            =  165278.1
                                              RMSE = root(RSS/df_r)  =  66.83544
                                              Procrustes = RSS/SS    =    0.3338
Translation c
```

	survey_x	survey_y
_cons	741.4458	435.6215

```
Rotation & reflection matrix A (orthogonal)
```

	survey_x	survey_y
speed_x	.9841521	-.1773266
speed_y	.1773266	.9841521

```
Dilation factor
             rho =    1.0000
Fit statistics by target variable
```

Statistics	survey_x	survey_y
SS	216310.2	278759.8
RSS	70385.78	94892.36
RMSE	61.68174	71.61925
Procrustes	.3253928	.340409
Corr_y_yhat	.9976669	.9985076

As expected, the optimal transformation without dilation implies a much weaker relation between the Speed and Survey maps; the Procrustes statistic has increased from 0.0040 to 0.3338. We conclude that we cannot adequately describe the correspondence between the maps if we ignore differences in scale.

◁

Is an orthogonal Procrustes analysis symmetric?

In examples 1 and 2, we transformed the Speed map to optimally match the modern Survey map. We could also have reversed the procedure, that is, transform the Survey map to match the Speed map.

▷ Example 3

Here we change the order of the Speed and Survey map in our call to `procrustes` from example 1.

```
. procrustes (speed_x speed_y) (survey_x survey_y)
Procrustes analysis (orthogonal)            Number of observations =        20
                                            Model df (df_m)        =         4
                                            Residual df (df_r)     =        36
                                            SS(target)             =  88862.75
                                            RSS(target)            =  354.2132
                                            RMSE = root(RSS/df_r)  =  3.136759
                                            Procrustes = RSS/SS    =    0.0040
Translation c
```

	speed_x	speed_y
_cons	-187.0142	-159.5801

```
Rotation & reflection matrix A (orthogonal)
```

	speed_x	speed_y
survey_x	.9841521	.1773266
survey_y	-.1773266	.9841521

```
Dilation factor
          rho =    0.4228
Fit statistics by target variable
```

Statistics	speed_x	speed_y
SS	41544.95	47317.8
RSS	218.3815	135.8317
RMSE	3.483146	2.747036
Procrustes	.0052565	.0028706
Corr_y_yhat	.9975074	.9986641

The implied transformations are similar but not identical. For instance, the product of estimated scale factors is $2.3556 \times 0.4228 = 0.9959$, which is close to 1 but not identical to 1—this is not due to roundoff error. Why do the results differ? Think about the analogy with regression analysis. The regression of Y on X and the regression of X on Y generally imply different relationships between the variables. In geometric terms, one minimizes the sum of squares of the "vertical" distances between the data point and the regression line, whereas the other minimizes the "horizontal" distances. The implied regression lines are the same if the variance in X and Y are the same. Even if this does not hold, the proportion of explained variance R^2 in both regressions is the same. In Procrustes analysis, an analogous relationship holds between the analyses "Speed = transformed(Survey) + E" and "Survey = transformed(Speed) + E". Both analyses yield the same Procrustes statistic. The implied analyses are equivalent (that is, the implied transformation in one analysis is the mathematical inverse of the transformation in the other analysis) only if the Speed and Survey data are scaled so that the trace of the associated covariance matrices is the same.

◁

Other transformations

A Procrustes analysis can also be applied with other classes of transformations. Browne (1967) analyzed Procrustes analyses with oblique rotations. Cramer (1974) and ten Berge and Nevels (1977) identified and solved some problems in Browne's solution (but still ignore the problem that the derived oblique rotations are not necessarily orientation preserving). `procrustes` supports oblique transformations. `procrustes` also allows dilation; see *Methods and formulas*.

▷ Example 4

Even though the orthogonal Procrustes analysis of example 1 demonstrated a similarity between the two configurations assuming an orthogonal transformation, we now investigate what happens with an oblique transformation.

```
. procrustes (survey_x survey_y) (speed_x speed_y), trans(oblique)
Procrustes analysis (oblique)                Number of observations =        20
                                             Model df (df_m)        =         5
                                             Residual df (df_r)     =        35
                                             SS(target)             =    495070
                                             RSS(target)            =  1967.854
                                             RMSE = root(RSS/df_r)  =  7.498294
                                             Procrustes = RSS/SS    =    0.0040
Translation c

             |   survey_x    survey_y
-------------+-----------------------
       _cons |   503.0093    292.4346
-------------+-----------------------

Rotation & reflection matrix A (oblique)

             |   survey_x    survey_y
-------------+-----------------------
     speed_x |   .9835969   -.1737553
     speed_y |   .1803803    .9847889
-------------+-----------------------

Dilation factor
          rho =    2.3562
Fit statistics by target variable

  Statistics |   survey_x    survey_y
-------------+-----------------------
          SS |   216310.2    278759.8
         RSS |   1080.677    887.1769
        RMSE |   7.858307      7.1201
  Procrustes |    .004996    .0031826
  Corr_y_yhat|   .9976685    .9985163
-------------+-----------------------
```

We see that the optimal oblique transformation is almost orthogonal; the columns of the oblique rotation and reflection matrix are almost perpendicular. The dilation factor and translation vector hardly differ from the orthogonal case shown in example 1. Finally, we see that the residual sum of squares decreased little, namely, from 1,973.4 to 1,967.9.

◁

Procrustes analysis can be interpreted as multivariate regression $\mathbf{Y} = \mathbf{c} + \mathbf{x}\mathbf{B} + \mathbf{e}$ in which some nonlinear restriction is applied to the coefficients $\mathbf{B}$. Procrustes analysis assumes $\mathbf{B} = \rho\mathbf{A}$ with $\mathbf{A}$

assumed to be orthogonal or $\mathbf{A}$ assumed to be oblique. The intercepts of the multivariate regression are, of course, the translation of the Procrustean transform. In contrast to multivariate regression, it is assumed that the distribution of the residuals $\mathbf{e}$ is spherical; that is, all that is assumed is that $\text{var}(\mathbf{e}) = \sigma^2\mathbf{I}$. This assumption affects standard errors, not the estimated coefficients. Multivariate regression serves as a useful baseline to gauge the extent to which the Procrustean analysis is appropriate. procrustes supports the transform(unrestricted) option and displays the fitted model in a format comparable to Procrustes analysis.

▷ Example 5

We demonstrate with Speed's map data.

```
. procrustes (survey_x survey_y) (speed_x speed_y), trans(unrestricted)
Procrustes analysis (unrestricted)              Number of observations =        20
                                                Model df (df_m)        =         6
                                                Residual df (df_r)     =        34
                                                SS(target)             =    495070
                                                RSS(target)            =  1833.435
                                                RMSE = root(RSS/df_r)  =  7.343334
                                                Procrustes = RSS/SS    =    0.0037
Translation c
```

	survey_x	survey_y
_cons	510.8028	288.243

```
Rotation & reflection matrix A (unrestricted)
```

	survey_x	survey_y
speed_x	2.27584	-.4129564
speed_y	.4147244	2.355725

```
Fit statistics by target variable
```

Statistics	survey_x	survey_y
SS	216310.2	278759.8
RSS	1007.14	826.2953
RMSE	7.696981	6.971772
Procrustes	.004656	.0029642
Corr_y_yhat	.9976693	.9985168

Because we already saw that there is almost no room to improve on the orthogonal Procrustes transform with this particular dataset, dropping the restrictions on the coefficients hardly improves the fit. For instance, the residual sum of squares further decreases from 1,967.9 in the oblique case to 1,833.4 in the unrestricted case, with only a small reduction in the value of the Procrustes statistic. ◁

Saved results

`procrustes` saves the following in `e()`:

Scalars

`e(N)`	number of observations
`e(rho)`	dilation factor
`e(P)`	Procrustes statistic
`e(ss)`	total sum of squares, summed over all y variables
`e(rss)`	residual sum of squares, summed over all y variables
`e(rmse)`	root mean squared error
`e(urmse)`	root mean squared error (unadjusted for # of estimated parameters)
`e(df_m)`	model degrees of freedom
`e(df_r)`	residual degrees of freedom
`e(ny)`	number of y variables (target variables)

Macros

`e(cmd)`	`procrustes`
`e(cmdline)`	command as typed
`e(ylist)`	y variables (target variables)
`e(xlist)`	x variables (source variables)
`e(transform)`	`orthogonal`, `oblique`, or `unrestricted`
`e(uniqueA)`	`1` if rotation is unique, `0` otherwise
`e(wtype)`	weight type
`e(wexp)`	weight expression
`e(properties)`	`nob noV`
`e(estat_cmd)`	program used to implement `estat`
`e(predict)`	program used to implement `predict`
`e(marginsnotok)`	predictions disallowed by `margins`

Matrices

`e(c)`	translation vector
`e(A)`	orthogonal transformation matrix
`e(ystats)`	matrix containing fit statistics

Functions

`e(sample)`	marks estimation sample

Methods and formulas

`procrustes` is implemented as an ado-file.

Methods and formulas are presented under the following headings:

Introduction
Orthogonal transformations
Oblique transformations
Unrestricted transformations
Reported statistics

Introduction

A Procrustes analysis is accomplished by solving a matrix minimization problem

$$\text{Minimize } |\,\mathbf{Y} - (\mathbf{1c}' + \rho\,\mathbf{X}\,\mathbf{A})\,|$$

with respect to $\mathbf{A}$, $\mathbf{c}$, and ρ. $\mathbf{A}$ is a matrix representing a linear transformation, $\rho > 0$ is a scalar called the "dilation factor", $\mathbf{c}$ is a translation (row-) vector, and $|.|$ is the Frobenius (or L2) norm. Three classes of transformations are available in `procrustes`: orthogonal, oblique, and unrestricted. The orthogonal class consists of all orthonormal matrices $\mathbf{A}$, that is, all square matrices that satisfy

$\mathbf{A}'\mathbf{A} = \mathbf{I}$, representing orthogonal norm-preserving rotations and reflections. The oblique class comprises all normal matrices $\mathbf{A}$, characterized by $\text{diag}(\mathbf{A}'\mathbf{A}) = \mathbf{1}$. Oblique transformations preserve the length of vectors but not the angles between vectors—orthogonal vectors will generally not remain orthogonal under oblique transformations. Finally, the unrestricted class consists of all conformable regular matrices $\mathbf{A}$.

Define $\widetilde{\mathbf{Y}}$ and $\widetilde{\mathbf{X}}$ as the centered $\mathbf{Y}$ and $\mathbf{X}$, respectively, if a constant $\mathbf{c}$ is included in the analysis and as the uncentered $\mathbf{Y}$ and $\mathbf{X}$ otherwise.

The derivation of the optimal $\mathbf{A}$ obviously differs for the three classes of transformations.

Orthogonal transformations

The solution for the orthonormal case can be expressed in terms of the singular value decomposition of $\widetilde{\mathbf{Y}}'\widetilde{\mathbf{X}}$,

$$\widetilde{\mathbf{Y}}'\widetilde{\mathbf{X}} = \mathbf{U}\mathbf{\Lambda}\mathbf{V}'$$

where $\mathbf{U}'\mathbf{U} = \mathbf{V}'\mathbf{V} = \mathbf{I}$. Then

$$\widehat{\mathbf{A}} = \mathbf{V}\mathbf{U}'$$

$\widehat{\mathbf{A}}$ is the same whether or not scaling is required, that is, whether ρ is a free parameter or a fixed parameter. When ρ is a free parameter, the optimal ρ is

$$\widehat{\rho} = \frac{\text{trace}(\widehat{\mathbf{A}}\widetilde{\mathbf{Y}}'\widetilde{\mathbf{X}})}{\text{trace}(\widetilde{\mathbf{X}}'\widetilde{\mathbf{X}})}$$

See ten Berge (1977) for a modern and elementary derivation; see Mardia, Kent, and Bibby (1979) for a derivation using matrix differential calculus.

Oblique transformations

Improving on earlier studies by Browne (1967) and Cramer (1974), ten Berge and Nevels (1977) provide a full algorithm to compute the optimal oblique rotation without dilation, that is, with uniform scaling $\rho = 1$. In contrast to the orthogonal case, the optimal oblique rotation $\widehat{\mathbf{A}}$ depends on ρ. To the best of our knowledge, this case has not been treated in the literature (J. M. F. ten Berge, 2004, pers. comm.). However, an "alternating least squares" extension of the ten Berge and Nevels (1977) algorithm deals with this case.

For each iteration, step (a) follows ten Berge and Nevels (1977) for calculating $\widetilde{\mathbf{Y}}$ and $\widehat{\rho}\widetilde{\mathbf{X}}$. In step (b) of an iteration, ρ is optimized, keeping $\widehat{\mathbf{A}}$ fixed, with solution

$$\widehat{\rho} = \frac{\text{trace}(\widehat{\mathbf{A}}\widetilde{\mathbf{Y}}'\widetilde{\mathbf{X}})}{\text{trace}(\widetilde{\mathbf{X}}'\widetilde{\mathbf{X}}\widehat{\mathbf{A}}\widehat{\mathbf{A}}')}$$

Iteration continues while the relative decrease in the residual sum of squares is large enough. This algorithm is ensured to yield a local optimum of the residual sum of squares as the RSS decreases both when updating the rotation $\mathbf{A}$ and when updating the dilation factor ρ. Beware that the algorithm is not guaranteed to find the global minimum.

Unrestricted transformations

In the unrestricted solution, the dilation factor ρ is fixed at 1. The computation of the Procrustes transformation is obviously equivalent to the least-squares solution of multivariate regression

$$\widehat{\mathbf{A}} = (\widetilde{\mathbf{X}}'\widetilde{\mathbf{X}})^{-1}\widetilde{\mathbf{X}}'\widetilde{\mathbf{Y}}$$

Given $\widehat{\mathbf{A}}$ and $\widehat{\rho}$, the optimal translation $\widehat{\mathbf{c}}$ can be written as

$$\widehat{\mathbf{c}} = \mathbf{Y}'\mathbf{1} - \widehat{\rho}\widehat{\mathbf{A}}\mathbf{X}$$

If the constant is suppressed, $\mathbf{c}$ is simply set to $\mathbf{0}$.

Reported statistics

`procrustes` computes and displays the following statistics for each target variable separately and globally by adding the appropriate sums of squares over all target variables. The predicted values $\widehat{\mathbf{Y}}$ for $\mathbf{Y}$ are defined as

$$\widehat{\mathbf{Y}} = \mathbf{1}\widehat{\mathbf{c}}' + \widehat{\rho}\mathbf{X}\widehat{\mathbf{A}}$$

The Procrustes statistic, P, is a scaled version of the squared distance of $\mathbf{Y}$:

$$P = \text{RSS}/\text{SS}$$

where

$$\text{RSS} = \text{trace}((\mathbf{Y} - \widehat{\mathbf{Y}})(\mathbf{Y} - \widehat{\mathbf{Y}})')$$

$$\text{SS} = \text{trace}(\ddot{\mathbf{Y}}'\ddot{\mathbf{Y}})$$

Note that $0 \leq P \leq 1$, and a small value of P means that $\mathbf{Y}$ is close to the transformed value of $\mathbf{X}$, that is, the $\mathbf{X}$ and $\mathbf{Y}$ configurations are similar. In the literature, this statistic is often denoted by R^2. It is easy to confuse this with the R^2 statistic in a regression context, which is actually $1 - P$.

A measure for the size of the residuals is the root mean squared error,

$$\text{RMSE} = \sqrt{\text{RSS}/\text{df}_{\text{r}}}$$

Here df_{r} are $Nn_y - \text{df}_{\text{m}}$, with $\text{df}_{\text{m}} = n_y n_x + n_y + 1 - k$, and with N the number of observations, n_y and n_x the number of target variables and source variables, respectively, and k, the number of restrictions, defined as

orthogonal:	$k = n_x(n_x - 1)/2$
oblique:	$k = n_y$
unrestricted:	$k = 1$

`procrustes` computes RMSE(j) for target variable y_j as

$$\text{RMSE}(j) = \sqrt{\text{RSS}(j)/(\text{df}_{\text{r}}/n_y)}$$

Finally, `procrustes` computes the Pearson correlation between y_j and $\widehat{y}_j$. For the unrestricted transformation, this is just the square root of the explained variance $1 - P(j)$, where $P(j) = \text{RSS}(j)/\text{SS}$. For the orthogonal and oblique transformation, this relationship does not hold.

References

Browne, M. W. 1967. On oblique Procrustes rotation. *Psychometrika* 32: 125–132.

Cox, T. F., and M. A. A. Cox. 2001. *Multidimensional Scaling*. 2nd ed. Boca Raton, FL: Chapman & Hall/CRC.

Cramer, E. M. 1974. On Browne's solution for oblique Procrustes rotation. *Psychometrika* 39: 159–163.

Gower, J. C., and G. B. Dijksterhuis. 2004. *Procrustes Problems*. Oxford: Oxford University Press.

Green, B. F. 1952. The orthogonal approximation of an oblique structure in factor analysis. *Psychometrika* 17: 429–440.

Hurley, J. R., and R. B. Cattell. 1962. The Procrustes program: Producing direct rotation to test a hypothesized factor structure. *Behavioral Science* 7: 258–262.

Mardia, K. V., J. T. Kent, and J. M. Bibby. 1979. *Multivariate Analysis*. London: Academic Press.

Mosier, C. I. 1939. Determining a simple structure when loadings for certain tests are known. *Psychometrika* 4: 149–162.

Peen, C. 2007. Old towns of England. http://www.oldtowns.co.uk.

ten Berge, J. M. F. 1977. Orthogonal Procrustes rotation for two or more matrices. *Psychometrika* 42: 267–276.

ten Berge, J. M. F., and K. Nevels. 1977. A general solution to Mosier's oblique Procrustes problem. *Psychometrika* 42: 593–600.

Also see

[MV] **procrustes postestimation** — Postestimation tools for procrustes

[MV] **ca** — Simple correspondence analysis

[MV] **pca** — Principal component analysis

[MV] **rotate** — Orthogonal and oblique rotations after factor and pca

[R] **mvreg** — Multivariate regression

[U] **20 Estimation and postestimation commands**

Title

procrustes postestimation — Postestimation tools for procrustes

Description

The following postestimation commands are of special interest after `procrustes`:

Command	Description
`estat compare`	fit statistics for orthogonal, oblique, and unrestricted transformations
`estat mvreg`	display multivariate regression resembling unrestricted transformation
`estat summarize`	display summary statistics over the estimation sample
`procoverlay`	produce a Procrustes overlay graph

For information about these commands, see below.

The following standard postestimation commands are also available:

Command	Description
* `estimates`	cataloging estimation results
`predict`	compute fitted values and residuals

* All `estimates` subcommands except `table` and `stats` are available; see [R] **estimates**.

See the corresponding entries in the *Base Reference Manual* for details.

Special-interest postestimation commands

`estat compare` displays a table with fit statistics of the three transformations provided by `procrustes`: `orthogonal`, `oblique`, and `unrestricted`. The two additional procrustes analyses are performed on the same sample as the original procrustes analysis and with the same options. F tests comparing the models are provided.

`estat mvreg` produces the `mvreg` (see [R] **mvreg**) output related to the unrestricted Procrustes analysis (the `transform(unrestricted)` option of `procrustes`).

`estat summarize` displays summary statistics over the estimation sample of the target and source variables ($varlist_y$ and $varlist_x$).

`procoverlay` displays a plot of the target variables overlaid with the fitted values derived from the source variables. If there are more than two target variables, multiple plots are shown in one graph.

Syntax for predict

predict [*type*] {*stub*∗ | *newvarlist*} [*if*] [*in*] , [*statistic*]

statistic	Description
Main	
fitted	fitted values $\mathbf{1}\,\mathbf{c}' + \rho\,\mathbf{X}\,\mathbf{A}$; the default (specify $\#_y$ vars)
residuals	unstandardized residuals (specify $\#_y$ vars)
q	residual sum of squares over the target variables (specify one var)

These statistics are available both in and out of sample; type `predict ... if e(sample) ...` if wanted only for the estimation sample.

Menu

Statistics > Postestimation > Predictions, residuals, etc.

Options for predict

Main

`fitted`, the default, computes fitted values, that is, the least-squares approximations of the target ($varlist_y$) variables. You must specify the same number of new variables as there are target variables.

`residuals` computes the raw (unstandardized) residuals for each target ($varlist_y$) variable. You must specify the same number of new variables as there are target variables.

`q` computes the residual sum of squares over all variables, that is, the squared Euclidean distance between the target and transformed source points. Specify one new variable.

Syntax for estat

Table of fit statistics

estat compare [, detail]

Comparison of `mvreg` *and* `procrustes` *output*

estat mvreg [, *mvreg_options*]

Display summary statistics

estat summarize [, labels noheader noweights]

Menu

Statistics > Postestimation > Reports and statistics

Options for estat

`detail`, an option with `estat compare`, displays the standard `procrustes` output for the two additional transformations.

mvreg_options, allowed with `estat mvreg`, are any of the options allowed by `mvreg`; see [R] **mvreg**. The constant is already suppressed if the Procrustes analysis suppressed it.

`labels`, `noheader`, and `noweights` are the same as for the generic `estat summarize` command; see [R] **estat**.

Syntax for procoverlay

`procoverlay` [*if*] [*in*] [, *procoverlay_options*]

procoverlay_options	Description
Main	
`autoaspect`	adjust aspect ratio on the basis of the data; default aspect ratio is 1
`targetopts(`*target_opts*`)`	affect the rendition of the target
`sourceopts(`*source_opts*`)`	affect the rendition of the source
Y axis, X axis, Titles, Legend, Overall	
twoway_options	any options other than `by()` documented in [G-3] ***twoway_options***
By	
`byopts(`*by_option*`)`	affect the rendition of combined graphs

target_opts	Description
Main	
`nolabel`	removes the default observation label from the target
marker_options	change look of markers (color, size, etc.)
marker_label_options	change look or position of marker labels

source_opts	Description
Main	
`nolabel`	removes the default observation label from the source
marker_options	change look of markers (color, size, etc.)
marker_label_options	change look or position of marker labels

Menu

Statistics > Multivariate analysis > Procrustes overlay graph

Options for procoverlay

Main

`autoaspect` specifies that the aspect ratio be automatically adjusted based on the range of the data to be plotted. This option can make some `procoverlay` plots more readable. By default, `procoverlay` uses an aspect ratio of one, producing a square plot.

As an alternative to `autoaspect`, the *twoway_option* `aspectratio()` can be used to override the default aspect ratio. `procoverlay` accepts the `aspectratio()` option as a suggestion only and will override it when necessary to produce plots with balanced axes, that is, where distance on the x axis equals distance on the y axis.

twoway_options, such as `xlabel()`, `xscale()`, `ylabel()`, and `yscale()`, should be used with caution. These *axis_options* are accepted but may have unintended side effects on the aspect ratio. See [G-3] ***twoway_options***.

`targetopts(`*target_opts*`)` affects the rendition of the target plot. The following *target_opts* are allowed:

`nolabel` removes the default target observation label from the graph.

marker_options affect the rendition of markers drawn at the plotted points, including their shape, size, color, and outline; see [G-3] ***marker_options***.

marker_label_options specify if and how the markers are to be labeled; see [G-3] ***marker_label_options***.

`sourceopts(`*source_opts*`)` affects the rendition of the source plot. The following *source_opts* are allowed:

`nolabel` removes the default source observation label from the graph.

marker_options affect the rendition of markers drawn at the plotted points, including their shape, size, color, and outline; see [G-3] ***marker_options***.

marker_label_options specify if and how the markers are to be labeled; see [G-3] ***marker_label_options***.

Y axis, X axis, Titles, Legend, Overall

twoway_options are any of the options documented in [G-3] ***twoway_options***, excluding `by()`. These include options for titling the graph (see [G-3] ***title_options***) and for saving the graph to disk (see [G-3] ***saving_option***). See `autoaspect` above for a warning against using options such as `xlabel()`, `xscale()`, `ylabel()`, and `yscale()`.

By

`byopts(`*by_option*`)` is documented in [G-3] ***by_option***. This option affects the appearance of the combined graph and is ignored, unless there are more than two target variables specified in `procrustes`.

Remarks

The examples in [MV] **procrustes** demonstrated a Procrustes transformation of a historical map, produced by John Speed in 1610, to a modern map. Here we demonstrate the use of `procrustes` postestimation tools in assessing the accuracy of Speed's map. Example 1 of [MV] **procrustes** performed the following analysis:

```
. use http://www.stata-press.com/data/r12/speed_survey
(Data on Speed's Worcestershire map (1610))

. procrustes (survey_x survey_y) (speed_x speed_y)
  (output omitted )
```

See example 1 of [MV] **procrustes**. The following examples are based on this `procrustes` analysis.

▷ Example 1

Did John Speed get the coordinates of the towns right—up to the location, scale, and orientation of his map relative to the modern map? In example 1 of [MV] **procrustes**, we demonstrated how the optimal transformation from the historical coordinates to the modern (true) coordinates can be estimated by `procrustes`.

It is possible to "predict" the configuration of 20 cities on Speed's historical map, optimally transformed (rotated, dilated, and translated) to approximate the true configuration. `predict` with the `fitted` option expects the same number of variables as the number of target (dependent) variables (`survey_x` and `survey_y`).

```
. predict fitted_x fitted_y
(fitted assumed)
```

We omitted the `fitted` option because it is the default.

It is often useful to also compute the (squared) distance between the true location and the transformed location of the historical map. This can be seen as a quality measure—the larger the value, the more Speed erred in the location of the respective town.

```
. predict q, q
```

We now list the target data (`survey_x` and `survey_y`, the values from the modern map), the fitted values (`fitted_x` and `fitted_y`, produced by `predict`), and the squared distance between them (`q`, produced by `predict` with the `q` option).

```
. list name survey_x survey_y fitted_x fitted_y q, sep(0) noobs
```

name	survey_x	survey_y	fitted_x	fitted_y	q
Alve	1027	725	1037.117	702.9464	588.7149
Arro	1083	565	1071.682	562.6791	133.4802
Astl	787	677	783.0652	674.5216	21.62482
Beck	976	358	978.8665	366.3761	78.37637
Beng	1045	435	1055.245	431.6015	116.51
Crad	736	471	725.8594	476.5895	134.075
Droi	893	633	890.5839	633.6066	6.205747
Ecki	922	414	929.4932	411.1757	64.12465
Eves	1037	437	1036.887	449.2707	150.5827
Hall	828	579	825.1494	575.9836	17.22464
Hanb	944	637	954.6189	643.6107	156.4629
Inkb	1016	573	1004.869	577.1111	140.7917
Kemp	848	490	845.7215	490.8959	5.994327
Kidd	826	762	836.8665	760.5699	120.1264
Mart	756	598	745.2623	597.5585	115.4937
Stud	1074	632	1072.622	634.3164	7.264294
Tewk	891	324	898.4571	318.632	84.42448
UpSn	943	544	939.3932	545.8247	16.33858
Upto	852	403	853.449	400.9419	6.335171
Worc	850	545	848.7917	547.7881	9.233305

We see that Speed especially erred in the location of Alvechurch—it is off by no less than $\sqrt{588} = 24$ miles, whereas the average error is about 8 miles. In a serious analysis of this dataset, we would check the data on Alvechurch, and, if we found it to be in order, consider whether we should actually drop Alvechurch from the analysis. In this illustration, we ignore this potential problem.

◁

▷ Example 2

Although the numerical information convinces us that Speed's map is generally accurate, a plot will convey this message more convincingly. `procoverlay` produces a plot that contains the target (survey) coordinates and the Procrustes-transformed historical coordinates. We could just type

```
. procoverlay
```

However, we decide to set several options to produce a presentation-quality graph. The suboption `mlabel()` of `target()` (or of `source()`) adds labels, identifying the towns. Because the target and source points are so close, there can be no confusing how they are matched. Displaying the labels twice in the plot is not helpful for this dataset. Therefore, we choose to label the target points, but not the source points using the `nolabel` suboption of `source()`. We preserve the equivalence of the x and y scale while using as much of the graphing region as possible with the `autoaspect` option. The `span` suboption of `title()` allows the long title to extend beyond the graph region if needed. We override the default legend by using the `legend()` option.

```
. procoverlay, target(mlabel(name)) source(nolabel) autoaspect
> title(Historic map of 20 towns and villages in Worcestershire, span)
> subtitle(overlaid with actual positions)
> legend(label(1 historic map) label(2 actual position))
```

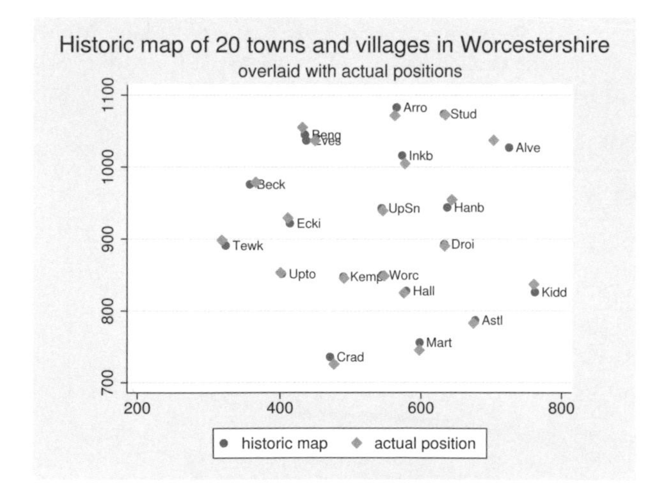

◁

▷ Example 3

`estat` offers three specific facilities after `procrustes`. These can all be seen as convenience tools that accomplish simple analyses, ensuring that the same variables and the same observations are used as in the Procrustes analysis.

The variables involved in the Procrustes analysis can be summarized over the estimation sample, for instance, in order to gauge differences in scales and location of the target and source variables.

```
. estat summarize
Estimation sample procrustes                    Number of obs =      20
```

Variable	Mean	Std. Dev.	Min	Max
target				
survey_x	916.7	106.6993	736	1083
survey_y	540.1	121.1262	324	762
source				
speed_x	153.95	46.76084	78	220
speed_y	133.9	49.90401	40	220

From the summarization, the two maps have different origins and scale.

As pointed out in [MV] **procrustes**, orthogonal and oblique Procrustes analyses can be thought of as special cases of multivariate regression (see [R] **mvreg**), subject to nonlinear restrictions on the coefficient matrix. Comparing the Procrustes statistics and the transformations for each of the three classes of transformations is helpful in selecting a transformation. The `compare` subcommand of `estat` provides summary information for the optimal transformations in each of the three classes.

```
. estat compare
Summary statistics for three transformations
```

	Procrustes	df_m	df_r	rmse
orthogonal	0.0040	4	36	7.403797
oblique	0.0040	5	35	7.498294
unrestricted	0.0037	6	34	7.343334

```
(F tests comparing the models suppressed)
```

The Procrustes statistic is ensured to decrease (not increase) from orthogonal to oblique to unrestricted because the associated classes of transformations are getting less restrictive. The model degrees of freedom (`df_m`) of the three transformation classes are the dimension of the classes, that is, the number of "free parameters". For instance, with orthogonal transformations between two source and two target variables, there is 1 degree of freedom for the rotation (representing the rotation angle), 2 degrees of freedom for the translation, and 1 degree of freedom for dilation (uniform scaling), that is, four in total. The residual degrees of freedom (df_r) are the number of observations (number of target variables times the number of observations) minus the model degrees of freedom. The root mean squared error RMSE, defined as

$$\mathrm{RMSE} = \sqrt{\frac{\mathrm{RSS}}{\mathrm{df}_r}}$$

does not, unlike the Procrustes statistic, surely become smaller with the less restrictive models. In this example, in fact, the RMSE of the orthogonal transformation is smaller than that of the oblique transformation. This indicates that the additional degree of freedom allowing for skew rotations does not produce a closer fit. In this example, we see little reason to relax orthogonal transformations; very little is gained in terms of the Procrustes statistic (an illness-of-fit measure) or the RMSE. In this interpretation, we used our intuition to guide us whether a difference in fit is substantively and statistically meaningful—formal significance tests are not provided.

Finally, the unrestricted transformation can be estimated with procrustes ..., transform(unrestricted). This analysis is related to a multivariate regression with the target variables as the dependent variables and the source variables as the independent variables. Although the unrestricted Procrustes analysis assumes spherical (uncorrelated homoskedastic) residuals, this restrictive assumption is not made in multivariate regression as estimated by the mvreg command. The comparable multivariate regression over the same estimation sample can be viewed simply by typing

```
. estat mvreg
Multivariate regression, similar to "procrustes ..., transform(unrestricted)"
Equation             Obs  Parms        RMSE    "R-sq"          F        P
--------------------------------------------------------------------------
survey_x              20      3    7.696981    0.9953   1817.102   0.0000
survey_y              20      3    6.971772    0.9970   2859.068   0.0000

------------------------------------------------------------------------------
             |      Coef.   Std. Err.      t    P>|t|     [95% Conf. Interval]
-------------+----------------------------------------------------------------
survey_x     |
     speed_x |    2.27584   .0379369    59.99   0.000       2.1958     2.35588
     speed_y |   .4147244   .0355475    11.67   0.000     .3397257     .489723
       _cons |   510.8028   8.065519    63.33   0.000     493.7861    527.8196
-------------+----------------------------------------------------------------
survey_y     |
     speed_x |  -.4129564   .0343625   -12.02   0.000     -.485455   -.3404579
     speed_y |   2.355725   .0321982    73.16   0.000     2.287793    2.423658
       _cons |    288.243   7.305587    39.46   0.000     272.8296    303.6564
------------------------------------------------------------------------------
```

This analysis is seen as postestimation after a Procrustes analysis, so it does not change the "last estimation results". We may still replay procrustes and use other procrustes postestimation commands.

◁

Saved results

estat compare after procrustes saves the following in r():

Matrices

r(cstat)	Procrustes statistics, degrees of freedom, and RMSEs
r(fstat)	F statistics, degrees of freedom, and p-values

estat mvreg does not return results.

estat summarize after procrustes saves the following in r():

Matrices

r(stats)	means, standard deviations, minimums, and maximums

Methods and formulas

All postestimation commands listed above are implemented as ado-files.

The predicted values for the jth variable are defined as

$$\widehat{y}_j = \widehat{c}_j + \widehat{\rho}\,\mathbf{X}\,\widehat{\mathbf{A}}[.,j]$$

The residual for y_j is simply $y_j - \widehat{y}_j$. The "rowwise" quality q of the approximation is defined as the residual sum of squares:

$$q = \sum_j (y_j - \widehat{y}_j)^2$$

The entries of the summary table produced by `estat compare` are described in *Methods and formulas* of [MV] **procrustes**. The F tests produced by `estat compare` are similar to standard nested model tests in linear models.

References

See *References* in [MV] **procrustes**.

Also see

[MV] **procrustes** — Procrustes transformation

[R] **mvreg** — Multivariate regression

Title

rotate — Orthogonal and oblique rotations after factor and pca

Syntax

`rotate` [, *options*]

`rotate, clear`

options	Description
Main	
`orthogonal`	restrict to orthogonal rotations; the default, except with `promax()`
`oblique`	allow oblique rotations
rotation_methods	rotation criterion
`normalize`	rotate Kaiser normalized matrix
`factors(#)`	rotate # factors or components; default is to rotate all
`components(#)`	synonym for `factors()`
Reporting	
`blanks(#)`	display loadings as blanks when |loading| < #; default is `blanks(0)`
`detail`	show `rotatemat` output; seldom used
`format(%fmt)`	display format for matrices; default is `format(%9.5f)`
`noloading`	suppress display of rotated loadings
`norotation`	suppress display of rotation matrix
Optimization	
optimize_options	control the maximization process; seldom used

rotation_methods	Description
* varimax	varimax (orthogonal only); the default
vgpf	varimax via the GPF algorithm (orthogonal only)
quartimax	quartimax (orthogonal only)
equamax	equamax (orthogonal only)
parsimax	parsimax (orthogonal only)
entropy	minimum entropy (orthogonal only)
tandem1	Comrey's tandem 1 principle (orthogonal only)
tandem2	Comrey's tandem 2 principle (orthogonal only)
* promax[(#)]	promax power # (implies oblique); default is promax(3)
oblimin[(#)]	oblimin with $\gamma = \#$; default is oblimin(0)
cf(#)	Crawford–Ferguson family with $\kappa = \#$, $0 \leq \# \leq 1$
bentler	Bentler's invariant pattern simplicity
oblimax	oblimax
quartimin	quartimin
target(*Tg*)	rotate toward matrix *Tg*
partial(*Tg W*)	rotate toward matrix *Tg*, weighted by matrix *W*

* varimax and promax ignore all *optimize_options*.

Menu

Statistics > Multivariate analysis > Factor and principal component analysis > Postestimation > Rotate loadings

Description

rotate performs a rotation of the loading matrix after factor, factormat, pca, or pcamat; see [MV] **factor** and [MV] **pca**. Many rotation criteria (such as varimax and oblimin) are available that can be applied with respect to the orthogonal and/or oblique class of rotations. rotate stores in e() object of the estimation command in fields e(r_*name*). For instance, e(r_L) will contain the rotated loadings.

rotate, clear removes the rotation results from the estimation results.

If you want to rotate a given matrix, see [MV] **rotatemat**. Actually, rotate is implemented using rotatemat.

If you want a Procrustes rotation, which rotates variables optimally toward other variables, see [MV] **procrustes**.

Options

Main

orthogonal specifies that an orthogonal rotation be applied. This is the default.

See *Rotation criteria* below for details on the *rotation_methods* available with orthogonal.

`oblique` specifies that an oblique rotation be applied. This often yields more interpretable factors with a simpler structure than that obtained with an orthogonal rotation. In many applications (for example, after `factor` and `pca`) the factors before rotation are orthogonal (uncorrelated), whereas the oblique rotated factors are correlated.

See *Rotation criteria* below for details on the *rotation_methods* available with `oblique`.

`clear` specifies that rotation results be cleared (removed) from the last estimation command. `clear` may not be combined with any other option.

`rotate` stores its results within the `e()` results of `pca` and `factor`, overwriting any previous rotation results. Postestimation commands such as `predict` operate on the last rotated results, if any, instead of the unrotated results, and allow you to specify `norotated` to use the unrotated results. The `clear` option of `rotate` allows you to remove the rotation results from `e()`, thus freeing you from having to specify `norotated` for the postestimation commands.

`normalize` requests that the rotation be applied to the Kaiser normalization (Horst 1965) of the matrix **A**, so that the rowwise sums of squares equal 1. Kaiser normalization applies to the rotated columns only (see the `factors()` option below).

`factors(#)`, and synonym `components(#)`, specifies the number of factors or components (columns of the loading matrix) to be rotated, counted "from the left", that is, with the lowest column index. The other columns are left unrotated. All columns are rotated by default.

Reporting

`blanks(#)` shows blanks for loadings with absolute values smaller than #.

`detail` displays the `rotatemat` output; seldom used.

`format(%fmt)` specifies the display format for matrices. The default is `format(%9.5f)`.

`noloading` suppresses the display of the rotated loadings.

`norotation` suppresses the display of the optimal rotation matrix.

Optimization

optimize_options are seldom used; see [MV] **rotatemat**.

Rotation criteria

In the descriptions below, the matrix to be rotated is denoted as **A**, p denotes the number of rows of **A**, and f denotes the number of columns of **A** (factors or components). If **A** is a loading matrix from `factor` or `pca`, p is the number of variables, and f is the number of factors or components.

Criteria suitable only for orthogonal rotations

`varimax` and `vgpf` apply the orthogonal varimax rotation (Kaiser 1958). `varimax` maximizes the variance of the squared loadings within factors (columns of **A**). It is equivalent to `cf(1/p)` and to `oblimin(1)`. `varimax`, the most popular rotation, is implemented with a dedicated fast algorithm and ignores all *optimize_options*. Specify `vgpf` to switch to the general GPF algorithm used for the other criteria.

`quartimax` uses the quartimax criterion (Harman 1976). `quartimax` maximizes the variance of the squared loadings within the variables (rows of **A**). For orthogonal rotations, `quartimax` is equivalent to `cf(0)` and to `oblimax`.

`equamax` specifies the orthogonal equamax rotation. `equamax` maximizes a weighted sum of the `varimax` and `quartimax` criteria, reflecting a concern for simple structure within variables (rows of **A**) as well as within factors (columns of **A**). `equamax` is equivalent to `oblimin(`p`/2)` and `cf(#)`, where $\# = f/(2p)$.

`parsimax` specifies the orthogonal parsimax rotation. `parsimax` is equivalent to `cf(#)`, where $\# = (f-1)/(p+f-2)$.

`entropy` applies the minimum entropy rotation criterion (Jennrich 2004).

`tandem1` specifies that the first principle of Comrey's tandem be applied. According to Comrey (1967), this principle should be used to judge which "small" factors should be dropped.

`tandem2` specifies that the second principle of Comrey's tandem be applied. According to Comrey (1967), `tandem2` should be used for "polishing".

Criteria suitable only for oblique rotations

`promax`[`(#)`] specifies the oblique promax rotation. The optional argument specifies the `promax` power. Not specifying the argument is equivalent to specifying `promax(3)`. Values smaller than 4 are recommended, but the choice is yours. Larger `promax` powers simplify the loadings (generate numbers closer to zero and one) but at the cost of additional correlation between factors. Choosing a value is a matter of trial and error, but most sources find values in excess of 4 undesirable in practice. The power must be greater than 1 but is not restricted to integers.

Promax rotation is an oblique rotation method that was developed before the "analytical methods" (based on criterion optimization) became computationally feasible. Promax rotation comprises an oblique Procrustean rotation of the original loadings **A** toward the elementwise #-power of the orthogonal varimax rotation of **A**.

Criteria suitable for orthogonal and oblique rotations

`oblimin`[`(#)`] specifies that the oblimin criterion with $\gamma = \#$ be used. When restricted to orthogonal transformations, the `oblimin()` family is equivalent to the orthomax criterion function. Special cases of `oblimin()` include

γ	Special case
0	quartimax / quartimin
1/2	biquartimax / biquartimin
1	varimax / covarimin
$p/2$	equamax

p = number of rows of **A**.

γ defaults to zero. Jennrich (1979) recommends $\gamma \leq 0$ for oblique rotations. For $\gamma > 0$, it is possible that optimal oblique rotations do not exist; the iterative procedure used to compute the solution will wander off to a degenerate solution.

`cf(#)` specifies that a criterion from the Crawford–Ferguson (1970) family be used with $\kappa = \#$. `cf(`κ`)` can be seen as $(1-\kappa)\mathrm{cf}_1(\mathbf{A}) + (\kappa)\mathrm{cf}_2(\mathbf{A})$, where $\mathrm{cf}_1(\mathbf{A})$ is a measure of row parsimony and $\mathrm{cf}_2(\mathbf{A})$ is a measure of column parsimony. $\mathrm{cf}_1(\mathbf{A})$ attains its greatest lower bound when no row of **A** has more than one nonzero element, whereas $\mathrm{cf}_2(\mathbf{A})$ reaches zero if no column of **A** has more than one nonzero element.

For orthogonal rotations, the Crawford–Ferguson family is equivalent to the `oblimin()` family. For orthogonal rotations, special cases include the following:

κ	Special case
0	quartimax / quartimin
$1/p$	varimax / covarimin
$f/(2p)$	equamax
$(f-1)/(p+f-2)$	parsimax
1	factor parsimony

p = number of rows of **A**.
f = number of columns of **A**.

`bentler` specifies that the "invariant pattern simplicity" criterion (Bentler 1977) be used.

`oblimax` specifies the oblimax criterion. `oblimax` maximizes the number of high and low loadings. `oblimax` is equivalent to `quartimax` for orthogonal rotations.

`quartimin` specifies that the quartimin criterion be used. For orthogonal rotations, `quartimin` is equivalent to `quartimax`.

`target(`*Tg*`)` specifies that **A** be rotated as near as possible to the conformable matrix *Tg*. Nearness is expressed by the Frobenius matrix norm.

`partial(`*Tg W*`)` specifies that **A** be rotated as near as possible to the conformable matrix *Tg*. Nearness is expressed by a weighted (by *W*) Frobenius matrix norm. *W* should be nonnegative and usually is zero–one valued, with ones identifying the target values to be reproduced as closely as possible by the factor loadings, whereas zeros identify loadings to remain unrestricted.

Remarks

Remarks are presented under the following headings:

Orthogonal rotations
Oblique rotations
Other types of rotation

In this entry, we focus primarily on the rotation of factor loading matrices in factor analysis. `rotate` may also be used after `pca`, with the same syntax. We advise caution in the interpretation of rotated loadings in principal component analysis because some of the optimality properties of principal components are not preserved under rotation. See [MV] **pca postestimation** for more discussion of this point.

Orthogonal rotations

The interpretation of a factor analytical solution is not always easy—an understatement, many will agree. This is due partly to the standard way in which the inherent indeterminacy of factor analysis is resolved. Orthogonal transformations of the common factors and the associated factor loadings are possible without affecting the reconstructed (fitted) correlation matrix and preserving the property that common factors are uncorrelated. This gives considerable freedom in selecting an orthogonal rotation to facilitate the interpretation of the factor loadings. Thurstone (1935) offered criteria for a "simple structure" required for a psychologically meaningful factor solution. These informal criteria for interpretation were then formalized into formal rotation criteria, for example, Harman (1976) and Gorsuch (1983).

▷ Example 1

We illustrate `rotate` by using a factor analysis of the correlation matrix of eight physical variables (height, arm span, length of forearm, length of lower leg, weight, bitrochanteric diameter, chest girth, and chest width) of 305 girls.

```
. matrix input R = ( 1000  846  805  859  473  398  301  382 \
>                     846 1000  881  826  376  326  277  415 \
>                     805  881 1000  801  380  319  237  345 \
>                     859  826  801 1000  436  329  327  365 \
>                     473  376  380  436 1000  762  730  629 \
>                     398  326  319  329  762 1000  583  577 \
>                     301  277  237  327  730  583 1000  539 \
>                     382  415  345  365  629  577  539 1000 )

. matrix R = R/1000

. matrix colnames R = height  arm_span  fore_arm  lower_leg
>                     weight  bitrod    ch_girth  ch_width

. matrix rownames R = height  arm_span  fore_arm  lower_leg
>                     weight  bitrod    ch_girth  ch_width

. matlist R, border format(%7.3f)
```

	height	arm_s~n	fore_~m	lower~g	weight	bitrod	ch_gi~h	ch_wi~h
height	1.000							
arm_span	0.846	1.000						
fore_arm	0.805	0.881	1.000					
lower_leg	0.859	0.826	0.801	1.000				
weight	0.473	0.376	0.380	0.436	1.000			
bitrod	0.398	0.326	0.319	0.329	0.762	1.000		
ch_girth	0.301	0.277	0.237	0.327	0.730	0.583	1.000	
ch_width	0.382	0.415	0.345	0.365	0.629	0.577	0.539	1.000

We extract two common factors with the iterated principal-factor method. See the description of `factormat` in [MV] **factor** for details on running a factor analysis on a Stata matrix rather than on a dataset.

```
. factormat R, n(305) fac(2) ipf
(obs=305)

Factor analysis/correlation                    Number of obs    =      305
    Method: iterated principal factors         Retained factors =        2
    Rotation: (unrotated)                      Number of params =       15
```

Factor	Eigenvalue	Difference	Proportion	Cumulative
Factor1	4.44901	2.93878	0.7466	0.7466
Factor2	1.51023	1.40850	0.2534	1.0000
Factor3	0.10173	0.04705	0.0171	1.0171
Factor4	0.05468	0.03944	0.0092	1.0263
Factor5	0.01524	0.05228	0.0026	1.0288
Factor6	-0.03703	0.02321	-0.0062	1.0226
Factor7	-0.06025	0.01415	-0.0101	1.0125
Factor8	-0.07440	.	-0.0125	1.0000

```
LR test: independent vs. saturated:  chi2(28) = 2092.68 Prob>chi2 = 0.0000
```

Factor loadings (pattern matrix) and unique variances

Variable	Factor1	Factor2	Uniqueness
height	0.8560	-0.3244	0.1620
arm_span	0.8482	-0.4115	0.1112
fore_arm	0.8082	-0.4090	0.1795
lower_leg	0.8309	-0.3424	0.1923
weight	0.7503	0.5712	0.1108
bitrod	0.6307	0.4922	0.3600
ch_girth	0.5687	0.5096	0.4169
ch_width	0.6074	0.3507	0.5081

The default factor solution is rather poor from the perspective of a "simple structure", namely, that variables should have high loadings on few (one) factors and factors should ideally have only low and high values. A plot of the loadings is illuminating.

```
. loadingplot, xlab(0(.2)1) ylab(-.4(.2).6) aspect(1) yline(0) xline(0)
```

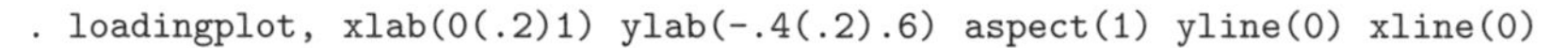

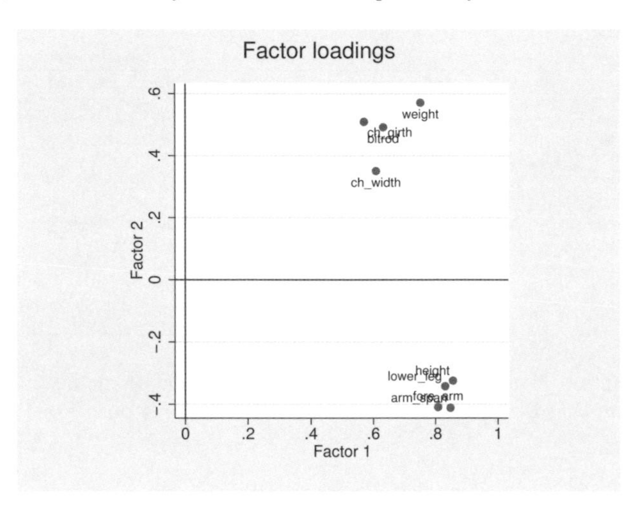

There are two groups of variables. We would like to see one group of variables close to one axis and the other group of variables close to the other axis. Turning the plot by about 45 degrees counterclockwise should make this possible and offer a much "simpler" structure. This is what the `rotate` command accomplishes.

```
. rotate

Factor analysis/correlation                  Number of obs    =        305
    Method: iterated principal factors       Retained factors =          2
    Rotation: orthogonal varimax (Kaiser off)  Number of params =       15
```

Factor	Variance	Difference	Proportion	Cumulative
Factor1	3.39957	0.83989	0.5705	0.5705
Factor2	2.55968	.	0.4295	1.0000

LR test: independent vs. saturated: chi2(28) = 2092.68 Prob>chi2 = 0.0000

```
Rotated factor loadings (pattern matrix) and unique variances
```

Variable	Factor1	Factor2	Uniqueness
height	0.8802	0.2514	0.1620
arm_span	0.9260	0.1770	0.1112
fore_arm	0.8924	0.1550	0.1795
lower_leg	0.8708	0.2220	0.1923
weight	0.2603	0.9064	0.1108
bitrod	0.2116	0.7715	0.3600
ch_girth	0.1515	0.7484	0.4169
ch_width	0.2774	0.6442	0.5081

```
Factor rotation matrix
```

	Factor1	Factor2
Factor1	0.8018	0.5976
Factor2	-0.5976	0.8018

See [MV] **factor** for the interpretation of the first panel. Here we will focus on the second and third panel. The rotated factor loadings satisfy

$$\text{Factor1}_{\text{rotated}} = 0.8018 \times \text{Factor1}_{\text{unrotated}} - 0.5976 \times \text{Factor2}_{\text{unrotated}}$$

$$\text{Factor2}_{\text{rotated}} = 0.5976 \times \text{Factor1}_{\text{unrotated}} + 0.8018 \times \text{Factor2}_{\text{unrotated}}$$

The uniqueness—the variance of the specific factors—is not affected, because we are changing only the coordinates in common factor space. The purpose of rotation is to make factor loadings easier to interpret. The first factor loads high on the first four variables and low on the last four variables; for the second factor, the roles are reversed. This is really a simple structure according to Thurstone's criteria. This is clear in the plot of the factor loadings.

```
. loadingplot, xlab(0(.2)1) ylab(0(.2)1) aspect(1) yline(0) xline(0)
```

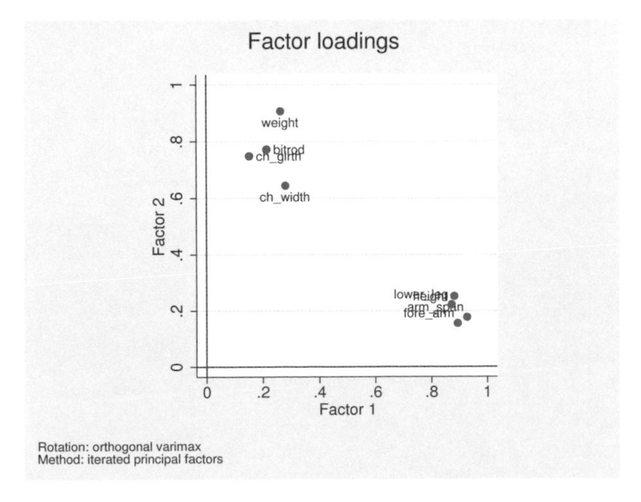

`rotate` provides several different rotations. You may make your intention clearer by typing the command as

```
. rotate, orthogonal varimax
```
(output omitted)

rotate defaults to orthogonal (angle and length preserving) rotations of the axes; thus, orthogonal may be omitted. The default rotation method is varimax, probably the most popular method. We warn that the varimax rotation is not appropriate if you expect a general factor contributing to all variables (see also Gorsuch 1983, chap. 9). In such a case you could, for instance, consider a quartimax rotation.

◁

▷ Example 2

rotate has performed what is known as "raw varimax", rotating the axes to maximize the sum of the variance of the squared loadings in the columns—the variance in a column is large if it comprises small and large (in the absolute sense) values. In rotating the axes, rows with large initial loadings—that is, with high communalities—have more influence than rows with only small values. Kaiser suggested that in the computation of the optimal rotation, all rows should have the same weight. This is usually known as the Kaiser normalization and sometimes known as the Horst normalization (Horst 1965). The option normalize applies this normalization method for rotation.

```
. rotate, normalize
Factor analysis/correlation                        Number of obs    =       305
    Method: iterated principal factors             Retained factors =         2
    Rotation: orthogonal varimax (Kaiser on)       Number of params =        15

    --------------------------------------------------------------------------
         Factor  |     Variance   Difference        Proportion   Cumulative
    -------------+------------------------------------------------------------
        Factor1  |      3.31500      0.67075            0.5563       0.5563
        Factor2  |      2.64425            .            0.4437       1.0000
    --------------------------------------------------------------------------
    LR test: independent vs. saturated:  chi2(28) = 2092.68 Prob>chi2 = 0.0000

Rotated factor loadings (pattern matrix) and unique variances

    -------------------------------------------------
        Variable |  Factor1   Factor2 |   Uniqueness
    -------------+--------------------+--------------
          height |   0.8724    0.2775 |      0.1620
        arm_span |   0.9203    0.2045 |      0.1112
        fore_arm |   0.8874    0.1815 |      0.1795
       lower_leg |   0.8638    0.2478 |      0.1923
          weight |   0.2332    0.9137 |      0.1108
          bitrod |   0.1885    0.7775 |      0.3600
        ch_girth |   0.1292    0.7526 |      0.4169
        ch_width |   0.2581    0.6522 |      0.5081
    -------------------------------------------------

Factor rotation matrix

    --------------------------------
                 | Factor1  Factor2
    -------------+------------------
         Factor1 |  0.7837   0.6212
         Factor2 | -0.6212   0.7837
    --------------------------------
```

Here the raw and normalized varimax rotated loadings are not much different.

◁

In the first example, loadingplot after rotate showed the rotated loadings, not the unrotated loadings. How can this be? Remember that Stata estimation commands store their results in e(), which we can list using ereturn list.

```
. ereturn list
scalars:
                  e(f) =  2
                  e(N) =  305
               e(df_m) =  15
               e(df_r) =  13
             e(chi2_i) =  2092.68137837692
               e(df_i) =  28
                e(p_i) =  0
              e(evsum) =  5.95922412962743
                e(r_f) =  2
macros:
    e(r_normalization) : "kaiser"
            e(r_class) : "orthogonal"
        e(r_criterion) : "varimax"
           e(r_ctitle) : "varimax"
            e(cmdline) : "factormat R, n(305) fac(2) ipf"
                e(cmd) : "factor"
       e(marginsnotok) : "_ALL"
         e(properties) : "nob noV eigen"
              e(title) : "Factor analysis"
            e(predict) : "factor_p"
          e(estat_cmd) : "factor_estat"
         e(rotate_cmd) : "factor_rotate"
            e(factors) : "factors(2)"
             e(mtitle) : "iterated principal factors"
             e(method) : "ipf"
         e(matrixname) : "R"
matrices:
               e(r_Ev) :  1 x 2
              e(r_Phi) :  2 x 2
                e(r_T) :  2 x 2
                e(r_L) :  8 x 2
                  e(C) :  8 x 8
                e(Phi) :  2 x 2
                  e(L) :  8 x 2
                e(Psi) :  1 x 8
                 e(Ev) :  1 x 8
```

When you replay an estimation command, it simply knows where to look, so that it can redisplay the output. `rotate` does something that few other postestimation commands are allowed to do: it adds information to the estimation results computed by `factor` or `pca`. But to avoid confusion, it writes in `e()` fields with the prefix `r_`. For instance, the matrix `e(r_L)` contains the rotated loadings.

If you replay `factor` after `rotate`, `factor` will display the rotated results. And this is what all `factor` and `pca` postestimation commands do. For instance, if you `predict` after `rotate`, `predict` will use the rotated results. Of course, it is still possible to operate on the unrotated results. `factor, norotated` replays the unrotated results. `predict` with the `norotated` option computes the factor scores for the unrotated results.

`rotate` stores information only about the most recent rotation, overwriting any information from the previous rotation. If you need the previous results again, run `rotate` with the respective options again; you do not need to run `factor` again. It is also possible to use `estimates store` to store estimation results for different rotations, which you may later restore and replay at will. See [R] **estimates store** for details.

If you no longer need the rotation results, you may type

```
. rotate, clear
```

to clean up the rotation result and return the `factor` results back to their pristine state (as if `rotate` had never been called).

▷ Example 3

`rotate` provides many more orthogonal rotations. Previously we stated that the varimax rotation can be thought of as the rotation that maximizes the varimax criterion, namely, the variance of the squared loadings summed over the columns. A column of loadings with a high variance tends to contain a series of large values and a series of low values, achieving the simplicity aim of factor analytic interpretation. The other types of rotation simply maximize other concepts of simplicity. For instance, the `quartimax` rotation aims at rowwise simplicity—preferably, the loadings within variables fall into a grouping of a few large ones and a few small ones, using again the variance in squared loadings as the criterion to be maximized.

```
. rotate, quartimax normalize
Factor analysis/correlation                        Number of obs    =       305
    Method: iterated principal factors             Retained factors =         2
    Rotation: orthogonal quartimax (Kaiser on)     Number of params =        15
```

Factor	Variance	Difference	Proportion	Cumulative
Factor1	3.32371	0.68818	0.5577	0.5577
Factor2	2.63553	.	0.4423	1.0000

```
    LR test: independent vs. saturated:  chi2(28) = 2092.68 Prob>chi2 = 0.0000
Rotated factor loadings (pattern matrix) and unique variances
```

Variable	Factor1	Factor2	Uniqueness
height	0.8732	0.2749	0.1620
arm_span	0.9210	0.2017	0.1112
fore_arm	0.8880	0.1788	0.1795
lower_leg	0.8646	0.2452	0.1923
weight	0.2360	0.9130	0.1108
bitrod	0.1909	0.7769	0.3600
ch_girth	0.1315	0.7522	0.4169
ch_width	0.2601	0.6514	0.5081

```
Factor rotation matrix
```

	Factor1	Factor2
Factor1	0.7855	0.6188
Factor2	-0.6188	0.7855

Here the quartimax and the varimax rotated results are rather similar. This need not be the case—varimax focuses on simplicity within columns (factors) and quartimax within rows (variables). It is possible to compromise, rotating to strive for a weighted sum of row simplicity and column simplicity. This is known as the orthogonal oblimin criterion; in the orthogonal case, `oblimin()` is equivalent to the Crawford–Ferguson (option `cf()`) family and to the orthomax family. These are parameterized families of criteria with, for instance, the following special cases:

`oblimin(0)`	quartimax rotation
`oblimin(0.5)`	biquartimax rotation
`oblimin(1)`	varimax rotation

```
. rotate, oblimin(0.5) normalize
Factor analysis/correlation                        Number of obs    =       305
    Method: iterated principal factors             Retained factors =         2
    Rotation: orthogonal oblimin (Kaiser on)       Number of params =        15
```

Factor	Variance	Difference	Proportion	Cumulative
Factor1	3.31854	0.67783	0.5569	0.5569
Factor2	2.64071	.	0.4431	1.0000

```
    LR test: independent vs. saturated:  chi2(28) = 2092.68 Prob>chi2 = 0.0000
Rotated factor loadings (pattern matrix) and unique variances
```

Variable	Factor1	Factor2	Uniqueness
height	0.8727	0.2764	0.1620
arm_span	0.9206	0.2033	0.1112
fore_arm	0.8877	0.1804	0.1795
lower_leg	0.8642	0.2468	0.1923
weight	0.2343	0.9134	0.1108
bitrod	0.1895	0.7772	0.3600
ch_girth	0.1301	0.7525	0.4169
ch_width	0.2589	0.6518	0.5081

```
Factor rotation matrix
```

	Factor1	Factor2
Factor1	0.7844	0.6202
Factor2	-0.6202	0.7844

Because the varimax and orthomax rotation are relatively close, the factor loadings resulting from an optimal rotation of a compromise criterion are close as well.

The orthogonal quartimax rotation may be obtained in different ways, namely, directly or by the appropriate member of the `oblimin()` or `cf()` families:

```
. rotate, quartimax
  (output omitted)
. rotate, oblimin(0)
  (output omitted)
. rotate, cf(0)
  (output omitted)
```

◁

❑ Technical note

The orthogonal varimax rotation also belongs to the oblimin and Crawford–Ferguson families.

```
. rotate, varimax
  (output omitted)
. rotate, oblimin(1)
  (output omitted)
. rotate, cf(0.125)
  (output omitted)
```

(The $0.125 = 1/8$ above is 1 divided by the number of variables.) All three produce the orthogonal varimax rotation. (There is actually a fourth way, namely `rotate, vgpf`.) There is, however, a subtle difference in algorithms used. The varimax rotation as specified by the `varimax` option (which is also the default) is computed by the classic algorithm of cycling through rotations of two factors at a time. The other ways use the general "gradient projection" algorithm proposed by Jennrich; see [MV] **rotatemat** for more information.

❏

Oblique rotations

In addition to orthogonal rotations, oblique rotations are also available.

▷ Example 4

The rotation methods that we have discussed so far are all orthogonal: the angles between the axes are unchanged, so the rotated factors are uncorrelated.

Returning to our original factor analysis,

```
. factormat R, n(305) fac(2) ipf
  (output omitted )
```

we examine the correlation matrix of the common factors,

```
. estat common
Correlation matrix of the common factors
```

Factors	Factor1	Factor2
Factor1	1	
Factor2	0	1

and see that they are uncorrelated.

The indeterminacy in the factor analytic model, however, allows us to consider other transformations of the common factors, namely, oblique rotations. These are rotations of the axes that preserve the norms of the rows of the loadings but not the angles between the axes or the angles between the rows. There are advantages and disadvantages of oblique rotations. See, for instance, Gorsuch (1983, chap. 9). In many substantive theories, there seems little reason to impose the restriction that the common factors be uncorrelated. The additional freedom in choosing the axes generally leads to more easily interpretable factors, sometimes to a great extent. However, although most researchers are willing to accept mildly correlated factors, they would prefer to use fewer of such factors.

`rotate` provides an extensive menu of oblique rotations; with a few exceptions, criteria suitable for orthogonal rotations are also suitable for oblique rotation. Again oblique rotation can be conceived of as maximizing some "simplicity" criterion. We illustrate with the oblimin oblique rotation.

```
. rotate, oblimin oblique normalize
Factor analysis/correlation                        Number of obs    =      305
    Method: iterated principal factors             Retained factors =        2
    Rotation: oblique oblimin (Kaiser on)          Number of params =       15

        Factor  |     Variance   Proportion    Rotated factors are correlated
    ------------+-----------------------------------------------------------
       Factor1  |      3.95010       0.6629
       Factor2  |      3.35832       0.5635
    ------------------------------------------------------------------------
    LR: independence vs saturated:  Chi2( 28) =  2092.68, Prob > chi2 = 0.0000
Rotated factor loadings (pattern matrix) and unique variances

        Variable |  Factor1   Factor2 |   Uniqueness
    -------------+--------------------+--------------
          height |   0.8831    0.0648 |      0.1620
        arm_span |   0.9560   -0.0288 |      0.1112
        fore_arm |   0.9262   -0.0450 |      0.1795
       lower_leg |   0.8819    0.0344 |      0.1923
          weight |   0.0047    0.9408 |      0.1108
          bitrod |  -0.0069    0.8032 |      0.3600
        ch_girth |  -0.0653    0.7923 |      0.4169
        ch_width |   0.1042    0.6462 |      0.5081
    -----------------------------------------------

Factor rotation matrix

                 |  Factor1  Factor2
    -------------+------------------
         Factor1 |   0.9112   0.7930
         Factor2 |  -0.4120   0.6092
    --------------------------------
```

The oblique rotation yields a much "simpler" structure in the Thurstone (1935) sense than that of the orthogonal rotations. This time, the common factors are moderately correlated.

```
. estat common
Correlation matrix of the Oblimin(0) rotated common factors

         Factors |  Factor1   Factor2
    -------------+--------------------
         Factor1 |        1
         Factor2 |    .4716         1
    ----------------------------------
```

◁

❑ Technical note

The numerical maximization of a simplicity criterion with respect to the class of orthogonal or oblique rotations proceeds in a stepwise method, making small improvements from an initial guess, until no more small improvements are possible. Such a procedure is not guaranteed to converge to the global optimum but to a local optimum instead. In practice, we experience few such problems. To some extent, this is because we have a reasonable starting value using the unrotated factors or loadings. As a safeguard, Stata starts the improvement from multiple initial positions chosen at random from the classes of orthonormal and normal rotation matrices. If the maximization procedure converges to the same criterion value at each trial, we may be reasonably certain that we have found the global optimum. Let us illustrate.

```
. set seed 123
. rotate, oblimin oblique normalize protect(10)
Trial   1 : min criterion   .0181657
Trial   2 : min criterion   46234.38
Trial   3 : min criterion   .0181657
Trial   4 : min criterion   .0181657
Trial   5 : min criterion   .0181657
Trial   6 : min criterion   .0181657
Trial   7 : min criterion   .0181657
Trial   8 : min criterion   1769.989
Trial   9 : min criterion   250205.6
Trial  10 : min criterion   .0181657
Factor analysis/correlation                        Number of obs    =      305
    Method: iterated principal factors             Retained factors =        2
    Rotation: oblique oblimin (Kaiser on)          Number of params =       15

    --------------------------------------------------------------------------
         Factor  |     Variance   Proportion    Rotated factors are correlated
    -------------+------------------------------------------------------------
        Factor1  |      3.95010       0.6629
        Factor2  |      3.35832       0.5635
    --------------------------------------------------------------------------
    LR test: independent vs. saturated:  chi2(28) = 2092.68 Prob>chi2 = 0.0000
Rotated factor loadings (pattern matrix) and unique variances

    -------------------------------------------------
        Variable |  Factor1   Factor2 |   Uniqueness
    -------------+--------------------+--------------
          height |   0.8831    0.0648 |      0.1620
        arm_span |   0.9560   -0.0288 |      0.1112
        fore_arm |   0.9262   -0.0450 |      0.1795
       lower_leg |   0.8819    0.0344 |      0.1923
          weight |   0.0047    0.9408 |      0.1108
          bitrod |  -0.0069    0.8032 |      0.3600
        ch_girth |  -0.0653    0.7923 |      0.4169
        ch_width |   0.1042    0.6462 |      0.5081
    -------------------------------------------------
Factor rotation matrix

    ----------------------------------
                 |  Factor1  Factor2
    -------------+--------------------
         Factor1 |   0.9112   0.7930
         Factor2 |  -0.4120   0.6092
    ----------------------------------
```

Here three of the random trials converged to distinct rotations from the rest. Specifying options `log` and `trace` would demonstrate that in these cases, the initial configurations were so far off that no improvements could be found. In a real application, we would probably rerun `rotate` with more trials, say, `protect(50)`, for more reassurance.

❑

❑ Technical note

There is another but almost trivial source of nonuniqueness. All simplicity criteria supported by `rotate` and `rotatemat` are invariant with respect to permutations of the rows and of the columns. Also, the signs of rotated loadings are undefined. `rotatemat`, the computational engine of `rotate`, makes sure that all columns have a positive orientation, that is, have a positive sum. `rotate`, after `factor` and `pca`, also sorts the columns into decreasing order of explained variance.

❑

Other types of rotation

`rotate` supports a few rotation methods that do not fit into the scheme of “simplicity maximization”. The first is known as the target rotation, which seeks to rotate the factor loading matrix to approximate as much as possible a target matrix of the same size as the factor loading matrix.

▷ Example 5

We continue with our same example. If we had expected a factor loading structure in which the first group of four variables would load especially high on the first factor and the second group of four variables on the second factor, we could have set up the following target matrix.

```
. matrix W = ( 1,0 \ 1,0 \ 1,0 \ 1,0 \ 0,1 \ 0,1 \ 0,1 \ 0,1 )

. matrix list W

W[8,2]
    c1  c2
r1   1   0
r2   1   0
r3   1   0
r4   1   0
r5   0   1
r6   0   1
r7   0   1
r8   0   1
```

It is also possible to request an orthogonal or oblique rotation toward the target **W**.

```
. rotate, target(W) normalize

Factor analysis/correlation                        Number of obs    =      305
    Method: iterated principal factors             Retained factors =        2
    Rotation: orthogonal target (Kaiser on)        Number of params =       15

    ---------------------------------------------------------------------------
         Factor  |     Variance   Difference        Proportion   Cumulative
    -------------+-------------------------------------------------------------
        Factor1  |      3.30616      0.65307            0.5548       0.5548
        Factor2  |      2.65309            .            0.4452       1.0000
    ---------------------------------------------------------------------------
    LR test: independent vs. saturated:  chi2(28) = 2092.68 Prob>chi2 = 0.0000

Rotated factor loadings (pattern matrix) and unique variances

    -----------------------------------------------------
        Variable |  Factor1   Factor2 |   Uniqueness
    -------------+--------------------+------------------
          height |   0.8715    0.2802 |      0.1620
        arm_span |   0.9197    0.2073 |      0.1112
        fore_arm |   0.8869    0.1843 |      0.1795
       lower_leg |   0.8631    0.2505 |      0.1923
          weight |   0.2304    0.9144 |      0.1108
         bitrod  |   0.1861    0.7780 |      0.3600
        ch_girth |   0.1268    0.7530 |      0.4160
        ch_width |   0.2561    0.6530 |      0.5081
    -----------------------------------------------------

Factor rotation matrix

    --------------------------------
                 | Factor1  Factor2
    -------------+------------------
         Factor1 |  0.7817   0.6236
         Factor2 | -0.6236   0.7817
    --------------------------------
```

With this target matrix, the result is not far different from the varimax and other orthogonal rotations.

◁

▷ Example 6

For our last example, we return to the early days of factor analysis, the time before fast computing. Analytical methods for orthogonal rotation, such as varimax, were developed relatively early. Analogous methods for oblique rotations proved more complicated. Hendrickson and White (1964) proposed a computationally simple method to obtain an oblique rotation that comprises an oblique Procrustes rotation of the factor loadings toward a signed power of the varimax rotation of the factor loadings. The promax method has one parameter, the power to which the varimax loadings are raised. Larger promax powers simplify the factor loadings (that is, generate more zeros and ones) at the cost of more correlation between the common factors. Generally, we recommend that you keep the power in the range (1,4] and not restricted to integers. Specifying `promax` is equivalent to `promax(3)`.

```
. rotate, promax normalize
Factor analysis/correlation                        Number of obs    =      305
    Method: iterated principal factors             Retained factors =        2
    Rotation: oblique promax (Kaiser on)           Number of params =       15

    --------------------------------------------------------------------------
         Factor  |     Variance   Proportion    Rotated factors are correlated
    -------------+------------------------------------------------------------
        Factor1  |      3.92727       0.6590
        Factor2  |      3.31295       0.5559
    --------------------------------------------------------------------------
    LR test: independent vs. saturated:  chi2(28) = 2092.68 Prob>chi2 = 0.0000

Rotated factor loadings (pattern matrix) and unique variances

    -------------------------------------------------
        Variable |  Factor1   Factor2 |   Uniqueness 
    -------------+--------------------+--------------
          height |   0.8797    0.0744 |      0.1620  
        arm_span |   0.9505   -0.0176 |      0.1112  
        fore_arm |   0.9205   -0.0340 |      0.1795  
       lower_leg |   0.8780    0.0443 |      0.1923  
          weight |   0.0214    0.9332 |      0.1108  
          bitrod |   0.0074    0.7966 |      0.3600  
        ch_girth |  -0.0509    0.7851 |      0.4169  
        ch_width |   0.1152    0.6422 |      0.5081  
    -------------------------------------------------

Factor rotation matrix

    --------------------------------
                 |  Factor1  Factor2 
    -------------+------------------
         Factor1 |   0.9069   0.7832 
         Factor2 |  -0.4214   0.6218 
    --------------------------------
```

In this simple two-factor example, the promax solution is similar to the oblique oblimin solution.

◁

Saved results

rotate adds saved results named e(r_*name*) to the saved results that were already defined by factor or pca.

rotate adds to the following results:

Scalars

e(r_f)	number of factors/components in rotated solution
e(r_fmin)	rotation criterion value

Macros

e(r_class)	orthogonal or oblique
e(r_criterion)	rotation criterion
e(r_ctitle)	title for rotation
e(r_normalization)	kaiser or none

Matrices

e(r_L)	rotated loadings
e(r_T)	rotation
e(r_Phi)	correlations between common factors (after factor only)
e(r_Ev)	explained variance by common factors (factor) or rotated components (pca)

The factors/components in the rotated solution are in decreasing order of e(r_Ev).

❑ Technical note

The rest of this section contains information of interest to programmers who want to provide rotate support to other estimation commands. Similar to other postestimation commands, such as estat and predict, rotate invokes a handler command. The name of this command is extracted from the field e(rotate_cmd). The estimation command *cmd* should set this field appropriately. For instance, pca sets the macro e(rotate_cmd) to pca_rotate. The command pca_rotate implements rotation after pca and pcamat, using rotatemat as the computational engine. pca_rotate does not display output itself; it relies on pca to do so.

For consistent behavior for end users and programmers alike, we recommend that the estimation command *cmd*, the driver commands, and other postestimation commands adhere to the following guidelines:

Driver command

- The rotate driver command for *cmd* should be named *cmd*_rotate.
- *cmd*_rotate should be an e-class command, that is, returning in e().
- Make sure that *cmd*_rotate is invoked after the correct estimation command (for example, if "`e(cmd)'" != "pca" ...).
- Allow at least the option detail and any option available to rotatemat.
- Extract from e() the matrix you want to rotate; invoke rotatemat on the matrix; and run this command quietly (that is, suppress all output) unless the option detail was specified.
- Extract the r() objects returned by rotatemat; see *Methods and formulas* of [MV] **rotatemat** for details.
- Compute derived results needed for your estimator.
- Store in e() fields (macros, scalars, matrices) named r_*name*, adding to the existing e() fields.

  Store the macros returned by rotatemat under the same named prefixed with r_. In particular, the macro e(r_criterion) should be set to the name of the rotation criterion returned by rotatemat as r(criterion). Other commands can check this field to find out whether rotation results are available.

We suggest that only the most recent rotation results be stored, overwriting any existing `e(r_*)` results. The programmer command `_rotate_clear` clears any existing `r_*` fields from `e()`.

- Display the rotation results by replaying *cmd*.

Estimation command cmd

- In *cmd*, define `e(rotate_cmd)` to *cmd*`_rotate`.
- *cmd* should be able to display the rotated results and should default to do so if rotated results are available. Include an option `noROTated` to display the unrotated results.
- You may use the programmer command `_rotate_text` to obtain a standard descriptive text for the rotation method.

Other postestimation commands

- Other postestimation commands after *cmd* should operate on the rotated results whenever they are appropriate and available, unless the option `noROTated` specifies otherwise.
- Mention that you operate on the unrotated results only if rotated results are available, but the user or you as the programmer decided not to use them.

❑

Methods and formulas

`rotate` is implemented as an ado-file.

See *Methods and formulas* of [MV] **rotatemat**.

Henry Felix Kaiser (1927–1992) was born in Morristown, New Jersey, and educated in California, where he earned degrees at Berkeley in between periods of naval service during and after World War II. A specialist in psychological and educational statistics and measurement, Kaiser worked at the Universities of Illinois and Wisconsin before returning to Berkeley in 1968. He made several contributions to factor analysis, including varimax rotation (the subject of his PhD) and a measure for assessing sampling adequacy. Kaiser is remembered as an eccentric who spray-painted his shoes in unusual colors and listed ES (Eagle Scout) as his highest degree.

References

Bentler, P. M. 1977. Factor simplicity index and transformations. *Psychometrika* 42: 277–295.

Comrey, A. L. 1967. Tandem criteria for analytic rotation in factor analysis. *Psychometrika* 32: 277–295.

Crawford, C. B., and G. A. Ferguson. 1970. A general rotation criterion and its use in orthogonal rotation. *Psychometrika* 35: 321–332.

Gorsuch, R. L. 1983. *Factor Analysis*. 2nd ed. Hillsdale, NJ: Lawrence Erlbaum.

Harman, H. H. 1976. *Modern Factor Analysis*. 3rd ed. Chicago: University of Chicago Press.

Hendrickson, A. E., and P. O. White. 1964. Promax: A quick method for rotation to oblique simple structure. *British Journal of Statistical Psychology* 17: 65–70.

Horst, P. 1965. *Factor Analysis of Data Matrices*. New York: Holt, Rinehart & Winston.

Jennrich, R. I. 1979. Admissible values of γ in direct oblimin rotation. *Psychometrika* 44: 173–177.

——. 2004. Rotation to simple loadings using component loss functions: The orthogonal case. *Psychometrika* 69: 257–273.

Jensen, A. R., and M. Wilson. 1994. Henry Felix Kaiser (1927–1992). *American Psychologist* 49: 1085.

Kaiser, H. F. 1958. The varimax criterion for analytic rotation in factor analysis. *Psychometrika* 23: 187–200.

Mulaik, S. A. 1992. Henry Felix Kaiser 1927–1992. *Multivariate Behavioral Research* 27: 159–171.

Thurstone, L. L. 1935. *The Vectors of Mind: Multiple-Factor Analysis for the Isolation of Primary Traits*. Chicago: University of Chicago Press.

Also see *References* in [MV] **rotatemat**.

Also see

[MV] **factor** — Factor analysis

[MV] **factor postestimation** — Postestimation tools for factor and factormat

[MV] **pca** — Principal component analysis

[MV] **pca postestimation** — Postestimation tools for pca and pcamat

[MV] **procrustes** — Procrustes transformation

[MV] **rotatemat** — Orthogonal and oblique rotations of a Stata matrix

Title

rotatemat — Orthogonal and oblique rotations of a Stata matrix

Syntax

`rotatemat` *matrix_L* [, *options*]

options	Description
Main	
`orthogonal`	restrict to orthogonal rotations; the default, except with `promax()`
`oblique`	allow oblique rotations
rotation_methods	rotation criterion
`normalize`	rotate Kaiser normalized matrix
Reporting	
`format(%`*fmt*`)`	display format for matrices; default is `format(%9.5f)`
`blanks(`*#*`)`	display loadings as blanks when $\lvert\text{loading}\rvert < \#$; default is `blanks(0)`
`nodisplay`	suppress all output except log and trace
`noloading`	suppress display of rotated loadings
`norotation`	suppress display of rotation matrix
`matname(`*string*`)`	descriptive label of the matrix to be rotated
`colnames(`*string*`)`	descriptive name for columns of the matrix to be rotated
Optimization	
optimize_options	control the optimization process; seldom used

rotation_methods	Description
*varimax	varimax (orthogonal only); the default
vgpf	varimax via the GPF algorithm (orthogonal only)
quartimax	quartimax (orthogonal only)
equamax	equamax (orthogonal only)
parsimax	parsimax (orthogonal only)
entropy	minimum entropy (orthogonal only)
tandem1	Comrey's tandem 1 principle (orthogonal only)
tandem2	Comrey's tandem 2 principle (orthogonal only)
*promax[(#)]	promax power # (implies oblique); default is promax(3)
oblimin[(#)]	oblimin with $\gamma = \#$; default is oblimin(0)
cf(#)	Crawford–Ferguson family with $\kappa = \#$, $0 \leq \# \leq 1$
bentler	Bentler's invariant pattern simplicity
oblimax	oblimax
quartimin	quartimin
target(*Tg*)	rotate toward matrix *Tg*
partial(*Tg W*)	rotate toward matrix *Tg*, weighted by matrix *W*

* varimax and promax ignore all *optimize_options*.

Menu

Statistics > Multivariate analysis > Orthogonal and oblique rotations of a matrix

Description

rotatemat applies a linear transformation $\mathbf{T}$ to the matrix *matrix_L*, which we will call $\mathbf{A}$, so that the result $c(\mathbf{A}(\mathbf{T}')^{-1})$ minimizes some criterion function $c()$ over all matrices $\mathbf{T}$ in a class of feasible transformations. Two classes are supported: orthogonal (orthonormal) and oblique. Orthonormal rotations comprise all orthonormal matrices $\mathbf{T}$, such that $\mathbf{T}'\mathbf{T} = \mathbf{T}\mathbf{T}' = \mathbf{I}$; here $\mathbf{A}(\mathbf{T}')^{-1}$ simplifies to $\mathbf{AT}$. Oblique rotations are characterized by $\text{diag}(\mathbf{T}'\mathbf{T}) = \mathbf{1}$. A wide variety of criteria $c()$ is available, representing different ways to measure the "simplicity" of a matrix. Most of these criteria can be applied with both orthogonal and oblique rotations.

If you are interested in rotation after factor, factormat, pca, or pcamat, see [MV] **factor postestimation**, [MV] **pca postestimation**, and the general description of rotate as a postestimation facility in [MV] **rotate**.

This entry describes the computation engine for orthogonal and oblique transformations of Stata matrices. This command may be used directly on any Stata matrix.

Options

Main

orthogonal specifies that an orthogonal rotation be applied. This is the default.

See *Rotation criteria* below for details on the *rotation_methods* available with orthogonal.

`oblique` specifies that an oblique rotation be applied. This often yields more interpretable factors with a simpler structure than that obtained with an orthogonal rotation. In many applications (for example, after `factor` and `pca`), the factors before rotation are orthogonal (uncorrelated), whereas the oblique rotated factors are correlated.

See *Rotation criteria* below for details on the *rotation_methods* available with `oblique`.

`normalize` requests that the rotation be applied to the Kaiser normalization (Horst 1965) of the matrix **A** so that the rowwise sums of squares equal 1.

Reporting

`format(%`*fmt*`)` specifies the display format for matrices. The default is `format(%9.5f)`.

`blanks(`*#*`)` specifies that small values of the rotated matrix—that is, those elements of $\mathbf{A}(\mathbf{T}')^{-1}$ that are less than # in absolute value—are displayed as spaces.

`nodisplay` suppresses all output except the log and trace.

`noloading` suppresses the display of the rotated loadings.

`norotation` suppresses the display of the optimal rotation matrix.

`matname(`*string*`)` is a rarely used output option; it specifies a descriptive label of the matrix to be rotated.

`colnames(`*string*`)` is a rarely used output option; it specifies a descriptive name to refer to the columns of the matrix to be rotated. For instance, `colnames(components)` specifies that the output label the columns as "components". The default is "factors".

Optimization

optimize_options control the iterative optimization process. These options are seldom used.

`iterate(`*#*`)` is a rarely used option; it specifies the maximum number of iterations. The default is `iterate(1000)`.

`log` specifies that an iteration log be displayed.

`trace` is a rarely used option; it specifies that the rotation be displayed at each iteration.

`tolerance(`*#*`)` is one of three criteria for declaring convergence and is rarely used. The `tolerance()` convergence criterion is satisfied when the relative change in the rotation matrix **T** from one iteration to the next is less than or equal to #. The default is `tolerance(1e-6)`.

`gtolerance(`*#*`)` is one of three criteria for declaring convergence and is rarely used. The `gtolerance()` convergence criterion is satisfied when the Frobenius norm of the gradient of the criterion function $c()$ projected on the manifold of orthogonal matrices or of normal matrices is less than or equal to #. The default is `gtolerance(1e-6)`.

`ltolerance(`*#*`)` is one of three criteria for declaring convergence and is rarely used. The `ltolerance()` convergence criterion is satisfied when the relative change in the minimization criterion $c()$ from one iteration to the next is less than or equal to #. The default is `ltolerance(1e-6)`.

`protect(`*#*`)` requests that # optimizations with random starting values be performed and that the best of the solutions be reported. The output also indicates whether all starting values converged to the same solution. When specified with a large number, such as `protect(50)`, this provides reasonable assurance that the solution found is the global maximum and not just a local maximum. If `trace` is also specified, the rotation matrix and rotation criterion value of each optimization will be reported.

maxstep(*#*) is a rarely used option; it specifies the maximum number of step-size halvings. The default is maxstep(20).

init(*matname*) is a rarely used option; it specifies the initial rotation matrix. *matname* should be square and regular (nonsingular) and have the same number of columns as the matrix *matrix_L* to be rotated. It should be orthogonal ($\mathbf{T'T} = \mathbf{TT'} = \mathbf{I}$) or normal ($\text{diag}(\mathbf{T'T}) = \mathbf{1}$), depending on whether orthogonal or oblique rotations are performed. init() cannot be combined with random. If neither init() nor random is specified, the identity matrix is used as the initial rotation.

random is a rarely used option; it specifies that a random orthogonal or random normal matrix be used as the initial rotation matrix. random cannot be combined with init(). If neither init() nor random is specified, the identity matrix is used as the initial rotation.

Rotation criteria

In the descriptions below, the matrix to be rotated is denoted as $\mathbf{A}$, p denotes the number of rows of $\mathbf{A}$, and f denotes the number of columns of $\mathbf{A}$ (factors or components). If $\mathbf{A}$ is a loading matrix from factor or pca, p is the number of variables and f is the number of factors or components.

Criteria suitable only for orthogonal rotations

varimax and vgpf apply the orthogonal varimax rotation (Kaiser 1958). varimax maximizes the variance of the squared loadings within factors (columns of $\mathbf{A}$). It is equivalent to cf(*1/p*) and to oblimin(1). varimax, the most popular rotation, is implemented with a dedicated fast algorithm and ignores all *optimize_options*. Specify vgpf to switch to the general GPF algorithm used for the other criteria.

quartimax uses the quartimax criterion (Harman 1976). quartimax maximizes the variance of the squared loadings within the variables (rows of $\mathbf{A}$). For orthogonal rotations, quartimax is equivalent to cf(0) and to oblimax.

equamax specifies the orthogonal equamax rotation. equamax maximizes a weighted sum of the varimax and quartimax criteria, reflecting a concern for simple structure within variables (rows of $\mathbf{A}$) as well as within factors (columns of $\mathbf{A}$). equamax is equivalent to oblimin(*p/2*) and cf(*#*), where $\# = f/(2p)$.

parsimax specifies the orthogonal parsimax rotation. parsimax is equivalent to cf(*#*), where $\# = (f-1)/(p+f-2)$.

entropy applies the minimum entropy rotation criterion (Jennrich 2004).

tandem1 specifies that the first principle of Comrey's tandem be applied. According to Comrey (1967), this principle should be used to judge which "small" factors be dropped.

tandem2 specifies that the second principle of Comrey's tandem be applied. According to Comrey (1967), tandem2 should be used for "polishing".

Criteria suitable only for oblique rotations

promax[(*#*)] specifies the oblique promax rotation. The optional argument specifies the promax power. Not specifying the argument is equivalent to specifying promax(3). Values less than 4 are recommended, but the choice is yours. Larger promax powers simplify the loadings (generate numbers closer to zero and one) but at the cost of additional correlation between factors. Choosing a value is a matter of trial and error, but most sources find values in excess of 4 undesirable in practice. The power must be greater than 1 but is not restricted to integers.

Promax rotation is an oblique rotation method that was developed before the "analytical methods" (based on criterion optimization) became computationally feasible. Promax rotation comprises an oblique Procrustean rotation of the original loadings $\mathbf{A}$ toward the elementwise #-power of the orthogonal varimax rotation of $\mathbf{A}$.

Criteria suitable for orthogonal and oblique rotations

`oblimin`[`(`*#*`)`] specifies that the oblimin criterion with $\gamma = \#$ be used. When restricted to orthogonal transformations, the `oblimin()` family is equivalent to the orthomax criterion function. Special cases of `oblimin()` include

γ	Special case
0	quartimax / quartimin
1/2	biquartimax / biquartimin
1	varimax / covarimin
$p/2$	equamax

p = number of rows of $\mathbf{A}$.

γ defaults to zero. Jennrich (1979) recommends $\gamma \leq 0$ for oblique rotations. For $\gamma > 0$, it is possible that optimal oblique rotations do not exist; the iterative procedure used to compute the solution will wander off to a degenerate solution.

`cf(`*#*`)` specifies that a criterion from the Crawford–Ferguson (1970) family be used with $\kappa = \#$. `cf(`κ`)` can be seen as $(1-\kappa)\text{cf}_1(\mathbf{A}) + (\kappa)\text{cf}_2(\mathbf{A})$, where $\text{cf}_1(\mathbf{A})$ is a measure of row parsimony and $\text{cf}_2(\mathbf{A})$ is a measure of column parsimony. $\text{cf}_1(\mathbf{A})$ attains its greatest lower bound when no row of $\mathbf{A}$ has more than one nonzero element, whereas $\text{cf}_2(\mathbf{A})$ reaches zero if no column of $\mathbf{A}$ has more than one nonzero element.

For orthogonal rotations, the Crawford–Ferguson family is equivalent to the `oblimin()` family. For orthogonal rotations, special cases include the following:

κ	Special case
0	quartimax / quartimin
$1/p$	varimax / covarimin
$f/(2p)$	equamax
$(f-1)/(p+f-2)$	parsimax
1	factor parsimony

p = number of rows of $\mathbf{A}$.
f = number of columns of $\mathbf{A}$.

`bentler` specifies that the "invariant pattern simplicity" criterion (Bentler 1977) be used.

`oblimax` specifies the oblimax criterion, which maximizes the number of high and low loadings. `oblimax` is equivalent to `quartimax` for orthogonal rotations.

`quartimin` specifies that the quartimin criterion be used. For orthogonal rotations, `quartimin` is equivalent to `quartimax`.

`target(`*Tg*`)` specifies that $\mathbf{A}$ be rotated as near as possible to the conformable matrix *Tg*. Nearness is expressed by the Frobenius matrix norm.

`partial(`*Tg W*`)` specifies that $\mathbf{A}$ be rotated as near as possible to the conformable matrix *Tg*. Nearness is expressed by a weighted (by *W*) Frobenius matrix norm. *W* should be nonnegative and usually is zero–one valued, with ones identifying the target values to be reproduced as closely as possible by the factor loadings, whereas zeros identify loadings to remain unrestricted.

Remarks

Remarks are presented under the following headings:

Introduction
Orthogonal rotations
Oblique rotations
Promax rotation

Introduction

For an introduction to rotation, see Harman (1976) and Gorsuch (1983).

All supported rotation criteria are invariant with respect to permutations of the columns and change of signs of the columns. `rotatemat` returns the solution with positive column sums and with columns sorted by the L2 norm; columns are ordered with respect to the L1 norm if the columns have the same L2 norm.

A factor analysis of 24 psychological tests on 145 seventh- and eighth-grade school children with four retained factors is used for illustration. Factors were extracted with maximum likelihood. The loadings are reported by Harman (1976). We enter the factor loadings as a Stata matrix with 24 rows and four columns. For more information, we add full descriptive labels as comments and short labels as row names.

```
. matrix input L = (
    601   019   388   221 \          Visual perception
    372  -025   252   132 \          Cubes
    413  -117   388   144 \          Paper form board
    487  -100   254   192 \          Flags
    691  -304  -279   035 \          General information
    690  -409  -200  -076 \          Paragraph comprehension
    677  -409  -292   084 \          Sentence completion
    674  -189  -099   122 \          Word classification
    697  -454  -212  -080 \          Word meaning
    476   534  -486   092 \          Addition
    558   332  -142  -090 \          Code
    472   508  -139   256 \          Counting dots
    602   244   028   295 \          Straight-curved capitals
    423   058   015  -415 \          Word recognition
    394   089   097  -362 \          Number recognition
    510   095   347  -249 \          Figure recognition
    466   197  -004  -381 \          Object-number
    515   312   152  -147 \          Number-figure
    443   089   109  -150 \          Figure-word
    614  -118   126  -038 \          Deduction
    589   227   057   123 \          Numerical puzzles
    608  -107   127  -038 \          Problem reasoning
    687  -044   138   098 \          Series completion
    651   177  -212  -017 )          Arithmetic problems

. matrix colnames L = F1 F2 F3 F4

. matrix rownames L = visual    cubes     board
                      flags     general   paragraph
                      sentence  wordclas  wordmean
                      add       code      dots
                      capitals  wordrec   numbrec
                      figrec    obj-num   num-fig
                      fig-word  deduct    numpuzz
                      reason    series    arith

. matrix L = L/1000
```

Thus using `rotatemat`, we can study various rotations of `L` without access to the full data or the correlation matrix.

Orthogonal rotations

We can rotate the matrix `L` according to an extensive list of criteria, including orthogonal rotations.

▷ Example 1

The default rotation, orthogonal varimax, is probably the most popular method:

```
. rotatemat L, format(%6.3f)
Rotation of L[24,4]

    Criterion                  varimax
    Rotation class             orthogonal
    Kaiser normalization       off

Rotated factors
```

	F1	F2	F3	F4
visual	0.247	0.151	0.679	0.128
cubes	0.171	0.060	0.425	0.078
board	0.206	-0.049	0.549	0.097
flags	0.295	0.068	0.504	0.050
general	0.765	0.214	0.117	0.067
paragraph	0.802	0.074	0.122	0.160
sentence	0.826	0.148	0.117	-0.008
wordclas	0.612	0.230	0.290	0.061
wordmean	0.840	0.049	0.112	0.152
add	0.166	0.846	-0.076	0.082
code	0.222	0.533	0.134	0.313
dots	0.048	0.705	0.257	0.025
capitals	0.240	0.500	0.450	0.020
wordrec	0.249	0.124	0.032	0.526
numbrec	0.178	0.109	0.106	0.499
figrec	0.158	0.076	0.401	0.510
obj-num	0.197	0.262	0.060	0.539
num-fig	0.096	0.352	0.311	0.422
fig-word	0.204	0.175	0.232	0.336
deduct	0.443	0.115	0.365	0.255
numpuzz	0.233	0.428	0.389	0.169
reason	0.432	0.120	0.363	0.256
series	0.440	0.228	0.472	0.184
arith	0.409	0.509	0.150	0.228

```
Orthogonal rotation
```

	F1	F2	F3	F4
F1	0.677	0.438	0.475	0.352
F2	-0.632	0.737	0.049	0.232
F3	-0.376	-0.458	0.760	0.268
F4	-0.011	0.234	0.441	-0.866

◁

The varimax rotation $\mathbf{T}$ of $\mathbf{A}$ maximizes the (raw) varimax criterion over all orthogonal $\mathbf{T}$, which for $p \times f$ matrices is defined as (Harman 1976)

$$c_{\text{varimax}}(\mathbf{A}) = \frac{1}{p}\sum_{j=1}^{f}\left\{\left(\sum_{i=1}^{p} A_{ij}^4\right) - \frac{1}{p}\left(\sum_{i=1}^{p} A_{ij}^2\right)^2\right\}$$

The criterion $c_{\text{varimax}}(\mathbf{A})$ can be interpreted as the sum over the columns of the variances of the squares of the loadings A_{ij}. A column with large variance will typically consist of many small values and a few large values. Achieving such "simple" columnwise distributions is often helpful for interpretation.

❑ Technical note

The raw varimax criterion as defined here has been criticized because it weights variables by the size of their loadings, that is, by their communalities. This is often not desirable. A common rotation strategy is to weight all rows equally by rescaling to the same rowwise sum of squared loadings. This is known as the Kaiser normalization. You may request this normalized solution with the `normalize` option. The default in `rotatemat` and in `rotate` (see [MV] **rotate**) is not to normalize.

❑

Many other criteria for the rotation of matrices have been proposed and studied in the literature. Most of these criteria can be stated in terms of a "simplicity function". For instance, quartimax rotation (Carroll 1953) seeks to achieve interpretation within rows—in a factor analytic setup, this means that variables should have a high loading on a few factors and a low loading on the other factors. The quartimax criterion is defined as (Harman 1976)

$$c_{\text{quartimax}}(\mathbf{A}) = \left(\frac{1}{pf}\sum_{i=1}^{p}\sum_{j=1}^{f} A_{ij}^4\right) - \left(\frac{1}{pf}\sum_{i=1}^{p}\sum_{j=1}^{f} A_{ij}^2\right)^2$$

▷ Example 2

We display the quartimax solution, use blanks to represent loadings with absolute values smaller than 0.3, and suppress the display of the rotation matrix.

```
. rotatemat L, quartimax format(%6.3f) norotation blanks(0.3)
Rotation of L[24,4]

    Criterion                 quartimax
    Rotation class            orthogonal
    Kaiser normalization      off
    Criterion value           -1.032898
    Number of iterations      35

Rotated factors (blanks represent abs()<.3)
```

	F1	F2	F3	F4
visual	0.374		0.630	
cubes			0.393	
board			0.513	
flags	0.379		0.450	
general	0.791			
paragraph	0.827			
sentence	0.838			
wordclas	0.669			
wordmean	0.860			
add		0.829		
code	0.316	0.521		
dots		0.701		
capitals	0.348	0.482	0.393	
wordrec	0.316			0.492
numbrec				0.469
figrec			0.382	0.470
obj-num				0.503
num-fig		0.357		0.383
fig-word				
deduct	0.528			
numpuzz	0.342	0.414	0.340	
reason	0.517			
series	0.543		0.395	
arith	0.490	0.478		

◁

Some of the criteria supported by `rotatemat` are defined as one-parameter families. The oblimin(γ) criterion and the Crawford and Ferguson cf(κ) criterion families contain the varimax and quartimax criteria as special cases; that is, they can be obtained by certain values of γ and κ, respectively. Intermediate parameter values provide compromises between varimax's aim of column simplification and quartimax's aim of row simplification. Varimax and quartimax are equivalent to oblimin(1) and oblimin(0), respectively. A compromise, oblimin(0.5), is also known as biquartimax.

▷ Example 3

Because the varimax and quartimax solutions are so close for our matrix L, the biquartimax compromise will also be rather close.

```
. rotatemat L, oblimin(0.5) format(%6.3f) norotation
  (output omitted)
```

◁

❑ Technical note

You may have noticed a difference between the output of rotatemat in the default case or equivalently when we type

```
. rotatemat L, varimax
```

and in other cases. In the default case, no mention is made of the criterion value and the number of iterations. rotatemat uses a fast special algorithm for this most common case, whereas for other rotations it uses a general gradient projection algorithm (GPF) proposed by Jennrich (2001, 2002); see also Bernaards and Jennrich (2005). The general algorithm is used to obtain the varimax rotation if you specify the option vgpf rather than varimax.

❑

The rotations we have illustrated are orthogonal—the lengths of the rows and the angles between the rows are not affected by the rotations. We may verify—we do not show this in the manual to conserve paper—that after an orthogonal rotation of L

```
. matlist L*L'
```

and

```
. matlist r(AT)*r(AT)'
```

return the same 24 by 24 matrix, whereas

```
. matlist r(T)*r(T)'
```

and

```
. matlist r(T)'*r(T)
```

both return a 2×2 identity matrix. rotatemat returns in r(AT) the rotated matrix and in r(T) the rotation matrix.

Oblique rotations

rotatemat provides a second class of rotations: oblique rotations. These rotations maintain the norms of the rows of the matrix but not their inner products. In geometric terms, interpreting the rows of the matrix to be rotated as vectors, both the orthogonal and the oblique rotations maintain the lengths of the vectors. Under orthogonal transformations, the angles between the vectors are also left unchanged—these transformations comprise true reorientations in space and reflections. Oblique rotations do not conserve angles between vectors. If the vectors are orthogonal before rotations—as will be the case if we are rotating factor or component loading matrices—this will no longer be the case after the rotation. The "freedom" to select angles between the rows allows oblique rotations to generate simpler loading structures than the orthogonal rotations—sometimes much simpler. In a factor analytic setting, the disadvantage is, however, that the rotated factors are correlated.

rotatemat can obtain oblique rotations for most of the criteria that are available for orthogonal rotations; some of the criteria (such as the entropy criterion) are available only for the orthogonal case.

▷ Example 4

We illustrate with the psychological tests matrix L and apply the oblique oblimin criterion.

```
. rotatemat L, oblimin oblique format(%6.3f) blanks(0.3)
Rotation of L[24,4]

    Criterion                 oblimin(0)
    Rotation class            oblique
    Kaiser normalization      off
    Criterion value           .1957363
    Number of iterations      78

Rotated factors (blanks represent abs()<.3)
```

	F1	F2	F3	F4
visual		0.686		
cubes		0.430		
board		0.564		
flags		0.507		
general	0.771			
paragraph	0.808			
sentence	0.865			
wordclas	0.560			
wordmean	0.857			
add			0.864	
code			0.460	0.305
dots			0.701	
capitals		0.437	0.442	
wordrec				0.571
numbrec				0.543
figrec		0.314		0.540
obj-num				0.584
num-fig				0.438
fig-word				0.341
deduct	0.325			
numpuzz		0.344	0.347	
reason	0.311			
series		0.417		
arith			0.428	

```
Oblique rotation
```

	F1	F2	F3	F4
F1	0.823	0.715	0.584	0.699
F2	-0.483	0.019	0.651	0.213
F3	-0.299	0.587	-0.435	0.207
F4	-0.006	0.379	0.213	-0.651

The option `oblique` requested an oblique rotation rather than the default `orthogonal`. You may verify that `r(AT)` equals `L * inv(r(T)')` within reasonable roundoff with

```
. matlist r(AT) - L * inv(r(T)')
  (output omitted)
```

The correlation between the rotated dimensions is easily obtained.

```
. matlist r(T)' * r(T)

             |        F1         F2         F3         F4
-------------+--------------------------------------------
          F1 |         1
          F2 |  .4026978          1
          F3 |   .294928   .2555824          1
          F4 |  .4146879   .3784689   .3183115          1
```

◁

Promax rotation

`rotatemat` also offers promax rotation.

▷ Example 5

We use the matrix L to illustrate promax rotation.

```
. rotatemat L, promax blanks(0.3) format(%6.3f)
Rotation of L[24,4]

    Criterion                promax(3)
    Rotation class           oblique
    Kaiser normalization     off

Rotated factors (blanks represent abs()<.3)

    ----------------------------------------------
                 |    F1      F2      F3      F4
    -------------+--------------------------------
          visual |          0.775
           cubes |          0.487
           board |          0.647
           flags |          0.572
         general | 0.786
       paragraph | 0.825
        sentence | 0.888
        wordclas | 0.543
        wordmean | 0.878
             add |                  0.921
            code |                  0.466
            dots |                  0.728
        capitals |          0.468   0.441
         wordrec |                          0.606
        numbrec  |                          0.570
          figrec |          0.364           0.539
         obj-num |                          0.610
         num-fig |                          0.425
        fig-word |                          0.337
          deduct |          0.323
         numpuzz |          0.369   0.336
          reason |          0.322
          series |          0.462
           arith |                  0.436
    ----------------------------------------------
```

```
Oblique rotation
```

	F1	F2	F3	F4
F1	0.841	0.829	0.663	0.743
F2	-0.462	0.020	0.614	0.215
F3	-0.282	0.478	-0.386	0.159
F4	-0.012	0.290	0.184	-0.614

The correlation between the rotated dimensions can be obtained as

```
. matlist r(T)' * r(T)
```

	F1	F2	F3	F4
F1	1			
F2	.5491588	1		
F3	.3807942	.4302401	1	
F4	.4877064	.5178414	.4505817	1

◁

Saved results

`rotatemat` saves the following in `r()`:

Scalars

`r(f)`	criterion value
`r(iter)`	number of GPF iterations
`r(rc)`	return code
`r(nnconv)`	number of nonconvergent trials; `protect()` only

Macros

`r(cmd)`	`rotatemat`
`r(ctitle)`	descriptive label of rotation method
`r(ctitle12)`	version of `r(ctitle)` at most 12 characters long
`r(criterion)`	criterion name (e.g., `oblimin`)
`r(class)`	`orthogonal` or `oblique`
`r(normalization)`	`kaiser` or `none`
`r(carg)`	criterion argument

Matrices

`r(T)`	optimal transformation $\mathbf{T}$
`r(AT)`	optimal $\mathbf{AT} = \mathbf{A}(\mathbf{T}')^{-1}$
`r(fmin)`	minimums found; `protect()` only

Methods and formulas

`rotatemat` is implemented as an ado-file.

`rotatemat` minimizes a scalar-valued criterion function $c(\mathbf{AT})$ with respect to the set of orthogonal matrices $\mathbf{T}'\mathbf{T} = \mathbf{I}$, or $c(\mathbf{A}(\mathbf{T}')^{-1})$ with respect to the normal matrix, $\text{diag}(\mathbf{T}'\mathbf{T}) = \mathbf{1}$. For orthonormal $\mathbf{T}$, $\mathbf{T} = (\mathbf{T}')^{-1}$.

The rotation criteria can be conveniently written in terms of scalar-valued functions; see Bernaards and Jennrich (2005). Define the inner product $\langle \mathbf{A}, \mathbf{B} \rangle = \text{trace}(\mathbf{A}'\mathbf{B})$. $|\mathbf{A}| = \sqrt{\langle \mathbf{A}, \mathbf{A} \rangle}$ is called the Frobenius norm of the matrix $\mathbf{A}$. Let $\boldsymbol{\Lambda}$ be a $p \times k$ matrix. Denote by $\mathbf{X}^2$ the direct product $\mathbf{X} \cdot \mathbf{X}$. See Harman (1976) for information on many of the rotation criteria and references to the authors originally proposing the criteria. Sometimes we list an alternative reference. Our notation is similar to that of Bernaards and Jennrich (2005).

`rotatemat` uses the iterative "gradient projection algorithm" (Jennrich 2001, 2002) for the optimization of the criterion over the permissible transformations. Different versions are provided for optimal orthogonal and oblique rotations; see Bernaards and Jennrich (2005).

Varimax (orthogonal only)

Varimax is equivalent to oblimin with $\gamma = 1$ or to the Crawford–Ferguson family with $\kappa = 1/p$; see below.

Quartimax (orthogonal only)

$$c(\mathbf{\Lambda}) = \sum_i \sum_r \lambda_{ir}^4 = -\frac{1}{4} \left\langle \mathbf{\Lambda}^2, \mathbf{\Lambda}^2 \right\rangle$$

Equamax (orthogonal only)

Equamax is equivalent to oblimin with $\gamma = p/2$ or to the Crawford–Ferguson family with $\kappa = f/(2p)$; see below.

Parsimax (orthogonal only)

Parsimax is equivalent to the Crawford–Ferguson family with $\kappa = (f-1)/(p+f-2)$; see below.

Entropy (orthogonal only); see Jennrich (2004)

$$c(\mathbf{\Lambda}) = -\frac{1}{2} \left\langle \mathbf{\Lambda}^2, \log \mathbf{\Lambda}^2 \right\rangle$$

Tandem principal 1 (orthogonal only); see Comrey (1967)

$$c(\mathbf{\Lambda}) = -\left\langle \mathbf{\Lambda}^2, (\mathbf{\Lambda}\mathbf{\Lambda}')^2 \mathbf{\Lambda}^2 \right\rangle$$

Tandem principal 2 (orthogonal only); see Comrey (1967)

$$c(\mathbf{\Lambda}) = \left\langle \mathbf{\Lambda}^2, \{\mathbf{1}\mathbf{1}' - (\mathbf{\Lambda}\mathbf{\Lambda}')^2\} \mathbf{\Lambda}^2 \right\rangle$$

Promax (oblique only)

Promax does not fit in the maximizing-of-a-simplicity-criterion framework that is at the core of `rotatemat`. The promax method (Hendrickson and White 1964) was proposed before computing power became widely available. The promax rotation comprises three steps:

1. Perform an orthogonal rotation on $\mathbf{A}$; `rotatemat` uses varimax.
2. Raise the elements of the rotated matrix to some power, preserving the signs of the elements. Typically, the power is taken from the range [2,4]. This operation is meant to distinguish more clearly between small and large values.
3. The matrix from step 2 is used as the target for an oblique Procrustean rotation from the original matrix $\mathbf{A}$. The method to compute this rotation in promax is different from the method in the `procrustes` command (see [MV] **procrustes**). The latter produces the real least-squares oblique rotation; promax uses an approximation.

Oblimin; see Jennrich (1979)

$$c(\mathbf{\Lambda}) = \frac{1}{4}\left\langle \mathbf{\Lambda}^2, \{\mathbf{I} - (\gamma/p)\mathbf{1}\mathbf{1}'\}\mathbf{\Lambda}^2(\mathbf{1}\mathbf{1}' - \mathbf{I})\right\rangle$$

Orthomax and oblimin are equivalent when restricted to orthogonal rotations. Special cases of `oblimin()` include the following:

γ	Special case
0	quartimin
$1/2$	biquartimin
$p/2$	equamax
1	varimax

Crawford and Ferguson (1970) family

$$c(\mathbf{\Lambda}) = \frac{1-\kappa}{4}\left\langle \mathbf{\Lambda}^2, \mathbf{\Lambda}^2(\mathbf{1}\mathbf{1}' - \mathbf{I})\right\rangle + \frac{\kappa}{4}\left\langle \mathbf{\Lambda}^2, (\mathbf{1}\mathbf{1}' - \mathbf{I})\mathbf{\Lambda}^2\right\rangle$$

When restricted to orthogonal transformations, `cf()` and `oblimin()` are in fact equivalent. Special cases of `cf()` include the following:

κ	Special case
0	quartimax
$1/p$	varimax
$f/(2p)$	equamax
$(f-1)/(p+f-2)$	parsimax
1	factor parsimony

Bentler's invariant pattern simplicity; see Bentler (1977)

$$c(\mathbf{\Lambda}) = \log[\det\{(\mathbf{\Lambda}^2)'\mathbf{\Lambda}^2\}] - \log(\det[\operatorname{diag}\{(\mathbf{\Lambda}^2)'\mathbf{\Lambda}^2\}])$$

Oblimax

$$c(\mathbf{\Lambda}) = -\log(\langle \mathbf{\Lambda}^2, \mathbf{\Lambda}^2 \rangle) + 2\log(\langle \mathbf{\Lambda}, \mathbf{\Lambda} \rangle)$$

For orthogonal transformations, oblimax is equivalent to quartimax; see above.

Quartimin

$$c(\mathbf{\Lambda}) = \sum_{r \neq s} \sum_i \lambda_{ir}^2 \lambda_{is}^2 = -\frac{1}{4} \langle \mathbf{\Lambda}^2, \mathbf{\Lambda}^2(\mathbf{11}' - \mathbf{I}) \rangle$$

Target

$$c(\mathbf{\Lambda}) = \frac{1}{2} |\mathbf{\Lambda} - \mathbf{H}|^2$$

for given target matrix $\mathbf{H}$.

Partially specified target

$$c(\mathbf{\Lambda}) = |\mathbf{W} \cdot (\mathbf{\Lambda} - \mathbf{H})|^2$$

for given target matrix $\mathbf{H}$, nonnegative weighting matrix $\mathbf{W}$ (usually zero–one valued) and with $\cdot$ denoting the direct product.

References

Bentler, P. M. 1977. Factor simplicity index and transformations. *Psychometrika* 42: 277–295.

Bernaards, C. A., and R. I. Jennrich. 2005. Gradient projection algorithms and software for arbitrary rotation criteria in factor analysis. *Educational and Psychological Measurement* 65: 676–696.

Carroll, J. B. 1953. An analytical solution for approximating simple structure in factor analysis. *Psychometrika* 18: 23–38.

Comrey, A. L. 1967. Tandem criteria for analytic rotation in factor analysis. *Psychometrika* 32: 277–295.

Crawford, C. B., and G. A. Ferguson. 1970. A general rotation criterion and its use in orthogonal rotation. *Psychometrika* 35: 321–332.

Gorsuch, R. L. 1983. *Factor Analysis*. 2nd ed. Hillsdale, NJ: Lawrence Erlbaum.

Harman, H. H. 1976. *Modern Factor Analysis*. 3rd ed. Chicago: University of Chicago Press.

Hendrickson, A. E., and P. O. White. 1964. Promax: A quick method for rotation to oblique simple structure. *British Journal of Statistical Psychology* 17: 65–70.

Horst, P. 1965. *Factor Analysis of Data Matrices*. New York: Holt, Rinehart & Winston.

Jennrich, R. I. 1979. Admissible values of γ in direct oblimin rotation. *Psychometrika* 44: 173–177.

——. 2001. A simple general procedure for orthogonal rotation. *Psychometrika* 66: 289–306.

——. 2002. A simple general method for oblique rotation. *Psychometrika* 67: 7–20.

——. 2004. Rotation to simple loadings using component loss functions: The orthogonal case. *Psychometrika* 69: 257–273.

Kaiser, H. F. 1958. The varimax criterion for analytic rotation in factor analysis. *Psychometrika* 23: 187–200.

Also see

[MV] **rotate** — Orthogonal and oblique rotations after factor and pca

[MV] **procrustes** — Procrustes transformation

Title

scoreplot — Score and loading plots

Syntax

Plot score variables

scoreplot [*if*] [*in*] [, *scoreplot_options*]

Plot the loadings (factors, components, or discriminant functions)

loadingplot [, *loadingplot_options*]

scoreplot_options	Description
Main	
factors(#)	number of factors/scores to be plotted; default is factors(2)
components(#)	synonym for factors()
norotated	use unrotated factors or scores, even if rotated results exist
matrix	graph as a matrix plot, available only when factors(2) is specified; default is a scatterplot
combined	graph as a combined plot, available when factors(# > 2); default is a matrix plot
half	graph lower half only; allowed only with matrix
graph_matrix_options	affect the rendition of the matrix graph
combine_options	affect the rendition of the combined graph
scoreopt(*predict_opts*)	options for predict generating score variables
marker_options	change look of markers (color, size, etc.)
marker_label_options	change look or position of marker labels
Y axis, X axis, Titles, Overall	
twoway_options	any options other than by() documented in [G-3] ***twoway_options***

loadingplot_options	Description
Main	
factors(#)	number of factors/scores to be plotted; default is factors(2)
components(#)	synonym for factors()
norotated	use unrotated factors or scores, even if rotated results exist
matrix	graph as a matrix plot, available only when factors(2) is specified; default is a scatterplot
combined	graph as a combined plot, available when factors(# > 2); default is a matrix plot
half	graph lower half only; allowed only with matrix
graph_matrix_options	affect the rendition of the matrix graph
combine_options	affect the rendition of the combined graph
maxlength(#)	abbreviate variable names to # characters; default is maxlength(12)
marker_options	change look of markers (color, size, etc.)
marker_label_options	change look or position of marker labels
Y axis, X axis, Titles, Overall	
twoway_options	any options other than by() documented in [G-3] ***twoway_options***

Menu

scoreplot

Statistics > Multivariate analysis > Factor and principal component analysis > Postestimation > Score variables plot

loadingplot

Statistics > Multivariate analysis > Factor and principal component analysis > Postestimation > Loading plot

Description

scoreplot produces scatterplots of the score variables after factor, factormat, pca, or pcamat, and scatterplots of the discriminant score variables after discrim lda or candisc.

loadingplot produces scatterplots of the loadings (factors or components) after factor, factormat, pca, or pcamat, and the standardized discriminant function loadings after discrim lda or candisc.

Options

Main

factors(#) produces plots for all combinations of score variables up to #. # should not exceed the number of retained factors (components or discriminant functions) and defaults to 2. components() is a synonym. No plot is produced with factors(1).

norotated uses unrotated results, even when rotated results are available. The default is to use rotated results if they are available. norotated is ignored if rotated results are not available.

matrix specifies that plots be produced using graph matrix; see [G-2] **graph matrix**. This is the default when three or more factors are specified. This option may not be used with combined.

`combined` specifies that plots be produced using `graph combine`; see [G-2] **graph combine**. This option may not be used with `matrix`.

`half` specifies that only the lower half of the matrix be graphed. `half` can be specified only with the `matrix` option.

graph_matrix_options affect the rendition of the matrix plot; see [G-2] **graph matrix**.

combine_options affect the rendition of the combined plot; see [G-2] **graph combine**. *combine_options* may not be specified unless `factors()` is greater than 2.

`scoreopt(`*predict_opts*`)`, an option used with `scoreplot`, specifies options for `predict` to generate the score variables. For example, after `factor`, `scoreopt(bartlett)` specifies that Bartlett scoring be applied.

`maxlength(`*#*`)`, an option used with `loadingplot`, causes the variable names (used as point markers) to be abbreviated to # characters. The `abbrev()` function performs the abbreviation, and if # is less than 5, it is treated as 5; see *String functions* in [D] **functions**.

marker_options affect the rendition of markers drawn at the plotted points, including their shape, size, color, and outline; see [G-3] ***marker_options***.

marker_label_options specify if and how the markers are to be labeled; see [G-3] ***marker_label_options***.

Y axis, X axis, Titles, Overall

twoway_options are any of the options documented in [G-3] ***twoway_options***, excluding `by()`. These include options for titling the graph (see [G-3] ***title_options***) and for saving the graph to disk (see [G-3] ***saving_option***).

Remarks

One of the main results from a principal component analysis, factor analysis, or a linear discriminant analysis is a set of eigenvectors that are called components, factors, or linear discriminant functions. These are saved in what is called a loading matrix. `pca`, `pcamat`, `factor`, and `factormat` save the loading matrix in `e(L)`. If there were p variables involved in the PCA or factor analysis, and f components or factors were retained, there will be p rows and f columns in the resulting loading matrix. `discrim lda` and `candisc` save the standardized canonical discriminant function coefficients or loadings in `e(L_std)`.

The columns of the loading matrix are in order of importance. For instance, with PCA, the first column of the loading matrix is the component that accounts for the most variance, the second column accounts for the next most variance, and so on.

In a loading plot, the values from one column of the loading matrix are plotted against the values from another column of the loading matrix. Of most interest is the plot of the first and second columns (the first and second components, factors, or discriminant functions), and this is what `loadingplot` produces by default. The rows of the loading matrix provide the points to be graphed. Variable names are automatically used as the marker labels for these points.

▷ Example 1

We use the Renaissance painters' data introduced in example 2 of [MV] **biplot**. There are four attribute variables recorded for 10 painters. We examine the first two principal component loadings for this dataset.

```
. use http://www.stata-press.com/data/r12/renpainters
(Scores by Roger de Piles for Renaissance Painters)
. pca composition drawing colour expression
  (output omitted)
. loadingplot
```

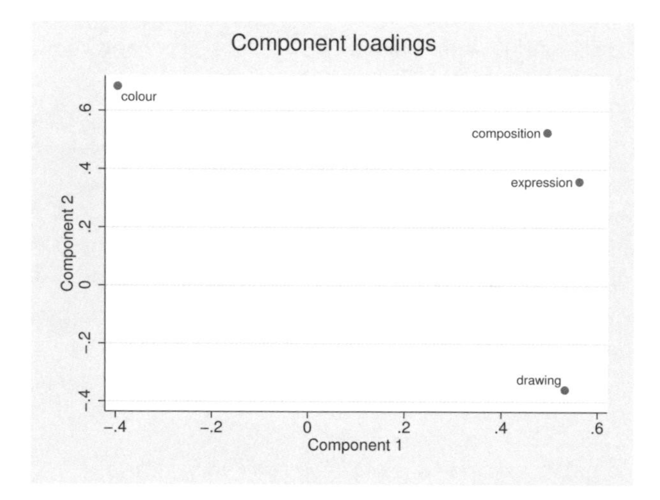

From the first component, we see that color (or colour, if you prefer) is separated from the other three attributes (variables): composition, expression, and drawing. From the second component, the difference in drawing stands out.

◁

Score plots approach the view of the loading matrix from the perspective of the observations. `predict` after `pca` and `factor` produces scores; see [MV] **pca postestimation** and [MV] **factor postestimation**. `predict` after `discrim lda` and `candisc` can request discriminant function scores; see [MV] **discrim lda postestimation**. A score for an observation from a particular column of the loading matrix is obtained as the linear combination of that observation's data by using the coefficients found in the loading. From the raw variables, the linear combinations described in the columns of the loading matrix are applied to generate new component or factor score variables. A score plot graphs one score variable against another. By default, `scoreplot` graphs the scores generated from the first and second columns of the loading matrix (the first two components or factors).

▷ Example 2

We continue with the PCA of the Renaissance painters.

```
. scoreplot
```

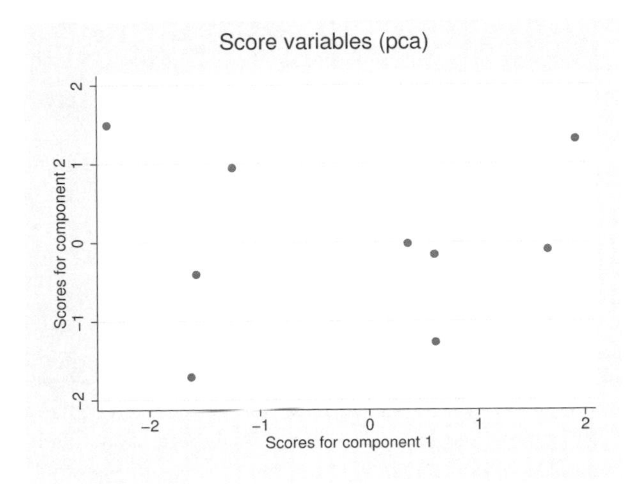

Unlike `loadingplot`, which can use the variable names as labels, `scoreplot` does not automatically know how to label the points after `factor` and `pca`. The graph above is not helpful. The *marker_label_option* `mlabel()` takes care of this.

```
. scoreplot, mlabel(painter) aspect(1) xlabel(-2(1)3) ylabel(-2(1)3)
> title(Renaissance painters)
```

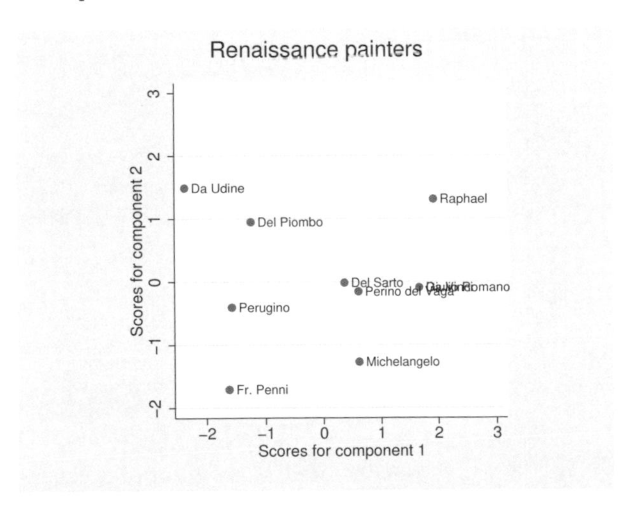

We added a few other options to improve the graph. We extended the axes to include the value 3 so that there would be room for the marker labels, imposed an aspect ratio of 1, and added a title.

The score plot gives us a feeling for the similarities and differences between the painters. Da Udine is in an opposite corner from Michelangelo. The other two corners have Fr. Penni and Raphael.

You could refer to the loading plot and compare it with this score plot to see that the corner with Da Udine is most associated with the `colour` variable, whereas the corner with Michelangelo is best associated with `drawing`. Raphael is in the corner where the variables `composition` and `expression` are predominant.

If you like to make these kinds of associations between the variables and the observations, you will enjoy using the `biplot` command. It provides a joint view of the variables and observations; see [MV] **biplot**.

◁

After a rotation, the rotated factor or component loading matrix is saved in `e(r_L)`. By default, both `loadingplot` and `scoreplot` work with the rotated loadings if they are available. The `norotated` option allows you to obtain the graphs for the unrotated loadings and scores.

You can also request a matrix or combined graph for all combinations of the first several components or factors with the `components()` option (or the alias `factors()`).

▷ Example 3

Even though the results from our initial look at the principal components of the Renaissance painters seem clear enough, we continue our demonstration by showing the score plots for the first three components after a rotation. See [MV] **rotate** for information on rotation in general, and see [MV] **pca postestimation** for specific guidance and warnings concerning rotation after a PCA.

```
. rotate
  (output omitted )
. scoreplot, mlabel(painter) components(3) combined aspect(.8)
> xlabel(-2(1)3) ylabel(-2(1)2)
```

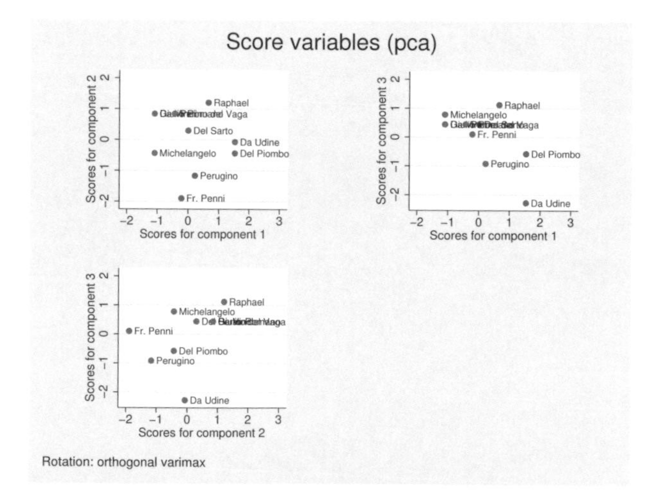

By default, the rotation information was included as a note in the graph. We specified `components(3)` to obtain all paired plots between the first three components. The `combined` option selected the combined view instead of the default matrix view of the graphs. The `aspect()`, `xlabel()`, and `ylabel()` options provide reasonable use of the graphing region while providing the same scale in the horizontal and vertical directions.

◁

As the number of `factors()` or `components()` increases, the graphing area for each plot gets smaller. Although the default matrix view (option `matrix`) may be the most natural, the combined view (option `combined`) displays half as many graphs. However, the combined view uses more space for the labeling of axes than the matrix view. Regardless of the choice, with many requested factors or components, the graphs become too small to be of any use. In `loadingplot`, the `maxlength()`

option will trim the variable name marker labels that are automatically included. This may help reduce overlap when multiple small graphs are shown. You can go further and remove these marker labels by using the `mlabel("")` option.

Other examples of `loadingplot` and `scoreplot` are found in [MV] **pca postestimation**, [MV] **factor postestimation**, [MV] **discrim lda**, and [MV] **discrim lda postestimation**.

Methods and formulas

`scoreplot` and `loadingplot` are implemented as ado-files.

Also see

[MV] **candisc** — Canonical linear discriminant analysis

[MV] **discrim lda** — Linear discriminant analysis

[MV] **discrim lda postestimation** — Postestimation tools for discrim lda

[MV] **factor** — Factor analysis

[MV] **factor postestimation** — Postestimation tools for factor and factormat

[MV] **pca** — Principal component analysis

[MV] **pca postestimation** — Postestimation tools for pca and pcamat

[MV] **screeplot** — Scree plot

Title

screeplot — Scree plot

Syntax

`screeplot` [*eigvals*] [, *options*]

`scree` is a synonym for `screeplot`.

options	Description
Main	
`neigen(#)`	graph only largest # eigenvalues; default is to plot all eigenvalues
Mean	
`mean`	graph horizontal line at the mean of the eigenvalues
`meanlopts(`*cline_options*`)`	affect rendition of the mean line
CI	
`ci`[`(`*ci_options*`)`]	graph confidence intervals (after `pca` only); `ci` is a synonym for `ci(asymptotic)`
Plot	
cline_options	affect rendition of the lines connecting points
marker_options	change look of markers (color, size, etc.)
Add plots	
`addplot(`*plot*`)`	add other plots to the generated graph
Y axis, X axis, Titles, Legend, Overall	
twoway_options	any options other than `by()` documented in [G-3] ***twoway_options***

ci_options	Description
`asymptotic`	compute asymptotic confidence intervals; the default
`heteroskedastic`	compute heteroskedastic bootstrap confidence intervals
`homoskedastic`	compute homoskedastic bootstrap confidence intervals
area_options	affect the rendition of the confidence bands
`table`	produce a table of confidence intervals
`level(#)`	set confidence level; default is `level(95)`
`reps(#)`	number of bootstrap simulations; default is `reps(200)`
`seed(`*str*`)`	random-number seed used for the bootstrap simulations

Menu

Statistics > Multivariate analysis > Factor and principal component analysis > Postestimation > Scree plot of eigenvalues

Description

screeplot produces a scree plot of the eigenvalues of a covariance or correlation matrix.

screeplot automatically obtains the eigenvalues after estimation commands that have eigen as one of their e(properties) and that store the eigenvalues in the matrix e(Ev). These commands include candisc, discrim lda, factor, factormat, pca, and pcamat; see [MV] **candisc**, [MV] **discrim lda**, [MV] **factor**, and [MV] **pca**. screeplot also works automatically to plot singular values after ca and camat, canonical correlations after canon, and eigenvalues after manova, mca, mds, mdsmat, and mdslong; see [MV] **ca**, [MV] **canon**, [MV] **manova**, [MV] **mca**, [MV] **mds**, [MV] **mdsmat**, and [MV] **mdslong**.

screeplot lets you obtain a scree plot in other cases by directly specifying *eigvals*, a vector containing the eigenvalues.

Options

Main

neigen(*#*) specifies the number of eigenvalues to plot. The default is to plot all eigenvalues.

Mean

mean displays a horizontal line at the mean of the eigenvalues.

meanlopts(*cline_options*) affects the rendition of the mean reference line added using the mean option; see [G-3] ***cline_options***.

CI

ci[(*ci_options*)] displays confidence intervals for the eigenvalues. The option ci is a synonym for ci(asymptotic). The following methods for estimating confidence intervals are available:

ci(asymptotic) specifies the asymptotic distribution of the eigenvalues of a central Wishart distribution, the distribution of the covariance matrix of a sample from a multivariate normal distribution. The asymptotic theory applied to correlation matrices is not fully correct, probably giving confidence intervals that are somewhat too narrow.

ci(heteroskedastic) specifies a parametric bootstrap by using the percentile method and assuming that the eigenvalues are from a matrix that is multivariate normal with the same eigenvalues as observed.

ci(homoskedastic) specifies a parametric bootstrap by using the percentile method and assuming that the eigenvalues are from a matrix that is multivariate normal with all eigenvalues equal to the mean of the observed eigenvalues. For a PCA of a correlation matrix, this mean is 1.

ci(*area_options*) affects the rendition of the confidence bands; see [G-3] ***area_options***.

ci(table) produces a table with the confidence intervals.

ci(level(*#*)) specifies the confidence level, as a percentage, for confidence intervals. The default is level(95) or as set by set level; see [U] **20.7 Specifying the width of confidence intervals**.

ci(reps(*#*)) specifies the number of simulations to be performed for estimating the confidence intervals. This option is valid only when heteroskedastic or homoskedastic is specified. The default is reps(200).

`ci(seed(`*str*`))` sets the random-number seed used for the parametric bootstrap. Setting the seed makes sure that results are reproducible. See `set seed` in [R] **set seed**. This option is valid only when `heteroskedastic` or `homoskedastic` is specified.

The confidence intervals are not adjusted for "simultaneous inference".

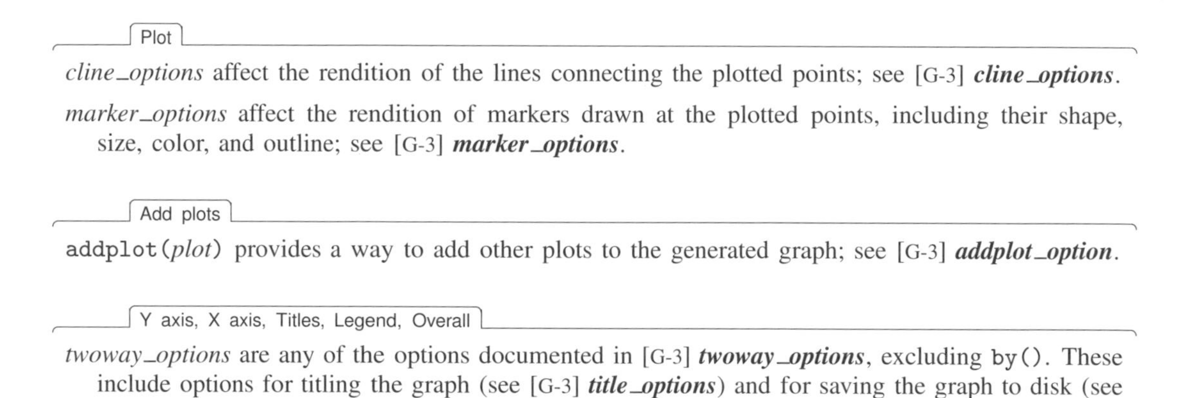

Plot

cline_options affect the rendition of the lines connecting the plotted points; see [G-3] ***cline_options***.

marker_options affect the rendition of markers drawn at the plotted points, including their shape, size, color, and outline; see [G-3] ***marker_options***.

Add plots

`addplot(`*plot*`)` provides a way to add other plots to the generated graph; see [G-3] ***addplot_option***.

Y axis, X axis, Titles, Legend, Overall

twoway_options are any of the options documented in [G-3] ***twoway_options***, excluding `by()`. These include options for titling the graph (see [G-3] ***title_options***) and for saving the graph to disk (see [G-3] ***saving_option***).

Remarks

Cattell (1966) introduced scree plots, which are visual tools used to help determine the number of important components or factors in multivariate settings, such as principal component analysis and factor analysis; see [MV] **pca** and [MV] **factor**. The scree plot is examined for a natural break between the large eigenvalues and the remaining small eigenvalues. The word "scree" is used in reference to the appearance of the large eigenvalues as the hill and the small eigenvalues as the debris of loose rocks at the bottom of the hill. Examples of scree plots can be found in most books that discuss principal component analysis or factor analysis, including Rabe-Hesketh and Everitt (2007), Hamilton (1992, 249–288), Rencher (2002), and Hamilton (2009, chap. 12).

▷ Example 1

Multivariate commands, such as `pca` and `factor` (see [MV] **pca** and [MV] **factor**), produce eigenvalues and eigenvectors. The `screeplot` command graphs the eigenvalues, so you can decide how many components or factors to retain.

We demonstrate scree plots after a principal component analysis. Say that we have been hired by the restaurant industry to study expenditures on eating and drinking. We have data on 898 U.S. cities:

```
. use http://www.stata-press.com/data/r12/emd
(1980 City Data)
. pca ln_eat - hhsize
  (output omitted )
```

```
. screeplot
```

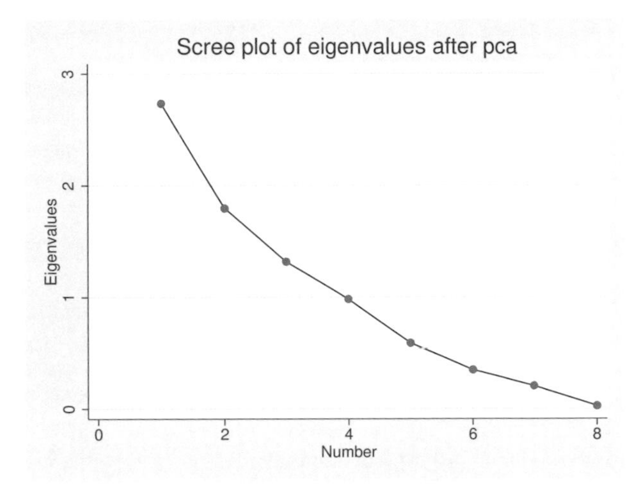

This scree plot does not suggest a natural break between high and low eigenvalues.

We render this same scree plot with the addition of confidence bands by using the `ci()` option. The `asymptotic` suboption selects confidence intervals that are based on the assumption of asymptotic normality. Because the asymptotic theory applied to correlation matrices is not fully correct, we also use the `level()` suboption to adjust the confidence interval. The `table` suboption displays a table of the eigenvalues and lower and upper confidence-interval values.

```
. scree, ci(asympt level(90) table)
(caution is advised in interpreting an asymptotic theory-based confidence
 interval of eigenvalues of a correlation matrix)
```

	eigval	low	high
Comp1	2.733531	2.497578	2.991777
Comp2	1.795623	1.640628	1.965261
Comp3	1.318192	1.204408	1.442725
Comp4	.9829996	.8981487	1.075867
Comp5	.5907632	.5397695	.6465744
Comp6	.3486276	.3185346	.3815635
Comp7	.2052582	.1875407	.2246496
Comp8	.0250056	.0228472	.027368

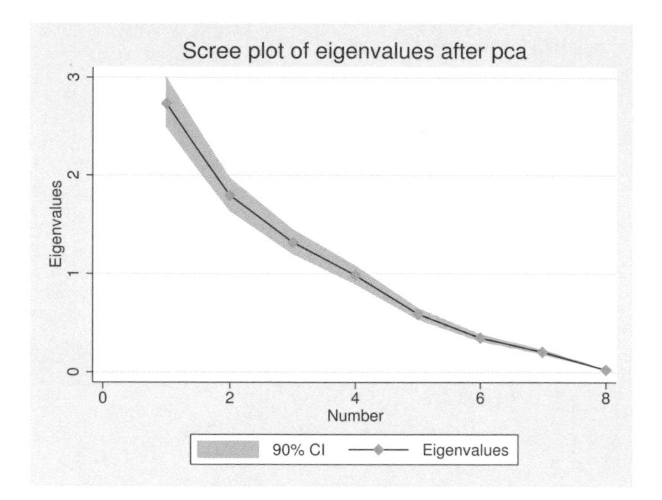

`screeplot` warned us about using asymptotic confidence intervals with eigenvalues based on a correlation matrix. `screeplot` knew that the eigenvalues were based on a correlation matrix instead of a covariance matrix by examining the information available in the `e()` results from the `pca` that we ran.

Instead of displaying an asymptotic confidence band, we now display a parametric bootstrap confidence interval. We select the `heteroskedastic` suboption of `ci()` that allows for unequal eigenvalues. The `reps()` and `seed()` suboptions control the execution of the bootstrap. The `mean` option provides a horizontal line at the mean of the eigenvalues, here at 1 because we are dealing with a principal component analysis performed on a correlation matrix.

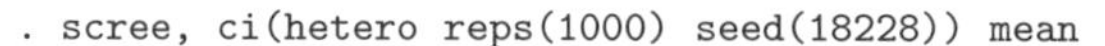

```
. scree, ci(hetero reps(1000) seed(18228)) mean
```

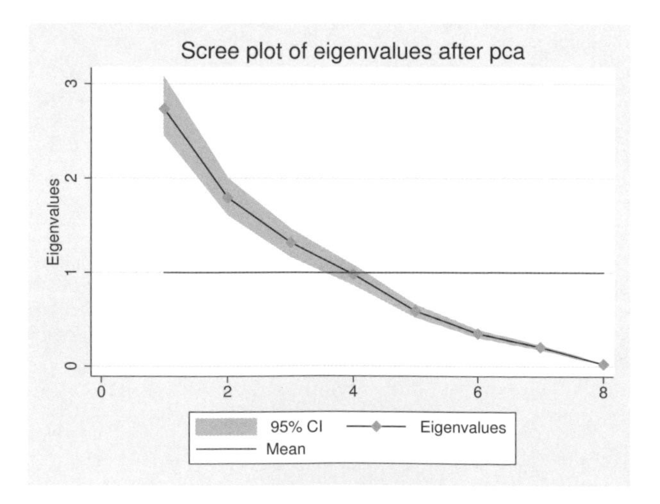

Our final scree plot switches to computing the bootstrap confidence intervals on the basis of the assumption that the eigenvalues are equal to the mean of the observed eigenvalues (using the `homoskedastic` suboption of `ci()`). We again set the seed by using the `seed()` suboption.

```
. scree, ci(homo seed(56227))
```

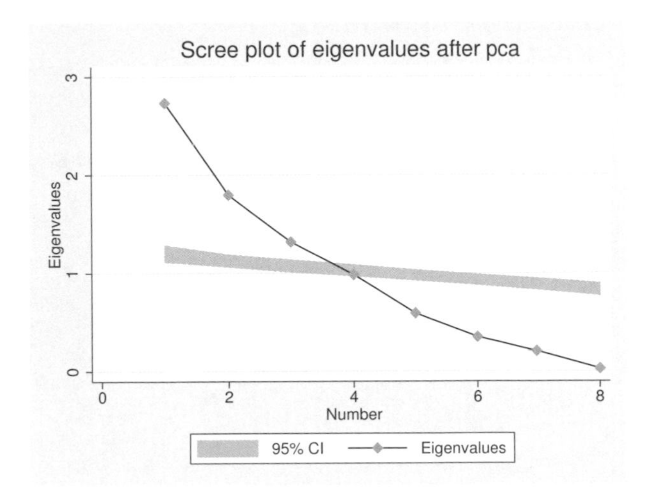

The graph indicates that our data do not support using the homoskedastic assumption.

◁

Saved results

`screeplot` saves the following in `r()`:

Scalars

`r(level)`	confidence level for confidence intervals

Macros

`r(Ctype)`	`correlation` or `covariance`
`r(ci)`	method for estimating confidence interval
`r(seed)`	random-number seed used for parametric bootstrap

Matrices

`r(ci)`	confidence intervals
`r(eigvals)`	eigenvalues

Scree is an English word with Old Norse origins. In geomorphology, scree denotes an accumulation of loose, coarse, usually angular debris at the foot of steep rock slopes (Luckman 2004). The accumulation must be sufficiently thick to be independent of the underlying slope; thus thin veneers of rock particles do not qualify. Screes are themselves quite steep; angles between 30° and 40° are common.

Outside geomorphology, and, notably, in some discussions of scree plots, the term is commonly misapplied to imply deposits on much gentler slopes, or those gentler slopes themselves, and by extension parts of curves thought similar to either.

However appropriate in general or in particular, the term *scree plot* is so widely used in multivariate statistics that it is echoed in the name of this Stata command.

Methods and formulas

`screeplot` is implemented as an ado-file.

References

Cattell, R. B. 1966. The scree test for the number of factors. *Multivariate Behavioral Research* 1: 245–276.

Hamilton, L. C. 1992. *Regression with Graphics: A Second Course in Applied Statistics.* Belmont, CA: Duxbury.

——. 2009. *Statistics with Stata (Updated for Version 10).* Belmont, CA: Brooks/Cole.

Luckman, B. 2004. Scree. In *Encyclopedia of Geomorphology*, ed. A. S. Goudie, 915–917. London: Routledge.

Rabe-Hesketh, S., and B. S. Everitt. 2007. *A Handbook of Statistical Analyses Using Stata.* 4th ed. Boca Raton, FL: Chapman & Hall/CRC.

Rencher, A. C. 2002. *Methods of Multivariate Analysis.* 2nd ed. New York: Wiley.

Also see

[MV] **factor** — Factor analysis

[MV] **pca** — Principal component analysis

[MV] **mds** — Multidimensional scaling for two-way data

Glossary

agglomerative hierarchical clustering methods. Agglomerative hierarchical clustering methods are bottom-up methods for hierarchical clustering. Each observation begins in a separate group. The closest pair of groups is agglomerated or merged in each iteration until all of the data is in one cluster. This process creates a hierarchy of clusters. Contrast to *divisive hierarchical clustering methods*.

anti-image correlation matrix or **anti-image covariance matrix**. The image of a variable is defined as that part which is predictable by regressing each variable on all the other variables; hence, the anti-image is the part of the variable that cannot be predicted. The anti-image correlation matrix $\mathbf{A}$ is a matrix of the negatives of the partial correlations among variables. Partial correlations represent the degree to which the factors explain each other in the results. The diagonal of the anti-image correlation matrix is the KMO measure of sampling adequacy for the individual variables. Variables with small values should be eliminated from the analysis. The anti-image covariance matrix $\mathbf{C}$ contains the negatives of the partial covariances and has one minus the squared multiple correlations in the principal diagonal. Most of the off-diagonal elements should be small in both anti-image matrices in a good factor model. Both anti-image matrices can be calculated from the inverse of the correlation matrix $\mathbf{R}$ via

$$\mathbf{A} = \{\text{diag}(\mathbf{R})\}^{-1}\mathbf{R}\{\text{diag}(\mathbf{R})\}^{-1}$$
$$\mathbf{C} = \{\text{diag}(\mathbf{R})\}^{-1/2}\mathbf{R}\{\text{diag}(\mathbf{R})\}^{-1/2}$$

Also see *KMO*.

average-linkage clustering. Average-linkage clustering is a hierarchical clustering method that uses the average proximity of observations between groups as the proximity measure between the two groups.

Bayes' theorem. Bayes' theorem states that the probability of an event, A, conditional on another event, B, is generally different from the probability of B conditional on A, although the two are related. Bayes' theorem is that

$$P(A|B) = \frac{P(B|A)P(A)}{P(B)}$$

where $P(A)$ is the marginal probability of A, and $P(A|B)$ is the conditional probability of A given B, and likewise for $P(B)$ and $P(B|A)$.

Bentler's invariant pattern simplicity rotation. Bentler's (1977) rotation maximizes the invariant pattern simplicity. It is an oblique rotation that minimizes the criterion function

$$c(\mathbf{\Lambda}) = -\log[|(\mathbf{\Lambda}^2)'\mathbf{\Lambda}^2|] + \log[|\text{diag}\{(\mathbf{\Lambda}^2)'\mathbf{\Lambda}^2\}|]$$

See *Crawford–Ferguson rotation* for a definition of $\mathbf{\Lambda}$. Also see *oblique rotation*.

between matrix and **within matrix**. The between and within matrices are SSCP matrices that measure the spread between groups and within groups, respectively. These matrices are used in multivariate analysis of variance and related hypothesis tests: Wilk's lambda, Roy's largest root, Lawley–Hotelling trace, and Pillai's trace.

Here we have k independent random samples of size n. The between matrix $\mathbf{H}$ is given by

$$\mathbf{H} = n\sum_{i=1}^{k}(\overline{\mathbf{y}}_{i\bullet} - \overline{\mathbf{y}}_{\bullet\bullet})(\overline{\mathbf{y}}_{i\bullet} - \overline{\mathbf{y}}_{\bullet\bullet})' = \sum_{i=1}^{k}\frac{1}{n}\mathbf{y}_{i\bullet}\mathbf{y}'_{i\bullet} - \frac{1}{kn}\mathbf{y}_{\bullet\bullet}\mathbf{y}'_{\bullet\bullet}$$

The within matrix $\mathbf{E}$ is defined as

$$\mathbf{E} = \sum_{i=1}^{k}\sum_{j=1}^{n}(\mathbf{y}_{ij} - \overline{\mathbf{y}}_{i\bullet})(\mathbf{y}_{ij} - \mathbf{y}_{i\bullet})' = \sum_{i=1}^{k}\sum_{j=1}^{n}\mathbf{y}_{ij}\mathbf{y}'_{ij} - \sum_{i=1}^{k}\frac{1}{n}\mathbf{y}_{i\bullet}\mathbf{y}'_{i\bullet}$$

Also see *SSCP matrix*.

biplot. A biplot is a scatterplot which represents both observations and variables simultaneously. There are many different biplots; variables in biplots are usually represented by arrows and observations are usually represented by points.

biquartimax rotation or **biquartimin rotation**. Biquartimax rotation and biquartimin rotation are synonyms. They put equal weight on the varimax and quartimax criteria, simplifying the columns and rows of the matrix. This is an oblique rotation equivalent to an oblimin rotation with $\gamma = 0.5$. Also see *varimax rotation*, *quartimax rotation*, and *oblimin rotation*.

boundary solution or **Heywood solution**. See *Heywood case*.

CA. Correspondence analysis (CA) gives a geometric representation of the rows and columns of a two-way frequency table. The geometric representation is helpful in understanding the similarities between the categories of variables and associations between variables. CA is calculated by singular value decomposition. Also see *singular value decomposition*.

canonical correlation analysis. Canonical correlation analysis attempts to describe the relationships between two sets of variables by finding linear combinations of each so that the correlation between the linear combinations is maximized.

canonical discriminant analysis. Canonical linear discriminant analysis is LDA where describing how groups are separated is of primary interest. Also see *LDA*.

canonical loadings. The canonical loadings are coefficients of canonical linear discriminant functions. Also see *canonical discriminant analysis* and *loading*.

canonical variate set. The canonical variate set is a linear combination or weighted sum of variables obtained from canonical correlation analysis. Two sets of variables are analyzed in canonical correlation analysis. The first canonical variate of the first variable set is the linear combination in standardized form that has maximal correlation with the first canonical variate from the second variable set. The subsequent canonical variates are uncorrelated to the previous and have maximal correlation under that constraint.

centered data. Centered data has zero mean. You can center data $\mathbf{x}$ by taking $\mathbf{x} - \overline{\mathbf{x}}$.

centroid-linkage clustering. Centroid-linkage clustering is a hierarchical clustering method that computes the proximity between two groups as the proximity between the group means.

classical scaling. Classical scaling is a method of performing MDS via an eigen decomposition. This is contrasted to modern MDS, which is achieved via the minimization of a loss function. Also see *MDS* and *modern scaling*.

classification. Classification is the act of allocating or classifying observations to groups as part of discriminant analysis. In some sources, classification is synonymous with cluster analysis.

classification function. Classification functions can be obtained after LDA or QDA. They are functions based on Mahalanobis distance for classifying observations to the groups. See *discriminant function* for an alternative. Also see *LDA* and *QDA*.

classification table. A classification table, also known as a confusion matrix, gives the count of observations from each group that are classified into each of the groups as part of a discriminant analysis. The element at (i, j) gives the number of observations that belong to the ith group but were classified into the jth group. High counts are expected on the diagonal of the table where observations are correctly classified, and small values are expected off the diagonal. The columns of the matrix are categories of the predicted classification; the rows represent the actual group membership.

cluster analysis. Cluster analysis is a method for determining natural groupings or clusters of observations.

cluster tree. See *dendrogram*.

clustering. See *cluster analysis*.

common factors. Common factors are found by factor analysis. They linearly reconstruct the original variables. In factor analysis, reconstruction is defined in terms of prediction of the correlation matrix of the original variables.

communality. Communality is the proportion of a variable's variance explained by the common factors in factor analysis. It is also "1 − uniqueness". Also see *uniqueness*.

complete-linkage clustering. Complete-linkage clustering is a hierarchical clustering method that uses the farthest pair of observations between two groups to determine the proximity of the two groups.

component scores. Component scores are calculated after PCA. Component scores are the coordinates of the original variables in the space of principal components.

Comrey's tandem 1 and 2 rotations. Comrey (1967) describes two rotations, the first (tandem 1) to judge which "small" factors should be dropped, the second (tandem 2) for "polishing".

Tandem principal 1 minimizes the criterion

$$c(\mathbf{\Lambda}) = \langle \mathbf{\Lambda}^2, (\mathbf{\Lambda}\mathbf{\Lambda}')^2\mathbf{\Lambda}^2 \rangle$$

Tandem principal 2 minimizes the criterion

$$c(\mathbf{\Lambda}) = \langle \mathbf{\Lambda}^2, \{\mathbf{1}\mathbf{1}' - (\mathbf{\Lambda}\mathbf{\Lambda}')^2\}\mathbf{\Lambda}^2 \rangle$$

See *Crawford–Ferguson rotation* for a definition of $\mathbf{\Lambda}$.

configuration. The configuration in MDS is a representation in a low-dimensional (usually 2-dimensional) space with distances in the low-dimensional space approximating the dissimilarities or disparities in high-dimensional space. Also see *MDS*, *dissimilarity*, and *disparity*.

configuration plot. A configuration plot after MDS is a (usually 2-dimensional) plot of labeled points showing the low-dimensional approximation to the dissimilarities or disparities in high-dimensional space. Also see *MDS*, *dissimilarity*, and *disparity*.

confusion matrix. A confusion matrix is a synonym for a classification table after discriminant analysis. See *classification table*.

contrast or **contrasts**. In ANOVA, a contrast in k population means is defined as a linear combination

$$\delta = c_1\mu_1 + c_2\mu_2 + \cdots + c_k\mu_k$$

where the coefficients satisfy

$$\sum_{i=1}^{k} c_i = 0$$

In the multivariate setting (MANOVA), a contrast in k population mean vectors is defined as

$$\boldsymbol{\delta} = c_1\boldsymbol{\mu}_1 + c_2\boldsymbol{\mu}_2 + \cdots c_k\boldsymbol{\mu}_k$$

where the coefficients again satisfy

$$\sum_{i=1}^{k} c_i = 0$$

The univariate hypothesis $\delta = 0$ may be tested with `contrast` (or `test`) after ANOVA. The multivariate hypothesis $\boldsymbol{\delta} = 0$ may be tested with `manovatest` after MANOVA.

correspondence analysis. See *CA*.

correspondence analysis projection. A correspondence analysis projection is a line plot of the row and column coordinates after CA. The goal of this graph is to show the ordering of row and column categories on each principal dimension of the analysis. Each principal dimension is represented by a vertical line; markers are plotted on the lines where the row and column categories project onto the dimensions. Also see *CA*.

costs. Costs in discriminant analysis are the cost of misclassifying observations.

covarimin rotation. Covarimin rotation is an orthogonal rotation equivalent to varimax. Also see *varimax rotation*.

Crawford–Ferguson rotation. Crawford–Ferguson (1970) rotation is a general oblique rotation with several interesting special cases.

Special cases of the Crawford–Ferguson rotation include

κ	Special case
0	quartimax / quartimin
$1/p$	varimax / covarimin
$f/(2p)$	equamax
$(f-1)/(p+f-2)$	parsimax
1	factor parsimony

p = number of rows of $\mathbf{A}$.
f = number of columns of $\mathbf{A}$.

Where $\mathbf{A}$ is the matrix to be rotated, $\mathbf{T}$ is the rotation and $\boldsymbol{\Lambda} = \mathbf{AT}$. The Crawford–Ferguson rotation is achieved by minimizing the criterion

$$c(\boldsymbol{\Lambda}) = \frac{1-\kappa}{4}\left\langle \boldsymbol{\Lambda}^2, \boldsymbol{\Lambda}^2(\mathbf{11}' - \mathbf{I})\right\rangle + \frac{\kappa}{4}\left\langle \boldsymbol{\Lambda}^2, (\mathbf{11}' - \mathbf{I})\boldsymbol{\Lambda}^2\right\rangle$$

Also see *oblique rotation*.

crossed variables or **stacked variables**. In CA and MCA crossed categorical variables may be formed from the interactions of two or more existing categorical variables. Variables that contain these interactions are called crossed or stacked variables.

crossing variables or **stacking variables**. In CA and MCA, crossing or stacking variables are the existing categorical variables whose interactions make up a crossed or stacked variable.

curse of dimensionality. The curse of dimensionality is a term coined by Richard Bellman (1961) to describe the problem caused by the exponential increase in size associated with adding extra dimensions to a mathematical space. On the unit interval, 10 evenly spaced points suffice to sample with no more distance than 0.1 between them; however a unit square requires 100 points, and a unit cube requires 1000 points. Many multivariate statistical procedures suffer from the curse of dimensionality. Adding variables to an analysis without adding sufficient observations can lead to imprecision.

dendrogram or **cluster tree**. A dendrogram or cluster tree graphically presents information about how observations are grouped together at various levels of (dis)similarity in hierarchical cluster analysis. At the bottom of the dendrogram, each observation is considered its own cluster. Vertical lines extend up for each observation, and at various (dis)similarity values, these lines are connected to the lines from other observations with a horizontal line. The observations continue to combine until, at the top of the dendrogram, all observations are grouped together. Also see *hierarchical clustering*.

dilation. A dilation stretches or shrinks distances in Procrustes rotation.

dimension. A dimension is a parameter or measurement required to define a characteristic of an object or observation. Dimensions are the variables in the dataset. Weight, height, age, blood pressure, and drug dose are examples of dimensions in health data. Number of employees, gross income, net income, tax, and year are examples of dimensions in data about companies.

discriminant analysis. Discriminant analysis is used to describe the differences between groups and to exploit those differences when allocating (classifying) observations of unknown group membership. Discriminant analysis is also called classification in many references.

discriminant function. Discriminant functions are formed from the eigenvectors from Fisher's approach to LDA. See *LDA*. See *classification function* for an alternative.

discriminating variables. Discriminating variables in a discriminant analysis are analyzed to determine differences between groups where group membership is known. These differences between groups are then exploited when classifying observations to the groups.

disparity. Disparities are transformed dissimilarities, that is, dissimilarity values transformed by some function. The class of functions to transform dissimilarities to disparities may either be (1) a class of metric, or known functions such as linear functions or power functions that can be parameterized by real scalars or (2) a class of more general (nonmetric) functions, such as any monotonic function. Disparities are used in MDS. Also see *dissimilarity*, *MDS*, *metric scaling*, and *nonmetric scaling*.

dissimilarity, **dissimilarity matrix**, and **dissimilarity measure**. Dissimilarity or a dissimilarity measure is a quantification of the difference between two things, such as observations or variables or groups of observations or a method for quantifying that difference. A dissimilarity matrix is a matrix containing dissimilarity measurements. Euclidean distance is one example of a dissimilarity measure. Contrast to *similarity*. Also see *proximity* and *Euclidean distance*.

divisive hierarchical clustering methods. Divisive hierarchical clustering methods are top-down methods for hierarchical clustering. All the data begins as a part of one large cluster; with each iteration, a cluster is broken into two to create two new clusters. At the first iteration there are two clusters, then three, and so on. Divisive methods are very computationally expensive. Contrast to *agglomerative hierarchical clustering methods*.

eigenvalue. An eigenvalue is the scale factor by which an eigenvector is multiplied. For many multivariate techniques, the size of an eigenvalue indicates the importance of the corresponding eigenvector. Also see *eigenvector*.

eigenvector. An eigenvector of a linear transformation is a nonzero vector that is either left unaffected or simply multiplied by a scale factor after the transformation.

Here $\mathbf{x}$ is an eigenvector of linear transformation $\mathbf{A}$ with eigenvalue λ:

$$\mathbf{A}\mathbf{x} = \lambda\mathbf{x}$$

For many multivariate techniques, eigenvectors form the basis for analysis and interpretation. Also see *loading*.

equamax rotation. Equamax rotation is an orthogonal rotation whose criterion is a weighted sum of the varimax and quartimax criteria. Equamax reflects a concern for simple structure within the rows and columns of the matrix. It is equivalent to oblimin with $\gamma = p/2$, or to the Crawford–Ferguson family with $\kappa = f/2p$, where p is the number of rows of the matrix to be rotated, and f is the number of columns. Also see *orthogonal rotation*, *varimax rotation*, *quartimax rotation*, *oblimin rotation*, and *Crawford–Ferguson rotation*.

Euclidean distance. The Euclidean distance between two observations is the distance one would measure with a ruler. The distance between vector $\mathbf{P} = (P_1, P_2, \dots, P_n)$ and $\mathbf{Q} = (Q_1, Q_2, \dots, Q_n)$ is given by

$$D(\mathbf{P}, \mathbf{Q}) = \sqrt{(P_1 - Q_1)^2 + (P_2 - Q_2)^2 + \cdots + (P_n - Q_n)^2} = \sqrt{\sum_{i=1}^{n} (P_i - Q_i)^2}$$

factor. A factor is an unobserved random variable that is thought to explain variability among observed random variables.

factor analysis. Factor analysis is a statistical technique used to explain variability among observed random variables in terms of fewer unobserved random variables called factors. The observed variables are then linear combinations of the factors plus error terms.

If the correlation matrix of the observed variables is $\mathbf{R}$, then $\mathbf{R}$ is decomposed by factor analysis as

$$\mathbf{R} = \mathbf{\Lambda}\mathbf{\Phi}\mathbf{\Lambda}' + \mathbf{\Psi}$$

$\mathbf{\Lambda}$ is the loading matrix, and $\mathbf{\Psi}$ contains the specific variances, for example, the variance specific to the variable not explained by the factors. The default unrotated form assumes uncorrelated common factors, $\mathbf{\Phi} = \mathbf{I}$.

factor loading plot. A factor loading plot produces a scatter plot of the factor loadings after factor analysis.

factor loadings. Factor loadings are the regression coefficients which multiply the factors to produce the observed variables in the factor analysis.

factor parsimony. Factor parsimony is an oblique rotation, which maximizes the column simplicity of the matrix. It is equivalent to a Crawford–Ferguson rotation with $\kappa = 1$. Also see *oblique rotation* and *Crawford–Ferguson rotation*.

factor scores. Factor scores are computed after factor analysis. Factor scores are the coordinates of the original variables, $\mathbf{x}$, in the space of the factors. The two types of scoring are regression scoring (Thomson 1951) and Bartlett (1937, 1938) scoring.

Using the symbols defined in *factor analysis*, the formula for regression scoring is

$$\widehat{\mathbf{f}} = \mathbf{\Lambda}'\mathbf{R}^{-1}\mathbf{x}$$

In the case of oblique rotation the formula becomes

$$\widehat{\mathbf{f}} = \mathbf{\Phi}\mathbf{\Lambda}'\mathbf{R}^{-1}\mathbf{x}$$

The formula for Bartlett scoring is

$$\widehat{\mathbf{f}} = \mathbf{\Gamma}^{-1}\mathbf{\Lambda}'\mathbf{\Psi}^{-1}\mathbf{x}$$

where

$$\mathbf{\Gamma} = \mathbf{\Lambda}'\mathbf{\Psi}^{-1}\mathbf{\Lambda}$$

Also see *factor analysis*.

Heywood case or **Heywood solution**. A Heywood case can appear in factor analysis output; this indicates that a boundary solution, called a Heywood solution, was produced. The geometric assumptions underlying the likelihood-ratio test are violated, though the test may be useful if interpreted cautiously.

hierarchical clustering and **hierarchical clustering methods**. In hierarchical clustering, the data is placed into clusters via iterative steps. Contrast to *partition clustering*. Also see *agglomerative hierarchical clustering methods* and *divisive hierarchical clustering methods*.

Hotelling's T-squared generalized means test. Hotelling's T-squared generalized means test is a multivariate test that reduces to a standard t test if only one variable is specified. It tests whether one set of means is zero or if two sets of means are equal.

inertia. In CA, the inertia is related to the definition in applied mathematics of "moment of inertia", which is the integral of the mass times the squared distance to the centroid. Inertia is defined as the total Pearson chi-squared for the two-way table divided by the total number of observations, or the sum of the squared singular values found in the singular value decomposition.

$$\text{total inertia} = \frac{1}{n}\chi^2 = \sum_k \lambda_k^2$$

In MCA, the inertia is defined analogously. In the case of the indicator or Burt matrix approach, it is given by the formula

$$\text{total inertia} = \left(\frac{q}{q-1}\right)\sum \phi_t^2 - \frac{(J-q)}{q^2}$$

where q is the number of active variables, J is the number of categories and ϕ_t is the tth (unadjusted) eigenvalue of the eigen decomposition. In JCA the total inertia of the modified Burt matrix is defined as the sum of the inertias of the off-diagonal blocks. Also see *CA* and *MCA*.

iterated principal-factor method. The iterated principal-factor method is a method for performing factor analysis in which the communalities $\widehat{h}_i^2$ are estimated iteratively from the loadings in $\widehat{\mathbf{\Lambda}}$ using

$$\widehat{h}_i^2 = \sum_{j=1}^{m} \widehat{\lambda}_{ij}^2$$

Also see *factor analysis* and *communality*.

JCA. An acronym for joint correspondence analysis; see *MCA*.

joint correspondence analysis. See *MCA*.

Kaiser–Meyer–Olkin measure of sampling adequacy. See *KMO*.

kmeans. Kmeans is a method for performing partition cluster analysis. The user specifies the number of clusters, k, to create using an iterative process. Each observation is assigned to the group whose mean is closest, and then based on that categorization, new group means are determined. These steps continue until no observations change groups. The algorithm begins with k seed values, which act as the k group means. There are many ways to specify the beginning seed values. Also see *partition clustering*.

kmedians. Kmedians is a variation of kmeans. The same process is performed, except that medians instead of means are computed to represent the group centers at each step. Also see *kmeans* and *partition clustering*.

KMO. The Kaiser–Meyer–Olkin (KMO) measure of sampling adequacy takes values between 0 and 1, with small values meaning that the variables have too little in common to warrant a factor analysis or PCA. Historically, the following labels have been given to values of KMO (Kaiser 1974):

0.00 to 0.49	unacceptable
0.50 to 0.59	miserable
0.60 to 0.69	mediocre
0.70 to 0.79	middling
0.80 to 0.89	meritorious
0.90 to 1.00	marvelous

KNN. kth-nearest-neighbor (KNN) discriminant analysis is a nonparametric discrimination method based on the k nearest neighbors of each observation. Both continuous and binary data can be handled through the different similarity and dissimilarity measures. KNN analysis can distinguish irregular-shaped groups, including groups with multiple modes. Also see *discriminant analysis* and *nonparametric methods*.

Kruskal stress. The Kruskal stress measure (Kruskal 1964; Cox and Cox 2001, 63) used in MDS is given by

$$\text{Kruskal}(\widehat{\mathbf{D}}, \mathbf{E}) = \left\{ \frac{\sum (E_{ij} - \widehat{D}_{ij})^2}{\sum E_{ij}^2} \right\}^{1/2}$$

where D_{ij} is the dissimilarity between objects i and j, $1 \leq i, j \leq n$, and $\widehat{D}_{ij}$ is the disparity, that is, the transformed dissimilarity, and E_{ij} is the Euclidean distance between rows i and j of the matching configuration. Kruskal stress is an example of a loss function in modern MDS. After classical MDS, `estat stress` gives the Kruskal stress. Also see *classical scaling*, *MDS*, and *stress*.

kth nearest neighbor. See *KNN*.

Lawley–Hotelling trace. The Lawley–Hotelling trace is a test statistic for the hypothesis test $H_0 : \boldsymbol{\mu}_1 = \boldsymbol{\mu}_2 = \cdots = \boldsymbol{\mu}_k$ based on the eigenvalues $\lambda_1, \lambda_2, \ldots, \lambda_s$ of $\mathbf{E}^{-1}\mathbf{H}$. It is defined as

$$U^{(s)} = \text{trace}(\mathbf{E}^{-1}\mathbf{H}) = \sum_{i=1}^{s} \lambda_i$$

where $\mathbf{H}$ is the between matrix and $\mathbf{E}$ is the within matrix, see *between matrix*.

LDA. Linear discriminant analysis (LDA) is a parametric form of discriminant analysis. In Fisher's (1936) approach to LDA, linear combinations of the discriminating variables that provide maximal separation between the groups. The Mahalanobis (1936) formulation of LDA assumes that the observations come from a multivariate normal distribution with equal covariance matrices. Also see *discriminant analysis* and *parametric methods*.

linear discriminant analysis. See *LDA*.

linkage. In cluster analysis, the linkage refers to the measure of proximity between groups or clusters.

loading. A loading is a coefficient or weight in a linear transformation. Loadings play an important role in many multivariate techniques, including factor analysis, PCA, MANOVA, LDA, and canonical correlations. In some settings, the loadings are of primary interest and are examined for interpretability. For many multivariate techniques, loadings are based on an eigenanalysis of a correlation or covariance matrix. Also see *eigenvector*.

loading plot. A loading plot is a scatter plot of the loadings after LDA, factor analysis or PCA.

logistic discriminant analysis. Logistic discriminant analysis is a form of discriminant analysis based on the assumption that the likelihood ratios of the groups have an exponential form. Multinomial logistic regression provides the basis for logistic discriminant analysis. Because multinomial logistic regression can handle binary and continuous regressors, logistic discriminant analysis is also appropriate for binary and continuous discriminating variables. Also see *discriminant analysis*.

LOO. LOO is an acronym for leave one out. In discriminant analysis, classification of an observation while leaving it out of the estimation sample is done to check the robustness of the analysis. Also see *discriminant analysis*.

loss. Modern MDS is performed by minimizing a loss function, also called a loss criterion. The loss quantifies the difference between the disparities and the Euclidean distances.

Loss functions include Kruskal's stress and its square, both normalized with either disparities or distances, the strain criterion which is equivalent to classical metric scaling when the disparities equal the dissimilarities, and the Sammon (1969) mapping criterion which is the sum of the scaled, squared differences between the distances and the disparities, normalized by the sum of the disparities.

Also see *MDS*, *Kruskal stress*, *classical scaling*, and *disparity*.

Mahalanobis distance. The Mahalanobis distance measure is a scale-invariant way of measuring distance. It takes into account the correlations of the dataset.

Mahalanobis transformation. The Mahalanobis transformation takes a Cholesky factorization of the inverse of the covariance matrix $\mathbf{S}^{-1}$ in the formula for Mahalanobis distance and uses it to transform the data. If we have the Cholesky factorization $\mathbf{S}^{-1} = \mathbf{L}'\mathbf{L}$, then the Mahalanobis transformation of $\mathbf{x}$ is $\mathbf{z} = \mathbf{L}\mathbf{x}$, and $\mathbf{z}'\mathbf{z} = D^2_M(\mathbf{x})$.

MANCOVA. MANCOVA is multivariate analysis of covariance. See *MANOVA*.

MANOVA. MANOVA is multivariate analysis of variance; it is used to test hypotheses about means. Four multivariate statistics are commonly computed in MANOVA: Wilk's lambda, Pillai's trace, Lawley–Hotelling trace, and Roy's largest root. Also see *Wilk's lambda*, *Pillai's trace*, *Lawley–Hotelling trace*, and *Roy's largest root*.

mass. In CA and MCA, the mass is the marginal probability. The sum of the mass over the active row or column categories equals 1.

matching coefficient. The matching similarity coefficient is used to compare two binary variables. If a is the number of observations that both have value 1, and d is the number of observations that both have value 0, and b, c are the number of $(1, 0)$ and $(0, 1)$ observations, respectively, then the matching coefficient is given by

$$\frac{a+d}{a+b+c+d}$$

Also see *similarity measure*.

matching configuration. In MDS, the matching configuration is the low dimensional configuration whose distances approximate the high-dimensional dissimilarities or disparities. Also see *MDS*, *dissimilarity*, and *disparity*.

matching configuration plot. After MDS, this is a scatter plot of the matching configuration.

maximum likelihood factor method. The maximum likelihood factor method is a method for performing factor analysis that assumes multivariate normal observations. It maximizes the determinant of the partial correlation matrix; thus, this solution is also meaningful as a descriptive method for nonnormal data. Also see *factor analysis*.

MCA. Multiple correspondence analysis (MCA) and joint correspondence analysis (JCA) are methods for analyzing observations on categorical variables. MCA and JCA analyze a multiway table and are usually viewed as an extension of CA. Also see *CA*.

MDS. Multidimensional scaling (MDS) is a dimension-reduction and visualization technique. Dissimilarities (for instance, Euclidean distances) between observations in a high-dimensional space are represented in a lower-dimensional space which is typically two dimensions so that the Euclidean distance in the lower-dimensional space approximates in some sense the dissimilarities in the higher-dimensional space. Often the higher-dimensional dissimilarities are first transformed to disparities, and the disparities are then approximated by the distances in the lower-dimensional space. Also see *dissimilarity*, *disparity*, *classical scaling*, *loss*, *modern scaling*, *metric scaling*, and *nonmetric scaling*.

MDS configuration plot. See *configuration plot*.

measure. A measure is a quantity representing the proximity between objects or method for determining the proximity between objects. Also see *proximity*.

median-linkage clustering. Median-linkage clustering is a hierarchical clustering method that uses the distance between the medians of two groups to determine the similarity or dissimilarity of the two groups. Also see *cluster analysis* and *agglomerative hierarchical clustering methods*.

metric scaling. Metric scaling is a type of MDS, in which the dissimilarities are transformed to disparities via a class of known functions. This is contrasted to *nonmetric scaling* Also see *MDS*.

minimum entropy rotation. The minimum entropy rotation is an orthogonal rotation achieved by minimizing the deviation from uniformity (entropy). The minimum entropy criterion (Jennrich 2004) is

$$c(\mathbf{\Lambda}) = -\frac{1}{2}\left\langle \mathbf{\Lambda}^2, \log \mathbf{\Lambda}^2 \right\rangle$$

See *Crawford–Ferguson rotation* for a definition of $\mathbf{\Lambda}$. Also see *orthogonal rotation*.

misclassification rate. The misclassification rate calculated after discriminant analysis is, in its simplest form, the fraction of observations incorrectly classified. See *discriminant analysis*.

modern scaling. Modern scaling is a form of MDS that is achieved via the minimization of a loss function that compares the disparities (transformed dissimilarities) in the higher-dimensional space and the distances in the lower-dimensional space. Contrast to *classical scaling*. Also see *dissimilarity*, *disparity*, *MDS*, and *loss*.

multidimensional scaling. See *MDS*.

multiple correspondence analysis. See *MCA*.

multivariate regression. Multivariate regression is a method of estimating a linear (matrix) model

$$\mathbf{Y} = \mathbf{XB} + \boldsymbol{\Xi}$$

Multivariate regression is estimated by least-squares regression, and it can be used to test hypotheses, much like MANOVA.

nearest neighbor. See *KNN*.

nonmetric scaling. Nonmetric scaling is a type of modern MDS in which the dissimilarities may be transformed to disparities via any monotonic function as opposed to a class of known functions. Contrast to *metric scaling*. Also see *MDS*, *dissimilarity*, *disparity*, and *modern scaling*.

nonparametric methods. Nonparametric statistical methods, such as KNN discriminant analysis, do not assume the population fits any parameterized distribution.

normalization. Normalization presents information in a standard form for interpretation. In CA the row and column coordinates can be normalized in different ways depending on how one wishes to interpret the data. Normalization is also used in rotation, MDS, and MCA.

oblimax rotation. Oblimax rotation is a method for oblique rotation which maximizes the number of high and low loadings. When restricted to orthogonal rotation, oblimax is equivalent to quartimax rotation. Oblimax minimizes the oblimax criterion

$$c(\boldsymbol{\Lambda}) = -\log(\langle \boldsymbol{\Lambda}^2, \boldsymbol{\Lambda}^2 \rangle) + 2\log(\langle \boldsymbol{\Lambda}, \boldsymbol{\Lambda} \rangle)$$

See *Crawford–Ferguson rotation* for a definition of $\boldsymbol{\Lambda}$. Also see *oblique rotation*, *orthogonal rotation*, and *quartimax rotation*.

oblimin rotation. Oblimin rotation is a general method for oblique rotation, achieved by minimizing the oblimin criterion

$$c(\boldsymbol{\Lambda}) = \frac{1}{4}\langle \boldsymbol{\Lambda}^2, \{\mathbf{I} - (\gamma/p)\mathbf{11}'\}\boldsymbol{\Lambda}^2(\mathbf{11}' - \mathbf{I})\rangle$$

Oblimin has several interesting special cases:

γ	Special case
0	quartimax / quartimin
$1/2$	biquartimax / biquartimin
1	varimax / covarimin
$p/2$	equamax

p = number of rows of $\mathbf{A}$.

See *Crawford–Ferguson rotation* for a definition of $\boldsymbol{\Lambda}$ and $\mathbf{A}$. Also see *oblique rotation*.

oblique rotation or **oblique transformation**. An oblique rotation maintains the norms of the rows of the matrix but not their inner products. In geometric terms, this maintains the lengths of vectors, but not the angles between them. In contrast, in orthogonal rotation, both are preserved.

ordination. Ordination is the ordering of a set of data points with respect to one or more axes. MDS is a form of ordination.

orthogonal rotation or **orthogonal transformation**. Orthogonal rotation maintains both the norms of the rows of the matrix and also inner products of the rows of the matrix. In geometric terms, this maintains both the lengths of vectors and the angles between them. In contrast, oblique rotation maintains only the norms, that is, the lengths of vectors.

parametric methods. Parametric statistical methods such as LDA and QDA, assume the population fits a parameterized distribution. For example, for LDA we assume the groups are multivariate normal with equal covariance matrices.

parsimax rotation. Parsimax rotation is an orthogonal rotation that balances complexity between the rows and the columns. It is equivalent to the Crawford–Ferguson family with $\kappa = (f-1)/(p+f-2)$, where p is the number of rows of the original matrix, and f is the number of columns. See *orthogonal rotation* and *Crawford–Ferguson rotation*.

partially specified target rotation. Partially specified target rotation minimizes the criterion

$$c(\mathbf{\Lambda}) = \|\mathbf{W} \otimes (\mathbf{\Lambda} - \mathbf{H})\|^2$$

for a given target matrix $\mathbf{H}$ and a nonnegative weighting matrix $\mathbf{W}$ (usually zero–one valued). See *Crawford–Ferguson rotation* for a definition of $\mathbf{\Lambda}$.

partition clustering and **partition cluster-analysis methods**. Partition clustering methods break the observations into a distinct number of nonoverlapping groups. This is accomplished in one step, unlike hierarchical cluster-analysis methods, in which an iterative procedure is used. Consequently, this method is quicker and will allow larger datasets than the hierarchical clustering methods. Contrast to *hierarchical clustering*. Also see *kmeans* and *kmedians*.

PCA. Principal component analysis (PCA) is a statistical technique used for data reduction. The leading eigenvectors from the eigen decomposition of the correlation or the covariance matrix of the variables describe a series of uncorrelated linear combinations of the variables that contain most of the variance. In addition to data reduction, the eigenvectors from a PCA are often inspected to learn more about the underlying structure of the data.

Pillai's trace. Pillai's trace is a test statistic for the hypothesis test $H_0 : \boldsymbol{\mu}_1 = \boldsymbol{\mu}_2 = \cdots = \boldsymbol{\mu}_k$ based on the eigenvalues $\lambda_1, \ldots, \lambda_s$ of $\mathbf{E}^{-1}\mathbf{H}$. It is defined as

$$V^{(s)} = \text{trace}[(\mathbf{E} + \mathbf{H})^{-1}\mathbf{H}] = \sum_{i=1}^{s} \frac{\lambda_i}{1 + \lambda_i}$$

where $\mathbf{H}$ is the between matrix and $\mathbf{E}$ is the within matrix. See *between matrix*.

posterior probabilities. After discriminant analysis, the posterior probabilities are the probabilities of a given observation being assigned to each of the groups based on the prior probabilities, the training data, and the particular discriminant model. Contrast to *prior probabilities*.

principal component analysis. See *PCA*.

principal factor method. The principal factor method is a method for factor analysis in which the factor loadings, sometimes called factor patterns, are computed using the squared multiple correlations as estimates of the communality. Also see *factor analysis* and *communality*.

prior probabilities Prior probabilities in discriminant analysis are the probabilities of an observation belonging to a group before the discriminant analysis is performed. Prior probabilities are often based on the prevalence of the groups in the population as a whole. Contrast to *posterior probabilities*.

Procrustes rotation. A Procrustes rotation is an orthogonal or oblique transformation, that is, a restricted Procrustes transformation without translation or dilation (uniform scaling).

Procrustes transformation. The goal of Procrustes transformation is to transform the source matrix $\mathbf{X}$ to be as close as possible to the target $\mathbf{Y}$. The permitted transformations are any combination of dilation (uniform scaling), rotation and reflection (that is, orthogonal or oblique transformations), and translation. Closeness is measured by residual sum of squares. In some cases, unrestricted Procrustes transformation is desired; this allows the data to be transformed not just by orthogonal or oblique rotations, but by all conformable regular matrices $\mathbf{A}$. Unrestricted Procrustes transformation is equivalent to a multivariate regression.

The name comes from Procrustes of Greek mythology; Procrustes invited guests to try his iron bed. If the guest was too tall for the bed, Procrustes would amputate the guest's feet, and if the guest was too short, he would stretching the guest out on a rack.

Also see *orthogonal rotation*, *oblique rotation*, *dilation*, and *multivariate regression*.

promax rotation. Promax power rotation is an oblique rotation. It does not fit in the minimizing-a-criterion framework that is at the core of most other rotations. The promax method (Hendrickson and White 1964) was proposed before computing power became widely available. The promax rotation consists of three steps:

1. Perform an orthogonal rotation.
2. Raise the elements of the rotated matrix to some power, preserving the sign of the elements. Typically the power is in the range $2 \leq \text{power} \leq 4$. This operation is meant to distinguish clearly between small and large values.
3. The matrix from step two is used as the target for an oblique Procrustean rotation from the original matrix.

proximity, **proximity matrix**, and **proximity measure**. Proximity or a proximity measure means the nearness or farness of two things, such as observations or variables or groups of observations or a method for quantifying the nearness or farness between two things. A proximity is measured by a similarity or dissimilarity. A proximity matrix is a matrix of proximities. Also see *similarity* and *dissimilarity*.

QDA. Quadratic discriminant analysis (QDA) is a parametric form of discriminant analysis and is a generalization of LDA. Like LDA, QDA assumes that the observations come from a multivariate normal distribution, but unlike LDA, the groups are not assumed to have equal covariance matrices. Also see *discriminant analysis*, *LDA*, and *parametric methods*.

quadratic discriminant analysis. See *QDA*.

quartimax rotation. Quartimax rotation maximizes the variance of the squared loadings within the rows of the matrix. It is an orthogonal rotation that is equivalent to minimizing the criterion

$$c(\mathbf{\Lambda}) = \sum_i \sum_r \lambda_{ir}^4 = -\frac{1}{4}\langle \mathbf{\Lambda}^2, \mathbf{\Lambda}^2 \rangle$$

See *Crawford–Ferguson rotation* for a definition of $\mathbf{\Lambda}$.

quartimin rotation. Quartimin rotation is an oblique rotation that is equivalent to quartimax rotation when quartimin is restricted to orthogonal rotations. Quartimin is equivalent to oblimin rotation with $\gamma = 0$. Also see *quartimax rotation*, *oblique rotation*, *orthogonal rotation*, and *oblimin rotation*.

reflection. A reflection is an orientation reversing orthogonal transformation, that is, a transformation that involves negating coordinates in one or more dimensions. A reflection is a Procrustes transformation.

repeated measures. Repeated measures data have repeated measurements for the subjects over some dimension, such as time—for example test scores at the start, midway, and end of the class. The repeated observations are typically not independent. Repeated-measures ANOVA is one approach for analyzing repeated measures data, and MANOVA is another. Also see *sphericity*.

rotation. A rotation is an orientation preserving orthogonal transformation. A rotation is a Procrustes transformation.

Roy's largest root. Roy's largest root test is a test statistic for the hypothesis test $H_0 : \boldsymbol{\mu}_1 = \cdots = \boldsymbol{\mu}_k$ based on the largest eigenvalue of $\mathbf{E}^{-1}\mathbf{H}$. It is defined as

$$\theta = \frac{\lambda_1}{1 + \lambda_1}$$

Here $\mathbf{H}$ is the between matrix, and $\mathbf{E}$ is the within matrix. See *between matrix*.

Sammon mapping criterion. The Sammon (1969) mapping criterion is a loss criterion used with MDS; it is the sum of the scaled, squared differences between the distances and the disparities, normalized by the sum of the disparities. Also see *MDS*, *modern scaling*, and *loss*.

score. A score for an observation after factor analysis, PCA, or LDA is derived from a column of the loading matrix and is obtained as the linear combination of that observation's data by using the coefficients found in the loading.

score plot. A score plot produces scatterplots of the score variables after factor analysis, PCA, or LDA.

scree plot. A scree plot is a plot of eigenvalues or singular values ordered from greatest to least after an eigen decomposition or singular value decomposition. Scree plots help determine the number of factors or components in an eigen analysis. Scree is the accumulation of loose stones or rocky debris lying on a slope or at the base of a hill or cliff; this plot is called a scree plot because it looks like a scree slope. The goal is to determine the point where the mountain gives way to the fallen rock.

Shepard diagram. A Shepard diagram after MDS is a 2-dimensional plot of high-dimensional dissimilarities or disparities versus the resulting low-dimensional distances. Also see *MDS*.

similarity, **similarity matrix**, and **similarity measure**. A similarity or a similarity measure is a quantification of how alike two things are, such as observations or variables or groups of observations, or a method for quantifying that alikeness. A similarity matrix is a matrix containing similarity measurements. The matching coefficient is one example of a similarity measure. Contrast to *dissimilarity*. Also see *proximity* and *matching coefficient*.

single-linkage clustering. Single-linkage clustering is a hierarchical clustering method that computes the proximity between two groups as the proximity between the closest pair of observations between the two groups.

singular value decomposition. A singular value decomposition (SVD) is a factorization of a rectangular matrix. It says that if $\mathbf{M}$ is an $m \times n$ matrix, there exists a factorization of the form

$$\mathbf{M} = \mathbf{U}\boldsymbol{\Sigma}\mathbf{V}^*$$

where $\mathbf{U}$ is an $m \times m$ unitary matrix, $\boldsymbol{\Sigma}$ is an $m \times n$ matrix with nonnegative numbers on the diagonal and zeros off the diagonal, and $\mathbf{V}^*$ is the conjugate transpose of $\mathbf{V}$, an $n \times n$ unitary matrix. If $\mathbf{M}$ is a real matrix, then so is $\mathbf{V}$, and $\mathbf{V}^* = \mathbf{V}'$.

sphericity. Sphericity is the state or condition of being a sphere. In repeated measures ANOVA, sphericity concerns the equality of variance in the difference between successive levels of the repeated measure. The multivariate alternative to ANOVA, called MANOVA, does not require the assumption of sphericity. Also see *repeated measures*.

SSCP matrix. SSCP is an acronym for the sums of squares and cross products. Also see *between matrix*.

stacked variables. See *crossed variables*.

stacking variables. See *crossing variables*.

standardized data. Standardized data has a mean of zero and a standard deviation of one. You can standardize data $\mathbf{x}$ by taking $(\mathbf{x} - \overline{\mathbf{x}})/\sigma$, where σ is the standard deviation of the data.

stopping rules. Stopping rules for hierarchical cluster analysis are used to determine the number of clusters. A stopping-rule value (also called an index) is computed for each cluster solution, that is, at each level of the hierarchy in hierarchical cluster analysis. Also see *hierarchical clustering*.

stress. See *Kruskal stress* and *loss*.

structure. Structure, as in factor structure, is the correlations between the variables and the common factors after factor analysis. Structure matrices are available after factor analysis and LDA. Also see *factor analysis* and *LDA*.

supplementary rows or columns or **supplementary variables**. Supplementary rows or columns can be included in CA, and supplementary variables can be included in MCA. They do not affect the CA or MCA solution, but they are included in plots and tables with statistics of the corresponding row or column points. Also see *CA* and *MCA*.

SVD. See *singular value decomposition*.

target rotation. Target rotation minimizes the criterion

$$c(\mathbf{\Lambda}) = \frac{1}{2}||\mathbf{\Lambda} - \mathbf{H}||^2$$

for a given target matrix $\mathbf{H}$.

See *Crawford–Ferguson rotation* for a definition of $\mathbf{\Lambda}$.

taxonomy. Taxonomy is the study of the general principles of scientific classification. It also denotes classification, especially the classification of plants and animals according to their natural relationships. Cluster analysis is a tool used in creating a taxonomy and is synonymous with numerical taxonomy. Also see *cluster analysis*.

tetrachoric correlation. A tetrachoric correlation estimates the correlation coefficients of binary variables by assuming a latent bivariate normal distribution for each pair of variables, with a threshold model for manifest variables.

ties. After discriminant analysis, ties in classification occur when two or more posterior probabilities are equal for an observation. They are most common with KNN discriminant analysis.

total inertia or **total principal inertia**. The total (principal) inertia in CA and MCA is the sum of the principal inertias. In CA, total inertia is the Pearson χ^2/n. In CA, the principal inertias are the singular values; in MCA the principal inertias are the eigenvalues. Also see *CA* and *MCA*.

uniqueness. In factor analysis, the uniqueness is the percentage of a variable's variance that is not explained by the common factors. It is also "1 − communality". Also see *communality*.

unrestricted transformation. An unrestricted transformation is a Procrustes transformation that allows the data to be transformed, not just by orthogonal and oblique rotations, but by all conformable regular matrices. This is equivalent to a multivariate regression. Also see *Procrustes transformation* and *multivariate regression*.

varimax rotation. Varimax rotation maximizes the variance of the squared loadings within the columns of the matrix. It is an orthogonal rotation equivalent to oblimin with $\gamma = 1$ or to the Crawford–Ferguson family with $\kappa = 1/p$, where p is the number of rows of the matrix to be rotated. Also see *orthogonal rotation*, *oblimin rotation*, and *Crawford–Ferguson rotation*.

Ward's linkage clustering. Ward's-linkage clustering is a hierarchical clustering method that joins the two groups resulting in the minimum increase in the error sum of squares.

weighted-average linkage clustering. Weighted-average linkage clustering is a hierarchical clustering method that uses the weighted average similarity or dissimilarity of the two groups as the measure between the two groups.

Wilk's lambda. Wilk's lambda is a test statistic for the hypothesis test $H_0 : \boldsymbol{\mu}_1 = \boldsymbol{\mu}_2 = \cdots = \boldsymbol{\mu}_k$ based on the eigenvalues $\lambda_1, \ldots, \lambda_s$ of $\mathbf{E}^{-1}\mathbf{H}$. It is defined as

$$\Lambda = \frac{|\mathbf{E}|}{|\mathbf{E} + \mathbf{H}|} = \prod_{i=1}^{s} \frac{1}{1 + \lambda_i}$$

where $\mathbf{H}$ is the between matrix and $\mathbf{E}$ is the within matrix. See *between matrix*.

Wishart distribution. The Wishart distribution is a family of probability distributions for nonnegative-definite matrix-valued random variables ("random matrices"). These distributions are of great importance in the estimation of covariance matrices in multivariate statistics.

within matrix. See *between matrix*.

References

Bartlett, M. S. 1937. The statistical conception of mental factors. *British Journal of Psychology* 28: 97–104.

——. 1938. Methods of estimating mental factors. *Nature, London* 141: 609–610.

Bellman, R. E. 1961. *Adaptive Control Processes*. Princeton, NJ: Princeton University Press.

Bentler, P. M. 1977. Factor simplicity index and transformations. *Psychometrika* 42: 277–295.

Comrey, A. L. 1967. Tandem criteria for analytic rotation in factor analysis. *Psychometrika* 32: 277–295.

Cox, T. F., and M. A. A. Cox. 2001. *Multidimensional Scaling*. 2nd ed. Boca Raton, FL: Chapman & Hall/CRC.

Crawford, C. B., and G. A. Ferguson. 1970. A general rotation criterion and its use in orthogonal rotation. *Psychometrika* 35: 321–332.

Fisher, R. A. 1936. The use of multiple measurements in taxonomic problems. *Annals of Eugenics* 7: 179–188.

Hendrickson, A. E., and P. O. White. 1964. Promax: A quick method for rotation to oblique simple structure. *British Journal of Statistical Psychology* 17: 65–70.

Jennrich, R. I. 1979. Admissible values of γ in direct oblimin rotation. *Psychometrika* 44: 173–177.

——. 2004. Rotation to simple loadings using component loss functions: The orthogonal case. *Psychometrika* 69: 257–273.

Kaiser, H. F. 1974. An index of factor simplicity. *Psychometrika* 39: 31–36.

Kruskal, J. B. 1964. Multidimensional scaling by optimizing goodness of fit to a nonmetric hypothesis. *Psychometrika* 29: 1–27.

Mahalanobis, P. C. 1936. On the generalized distance in statistics. *National Institute of Science of India* 12: 49–55.

Sammon, J. W., Jr. 1969. A nonlinear mapping for data structure analysis. *IEEE Transactions on Computers* 18: 401–409.

Thomson, G. H. 1951. *The Factorial Analysis of Human Ability*. London: University of London Press.

Subject and author index

This is the subject and author index for the *Multivariate Statistics Reference Manual*. Readers interested in other topics should see the combined subject index (and the combined author index) in the *Quick Reference and Index*.

Semicolons set off the most important entries from the rest. Sometimes no entry will be set off with semicolons, meaning that all entries are equally important.

A

B

C

D

F

G

H

I

J

K

L

N

O

P

Q

T

U

V

W

Y

Z